Biomaterials Associated Infection

T. Fintan Moriarty • Sebastian A.J. Zaat
Henk J. Busscher
Editors

Biomaterials Associated Infection

Immunological Aspects and Antimicrobial Strategies

Springer

Editors
T. Fintan Moriarty
AO Research Institute Davos
Davos Platz
Switzerland

Henk J. Busscher
Department of Biomedical Engineering
University Medical Center Groningen
Groningen
Netherlands

Sebastian A.J. Zaat
Department of Medical Microbiology
Center for Infection and Immunity
Amsterdam (CINIMA)
University of Amsterdam
Amsterdam, The Netherlands

ISBN 978-1-4899-9374-8 ISBN 978-1-4614-1031-7 (eBook)
DOI 10.1007/978-1-4614-1031-7
Springer New York Heidelberg Dordrecht London

Softcover reprint of the hardcover 1st edition 2013

Printed on acid-free paper

Springer is part of Springer Science+Business Media (www.springer.com)

Preface

The use of biomaterials in modern medicine has greatly contributed to patient care allowing less-invasive patient monitoring, improvements in the administration of medicines, and enhanced patient mobility contributing to improvements in quality of life. Relatively early in the evolution of medical devices, however, it was realized that the use of implanted biomaterials also results in an increased susceptibility to infection. This increase in infection susceptibility appears to be present across all biomaterials classes, regardless of form or function and has remained to this day as one of the most common, yet unresolved problems associated with the use of implanted biomaterials. With an aging society, and increasing use of biomaterials to ensure restoration of function and quality of life, the problem of biomaterials associated infection (BAI) may be expected to increase in the coming decades. In this book we aim to provide the reader with an overview of the problem of BAI with a particular emphasis on the pathogenesis of BAI and opportunities available through biomaterials research for reducing its incidence.

Medical devices and the biomaterials from which they are composed are clearly central players in the pathogenesis of BAI. The presence of an implanted biomaterial causes a local defect in host immune defenses that contributes to the failure of the host to efficiently clear contaminating bacteria in a certain percentage of cases. The processes of BAI then progress from bacterial adhesion to the biomaterial and colonization of adjacent tissues through to formation of an antibiotic recalcitrant bacterial biofilm. A better understanding of this process from microbiological, pharmacological, immunological, and biomaterials science perspectives is required if significant reduction in BAI incidence is to be achieved.

One area in which considerable research has been performed is that of bacterial virulence and the pathogenesis of BAI. This theme is discussed in depth in the first part of the book, with a number of chapters dedicated to the pathogenic process of BAI for the most common causative microorganisms *Staphylococcus aureus* and *Staphylococcus epidermidis,* as well as a focus upon the increasingly acknowledged opportunistic pathogen in BAI, *Propionibacterium acnes.* There is also a growing body of data available on the host response to biomaterials, although a clearer understanding of the interplay between the local immune activation resulting from the biomaterials and subsequent impact upon the contaminating bacteria are required if we are to fully elucidate the pathogenic process of BAI and develop successful interventions. The immune response to biomaterials is also discussed in the book with contributions discussing the role of the foreign body response and dendritic cell activation by biomaterials.

Biomaterials research presents the single largest target, with most potential for clinical impact, in achieving reductions in BAI. Modern biomaterials science has provided a vast array of modification and activation strategies to impart anti-infective properties upon biomaterials. In this book some of the most commonly applied approaches to anti-infective biomaterials design are discussed, such as antibiotic-loaded materials and coatings, anti-adhesive surfaces, surface-functionalization with silver, and also a more novel and contemporary approach, prevention of infection by medical-grade honey. Furthermore, with clinicians and insurance providers demanding reductions in BAI incidence, research into new anti-infective biomaterials and the clinical introduction of devices with anti-infective properties represent the front line in modern translational research. Ultimately, the most promising anti-infective biomaterials solutions must be tested in the clinics and hospital wards, where treatment outcomes and efficacy information may be collated. Clinical data is already available for some antimicrobial devices such as catheters, endotracheal tubes, periodontal implants, and orthopedic devices. The case histories of several of these devices are presented in this book together with a frank discussion of their impact on reducing or preventing BAI and potential future use based on the latest clinical data. Careful attention to, and continuous observation of, the clinical results associated with the use of antimicrobial devices is expected to provide the necessary direction for future biomaterials research. In this book we have aimed to provide an overview of the most crucial parameters in both preclinical and clinical fields allowing the reader to obtain a complete understanding of the problem of BAI from all of the most important perspectives.

The Editors

April 2012

Contents

Contributors

James M. Anderson Department of Pathology, Case Western Reserve University, Cleveland, OH, USA

Javier Ariza Department of Infectious Diseases, IDIBELL, Hospital Universitari Bellvitge, Barcelona, Spain

Georgios N. Belibasakis Faculty of Medicine, Oral Microbiology and Immunology, Institute of Oral Biology, Center of Dental Medicine, University of Zürich, Zürich, Switzerland

Nagihan Bostanci Faculty of Medicine, Oral Translational Research, Institute of Oral Biology, Center of Dental Medicine, University of Zürich, Zürich, Switzerland

Amanda E. Brooks Department of Pharmaceutics and Pharmaceutical Chemistry, University of Utah, Salt Lake City, UT, USA

Benjamin D. Brooks Department of Pharmaceutics and Pharmaceutical Chemistry, University of Utah, Salt Lake City, UT, USA

Henk J. Busscher Department of Biomedical Engineering, University Medical Center Groningen, Groningen, The Netherlands

Lorenzo Calabro AO Research Institute Davos, Davos, Switzerland, Queensland University of Technology, Brisbane, Australia

Rabih O. Darouiche Center for Prostheses Infection and Infectious Disease Section, Michael E. DeBakey Veterans Affairs Medical Center, and Baylor College of Medicine, Houston, TX, USA

Joana F. da Silva Domingues Department of Biomedical Engineering, University Medical Center Groningen, Groningen and University of Groningen, Groningen, The Netherlands

Angharad P. Davies Medical Microbiology and Infectious Diseases, Institute of Life Science, College of Medicine, Swansea University, Swansea, UK

Roger Feneley North Bristol NHS Trust, Southmead Hospital, Bristol, UK

Abhay D. Gahukamble AO Research Institute Davos, Davos Platz, Switzerland

Friedrich Götz Department of Microbial Genetics, Faculty of Biology, University of Tübingen, Tübingen, Germany

David W. Grainger Department of Pharmaceutics and Pharmaceutical Chemistry, University of Utah, Salt Lake City, UT, USA

Department of Bioengineering, University of Utah, Salt Lake City, UT, USA

Hans J. Griesser Ian Wark Research Institute, University of South Australia, Mawson Lakes, SA, Australia

Stefani S. Griesser Ian Wark Research Institute, Mawson Institute, University of South Australia, Mawson Lakes, SA, Australia

Llinos G. Harris Medical Microbiology and Infectious Diseases, Institute of Life Science, College of Medicine, Swansea University, Swansea, UK

Christine Heilmann Institute for Medical Microbiology, The Interdisciplinary Center for Clinical Research (IZKF), University Hospital of Münster, Münster, Germany

Rose Jeeves Medical Microbiology and Infectious Diseases, Institute of Life Science, College of Medicine, Swansea University, Swansea, UK

E.T. Kang Department of Chemical and Biomolecular Engineering, National University of Singapore, Kent Ridge, Singapore

Johannes K.-M. Knobloch Institut für Medizinische Mikrobiologie und Hygiene, Universität Lübeck, Lübeck, Germany

Marin Kollef Washington University School of Medicine, St. Louis, MO, USA

Roel Kuijer Department of Biomedical Engineering, University Medical Center Groningen, Groningen and University of Groningen, Groningen, The Netherlands

Paul H.S. Kwakman Department of Medical Microbiology, Center for Infection and Immunity Amsterdam (CINIMA), University of Amsterdam, Amsterdam, The Netherlands

J.A. Loontjens University of Groningen, Groningen, The Netherlands

Jaime Lora-Tamayo Department of Infectious Diseases, IDIBELL, Hospital Universitari Bellvitge, Barcelona, Spain

Cameron Lutton AO Research Institute Davos, Davos, Switzerland

Dietrich Mack Medical Microbiology and Infectious Diseases, Institute of Life Science, College of Medicine, Swansea University, Swansea, UK

Andrew McDowell Centre for Infection and Immunity, School of Medicine, Dentistry and Biomedical Sciences, Queen's University Belfast, Belfast, UK

T. Fintan Moriarty AO Research Institute Davos, Davos Platz, Switzerland

Oscar Murillo Department of Infectious Diseases, IDIBELL, Hospital Universitari Bellvitge, Barcelona, Spain

K.G. Neoh Department of Chemical and Biomolecular Engineering, National University of Singapore, Kent Ridge, Singapore

Ben Pascoe Medical Microbiology and Infectious Diseases, Institute of Life Science, College of Medicine, Swansea University, Swansea, UK

Jasmine D. Patel Exponent, Philadelphia, PA, USA

Sheila Patrick Centre for Infection and Immunity, School of Medicine, Dentistry and Biomedical Sciences, Queen's University Belfast, Belfast, UK

G. Rutger Persson University of Kristianstad, Kristianstad, SwedenUniversity of Washington, Seattle, WA, USAUniversity of Bern, Bern, Switzerland

Neil Poulter Ian Wark Research Institute, Mawson Institute, University of South Australia, Mawson Lakes, SA, Australia

Stefan Renvert University of Kristianstad, Kristianstad, SwedenTrinity College, Dublin, Ireland

Blekinge Tekniska Högskola, Karlskrona, Sweden

R. Geoff Richards AO Research Institute Davos, Davos, Switzerland

Holger Rohde Institut für Medizinische Mikrobiologie, Virologie und Hygiene, Zentrum für Klinische Pathologie, Universitäts-Klinikum Hamburg-Eppendorf, Hamburg, Germany

Gerhard Schmidmaier Department of Orthopedics, Trauma Surgery and Paraplegiology, Research Center for Experimental Orthopedics, Heidelberg University Hospital, Heidelberg, Germany

Antonio S. Sechi Department of Cell Biology, Institute of Biomedical Engineering, Universitätsklinikum Aachen, Rheinisch-Westfälische Technische Hochschule (RWTH), Aachen, Germany

Ahmed Fouad Seif El Din AO Research Institute Davos, Davos, Switzerland

Orthopaedic Surgery and Trauma Department, Kasr Al Ainy Hospital, Cairo, Egypt

Behnaz Shokouhi Department of Cell Biology, Institute of Biomedical Engineering, Universitätsklinikum Aachen, Rheinisch-Westfälische Technische Hochschule (RWTH), Aachen, Germany

Z.L. Shi Department of Chemical and Biomolecular Engineering, National University of Singapore, Kent Ridge, Singapore

David Stickler Cardiff School of Biosciences, Cardiff University, Cardiff, Wales, UK

Guruprakash Subbiahdoss Department of Biomedical Engineering, University Medical Center Groningen, Groningen and University of Groningen, Groningen, The Netherlands

Thomas Thurnheer Faculty of Medicine, Oral Microbiology and Immunology, Institute of Oral Biology, Center of Dental Medicine, University of Zürich, Zürich, Switzerland

Andrej Trampuz Infectious Diseases Service, Department of Medicine, University Hospital (CHUV) and University of Lausanne, Lausanne, Switzerland

Henny C. van der Mei Department of Biomedical Engineering, University Medical Center Groningen, Groningen and University of Groningen, Groningen, The Netherlands

Krasimir Vasilev Mawson Institute, University of South Australia, Mawson Lakes, SA, Australia

Thomas S. Wilkinson Medical Microbiology and Infectious Diseases, Institute of Life Science, College of Medicine, Swansea University, Swansea, UK

Sebastian A.J. Zaat Department of Medical Microbiology, Center for Infection and Immunity Amsterdam (CINIMA), University of Amsterdam, Amsterdam, The Netherlands

Werner Zimmerli Basel University Medical Clinic, Kantonsspital, Liestal, Switzerland

Part I
The Pathogenesis of Medical Device-Associated Infection

Chapter 1
Biomaterial-Associated Infection: A Perspective from the Clinic

Werner Zimmerli and Andrej Trampuz

Abstract Implant-associated infections are caused by microorganisms which adhere to the device and form biofilms. Medical devices are highly susceptible to bacterial and fungal infections. The minimal abscess-producing inoculum is >10,000-fold lower in the presence than in the absence of a foreign body. This is mainly due to a local granulocyte defect. Not only abiotic, but also biological (devitalized) implants are prone to infection. Among the clinically most important devices are prosthetic joints. These implants are at life-long risk for infection. During the first 2 years after implantation, the majority of the infections are exogenous; later, they are mainly caused by the hematogenous route. The treatment goal is complete eradication of infection, freedom of pain, and correct function of the joint. Reaching this goal requires rapid diagnosis and a rational treatment strategy including adequate surgery (debridement, one-stage or two-stage exchange) combined with prolonged antibiotic therapy. Surface adhering biofilms are highly resistant to host defense and antimicrobial agents. According to animal experiments and clinical studies, rifampin is more efficacious against surface adhering staphylococci than other agents. In view of the limited efficacy of treatment, novel preventive options are required such as implant coating or quorum sensing inhibitors.

W. Zimmerli, M.D. (✉)
Basel University Medical Clinic, Kantonsspital,
Rheinstrasse 26, Liestal CH-4410, Switzerland
e-mail: werner.zimmerli@unibas.ch

A. Trampuz
Infectious Diseases Service, Department of Medicine, University Hospital (CHUV) and
University of Lausanne, Lausanne, Switzerland

T.F. Moriarty et al. (eds.), *Biomaterials Associated Infection: Immunological Aspects and Antimicrobial Strategies*, DOI 10.1007/978-1-4614-1031-7_1,

1.1 Introduction

Biomaterial-associated infections are typically caused by microorganisms that grow in biofilms [1–3]. In contrast to infections caused by free-floating (planktonic) bacteria, biofilm infections have the following distinct characteristics: (1) bacteria are adhering to a surface, (2) bacterial clusters are encased in a matrix, (3) infection is confined to a local site, (4) adherent bacteria may escape routine diagnostic procedures, (5) infection generally persists despite susceptibility of planktonic bacteria to the antimicrobial agent, and (6) host defenses are unable to eradicate microorganisms, i.e., spontaneous cure does not occur [4, 5]. Many types of chronic infections, especially posttraumatic osteomyelitis, chronic prostatitis, chronic wounds, and pulmonary infections in patients with cystic fibrosis are caused by bacterial biofilms. These infections are all characterized by a chronic disease course due to biofilm formation. However, in implant-associated infections, no chronic course is required for the biofilm formation. Adherent bacterial clusters can be observed within a few hours after inoculation, as shown in animal models of foreign-body infection [6, 7].

During the past decades, medical devices have gained growing importance due to considerable medical and technical progress [8, 9]. In particular, biocompatibility, functionality, and durability of various implants have been considerably improved. Therefore, many implanted devices can properly function for decades. The drawback of this technological advance is the high susceptibility of implants to infection. In addition, such infections chronically persist, if no comprehensive treatment concept is applied. This concept consists of appropriate surgical intervention and long-term antimicrobial therapy directed against biofilms. Whereas acute (planktonic) infection generally needs only one single antimicrobial treatment course to eradicate microorganisms, chronic (biofilm) infection may require sophisticated diagnostic procedures, long-term antimicrobial therapy and repetitive surgical interventions. In the case of orthopedic device-associated biofilm infection, cure of infection should be aimed at in the first attempt, because each treatment failure results in a worse functional result [10, 11]. Thus, the prerequisite for correct treatment of device associated infection is a rational concept for the optimal surgical and antimicrobial therapy [10–12].

1.2 Type of Implants

The introduction of implantable devices represents a major advance of modern human medicine. Implants are used to improve impaired function, replace missing anatomic structure, or optimize physical appearance [8]. They significantly improve the quality of life in children (e.g., neurosurgical shunts, spinal implants for correction of scoliosis), adults (e.g., dental implants, artificial heart valves, breast prostheses) and in the elderly (joint prostheses, fracture-fixation devices, cardiac pacemakers).

Medical devices are made out of abiotic materials such as metals, polymers (e.g., silicone), but may also contain biological materials such as devitalized bone, blood vessels, muscle fascia from autologous (venous bypass), allogeneic (processed bone), or xenogeneic sources (e.g., porcine heart valve). Regardless of the type of material (synthetic or devitalized biological origin), it represents foreign material, and is therefore labeled as a biomaterial. Thus, the affix "bio" does not refer to a biological origin, but to the fact that "foreign" material is implanted in a living organism. Interestingly, neither synthetic nor devitalized biological devices are rejected by the body, despite the fact that the host reacts to such implants in different ways depending upon the biocompatibility of the device [13, 14]. However, no implant is completely inert after implantation.

Medical prostheses can be used either as extracorporeal devices (e.g., limb prosthesis after amputation), intracorporeal (indwelling) devices (e.g., vascular prosthesis), or as implants that cross the anatomic cutaneous or mucosal barriers (e.g., dental implants). Table 1.1 summarizes the most important implanted devices which are used in clinical medicine. Intracorporeal implants can be classified according to their localization as intravascular or extravascular devices. The two different types of devices interact very differently with the host. Whereas intravascular implants mainly interact with coagulation factors and circulating blood cells, extravascular implants interact with surrounding tissue, interstitial fluid and attracted phagocytes. Permanent implants cannot be removed without compromising the replaced function. Therefore, the primary goal is to prevent implant failure due to mechanical reasons or infection. In this chapter, we limit the detailed discussion to joint prostheses.

1.2.1 Intravascular Devices

Among the most common intravascular devices are arteriovenous shunts which are required in patients with chronic renal failure undergoing hemodialysis [15]. These implants are typically made of synthetic material (expanded polytetrafluoroethylene, ePTFE). With such catheters, 3–4 infections occur per 1,000 dialysis sessions [16]. In addition, vascular prostheses are used for revascularization in case of arterial occlusive disease [17]. The average infection rate of vascular grafts is about 4 %, with lower rates of aortic and higher rates of peripheral grafts [8]. A more recent technique for the treatment of vascular obstruction is the placing of intravascular stents in diseased vessels, which are mainly used in interventional cardiology [18, 19]. They are also used to improve blood flow in obstructed or damaged (e.g., aneurysm) large vessels such as the aorta or the carotid artery [20]. Stent associated infections are very rare [21].

After implantation, vascular prostheses are covered by a new cell layer which is called neointima. Experimentally, neointimal formation on the inner side of the vascular prosthesis plays a crucial role for reduction of its susceptibility to infection. In a dog experimental model, the rate of prosthetic graft-related infection induced by

Table 1.1 Implantable devices currently used in medicine

Anatomic site/surgical field	Implant type
Bone and joint	Prosthetic joint (hip, knee, shoulder, ankle, elbow, other)
	Internal fixation devices (nail, plate, and screw)
	External fixation (pin)
	Spine implants (spondylodesis, discus implant)
	Bone cement (for temporary or definitive joint fixation or after vertebral fracture)
	Bone substitutes (for bone defects), bone matrix putties/gels and osteoconductive scaffolds
	Kirschner wires (finger/toe osteosynthesis) Sternum fixation (cerclage)
Plastic/reconstructive	Breast, nose, chin, testis
	Penile prosthesis
	Skin regeneration (dermal regeneration template)
Cardiovascular	Artificial/mechanical heart valve
	Vascular prostheses
	Electrophysiologic cardiac devices (cardiac pacemaker, implantable cardioverter/defibrillator)
	Left ventricular assist device (artificial heart)
	Venous access devices with subcutaneous reservoir
	Intravascular stents
	Permanent intravascular catheters (tunneled venous catheters, hemodialysis shunts)
Neurosurgical	Internal CSF drainage devices (ventriculoperitoneal/-atrial, lumbar)
	External cerebrospinal fluid drainage devices
	Intracranial pressure and brain tissue oxygen monitoring devices
	Deep brain stimulator
	Intrathecal opiate pump
	Nerve stimulator
	Cranial closure (duraplastic, mesh, plates)
	Dura repair for cerebrospinal fluid leakage (collagen and other patches)
	Bone repair after craniotomy (bone cement)
	Absorbable collagen implant (to support tendon and nerve injury)
Ophthalmology	Implanted lenses
	Retinal implant
Ear/nose/throat	Cochlear implant
Visceral surgery	Abdominal mesh graft (e.g., for hernioplastic)
Urology	Ureteric catheter (double-J)
	Permanent urinary catheter
	Suprapubic catheter
Gynecology	Intrauterine device
Dental medicine	Dental implants
Other	Insulin pump
	Peritoneal dialysis catheter

experimental bacteremia dropped from 100 % during the first 2 weeks after surgery to 30 % 1 year later [22]. The correlation between neointimal integrity and protection against hematogenous seeding illustrates the importance of the foreign surface for bacterial adherence [22, 23].

Artificial heart valves are used to correct structural valvular defects such as stenosis or regurgitation [24, 25]. According to a randomized prospective trial comparing mechanical heart valves with porcine bioprostheses, not only synthetic but also biological devices of devascularized tissue (e.g., porcine bioprosthesis) increase the susceptibility to endocarditis as compared to native valves [26, 27]. This indicates that avascular biological material obviously behaves like synthetic material in terms of susceptibility to infection.

Implantable electrophysiologic cardiac devices are increasingly used for prevention and therapy of cardiac arrhythmias and heart failure. Cardiac pacemakers are commonly used in patients with atrioventricular conduction block, sick sinus syndrome and sinus bradycardia, whereas implantable cardioverter/defibrillators target primarily patients with heart failure after myocardial infarction and those having experienced a life-threatening ventricular arrhythmia. Cardiac resynchronisation therapy ameliorates cardiac function and reduces the mortality in patients with heart failure and improved exercise capacity. Infection of electrophysiological cardiac devices is a rare, but serious and potentially life-threatening complication, ranging between 0.7 and 1.6 % [28]. The infection may involve the generator pocket, the leads (with or without the endocardium) or both components. Pacemaker endocarditis usually requires complete removal of the device and replacement at another anatomic location, if it is still needed. As temporary solution, a temporary epicardial (extravascular) pacing system can be used.

1.2.2 Extravascular Devices

These implants are localized in different compartments of the host and have no direct interaction with the circulating blood. Orthopedic devices, such as prosthetic joints and internal fixation devices (nails, plates, cerclage, and screws), are the most important implants in human medicine [10, 29–32]. In addition, other foreign materials such as bone grafts (devitalized bone) and bone substitutes (e.g., calcium phosphates) are often used in orthopedic surgery for filling bone defects [33]. Despite the fact that joint replacement is a so-called clean procedure [34], implant-associated infections are observed in 0.5–1 % (hip or knee arthroplasty) to >5 % (elbow or ankle arthroplasty) of primary implant surgeries [10, 35]. In general, all types of revision surgery increase the risk for wound infection due to less favorable soft tissue conditions, more extensive tissue damage, and longer operation times. In addition to intraoperative infections, implanted devices are at life-long risk of hematogenous or lymphogenous seeding during bacteremia or local skin infection. In general, mainly high-virulent organisms may cause hematogenous infections, originating from skin/soft tissue, respiratory and urinary tract infections. The risk for hematogenous infection is estimated between 0.25 and 0.5 % per year per prosthesis, despite the fact that large differences exist between the type of organism (e.g., prosthetic devices are more prone to gram-positive cocci than to gram-negative rods) and type of prosthesis involved (e.g., large weight-bearing joints are more susceptible to infection than small non-weight-bearing joints). Secondary hematogenous infection may occur not only in patients with intravascular implants, but also in those with extravascular devices, such as prosthetic joints (see below). Furthermore, implants can get infected "*per continuitatem*" through adjacent focus of skin or soft tissue infection. These clinical observations demonstrate an acquired local host defense defect around the implant (see below).

1.3 Pathogenesis

1.3.1 Role of the Host

Implanted devices are susceptible to virtually all types of bacteria and fungi [8, 36, 37]. Apparently, innate host defense mechanisms do not efficiently clear invading microorganisms [38–41]. The first line of defense against skin-borne microorganisms such as staphylococci and streptococci is phagocytosis by granulocytes or mononuclear phagocytes. Killing of bacteria by these cells depends on efficient phagocytosis and intracellular killing. For rapid ingestion of bacteria, opsonization of the microorganisms is essential. This involves nonspecific (complement) and in some microorganisms specific soluble components (antibodies), as well as the corresponding receptors on the phagocytes. The process of phagocytosis involves chemotaxis, cell adherence to the microorganism, ingestion, killing, and digestion [42]. If this complex process is impaired at a given level, there is an enhanced risk of microbial persistence and therefore infection [43].

1.3.2 Interaction of Implant with Granulocytes

In view of the susceptibility of implants to infection, various possible mechanisms for the impaired bacterial clearance have been hypothesized. Infections around an implant never heal spontaneously, except after extrusion of the foreign body [6, 39, 41, 44]. This indicates that granulocytes around an implant are incompetent, or alternatively, bacteria within a biofilm resist phagocytosis, or both. Indeed, we could show that the interaction of granulocytes with the nonphagocytosable surface of the foreign body results in an impaired function [39, 40]. Specifically, ingestion of opsonized particles, killing of *Staphylococcus aureus* and superoxide production is impaired in granulocytes accumulating around an implanted device [39, 40, 45]. In addition, these granulocytes are partially degranulated, indicating exocytosis on the surface of the foreign body [40]. Similarly, the presence of wear particles also impairs the antibacterial activity of neutrophil granulocytes [46, 47]. This suggests that wear particles (ultrahigh molecular weight polyethylene, UHWP) may contribute to the increased susceptibility of loose prostheses to bacterial infection. Taken together, these experimental observations explain the paradox of microbial persistence in the presence of abundant granulocytes around the implant.

1.3.3 Interaction of Implant with Complement

Contact of blood or other body fluids with biomaterials leads to complement activation [48–50]. This results in tissue inflammation, because of the production of chemotactic peptides (C3a and C5a) [51]. Indeed, Tang et al. [52] described a close relationship between surface-mediated complement activation *in vitro* and accumulation of phagocytes on polymer surfaces in a mouse model. In decomplemented mice there was no phagocyte accumulation around intraperitoneal implants, indicating the role of complement activation on inflammation *in vivo*. Wear particles have been shown to activate complement, leading to increased concentrations of C3a [53]. Therefore, it can be speculated that inflammation around ultrahigh molecular weight polymer particles lead to local inflammation and thereby to loosening of the prosthesis.

1.4 Role of the Microorganism

In human and animal experiments, it has been shown that the presence of a foreign body decreases the minimal abscess-forming dose at least 10,000-fold [39, 54, 55]. This highly increased susceptibility of implants to infection can be explained by the described compromised local host defense. Evading microorganisms rapidly adhere to the implant, where they resist elimination by host defense mechanisms [56]. Biofilm formation has been extensively studied not only in the context of device associated infection, but also in many types of chronic infections such as cystic fibrosis [5, 57, 58]. Biofilm formation is discussed in detail in Chaps 2 and 3 of this book. In brief, a biofilm is a community of bacteria encased in exopolysaccharides surrounding the microorganism (Fig. 1.1). This protective matrix protects bacteria from host immunity and antimicrobial agents [3, 59].

1.5 Periprosthetic Joint Infection

With aging of the general population, the prevalence of joint degenerative diseases is continuously rising. Joint replacement is one of the most successful surgical interventions alleviating pain, improving joint function, but also restoring general

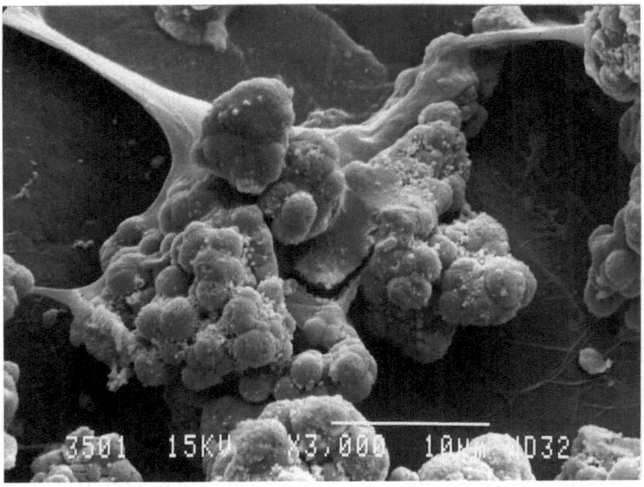

Fig. 1.1 SEM from the surface of an explanted hip cup from a patient with hematogenous periprosthetic infection due to *Samonella dublin* [59]. The amorphous conglomerates show the microorganisms covered by a slimy substance, corresponding to exopolysaccharides

mobility and personal independence. The number of implanted hip and knee prostheses increases exponentially, which is followed by an increase of revision surgeries several years later [60]. In addition, shoulder and ankle joint replacements have been increasingly performed during the last two decades [61, 62], but also other joint replacement surgeries are increasingly performed, such as elbow [63]. The number of arthroplasties and consequently also the absolute number of complications is steadily increasing. In our cohort of patients with primary hip arthroplasty, the most frequent local complications, that mostly required revision surgery, were hematomas (4 %), perioperative fracture (2.3 %), dislocation (2.2 %), paresis (2.1 %), and infection (0.4 %) [64]. Thus, periprosthetic joint infection is a rare complication compared to other postoperative complications. However, infections cause a considerable morbidity for the patient and accounts for a substantial proportion of health care expenditures [8, 65].

1.5.1 Definition of Periprosthetic Joint Infection

Periprosthetic joint infection can be diagnosed in a patient with clinical signs of infection and at least one of the following criteria: (a) growth of the same microorganism in at least two cultures of synovial fluid or periprosthetic tissue, (b) predominance of neutrophils in synovial fluid [66, 67], (c) purulence at the implant site, (d) acute inflammation on histopathological examination of periprosthetic tissue, or (e) presence of a sinus tract communicating with the prosthesis [10, 68, 69].

1.5.2 Classification and Microbiology

Periprosthetic joint infections can be classified either according to their route of infection (exogenous or hematogenous), or according to the time of appearance after implantation. In case of inoculation from the outside of the body, infection is labeled as exogenous. Such infections typically occur either during surgery or in the postoperative phase in patients with a secreting wound. Alternatively, it can be acquired through joint tap. Exogenous infections manifest themselves mainly during the first 2 years after implantation. As an exception, infection after arthocentesis may occur at any time after implantation. Such infections manifest themselves either acute (within days) when caused by *S. aureus* or subacute (within weeks) when caused by coagulase-negative staphylococci. Hematogenous infections are acquired via bloodstream at any time after surgery. In view of the impaired local host defense (see above), selective seeding on the implant is not surprising [45].I As a consequence, infections occurring after the second year are almost exclusively of hematogenous origin. The differentiation in exogenous versus hematogenous inoculation is sometimes not clear. Early postoperative infection is not always caused by the exogenous route. The risk for hematogenous infection is even higher early after surgery and decreases thereafter [65].

The interval between implantation and appearance of signs of periprosthetic infection allows inference regarding infecting agent and pathogenesis. Infections can be classified as "early", "delayed" and "late" [10]. Early infection occurs within the first 2 months, and delayed infection between the third and the 24th-month after surgery. Late infection is diagnosed >2 years after implantation. Early as well as delayed infections are typically of exogenous origin. The former type is mainly due to virulent microorganisms such as *S. aureus* or *Escherichia coli*, whereas the latter is typically caused by low-virulent bacteria such as coagulase-negative staphylococci or *Propionibacterium acnes*.

Late infections are a consequence of hematogenous or lymphogenous seeding. Symptoms may be dominated by symptoms of the primary focus, mainly skin/soft tissue, urinary tract, or respiratory tract infection [70]. However, in many patients no obvious primary focus can be detected, or the primary infection is already healed when local signs of periprosthetic infection appear. By far the most common microorganism causing hematogenous implant associated infection is *S. aureus*. In three independent clinical studies, the rate of periprosthetic joint infection after documented *S. aureus* sepsis has shown to be between 34 and 39 % [71–73]. This feature was also observed in the experimental setting showing that *S. aureus* bacteremia with a density of 1,000 colony forming units (CFU)/ml blood results in 42 % infection of extravascular implants in guinea pigs [74]. In contrast, occasional bacteremia induced by dental work does not jeopardize orthopedic implants, as recently demonstrated with a case–control study [75]. This is probably due to the low bacterial density during dental intervention [76].

Overall, the most common microorganisms causing periprosthetic infection are coagulase-negative staphylococci (30–43 %) and *S. aureus* (12–23 %), followed by streptococci (9–10 %), enterococci (3–7 %), gram-negative bacilli (3–6 %), and anaerobes (2–4 %) [65, 77, 78]. Polymicrobial infections are observed in up to >20 % [79], and no organism is isolated in 10–30 %, depending on the quality of the diagnostic procedure and preceding antimicrobial therapy.

1.5.3 Predisposing Factors and Risk Factors

The characteristics predisposing patients to periprosthetic joint infection have been described in many observational studies [80–84]. The major drawback in these studies is the lack of statistical rigor. None of these studies is a case–control study or a formal observational cohort study. As an exception, Berbari et al. [85] performed a solid case–control study. In this study, many previously reported risk factors have also been detected in the univariate analysis, but not confirmed in the multivariate analysis. Among them are rheumatoid arthritis, diabetes mellitus, prolonged duration of index hospitalization, wound problems (drainage, hematoma, and dehiscence), and decubitus ulcer. Significant risk factors in the multivariate analysis were a high comorbidity measured by the National Nosocomial Infection Surveillance System (NNIS) risk index score (OR 3.9), systemic malignancy (OR 3.1), prior joint arthroplasty (OR 2.0), and postoperative surgical site infection (35.9) [85]. Since surgical site infection increases the risk for implant associated infection almost 36-fold, the distinction between superficial and deep wound infection in patients with arthroplasties seems to be arbitrary.

We recently performed a case–control study to evaluate potential risk factors for periprosthetic ankle joint infection [86]. In a cohort of 26 patients with infection and 105 control patients after ankle arthroplasty, we found the following significant risk factors: Prior surgery at the index articulation ($p<0.05$), low functional activity prior to surgery ($p<0.01$), long duration of index surgery ($p<0.01$, revision surgery ($p=0.02$), persistent wound dehiscence ($p=0.01$), and secondary wound secretion ($p=0.02$).

1.5.4 Clinical Presentation and Diagnostic Workup

Rapid and accurate diagnosis of periprosthetic joint infection is crucial for optimal treatment. The clinical presentation depends on the above mentioned type of infection. Early infection is characterized by acute onset of arthralgia, effusion, erythema, and warmth at the implant site. Fever may not be present, especially if the patient is treated with pain relievers acting also as antipyretics. After prolonged duration, skin and soft tissue is damaged by cellulitis or even sinus tracts. As a rule, patients with delayed infection present with subtle signs and symptoms. Typically, the clinical picture simulates mechanical failure with persistent joint pain and early implant loosening. In case of late hematogenous infection, the patient may present with clinical signs of sepsis or more subtle and nonspecific with increasing joint pain and local signs of inflammation.

Sensitivity and/or specificity of laboratory inflammatory blood maker levels are low in case of periprosthetic joint infection. Berbari et al. [87] recently published a systematic review analyzing data from 3,909 revision total hip or knee arthroplasties. The prevalence of periprosthetic joint infection was 32.5 %. The pooled sensitivity was 45 % for leukocyte counts, 75 % for the erythrocyte sedimentation rate, 88 % for the C-reactive protein level, and 97 % for the interleukin-6 level. Pooled specificity for the same markers was 87, 70, 74, and 91 %. Thus, interleukin-6, which is not a routinely available test, has the highest accuracy, followed by C-reactive protein. In this review, variable cut-offs were taken from the included original studies. In a recent single-center study, the optimal cut-offs of C-reactive protein and erythrocyte sedimentation rate for detecting implant-associated infections have been analyzed [88]. With a C-reactive protein cut-off of 14.5 mg/L the sensitivity and specificity were 79 % and 88 %, respectively, in 297 patients with either aseptic failure ($n=215$) or implant-associated infection ($n=82$) after knee arthroplasty.

Analysis of synovial fluid has a better diagnostic value. Synovial fluid leukocyte count >1,700 per cubic millimeter or >65 % neutrophils had sensitivity for infection of 94 % and 97 %, respectively, and specificities of 88 % and 98 %, respectively to detect infection in patients without underlying inflammatory joint diseases who were more than 6 months from total knee arthroplasty [66]. In patients after total hip arthroplasty, a leukocyte count of 4,200/mm^3 had a sensitivity of 84 % and a specificity of 93 % to detect implant associated infection [67].

In addition to laboratory workup, imaging studies are used to diagnose infection. Plain radiographs lack sensitivity and specificity, but may be useful when studied serially over time after implantation in order to detect subperiosteal bone growth, migration of the implant, or osteolysis [89]. Radionuclide scans, CT-scans, single-photon emission computed tomography (SPECT), magnetic resonance imaging (MRI), and positron emission tomography (FDG-PET) are increasingly used in patients with suspected periprosthetic joint infection [90–92]. However, not all these techniques are available at each center, and the reported specificity is variable in different studies [93–95].

Histopathology can be used intraoperatively (frozen section) for differentiation between aseptic loosening and infection. This technique needs special competence of the pathologist and is therefore available in only a limited number of centers. Typically, >5 polymorphonuclear leukocytes per high-power field (40×) is defined as cut-off for infection. The sensitivity is good for high-virulence microorganisms (*S. aureus*, *Pseudomonas* spp., enterobacteriaceae). However, it is poor for low-virulence bacteria such as coagulase-negative staphylococci or *P. acnes* [96]. Therefore, frozen section is not useful in those patients in whom it would be needed most. However, the correlation of histopathology with tissue biopsy culture allows to better separate infection from contamination. At least three and optimally six periprosthetic tissue biopsies should be obtained for anaerobic and aerobic culture [97]. With a long incubation time, the yield of fastidious and slow growing bacteria improves [98]. Sonication of the hardware (modular parts, screws, prosthesis) is a novel technique which is not yet broadly available. Culture of sonication fluid improves the sensitivity especially in patients treated with antibiotics within a time-period of 2 weeks before surgery [99].

1.5.5 Treatment Principles

There is never spontaneous healing of an implant associated infection. The treatment goal is complete eradication of infection, freedom of pain, and correct function of the joint. In order to reach this goal, management should not be limited to antimicrobial therapy alone. The therapeutic approach is always interdisciplinary, including surgical revision and antimicrobial therapy. The cornerstone of successful treatment is early diagnosis, since surgical intervention is less invasive in patients with a short history of infection. This requires a high degree of suspicion and strict avoidance of empirical antimicrobial therapy in case of uncertain infection. The first treatment attempt is the most decisive in order to avoid tissue damage and loss of function. Therefore, a well-defined treatment strategy is required.

1.5.5.1 Surgical Treatment

Figure 1.2 shows a treatment algorithm for surgery in periprosthetic joint infection [10, 100]. Debridement with retention has a success rate >80 % [101–105], if the patients fulfill the following conditions: (1) stable implant, (2) pathogen with susceptibility to antimicrobial agents active against surface-adhering microorganisms, (3) absence of a sinus tract or an abscess, and (4) duration of symptoms of infection less than 3 weeks. One-stage exchange includes the removal and implantation of a new prosthesis during the same surgical procedure. Patients with intact or only slightly compromised soft tissue qualify for this procedure. A success rate of 86–100 % can be expected in appropriately selected patients [10, 100]. If resistant or difficult-to-treat microorganisms are causing the infection, such as multi-resistant microorganisms for which no drug with good oral bioavailability is obtainable, small-colony variants of staphylococci [106], enterococci, quinolone-resistant *Pseudomonas aeruginosa* or fungi, a two-stage revision should be preferred. Two-stage exchange includes removal of the prosthesis with implantation of a new prosthesis during a later surgical procedure. If no difficult-to-treat microorganisms are isolated, a short interval until reimplantation (2–4 weeks) and a temporary antimicrobial-impregnated bone cement spacer may be used. If difficult-to-treat microorganisms are isolated, a longer interval (8 weeks) is preferred. Whether a spacer influences the eradication of all microorganisms is still unknown. The two-stage procedure can be used for every patient and has a success rate generally exceeding 90 % [10]. However, the expenditure for the patient and the surgeon is higher than for other surgical options. Permanent removal of the device is usually reserved for patients with a high risk of reinfection (e.g., severely compromised host, active intravenous drug use) or when no functional improvement after reimplantation is expected. Alternatively, long-term antimicrobial suppression may be chosen, if the patient is inoperable, bedridden or debilitated. However, suppressive therapy only controls clinical symptoms rather than cures the infection. Therefore, the risk for relapse is high after stopping suppressive antimicrobials (Fig. 1.2).

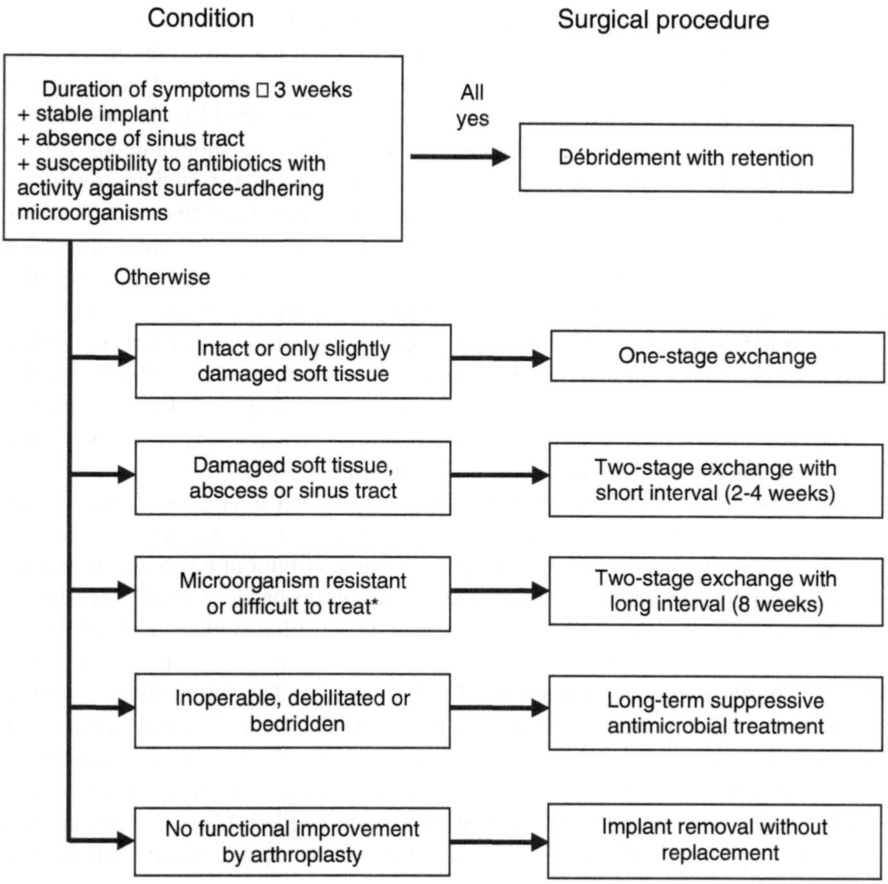

Fig. 1.2 Surgical treatment algorithm for prosthetic joint infections. Modified according to Trampuz and Zimmerli [100]. *Difficult-to-treat microorganisms include microorganisms resistant to antibiotics with good oral bioavailability, rifampin-resistant staphylococci, enterococci, quinolone-resistant gram-negative bacilli and fungi. (©2005 by EMH editions, Basel/Switzerland: original publication available on http://www.smw.ch)

1.5.5.2 Antimicrobial Treatment

Device-related infections often persist until the device is removed, even if the microorganism is highly susceptible to the antibiotic *in vitro*. Therefore, until recently it has been a dogma that infections around a foreign-body cannot be cured. Thus, there is obviously a discrepancy between the *in vitro* and *in vivo* efficacy of antimicrobial agents against such infections. Treatment results of clinical studies on device-associated infections are difficult to interpret because of the widely varying characteristics of the different patients. Therefore, there are almost no controlled studies [107]. In addition, if patients are treated without removal of the device, recurrence from persisting adherent microorganisms may occur within 2 years. Thus, a very long follow-up is needed in order to reach the hard endpoint of cure. We therefore developed a guinea-pig model to test not only local host-defense mechanisms, but also different antibiotic regimens [55, 108–113]. In this model, staphylococcal infection was established in the vicinity of subcutaneous tissue-cages (Teflon). After 1–3 days, a 4-day-course of antibiotics was started. Interestingly, the cure rate was extremely low with most antibiotics including vancomycin, teicoplanin, fluoroquinolones, linezolid, and daptomycin. In contrast, rifampin and a novel rifamycin were highly effective resulting in cure-rates up to 100 %, especially if given in combination with a second antibiotic in order to prevent emergence of resistance [55, 110–113]. The favorable properties of rifamycins correlate with a good *in vitro* activity on nongrowing and adherent staphylococci [108]. Similarly, only the antibiotics acting on stationary-phase gram-negative bacilli, namely, fluoroquinolones, were also active in the guinea-pig foreign-body infection model using *E. coli* or *Salmonella dublin* as test strains [59, 109].

Taken together, antibiotics against implant-associated infections should act on stationary-phase bacteria, on biofilm microorganisms, and should have a good bioavailability for long-term use on an outpatient basis. Rifampin has these properties against staphylococci and fluoroquinolones against gram-negative bacilli. The favorable properties of rifampin have been proven in a controlled clinical study [101].

Table 1.2 summarizes the choice of antimicrobial agents according to the pathogen. The initial treatment is generally by the iv-route. The initial therapy with a beta-lactam, a glycopeptide, or daptomycin is administered to lower the number of bacteria before starting oral therapy. This minimizes the risk for emergence of resistance during initial treatment when the bacterial density is highest. For long-term treatment, drugs with good oral bioavailability are needed. If such drugs are not available, outpatient iv-therapy with a once-daily drug may be a feasible alternative option [114].

1.6 Outlook

Biomaterial associated infection compromises the quality of life, has a high morbidity and is even associated with mortality. In addition, it has a high economic impact, since treatment of infection costs many times more than the primary implantation of the device. Therefore, major efforts should be invested to minimize susceptibility of implants to infection. Coating of the implant surface with antimicrobial substances such as antibiotics (e.g., vancomycin), antimicrobial peptides, or silver is an option. This strategy has been tested *in vitro* and in experimental models [115–117]. However, clinical data are still scarce, because in orthopedic surgery biomechanical research has traditionally been more important than the study of prevention of infection [118, 119]. An upcoming and promising technology

Table 1.2 Treatment of prosthetic joint infections (updated after Zimmerli et al. [10])

Microorganism	Antimicrobial agent	Dose	Route
Staphylococcus aureus or coagulase-negative staphylococci			
Methicillin-susceptible	Rifampin plus	450 mg every 12 h	PO/IV
	(flu)cloxacillin[a]	2 g every 6 h	IV
	For 2 weeks, followed by		
	Rifampin plus	450 mg every 12 h	PO
	Levofloxacin or	750 mg every 24 h to 500 mg every 12 h	PO
	Other fluoroquinolone	According to drug	PO
Methicillin-resistant	Rifampin plus	450 mg every 12 h	PO/IV
	Vancomycin[b] or	1 g every 12 h	IV
	Daptomycin	8–10 mg/kg 24 h	IV
	For 2 weeks, followed by		
	Rifampin plus	450 mg every 12 h	PO
	Levofloxacin or	750 mg every 24 h to 500 mg every 12 h	PO
	Other fluoroquinolone or	according to drug	PO
	Daptomycin or	8–10 mg/kg every 24 h	IV
	Teicoplanin or	400 mg every 24 h	IV/IM
	Fusidic acid or	500 mg every 8 h	PO
	Cotrimoxazole or	One forte tablet every 8 h	PO
	Minocycline	100 mg every 12 h	PO
Streptococcus spp.	Penicillin G or	Five million U every 6 h	IV
	Ceftriaxone	2 g every 24 h	IV
	For 4 weeks, followed by		
	Amoxicillin	750–1,000 mg every 8 h	PO
Enterococcus spp. (penicillin-susceptible)	Penicillin G or	Five million U every 6 h	IV
	Ampicillin or amoxicillin or	2 g every 4–6 h	IV
	Daptomycin plus	8–10 mg/kg every 24 h	IV
	Aminoglycoside[c]	According to drug	IV
	For 2–4 weeks, followed by		
	Amoxicillin	750–1,000 mg every 8 h	PO
Enterococcus spp. (penicillin-resistant)	Vancomycin or	1 g every 12 h	IV
	Daptomycin plus	8–12 mg/kg every 24 h	IV
	Aminoglycoside[c]	According to drug	IV
Enterobacteriaceae (quinolone-susceptible)	Ciprofloxacin	750 mg every 12 h	PO
Nonfermenters (e.g., *Pseudomonas aeruginosa*, *Acinetobacter* spp.)	Cefepime or ceftazidime plus	2 g every 6 h	IV
	aminoglycoside[c]	According to drug	IV
	For 2–4 weeks, followed by		
	Ciprofloxacin	750 mg every 12 h	PO
Anaerobes[d]	Clindamycin	600 mg every 6–8 h	IV
	For 2–4 weeks, followed by		
	Clindamycin	300 mg every 6 h	PO

(continued)

Table 1.2 (continued)

Microorganism	Antimicrobial agent	Dose	Route
Mixed infections (without methicillin-resistant staphylococci)	Amoxicillin/clavulanic acid	2.2 g every 8 h	IV
	or Piperacillin/tazobactam	4.5 g every 8 h	IV
	or		
	Imipenem	500 mg every 6 h	IV
	or		
	Meropenem	1 g every 8 h	IV
	For 2–4 weeks, followed by individual regimens according to antimicrobial susceptibility		

PO orally, *IV* intravenously, *IM* intramuscularly, *forte tablet* trimethoprim 160 mg plus sulfamethoxazole 800 mg

[a]In patients with delayed hypersensitivity, cefazolin (2 g every 8 h IV) can be administered. In patients with immediate hypersensitivity, penicillin should be replaced by vancomycin (1 g every 12 h IV)

[b]Vancomycin trough levels should be monitored before fourth dose and thereafter as appropriate. The target serum concentration of vancomycin should be 15–20 μg/ml. In case of increased vancomycin MIC (≥2 μg/ml), daptomycin is the preferred antibiotic of choice

[c]Aminoglycosides can be administered as single daily dose (except for enterococcal infections where it should be divided in 2–3 daily doses). For enterococci, aminoglycosides are used if they are susceptible to high-level resistance testing

[d]Alternatively, penicillin G (five million units every 6 h IV) or ceftriaxone (2 g every 24 h IV) can be used for gram-positive anaerobes (e.g., *Propionibacterium acnes*), and metronidazole (500 mg every 8 h IV or PO) for gram-negative anaerobes (e.g., *Bacteroides* spp.)

is the use of "Microsystems Technology," also known as MEMS (Micro-Electro-Mechanical Systems) [120]. This technique could potentially be used to perceive bacterial communication, also known as quorum sensing [121]. Quorum sensing is crucial for survival of biofilm bacteria. Early discovering of quorum sensing may allow very early diagnosis of implant associated infection. Interference with quorum sensing by release of antimicrobial agents upon sensing of an early biofilm may prevent clinical infection.

1.7 Conclusions

Any type of artificial implant is highly susceptible to infection, due to an inefficient host defense in the vicinity of an implanted foreign body. Microorganisms rapidly form biofilms on the artificial surface. Such infections resist host defense and most antimicrobial agents. In case of periprosthetic joint infections, successful management requires a combination of surgery and antibiotic treatment. Respecting the principles of a rational algorithm for orthopedic revision surgery results in success rates between 80 and 90 %. In view of the difficult treatment, future research should focus on prevention of implant associated infections.

References

1. Gristina AG. Biomaterial-centered infection: microbial adhesion versus tissue integration. Science. 1987;237:1588–95.
2. Costerton JW, Montanaro L, Arciola CR. Biofilm in implant infections: its production and regulation. Int J Artif Organs. 2005;28:1062–8.
3. Wolcott R, Dowd S. The role of biofilms: are we hitting the right target? Plast Reconstr Surg. 2011;127 Suppl 1:28S–35S.
4. Hall-Stoodley L, Stoodley P. Evolving concepts in biofilm infections. Cell Microbiol. 2009;11:1034–43.
5. Parsek MR, Singh PK. Bacterial biofilms: an emerging link to disease pathogenesis. Annu Rev Microbiol. 2003;57:677–701.
6. Zimmerli W. Tissue cage infection model. In: Zak O, Sande M, editors. Handbook of animal models of infection. London: Academic Press Ltd.; 1999. p. 409–17.
7. Furustrand U, Majic I, Belkhodja Zalila C, et al. Gentamicin improves the activity of daptomycin and vancomycin against Enterococcus faecalis *in vitro* and in an experimental foreign-body infection model. Antimicrob Agents Chemother. 2011;55:4821–7.
8. Darouiche RO. Treatment of infections associated with surgical implants. N Engl J Med. 2004;350:1422–9.
9. Zimmerli W, Trampuz A. Implant-associated infection. In: Bjarnsholt T, Hoiby N, Moser C, Jensen PO, editors. Biofilm infections. 1st ed. Heidelberg: Springer; 2011. p. 69–90.
10. Zimmerli W, Trampuz A, Ochsner PE. Prosthetic-joint infections. N Engl J Med. 2004;351:1645–54.
11. De Man FH, Sendi P, Zimmerli W, et al. Infectiological, functional, and radiographic outcome after revision for prosthetic hip infection according to a strict algorithm. Acta Orthop. 2011;82:27–34.
12. Zimmerli W. Infection and musculoskeletal conditions: prosthetic-joint-associated infections. Best Pract Res Clin Rheumatol. 2006;20:1045–63.
13. Anderson JM, McNally AK. Biocompatibility of implants: lymphocyte/macrophage interactions. Semin Immunopathol. 2011;33:221–33.
14. Anderson JM, Rodriguez A, Chang DT. Foreign body reaction to biomaterials. Semin Immunol. 2008;20:86–100.
15. Peck MK, Dusserre N, Zagalski K, et al. New biological solutions for hemodialysis access. J Vasc Access. 2011;12(3):185–92.
16. Hannah EL, Stevenson KB, Lowder CA, et al. Outbreak of hemodialysis vascular access site infections related to malfunctioning permanent tunneled catheters: making the case for active infection surveillance. Infect Control Hosp Epidemiol. 2002;23:538–41.
17. Chiesa R, Marone EM, Tshomba Y, et al. Aortobifemoral bypass grafting using expanded polytetrafluoroethylene stretch grafts in patients with occlusive atherosclerotic disease. Ann Vasc Surg. 2009;23:764–9.
18. Viola GM, Darouiche RO. Cardiovascular implantable device infections. Curr Infect Dis Rep. 2011;13:333–42.
19. Raber L, Wohlwend L, Wigger M, et al. Five-year clinical and angiographic outcomes of a randomized comparison of sirolimus-eluting and paclitaxel-eluting stents: results of the Sirolimus-Eluting Versus Paclitaxel-Eluting Stents for Coronary Revascularization LATE trial. Circulation. 2011;123:2819–28.
20. Grunwald IQ, Kuhn AL. Controversies around carotid stenting. Acta Chir Belg. 2011;111:63–7.
21. Antonios VS, Baddour LM. Intra-arterial device infections. Curr Infect Dis Rep. 2004;6:263–9.
22. Malone JM, Moore WS, Campagna G, et al. Bacteremic infectability of vascular grafts: the influence of pseudointimal integrity and duration of graft function. Surgery. 1975;78:211–6.
23. Moore WS, Malone JM, Keown K. Prosthetic arterial graft material. Influence on neointimal healing and bacteremic infectibility. Arch Surg. 1980;115:1379–83.

24. Smith CR, Leon MB, Mack MJ, et al. Transcatheter versus surgical aortic-valve replacement in high-risk patients. N Engl J Med. 2011;364:2187–98.
25. Hammermeister KE, Sethi GK, Henderson WG, et al. A comparison of outcomes in men 11 years after heart-valve replacement with a mechanical valve or bioprosthesis. Veterans Affairs Cooperative Study on Valvular Heart Disease. N Engl J Med. 1993;328:1289–96.
26. Hammermeister K, Sethi GK, Henderson WG, et al. Outcomes 15 years after valve replacement with a mechanical versus a bioprosthetic valve: final report of the Veterans Affairs randomized trial. J Am Coll Cardiol. 2000;36:1152–8.
27. Oxenham H, Bloomfield P, Wheatley DJ, et al. Twenty year comparison of a Bjork-Shiley mechanical heart valve with porcine bioprostheses. Heart. 2003;89:715–21.
28. Baddour LM, Epstein AE, Erickson CC, et al. Update on cardiovascular implantable electronic device infections and their management: a scientific statement from the American Heart Association. Circulation. 2010;121:458–77.
29. Del Pozo JL, Patel R. Clinical practice. Infection associated with prosthetic joints. N Engl J Med. 2009;361:787–94.
30. Ong KL, Kurtz SM, Lau E, et al. Prosthetic joint infection risk after total hip arthroplasty in the Medicare population. J Arthroplasty. 2009;24:105–9.
31. Lethaby A, Temple J, Santy J. Pin site care for preventing infections associated with external bone fixators and pins. Cochrane Database Syst Rev. 2008:CD004551.
32. Mueller ME, Allgoewer M, Willenegger H. Mass survey by a study group on osteosynthesis problems. Langenbecks Arch Klin Chir Ver Dtsch Z Chir. 1963;304:808–17.
33. Clauss M, Trampuz A, Borens O, et al. Biofilm formation on bone grafts and bone graft substitutes: comparison of different materials by a standard *in vitro* test and microcalorimetry. Acta Biomater. 2011;6:3791–7.
34. Evans RP. Current concepts for clean air and total joint arthroplasty: laminar airflow and ultraviolet radiation: a systematic review. Clin Orthop Relat Res. 2011;469:945–53.
35. Krenek L, Farng E, Zingmond D, et al. Complication and revision rates following total elbow arthroplasty. J Hand Surg Am. 2011;36:68–73.
36. Trampuz A, Zimmerli W. Diagnosis and treatment of implant-associated septic arthritis and osteomyelitis. Curr Infect Dis Rep. 2008;10:394–403.
37. Sendi P, Frei R, Maurer TB, et al. *Escherichia coli* variants in periprosthetic joint infection: diagnostic challenges with sessile bacteria and sonication. J Clin Microbiol. 2011;48:1720–5.
38. Vaudaux P, François P, Waldvogel F. Host factors predisposing to and influencing therapy of foreign body infections. In: Waldvogel FA, editor. Infections associated with undwelling medical devices. 3rd ed. Washington: ASM Press; 2000. p. 173–205
39. Zimmerli W, Waldvogel FA, Vaudaux P, et al. Pathogenesis of foreign body infection: description and characteristics of an animal model. J Infect Dis. 1982;146:487–97.
40. Zimmerli W, Lew PD, Waldvogel FA. Pathogenesis of foreign body infection. Evidence for a local granulocyte defect. J Clin Invest. 1984;73:1191–200.
41. Kristian SA, Birkenstock TA, Sauder U, et al. Biofilm formation induces C3a release and protects *Staphylococcus epidermidis* from IgG and complement deposition and from neutrophil-dependent killing. J Infect Dis. 2008;197:1028–35.
42. Verhoef J. The phagocytic process and the role of complement in host defense. J Chemother. 1991;3 Suppl 1:93–7.
43. Zimmerli W, Zarth A, Gratwohl A, et al. Neutrophil function and pyogenic infections in bone marrow transplant recipients. Blood. 1991;77:393–9.
44. Kristian SA, Golda T, Ferracin F, et al. The ability of biofilm formation does not influence virulence of *Staphylococcus aureus* and host response in a mouse tissue cage infection model. Microb Pathog. 2004;36:237–45.
45. Zimmerli W, Sendi P. Pathogenesis of implant-associated infection: the role of the host. Semin Immunopathol. 2011;33:295–306.
46. Bernard L, Vaudaux P, Merle C, et al. The inhibition of neutrophil antibacterial activity by ultra-high molecular weight polyethylene particles. Biomaterials. 2005;26:5552–7.

47. Hosman AH, van der Mei HC, Bulstra SK, et al. Effects of metal-on-metal wear on the host immune system and infection in hip arthroplasty. Acta Orthop. 2010;81:526–34.
48. Bohler J, Kramer P, Gotze O, et al. Leucocyte counts and complement activation during pump-driven and arteriovenous haemofiltration. Contrib Nephrol. 1983;36:15–25.
49. Hammerschmidt DE, Craddock PR, McCullough F, et al. Complement activation and pulmonary leukotasis during nylon fiber filtration leukapheresis. Blood. 1978;51:721–30.
50. Shepard AD, Gelfand JA, Callow AD, et al. Complement activation by synthetic vascular prostheses. J Vasc Surg. 1984;1:829–38.
51. Anderson JM. Inflammatory response to implants. ASAIO Trans. 1988;34:101–7.
52. Tang L, Liu L, Elwing HB. Complement activation and inflammation triggered by model biomaterial surfaces. J Biomed Mater Res. 1998;41:333–40.
53. Noordin S, Shortkroff S, Sledge CB, et al. Investigation of the activation of a human serum complement protein, C3, by orthopedic prosthetic particulates. Biomaterials. 2004;25:5347–52.
54. Elek SD, Conen PE. The virulence of *Staphylococcus pyogenes* for man; a study of the problems of wound infection. Br J Exp Pathol. 1957;38:573–86.
55. Widmer AF, Frei R, Rajacic Z, et al. Correlation between *in vivo* and *in vitro* efficacy of antimicrobial agents against foreign body infections. J Infect Dis. 1990;162:96–102.
56. Vaudaux PE, Zulian G, Huggler E, et al. Attachment of *Staphylococcus aureus* to polymethylmethacrylate increases its resistance to phagocytosis in foreign body infection. Infect Immun. 1985;50:472–7.
57. Borlee BR, Goldman AD, Murakami K, et al. Pseudomonas aeruginosa uses a cyclic-di-GMP-regulated adhesin to reinforce the biofilm extracellular matrix. Mol Microbiol. 2010;75: 827–42.
58. Lindsay D, von Holy A. Bacterial biofilms within the clinical setting: what healthcare professionals should know. J Hosp Infect. 2006;64:313–25.
59. Widmer AF, Colombo VE, Gachter A, et al. Salmonella infection in total hip replacement: tests to predict the outcome of antimicrobial therapy. Scand J Infect Dis. 1990;22:611–8.
60. Kurtz S, Ong K, Lau E, et al. Projections of primary and revision hip and knee arthroplasty in the United States from 2005 to 2030. J Bone Joint Surg Am. 2007;89:780–5.
61. Fevang BT, Lie SA, Havelin LI, et al. Risk factors for revision after shoulder arthroplasty: 1,825 shoulder arthroplasties from the Norwegian Arthroplasty Register. Acta Orthop. 2009;80:83–91.
62. Kim BS, Knupp M, Zwicky L, et al. Total ankle replacement in association with hindfoot fusion: outcome and complications. J Bone Joint Surg Br. 2010;92:1540–7.
63. Achermann Y, Vogt M, Spormann C, et al. Characteristics and outcome of 27 elbow periprosthetic joint infections: results from a 14-year cohort study of 358 elbow prostheses. Clin Microbiol Infect. 2011;17:432–8.
64. Lüem M. Patient population. In: Ochsner PE, editor. Total hip replacement: implantation technique and local complications. Berlin: Springer; 2003. p. 5–13.
65. Steckelberg JM, Osmon DR. Prosthetic joint infection. In: Waldvogel FA, ed. Infections associated with indwelling medical devices. 3rd ed. Washington, DC: ASM Press; 2000. p. 173–205.
66. Trampuz A, Hanssen AD, Osmon DR, et al. Synovial fluid leukocyte count and differential for the diagnosis of prosthetic knee infection. Am J Med. 2004;117:556–62.
67. Schinsky MF, Della Valle CJ, Sporer SM, et al. Perioperative testing for joint infection in patients undergoing revision total hip arthroplasty. J Bone Joint Surg Am. 2008;90:1869–75.
68. Osmon D, Berbari E, Berendt A, et al. Diagnosis and management of prosthetic joint infection: clinical practice guidelines by the infectious diseases society of America. Clin Infect Dis. 2012 (in press)
69. Parvizi J, Jacovides C, Zmistowski B, et al. Definition of periprosthetic joint infection: is there a consensus? Clin Orthop Relat Res. 2011;469(11):3022–30.
70. Maderazo EG, Judson S, Pasternak H. Late infections of total joint prostheses. A review and recommendations for prevention. Clin Orthop Relat Res. 1988;229:131–42.

71. Murdoch DR, Roberts SA, Fowler Jr VG, et al. Infection of orthopedic prostheses after *Staphylococcus aureus* bacteremia. Clin Infect Dis. 2001;32:647–49.
72. Lalani T, Chu VH, Grussemeyer CA, et al. Clinical outcomes and costs among patients with *Staphylococcus aureus* bacteremia and orthopedic device infections. Scand J Infect Dis. 2008;40:973–7.
73. Sendi P, Banderet F, Graber P, et al. Periprosthetic joint infection following *Staphylococcus aureus* bacteremia. J Infect. 2011;63:17–22.
74. Zimmerli W, Zak O, Vosbeck K. Experimental hematogenous infection of subcutaneously implanted foreign bodies. Scand J Infect Dis. 1985;17:303–10.
75. Berbari EF, Osmon DR, Carr A, et al. Dental procedures as risk factors for prosthetic hip or knee infection: a hospital-based prospective case-control study. Clin Infect Dis. 2010;50:8–16.
76. Zimmerli W, Sendi P. Antibiotics for prevention of periprosthetic joint infection following dentistry: time to focus on data. Clin Infect Dis. 2010;50:17–9.
77. Pandey R, Berendt AR, Athanasou NA. Histological and microbiological findings in non-infected and infected revision arthroplasty tissues. The OSIRIS Collaborative Study Group. Oxford Skeletal Infection Research and Intervention Service. Arch Orthop Trauma Surg. 2000;120:570–4.
78. Segawa H, Tsukayama DT, Kyle RF, et al. Infection after total knee arthroplasty. A retrospective study of the treatment of eighty-one infections. J Bone Joint Surg Am. 1999;81:1434–45.
79. Marculescu CE, Cantey JR. Polymicrobial prosthetic joint infections: risk factors and outcome. Clin Orthop Relat Res. 2008;466:1397–404.
80. Wilson MG, Kelley K, Thornhill TS. Infection as a complication of total knee-replacement arthroplasty. Risk factors and treatment in sixty-seven cases. J Bone Joint Surg Am. 1990;72:878–83.
81. Salvati EA, Robinson RP, Zeno SM, et al. Infection rates after 3175 total hip and total knee replacements performed with and without a horizontal unidirectional filtered air-flow system. J Bone Joint Surg Am. 1982;64:525–35.
82. Fitzgerald Jr RH, Nolan DR, Ilstrup DM, et al. Deep wound sepsis following total hip arthroplasty. J Bone Joint Surg Am. 1977;59:847–55.
83. Rand JA, Fitzgerald Jr RH. Diagnosis and management of the infected total knee arthroplasty. Orthop Clin North Am. 1989;20:201–10.
84. Wymenga AB, van Horn JR, Theeuwes A, et al. Perioperative factors associated with septic arthritis after arthroplasty. Prospective multicenter study of 362 knee and 2,651 hip operations. Acta Orthop Scand. 1992;63:665–71.
85. Berbari EF, Hanssen AD, Duffy MC, et al. Risk factors for prosthetic joint infection: case-control study. Clin Infect Dis. 1998;27:1247–54.
86. Kessler B, Sendi P, Graber P et al. Risk factors for periprosthetic ankle joint infection: a hospital-based case-control study. J Bone Joint Surg Am. 2012 (in revision).
87. Berbari E, Mabry T, Tsaras G, et al. Inflammatory blood laboratory levels as markers of prosthetic joint infection: a systematic review and meta-analysis. J Bone Joint Surg Am. 2010;92:2102–9.
88. Piper KE, Fernandez-Sampedro M, Steckelberg KE, et al. C-reactive protein, erythrocyte sedimentation rate and orthopedic implant infection. PLoS One. 2010;5:e9358.
89. Tigges S, Stiles RG, Roberson JR. Appearance of septic hip prostheses on plain radiographs. AJR. 1994;163:377–80.
90. van der Bruggen W, Bleeker-Rovers CP, Boerman OC, et al. PET and SPECT in osteomyelitis and prosthetic bone and joint infections: a systematic review. Semin Nucl Med. 2010;40:3–15.
91. Filippi L, Schillaci O. Usefulness of hybrid SPECT/CT in 99mTc-HMPAO-labeled leukocyte scintigraphy for bone and joint infections. J Nucl Med. 2006;47:1908–13.
92. Graute V, Feist M, Lehner S, et al. Detection of low-grade prosthetic joint infections using 99mTc-antigranulocyte SPECT/CT: initial clinical results. Eur J Nucl Med Mol Imaging. 2010;37:1751–9.
93. Strobel K, Stumpe KD. PET/CT in musculoskeletal infection. Semin Musculoskelet Radiol. 2007;11:353–64.

94. Kwee TC, Kwee RM, Alavi A. FDG-PET for diagnosing prosthetic joint infection: systematic review and metaanalysis. Eur J Nucl Med Mol Imaging. 2008;35:2122–32.
95. Love C, Marwin SE, Palestro CJ. Nuclear medicine and the infected joint replacement. Semin Nucl Med. 2009;39:66–78.
96. Bori G, Soriano A, Garcia S, et al. Low sensitivity of histology to predict the presence of microorganisms in suspected aseptic loosening of a joint prosthesis. Mod Pathol. 2006;19:874–7.
97. Atkins BL, Athanasou N, Deeks JJ, et al. Prospective evaluation of criteria for microbiological diagnosis of prosthetic-joint infection at revision arthroplasty. The OSIRIS Collaborative Study Group. J Clin Microbiol. 1998;36:2932–9.
98. Zappe B, Graf S, Ochsner PE, et al. Propionibacterium spp. in prosthetic joint infections: a diagnostic challenge. Arch Orthop Trauma Surg. 2008;128:1039–46.
99. Trampuz A, Piper KE, Jacobson MJ, et al. Sonication of removed hip and knee prostheses for diagnosis of infection. N Engl J Med. 2007;357:654–63.
100. Trampuz A, Zimmerli W. Prosthetic joint infections: update in diagnosis and treatment. Swiss Med Wkly. 2005;135:243–51.
101. Zimmerli W, Widmer AF, Blatter M, et al. Role of rifampin for treatment of orthopedic implant-related staphylococcal infections: a randomized controlled trial. Foreign-Body Infection (FBI) Study Group. JAMA. 1998;279:1537–41.
102. Byren I, Bejon P, Atkins BL, et al. One hundred and twelve infected arthroplasties treated with 'DAIR' (debridement, antibiotics and implant retention): antibiotic duration and outcome. J Antimicrob Chemother. 2009;63:1264–71.
103. Giulieri SG, Graber P, Ochsner PE, et al. Management of infection associated with total hip arthroplasty according to a treatment algorithm. Infection. 2004;32:222–8.
104. Laffer RR, Graber P, Ochsner PE, et al. Outcome of prosthetic knee-associated infection: evaluation of 40 consecutive episodes at a single centre. Clin Microbiol Infect. 2006;12:433–9.
105. Sendi P, Christensson B, Uckay I, et al. Group B streptococcus in prosthetic hip and knee joint-associated infections. J Hosp Infect. 2011;79:64–9.
106. Sendi P, Rohrbach M, Graber P, et al. *Staphylococcus aureus* small colony variants in prosthetic joint infection. Clin Infect Dis. 2006;43:961–7.
107. Stengel D, Bauwens K, Sehouli J, et al. Systematic review and meta-analysis of antibiotic therapy for bone and joint infections. Lancet Infect Dis. 2001;1:175–88.
108. Zimmerli W, Frei R, Widmer AF, et al. Microbiological tests to predict treatment outcome in experimental device-related infections due to *Staphylococcus aureus*. J Antimicrob Chemother. 1994;33:959–67.
109. Widmer AF, Wiestner A, Frei R, et al. Killing of nongrowing and adherent *Escherichia coli* determines drug efficacy in device-related infections. Antimicrob Agents Chemother. 1991;35:741–6.
110. Schwank S, Rajacic Z, Zimmerli W, et al. Impact of bacterial biofilm formation on *in vitro* and *in vivo* activities of antibiotics. Antimicrob Agents Chemother. 1998;42:895–8.
111. Baldoni D, Haschke M, Rajacic Z, et al. Linezolid alone or combined with rifampin against methicillin-resistant *Staphylococcus aureus* in experimental foreign-body infection. Antimicrob Agents Chemother. 2009;53:1142–8.
112. John AK, Baldoni D, Haschke M, et al. Efficacy of daptomycin in implant-associated infection due to methicillin-resistant *Staphylococcus aureus*: importance of combination with rifampin. Antimicrob Agents Chemother. 2009;53:2719–24.
113. Trampuz A, Murphy CK, Rothstein DM, et al. Efficacy of a novel rifamycin derivative, ABI-0043, against *Staphylococcus aureus* in an experimental model of foreign-body infection. Antimicrob Agents Chemother. 2007;51:2540–5.
114. Osmon DR, Berbari EF. Outpatient intravenous antimicrobial therapy for the practicing orthopaedic surgeon. Clin Orthop Relat Res. 2002;403:80–6.
115. Batoni G, Maisetta G, Brancatisano FL, et al. Use of antimicrobial peptides against microbial biofilms: advantages and limits. Curr Med Chem. 2010;18:256–79.
116. Gao G, Lange D, Hilpert K, et al. The biocompatibility and biofilm resistance of implant coatings based on hydrophilic polymer brushes conjugated with antimicrobial peptides. Biomaterials. 2011;32:3899–909.

117. Ketonis C, Barr S, Adams CS, et al. Bacterial colonization of bone allografts: establishment and effects of antibiotics. Clin Orthop Relat Res. 2010;468:2113–21.
118. Masse A, Bruno A, Bosetti M, et al. Prevention of pin track infection in external fixation with silver coated pins: clinical and microbiological results. J Biomed Mater Res. 2000;53:600–4.
119. Khare MD, Bukhari SS, Swann A, et al. Reduction of catheter-related colonisation by the use of a silver zeolite-impregnated central vascular catheter in adult critical care. J Infect. 2007; 54:146–50.
120. Wang CH, Lee GB. Automatic bio-sampling chips integrated with micro-pumps and micro-valves for disease detection. Biosens Bioelectron. 2005;21:419–25.
121. Coelho LR, Souza RR, Ferreira FA, et al. agr RNAIII divergently regulates glucose-induced biofilm formation in clinical isolates of *Staphylococcus aureus*. Microbiology. 2008; 154:3480–90.

Chapter 2
Staphylococcus epidermidis in Biomaterial-Associated Infections

Dietrich Mack, Angharad P. Davies, Llinos G. Harris, Rose Jeeves, Ben Pascoe, Johannes K.-M. Knobloch, Holger Rohde, and Thomas S. Wilkinson

Abstract Coagulase-negative staphylococci, mainly *Staphylococcus epidermidis*, are currently the most frequent cause of hospital acquired infections in the USA. Mostly, but not exclusively, *S. epidermidis* infections are linked to the use of implanted medical devices like central venous catheters, prosthetic joints and heart valves, pacemakers, cardiac assist devices, cerebrospinal fluid shunts, and intraocular lenses. As new molecular techniques reveal that *S. epidermidis* are by no means the most prominent bacteria of the skin and mucous membrane flora, the implication is that *S. epidermidis* has specific virulence factors, which transforms this commensal bacterial species into one of the most successful pathogens in modern medicine. A vast array of specific attachment factors for native and host protein-modified device surfaces and the ability to accumulate in adherent multilayered biofilms appear to be vital for the success of *S. epidermidis* as a pathogen. Biofilm formation contributes to the ability of the organism to withstand the host's innate and acquired immune defense mechanisms and to resist antimicrobial therapy, so that device removal is a regular feature for the treatment of *S. epidermidis* biomaterial-associated infection. Recent developments in the understanding of *S. epidermidis* virulence are reviewed in this chapter.

D. Mack, M.D., F.R.C. Path.(✉) • A.P. Davies • L.G. Harris • R. Jeeves • B. Pascoe • T.S. Wilkinson,
Medical Microbiology and Infectious Diseases, Institute of Life Science,
College of Medicine, Swansea University, Swansea SA2 8PP, UK
e-mail: D.Mack@swansea.ac.uk

J.K.-M. Knobloch
Institut für Medizinische Mikrobiologie und Hygiene, Universität zu Lübeck, Lübeck, Germany

H. Rohde
Institut für Medizinische Mikrobiologie, Virologie und Hygiene, Zentrum für Klinische Pathologie, Universitäts-Klinikum Hamburg-Eppendorf, Hamburg, Germany

T.F. Moriarty et al. (eds.), *Biomaterials Associated Infection: Immunological Aspects and Antimicrobial Strategies*, DOI 10.1007/978-1-4614-1031-7_2,

2.1 Introduction

Medical progress in recent decades is closely linked with the ever-increasing use of implanted biomedical devices, which allow monitoring of vital functions and administration of drugs to critically ill patients, and intermittently or permanently support or replace failing organ functions. These implants usually are highly beneficial to the individual patient; however, their use is linked to complications and one of the most important of these is infection. Biomaterial-associated infections have a number of interesting features, which make them special. First, they are most commonly caused by coagulase-negative staphylococci (mostly *Staphylococcus epidermidis*), *Staphylococcus aureus*, and also other bacteria of low virulence potential for otherwise healthy individuals, which reside on skin and mucous membranes. Second, they frequently take a chronic persistent course indicating that the innate and acquired immune system does not deal with the causative organisms in an effective way. And third, antimicrobial therapy is regularly unsuccessful due to phenotypic resistance of the causative organisms, while the isolated microbes are susceptible when tested under standard planktonic laboratory conditions. This makes removal of the device a frequent necessity. The phenomenon of phenotypic resistance is frequently attributed to impaired penetration of antimicrobials through the biofilm; however, it is more likely to be an altered metabolic state referred to as tolerance, occurring in bacterial persisters [1].

Currently, it is thought that the formation of adherent multilayered biofilms is of genuine importance in the pathogenesis of biomaterial-associated infections. Biofilms are complex consortia of adherent microorganisms encased in a polymeric matrix [2]. Biofilm formation can be separated into four distinct phases. First, primary attachment of cells to a surface; second, accumulation of microorganisms in multiple layers; followed third by maturation of the biofilm; and fourth, detachment of single cells or groups of cells, which may start the whole process of biofilm formation elsewhere [3].

Coagulase-negative staphylococci, in particular *S. epidermidis*, are prototypic bacteria forming biofilms important in infection.

2.2 Clinical Importance of Coagulase-Negative Staphylococci

The major risk factor for infection with coagulase-negative staphylococci is the presence of implanted biomedical devices like central venous catheters, prosthetic joints, fracture fixation devices, cardiac pacemakers and heart valves, artificial lenses, vascular grafts, mammary implants, and CSF-shunts [4–6]. In Germany alone more than 2.5 million of these biomedical devices are used annually [5]. A major complication of their use is infection, affecting up to 100,000 patients annually in Germany. Similar figures were reported for other industrialized countries like the USA [6] indicating that millions of patients are at risk worldwide. Additional patient-related risk factors for infection with coagulase-negative staphylococci are malignancy, chemotherapy, leukopenia, premature birth, care in an ICU, bone marrow transplantation, and immunosuppression for reasons such as polytrauma, HIV infection, and transplantation [4, 5, 7].

According to data reported to the National Health Care Safety Network, coagulase-negative staphylococci today are the most frequent cause of device- or surgery-associated hospital-acquired infection in the USA [8]. Coagulase-negative staphylococci were ranked first for all infections (15.3%) and for central line associated blood stream infections (CLABSI; 34.1%) and second for surgical site infections (13.7%) [8]. Similarly, coagulase-negative staphylococci were the most common cause of CVC-associated blood-stream infection (32.1%) reported from German ICUs to the hospital infection surveillance system KISS [9] and for CLABSI reported from ICUs to the provincial hospital infection surveillance program SPIN in Quebec, Canada (53%) [10]. Incidence of CLABSI varies from 0.9 to 1.9/1,000 central line days [9] and 1.67 to 4.4/1,000 central line days depending on type of ICU reporting [10]. In the 1990s 50,000–120,000 patients were estimated to contract nosocomial catheter-related blood stream infections annually in the USA [11, 12]. More recently CLABSI episodes associated with US ICUs were estimated to have decreased from 47,000 in 2001 to 18,000 in 2009, which was mainly associated with a decrease in *S. aureus* infections and a more modest decrease in infections with Gram-negative rods, *Candida* spp., and *Enterococcus* spp. [13]. In 2009 an estimated 23,000 CLABSIs occurred among patients from inpatient wards and in 2008 an estimated 37,000 CLABSIs occurred in patients receiving outpatient hemodialysis [13], indicating that coagulase-negative staphylococci are pathogens of major importance in CLABSI-associated morbidity. Up to 80% of coagulase-negative staphylococci isolated from blood cultures are *S. epidermidis* while the remainder represent other coagulase-negative staphylococcal species, most frequently *Staphylococcus haemolyticus*, *Staphylococcus hominis*, *Staphylococcus capitis*, *Staphylococcus saprophyticus*, *Staphylococcus warneri*, *Staphylococcus cohnii*, *Staphylococcus xylosus*, *Staphylococcus chromogenes*, *Staphylococcus schleiferi*, or *Staphylococcus lugdunensis* [5, 14]. CLABSI is associated with a significant mortality of 12–25% [13].

Coagulase-negative staphylococci (11%) are the third most common cause of infective endocarditis after *S. aureus* (31%) and viridans streptococci (17%) and the second most common cause of prosthetic valve associated infective endocarditis (17%) and other intracardiac devices (26%) after *S. aureus* with 23% and 35%, respectively [15]. Eighty-two percent of coagulase-negative staphylococci causing prosthetic valve infective endocarditis were *S. epidermidis* [16]. However, despite the predilection of coagulase-negative staphylococcal infection for implanted biomaterials it emerges that these organisms also cause severe cases of native valve infective endocarditis particularly in nondrug abusers [15]. Almost 25% of cases of native valve infective endocarditis due to staphylococci were caused by coagulase-negative staphylococci [15], with a mortality (19%) not statistically significantly different from that of *S. aureus* infective endocarditis [17].

Eighty percent of coagulase-negative staphylococcal isolates causing native-valve endocarditis were *S. epidermidis* [18]. *S. lugdunensis* is of particular importance in infective endocarditis with a high associated morbidity and mortality (50%) in excess of *S. aureus* [19].

Cardiac implantable electronic devices—pacemakers and implanted defibrillators—pose a risk of infection, which are caused by coagulase-negative staphylococci in 27% of cases, the second most common cause following *S. aureus* (34%) [20]. *S. epidermidis* (38%) was the cause of bloodstream infection related with cardiac assist devices followed by *S. aureus* (24%) [21, 22], while in another large series *S. aureus* (34%) was more frequent than coagulase-negative staphylococci (27%) [20].

Cerebrospinal fluid (CSF) shunts had a high infection rate of 11.7% per patient and 7.2% per procedure in a recent US survey of 7,071 children over a 5 year period, where the aggregated infection rates per hospital varied widely from 4.1 to 20.5% per patient and 2.5–12.3% per procedure [23]. A similarly wide range of hospital infection rates were reported from a Canadian CSF shunt infection surveillance system with a mean infection rate of 4.1% per procedure ranging from 0 to 11.1% in different hospitals [24]. Coagulase-negative staphylococci remain the most prominent cause of CSF-shunt infection, causing 37% of cases, followed by *S. aureus* (18%) in a series from Switzerland [25].

Prosthetic joint infections represent a major burden as the infected joint usually has to be removed, which causes additional major morbidity for the affected patients. The United Kingdom National Joint Registry reports that numbers of total hip and knee replacement procedures have increased from approximately 130,000 in 2005–2006 to almost 179,000 in 2010–2011 in England and Wales [26]. In 2010–2011 there were 7,833 revision hip arthroplasty procedures: 45% due to aseptic loosening and 13% due to infection; and 5,082 knee revision procedures: 33% due to aseptic loosening and 23% due to infection [26]. Deep infection rates reported from a single center in the UK were 0.57% for hip and 0.86% for knee replacements [27]. The Norwegian Joint Registry found a threefold increased risk for revision hip replacements due to deep infection in the time period from 1987 to 1992 as compared to 2003–2007. This increased risk for deep infections was particularly pronounced in uncemented hip replacements where a 5.3-fold risk was observed for the later time period as compared to the 1987–1992 period [28]. In a Medicare patient population, covered by the US government health insurance for people aged 65 or over, the incidence of infection in hip replacements was 1.63% within 2 years and 0.59% from years 2 to 10 [29], while in total knee replacements it was 1.55% within 2 years and 0.46% between years 2 and 10 [30]. In hip and knee replacement infections in the UK coagulase-negative staphylococci were the most frequently observed causative organisms in 36% and 49% of cases compared to 29% and 13% *S. aureus*, respectively [27, 31]. In another study of periprosthetic joint infections coagulase-negative staphylococci accounted for 54.5% of organisms, while *S. aureus* was third with 10.9% after *Propionibacterium acnes* with 13.6% [32]. In a study evaluating proven staphylococcal prosthetic hip and knee joint infections, 77%

S. epidermidis, 23% *S. aureus*, 6.3% *S. capitis*, 5.1% *S. lugdunensis*, and 1.3% *S. haemolyticus* and *S. warneri*, respectively, were isolated [33, 34].

2.3 Coagulase-Negative Staphylococci as Skin Colonizers

Culture-based studies have identified coagulase-negative staphylococci as the most frequently found organisms of the normal skin flora [35]. The most prevalent and persistent species are *S. epidermidis* and *S. hominis* [36, 37]. Quantitative studies indicated that staphylococci usually made up 50% of the bacteria isolated from the head, nares and axillae and 10–70% of those isolated from legs and arms [38]. *S. epidermidis* represented 90–100% of the staphylococci isolated from the nares of individuals who were not carriers of *S. aureus*, whilst in *S. aureus* carriers *S. epidermidis* still made up 10–20% [38]. A link was recently found between *S. epidermidis* expressing Esp, a *S. epidermidis* serine protease similar to *S. aureus* V8 serine protease, and *S. aureus* noncarrier state [39]. This is reported to be due to interference through Esp with *S. aureus* biofilm formation [39], which could explain the high carrier rate of *S. epidermidis* in non-*S. aureus* carriers. *S. epidermidis* comprises greater than 75% of staphylococci from axillae and head and usually 10–45% of those isolated from legs and arms [38].

More recently skin microbiota from humans have been studied using 16S rRNA gene sequencing with sampling of different skin sites [40]. Superficial forearm skin samples revealed as leading genera in all six individuals studied *Propionibacterium* (22% of all clones), *Corynebacterium* (19% of all clones), and *Staphylococcus* (11.1% of all clones) [41]. However, *S. epidermidis* represented only 2.9% of all clones, while *P. acnes* was represented by 20.6% of all clones [41]. The most abundant staphylococcal species was *S. caprae* (3.5%), followed by *S. epidermidis* (2.9%) and *S. hominis* (1.6%) [41]. Similarly, analyzing skin samples from the inner elbow of five human subjects *S. epidermidis* and *P. acnes* comprised <5% of all represented 16S rRNA gene sequences [42]. In a larger study of 20 distinct skin sites in ten different individuals *Staphylococcus* spp. represented 16.8% of sequences of 205 genera present with more than three sequences [43]. Dry skin sites, e.g., volar forearms, hypothenar palm, and buttock, comprised 4.8% *Staphylococcus* spp. sequences, sebaceous sites including, e.g., glabella, alar crease, external auditory canal, back of scalp, manubrium and back 16.2% *Staphylococcus* spp. sequences, and moist sites, e.g., nares, axilla, antecubital fossa, inguinal crease, popliteal fossa, and umbilicus 21.2% *Staphylococcus* spp. sequences, respectively [43].

In contrast to the understanding of coagulase-negative staphylococci as accidental pathogens [44] in prosthetic-device associated infections despite low virulence potential, simply owing to their great abundance and frequent contamination of the device during implantation [45–47], the newer in-depth understanding of the skin microbiota would suggest that coagulase-negative staphylococci, in particular *S. epidermidis*, have specific virulence factors which make them successful pathogens in the scenario of biomaterial-associated infections.

2.4 Biofilm Accumulation

Biofilm accumulation in *S. epidermidis* is affected by intercellular adhesins, the predominant one being polysaccharide intercellular adhesin (PIA) [48–51] (Fig. 2.1). However, not all *S. epidermidis* strains identified in biomaterial-related infections are able to produce PIA or possess the gene locus *icaADBC* encoding synthetic machinery for PIA synthesis [52, 53]. An alternative, PIA-independent accumulation mechanism was therefore suspected [33, 48–50, 54–56], and eventually discovered in accumulation associated protein, Aap [57–59], biofilm associated protein Bap [60], extracellular matrix binding protein Embp [61, 62], and *S. epidermidis* SesC [63]. Additional proteinaceous intercellular adhesins may also exist [33].

2.4.1 Polysaccharide Intercellular Adhesin

PIA synthesis is probably a general mechanism employed by a diverse group of Gram-positive and Gram-negative eubacteria for biofilm formation [64]. It is

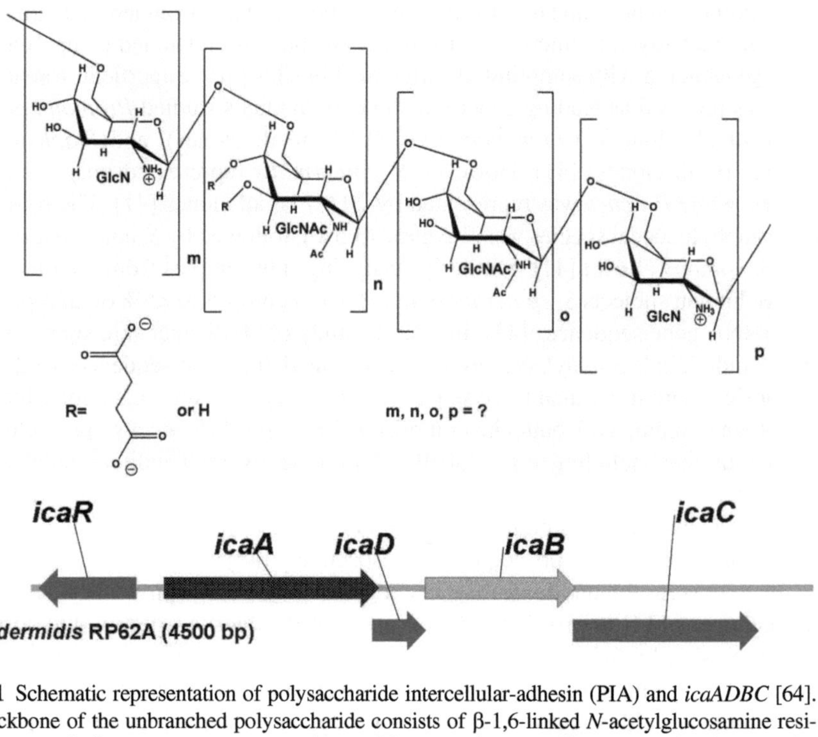

Fig. 2.1 Schematic representation of polysaccharide intercellular-adhesin (PIA) and *icaADBC* [64]. The backbone of the unbranched polysaccharide consists of β-1,6-linked *N*-acetylglucosamine residues. *m, n, o, p* indicate numerous sugar residues of the respective type. Chromosomal region containing *icaADBC* involved in PIA synthesis of *S. epidermidis* RP62A (4,500 bp). IcaAD have glycosyltransferase activity, IcaC is necessary for synthesis of full length PIA, and IcaB is the deacetylase [53, 116]. IcaR is a transcriptional regulator controlling in part *icaADBC* transcription [94]

certainly the major functional component for intercellular adhesion in *S. epidermidis* biofilms [48–50]. The *icaADBC* locus encodes enzymes for PIA synthesis in *S. epidermidis* [52, 53] and is also found in *S. aureus* [65, 66], *S. caprae* [67], and *S. lugdunensis* [68]: homologous DNA sequences have been detected in other coagulase-negative staphylococci [65, 69]. Other human pathogens such as *Escherichia coli* [70], *Aggregatibacter actinomycetemcomitans* [71], *Actinobacillus pleuropneumoniae* [71], *Yersinia pestis* [72–75], *Acinetobacter baumannii* [76], and *Bordetella spp.* [77, 78] have all been found to possess gene loci (*pgaABCD* or *hmsHFRS*) orthologous to *icaADBC,* which are involved in synthesis of polysaccharides used in biofilm formation.

2.4.1.1 Structure of PIA and Related Polysaccharides

The structural analysis of polysaccharides involved in biofilm formation of *S. epidermidis* was facilitated by the isolation of isogenic biofilm-negative transposon insertion mutants from biofilm-positive wild-type strains. Prior to that, difficulties had arisen with components from media used for bacterial culture being purified and analyzed, with misleading results ([79], for a review see also [80, 81]): however, comparison of extracts of wild-type and mutant bacteria circumvented this problem [49, 82]. The structural analysis of PIA and related polysaccharides from staphylococci and other bacteria like *Escherichia coli* has been recently reviewed [64, 83].

The structure of PIA was first described for biofilm-forming *S. epidermidis* 1457 and RP62A. PIA was extracted from the cells by sonication after the strains had been cultured in trypticase soy broth, which revealed the existence of both a major polysaccharide I (>80%), and a minor polysaccharide II (<20%) [82]. Polysaccharide I did not bind to Q-Sepharose, whilst polysaccharide II was moderately anionic. Chemical analyses and NMR spectroscopy have demonstrated that polysaccharide I is a linear homoglycan of β-1, 6-linked 2-amino-2-deoxy-D-glucopyranosyl residues. Approximately 80–85% of them are *N*-acetylated; the rest are non-*N*-acetylated and carry a positive charge. Cation exchange chromatography was used to identify molecular species with non-*N*-acetylated glucosaminyl residue content ranging between 2 and 26%. These were shown by chain cleavage by deamination with HNO_2 to be randomly distributed: glucosaminyl-rich sequences were somewhat predominant. Polysaccharide II, whilst structurally related to polysaccharide I, has a lower proportion of non-*N*-acetylated D-glucosaminyl residues. It was shown by colorimetric assay to contain ester-linked succinate and phosphate (see below), and is therefore anionic [49, 82]. Despite a high apparent molecular weight indicated by elution in the void volume of Sephadex G200 [49, 82] or Sephacryl S300 columns (C. Fischer and D. Mack, unpublished results), the ratio of reducing terminal sugar residues to total sugar residues was shown by methylation analysis to be 1:130, implying an average M_r of 30,000 for PIA polysaccharide chains [82]. This implies aggregation of PIA polysaccharide chains in solution. PIA was shown to function also as the hemagglutinin of *S. epidermidis* [84–86].

NMR spectroscopy and chemical analysis have been used for detailed analysis of polysaccharide produced by biofilm-producing *S. aureus* MN8m, which was referred to by Joyce et al. as *Staphylococcus aureus* exopolysaccharide (SAE) [87]. That study confirmed a basic structure of β-1,6-linked *N*-acetylglucosamine (GlcNAc) homopolymer, with presence of *O*-succinoylation at the 3- and 4-hydroxyl groups of the GlcNAc residues being revealed for the first time by means of NMR spectroscopy [87]. However, no phosphate was detected in SAE [87]. In comparison to PIA preparations from *S. epidermidis* 1457, which contained only 15–20% non-*N*-acetylated glucosamine residues, SAE from *S. aureus* MN8m had a higher proportion at 43% [82, 87]. The content of ester-linked succinate was comparable, at 8–9% for the *S. aureus* MN8m polysaccharide versus 6% for the *S. epidermidis* 1457 polysaccharide. Size-exclusion chromatography was used to reveal an apparent molecular weight of 346,000 for SAE. SAE was active in *in vitro* hemagglutination assays [87]. In conclusion, these studies demonstrate that *S. aureus* MN8m produces a polymer that is closely related both chemically and biologically to *S. epidermidis* PIA [82, 87].

The structure of PIA polysaccharide produced by *S. epidermidis* RP62A and *S. aureus* MN8m has now been independently investigated using a variety of different culture conditions prior to sonication for extraction of the polysaccharide [88]. Purification and NMR spectroscopy confirmed the basic structure of PIA and direct NMR-evidence for the presence of *O*-succinoylation in PIA from *S. epidermidis* RP62A was obtained [88]. The ratios of neutral and negatively charged PIA I and PIA II were similar to those previously observed for *S. epidermidis* 1457 [82]. It is worth noting, in particular, that Sadovskaya and colleagues compared the PIA and SAE extracted from *S. epidermidis* 1457 and *S. aureus* MN8m using identical conditions and NMR spectroscopy [88]. As well as similar elution profiles this revealed virtually identical ^{1}H NMR spectra. A higher proportion of charged groups, succinoyl residues and free amino groups in SAE was indicated by more pronounced peaks at 5.1, 2.7, and 3.0 ppm, in comparison to PIA from *S. epidermidis* RP62A [88]. No evidence for the presence of polysaccharide-bound phosphate was obtained when 0.9% NaCl was used instead of phosphate buffered saline in the purification process, with the implication that the initial observation of the presence of phosphate may have been due to carryover of buffer [82, 88].

PIA has also been referred to by other names including PS/A [89, 90], PNSG [66], PNAG [91], and SAE [87]. To summarize, PIA is a homoglycan of β-1,6-linked *N*-acetylglucosamine, with a fraction of free 2-amino groups (no *N*-acetylation) conferring positive charges and *O*-succinoyl ester residues conferring negative charges. In spite of variations in the degree of nonacetylated, free amino groups, *O*-succinoylation, and possibly molecular size, PIA, P/SA, SAE, and PNAG are now generally accepted to represent the same chemical entity [64, 81, 92, 93].

2.4.1.2 Biosynthesis of PIA and Related Polysaccharides

The production of PIA is controlled by the *icaADBC* locus of *S. epidermidis* [52], an operon with four open reading frames: *icaA*, *icaD*, *icaB*, and *icaC* [53]. *icaR* is found upstream of the *icaA* start codon and encodes a 185 amino acid (aa) *tetR*-type transcriptional regulator which is proposed to have a helix-turn-helix DNA binding motif at the *N*-terminus [94]. IcaR negatively controls *icaADBC* transcription, which is also influenced by TcaR [94, 95]. It is part of a regulatory network governing transcription of *icaADBC* including the alternative sigma factor σB, *sarA*, *sarX*, *sarZ*, *ygs*, *gpdS*, *spx*, ClpP, and quorum sensing regulator *luxS* [94, 96–108]. Regulation of *icaADBC* expression is beyond the scope of this article but is discussed in several reviews [44, 109–113]. It is known to differ significantly between *S. aureus* and *S. epidermidis* [114].

Production of a functionally active PIA molecule requires expression of all four *icaADBC* genes [93]. The process has been the subject of detailed study in recombinant strains of *S. carnosus* which expressed different combinations of the *icaADBC* genes and with UDP-GlcNAc as a sugar donor [53]. IcaA belongs to the glycosyltransferase 2 family. It is an integral membrane protein with 412 aa and 4 predicted transmembrane domains [52, 53, 115], and directs the synthesis of β-1,6-linked GlcNAc oligosaccharides of up to 20 GlcNAc units. IcaD is required for full activity of IcaA *in vitro*. It is a 101 aa integral membrane protein with two potential membrane spanning domains: it may be a chaperone directing folding and membrane insertion of IcaA and may act as a link between IcaA and IcaC [53]. Also essential for the synthesis of fully functioning PIA is IcaC, a 355 aa integral membrane protein with ten predicted transmembrane domains, which may be involved in externalization and elongation of the growing polysaccharide [53].

IcaB is a member of the polysaccharide deacetylase family, including, for example, chitin deacetylases or the chitooligosaccharide deacetylase NodB of *Rhizobium melioti*. In its mature form it is a 259 aa secreted protein with a predicted signal sequence, responsible for the de-*N*-acetylation of PIA, and crucial for PIA activity in biofilm formation and for virulence in *S. epidermidis* [116]. In Δ*icaB*-mutants, where the *icaB* gene has been deleted, PIA is poorly retained on the cell surface as it does not contain non-*N*-acetylated GlcNAc [116]. Similar results have been obtained for *S. aureus* [117].

In vitro membrane preparations of *icaADBC*-positive *S. epidermidis* 1457 grown in TSB lacking glucose generated production of immunoreactive PIA upon addition of UDP-GlcNAc. Bacteria grown *in vivo* under the same conditions did not synthesize PIA—an additional unknown glucose-dependent protein factor is required [118]. The biosynthetic pathway for *O*-succinoylation of PIA is currently uncertain.

2.4.2 *Proteinaceous Intercellular Adhesins in Biofilm Accumulation*

2.4.2.1 Accumulation Associated Protein

Accumulation associated protein Aap, previously described in RP62A derived biofilm negative mutant M7 [57], was discovered as an intercellular adhesin through analysis of a clinical *S. epidermidis* isolate 5179 from a patient with chronic ventriculoatrial CSF-shunt infection, who was bacteremic over a period of almost 2 months [119]. Despite clear clinical significance, all isolates of *S. epidermidis* from this patient were biofilm-negative *in vitro*, which was due to an insertion of IS257 in *icaA* of *S. epidermidis* 5179 [119]. Reports about precise excision of IS256 from *icaADBC* with restoration of biofilm formation [96] initiated long-term biofilm cultures of *S. epidermidis* 5179, which led to the isolation of stable biofilm-forming revertants [58]. However, IS257 was still in place in the revertants, which did not produce any PIA but still formed biofilms. Comparison of surface proteins of wild-type 5179 and biofilm-producing revertant 5179-R1 led to identification of an additional 140 kDa protein of the revertant as Aap [58]. Aap is anchored via a LPXTG motif and comprises two major domains: domain A, which may itself contain a lectin-like domain [120], and a repetitive domain B, composed of a variable number of 128 amino acid repeats [58]. In *S. epidermidis* removal of Aap N-terminal domain A through proteolytic processing leads to exposure of domain B, which gives Aap intercellular adhesive properties causing biofilm accumulation [58]. The proteolytic activation may be mediated either by staphylococcal exoproteases or by the host [58]. Aap has significant homology with cell wall protein SasG of *S. aureus* [121, 122], in which SasG-mediated biofilm accumulation has been recently demonstrated using the same mechanisms; however, a hypothesis of self-activation of SasG was put forward [123, 124]. Domain B repeats contain "G5 domains," which hypothetically were related to *N*-acetylglucosamine binding activity [120]. Recently, however, it was demonstrated that G5 domains of Aap are zinc (Zn^{2+})-dependent adhesion molecules incorporating 2–3 Zn^{2+} ions in the dimer interface [125]. Zinc chelation inhibits biofilm formation by *S. epidermidis* and *S. aureus* [125]. Tandem G5-domains associate in a modular fashion suggesting a Zinc-zipper mechanism for G5 domain-based intercellular adhesion. Clearly the ability to respond to host proteases by forming a biofilm is in the organism's favor and enables it to evade phagocytic clearance, as indicated by the less efficient killing of Aap-aggregated *S. epidermidis* 5179-R1 as compared to its biofilm-negative, nonaggregated wild-type [58, 126].

Aap is the essential component of fibrillar structures visible on *S. epidermidis* NCTC 11047 by electron microscopy [59], which resemble *S. epidermidis* surface proteins Ssp1 and Ssp2 [44, 127, 128]. Ssp1 and Ssp2 were discovered as proteins mediating adhesion to polystyrene, which was inhibited by Ssp1-specific monoclonal antibodies [127, 128].

Aap domain A mediates adhesion of *S. epidermidis* NCTC 11047 to human skin corneocytes, and may represent the primary function of Aap in colonizing *S. epidermidis* strains [129]. Monoclonal anti-Aap specific antibodies have been reported to inhibit biofilm formation of *S. epidermidis* RP62A, a prime example of a PIA-dependent biofilm forming *S. epidermidis* strain [130]. This is in line with the original observation of impaired Aap expression in biofilm-negative mutant M7 of RP62A [57]. Recently it was shown that three mAbs against the Aap C-terminal single B-repeat construct followed by the 79-aa half repeat (AapBrpt1.5) had differential effects on RP62A biofilm formation, depending on the specific epitopes

[131]. Biofilm inhibition and increased biofilm formation were observed with the respective mAbs, which also influenced exopolysaccharide synthesis including eDNA and PIA [131]. Proteases as contained in medicinal maggots excretion/secretions may specifically disperse Aap-dependent biofilms [132].

Aap is common in clinical isolates [33, 54, 133, 134], and Aap-dependent biofilm formation can be directly demonstrated in clinical isolates [33, 58], so it seems not unlikely that it is, like PIA, an important virulence factor.

2.4.2.2 Extracellular Matrix Binding Protein Embp

The intercellular adhesive properties of Embp were discovered in a clinically significant *S. epidermidis* 1585 isolate from blood cultures in a port-catheter infection, which was *icaADBC*- and *aap*-negative and did not form biofilm *in vitro* [62]. Long-term biofilm culture led to the isolation of a stable biofilm-positive variant 1585v, which was used to isolate isogenic biofilm-negative transposon mutants. Tn917 insertions were located in the distal gene of the giant 1 MDa Embp, which had been previously described as a fibronectin binding protein [61]. A translocation in 1585v led to overexpression of a 460 kDa truncated isoform of Embp necessary for biofilm formation [62]. Embp is a giant fibronectin-binding protein harboring 59 found in various architectures (FIVAR) and 38 protein G-related albumin-binding (GA) domains. The FIVAR domains of Embp mediate binding of *S. epidermidis* to solid-phase attached fibronectin, constituting the first step of biofilm formation on conditioned surfaces [62]. The binding site in fibronectin was assigned to the fibronectin domain type III12 [62]. Embp-mediated biofilm formation also protects *S. epidermidis* from phagocytosis by macrophages [62]. More detailed assessment of the antiphagocytic properties of Embp of *S. epidermidis* indicated that the aggregative state leads to decreased uptake and a diminished inflammatory response by J774A.1 macrophages [126]. Interestingly, culture in goat serum induced expression of Embp in biofilm-negative wild-type *S. epidermidis* 1585 and rendered it biofilm-positive, which probably explains its virulence in the initial *in vivo* infection [62].

2.4.2.3 Other Surface Proteins

S. aureus strains of bovine origin have been found to produce a surface protein known as biofilm associated protein (Bap) which supports biofilm formation [135]. It remains uncertain whether Bap is important in the attachment or the accumulative phases of biofilm formation. A homologous protein was found in a few (4/38) biofilm positive *S. epidermidis* strains from ovine or caprine mastitis [60], but not in strains from humans [33, 60]. Allelic replacement of Bap in those isolates led to a biofilm-negative phenotype indicating that indeed in these strains Bap was responsible for biofilm formation [60]. Recently, six human *S. epidermidis* isolates were shown to carry *bap*, but all but one of those isolates were also positive for *icaADBC* so that the role of Bap for a biofilm-positive phenotype in these strains requires further study [136]. Another surface protein, Bhp, which is related to Bap, is present in 10–19% of *S. epidermidis* isolates from human infections [54, 135]. Therefore, the significance of Bhp for *S. epidermidis* biofilm formation and pathogenesis remains to be determined.

SesC is an LPXTG motif-containing 68 kDa surface protein of *S. epidermidis* distantly related to clumping factor A of *S. aureus* and is expressed more strongly in biofilm-associated as compared to planktonic *S. epidermidis* 1457 and 10b cells *in vitro* and *in vivo* [63]. Rabbit anti-SesC inhibited biofilm formation of a number of *S. epidermidis* isolates *in vitro*, which may be related to changes in primary attachment to fibrinogen-coated surfaces in the presence of anti-SesC [63]. All of 105 *S. epidermidis* isolates were in possession of the *sesC* gene [63]. A specific role of SesC as an intercellular adhesin in biofilm accumulation remains to be demonstrated.

2.4.3 Importance of Biofilm Formation for Staphylococcal Virulence

Biofilm-positive *S. epidermidis* strains occur frequently in significant clinical infection (reviewed in ref. [137]). Therefore, the role of biofilm in virulence of coagulase-negative staphylococci, particularly *S. epidermidis,* has been widely discussed and studied. Initially, studies on the pathogenicity of clinical *S. epidermidis* isolates using *in vivo* foreign-body infection models, did not appear to demonstrate that biofilm-producing *S. epidermidis* clinical isolates were more virulent (for a review see ref. [5]). However, subsequent studies using isogenic pairs of biofilm-positive wild-type and biofilm-negative mutants were more conclusive. Biofilm-positive, PIA-producing *S. epidermidis* 1457 was more virulent than its isogenic biofilm-negative transductant 1457-M10 (containing a Tn917 insertion in *icaA* [85]) in a subcutaneous catheter infection model in mice and a central venous catheter infection model in rats [138–140]. A similar result was demonstrated with the same models using an independent isogenic strain pair obtained from biofilm-positive *S. epidermidis* O-47 [141]. An *atlE*-deletion mutant of this strain was also attenuated [141]. In a rat central venous catheter model, expression of *icaRADBC* in *icaADBC*-negative *S. epidermidis* strains also led to increased virulence [142]. A *Caenorhabditis elegans* infection model was used to study biofilm-positive *S. epidermidis* 9142. This strain killed the worms more rapidly than its isogenic *icaA*-insertion mutant 9142-M10 and accumulated to higher intestinal concentrations [143]. Virulence was restored to wild-type in the biofilm-negative mutant by complementation with cloned *icaADBC* [143]. In a collection of *S. epidermidis* infective endocarditis isolates PIA expression and pathogenicity for *C. elegans* was not closely associated [144].

The susceptibility of biofilm-positive, PIA-positive *S. epidermidis* 1457 to killing by antimicrobial peptides like LL-37, dermcidin, and human β-defensin 3 has been compared to that of its isogenic *icaA*-insertion mutant 1457-M10 [145]. The biofilm positive strain was less susceptible to killing and also displayed decreased phagocytosis and killing by polymorphonuclear granulocytes (PMNs). When *S. epidermidis* 1457 was grown as a biofilm and planktonically, the organism in a biofilm was less susceptible to phagocytic killing after opsonization with normal human serum. The planktonic cells were killed to the same extent as planktonic cells of the isogenic biofilm-negative *icaA*-insertion mutant [146]. Biofilm-positive wild-type bacteria preopsonized with normal human serum were more resistant to killing than the isogenic biofilm-negative cells [146], and killing was dependent on activation of classic or lectin-mediated complement pathways. Compared to the mutant, biofilm-positive *S. epidermidis* 1457 induced more C3a, but C3b and human IgG surface deposition was decreased. In an experiment which reproduced the previously observed virulence differences in the isogenic strain pair, bacteria from explanted infected catheters from a mouse catheter infection model, displayed ex vivo, a lower level of phagocytic killing by PMNs in the biofilm-positive wild-type bacteria probably due to inadequate opsonization of biofilm bacteria *in vivo* [146]. Biofilm-grown cells of *S. epidermidis* 9142 were killed less efficiently than planktonically grown cells opsonized with antibodies raised against de-*N*-acetylated PIA/PNAG [147, 148]. Anti-PIA/PNAG antibodies diffused into *S. epidermidis* biofilms sufficiently to allow opsonization, when observed by confocal microscopy [147, 149]. Irrespective of the mechanism of biofilm accumulation, PIA, Aap, or Embp, aggregated biofilm *S. epidermidis* cells induced a lower inflammatory response in macrophages than the dispersed biofilm forming bacteria or isogenic biofilm-negative mutants [126], which might indicate another mechanism of how *S. epidermidis* remains below the radar of the immune system in chronic infection.

Since *S. epidermidis* is an opportunistic pathogen, mechanisms of pathogenicity which are important in some types of device-related infection might be less crucial in others. For example, in the guinea-pig tissue cage model [150], there was no difference in virulence between a biofilm-positive wild-type *S. epidermidis* 1457 and its isogenic *icaA*-insertion mutant, and no difference between *icaADBC*-positive and -negative clinical isolates [151, 152]. Nonetheless PIA was expressed *in vivo* in the tissue cages, and when animals were infected with both strains at the same time, the wild-type out-competed the mutant [153]. This may be because phagocytes are severely impaired in tissue cages [154], masking the expected advantage of the wild-type. During chronic catheter infection in rats, *icaADBC* transcription is continuously down-regulated. This suggests that once synthesized, PIA might have a long half-time *in vivo*. This is supported by the lack, to date, of any detectable PIA degrading enzymatic activity in staphylococci, a situation unlike that in *A. actinomycetemcomitans* which produces dispersin B [153, 155–157].

S. epidermidis produces a number of proinflammatory peptides called phenol-soluble modulins (PSMs), which are produced in a strictly *agr*-controlled manner [44, 158–160]. PSM δ has strong cytolytic activity towards neutrophils, however, is expressed only at low levels by *S. epidermidis* 1457, in line with a low cytolytic activity of *S. epidermidis* [161]. PSM δ is expressed only at very low levels in *S. epidermidis* 1457 biofilms as compared to planktonic cells [162]. PSM ß peptides promote *S. epidermidis* biofilm structuring and detachments *in vitro* and dissemination of infection during catheter infection *in vivo*, thereby providing the first mechanism of biofilm dispersal in *S. epidermidis* [162].

2.5 Primary Attachment to Modified and Unmodified Surfaces

Primary attachment occurs by both specific and nonspecific mechanisms, and to the native surfaces of implanted biomaterials as well as to surfaces primed by adsorption of host-derived plasma proteins, extracellular matrix (ECM) proteins and coagulation products (platelets and thrombi) [163, 164] (Fig. 2.2).

The level of attachment is usually greatest to a native surface, in part aided by the hydrophobicity of the staphylococcal cell surface and the surface of the biomaterial [165–173].

The most important adhesion molecule for native surfaces is *S. epidermidis* autolysin AtlE. This was identified by transposon mutagenesis in a mutant defective in attachment to polystyrene but which could still attach to glass [52, 174]. An AtlE-defective mutant was attenuated in virulence in a rat central venous catheter infection model [141], and in fact AtlE seems to have a dual role in attachment, as it also binds specifically to the ECM prote*in vitro*nectin [174, 175]. The high-resolution structure of the amidase domain AmiE of AtlE of *S. epidermidis* has recently been elucidated [176].

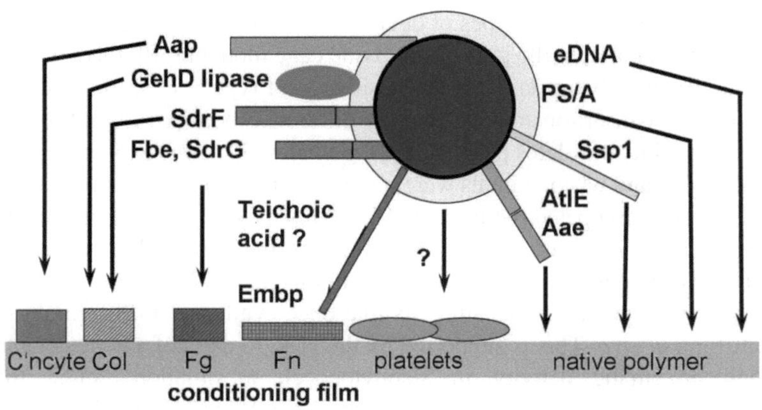

Fig. 2.2 Scheme of attachment mechanisms employed by *S. epidermidis* on native and conditioned polymer surfaces [64]. After implantation biomaterial surfaces are rapidly covered by a conditioning film composed of extracellular matrix proteins, thrombi, and activated platelets. *S. epidermidis* will attach to either native or conditioned surfaces employing a variety of different adhesive molecules. Fbe, SdrG *S. epidermidis* fibrinogen binding protein [182, 183], Embp *S. epidermidis* fibronectin binding protein [61, 62], Aap accumulation associated protein involved in corneocyte adhesion and biofilm accumulation [58, 129], AtlE *S. epidermidis* autolysin, Aae *S. epidermidis* autolysin/adhesin, Ssp1 staphylococcal surface protein 1 [127, 128], GehD *S. epidermidis* lipase, SdrF *S. epidermidis* collagen binding protein [192], PS/A *S. epidermidis* polysaccharide/adhesin, eDNA extracellular DNA, Col collagen, Fg fibrinogen, Fn fibronectin

The 220 kDa cell-wall associated protein Ssp1, which is organized in a fimbria-like structure, mediates attachment to polystyrene in an *S. epidermidis* strain [127, 128] but has not been further studied. However, more recent data raise the possibility that Ssp1 may be related to Aap, which also forms fibrillar appendages on the surface of *S. epidermidis* NCTC11047 [44, 59, 129].

A capsular polysaccharide adhesin, PS/A, of *S. epidermidis* RP62A mediates primary attachment to unmodified silastic catheter surfaces, and was later shown to be structurally very similar if not identical to polysaccharide intercellular adhesin (PIA) of *S. epidermidis* [81, 82, 87–91, 177].

Extracellular DNA (eDNA) released by *S. epidermidis* also contributes to biofilm formation [178, 179]. DNase I can degrade eDNA and thereby inhibit nascent biofilm formation as well as disintegrating preformed biofilms [178, 179]. However, analysis of primary attachment of DNAse I treated *S. epidermidis* cells indicated that eDNA acted predominantly in the early attachment phase of biofilm formation [179]. Release of eDNA from the staphylococci involved murein hydrolases like AtlE of *S. epidermidis* [179]. How eDNA release interacts with various other factors involved in *S. epidermidis* biofilm formation remains to be determined.

Specific binding to surface ECM proteins (such as fibronectin, fibrinogen, collagen, thrombospondin, and vitronectin) involves cell-wall associated adhesins known as MSCRAMMs (*m*icrobial *s*urface *c*omponents *r*ecognizing *a*dhesive *m*atrix *m*olecules) [180]. As well as the autolysin/adhesin AtlE, mentioned above, another *S. epidermidis* surface-associated autolysin, Aae, has been described, which also binds to vitronectin, though not to polystyrene [181]. The gene encoding the fibrinogen-binding protein Fbe/SdrG [182, 183] is common in clinical *S. epidermidis* isolates [33, 54, 182, 184], but its expression must be highly variable as many strains do not bind avidly to fibrinogen [182]. This may be related to the observation that SdrG/Fbe expression was only observed in *in vivo* grown *S. epidermidis* isolates, but not under *in vitro* conditions simulating *in vivo* conditions, i.e., growth in human serum [185]. Fbe-specific antibodies block adherence of *S. epidermidis* to fibrinogen-coated catheters [186] and attenuate infection in animal models [187, 188]; hence Fbe has potential as a vaccine candidate. In a rat model of intravascular catheter associated infection an *fbe*-deletion mutant was attenuated indicating that Fbe is a virulence factor in device-associated *S. epidermidis* infections [189].

The giant 1 MDa fibronectin-binding protein Embp [61] is also present in the majority of clinical *S. epidermidis* isolates [33, 54] but antibodies directed against it were not sufficiently opsonic to enhance phagocytosis [187]. Binding of *S. epidermidis* to fibronectin-modified surfaces is also enhanced by purified teichoic acids [190].

The lipase GehD is secreted by *S. epidermidis* and specifically binds collagen and promotes attachment to collagen coated surfaces [191]. More recently it was demonstrated that the LPXTG motif-containing *S. epidermidis* surface protein SdrF also mediates specific binding to type I collagen in a heterologous expression system in *Lactococcus lactis* [192]. Collagen binding occurs via the B domain of SdrF and both the $\alpha 1$ and $\alpha 2$ chains of type I collagen [192]. It was also found that a single B domain repeat of *S. epidermidis* 9491 retains the capacity to bind to type I collagen [192]. Using the *L. lactis* heterologous expression system and a murine infection model evidence was derived that SdrF may contribute to cardiac assist device driveline infections [193]. SdrF also mediated binding to unmodified Dacron surfaces covering drivelines [193]. In contrast *L. lactis* expressing GehD bound only weakly to driveline surfaces [193]. Anti-SdrF inhibited *S. epidermidis* 9491 binding in the *in vivo* model only by approx. 50% indicating that additional collagen binding factors may be involved [193].

2.6 Molecular Epidemiology of Mechanisms Used in Biofilm Formation in Biomaterial-Related Infection

Despite PIA production being a major mechanism of biofilm accumulation, strains of *S. epidermidis* lacking *icaADBC* may also produce biofilm, and biofilm-positive PIA-negative strains have been isolated in clinical contexts [33, 51, 54–56, 62, 119, 194–196].

Since the fundamental molecular basis of *S. epidermidis* biofilm formation has become apparent with the identification of PIA, *icaADBC*, and AtlE [52, 53, 82, 174] numerous studies have been undertaken in *S. epidermidis* strain collections from numerous types of biomaterial related infections, which investigated the presence of *icaADBC* and other genetic factors potentially involved in biofilm accumulation or primary attachment and eventually their correlation with a biofilm forming phenotype *in vitro* (Table 2.1). One of the drivers for these studies is the fact that a majority of clinical isolates of *S. epidermidis* and other coagulase-negative staphylococci represent contaminants, however, recognition of certain genetic traits might aid in resolving this dilemma of distinguishing true pathogens from contaminants. However, the rigor to differentiate contaminating from infecting isolates may have been different in various studies and some of the variation observed may be related to the various frequencies of included contaminants.

As CLABSI is the most frequent infection involving coagulase-negative staphylococci and *S. epidermidis*, many studies have focused on blood culture and catheter tip isolates. In early phenotypic studies it was recognized that biofilm-positive *S. epidermidis* occurs frequently in clinical significant infection, however, and that biofilm-negative strains also can cause infection (reviewed in ref. [137]). However, biofilm formation *in vitro* is due to variability between laboratories. Therefore, the detection of the presence of specific genetic determinants related to biofilm formation and virulence of *S. epidermidis* appears less cumbersome and better comparable between laboratories.

In the first clinical study after discovery of PIA and *icaADBC*, Ziebuhr and coworkers reported that 87% of *S. epidermidis* blood culture isolates were biofilm-positive and 85% were *icaADBC*-positive, while nasal control *S. epidermidis* isolates formed biofilm and were *icaADBC*-positive in 11% and 6%, respectively [51]. In an extension of this study 89.2% of nosocomial clinical isolates were *icaADBC*-positive and almost 50% belonged to multi locus sequence type ST27, whose representatives were all *icaADBC*-positive [197]. In a similar study 77% of clinically significant blood culture *S. epidermidis* isolates as opposed to 28% blood

Table 2.1 Molecular epidemiology of *icaADBC* in different biomaterial-related infections

Study	*icaADBC*-positive infection isolates % (*N*)	*icaADBC*-positive contaminant isolates % (*N*)	*icaADBC*-positive commensal control isolates % (*N*)	Comments
Bacteremia				
Ziebuhr 1997 [51]	85% (44/52)		6% (2/36)	Skin and mucosal strains from healthy volunteers
Frebourg 2000 [198]	77% (30/39)	28% (11/39)	38% (6/16)	Contaminants: only one positive BC; commensals: Isolates from hands of healthy volunteers
Vandecasteele 2003 [133]	68% (23/34)	88% (14/16)	52% (15/29)	Contaminants: CVC isolates from patients without bacteremia; commensals: skin isolates from healthy volunteers
CVC infection isolates				
Arciola 2001 [199]	48.5% (33/68)		0% (0/10)	Commensals: Nasal and mucosa from healthy volunteers
Klug 2003 [211]	90% (9/10)		25% (5/25)	Commensals: Hand skin isolates from healthy volunteers
Cafiso 2004 [202]	41% (11/27)			
Petrelli 2006 [201]	70% (47/67)			
Bacteremia in bone-marrow transplant patients				
Rohde 2004 [54]	94% (15/16)		80% (20/25)	Nasal isolates from BMT patients
			13% (2/15)	Nasal isolates from healthy volunteers
Ninin 2006 [203]	82% (31/38)	85% (60/71)		No significant difference between multiple or single positive BC
Bacteremia in neonatal ICU patients				
de Silva 2002 [204]	42% (5/12)	38% (15/39)	48% (12/25)	Sick baby skin isolates
			37% (17/46)	Well baby skin isolates
				Contaminants: BC isolates from babies without infection

(continued)

Table 2.1 (continued)

Study	*icaADBC*-positive infection isolates % (*N*)	*icaADBC*-positive contaminant isolates % (*N*)	*icaADBC*-positive commensal control isolates % (*N*)	Comments
Klingenberg 2005 [205]	61% (52/85)	52% (49/95)		Contaminants: BC isolates from babies without clinical signs of sepsis
Bradford 2006 [206]	67% (4/6)	63% (5/8)	20% (2/10)	Contaminants: BC isolates from babies without clinical signs of sepsis; commensals: Nasal strains from healthy adults
Prosthetic joint infections				
Galdbart 2000 [208]	81.5% (44/54)		17% (4/24)	Commensals: Hand skin flora strains from healthy volunteers
Arciola 2004 [209]	43% (51/120)			
Frank 2004 [195]	55% (17/31)	56% (5/9)		Contaminants: Arthroplasty-associated nonprosthetic joint-infection isolates
Chokr 2006 [55]	82% (31/38)			
Rohde 2007 [33]	62% (32/52)			All prosthetic joint infection isolates
	80% (12/15)			Knee prosthesis infection
	54% (20/37)			Hip prosthesis infection
Koskela 2009 [210]	50% (16/32)		33% (8/24)	Commensals: Skin and nasal isolates of healthy volunteers
CSF shunt infection				
Stevens 2008 [134]	79% (33/42)	31% (4/13)		Contaminants: CSF isolates deemed not clinical significant
Cardiac pacemaker infection				
Klug 2003 [211]	26% (7/27)		25% (5/25)	Commensals: Hand skin isolates from healthy volunteers

culture contaminant isolates carried *icaADBC* [198], while 68% of invasive blood culture *S. epidermidis* isolates, 88% of colonizing catheter tip isolates from asymptomatic patients, and 52% of skin isolates were *icaADBC*-positive, respectively, in a study by Vandecasteele and coworkers [133]. It seems reasonable to conclude that in blood culture isolates potentially PIA producing *S. epidermidis* isolates are highly prevalent highlighting a role of biofilm formation through this mechanism in pathogenesis.

This conclusion is partly supported by studies of *S. epidermidis* isolates from CVC associated infections where 48.5% of isolates were *icaADBC*-positive while all of ten skin isolates did not carry *icaADBC* [199]. In patients with clinical significant CVC-associated infection 70% of *S. epidermidis* isolates from CVC tips (>15 colonies per tip roll culture [200]) were *icaADBC*-positive [201]. In another study only 41% catheter tip isolates (>15 colonies per tip roll culture [200]) were *icaADBC*-positive [202]. The prevalence of *icaADBC*-positive isolates appears lower, however, isolates colonizing CVCs frequently do not cause CVC-associated bacteremia [133, 200] and this may explain the lower prevalence of *icaADBC* in strains from this specific biomaterial.

It seems pertinent to speculate that bacteremia isolates from bone marrow transplant patients may represent a different *S. epidermidis* population as many of these patients are severely immunosuppressed and often neutropenic and therefore virulence factors aiding in resistance to phagocytosis like biofilm formation and PIA production may be less important, which may lead to increased representation of *icaADBC*-negative skin isolates amongst infections in this patient population [126, 145, 146]. Eighty-three percent of *S. epidermidis* blood culture isolates from bacteremia in bone marrow transplant patients were *icaADBC*-positive, and there was no significant difference in *S. epidermidis* isolates in this regard from patients with only one positive blood culture (84.5%) as opposed to multiple positive blood cultures (81.6%) [203]. 60.5% of *icaADBC*-positive blood culture isolates produced biofilm, while 59% of *icaADBC*-negative isolates produced biofilm [203]. In a study with strict microbiological inclusion criteria in bone marrow transplant patients —> 2 clonally related or identical *S. epidermidis* isolates from independent blood cultures and signs of infection — all but one bacteremia episode were caused by isolates carrying *icaADBC*, while in commensal skin and nasal *S. epidermidis* isolates from uninfected patients and from healthy subjects were *icaADBC*-positive in 80% and 13%, respectively [54]. This would imply a shift in the commensal *S. epidermidis* populations from predominantly *icaADBC*-negative to predominantly *icaADBC*-positive in bone-marrow transplant patients, which may be caused through selection pressures in hospital. *Aap* was highly prevalent in all isolate groups in this study (87–94%) [54], but varied from 59% in invasive blood culture isolates, 88% in colonizing catheter isolates, and 38% skin isolates in Vandecasteele's study mentioned above [133].

In a study of blood culture isolates from a neonatal ICU 42% of clinical significant blood culture isolates, 38% of blood culture contaminants, 48% of skin isolates from sick babies and 37% of skin isolates from well babies carried *icaADBC* [204]. In a similar setting of a neonatal ICU 61% of clinical significant blood culture isolates as opposed to 52% of blood cultures contaminants were *icaADBC*-positive [205]. Also in neonatal ICU babies 66% invasive, 63% contaminant, and 20% nasal *S. epidermidis* isolates carried *icaADBC* [206]. Clearly, presence of *icaADBC* cannot be used to assess clinical significance of an isolate in this population. Isolation of *S. epidermidis* in bacteremia of premature neonates is difficult to assess clinically, especially as in many cases only a single positive blood culture is available, due to difficulties of obtaining blood. Therefore, many isolates may represent contaminants similar to *S. epidermidis* isolates from unselected adult blood cultures [207], which could in part explain the lower prevalence of *icaADBC* in *S. epidermidis* in this clinical situation.

In prosthetic joint infection 81.5% of *S. epidermidis* infection isolates carried *icaADBC*, while 16.7% skin flora isolates were positive for the operon [208]. In 120 *S. epidermidis* periprosthetic orthopedic infection isolates only 43% were *icaADBC* positive and 86.3% of those produced biofilm *in vitro* [209]. None of *icaADBC*-negative isolates from this study produced biofilm *in vitro* [209]. Similarly, Frank and coworkers found 55% prosthetic joint infection strains (≥2 positive cultures with the same organism) to carry *icaADBC*, a similar prevalence as in nonprosthetic joint infection arthroplasty-associated (only one isolate per episode; therefore possible contaminant) *S. epidermidis* strains (56%) [195]. In another study of isolates from orthopedic periprosthetic infections 81.6% *S. epidermidis* strains carried *icaADBC* [55]. In this study 45.2% of 31 *icaADBC*-positive *S. epidermidis* isolates produced biofilm, while none of the *icaADBC*-negative isolates were biofilm-positive [55]. Prevalence of *icaADBC* in prosthetic joint infection *S. epidermidis* isolates was 50% ($n=32$), while commensal control isolates had *icaADBC*-prevalence of 16.7% ($n=24$) [210].

In a study of clinical significant *S. epidermidis* hip and knee arthroplasty infection isolates prevalence of *icaADBC* was differentially distributed with 54 and 80% of isolates being positive [33]. This study had strict microbiological inclusion criteria for infections which were isolation of two clonally related or identical *S. epidermidis* strains from a joint fluid specimen and from intraoperative biopsies taken weeks to months later during revision surgery [33]. Interestingly, almost 70% of isolates formed biofilm *in vitro*; all biofilm-positive isolates from knee arthroplasty infections carried *icaADBC*, while amongst biofilm-positive hip arthroplasty infection strains only 63% were *icaADBC* positive [33], clearly indicating that other biofilm formation mechanisms in addition to PIA and *icaADBC* are of relevance in these infections. Direct evidence was obtained using specific hexosaminidase DspB and proteases that 50% of all *S. epidermidis* isolates used PIA as intercellular adhesin for biofilm formation, while 19.2% of biofilm-positive isolates used a proteinaceous intercellular adhesin (Fig. 2.3) [33]. For 9.6% of isolates Aap was the relevant intercellular adhesin as determined by biofilm inhibition with anti-Aap, while it is currently unresolved if Embp or other proteinaceous intercellular adhesins are active in the remaining biofilm-positive strains [33, 62]. Importantly, 30% of the clinically significant isolates were *in vitro* biofilm-negative. In four *icaADBC*-negative, *aap*-positive strains biofilm formation could be induced by limited proteolysis with trypsin, which led to Aap activation and Aap-dependent biofilm formation [58].

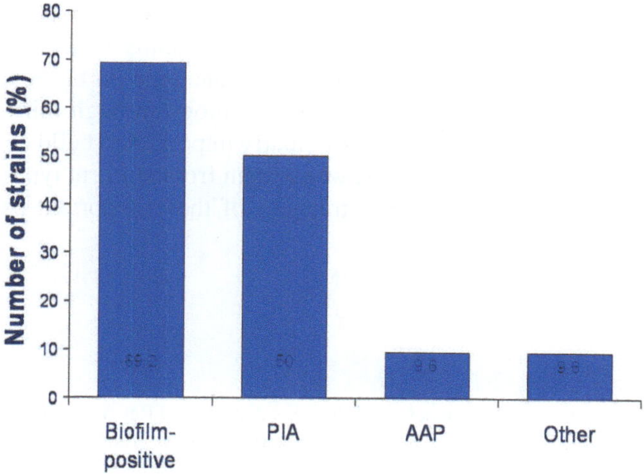

Fig. 2.3 Prevalence of biofilm formation and use of different mechanisms of intercellular adhesion in *S. epidermidis* from prosthetic hip and knee joint infections. Data compiled from Rohde et al. [33]

It should be noted that *in vitro* Embp expression is induced only in high concentrations of goat serum leading to biofilm formation of *S. epidermidis* 1585 [62], which could explain why many of the *S. epidermidis* isolates remained biofilm-negative under standard *in vitro* conditions [33].

In shunt-associated meningitis prevalence of *icaADBC* and *aap* in 42 *S. epidermidis* isolates was 78.6 and 47.6% while amongst 13 contaminants 30.8 and 46.2% were *icaADBC*- and *aap*-positive, respectively [134]. In pacemaker-associated infections prevalence of *icaADBC* was low (25.9%; $n=27$) similar to contaminant *S. epidermidis* isolates [211]. Prevalence of adhesion factors AtlE and Fbe was high at 89% in this study [211].

Molecular epidemiologic studies from our laboratory indicate that biofilm accumulation mechanisms are redundantly organized in *S. epidermidis*, the use of various mechanisms is differentially distributed in infection isolates from different types of device associated infections, e.g., CVC-associated bacteremia versus prosthetic hip or knee joint infections [33, 54, 58, 62]. From this work it is also valid to conclude that many *in vitro* biofilm-negative *S. epidermidis* isolates may produce biofilm due to mechanisms, which are not expressed under the respective *in vitro* conditions, but may be expressed *in vivo*. However, it cannot be excluded that some of these isolates do not form biofilm at all and are able to persist in periimplant tissues due to other mechanisms [212–214].

2.7 Conclusions

There is wide variation in the prevalence of *icaADBC* and biofilm formation in clinical molecular epidemiologic studies, which would be best explained on the one hand in the varying strictness of identification of clinical significance in different settings and the biologic fact that different biomedical-implants represent different types of infection selecting for *S. epidermidis* strains differentially equipped with virulence factors and molecules functional in biofilm formation. In light of the revolution of molecular epidemiology through third-generation full-genome sequencing it would be particularly important to build collections of clinically well-characterized isolates, which stem from uniform types of implant infections, to learn more about the strategies of these important pathogens of medical progress in the future.

References

1. Lewis K. Persister cells, dormancy and infectious disease. Nat Rev Microbiol. 2007;5:48–56.
2. Hall-Stoodley L, Costerton JW, Stoodley P. Bacterial biofilms: from the natural environment to infectious diseases. Nat Rev Microbiol. 2004;2:95–108.
3. Costerton JW, Stewart PS, Greenberg EP. Bacterial biofilms: a common cause of persistent infections. Science. 1999;284:1318–22.
4. Rupp ME, Archer GL. Coagulase-negative staphylococci: pathogens associated with medical progress. Clin Infect Dis. 1994;19:231–43.
5. Mack D, Horstkotte MA, Rohde H, Knobloch JKM. Coagulase-negative staphylococci. In: Pace JL, Rupp ME, Finch RG, editors. Biofilms, infection, and antimicrobial therapy. Boca Raton: CRC Press; 2006. p. 109–53. chap 7.
6. Darouiche RO. Device-associated infections: a macroproblem that starts with microadherence. Clin Infect Dis. 2001;33:1567–72.
7. Goldmann DA, Pier GB. Pathogenesis of infections related to intravascular catheterization. Clin Microbiol Rev. 1993;6:176–92.
8. Hidron AI, Edwards JR, Patel J, et al. NHSN annual update: antimicrobial-resistant pathogens associated with healthcare-associated infections: annual summary of data reported to the National Healthcare Safety Network at the Centers for Disease Control and Prevention, 2006–2007. Infect Control Hosp Epidemiol. 2008;29:996–1011.
9. Geffers C, Gastmeier P. Nosocomial infections and multidrug-resistant organisms in Germany: epidemiological data from KISS (the Hospital Infection Surveillance System). Dtsch Arztebl Int. 2011;108:87–93.
10. Fontela PS, Platt RW, Rocher I, et al. Surveillance provinciale des infections nosocomiales (SPIN) program: implementation of a mandatory surveillance program for central line-associated bloodstream infections. Am J Infect Control. 2011;39:329–35.
11. Mermel LA. Prevention of intravascular catheter-related infections. Ann Intern Med. 2000;132:391–402.
12. Raad II, Bodey GP. Infectious complications of indwelling vascular catheters. Clin Infect Dis. 1992;15:197–208.
13. Srinivasan A, Wise M, Bell M, et al. Vital signs: central line-associated blood stream infections–United States, 2001, 2008, and 2009. MMWR Morb Mortal Wkly Rep. 2011;60:243–8.
14. Jukes L, Mikhail J, Bome-Mannathoko N, et al. Rapid differentiation of *Staphylococcus aureus, Staphylococcus epidermidis* and other coagulase-negative staphylococci and methicillin susceptibility testing directly from growth positive blood cultures by multiplex real-time PCR. J Med Microbiol. 2010;59:1456–61.
15. Murdoch DR, Corey GR, Hoen B, et al. Clinical presentation, etiology, and outcome of infective endocarditis in the 21st century: the International Collaboration on Endocarditis-Prospective Cohort Study. Arch Intern Med. 2009;169:463–73.
16. Chu VH, Miro JM, Hoen B, et al. Coagulase-negative staphylococcal prosthetic valve endocarditis–a contemporary update based on the International Collaboration on Endocarditis: prospective cohort study. Heart. 2009;95:570–6.

17. Chu VH, Cabell CH, Abrutyn E, et al. Native valve endocarditis due to coagulase-negative staphylococci: report of 99 episodes from the International Collaboration on Endocarditis Merged Database. Clin Infect Dis. 2004;39:1527–30.
18. Chu VH, Woods CW, Miro JM, et al. Emergence of coagulase-negative staphylococci as a cause of native valve endocarditis. Clin Infect Dis. 2008;46:232–42.
19. Anguera I, Del Rio A, Miro JM, et al. Staphylococcus lugdunensis infective endocarditis: description of 10 cases and analysis of native valve, prosthetic valve, and pacemaker lead endocarditis clinical profiles. Heart. 2005;91(2):e10.
20. Viola GM, Awan LL, Darouiche RO. Nonstaphylococcal infections of cardiac implantable electronic devices. Circulation. 2010;121:2085–91.
21. Gordon RJ, Quagliarello B, Lowy FD. Ventricular assist device-related infections. Lancet Infect Dis. 2006;6:426–37.
22. Simon D, Fischer S, Grossman A, et al. Left ventricular assist device-related infection: treatment and outcome. Clin Infect Dis. 2005;40:1108–15.
23. Simon TD, Hall M, Riva-Cambrin J, et al. Infection rates following initial cerebrospinal fluid shunt placement across pediatric hospitals in the United States. Clinical article. J Neurosurg Pediatr. 2009;4:156–65.
24. Langley JM, Gravel D, Moore D, et al. Study of cerebrospinal fluid shunt-associated infections in the first year following placement, by the Canadian Nosocomial Infection Surveillance Program. Infect Control Hosp Epidemiol. 2009;30:285–8.
25. Conen A, Walti LN, Merlo A, et al. Characteristics and treatment outcome of cerebrospinal fluid shunt-associated infections in adults: a retrospective analysis over an 11-year period. Clin Infect Dis. 2008;47:73–82.
26. National Joint Registry England and Wales. Annual Report 2011. http://www.njrcentre.org.uk/NjrCentre/Portals/0/Documents/NJR%208th%20Annual%20Report%202011.pdf. Accessed 26 Sep 2011.
27. Phillips JE, Crane TP, Noy M, et al. The incidence of deep prosthetic infections in a specialist orthopaedic hospital: a 15-year prospective survey. J Bone Joint Surg Br. 2006;88:943–8.
28. Dale H, Hallan G, Hallan G, et al. Increasing risk of revision due to deep infection after hip arthroplasty. Acta Orthop. 2009;80:639–45.
29. Ong KL, Kurtz SM, Lau E, et al. Prosthetic joint infection risk after total hip arthroplasty in the Medicare population. J Arthroplasty. 2009;24:105–9.
30. Kurtz SM, Ong KL, Lau E, et al. Prosthetic joint infection risk after TKA in the Medicare population. Clin Orthop Relat Res. 2010;468:52–6.
31. Nickinson RS, Board TN, Gambhir AK, et al. The microbiology of the infected knee arthroplasty. Int Orthop. 2010;34:505–10.
32. Schafer P, Fink B, Sandow D, et al. Prolonged bacterial culture to identify late periprosthetic joint infection: a promising strategy. Clin Infect Dis. 2008;47:1403–9.
33. Rohde H, Burandt EC, Siemssen N, et al. Polysaccharide intercellular adhesin or protein factors in biofilm accumulation of *Staphylococcus epidermidis* and *Staphylococcus aureus* isolated from prosthetic hip and knee joint infections. Biomaterials. 2007;28:1711–20.
34. Harris LG, El-Bouri K, Johnston S, et al. Rapid identification of staphylococci from prosthetic joint infections using MALDI-TOF mass-spectrometry. Int J Artif Organs. 2010;33:568–74.
35. Roth RR, James WD. Microbial ecology of the skin. Annu Rev Microbiol. 1988;42:441–64.
36. Kloos WE. Natural populations of the genus *Staphylococcus*. Annu Rev Microbiol. 1980;34:559–92.
37. Carr DL, Kloos WE. Temporal study of the staphylococci and micrococci of normal infant skin. Appl Environ Microbiol. 1977;34:673–80.
38. Kloos WE, Musselwhite MS. Distribution and persistence of *Staphylococcus* and *Micrococcus* species and other aerobic bacteria on human skin. Appl Microbiol. 1975;30:381–5.
39. Iwase T, Uehara Y, Shinji H, et al. *Staphylococcus epidermidis* Esp inhibits *Staphylococcus aureus* biofilm formation and nasal colonization. Nature. 2010;465:346–9.
40. Grice EA, Segre JA. The skin microbiome. Nat Rev Microbiol. 2011;9:244–53.
41. Gao Z, Tseng CH, Pei Z, et al. Molecular analysis of human forearm superficial skin bacterial biota. Proc Natl Acad Sci U S A. 2007;104:2927–32.

42. Grice EA, Kong HH, Renaud G, et al. A diversity profile of the human skin microbiota. Genome Res. 2008;18:1043–50.
43. Grice EA, Kong HH, Conlan S, et al. Topographical and temporal diversity of the human skin microbiome. Science. 2009;324:1190–2.
44. Otto M. *Staphylococcus epidermidis* - the 'accidental' pathogen. Nat Rev Microbiol. 2009;7: 555–67.
45. Davis N, Curry A, Gambhir AK, et al. Intraoperative bacterial contamination in operations for joint replacement. J Bone Joint Surg Br. 1999;81:886–9 [see comments].
46. Byrne AM, Morris S, McCarthy T, et al. Outcome following deep wound contamination in cemented arthroplasty. Int Orthop. 2007;31:27–31.
47. Knobben BA, van Horn JR, van der Mei HC, et al. Evaluation of measures to decrease intra-operative bacterial contamination in orthopaedic implant surgery. J Hosp Infect. 2006;62:174–80.
48. Mack D, Siemssen N, Laufs R. Parallel induction by glucose of adherence and a polysaccharide antigen specific for plastic-adherent *Staphylococcus epidermidis*: evidence for functional relation to intercellular adhesion. Infect Immun. 1992;60:2048–57.
49. Mack D, Nedelmann M, Krokotsch A, et al. Characterization of transposon mutants of biofilm-producing *Staphylococcus epidermidis* impaired in the accumulative phase of biofilm production: genetic identification of a hexosamine-containing polysaccharide intercellular adhesin. Infect Immun. 1994;62:3244–53.
50. Mack D, Haeder M, Siemssen N, et al. Association of biofilm production of coagulase-negative staphylococci with expression of a specific polysaccharide intercellular adhesin. J Infect Dis. 1996;174:881–4.
51. Ziebuhr W, Heilmann C, Götz F, et al. Detection of the intercellular adhesion gene cluster (*ica*) and phase variation in *Staphylococcus epidermidis* blood culture strains and mucosal isolates. Infect Immun. 1997;65:890–6.
52. Heilmann C, Schweitzer O, Gerke C, et al. Molecular basis of intercellular adhesion in the biofilm-forming *Staphylococcus epidermidis*. Mol Microbiol. 1996;20:1083–91.
53. Gerke C, Kraft A, Süssmuth R, et al. Characterization of the N-acetylglucosaminyltransferase activity involved in the biosynthesis of the *Staphylococcus epidermidis* polysaccharide intercellular adhesin. J Biol Chem. 1998;273:18586–93.
54. Rohde H, Kalitzky M, Kröger N, et al. Detection of virulence-associated genes not useful for discriminating between invasive and commensal *Staphylococcus epidermidis* strains on a bone marrow transplant unit. J Clin Microbiol. 2004;42:5614–9.
55. Chokr A, Watier D, Eleaume H, et al. Correlation between biofilm formation and production of polysaccharide intercellular adhesin in clinical isolates of coagulase-negative staphylococci. Int J Med Microbiol. 2006;296:381–6.
56. Hennig S, Nyunt WS, Ziebuhr W. Spontaneous switch to PIA-independent biofilm formation in an *ica*-positive *Staphylococcus epidermidis* isolate. Int J Med Microbiol. 2007;297:117–22.
57. Hussain M, Herrmann M, von Eiff C, et al. A 140-kilodalton extracellular protein is essential for the accumulation of *Staphylococcus epidermidis* strains on surfaces. Infect Immun. 1997;65:519–24.
58. Rohde H, Burdelski C, Bartscht K, et al. Induction of *Staphylococcus epidermidis* biofilm formation via proteolytic processing of the accumulation-associated protein by staphylococcal and host proteases. Mol Microbiol. 2005;55:1883–95.
59. Banner MA, Cunniffe JG, Macintosh RL, et al. Localized tufts of fibrils on *Staphylococcus epidermidis* NCTC 11047 are comprised of the accumulation-associated protein. J Bacteriol. 2007;189:2793–804.
60. Tormo MA, Knecht E, Götz F, et al. Bap-dependent biofilm formation by pathogenic species of *Staphylococcus*: evidence of horizontal gene transfer? Microbiology. 2005;151:2465–75.
61. Williams RJ, Henderson B, Sharp LJ, et al. Identification of a fibronectin-binding protein from *Staphylococcus epidermidis*. Infect Immun. 2002;70:6805–10.
62. Christner M, Franke GC, Schommer NN, et al. The giant extracellular matrix-binding protein of *Staphylococcus epidermidis* mediates biofilm accumulation and attachment to fibronectin. Mol Microbiol. 2010;75:187–207.

63. Shahrooei M, Hira V, Stijlemans B, et al. Inhibition of *Staphylococcus epidermidis* biofilm formation by rabbit polyclonal antibodies against the SesC protein. Infect Immun. 2009;77:3670–8.
64. Mack D, Davies AP, Harris LG, et al. *Staphylococcus epidermidis* biofilms: functional molecules, relation to virulence, and vaccine potential. Top Curr Chem. 2009;288:157–82.
65. Cramton SE, Gerke C, Schnell NF, et al. The intercellular adhesion (*ica*) locus is present in *Staphylococcus aureus* and is required for biofilm formation. Infect Immun. 1999;67:5427–33.
66. McKenney D, Pouliot KL, Wang Y, et al. Broadly protective vaccine for *Staphylococcus aureus* based on an *in vivo*-expressed antigen. Science. 1999;284:1523–7.
67. Allignet J, Aubert S, Dyke KG, et al. *Staphylococcus caprae* strains carry determinants known to be involved in pathogenicity: a gene encoding an autolysin-binding fibronectin and the *ica* operon involved in biofilm formation. Infect Immun. 2001;69:712–8.
68. Frank KL, Patel R. Poly-N-acetylglucosamine is not a major component of the extracellular matrix in biofilms formed by *icaADBC*-positive *Staphylococcus lugdunensis* isolates. Infect Immun. 2007;75:4728–42.
69. Moretro T, Hermansen L, Holck AL, et al. Biofilm formation and the presence of the intercellular adhesion locus *ica* among staphylococci from food and food processing environments. Appl Environ Microbiol. 2003;69:5648–55.
70. Wang X, Preston III JF, Romeo T. The *pgaABCD* locus of *Escherichia coli* promotes the synthesis of a polysaccharide adhesin required for biofilm formation. J Bacteriol. 2004;186:2724–34.
71. Kaplan JB, Velliyagounder K, Ragunath C, et al. Genes involved in the synthesis and degradation of matrix polysaccharide in *Actinobacillus actinomycetemcomitans* and *Actinobacillus pleuropneumoniae* biofilms. J Bacteriol. 2004;186:8213–20.
72. Darby C, Hsu JW, Ghori N, et al. *Caenorhabditis elegans:* plague bacteria biofilm blocks food intake. Nature. 2002;417:243–4.
73. Itoh Y, Wang X, Hinnebusch BJ, et al. Depolymerization of beta-1, 6-N-acetyl-*d*-glucosamine disrupts the integrity of diverse bacterial biofilms. J Bacteriol. 2005;187:382–7.
74. Bobrov AG, Kirillina O, Forman S, et al. Insights into *Yersinia pestis* biofilm development: topology and co-interaction of Hms inner membrane proteins involved in exopolysaccharide production. Environ Microbiol. 2008;10:1419–32.
75. Bobrov AG, Kirillina O, Ryjenkov DA, et al. Systematic analysis of cyclic di-GMP signalling enzymes and their role in biofilm formation and virulence in *Yersinia pestis*. Mol Microbiol. 2011;79:533–51.
76. Choi AH, Slamti L, Avci FY, et al. The *pgaABCD* locus of *Acinetobacter baumannii* encodes the production of poly-beta-1-6-N-acetylglucosamine, which is critical for biofilm formation. J Bacteriol. 2009;191:5953–63.
77. Parise G, Mishra M, Itoh Y, et al. Role of a putative polysaccharide locus in *Bordetella* biofilm development. J Bacteriol. 2007;189:750–60.
78. Sloan GP, Love CF, Sukumar N, et al. The *Bordetella* Bps polysaccharide is critical for biofilm development in the mouse respiratory tract. J Bacteriol. 2007;189:8270–6.
79. Drewry DT, Galbraith L, Wilkinson BJ, et al. Staphylococcal slime: a cautionary tale. J Clin Microbiol. 1990;28:1292–6.
80. Mack D. Molecular mechanisms of *Staphylococcus epidermidis* biofilm formation. J Hosp Infect. 1999;43(Suppl):S113–25.
81. Maira-Litran T, Kropec A, Goldmann D, et al. Biologic properties and vaccine potential of the staphylococcal poly-N-acetyl glucosamine surface polysaccharide. Vaccine. 2004;22:872–9.
82. Mack D, Fischer W, Krokotsch A, et al. The intercellular adhesin involved in biofilm accumulation of *Staphylococcus epidermidis* is a linear beta-1,6-linked glucosaminoglycan: purification and structural analysis. J Bacteriol. 1996;178:175–83.
83. Jabbouri S, Sadovskaya I. Characteristics of the biofilm matrix and its role as a possible target for the detection and eradication of *Staphylococcus epidermidis* associated with medical implant infections. FEMS Immunol Med Microbiol. 2010;59:280–91.
84. Rupp ME, Archer GL. Hemagglutination and adherence to plastic by *Staphylococcus epidermidis*. Infect Immun. 1992;60:4322–7.

85. Mack D, Riedewald J, Rohde H, et al. Essential functional role of the polysaccharide intercellular adhesin of *Staphylococcus epidermidis* in hemagglutination. Infect Immun. 1999;67:1004–8.
86. Fey PD, Ulphani JS, Götz F, et al. Characterization of the relationship between polysaccharide intercellular adhesin and hemagglutination in *Staphylococcus epidermidis*. J Infect Dis. 1999;179:1561–4.
87. Joyce JG, Abeygunawardana C, Xu Q, et al. Isolation, structural characterization, and immunological evaluation of a high-molecular-weight exopolysaccharide from *Staphylococcus aureus*. Carbohydr Res. 2003;338:903–22.
88. Sadovskaya I, Vinogradov E, Flahaut S, et al. Extracellular carbohydrate-containing polymers of a model biofilm-producing strain, *Staphylococcus epidermidis* RP62A. Infect Immun. 2005;73:3007–17.
89. Tojo M, Yamashita N, Goldmann DA, et al. Isolation and characterization of a capsular polysaccharide adhesin from *Staphylococcus epidermidis*. J Infect Dis. 1988;157:713–22 [published erratum appears in J Infect Dis 1988 Jul;158(1):268].
90. McKenney D, Hübner J, Muller E, et al. The *ica* locus of *Staphylococcus epidermidis* encodes production of the capsular polysaccharide/adhesin. Infect Immun. 1998;66:4711–20.
91. Maira-Litran T, Kropec A, Abeygunawardana C, et al. Immunochemical properties of the staphylococcal poly-n-acetylglucosamine surface polysaccharide. Infect Immun. 2002;70: 4433–40.
92. Mack D, Rohde H, Harris LG, et al. Biofilm formation in medical device-related infection. Int J Artif Organs. 2006;29:343–59.
93. Götz F. *Staphylococcus* and biofilms. Mol Microbiol. 2002;43:1367–78.
94. Conlon KM, Humphreys H, O'Gara JP. *icaR* encodes a transcriptional repressor involved in environmental regulation of *ica* operon expression and biofilm formation in *Staphylococcus epidermidis*. J Bacteriol. 2002;184:4400–8.
95. Chang YM, Jeng WY, Ko TP, et al. Structural study of TcaR and its complexes with multiple antibiotics from *Staphylococcus epidermidis*. Proc Natl Acad Sci U S A. 2010;107:8617–22.
96. Ziebuhr W, Krimmer V, Rachid S, et al. A novel mechanism of phase variation of virulence in *Staphylococcus epidermidis*: evidence for control of the polysaccharide intercellular adhesin synthesis by alternating insertion and excision of the insertion sequence element IS256. Mol Microbiol. 1999;32:345–56.
97. Wang C, Li M, Dong D, et al. Role of ClpP in biofilm formation and virulence of *Staphylococcus epidermidis*. Microbes Infect. 2007;9:1376–83.
98. Wang L, Li M, Dong D, et al. SarZ is a key regulator of biofilm formation and virulence in *Staphylococcus epidermidis*. J Infect Dis. 2008;197:1254–62.
99. Knobloch JKM, Bartscht K, Sabottke A, et al. Biofilm formation by *Staphylococcus epidermidis* depends on functional RsbU, an activator of the *ΣB* operon: differential activation mechanisms due to ethanol and salt stress. J Bacteriol. 2001;183:2624–33.
100. Knobloch JKM, Jäger S, Horstkotte MA, et al. RsbU dependent regulation of *Staphylococcus epidermidis* biofilm formation is mediated via the alternative sigma factor σB by repression of the negative regulator gene *icaR*. Infect Immun. 2004;72:3838–48.
101. Jäger S, Jonas B, Pfanzelt D, et al. Regulation of biofilm formation by sigma B is a common mechanism in *Staphylococcus epidermidis* and is not mediated by transcriptional regulation of *sarA*. Int J Artif Organs. 2009;32:584–91.
102. Conlon KM, Humphreys H, O'Gara JP. Inactivations of *rsbU* and *sarA* by IS256 represent novel mechanisms of biofilm phenotypic variation in *Staphylococcus epidermidis*. J Bacteriol. 2004;186:6208–19.
103. Holland LM, O'Donnell ST, Ryjenkov DA, et al. A staphylococcal GGDEF domain protein regulates biofilm formation independently of c-di-GMP. J Bacteriol. 2008;190:5178–89.
104. Tormo MA, Marti M, Valle J, et al. SarA is an essential positive regulator of *Staphylococcus epidermidis* biofilm development. J Bacteriol. 2005;187:2348–56.
105. Xu L, Li H, Vuong C, et al. Role of the *luxS* quorum-sensing system in biofilm formation and virulence of *Staphylococcus epidermidis*. Infect Immun. 2006;74:488–96.

106. Rowe SE, Mahon V, Smith SG, et al. A novel role for SarX in *Staphylococcus epidermidis* biofilm regulation. Microbiology-Sgm. 2011;157:1042–9.
107. Wang X, Niu C, Sun G, et al. *ygs* is a novel gene that influences biofilm formation and the general stress response of *Staphylococcus epidermidis*. Infect Immun. 2011;79:1007–15.
108. Wang C, Fan J, Niu C, et al. Role of *spx* in biofilm formation of *Staphylococcus epidermidis*. FEMS Immunol Med Microbiol. 2010;59:152–60.
109. Mack D, Becker P, Chatterjee I, et al. Mechanisms of biofilm formation in *Staphylococcus epidermidis* and *Staphylococcus aureus*: functional molecules, regulatory circuits, and adaptive responses. Int J Med Microbiol. 2004;294:203–12.
110. Kong KF, Vuong C, Otto M. *Staphylococcus* quorum sensing in biofilm formation and infection. Int J Med Microbiol. 2006;296:133–9.
111. Mack D, Davies AP, Harris LG, et al. Microbial interactions in *Staphylococcus epidermidis* biofilms. Anal Bioanal Chem. 2007;387:399–408.
112. O'Gara JP. *ica* and beyond: biofilm mechanisms and regulation in *Staphylococcus epidermidis* and *Staphylococcus aureus*. FEMS Microbiol Lett. 2007;270:179–88.
113. Fey PD, Olson ME. Current concepts in biofilm formation of *Staphylococcus epidermidis*. Future Microbiol. 2010;5:917–33.
114. Knobloch JK, Horstkotte MA, Rohde H, et al. Evaluation of different detection methods of biofilm formation in *Staphylococcus aureus*. Med Microbiol Immunol. 2002;191:101–6.
115. Gill SR, Fouts DE, Archer GL, et al. Insights on evolution of virulence and resistance from the complete genome analysis of an early methicillin-resistant *Staphylococcus aureus* strain and a biofilm-producing methicillin-resistant *Staphylococcus epidermidis* strain. J Bacteriol. 2005;187:2426–38.
116. Vuong C, Kocianova S, Voyich JM, et al. A crucial role for exopolysaccharide modification in bacterial biofilm formation, immune evasion, and virulence. J Biol Chem. 2004;279: 54881–6.
117. Cerca N, Jefferson KK, Maira-Litran T, et al. Molecular basis for preferential protective efficacy of antibodies directed to the poorly acetylated form of staphylococcal poly-N-acetyl-beta-(1–6)-glucosamine. Infect Immun. 2007;75:3406–13.
118. Dobinsky S, Kiel K, Rohde H, et al. Glucose related dissociation between *icaADBC* transcription and biofilm expression by *Staphylococcus epidermidis*: evidence for an additional factor required for polysaccharide intercellular adhesin synthesis. J Bacteriol. 2003; 185:2879–86.
119. Rohde H, Knobloch JKM, Horstkotte MA, et al. Correlation of biofilm expression types of *Staphylococcus epidermidis* with polysaccharide intercellular adhesin synthesis: evidence for involvement of *icaADBC* genotype-independent factors. Med Microbiol Immunol. 2001; 190:105–12.
120. Bateman A, Holden MT, Yeats C. The G5 domain: a potential N-acetylglucosamine recognition domain involved in biofilm formation. Bioinformatics. 2005;21:1301–3.
121. Roche FM, Meehan M, Foster TJ. The *Staphylococcus aureus* surface protein SasG and its homologues promote bacterial adherence to human desquamated nasal epithelial cells. Microbiology. 2003;149:2759–67.
122. Roche FM, Massey R, Peacock SJ, et al. Characterization of novel LPXTG-containing proteins of *Staphylococcus aureus* identified from genome sequences. Microbiology. 2003;149:643–54.
123. Corrigan RM, Rigby D, Handley P, et al. The role of *Staphylococcus aureus* surface protein SasG in adherence and biofilm formation. Microbiology. 2007;153:2435–46.
124. Geoghegan JA, Corrigan RM, Gruszka DT, et al. Role of surface protein SasG in biofilm formation by *Staphylococcus aureus*. J Bacteriol. 2010;192:5663–73.
125. Conrady DG, Brescia CC, Horii K, et al. A zinc-dependent adhesion module is responsible for intercellular adhesion in staphylococcal biofilms. Proc Natl Acad Sci U S A. 2008;105: 19456–61.
126. Schommer NN, Christner M, Hentschke M, et al. *Staphylococcus epidermidis* uses distinct mechanisms of biofilm formation to interfere with phagocytosis and activation of mouse macrophage-like cells 774A.1. Infect Immun. 2011;79:2267–76.

127. Timmerman CP, Fleer A, Besnier JM, et al. Characterization of a proteinaceous adhesin of *Staphylococcus epidermidis* which mediates attachment to polystyrene. Infect Immun. 1991;59:4187–92.
128. Veenstra GJ, Cremers FF, van Dijk H, et al. Ultrastructural organization and regulation of a biomaterial adhesin of *Staphylococcus epidermidis*. J Bacteriol. 1996;178:537–41.
129. Macintosh RL, Brittan JL, Bhattacharya R, et al. The terminal a domain of the fibrillar accumulation-associated protein (Aap) of *Staphylococcus epidermidis* mediates adhesion to human corneocytes. J Bacteriol. 2009;191:7007–16.
130. Sun D, Accavitti MA, Bryers JD. Inhibition of biofilm formation by monoclonal antibodies against *Staphylococcus epidermidis* RP62A accumulation-associated protein. Clin Diagn Lab Immunol. 2005;12:93–100.
131. Hu J, Xu T, Zhu T, et al. Monoclonal antibodies against accumulation-associated protein affect eps biosynthesis and enhance bacterial accumulation of *Staphylococcus epidermidis*. PLoS One. 2011;6:e20918.
132. Harris LG, Bexfield A, Nigam Y, et al. Disruption of *Staphylococcus epidermidis* biofilms by medicinal maggot *Lucilia sericata* excretions/secretions. Int J Artif Organs. 2009;32:555–64.
133. Vandecasteele SJ, Peetermans WE, Merckx R, et al. Reliability of the *ica*, *aap* and *atlE* genes in the discrimination between invasive, colonizing and contaminant *Staphylococcus epidermidis* isolates in the diagnosis of catheter-related infections. Clin Microbiol Infect. 2003;9:114–9.
134. Stevens NT, Tharmabala M, Dillane T, et al. Biofilm and the role of the *ica* operon and *aap* in *Staphylococcus epidermidis* isolates causing neurosurgical meningitis. Clin Microbiol Infect. 2008;14:719–22.
135. Cucarella C, Solano C, Valle J, et al. Bap, a *Staphylococcus aureus* surface protein involved in biofilm formation. J Bacteriol. 2001;183:2888–96.
136. Potter A, Ceotto H, Giambiagi-Demarval M, et al. The gene *bap*, involved in biofilm production, is present in *Staphylococcus* spp. strains from nosocomial infections. J Microbiol. 2009;47:319–26.
137. Christensen GD, Baldassarri L, Simpson WA. Colonization of medical devices by coagulase-negative staphylococci. In: Bisno AL, Waldvogel FA, editors. Infections associated with indwelling medical devices. Washington, DC: American Society of Microbiology; 1994. p. 45–78. Chap 3.
138. Rupp ME, Ulphani JS, Fey PD, et al. Characterization of *Staphylococcus epidermidis* polysaccharide intercellular adhesin/hemagglutinin in the pathogenesis of intravascular catheter-associated infection in a rat model. Infect Immun. 1999;67:2656–9.
139. Rupp ME, Fey PD. *In vivo* models to evaluate adhesion and biofilm formation by *Staphylococcus epidermidis*. Methods Enzymol. 2001;336:206–15.
140. Rupp ME, Ulphani JS, Fey PD, et al. Characterization of the importance of polysaccharide intercellular adhesin/hemagglutinin of *Staphylococcus epidermidis* in the pathogenesis of biomaterial-based infection in a mouse foreign body infection model. Infect Immun. 1999;67:2627–32.
141. Rupp ME, Fey PD, Heilmann C, et al. Characterization of the importance of *Staphylococcus epidermidis* autolysin and polysaccharide intercellular adhesin in the pathogenesis of intravascular catheter-associated infection in a rat model. J Infect Dis. 2001;183:1038–42.
142. Li H, Xu L, Wang J, et al. Conversion of *Staphylococcus epidermidis* strains from commensal to invasive by expression of the *ica* locus encoding production of biofilm exopolysaccharide. Infect Immun. 2005;73:3188–91.
143. Begun J, Gaiani JM, Rohde H, et al. Staphylococcal biofilm exopolysaccharide protects against *Caenorhabditis elegans* immune defenses. PLoS Pathog. 2007;3:526–40.
144. Monk AB, Boundy S, Chu VH, et al. Analysis of the genotype and virulence of *Staphylococcus epidermidis* isolates from patients with infective endocarditis. Infect Immun. 2008;76:5127–32.
145. Vuong C, Voyich JM, Fischer ER, et al. Polysaccharide intercellular adhesin (PIA) protects *Staphylococcus epidermidis* against major components of the human innate immune system. Cell Microbiol. 2004;6:269–75.

146. Kristian SA, Birkenstock TA, Sauder U, et al. Biofilm formation induces C3a release and protects *Staphylococcus epidermidis* from IgG and complement deposition and from neutrophil-dependent killing. J Infect Dis. 2008;197:1028–35.
147. Cerca N, Jefferson KK, Oliveira R, et al. Comparative antibody-mediated phagocytosis of *Staphylococcus epidermidis* cells grown in a biofilm or in the planktonic state. Infect Immun. 2006;74:4849–55.
148. Maira-Litran T, Kropec A, Goldmann DA, et al. Comparative opsonic and protective activities of *Staphylococcus aureus* conjugate vaccines containing native or deacetylated Staphylococcal Poly-N-acetyl-beta-(1–6)-glucosamine. Infect Immun. 2005;73:6752–62.
149. Jäger S, Mack D, Rohde H, et al. Disintegration of *Staphylococcus epidermidis* biofilms under glucose-limiting conditions depends on the activity of the alternative sigma factor sigmaB. Appl Environ Microbiol. 2005;71:5577–81.
150. Zimmerli W, Waldvogel FA, Vaudaux P, et al. Pathogenesis of foreign body infection: description and characteristics of an animal model. J Infect Dis. 1982;146:487–97.
151. Francois P, Tu Quoc PH, Bisognano C, et al. Lack of biofilm contribution to bacterial colonisation in an experimental model of foreign body infection by *Staphylococcus aureus* and *Staphylococcus epidermidis*. FEMS Immunol Med Microbiol. 2003;35:135–40.
152. Chokr A, Leterme D, Watier D, et al. Neither the presence of *ica* locus, nor *in vitro*-biofilm formation ability is a crucial parameter for some *Staphylococcus epidermidis* strains to maintain an infection in a guinea pig tissue cage model. Microb Pathog. 2007;42:94–7.
153. Fluckiger U, Ulrich M, Steinhuber A, et al. Biofilm formation, *icaADBC* transcription, and polysaccharide intercellular adhesin synthesis by staphylococci in a device-related infection model. Infect Immun. 2005;73:1811–9.
154. Zimmerli W, Lew PD, Waldvogel FA. Pathogenesis of foreign body infection. Evidence for a local granulocyte defect. J Clin Invest. 1984;73:1191–200.
155. Vandecasteele SJ, Peetermans WE, Merckx R, et al. Expression of biofilm-associated genes in *Staphylococcus epidermidis* during *in vitro* and *in vivo* foreign body infections. J Infect Dis. 2003;188:730–7.
156. Kaplan JB, Ragunath C, Ramasubbu N, et al. Detachment of *Actinobacillus actinomycetemcomitans* biofilm cells by an endogenous beta-hexosaminidase activity. J Bacteriol. 2003;185:4693–8.
157. Kaplan JB, Ragunath C, Velliyagounder K, et al. Enzymatic detachment of *Staphylococcus epidermidis* biofilms. Antimicrob Agents Chemother. 2004;48:2633–6.
158. Mehlin C, Headley CM, Klebanoff SJ. An inflammatory polypeptide complex from *Staphylococcus epidermidis*: isolation and characterization. J Exp Med. 1999;189:907–18.
159. Vuong C, Durr M, Carmody AB, et al. Regulated expression of pathogen-associated molecular pattern molecules in *Staphylococcus epidermidis*: quorum-sensing determines pro-inflammatory capacity and production of phenol-soluble modulins. Cell Microbiol. 2004;6:753–9.
160. Yao Y, Sturdevant DE, Otto M. Genomewide analysis of gene expression in *Staphylococcus epidermidis* biofilms: insights into the pathophysiology of *S. epidermidis* biofilms and the role of phenol-soluble modulins in formation of biofilms. J Infect Dis. 2005;191:289–98.
161. Cheung GY, Rigby K, Wang R, et al. *Staphylococcus epidermidis* strategies to avoid killing by human neutrophils. PLoS Pathog. 2010;6:e1001133.
162. Wang R, Khan BA, Cheung GY, et al. *Staphylococcus epidermidis* surfactant peptides promote biofilm maturation and dissemination of biofilm-associated infection in mice. J Clin Invest. 2011;121:238–48.
163. Vaudaux PE, Lew DP, Waldvogel FA. Host factors predeposing to and influencing therapy of foreign body infections. In: Bisno AL, Waldvogel FA, editors. Infections associated with indwelling medical devices. Washington, DC: American Society for Microbiology; 1994. p. 1–29. Chap 1.
164. Mack D, Bartscht K, Dobinsky S, Horstkotte MA, Kiel K, Knobloch JKM, Schäfer P. Staphylococcal factors involved in adhesion and biofilm formation on biomaterials. In: An YH, Friedman RJ, editors. Handbook for studying bacterial adhesion: principles, methods, and applications. Totowa: Humana Press; 2000. p. 307–30. Chap 20.

165. Espersen F, Wilkinson BJ, Gahrn-Hansen B, et al. Attachment of staphylococci to silicone catheters *in vitro*. APMIS. 1990;98:471–8.
166. Hogt AH, Dankert J, Feijen J. Adhesion of *Staphylococcus epidermidis* and *Staphylococcus saprophyticus* to a hydrophobic biomaterial. J Gen Microbiol. 1985;131:2485–91.
167. Hogt AH, Dankert J, Feijen J. Adhesion of coagulase-negative staphylococci to methacrylate polymers and copolymers. J Biomed Mater Res. 1986;20:533–45.
168. Hogt AH, Dankert J, Hulstaert CE, et al. Cell surface characteristics of coagulase-negative staphylococci and their adherence to fluorinated poly(ethylenepropylene). Infect Immun. 1986;51:294–301.
169. Pascual A, Fleer A, Westerdaal NA, et al. Modulation of adherence of coagulase-negative staphylococci to teflon catheters *in vitro*. Eur J Clin Microbiol Infect Dis. 1986;5:518–22.
170. Muller E, Takeda S, Shiro H, et al. Occurrence of capsular polysaccharide/adhesin among clinical isolates of coagulase-negative staphylococci. J Infect Dis. 1993;168:1211–8.
171. Ludwicka A, Jansen B, Wadstrom T, et al. Attachment of staphylococci to various synthetic polymers. Zentralbl Bakteriol Mikrobiol Hyg A. 1984;256:479–89.
172. Mack D, Bartscht K, Fischer C, et al. Genetic and biochemical analysis of *Staphylococcus epidermidis* biofilm accumulation. Methods Enzymol. 2001;336:215–39.
173. Patel JD, Ebert M, Ward R, et al. *S. epidermidis* biofilm formation: effects of biomaterial surface chemistry and serum proteins. J Biomed Mater Res A. 2007;80:742–51.
174. Heilmann C, Hussain M, Peters G, et al. Evidence for autolysin-mediated primary attachment of *Staphylococcus epidermidis* to a polystyrene surface. Mol Microbiol. 1997;24:1013–24.
175. Li DQ, Lundberg F, Ljungh A. Characterization of vitronectin-binding proteins of *Staphylococcus epidermidis*. Curr Microbiol. 2001;42:361–7.
176. Zoll S, Patzold B, Schlag M, et al. Structural basis of cell wall cleavage by a staphylococcal autolysin. PLoS Pathog. 2010;6:e1000807.
177. Higashi JM, Wang IW, Shlaes DM, et al. Adhesion of *Staphylococcus epidermidis* and transposon mutant strains to hydrophobic polyethylene. J Biomed Mater Res. 1998;39:341–50.
178. Izano EA, Amarante MA, Kher WB, et al. Differential roles of poly-N-acetylglucosamine surface polysaccharide and extracellular DNA in *Staphylococcus aureus* and *Staphylococcus epidermidis* biofilms. Appl Environ Microbiol. 2008;74:470–6.
179. Qin Z, Ou Y, Yang L, et al. Role of autolysin-mediated DNA release in biofilm formation of *Staphylococcus epidermidis*. Microbiology. 2007;153:2083–92.
180. Patti JM, Allen BL, McGavin MJ, et al. MSCRAMM-mediated adherence of microorganisms to host tissues. Annu Rev Microbiol. 1994;48:585–617.
181. Heilmann C, Thumm G, Chhatwal GS, et al. Identification and characterization of a novel autolysin (Aae) with adhesive properties from *Staphylococcus epidermidis*. Microbiology. 2003;149:2769–78.
182. Nilsson M, Frykberg L, Flock JI, et al. A fibrinogen-binding protein of *Staphylococcus epidermidis*. Infect Immun. 1998;66:2666–73.
183. Hartford O, O'Brien L, Schofield K, et al. The Fbe (SdrG) protein of *Staphylococcus epidermidis* HB promotes bacterial adherence to fibrinogen. Microbiology. 2001;147:2545–52.
184. Arciola CR, Campoccia D, Gamberini S, et al. Presence of fibrinogen-binding adhesin gene in *Staphylococcus epidermidis* isolates from central venous catheters-associated and orthopaedic implant-associated infections. Biomaterials. 2004;25:4825–9.
185. Sellman BR, Timofeyeva Y, Nanra J, et al. Expression of *Staphylococcus epidermidis* SdrG increases following exposure to an *in vivo* environment. Infect Immun. 2008;76:2950–7.
186. Pei L, Flock JI. Functional study of antibodies against a fibrogenin-binding protein in *Staphylococcus epidermidis* adherence to polyethylene catheters. J Infect Dis. 2001;184:52–5.
187. Rennermalm A, Nilsson M, Flock JI. The fibrinogen binding protein of *Staphylococcus epidermidis* is a target for opsonic antibodies. Infect Immun. 2004;72:3081–3.
188. Vernachio JH, Bayer AS, Ames B, et al. Human immunoglobulin G recognizing fibrinogen-binding surface proteins is protective against both *Staphylococcus aureus* and *Staphylococcus epidermidis* infections *in vivo*. Antimicrob Agents Chemother. 2006;50:511–8.

189. Guo B, Zhao X, Shi Y, et al. Pathogenic implication of a fibrinogen-binding protein of *Staphylococcus epidermidis* in a rat model of intravascular-catheter-associated infection. Infect Immun. 2007;75:2991–5.
190. Hussain M, Heilmann C, Peters G, et al. Teichoic acid enhances adhesion of *Staphylococcus epidermidis* to immobilized fibronectin. Microb Pathog. 2001;31:261–70.
191. Bowden MG, Visai L, Longshaw CM, et al. Is the GehD lipase from *Staphylococcus epidermidis* a collagen binding adhesin? J Biol Chem. 2002;277:43017–23.
192. Arrecubieta C, Lee MH, Macey A, et al. SdrF, a *Staphylococcus epidermidis* surface protein, binds type I collagen. J Biol Chem. 2007;282:18767–76.
193. Arrecubieta C, Toba FA, von Bayern M, et al. SdrF, a *Staphylococcus epidermidis* surface protein, contributes to the initiation of ventricular assist device driveline-related infections. PLoS Pathog. 2009;5:e1000411.
194. Kogan G, Sadovskaya I, Chaignon P, et al. Biofilms of clinical strains of *Staphylococcus* that do not contain polysaccharide intercellular adhesin. FEMS Microbiol Lett. 2006; 255:11–6.
195. Frank KL, Hanssen AD, Patel R. *icaA* is not a useful diagnostic marker for prosthetic joint infection. J Clin Microbiol. 2004;42:4846–9.
196. Dice B, Stoodley P, Buchinsky F, et al. Biofilm formation by *ica*-positive and *ica*-negative strains of *Staphylococcus epidermidis in vitro*. Biofouling. 2009;25:367–75.
197. Kozitskaya S, Olson ME, Fey PD, et al. Clonal analysis of *Staphylococcus epidermidis* isolates carrying or lacking biofilm-mediating genes by multilocus sequence typing. J Clin Microbiol. 2005;43:4751–7.
198. Frebourg NB, Lefebvre S, Baert S, et al. PCR-Based assay for discrimination between invasive and contaminating *Staphylococcus epidermidis* strains. J Clin Microbiol. 2000; 38:877–80.
199. Arciola CR, Baldassarri L, Montanaro L. Presence of *icaA* and *icaD* genes and slime production in a collection of staphylococcal strains from catheter-associated infections. J Clin Microbiol. 2001;39:2151–6.
200. Maki DG, Weise CE, Sarafin HW. A semiquantitative culture method for identifying intravenous-catheter-related infection. N Engl J Med. 1977;296:1305–9.
201. Petrelli D, Zampaloni C, D'Ercole S, et al. Analysis of different genetic traits and their association with biofilm formation in *Staphylococcus epidermidis* isolates from central venous catheter infections. Eur J Clin Microbiol Infect Dis. 2006;25:773–81.
202. Cafiso V, Bertuccio T, Santagati M, et al. Presence of the *ica* operon in clinical isolates of *Staphylococcus epidermidis* and its role in biofilm production. Clin Microbiol Infect. 2004;10:1081–8.
203. Ninin E, Caroff N, Espaze E, et al. Assessment of *ica* operon carriage and biofilm production in *Staphylococcus epidermidis* isolates causing bacteraemia in bone marrow transplant recipients. Clin Microbiol Infect. 2006;12:446–52.
204. de Silva GD, Kantzanou M, Justice A, et al. The *ica* operon and biofilm production in coagulase-negative staphylococci associated with carriage and disease in a neonatal intensive care unit. J Clin Microbiol. 2002;40:382–8.
205. Klingenberg C, Aarag E, Ronnestad A, et al. Coagulase-negative staphylococcal sepsis in neonates. Association between antibiotic resistance, biofilm formation and the host inflammatory response. Pediatr Infect Dis J. 2005;24:817–22.
206. Bradford R, Abdul MR, Daley AJ, et al. Coagulase-negative staphylococci in very-low-birth-weight infants: inability of genetic markers to distinguish invasive strains from blood culture contaminants. Eur J Clin Microbiol Infect Dis. 2006;25:283–90.
207. Pien BC, Sundaram P, Raoof N, et al. The clinical and prognostic importance of positive blood cultures in adults. Am J Med. 2010;123:819–28.
208. Galdbart JO, Allignet J, Tung HS, et al. Screening for *Staphylococcus epidermidis* markers discriminating between skin-flora strains and those responsible for infections of joint prostheses. J Infect Dis. 2000;182:351–5.

209. Arciola CR, Campoccia D, Gamberini S, et al. Search for the insertion element IS256 within the *ica* locus of *Staphylococcus epidermidis* clinical isolates collected from biomaterial-associated infections. Biomaterials. 2004;25:4117–25.
210. Koskela A, Nilsdotter-Augustinsson A, Persson L, et al. Prevalence of the *ica* operon and insertion sequence IS256 among *Staphylococcus epidermidis* prosthetic joint infection isolates. Eur J Clin Microbiol Infect Dis. 2009;28:655–60.
211. Klug D, Wallet F, Kacet S, et al. Involvement of adherence and adhesion *Staphylococcus epidermidis* genes in pacemaker lead-associated infections. J Clin Microbiol. 2003; 41:3348–50.
212. Boelens JJ, Zaat SA, Meeldijk J, et al. Subcutaneous abscess formation around catheters induced by viable and nonviable *Staphylococcus epidermidis* as well as by small amounts of bacterial cell wall components. J Biomed Mater Res. 2000;50:546–56.
213. Broekhuizen CA, de Boer L, Schipper K, et al. Peri-implant tissue is an important niche for *Staphylococcus epidermidis* in experimental biomaterial-associated infection in mice. Infect Immun. 2007;75:1129–36.
214. Broekhuizen CA, Sta M, Vandenbroucke-Grauls CM, et al. Microscopic detection of viable *Staphylococcus epidermidis* in peri-implant tissue in experimental biomaterial-associated infection, identified by bromodeoxyuridine incorporation. Infect Immun. 2010;78:954–62.

Chapter 3
Staphylococcal Virulence Factors

Christine Heilmann and Friedrich Götz

Abstract The adherence of staphylococci to the unmodified or host factor-coated surface of a medical device is the first critical event in the establishment of a device-associated infection with these serious pathogens. *Staphylococcus aureus* in particular harbors a variety of proteinaceous and non-proteinaceous adhesins that mediate attachment to a multitude of host factors, such as extracellular matrix and plasma proteins and platelets, or intercellular adhesins, which are essential for biofilm accumulation. Proteinaceous adhesins may be classified in covalently surface-anchored proteins of the microbial surface components recognizing adhesive matrix molecules family or in proteins that are surface-associated by different means, such as ionic or hydrophobic interactions. Non-covalently surface-associated proteins include the autolysin/adhesins, proteins of the secretable expanded repertoire adhesive molecules family, or membrane-spanning proteins. Non-proteinaceous adhesins include the polysaccharide intercellular adhesin as well as lipoteichoic and wall teichoic acids. The features and functions of surface and surface-associated protein adhesins as well as of non-proteinaceous adhesins are discussed.

C. Heilmann (✉)
The Interdisciplinary Center for Clinical Research (IZKF), Institute for Medical Microbiology, Münster University Hospital of Münster, Domagkstraße 10, Münster 48149, Germany
e-mail: heilmac@uni-muenster.de

F. Götz
Department of Microbial Genetics, Faculty of Biology, University of Tübingen, Waldhäuser Straße 70/8, Tübingen 72076, Germany
e-mail: friedrich.goetz@uni-tuebingen.de

T.F. Moriarty et al. (eds.), *Biomaterials Associated Infection: Immunological Aspects and Antimicrobial Strategies*, DOI 10.1007/978-1-4614-1031-7_3,

Abbreviations

aa	Amino acids
CHAP	Cysteine, histidine-dependent amidohydrolase/peptidase
Clf	Clumping factor
Cn	Collagen
Eap	Extracellular adherence protein
Ebh	Extracellular matrix-binding protein homologue
ECM	Extracellular matrix
eDNA	Extracellular DNA
Efb	Extracellular fibrinogen-binding protein
Emp	Extracellular matrix protein-binding protein
Fg	Fibrinogen
Fn	Fibronectin
FnBP	Fibronectin-binding protein
LTA	Lipoteichoic acid
LysM	Lysine motif
Map	MHC class II analogous protein
MSCRAMM	Microbial surface component recognizing adhesive matrix molecules
PIA	Polysaccharide intercellular adhesin
PSM	Phenol-soluble modulin
SERAM	Secretable expanded repertoire adhesive molecules
TSP	Thrombospondin
Vn	Vitronectin
vWf	Von Willebrand factor
vWbp	Von Willebrand factor-binding protein
WTA	Wall teichoic acid

3.1 Introduction

The Gram-positive staphylococci are ubiquitous commensals of the skin and mucous membranes of humans and animals. In humans, *Staphylococcus aureus* and the coagulase-negative species *Staphylococcus epidermidis* are also among the leading causes of nosocomial infections [1]. Staphylococcal infections range from mild skin infections to serious and life-threatening syndromes, such as endocarditis, osteomyelitis, pneumonia, and sepsis [1–3]. Infections due to *S. epidermidis* typically are more subacute or even chronic and require an immune-compromised or predisposed host, i.e., patients with indwelling medical devices, like prosthetic heart valves and joints, artificial pacemakers, and intravascular catheters of various polymer materials (e.g., polyethylene, polyurethane, and silicon rubber) or metals (e.g., stainless steel, cobalt chrome, and titanium) [4, 5]. In contrast, the coagulase-positive *S. aureus* causes more acute infections associated with the colonization of the host tissue, such as endocarditis and osteomyelitis, but is also a common cause of medical

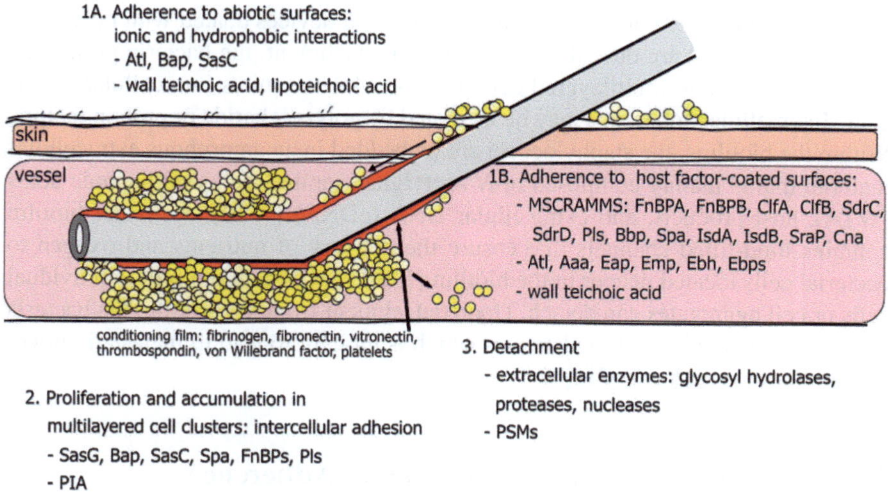

Fig. 3.1 Schematic model of mechanisms involved in the different phases of *Staphylococcus aureus* biofilm formation and detachment. *S. aureus* factors involved in the colonization of an unmodified or host factor-coated catheter surface are given

device-associated infections. The more virulent appearance of *S. aureus* is reflected by the production of numerous virulence factors, i.e., a variety of adhesins, extracellular enzymes, and toxins. In contrast, the capacity of *S. epidermidis* to produce adhesins and especially to secrete extracellular enzymes and toxins is much less pronounced.

The most critical pathogenicity factor in medical device-associated infections is the colonization of the foreign body by the bacteria and formation of a three-dimensional structure called biofilm. The presence of large adherent biofilms on explanted intravascular catheters has been demonstrated by scanning electron microscopy [6]. Microorganisms within a biofilm are protected against antimicrobial chemotherapy as well as against the host immune system. As a consequence, in the case of implant-associated infections the removal of the medical device frequently is necessary to eradicate the infection. In this chapter, the molecular mechanisms involved in staphylococcal (especially *S. aureus* and some aspects of *S. epidermidis*) colonization of a medical device, i.e., the different mechanisms involved in adherence to the unmodified or host factor-coated surface of the medical device and subsequent biofilm accumulation and detachment mechanisms, are discussed (summarized in Fig. 3.1).

3.2 Staphylococcal Biofilm Formation

To form a biofilm, staphylococci first attach to the surface of a medical device either directly or mediated by host products, such as extracellular matrix and plasma proteins or platelets, which are immobilized on the device surface shortly after its insertion. The factors involved either in direct adherence to the abiotic surfaces of medical devices (see Sect. 3.3) or to the surfaces of medical devices coated with host products (see Sect. 3.4) are described below. Upon attachment, the bacteria proliferate and accumulate into multilayered cell clusters, which requires intercellular adherence. Intercellular adherence may be mediated by polysaccharide or protein factors. Within the biofilms, the staphylococci are embedded in an amorphous extracellular material that is mainly composed of *N*-acetylglucosamine, cell wall teichoic acids (WTA), host products, and extracellular DNA (eDNA) [7–10]. A mature biofilm contains fluid-filled channels that ensure the delivery of nutrients and oxygen to bacterial cells located deeper in the biofilm [11]. From a mature biofilm, individual cells or cell aggregates can detach. Upon detachment from the biofilm, the bacteria may disseminate via the bloodstream, which is thought to lead to metastatic infection and/or development of sepsis.

3.3 Factors Involved in Staphylococcal Adherence to the Abiotic Surfaces of Medical Devices

In medical device-associated infection, the infection of the foreign material likely occurs by inoculation with only a few bacteria from the patient's skin or mucous membranes during implantation of the device. Microbial adherence to biomaterials largely depends on the nature of the polymer material and on the cell surface characteristics of the bacteria. The initial interactions are believed to occur via nonspecific physicochemical forces such as charge, van der Waal's forces, and hydrophobic interactions.

3.3.1 Covalently Linked Cell Surface Proteins (MSCRAMMs)

A covalently linked cell surface protein from *S. aureus*, the 239-kDa biofilm-associated protein Bap, is involved in attachment to a polystyrene surface and intercellular adhesion leading to biofilm formation (see also below) [12]. The structural features of Bap correspond to those of other microbial surface components recognizing adhesive matrix molecules (MSCRAMMs) (see below). The clinical significance of Bap is not clear, because it is apparently present in only 5 % of 350 bovine mastitis and absent from all human clinical *S. aureus* isolates tested so far [12]. However, a gene encoding a Bap-homologous protein, the 258 kDa Bhp, is present in the human clinical strain *S. epidermidis* RP62A [13].

3.3.2 Non-covalently Linked Surface-Associated Proteins: Autolysin/Adhesins

Initial adherence has also been attributed to non-covalently linked surface proteins. The 148-kDa *S. epidermidis* autolysin AtlE is a surface-associated component that mediates primary attachment of staphylococcal cells to a polystyrene surface [14] (see below). Possibly, the three repetitive sequences R1, R2, and R3 in the central part of the protein are involved in that adhesive function.

3.3.3 Non-proteinaceous Staphylococcal Adhesins: Wall Teichoic Acid and Lipoteichoic Acid

The *S. aureus* colonization of abiotic surfaces also depends on the charge of its teichoic acid. *S. aureus* teichoic acids are highly charged cell wall polymers, composed of alternating phosphate and ribitol (WTA) or glycerol (lipoteichoic acid; LTA) groups, which are substituted with D-alanine and *N*-acetylglucosamine. While the WTA is covalently linked to the peptidoglycan, the LTA is anchored in the outer leaflet of the cytoplasmic membrane via a glycolipid. The D-alanine–D-alanyl carrier protein ligase DltA is involved in the transfer of D-alanine into WTA. Thus, a *dltA* mutant lacks D-alanine in its WTA rendering it more negatively charged. The *dltA* mutant has a reduced capability to form biofilms due to a decreased initial attachment to polystyrene or glass, which is hydrophobic or negatively charged, respectively [15].

The glycolipid synthase YpfP is involved in the biosynthesis of LTA. An *S. aureus* SA113 *ypfP* mutant, which showed a markedly decreased production of LTA, revealed altered physicochemical properties and a reduced capacity to form a biofilm on a polystyrene surface [16]. Thus, LTA and its biosynthetic enzymes were proposed as potential targets in the development of novel anti-biofilm measures.

3.4 Factors Involved in Staphylococcal Adherence to the Surfaces of Medical Devices Coated with Host Products

Soon after insertion or implantation of a medical device, it becomes coated with extracellular matrix and plasma proteins and platelets. *S. aureus* is able to directly adhere to a multitude of plasma proteins and components of the extracellular matrix, such as fibronectin (Fn), fibrinogen (Fg), vitronectin (Vn), thrombospondin (TSP), bone sialoprotein, elastin, collagen (Cn), and von Willebrand factor (vWf) as well as platelets. Thus, all these host factors could serve as specific receptors for colonizing bacteria [17, 18]. Proteinaceous staphylococcal surface adhesins can either be covalently linked to the cell wall peptidoglycan or surface-associated by different means, such as ionic or hydrophobic interactions.

3.4.1 Covalently Linked Cell Surface Proteins (MSCRAMMs)

Most staphylococcal surface adhesins are covalently linked to the cell wall peptidoglycan and belong to the MSCRAMMs protein family [19]. Covalently linked

MSCRAMMs have a common overall organization including an N-terminal signal peptide, an exposed ligand-binding domain, which is often followed or interrupted by direct repeated sequences, a characteristic hydrophobic cell wall- and membrane-spanning domain, a C-terminal LPXTG motif responsible for cell-wall anchorage, and a positively charged tail. In most cases, cell-wall anchorage is mediated by a membrane-bound transpeptidase, called sortase (SrtA) that cleaves the peptide bond between the threonine and glycine of the LPXTG motif and covalently links the carboxyl group of threonine with the amino group of peptidoglycan cross-bridges [20–22]. A second iron-regulated sortase, SrtB, recognizes the motif NPQTN and is involved in cell surface anchorage of the iron-regulated surface determinant C (IsdC), which has been the only known substrate so far [23]. *S. aureus* genomes contain more than 20 genes encoding surface-anchored adhesins (most of them listed in Table 3.1). In contrast, there are only 12 genes encoding covalently linked MSCRAMMs in the *S. epidermidis* RP62A genome [24]. MSCRAMMs can bind to one or more host extracellular matrix and plasma proteins and a given host factor can be bound by more than one staphylococcal adhesin. However, neither the ligands nor the functions of all MSCRAMMs have been identified to date.

3.4.1.1 Fg-Binding Proteins (ClfA and ClfB) and the Sdr Protein Family

Fg-Binding Proteins: ClfA and ClfB

The binding of *S. aureus* to surface-immobilized Fg is believed to play a significant role in the development of biomaterial-associated infections [25]. Fg is a soluble 340-kDa glycoprotein and is composed of three pairs of nonidentical polypeptide chains (AαBβγ) that are arranged in a symmetrical dimeric structure. Fg is an important blood clotting factor and adhesion protein that mediates platelet adherence and aggregation. *S. aureus* produces several proteins that can specifically bind to Fg. Two of them are the covalently linked Fg-binding MSCRAMMs, clumping factors A and B (ClfA and ClfB), which are encoded by *clfA* and *clfB*, respectively [26, 27]. The Fg-binding MSCRAMM was first recognized as a factor that mediates cell clumping in the presence of human plasma and therefore was named clumping factor. *clfA* mutants not only failed to form clumps in soluble Fg, but also adhered poorly to surface-immobilized Fg [26]. In addition to the features common to all MSCRAMMs, ClfA and ClfB also possess the region R that consists of repeating serine–aspartate (SD) dipeptides, which is located between the wall-spanning domain and the ligand-binding domain A (Fig. 3.2). Thus, ClfA and ClfB are members of an MSCRAMM subfamily, the SD repeat-containing (Sdr) protein family (see below).

The predicted 92-kDa ClfA has a significant amino acid (aa) homology with the Fn-binding proteins (FnBPs, see below), especially in the N-terminal region A. The Fg-binding activity of ClfA was mapped to a 218-aa domain within the ~500-aa region A (residues 332–550) [28]. In a later study, the adjacent residues Glu^{526} and Val^{527} were identified as being important for the Fg-binding activity [29]. Despite similar

Table 3.1 *Staphylococcus aureus* (*Staphylococcus epidermidis*) genes encoding surface and surface-associated proteins involved in biofilm formation

Gene	Product	Ligand specificity/function	References
Covalently linked surface proteins involved in attachment			
bbp	Bbp	Binding to bone sialoprotein	[51]
Bap	Bap	Attachment to polystyrene	[12]
(bhp)	(Bhp)		[13]
clfA	ClfA	Binding to Fg; platelet aggregation	[26, 32]
clfB	ClfB	Binding to Fg, cytokeratin 10; platelet aggregation	[27, 32, 35]
cna	Cna	Binding to Cn	[160]
fnbA	FnBPA	Binding to Fn, Fg, elastin	[54, 57, 58]
fnbB	FnBPB	Binding to Fn, elastin	[55, 58]
isdA	IsdA	Binding to Fg, Fn, fetuin, hemoglobin, transferrin, hemin	[88, 89]
isdB	IsdB	Binding to hemoglobin, hemin, platelet integrin GPIIb/IIIa	[88, 92]
isdC	IsdC	Binding to hemin	[88]
isdH	IsdH	Binding to haptoglobulin, haptoglobulin–hemoglobin complex	[87]
sasA/sraP	SasA/SraP	Binding to platelets	[79]
sasC	SasC	Attachment to polystyrene	[130]
sdrC	SdrC	Binding to β-neurexin	[46]
sdrE	SdrE	Platelet aggregation	[32]
spa	Protein A/Spa	Binding to IgG, IgM, von Willebrand Factor, platelet receptor gC1qR	[81, 84, 85]
pls	Pls	Binding to glycolipids	[47]
Covalently linked surface proteins involved in biofilm accumulation			
Bap	Bap	Intercellular adhesion	[12]
(bhp)	(Bhp)		[13]
fnbA	FnBPA	Intercellular adhesion	[133]
fnbB	FnBPB	Intercellular adhesion	[133]
sasC	SasC	Intercellular adhesion	[130]
sasG	SasG	Intercellular adhesion	[123]
(aap)	(Aap)		[124]
spa	Protein A/Spa	Intercellular adhesion/bacterial cell aggregation	[131]
pls	Pls	Intercellular adhesion/bacterial cell aggregation	[48]
Non-covalently linked surface-associated proteins involved in attachment			
atl	Atl	Attachment to polystyrene, Fg, Fn, Vn	[96, 97]
(atlE)	(AtlE)		[14]
aaa (aae)	Aaa	Binding to Fg, Fn, Vn	[104]
	(Aae)		[105]
eap	Eap (Map, P70)	Binding to Fg, Fn, Vn, Cn, ICAM-1	[108, 111]
emp	Emp	Binding to Fg, Fn, Vn, Cn	[108]
ebh (embp)	Ebh (Embp)	Binding to Fn	[115]
			[116]
ebps	Ebps	Binding to elastin	[118]
Non-covalently linked surface-associated protein involved in biofilm accumulation			
(embp)	(Embp)	Intercellular adhesion	[134]

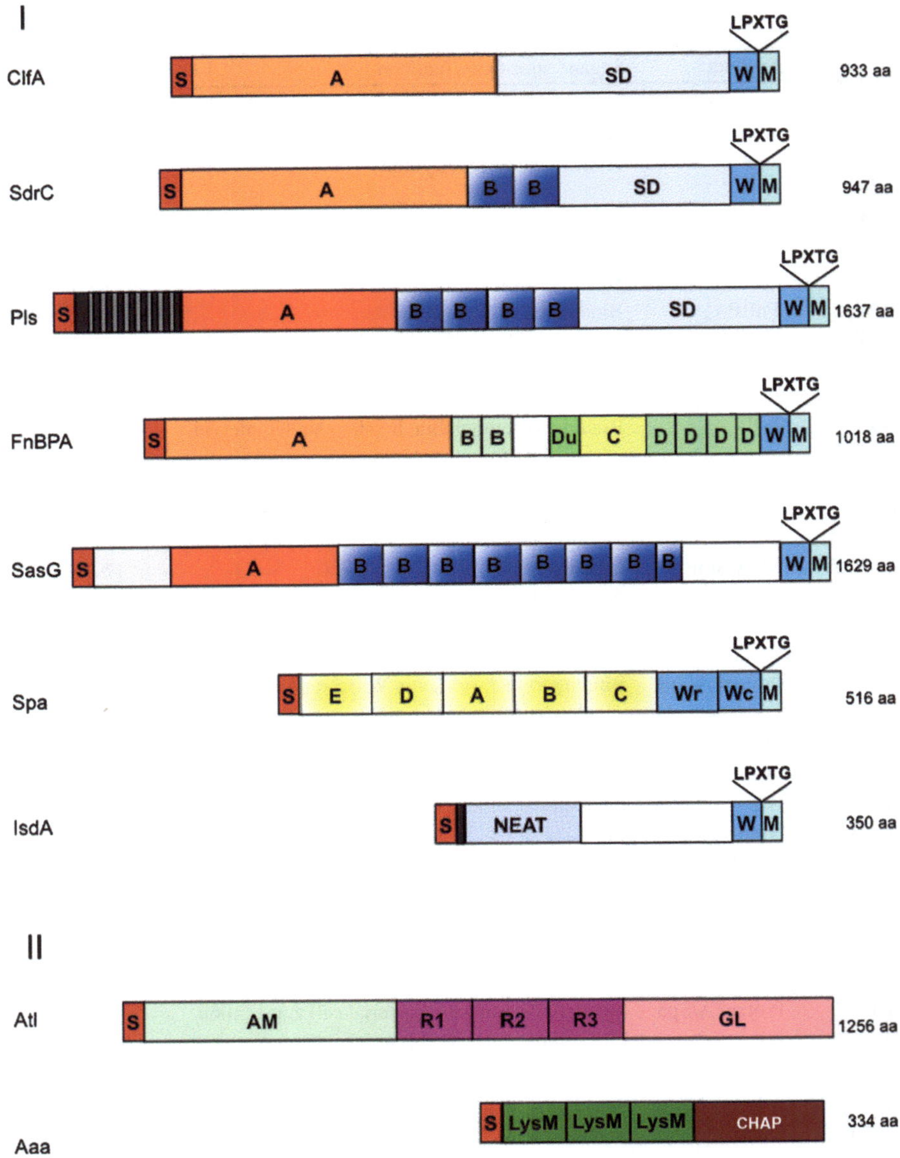

Fig. 3.2 Schematic model of selected *S. aureus* MSCRAMMs (I) and autolysin/adhesins (II). The positions of the signal sequences (S), the ligand-binding A domains (*A*), the B repeats (*B*), the serine–aspartate (SD) dipeptide repeats (or region R), the wall (W) and membrane (M)-spanning regions including positively charged residues, and the LPXTG motifs are shown. The domains are defined in the text. The A domains of ClfA, SdrC, and FnBPA are homologous. Moreover, the A domains of Pls and SasG as well as the B repeats of SdrC, Pls, and SasG are homologous. The sizes of the proteins may vary among different *S. aureus* strains. **I**: FnBPA: D, Fn-binding D repeats. C, non-repeated region with unknown function. Spa: E, D, A, B, C, IgG-binding domains, *Wr* is composed of an octapeptide repeat, and *Wc* is a non-repeated region. *IsdA*: ligand-binding near-iron transporter (NEAT) domain. **II**: Atl: *AM* amidase, *R* repeat, *GL* glucosaminidase. Aaa: *LysM* lysine motif, *CHAP* cysteine, histidine-dependent amidohydrolase/peptidase

functions and structural organization, the A domains of ClfA and the 124-kDa ClfB share only 26 % aa identity. ClfA exclusively binds to the C-terminus of the Fg γ-chain. In contrast, ClfB binds to the α- and β-chains of Fg [27, 30]. A recombinantly expressed and purified truncated ClfA protein comprising aa 221–550 (ClfA33) inhibits ADP-induced, Fg-dependent platelet aggregation in a concentration-dependent manner and moreover competes for platelet adhesion to immobilized Fg under shear stress indicating that the Fg-binding sites for the platelet receptor and ClfA overlap [30]. A later study indicated that ClfA also mediates direct binding to platelets by interacting with a 118-kDa platelet membrane receptor [31]. Furthermore, ClfA as well as ClfB are able to aggregate platelets, which are thought to play a role in the establishment of experimental endocarditis [32]. Platelet aggregation mediated by ClfA and ClfB occurs in an Fg-dependent manner. However, ClfA as well as ClfB can also cause platelet aggregation in an Fg-independent manner that requires IgG and complement deposition [33, 34].

ClfB binds also to human desquamated nasal epithelial cells and to cytokeratin 10, which was shown to be present on the surface of these cells, suggesting that ClfB may be an important factor in *S. aureus* nasal colonization [35]. A remarkable difference between ClfA and ClfB is their expression pattern with ClfB being only detectable on cells grown to the early exponential phase, but absent from cells from late exponential phase or stationary phase cultures [27].

The binding of ClfA as well as ClfB to Fg is regulated by the divalent cations Ca^{2+} and Mn^{2+} [27, 36]. Both cations inhibit ClfA-mediated clumping of *S. aureus* in the presence of soluble Fg as well as the interaction of a recombinant ClfA subdomain with a peptide resembling the C-terminal Fg γ-chain [36]. In agreement with this, ClfA harbors a putative EF-hand motif within the A region (aa 310–321) that mediates both the regulation by Ca^{2+} and ligand binding.

The ClfA region A is composed of three domains (N1, N2, and N3). The crystal structure of the Fg-binding segment (aa 221–559), which contains two of the domains (N2N3) demonstrates that each domain adopts an IgG-like fold [37]. Structural predictions suggested that the A domains of ClfB, FnBPA, and FnBPB also contain such IgG-like folds. Mutagenic substitution of the ClfA aa Tyr^{256}, Pro^{336}, Tyr^{338}, and Lys^{389}, which were proposed to contact the four C-terminal aa of the Fg γ-chain, resulted in proteins with no or strongly reduced affinity for Fg [37].

Consistent with its platelet-binding and -aggregation function, ClfA has been shown to be a virulence factor in a rabbit and rat infective endocarditis model as well as in a mouse model of septic arthritis [31, 38–40]. Moreover, active and passive immunization studies suggested ClfA as a suitable vaccine compound and from a structural model of the ClfA–Fg interaction, the potential for the design of novel anti-staphylococcal agents was proposed [40–42].

Sdr Protein Family: *S. aureus* SdrC, SdrD, SdrE, Pls

Other members of the *S. aureus* Sdr-protein family are SdrC, SdrD, SdrE, and Pls [43, 44]. The Sdr proteins are predicted to have a similar structural organization like ClfA and ClfB. Additionally, they contain two (SdrC), five (SdrD), three

(SdrE), and four (Pls) B repeats with 110–129 aa per repeat, which interconnect the regions A and R (Fig. 3.2). Like ClfA and ClfB, SdrE is able to induce platelet aggregation [32] and a function for SdrC and SdrD in adherence to human desquamated nasal epithelial cells could be established [45]. Moreover, by using the phage display technique, the N2N3 domain of SdrC was recently found to bind to β-neurexin [46].

Pls has surprisingly divergent functions. On one hand, Pls binds to cellular lipids and glycolipids and promotes bacterial cell–cell interaction as well as adherence to nasal epithelial cells [47, 48]. On the other hand, Pls prevents binding of *S. aureus* to immunoglobulin (Ig) G, soluble and immobilized Fn, as well as internalization by human host cells probably acting by steric hindrance [49]. In an animal model of septic arthritis, Pls has been found to be a virulence factor [50].

Sdr Protein Family: Bbp

The 97-kDa bone sialoprotein-binding protein Bbp has similarity with SdrC, SdrD, and SdrE [51]. Bbp was identified from *S. aureus* strains associated with bone and joint infections and found to specifically interact with bone sialoprotein, which is a glycoprotein of bone and dentine extracellular matrix [51]. Further studies indicated that Bbp is immunogenic and expressed during infection suggesting a relevant role in the pathogenesis of osteomyelitis [51].

Sdr Protein Family in Coagulase-Negative Staphylococcal Species

Less data on Fg-binding MSCRAMMs of coagulase-negative staphylococci are available. The *S. epidermidis* Fg-binding 119-kDa Fbe and the almost identical 97-kDa SdrG show significant similarities to the ClfA of *S. aureus* [52]. SdrG binds to the N-terminal region of the Fg β-chain [53]. Like with ClfA, the domain architecture with an IgG-like fold was also predicted from the crystal structure of the SdrG ligand-binding domain. Based on crystal structures of the SdrG N2N3 domains in complex with a synthetic peptide mimicking the target site in Fg, a dynamic mechanism for the binding of SdrG to Fg has been proposed, which has been termed "Dock, Lock, and Latch (DLL)" [53].

3.4.1.2 Fn-Binding Proteins (FnBPA and FnBPB)

Like the binding to Fg, the binding of *S. aureus* to surface-immobilized Fn is thought to be a relevant pathogenicity factor in device-associated infections [25]. Fn is a ~440 kDa glycoprotein and consists of two nearly identical monomers that are linked by a pair of disulfide bonds. Fn is a major component of blood plasma in its soluble form and a major component of the extracellular matrix in its insoluble form. Besides having other important functions, Fn plays a role in cell adhesion and wound healing. *S. aureus* produces two closely related covalently linked Fn-binding MSCRAMMs, FnBPA and FnBPB, which are encoded by *fnbA* and *fnbB*, respectively [54, 55]. The Fn-binding activity of both proteins has been localized to a C-terminally located and highly conserved repeat domain (D repeats) that is composed of an approximately 40-aa unit repeated four times (D1–D4) with the repeat D4 being incomplete (Fig. 3.2). A fifth repeat (Du) is located approximately 100 aa N-terminal to D1. In contrast to the *fnbA* or *fnbB* single mutants, only the *fnbA/fnbB* double-knockout mutant of *S. aureus* 8325–4 showed strongly reduced Fn-binding activity [56]. Complementation of the *fnbA/fnbB* double mutant with either plasmid-encoded *fnbA* or *fnbB* was able to fully restore the Fn-binding activity. This indicates that both proteins are expressed in *S. aureus* and contribute to the ability of strain 8325–4 to adhere to Fn-coated surfaces [56]. The approximately 500-aa N-terminal domains (A domain) of FnBPA and FnBPB share only 40 % sequence identity. The A domain of FnBPA exhibits substantial aa sequence identity to the A domains of the Fg-binding proteins ClfA, ClfB, and SdrG (see below) and has also been shown to contain Fg-binding activity [57]. Recombinant FnBPA seems to bind to the same binding site in the γ-chain of Fg-like ClfA and can compete with ClfA for binding to both immobilized and soluble Fg. Moreover, the A domains of both FnBPA and FnBPB can bind to immobilized elastin [58]. More precisely, Fg and elastin bind to the same region within the A domain of FnBPA [59]. Recently, the structural organization of FnBPA has been revised with the Fn-binding domain consisting of 11 tandem repeats [60]. Each repeat is predicted to interact with Fn by a tandem β-zipper mechanism [60, 61].

Most *S. aureus* strains carry both *fnb* genes [56, 62]. A study analyzing a larger collection of isolates from infected patients indicated that *S. aureus* strains isolated from orthopedic implant-associated infection showed a significantly greater adherence to Fn than isolates associated with nasal carriage, endocarditis, septic arthritis, or osteomyelitis, but there seems to be no difference in Fn-binding activity between the strains carrying one or two *fnb* genes [56, 62]. However, isolates associated with endocarditis, septic arthritis, or osteomyelitis were more likely to carry both *fnb* genes [62].

Besides being multifunctional adhesins, the FnBPs are efficient mediators of *S. aureus* internalization. Classically, staphylococci have been regarded to be extracellular pathogens. However, it is now widely accepted that *S. aureus* can also be internalized by a variety of human host cells, such as osteoblasts and endothelial and epithelial cells, and replicate intracellularly [63]. The FnBP-mediated mechanism of *S. aureus* internalization requires Fn as bridging molecule and the $\alpha_5\beta_1$ integrins as the host cell receptor resulting in signal transduction, tyrosine kinase activity, and cytoskeletal rearrangements [64–66]. The FnBPs also bind to the heat-shock protein 60 (Hsp60), which could act as a coreceptor in the FnBP-mediated uptake of *S. aureus* [67].

Although strain Newman possesses two *fnb* genes, it shows reduced Fn-binding activity and internalization rates. These conflicting results could be explained by the observation that strain Newman produces truncated versions of FnBPA and FnBPB due to point mutations that create a stop codon and thus FnBPs lacking the C-terminal anchor region [68].

There have been contrasting results regarding the importance of the FnBPs in virulence. Previous results suggested that the FnBPs do not play a significant role in infection, i.e., in the induction and/or propagation of endocarditis [69] or in the development of septic arthritis by *S. aureus* [70]. In a rat model of pneumonia, an *fnbA*/*fnbB* mutant even showed increased virulence compared to its wild-type strain [71]. However, expression of *fnbA* or *fnbB* in the nonpathogenic heterologous host *Staphylococcus carnosus* significantly increased adherence to intact endothelium *in vivo* [72] and *Lactococcus lactis* cells expressing *fnbA* revealed a considerably higher infectivity in an experimental rat endocarditis model in comparison with *L. lactis* cells harboring an empty vector [38]. The latter findings were further supported by the observation that FnBPA and FnBPB mediate adherence to platelets [73] and induce platelet activation [74]. Moreover, the FnBPs were found to play an important role in the induction of systemic inflammation [70]. Most recently, an increasing efficiency of host cell invasion and virulence in a murine sepsis model mediated by FnBPA has been associated with an increasing number of the 11 Fn-binding tandem repeats [75].

3.4.1.3 Collagen-Binding Adhesin

Another *S. aureus* MSCRAMM is the collagen (Cn)-binding protein Cna [76]. The Cn-binding site within the 133-kDa Cna is located in a region between the aa Asp^{209} and Tyr^{233}. Based on crystal structures from the subdomains N1 and N2 of Cna, each of them adopts an IgG-like fold, and in complex with a synthetic Cn-like triple helical peptide, a "collagen hug" model was proposed for the interaction of the multidomain Cna with its extended ropelike ligand [77]. In a rabbit model of soft contact lens-associated bacterial keratitis, Cna has been found to be a virulence factor [78].

3.4.1.4 Platelet-Binding Adhesin (SraP)

The serine-rich adhesin for platelets (SraP), which has been also termed SasA, is a glycoprotein with a calculated molecular mass of 227 kDa that is involved in adherence to platelets [79]. SraP contains an unusually long N-terminal signal peptide and two serine-rich repeat regions (srr1 and srr2) separated by a non-repeat region. A recombinant fragment of SraP consisting of the N-terminally located Srr1 and the non-repeat region was shown to directly bind to platelets, a trait that is thought to be an important pathogenicity factor in the development of infective endocarditis [79]. Consistently, in a rabbit model of endocarditis, an *sraP* mutant strain revealed significant lower bacterial counts within vegetations than the wild-type strain. Recently, the accessory Sec system that is encoded downstream of the *sraP* structural gene has been found to be required for the transport of SraP through the cytoplasmic membrane [80].

3.4.1.5 Protein A (Spa)

Protein A (Spa) was the first *S. aureus* surface protein identified and has been used as the model system to study the sortase-mediated cell wall anchoring [20, 81]. Protein A contains five N-terminally located approximately 60-aa-long tandem repeats (designated E, D, A, B, and C), which bind to the Fc portion of immunoglobulin G, leading to reduced antibody-mediated opsonization. Thus, the inhibition of phagocytosis has been proposed to be the basic contribution of protein A to virulence. The binding between these two molecules involves nine aa in the IgG fragment and 11 of the protein A repeat as determined by X-ray crystallography and NMR studies [82, 83]. However, it is now clear that its function in pathogenesis is more complex than previously thought. Protein A also binds to both soluble and immobilized vWF, a large multimeric glycoprotein that mediates platelet adhesion at sites of endothelial damage, which is thought to play a role in endovascular infection [84].

Protein A has also been identified to directly interact with platelets via binding to the platelet receptor gC1qR/p33 [85]. Moreover, protein A can bind to TNFR1, which is a receptor for tumor-necrosis factor-α (TNF-α) widely distributed on the airway epithelium, and stimulate an inflammatory response in airway epithelial cells. Thus, the protein A-TNFR1 signaling pathway is thought to have a central function in the pathogenesis of staphylococcal pneumonia [86].

3.4.1.6 Iron-Regulated Surface Determinants (IsdA, IsdB, IsdC, IsdH)

Iron-regulated surface determinants include IsdA, IsdB, IsdC, and IsdH. The transcription of *isdA*, *isdB*, *isdC*, and *isdH* is regulated by environmental iron and the iron-responsive regulator Fur [87–89]. Each of the Isd proteins contains one to three near-iron transporter (NEAT) domains and binds to one or more iron-containing proteins, such as transferrin, hemin, or hemoglobin. Thus, a function of these proteins in iron acquisition was suggested, but could not be proven to date. The NEAT domains in IsdA and IsdH are responsible for the binding to their ligands. IsdA binds to a wide range of host products, such as Fg, Fn, hemin, hemoglobulin, transferrin, and human desquamated nasal epithelial cells [45, 89]. However, a physiologically relevant binding of IsdA to Fg and Fn can only be observed when *S. aureus* is grown under iron-limited conditions, such as *in vitro* in serum or *in vivo*, favoring the expression of IsdA [89]. Moreover, IsdA was recently found to bind to proteins of the cornified envelope of human desquamated epithelial cells, such as involucrin, loricrin, and cytokeratin 10, which are thought to be the predominant ligands in the ecological niche of *S. aureus*, further delineating the importance of IsdA in nasal colonization [45, 89, 90]. IsdH binds to haptoglobin and haptoglobin–hemoglobin complexes [87]. Differences in the ligand-binding specificity of IsdA and IsdH might be reflected by the pronounced sequence differences among their NEAT domains [89]. Immunization with purified IsdA and IsdH resulted in reduced nasal colonization in the cotton rat model [91].

IsdB was found to be involved in platelet adhesion and aggregation: when *S. aureus* was grown under iron-limited conditions, mutants deficient in IsdB, but not IsdA or IsdH, were unable to adhere to or aggregate platelets. The platelet integrin GPIIb/IIIa was identified as the platelet receptor for IsdB and the direct interaction of these proteins could be demonstrated by surface plasmon resonance [92]. Immunization with IsdB protects against staphylococcal sepsis in different animal models [93].

3.4.2 Non-covalently Linked Surface-Associated Proteins

3.4.2.1 Autolysin/Adhesins

Another class of staphylococcal adhesins is represented by the autolysin/adhesins first described by us and others [14, 94] (Fig. 3.2). These non-covalently bound proteins are associated to the surface by ionic or hydrophobic interactions and have both enzymatic (peptidoglycan-hydrolytic) and adhesive functions. In general, peptidoglycan hydrolases or autolysins are thought to play important roles in cell-wall turnover, cell division, cell separation, and antibiotic-induced lysis of bacteria. Using transposon mutagenesis, the 148-kDa autolysin AtlE of *S. epidermidis* was identified as a surface-associated component, which mediates attachment to polystyrene, biofilm formation, and adherence to vitronectin (Vn) [14]. In a rat central venous catheter infection model, the *atlE* mutant was attenuated compared to the wild type [95]. The homologous 137-kDa autolysin Atl from *S. aureus* also mediates attachment to polystyrene and biofilm formation [96]. Moreover, we recently found that Atl additionally binds to Fg, Fn, Vn, and human endothelial cells [97]. AtlE and Atl show the same structural organization and are proteolytically cleaved into two bacteriolytically active domains, an N-terminal amidase and a C-terminal glucosaminidase [14, 96] (Fig. 3.2). Recently, the X-ray structure of the amidase revealing its active site has been resolved [98]. The bacteriolytically active domains are interconnected by three direct repeated sequences (R1, R2, and R3), which are involved in binding to peptidoglycan. Each repeat consists of approximately 170 aa with two glycine–tryptophan (GW)-dipeptide motifs, one in the central part and the other one at the end of each repeat. The amidase contains two repeats (R1 and R2) in its C-terminal portion and the glucosaminidase contains one repeat (R3) in its N-terminal portion. GW-containing repeats have been previously characterized from surface proteins of *Listeria monocytogenes*, such as InlB and Ami. InlB promotes bacterial entry into a variety of cultured cell lines and Ami contributes to the adherence of *L. monocytogenes* to eukaryotic cells [99, 100].

Recently, we could establish an analogous function for Atl/AtlE in staphylococcal internalization by endothelial cells that involves the 70-kDa heat-shock cognate protein Hsc70, which has been reported to be present on the surface of several types of cells [101] as the host cell receptor [97]. While this novel Atl- or

AtlE-mediated internalization mechanism may represent a "backup"-mechanism in *S. aureus* internalization, it may represent the major or even sole mechanism involved in the internalization of coagulase-negative staphylococci.

The homologous autolysin/adhesins from other coagulase-negative staphylococcal species, such as Aas from *Staphylococcus saprophyticus* and AtlC from *Staphylococcus caprae*, have been shown to bind to Fn [94, 102]. Aas also binds to human uroepithelial cells and agglutinates sheep erythrocytes [103].

Further multifunctional autolysin/adhesins include the Aaa from *S. aureus* and the homologous Aae from *S. epidermidis*. Aaa and Aae both have bacteriolytic activities and bind to Fg, Fn, and Vn in a dose-dependent and saturable fashion and with high affinity [104, 105]. Aaa and Aae also have a modular structure with three N-terminally located lysin motif (LysM) domains and a C-terminally located repetitive sequence with similarity to a domain called cysteine, histidine-dependent amidohydrolase/peptidase (CHAP) [106, 107] (Fig. 3.2). Both domains are found in a variety of bacteriolytically active proteins. LysM domains have been reported to confer peptidoglycan binding, while CHAP domains were found to have hydrolytic activities.

3.4.2.2 SERAMs (Eap/Map/P70, Emp)

Further examples of non-covalently associated surface proteins of *S. aureus* are secreted proteins that bind back to the bacterial cell surface by so far unknown mechanisms and have a broad spectrum of ligands; therefore they were termed secretable expanded repertoire adhesive molecules (SERAMs) [108]. Among the SERAMs are the 60–72 kDa extracellular adherence protein Eap (also designated as Map, "major histocompatibility complex (MHC) class II analogous protein" or P70) and the 40 kDa extracellular matrix and plasma-binding protein Emp, which both bind to various components of the extracellular matrix, such as Fg, Fn, or Vn [109, 110]. Map/Eap contains six repeated domains of 110 aa, with a subdomain of 31 aa sharing significant homology to a segment in the peptide-binding groove of the β-chain of the MHC class II proteins from different mammalian species [111]. *S. aureus* strains that do not express Eap revealed a lower extent to colonize and invade host tissue and mutants defective in *emp* showed reduced attachment to immobilized Fg and Fn [108]. Additional functions of Eap include the direct interaction with the host adhesive protein intercellular adhesion molecule 1 (ICAM-1), inhibition of neutrophil binding to and transmigration through the endothelium, and a decrease in phagocytic activity [108, 112]. Thus, Eap is a potent anti-inflammatory factor by the inhibition of the recruitment of host leukocytes. More recently, a further function for Eap as a potent angiostatic agent was described [113].

No homologues of the *S. aureus* extracellular matrix and plasma-binding protein Emp and the extracellular adherence protein Eap could be detected in *S. epidermidis* [109, 114].

3.4.2.3 Membrane-Spanning Proteins: ECM-Binding Protein Homologue and Elastin-Binding Protein

Staphylococcal surface-associated proteins that are non-covalently anchored to the cell surface include also the giant 1.1-mDa Fn-binding protein Ebh of *S. aureus* and the homologous ECM-binding protein Embp of *S. epidermidis* [115, 116], whose genes are by far the largest of the *S. aureus* and *S. epidermidis* genomes, respectively. The Fn-binding sites of Ebh and Embp seem to be unrelated to those of the *S. aureus* FnBPs. Ebh appears to be anchored to the cell surface by a C-terminally located membrane-spanning domain that is followed by a repeat region containing positively charged aa, which is predicted to be located intracellularly [117]. Ebps is another protein with a membrane anchor that binds elastin, which is a major component of the extracellular matrix [118].

3.4.3 Non-proteinaceous Staphylococcal Adhesins: WTA

The *S. aureus* WTA (see above) mediates adherence to human nasal epithelial cells and is involved in nasal colonization, which is considered a major risk factor for serious *S. aureus* infections [119]. Moreover, a WTA-deficient mutant (Δ*tagO*) showed decreased adherence to human endothelial cells especially under flow conditions and was attenuated in a rabbit model of infective endocarditis [120]. In *S. epidermidis*, it has been shown that the WTA is involved in the adherence to Fn [121].

3.5 Factors Involved in Intercellular Adhesion and Biofilm Accumulation

After successful attachment to a surface, bacteria proliferate and accumulate in multilayered cell clusters, which requires intercellular adhesion. Staphylococcal biofilm accumulation can be mediated by protein as well as polysaccharide factors.

3.5.1 Covalently Linked Cell Surface Proteins (MSCRAMMs)

Recently, several *S. aureus* surface-anchored proteins have been associated with the accumulation phase of biofilm formation and intercellular adherence. However, the first protein reported to be involved in the accumulative growth during biofilm formation was the 220-kDa accumulation-associated protein Aap from *S. epidermidis* [122]. Aap is highly homologous to the *S. aureus* surface protein SasG, which also mediates biofilm accumulation [123]. The function of Aap in the accumulation process was speculated to be the anchoring of the polysaccharide intercellular adhesin (PIA; see below) to the cell surface. However recently, it was shown that Aap is able to mediate intercellular adhesion and biofilm accumulation in a completely PIA-independent background. Intercellular adhesion is mediated by a repeat domain B, which becomes active only after proteolytic cleavage of the N-terminal A-domain [124]. The B repeats of Aap (also known as G5 domains) were found to be zinc-dependent adhesion modules and a "zinc zipper" mechanism was suggested for G5 domain-based intercellular adhesion in Aap- or SasG-mediated biofilm accumulation [125]. Antibodies against Aap of *S. epidermidis* inhibit biofilm formation to some extent, indicating that Aap may be a vaccine candidate to lower somewhat *S. epidermidis* biofilm infection [126]. A zinc-dependent dimerization of recombinant B repeats was also observed with SasG [127]. SasG not only mediates bacterial intercellular adhesion but also promotes binding to nasal epithelial cells [48]. Moreover like with Pls, which has sequence similarities with SasG and Aap, the expression of SasG masked the ability of the *S. aureus* cells to adhere to Fg, Fn, cytokeratin 10, and IgG. SasG-mediated binding to nasal epithelial cells was proposed to compensate for masking the ability of ClfB to bind to cytokeratin 10 [123].

While intercellular adhesion is mediated by the SasG B repeats with at least five of the eight repeats being required, binding to the nasal epithelial cells is carried out by the N-terminal region A. The functions of SasG and Aap might be explained by their fibrillar structure that was recently observed by transmission electron microscopy [123, 128].

The 239-kDa biofilm-associated protein Bap was the first protein reported to be involved in *S. aureus* biofilm formation. Bap mediates *S. aureus* attachment to a polystyrene surface and intercellular adhesion leading to biofilm accumulation [12]. Moreover, Bap was found to prevent adherence to Fg, Fn, and host tissue (sheep mammary glands) as well as cellular internalization [129].

Recently, we reported that the *S. aureus* surface protein C (SasC) mediates attachment to polystyrene and biofilm accumulation, but not adherence to Fg, TSP-1, vWf, or platelets [130]. SasC has an N-terminal domain, which confers intercellular adhesion, followed by an extended repeat region consisting of 17 repeats with similarity to the domain of unknown function 1542 (DUF 1542).

Furthermore, a function for the multifactorial virulence factor protein A (Spa) in biofilm development was detected, when expressed at high levels due to a mutation in the accessory gene regulator *agr* [131]. A role for Spa in the development of biofilm-associated infections was suggested in a murine model of subcutaneous catheter infection [131].

Additionally, FnBPA and FnBPB were found to be important for biofilm accumulation of methicillin-resistant *S. aureus* (MRSA) strains with the A domain of FnBPA, but not the Fn-binding domain being responsible for the biofilm accumulation [132]. Accordingly, the FnBPs play a significant role in a catheter-associated murine infection model [132, 133].

3.5.2 Non-covalently Linked Surface-Associated Proteins: Embp

The giant *S. epidermidis* protein Embp (see above) was not only found to bind Fn but recently also demonstrated to mediate biofilm accumulation [134]. Embp harbors 59 so-called found in various architectures (FIVAR) and 38 protein G-related albumin-binding (GA) domains. Further studies indicated that the FIVAR domains of Embp mediate the binding of *S. epidermidis* to immobilized Fn. Moreover, the Embp-mediated biofilm formation protected *S. epidermidis* from phagocytosis as detected by an *in vitro* phagocytosis assay [134]. A function of the homologous Ebh of *S. aureus* in biofilm accumulation is likely, but has not been established yet.

3.5.3 Non-proteinaceous Staphylococcal Adhesins

3.5.3.1 Polysaccharide Intercellular Adhesin

S. epidermidis transposon mutants not able to accumulate in multilayered cell clusters lack a specific polysaccharide antigen referred to as PIA [135, 136]. Purification and structural analysis of PIA revealed that it is a linear β-1,6-linked *N*-acetylglucosaminoglycan with 15–20 % of the *N*-acetylglucosaminyl residues being non-*N*-acetylated [8]. Later, it was found that PIA (also called poly-*N*-acetylglucosamine [PNAG]) is also produced by *S. aureus* [137, 138].

The partial deacetylation of 15–20 % of the *N*-acetylglucosaminyl residues renders the polysaccharide positively charged, which determines its biological activity. Possibly, it functions as an intercellular adhesin by electrostatically attracting the negatively charged teichoic acid at the bacterial cell surface. The structure of PIA so far is unique. However, PIA-mediated biofilm formation might represent a common principle, because PIA-related structures have also been identified to play a role in the biofilm formation of other pathogenic bacteria, such as the gram-negative *Escherichia coli* and *Actinobacillus actinomycetemcomitans* [139].

PIA is produced by the gene products encoded by the *icaADBC* operon. The *icaADBC* operon was first identified in *S. epidermidis* and is also present in *S. aureus* and other staphylococcal species [136, 137]. The *N*-acetylglucosaminyltransferase activity is carried out by IcaA, which requires IcaD for full activity. With its transmembrane helices, IcaC very likely is an integral membrane protein that putatively transports the *N*-acetylglucosamine oligomers across the membrane [140]. IcaB is mainly found in the culture supernatant and deacetylates PIA [140, 141].

The importance of PIA as a pathogenicity factor has been confirmed in various foreign-body animal infection models with different *S. epidermidis icaADBC* mutants [95, 142]. However, in *S. aureus* conflicting results were obtained: PIA production did not increase the capacity to induce persistent infections in a tissue cage model [143]. A study investigating the pathogenic properties of *S. epidermidis* strains obtained from polymer-associated septicemic disease compared with saprophytic skin and mucosal isolates demonstrated a strong correlation of biofilm formation and presence of the *ica* gene cluster essentially associated with disease isolates [144].

3.5.3.2 Extracellular DNA

Another polymeric molecule, eDNA, has been identified as an important component of the biofilm matrix of several bacterial species, such as *Streptococcus pneumoniae*, *Pseudomonas aeruginosa*, and *Enterococcus faecalis* [145–147]. Although it does not seem to mediate biofilm accumulation by itself, it contributes to *S. aureus* biofilm development [10]. DNA is a negatively charged molecule that upon its release could interact with the positively charged extracellular polymer PIA, thus acting as an additional "glue."

3.6 Factors Involved in Biofilm Detachment

Biofilm detachment may lead to the dissemination of a staphylococcal infection and thus to colonization of new sites and metastatic infection.

3.6.1 Extracellular Enzymes

Factors involved in biofilm detachment may include enzymatic activities that lead to the disintegration of the "glue." Depending on the nature of the substance that mediates the "stickiness," enzymatic activities like glycosyl hydrolases that would degrade PIA, proteases that would degrade protein components (such as Aap/SasG or Bap/Bhp), or nucleases that would degrade eDNA might be involved. Indeed, the gram-negative periodontal pathogen *A. actinomycetemcomitans* produces dispersin B, which is a soluble glycosyl hydrolase that degrades the self-synthesized extracellular polysaccharide PGA. Like PIA, PGA is a linear polymer of β [1, 6]-linked *N*-acetylglucosamine residues [139]. Dispersin B is also able to dissolve biofilms of clinical *S. epidermidis* strains by hydrolyzing the glycosidic linkages of PIA [139, 148]. However, the *S. aureus* and *S. epidermidis* genomes do not seem to encode analogous enzymatic activities.

eDNA has been shown to be an important component of the *S. aureus* biofilm matrix (see above) [10]. Accordingly, the addition of DNase I inhibits biofilm formation of *S. aureus* and promotes the detachment of preformed *S. aureus* biofilms [149]. Therefore, it may be speculated that the activity of an extracellular *S. aureus* nuclease would also contribute to biofilm detachment. In contrast to *S. aureus*, DNase I only slightly inhibits biofilm formation in *S. epidermidis*, but does not promote the detachment of preformed biofilms. Thus, in *S. epidermidis* eDNA seems to effect initial attachment to a surface, rather than biofilm accumulation and detachment [149].

Several studies indicate that the biofilm matrix of a significant proportion of biofilm-forming staphylococcal strains mainly contained teichoic acid and proteins, but not PIA [150, 151]. In this case, protease treatment disintegrated the biofilms, although sometimes only partially [151, 152]. At least in *S. aureus*, protease-mediated biofilm detachment is dependent on a functional *agr* quorum sensing system [153].

3.6.2 Phenol-Soluble Modulins

Another strategy leading to biofilm detachment involves the production and release of small peptides called phenol-soluble modulins (PSMs). PSMs have first been described as proinflammatory agents in *S. epidermidis* [154]. A first hint that PSMs are involved in preventing biofilm formation came from the investigation of a large collection of clinical *S. aureus* isolates with respect to their *agr* phenotype (production of delta-toxin) and the ability to form a biofilm. The effector molecule of the staphylococcal *agr* quorum sensing system is RNAIII. The gene for RNAIII also encodes the delta-toxin (*hld*) [155]. It turned out that predominantly the delta-toxin-negative strains and consequently *agr*-defective strains formed a biofilm, demonstrating a profound impact of *agr* on biofilm formation [156]. The observed effect seemed to be caused by the surfactant properties of delta-toxin, which also belongs to the PSMs. Externally added delta-toxin to biofilm-positive strains suppressed biofilm formation.

According to their length, the PSMs can be subdivided into two classes: α-type peptides (PSMα) have a length of approximately 20 aa and β-type peptides (PSMβ1, PSMβ2, PSMβ3) are between 40 and 45 aa in length [157]. PSMγ (delta-toxin) and the recently discovered PSMδ belong to the α-type peptides [158]. All PSMs have a pronounced amphipathic α-helical character and thus are supposed to have a surfactant-like effect, which might be responsible for their role in biofilm detachment [159]. A genome-wide analysis of gene expression and determination of the concentrations of PSM peptides by HPLC-MS in *S. epidermidis* biofilms revealed that *S. epidermidis* down-regulates the expression of the PSM peptide genes in biofilms, which is consistent with lower concentrations of PSM peptides in biofilms compared with planktonically grown cultures [159]. Moreover, a PSM β operon deletion strain formed a more compact biofilm compared to its isogenic wild-type strain [157]. PSM homologues also occur in *S. aureus* and other staphylococci [157].

3.7 Conclusion

In conclusion, the pathogenesis of medical device-associated infection with staphylococci is characterized by their ability to colonize the medical device by the formation of a biofilm. Staphylococci employ a variety of adhesion mechanisms that mediate attachment either to the abiotic surface of the medical device or to host products, such as extracellular matrix and plasma proteins, immobilized on the device's surface shortly after its insertion. Upon adherence, the bacteria proliferate and accumulate into multilayered cell clusters, which requires intercellular adhesive mechanisms finally leading to the formation of a multilayered biofilm in which the bacteria are protected against the host immune system as well as against antibiotic therapy. From a mature biofilm, staphylococcal cells can detach, which may lead to the dissemination of a staphylococcal infection and thus to colonization of new sites and metastatic infection. Thus, understanding of the molecular mechanisms involved in staphylococcal biofilm formation and detachment is required to develop necessary strategies to prevent or combat medical device-associated infections with these serious pathogens.

References

1. Karlowsky JA, Jones ME, Draghi DC, Thornsberry C, Sahm DF, Volturo GA. Prevalence and antimicrobial susceptibilities of bacteria isolated from blood cultures of hospitalized patients in the United States in 2002. Ann Clin Microbiol Antimicrob. 2004;3:7.
2. Lowy FD. *Staphylococcus aureus* infections. N Engl J Med. 1998;339(8):520–32.
3. Ziebuhr W. *Staphylococcus aureus* and *Staphylococcus epidermidis*: emerging pathogens in nosocomial infections. Contrib Microbiol. 2001;8:102–7.
4. Götz F, Peters G. Colonization of medical devices by coagulase-negative staphylococci. In: Waldvogel FA, Bisno AL, editors. Infections associated with indwelling medical devices. 3rd ed. Washington, DC: ASM; 2000. p. 55–88.
5. Lentino JR. Prosthetic joint infections: bane of orthopedists, challenge for infectious disease specialists. Clin Infect Dis. 2003;36(9):1157–61.
6. Peters G, Locci R, Pulverer G. Microbial colonization of prosthetic devices. II. Scanning electron microscopy of naturally infected intravenous catheters. Zentralbl Bakteriol Mikrobiol Hyg B. 1981;173(5):293–9.
7. Baldassarri L, Donnelli G, Gelosia A, Voglino MC, Simpson AW, Christensen GD. Purification and characterization of the staphylococcal slime-associated antigen and its occurrence among *Staphylococcus epidermidis* clinical isolates. Infect Immun. 1996;64(8):3410–5.
8. Mack D, Fischer W, Krokotsch A, Leopold K, Hartmann R, Egge H, et al. The intercellular adhesin involved in biofilm accumulation of *Staphylococcus epidermidis* is a linear beta-1,6-linked glucosaminoglycan: purification and structural analysis. J Bacteriol. 1996;178(1):175–83.
9. Hussain M, Wilcox MH, White PJ. The slime of coagulase-negative staphylococci: biochemistry and relation to adherence. FEMS Microbiol Rev. 1993;10(3–4):191–207.
10. Rice KC, Mann EE, Endres JL, Weiss EC, Cassat JE, Smeltzer MS, et al. The *cidA* murein hydrolase regulator contributes to DNA release and biofilm development in *Staphylococcus aureus*. Proc Natl Acad Sci U S A. 2007;104(19):8113–8.
11. Costerton JW, Stewart PS, Greenberg EP. Bacterial biofilms: a common cause of persistent infections. Science. 1999;284(5418):1318–22.
12. Cucarella C, Solano C, Valle J, Amorena B, Lasa I, Penades JR. Bap, a *Staphylococcus aureus* surface protein involved in biofilm formation. J Bacteriol. 2001;183(9):2888–96.
13. Bowden MG, Visai L, Longshaw CM, Holland KT, Speziale P, Höök M. Is the GehD lipase from *Staphylococcus epidermidis* a collagen binding adhesin? J Biol Chem. 2002;277(45):43017–23.
14. Heilmann C, Hussain M, Peters G, Götz F. Evidence for autolysin-mediated primary attachment of *Staphylococcus epidermidis* to a polystyrene surface. Mol Microbiol. 1997;24(5):1013–24.
15. Gross M, Cramton SE, Götz F, Peschel A. Key role of teichoic acid net charge in *Staphylococcus aureus* colonization of artificial surfaces. Infect Immun. 2001;69:3423–6.
16. Fedtke I, Mader D, Kohler T, Moll H, Nicholson G, Biswas R, et al. A *Staphylococcus aureus* *ypfP* mutant with strongly reduced lipoteichoic acid (LTA) content: LTA governs bacterial surface properties and autolysin activity. Mol Microbiol. 2007;65(4):1078–91.

17. Herrmann M, Lai QJ, Albrecht RM, Mosher DF, Proctor RA. Adhesion of *Staphylococcus aureus* to surface-bound platelets: role of fibrinogen/fibrin and platelet integrins. J Infect Dis. 1993;167(2):312–22.
18. Herrmann M, Hartleib J, Kehrel B, Montgomery RR, Sixma JJ, Peters G. Interaction of von Willebrand factor with *Staphylococcus aureus*. J Infect Dis. 1997;176(4):984–91.
19. Patti JM, Allen BL, McGavin MJ, Höök M. MSCRAMM-mediated adherence of microorganisms to host tissues. Annu Rev Microbiol. 1994;48:585–617.
20. Schneewind O, Mihaylova-Petkov D, Model P. Cell wall sorting signals in surface proteins of gram-positive bacteria. EMBO J. 1993;12(12):4803–11.
21. Ton-That H, Liu G, Mazmanian SK, Faull KF, Schneewind O. Purification and characterization of sortase, the transpeptidase that cleaves surface proteins of *Staphylococcus aureus* at the LPXTG motif. Proc Natl Acad Sci U S A. 1999;96(22):12424–9.
22. Mazmanian SK, Liu G, Ton-That H, Schneewind O. *Staphylococcus aureus* sortase, an enzyme that anchors surface proteins to the cell wall. Science. 1999;285(5428):760–3.
23. Marraffini LA, Schneewind O. Anchor structure of staphylococcal surface proteins. V. Anchor structure of the sortase B substrate IsdC. J Biol Chem. 2005;280(16):16263–71.
24. Gill SR, Fouts DE, Archer GL, Mongodin EF, Deboy RT, Ravel J, et al. Insights on evolution of virulence and resistance from the complete genome analysis of an early methicillin-resistant *Staphylococcus aureus* strain and a biofilm-producing methicillin-resistant *Staphylococcus epidermidis* strain. J Bacteriol. 2005;187(7):2426–38.
25. Herrmann M, Vaudaux PE, Pittet D, Auckenthaler R, Lew PD, Schumacher-Perdreau F, et al. Fibronectin, fibrinogen, and laminin act as mediators of adherence of clinical staphylococcal isolates to foreign material. J Infect Dis. 1988;158(4):693–701.
26. McDevitt D, Francois P, Vaudaux P, Foster TJ. Molecular characterization of the clumping factor (fibrinogen receptor) of *Staphylococcus aureus*. Mol Microbiol. 1994;11(2):237–48.
27. Ni Eidhin D, Perkins S, Francois P, Vaudaux P, Höök M, Foster TJ. Clumping factor B (ClfB), a new surface located fibrinogen binding adhesin of *Staphylococcus aureus*. Mol Microbiol. 1998;30(2):245–57.
28. McDevitt D, Francois P, Vaudaux P, Foster TJ. Identification of the ligand-binding domain of the surface-located fibrinogen receptor (clumping factor) of *Staphylococcus aureus*. Mol Microbiol. 1995;16(5):895–907.
29. Hartford OM, Wann ER, Höök M, Foster TJ. Identification of residues in the *Staphylococcus aureus* fibrinogen-binding MSCRAMM clumping factor A (ClfA) that are important for ligand binding. J Biol Chem. 2001;276(4):2466–73.
30. McDevitt D, Nanavaty T, House-Pompeo K, Bell E, Turner N, McIntire L, et al. Characterization of the interaction between the *Staphylococcus aureus* clumping factor (ClfA) and fibrinogen. Eur J Biochem. 1997;247(1):416–24.
31. Siboo IR, Cheung AL, Bayer AS, Sullam PM. Clumping factor a mediates binding of *Staphylococcus aureus* to human platelets. Infect Immun. 2001;69(5):3120–7.
32. O'Brien L, Kerrigan SW, Kaw G, Hogan M, Penades J, Litt D, et al. Multiple mechanisms for the activation of human platelet aggregation by *Staphylococcus aureus*: roles for the clumping factors ClfA and ClfB, the serine-aspartate repeat protein SdrE and protein A. Mol Microbiol. 2002;44(4):1033–44.
33. Loughman A, Fitzgerald JR, Brennan MP, Higgins J, Downer R, Cox D, et al. Roles for fibrinogen, immunoglobulin and complement in platelet activation promoted by *Staphylococcus aureus* clumping factor A. Mol Microbiol. 2005;57(3):804–18.
34. Miajlovic H, Loughman A, Brennan M, Cox D, Foster TJ. Both complement- and fibrinogen-dependent mechanisms contribute to platelet aggregation mediated by *Staphylococcus aureus* clumping factor B. Infect Immun. 2007;75(7):3335–43.
35. O'Brien LM, Walsh EJ, Massey RC, Peacock SJ, Foster TJ. *Staphylococcus aureus* clumping factor B (ClfB) promotes adherence to human type I cytokeratin 10: implications for nasal colonization. Cell Microbiol. 2002;4(11):759–70.
36. O'Connell DP, Nanavaty T, McDevitt D, Gurusiddappa S, Höök M, Foster TJ. The fibrinogen-binding MSCRAMM (clumping factor) of *Staphylococcus aureus* has a Ca2+–dependent inhibitory site. J Biol Chem. 1998;273(12):6821–9.

37. Deivanayagam CC, Wann ER, Chen W, Carson M, Rajashankar KR, Höök M, et al. A novel variant of the immunoglobulin fold in surface adhesins of *Staphylococcus aureus*: crystal structure of the fibrinogen-binding MSCRAMM, clumping factor A. EMBO J. 2002;21 (24):6660–72.
38. Que YA, Francois P, Haefliger J-A, Entenza J-M, Vaudaux P, Moreillon P. Reassessing the role of *Staphylococcus aureus* clumping factor and fibronectin-binding protein by expression in *Lactococcus lactis*. Infect Immun. 2001;69(10):6296–302.
39. Sullam PM, Bayer AS, Foss WM, Cheung AL. Diminished platelet binding *in vitro* by *Staphylococcus aureus* is associated with reduced virulence in a rabbit model of infective endocarditis. Infect Immun. 1996;64(12):4915–21.
40. Josefsson E, Hartford O, O'Brien L, Patti JM, Foster T. Protection against experimental *Staphylococcus aureus* arthritis by vaccination with clumping factor A, a novel virulence determinant. J Infect Dis. 2001;184(12):1572–80.
41. Hall AE, Domanski PJ, Patel PR, Vernachio JH, Syribeys PJ, Gorovits EL, et al. Characterization of a protective monoclonal antibody recognizing *Staphylococcus aureus* MSCRAMM protein clumping factor A. Infect Immun. 2003;71(12):6864–70.
42. Ganesh VK, Rivera JJ, Smeds E, Ko YP, Bowden MG, Wann ER, et al. A structural model of the *Staphylococcus aureus* ClfA-fibrinogen interaction opens new avenues for the design of anti-staphylococcal therapeutics. PLoS Pathog. 2008;4(11):e1000226.
43. Josefsson E, McCrea KW, Ni Eidhin D, O'Connell D, Cox J, Hook M, et al. Three new members of the serine-aspartate repeat protein multigene family of Staphylococcus aureus. Microbiology. 1998;144(Pt 12):3387–95.
44. Savolainen K, Paulin L, Westerlund-Wikstrom B, Foster TJ, Korhonen TK, Kuusela P. Expression of *pls*, a gene closely associated with the *mecA* gene of methicillin-resistant *Staphylococcus aureus*, prevents bacterial adhesion in vitro. Infect Immun. 2001;69(5):3013–20.
45. Corrigan RM, Miajlovic H, Foster TJ. Surface proteins that promote adherence of *Staphylococcus aureus* to human desquamated nasal epithelial cells. BMC Microbiol. 2009;9:22.
46. Barbu EM, Ganesh VK, Gurusiddappa S, Mackenzie RC, Foster TJ, Sudhof TC, et al. Beta-Neurexin is a ligand for the *Staphylococcus aureus* MSCRAMM SdrC. PLoS Pathog. 2010;6(1):e1000726.
47. Huesca M, Peralta R, Sauder DN, Simor AE, McGavin MJ. Adhesion and virulence properties of epidemic Canadian methicillin-resistant *Staphylococcus aureus* strain 1: identification of novel adhesion functions associated with plasmin-sensitive surface protein. J Infect Dis. 2002;185(9):1285–96.
48. Roche FM, Meehan M, Foster TJ. The *Staphylococcus aureus* surface protein SasG and its homologues promote bacterial adherence to human desquamated nasal epithelial cells. Microbiology. 2003;149(Pt 10):2759–67.
49. Hussain M, Schafer D, Juuti KM, Peters G, Haslinger-Löffler B, Kuusela PI, et al. Expression of Pls (plasmin sensitive) in *Staphylococcus aureus* negative for pls reduces adherence and cellular invasion and acts by steric hindrance. J Infect Dis. 2009;200(1):107–17.
50. Josefsson E, Juuti K, Bokarewa M, Kuusela P. The surface protein Pls of methicillin-resistant *Staphylococcus aureus* is a virulence factor in septic arthritis. Infect Immun. 2005;73 (5):2812–7.
51. Tung H, Guss B, Hellman U, Persson L, Rubin K, Ryden C. A bone sialoprotein-binding protein from *Staphylococcus aureus*: a member of the staphylococcal Sdr family. Biochem J. 2000;345(Pt 3):611–9.
52. McCrea KW, Hartford O, Davis S, Eidhin DN, Lina G, Speziale P, et al. The serine-aspartate repeat (Sdr) protein family in *Staphylococcus epidermidis*. Microbiology. 2000;46(Pt 7):1535–46.
53. Ponnuraj K, Bowden MG, Davis S, Gurusiddappa S, Moore D, Choe D, et al. A "dock, lock, and latch" structural model for a staphylococcal adhesin binding to fibrinogen. Cell. 2003;115(2):217–28.
54. Signas C, Raucci G, Jönsson K, Lindgren PE, Anantharamaiah GM, Höök M, et al. Nucleotide sequence of the gene for a fibronectin-binding protein from *Staphylococcus aureus*: use of this peptide sequence in the synthesis of biologically active peptides. Proc Natl Acad Sci U S A. 1989;86(2):699–703.

55. Jönsson K, Signas C, Müller HP, Lindberg M. Two different genes encode fibronectin binding proteins in Staphylococcus aureus. The complete nucleotide sequence and characterization of the second gene. Eur J Biochem. 1991;202(3):1041–8.
56. Greene C, McDevitt D, Francois P, Vaudaux PE, Lew DP, Foster TJ. Adhesion properties of mutants of *Staphylococcus aureus* defective in fibronectin-binding proteins and studies on the expression of *fnb* genes. Mol Microbiol. 1995;17(6):1143–52.
57. Wann ER, Gurusiddappa S, Höök M. The fibronectin-binding MSCRAMM FnbpA of *Staphylococcus aureus* is a bifunctional protein that also binds to fibrinogen. J Biol Chem. 2000;275(18):13863–71.
58. Roche FM, Downer R, Keane F, Speziale P, Park PW, Foster TJ. The N-terminal A domain of fibronectin-binding proteins A and B promotes adhesion of *Staphylococcus aureus* to elastin. J Biol Chem. 2004;279(37):38433–40.
59. Keane FM, Loughman A, Valtulina V, Brennan M, Speziale P, Foster TJ. Fibrinogen and elastin bind to the same region within the A domain of fibronectin binding protein A, an MSCRAMM of *Staphylococcus aureus*. Mol Microbiol. 2007;63(3):711–23.
60. Schwarz-Linek U, Werner JM, Pickford AR, Gurusiddappa S, Kim JH, Pilka ES, et al. Pathogenic bacteria attach to human fibronectin through a tandem beta-zipper. Nature. 2003;423:177–81.
61. Pilka ES, Werner JM, Schwarz-Linek U, Pickford AR, Meenan NA, Campbell ID, et al. Structural insight into binding of *Staphylococcus aureus* to human fibronectin. FEBS Lett. 2006;580(1):273–7.
62. Peacock SJ, Day NP, Thomas MG, Berendt AR, Foster TJ. Clinical isolates of *Staphylococcus aureus* exhibit diversity in *fnb* genes and adhesion to human fibronectin. J Infect. 2000;41(1):23–31.
63. Frees D, Chastanet A, Qazi S, Sorensen K, Hill P, Msadek T, et al. Clp ATPases are required for stress tolerance, intracellular replication and biofilm formation in *Staphylococcus aureus*. Mol Microbiol. 2004;54(5):1445–62.
64. Sinha B, Francois PP, Nüsse O, Foti M, Hartford OM, Vaudaux P, et al. Fibronectin-binding protein acts as *Staphylococcus aureus* invasin via fibronectin bridging to integrin alpha-5beta1. Cell Microbiol. 1999;1(2):101–17.
65. Dziewanowska K, Patti JM, Deobald CF, Bayles KW, Trumble WR, Bohach GA. Fibronectin binding protein and host cell tyrosine kinase are required for internalization of *Staphylococcus aureus* by epithelial cells. Infect Immun. 1999;67(9):4673–8.
66. Fowler T, Wann ER, Joh D, Johansson S, Foster TJ, Höök M. Cellular invasion by *Staphylococcus aureus* involves a fibronectin bridge between the bacterial fibronectin-binding MSCRAMMs and host cell beta1 integrins. Eur J Cell Biol. 2000;79(10):672–9.
67. Dziewanowska K, Carson AR, Patti JM, Deobald CF, Bayles KW, Bohach GA. Staphylococcal fibronectin binding protein interacts with heat shock protein 60 and integrins: role in internalization by epithelial cells. Infect Immun. 2000;68(11):6321–8.
68. Grundmeier M, Hussain M, Becker P, Heilmann C, Peters G, Sinha B. Truncation of fibronectin-binding proteins in *Staphylococcus aureus* strain Newman leads to deficient adherence and host cell invasion due to loss of the cell wall anchor function. Infect Immun. 2004;72(12):7155–63.
69. Flock JI, Hienz SA, Heimdahl A, Schennings T. Reconsideration of the role of fibronectin binding in endocarditis caused by *Staphylococcus aureus*. Infect Immun. 1996;64(5):1876–8.
70. Palmqvist N, Foster T, Fitzgerald JR, Josefsson E, Tarkowski A. Fibronectin-binding proteins and fibrinogen-binding clumping factors play distinct roles in staphylococcal arthritis and systemic inflammation. J Infect Dis. 2005;191(5):791–8.
71. McElroy MC, Cain DJ, Tyrrell C, Foster TJ, Haslett C. Increased virulence of a fibronectin-binding protein mutant of Staphylococcus aureus in a rat model of pneumonia. Infect Immun. 2002;70(7):3865–73.
72. Kerdudou S, Laschke MW, Sinha B, Preissner KT, Menger MD, Herrmann M. Fibronectin binding proteins contribute to the adherence of *Staphylococcus aureus* to intact endothelium *in vivo*. Thromb Haemost. 2006;96(2):183–9.

73. Heilmann C, Niemann S, Sinha B, Herrmann M, Kehrel BE, Peters G. *Staphylococcus aureus* fibronectin-binding protein (FnBP)-mediated adherence to platelets, and aggregation of platelets induced by FnBPA but not by FnBPB. J Infect Dis. 2004;190(2):321–9.
74. Fitzgerald JR, Loughman A, Keane F, Brennan M, Knobel M, Higgins J, et al. Fibronectin-binding proteins of *Staphylococcus aureus* mediate activation of human platelets via fibrinogen and fibronectin bridges to integrin GPIIb/IIIa and IgG binding to the FcgammaRIIa receptor. Mol Microbiol. 2006;59(1):212–30.
75. Edwards AM, Potts JR, Josefsson E, Massey RC. *Staphylococcus aureus* host cell invasion and virulence in sepsis is facilitated by the multiple repeats within FnBPA. PLoS Pathog. 2010;6(6):e1000964.
76. Patti JM, House-Pompeo K, Boles JO, Garza N, Gurusiddappa S, Höök M. Critical residues in the ligand-binding site of the *Staphylococcus aureus* collagen-binding adhesin (MSCRAMM). J Biol Chem. 1995;270(20):12005–11.
77. Zong Y, Xu Y, Liang X, Keene DR, Höök A, Gurusiddappa S, et al. A 'Collagen Hug' model for *Staphylococcus aureus* CNA binding to collagen. EMBO J. 2005;24(24):4224–36.
78. Rhem MN, Lech EM, Patti JM, McDevitt D, Höök M, Jones DB, et al. The collagen-binding adhesin is a virulence factor in *Staphylococcus aureus* keratitis. Infect Immun. 2000; 68(6):3776–9.
79. Siboo IR, Chambers HF, Sullam PM. Role of SraP, a Serine-Rich Surface Protein of *Staphylococcus aureus*, in binding to human platelets. Infect Immun. 2005;73(4): 2273–80.
80. Siboo IR, Chaffin DO, Rubens CE, Sullam PM. Characterization of the accessory Sec system of *Staphylococcus aureus*. J Bacteriol. 2008;190(18):6188–96.
81. Uhlen M, Guss B, Nilsson B, Gatenbeck S, Philipson L, Lindberg M. Complete sequence of the staphylococcal gene encoding protein A. A gene evolved through multiple duplications. J Biol Chem. 1984;259(3):1695–702.
82. Deisenhofer J. Crystallographic refinement and atomic models of a human Fc fragment and its complex with fragment B of protein A from *Staphylococcus aureus* at 2.9- and 2.8-A resolution. Biochemistry. 1981;20(9):2361–70.
83. Gouda H, Shiraishi M, Takahashi H, Kato K, Torigoe H, Arata Y, et al. NMR study of the interaction between the B domain of staphylococcal protein A and the Fc portion of immunoglobulin G. Biochemistry. 1998;37(1):129–36.
84. Hartleib J, Kohler N, Dickinson RB, Chhatwal GS, Sixma JJ, Hartford OM, et al. Protein A is the von Willebrand factor binding protein on *Staphylococcus aureus*. Blood. 2000;96(6):2149–56.
85. Nguyen T, Ghebrehiwet B, Peerschke EIB. *Staphylococcus aureus* Protein A recognizes platelet gC1qR/p33: a novel mechanism for staphylococcal interactions with platelets. Infect Immun. 2000;68:2061–8.
86. Gomez MI, Lee A, Reddy B, Muir A, Soong G, Pitt A, et al. *Staphylococcus aureus* protein A induces airway epithelial inflammatory responses by activating TNFR1. Nat Med. 2004;10(8):842–8.
87. Dryla A, Gelbmann D, von Gabain A, Nagy E. Identification of a novel iron regulated staphylococcal surface protein with haptoglobin-haemoglobin binding activity. Mol Microbiol. 2003;49(1):37–53.
88. Mazmanian SK, Skaar EP, Gaspar AH, Humayun M, Gornicki P, Jelenska J, et al. Passage of heme-iron across the envelope of *Staphylococcus aureus*. Science. 2003;99(5608):906–9.
89. Clarke SR, Wiltshire MD, Foster SJ. IsdA of *Staphylococcus aureus* is a broad spectrum, iron-regulated adhesin. Mol Microbiol. 2004;51(5):1509–19.
90. Clarke SR, Andre G, Walsh EJ, Dufrene YF, Foster TJ, Foster SJ. Iron-regulated surface determinant protein A mediates adhesion of *Staphylococcus aureus* to human corneocyte envelope proteins. Infect Immun. 2009;77(6):2408–16.
91. Clarke SR, Brummell KJ, Horsburgh MJ, McDowell PW, Mohamad SA, Stapleton MR, et al. Identification of in vivo-expressed antigens of *Staphylococcus aureus* and their use in vaccinations for protection against nasal carriage. J Infect Dis. 2006;193(8):1098–108.

92. Miajlovic H, Zapotoczna M, Geoghegan JA, Kerrigan SW, Speziale P, Foster TJ. Direct interaction of iron-regulated surface determinant IsdB of *Staphylococcus aureus* with the GPIIb/IIIa receptor on platelets. Microbiology. 2010;156(Pt 3):920–8.
93. Kuklin NA, Clark DJ, Secore S, Cook J, Cope LD, McNeely T, et al. A novel *Staphylococcus aureus* vaccine: iron surface determinant B induces rapid antibody responses in rhesus macaques and specific increased survival in a murine *S. aureus* sepsis model. Infect Immun. 2006;74(4):2215–23.
94. Hell W, Meyer HG, Gatermann SG. Cloning of *aas*, a gene encoding a *Staphylococcus saprophyticus* surface protein with adhesive and autolytic properties. Mol Microbiol. 1998;29(3):871–81.
95. Rupp ME, Fey PD, Heilmann C, Gotz F. Characterization of the importance of Staphylococcus epidermidis autolysin and polysaccharide intercellular adhesin in the pathogenesis of intravascular catheter-associated infection in a rat model. J Infect Dis. 2001;183(7):1038–42.
96. Biswas R, Voggu L, Simon UK, Hentschel P, Thumm G, Götz F. Activity of the major staphylococcal autolysin Atl. FEMS Microbiol Lett. 2006;259(2):260–8.
97. Hirschhausen N, Schlesier T, Schmidt MA, Götz F, Peters G, Heilmann C. A novel staphylococcal internalization mechanism involves the major autolysin Atl and heat shock cognate protein Hsc70 as host cell receptor. Cell Microbiol. 2010;12:1746–64.
98. Zoll S, Patzold B, Schlag M, Götz F, Kalbacher H, Stehle T. Structural basis of cell wall cleavage by a staphylococcal autolysin. PLoS Pathog. 2010;6(3):e1000807.
99. Braun L, Ohayon H, Cossart P. The InIB protein of *Listeria monocytogenes* is sufficient to promote entry into mammalian cells. Mol Microbiol. 1998;27(5):1077–87.
100. Milohanic E, Jonquieres R, Cossart P, Berche P, Gaillard J-L. The autolysin Ami contributes to the adhesion of *Listeria monocytogenes* to eukaryotic cells via its cell wall anchor. Mol Microbiol. 2001;39:1212–24.
101. Multhoff G. Heat shock protein 70 (Hsp70): membrane location, export and immunological relevance. Methods. 2007;43(3):229–37.
102. Allignet J, Aubert S, Dyke KG, El Solh N. *Staphylococcus caprae* strains carry determinants known to be involved in pathogenicity: a gene encoding an autolysin-binding fibronectin and the *ica* operon involved in biofilm formation. Infect Immun. 2001;69(2):712–8.
103. Meyer HG, Wengler-Becker U, Gatermann SG. The hemagglutinin of *Staphylococcus saprophyticus* is a major adhesin for uroepithelial cells. Infect Immun. 1996;64(9):3893–6.
104. Heilmann C, Hartleib J, Hussain M, Peters G. The multifunctional *Staphylococcus aureus* autolysin Aaa mediates adherence to immobilized fibrinogen and fibronectin. Infect Immun. 2005;73(8):4793–802.
105. Heilmann C, Thumm G, Chhatwal GS, Hartleib J, Uekötter A, Peters G. Identification and characterization of a novel autolysin (Aae) with adhesive properties from *Staphylococcus epidermidis*. Microbiology. 2003;149:2769–78.
106. Bateman A, Rawlings ND. The CHAP domain: a large family of amidases including GSP amidase and peptidoglycan hydrolases. Trends Biochem Sci. 2003;28(5):234–7.
107. Buist G, Steen A, Kok J, Kuipers OP. LysM, a widely distributed protein motif for binding to (peptido)glycans. Mol Microbiol. 2008;68(4):838–47.
108. Chavakis T, Wiechmann K, Preissner KT, Herrmann M. *Staphylococcus aureus* interactions with the endothelium: the role of bacterial "secretable expanded repertoire adhesive molecules" (SERAM) in disturbing host defense systems. Thromb Haemost. 2005;94(2):278–85.
109. Hussain M, Becker K, von Eiff C, Schrenzel J, Peters G, Herrmann M. Identification and characterization of a novel 38.5-kilodalton cell surface protein of *Staphylococcus aureus* with extended-spectrum binding activity for extracellular matrix and plasma proteins. J Bacteriol. 2001;183(23):6778–86.
110. McGavin MH, Krajewska-Pietrasik D, Ryden C, Höök M. Identification of a *Staphylococcus aureus* extracellular matrix-binding protein with broad specificity. Infect Immun. 1993;61(6):2479–85.
111. Jönsson K, McDevitt D, McGavin MH, Patti JM, Höök M. *Staphylococcus aureus* expresses a major histocompatibility complex class II analog. J Biol Chem. 1995;270(37):21457–60.

112. Chavakis T, Hussain M, Kanse SM, Peters G, Bretzel RG, Flock JI, et al. *Staphylococcus aureus* extracellular adherence protein serves as anti-inflammatory factor by inhibiting the recruitment of host leukocytes. Nat Med. 2002;8(7):687–93.
113. Sobke AC, Selimovic D, Orlova V, Hassan M, Chavakis T, Athanasopoulos AN, et al. The extracellular adherence protein from *Staphylococcus aureus* abrogates angiogenic responses of endothelial cells by blocking Ras activation. FASEB J. 2006;20(14):2621–3.
114. Hussain M, von Eiff C, Sinha B, Joost I, Herrmann M, Peters G, et al. eap Gene as novel target for specific identification of *Staphylococcus aureus*. J Clin Microbiol. 2008;46(2):470–6.
115. Clarke SR, Harris LG, Richards RG, Foster SJ. Analysis of Ebh, a 1.1-megadalton cell wall-associated fibronectin-binding protein of *Staphylococcus aureus*. Infect Immun. 2002; 70(12):6680–7.
116. Williams RJ, Henderson B, Sharp LJ, Nair SP. Identification of a fibronectin-binding protein from *Staphylococcus epidermidis*. Infect Immun. 2002;70(12):6805–10.
117. Kuroda M, Tanaka Y, Aoki R, Shu D, Tsumoto K, Ohta T. *Staphylococcus aureus* giant protein Ebh is involved in tolerance to transient hyperosmotic pressure. Biochem Biophys Res Commun. 2008;374(2):237–41.
118. Downer R, Roche F, Park PW, Mecham RP, Foster TJ. The elastin-binding protein of *Staphylococcus aureus* (EbpS) is expressed at the cell surface as an integral membrane protein and not as a cell wall-associated protein. J Biol Chem. 2002;277(1):243–50.
119. Weidenmaier C, Kokai-Kun JF, Kristian SA, Chanturiya T, Kalbacher H, Gross M, et al. Role of teichoic acids in *Staphylococcus aureus* nasal colonization, a major risk factor in nosocomial infections. Nat Med. 2004;10(3):243–5.
120. Weidenmaier C, Peschel A, Xiong YQ, Kristian SA, Dietz K, Yeaman MR, et al. Lack of wall teichoic acids in *Staphylococcus aureus* leads to reduced interactions with endothelial cells and to attenuated virulence in a rabbit model of endocarditis. J Infect Dis. 2005;191(10):1771–7.
121. Hussain M, Heilmann C, Peters G, Herrmann M. Teichoic acid enhances adhesion of *Staphylococcus epidermidis* to immobilized fibronectin. Microb Pathog. 2001;31(6):261–70.
122. Hussain M, Herrmann M, von Eiff C, Perdreau-Remington F, Peters G. A 140-kilodalton extracellular protein is essential for the accumulation of *Staphylococcus epidermidis* strains on surfaces. Infect Immun. 1997;65(2):519–24.
123. Corrigan RM, Rigby D, Handley P, Foster TJ. The role of *Staphylococcus aureus* surface protein SasG in adherence and biofilm formation. Microbiology. 2007;153(Pt 8):2435–46.
124. Rohde H, Burdelski C, Bartscht K, Hussain M, Buck F, Horstkotte MA, et al. Induction of *Staphylococcus epidermidis* biofilm formation via proteolytic processing of the accumulation-associated protein by staphylococcal and host proteases. Mol Microbiol. 2005;55(6):1883–95.
125. Conrady DG, Brescia CC, Horii K, Weiss AA, Hassett DJ, Herr AB. A zinc-dependent adhesion module is responsible for intercellular adhesion in staphylococcal biofilms. Proc Natl Acad Sci U S A. 2008;105(49):19456–61.
126. Sun D, Accavitti MA, Bryers JD. Inhibition of biofilm formation by monoclonal antibodies against *Staphylococcus epidermidis* RP62A accumulation-associated protein. Clin Diagn Lab Immunol. 2005;12(1):93–100.
127. Geoghegan JA, Corrigan RM, Gruszka DT, Speziale P, O'Gara JP, Potts JR, et al. Role of surface protein SasG in biofilm formation by *Staphylococcus aureus*. J Bacteriol. 2010;192: 5663–73.
128. Banner MA, Cunniffe JG, Macintosh RL, Foster TJ, Rohde H, Mack D, et al. Localized tufts of fibrils on *Staphylococcus epidermidis* NCTC 11047 are comprised of the accumulation-associated protein. J Bacteriol. 2007;189(7):2793–804.
129. Cucarella C, Tormo MA, Knecht E, Amorena B, Lasa I, Foster TJ, et al. Expression of the biofilm-associated protein interferes with host protein receptors of *Staphylococcus aureus* and alters the infective process. Infect Immun. 2002;70(6):3180–6.
130. Schroeder K, Jularic M, Horsburgh SM, Hirschhausen N, Neumann C, Bertling A, et al. Molecular characterization of a novel *Staphylococcus aureus* surface protein (SasC) involved in cell aggregation and biofilm accumulation. PLoS One. 2009;4(10):e7567.

131. Merino N, Toledo-Arana A, Vergara-Irigaray M, Valle J, Solano C, Calvo E, et al. Protein A-mediated multicellular behavior in *Staphylococcus aureus*. J Bacteriol. 2009;191(3):832–43.
132. O'Neill E, Pozzi C, Houston P, Humphreys H, Robinson DA, Loughman A, et al. A novel *Staphylococcus aureus* biofilm phenotype mediated by the fibronectin-binding proteins, FnBPA and FnBPB. J Bacteriol. 2008;190(11):3835–50.
133. Vergara-Irigaray M, Valle J, Merino N, Latasa C, Garcia B, de Ruiz Los Mozos I, et al. Relevant role of fibronectin-binding proteins in Staphylococcus aureus biofilm-associated foreign-body infections. Infect Immun. 2009;77(9):3978–91.
134. Christner M, Franke G, Schommer N, Wendt U, Wegert K, Pehle P, et al. The giant extracellular matrix binding protein of *Staphylococcus epidermidis* mediates biofilm accumulation and attachment to fibronectin. Mol Microbiol. 2010;75(1):187–207.
135. Mack D, Nedelmann M, Krokotsch A, Schwarzkopf A, Heesemann J, Laufs R. Characterization of transposon mutants of biofilm-producing *Staphylococcus epidermidis* impaired in the accumulative phase of biofilm production: genetic identification of a hexosamine-containing polysaccharide intercellular adhesin. Infect Immun. 1994;62(8):3244–53.
136. Heilmann C, Schweitzer O, Gerke C, Vanittanakom N, Mack D, Götz F. Molecular basis of intercellular adhesion in the biofilm-forming *Staphylococcus epidermidis*. Mol Microbiol. 1996;20(5):1083–91.
137. Cramton SE, Gerke C, Schnell NF, Nichols WW, Götz F. The intercellular adhesion (*ica*) locus is present in *Staphylococcus aureus* and is required for biofilm formation. Infect Immun. 1999;67(10):5427–33.
138. Maira-Litran T, Kropec A, Abeygunawardana C, Joyce J, Mark G, Goldmann DA, et al. Immunochemical properties of the staphylococcal poly-N-acetylglucosamine surface polysaccharide. Infect Immun. 2002;70(8):4433–40.
139. Kaplan JB, Velliyagounder K, Ragunath C, Rohde H, Mack D, Knobloch JK, et al. Genes involved in the synthesis and degradation of matrix polysaccharide in *Actinobacillus actinomycetemcomitans* and *Actinobacillus pleuropneumoniae* biofilms. J Bacteriol. 2004;186(24): 8213–20.
140. Gerke C, Kraft A, Sussmuth R, Schweitzer O, Götz F. Characterization of the N-acetylglucosaminyltransferase activity involved in the biosynthesis of the *Staphylococcus epidermidis* polysaccharide intercellular adhesin. J Biol Chem. 1998;273(29):18586–93.
141. Vuong C, Kocianova S, Voyich JM, Yao Y, Fischer ER, DeLeo FR, et al. A crucial role for exopolysaccharide modification in bacterial biofilm formation, immune evasion, and virulence. J Biol Chem. 2004;279(52):54881–6.
142. Rupp ME, Ulphani JS, Fey PD, Bartscht K, Mack D. Characterization of the importance of polysaccharide intercellular adhesin/hemagglutinin of *Staphylococcus epidermidis* in the pathogenesis of biomaterial-based infection in a mouse foreign body infection model. Infect Immun. 1999;67(5):2627–32.
143. Kristian SA, Golda T, Ferracin F, Cramton SE, Neumeister B, Peschel A, et al. The ability of biofilm formation does not influence virulence of *Staphylococcus aureus* and host response in a mouse tissue cage infection model. Microb Pathog. 2004;36(5):237–45.
144. Ziebuhr W, Heilmann C, Götz F, Meyer P, Wilms K, Straube E, et al. Detection of the intercellular adhesion gene cluster (*ica*) and phase variation in *Staphylococcus epidermidis* blood culture strains and mucosal isolates. Infect Immun. 1997;65(3):890–6.
145. Hall-Stoodley L, Nistico L, Sambanthamoorthy K, Dice B, Nguyen D, Mershon WJ, et al. Characterization of biofilm matrix, degradation by DNase treatment and evidence of capsule downregulation in *Streptococcus pneumoniae* clinical isolates. BMC Microbiol. 2008;8:173.
146. Allesen-Holm M, Barken KB, Yang L, Klausen M, Webb JS, Kjelleberg S, et al. A characterization of DNA release in *Pseudomonas aeruginosa* cultures and biofilms. Mol Microbiol. 2006;59(4):1114–28.
147. Thomas VC, Thurlow LR, Boyle D, Hancock LE. Regulation of autolysis-dependent extracellular DNA release by *Enterococcus faecalis* extracellular proteases influences biofilm development. J Bacteriol. 2008;190(16):5690–8.

148. Itoh Y, Wang X, Hinnebusch BJ, Preston 3rd JF, Romeo T. Depolymerization of beta-1,6-N-acetyl-D-glucosamine disrupts the integrity of diverse bacterial biofilms. J Bacteriol. 2005; 187(1):382–7.
149. Izano EA, Amarante MA, Kher WB, Kaplan JB. Differential roles of poly-N-acetylglucosamine surface polysaccharide and extracellular DNA in *Staphylococcus aureus* and *Staphylococcus epidermidis* biofilms. Appl Environ Microbiol. 2008;74(2):470–6.
150. Kogan G, Sadovskaya I, Chaignon P, Chokr A, Jabbouri S. Biofilms of clinical strains of *Staphylococcus* that do not contain polysaccharide intercellular adhesin. FEMS Microbiol Lett. 2006;255(1):11–6.
151. Rohde H, Burandt EC, Siemssen N, Frommelt L, Burdelski C, Wurster S, et al. Polysaccharide intercellular adhesin or protein factors in biofilm accumulation of *Staphylococcus epidermidis* and *Staphylococcus aureus* isolated from prosthetic hip and knee joint infections. Biomaterials. 2007;28(9):1711–20.
152. Chaignon P, Sadovskaya I, Ragunah C, Ramasubbu N, Kaplan JB, Jabbouri S. Susceptibility of staphylococcal biofilms to enzymatic treatments depends on their chemical composition. Appl Microbiol Biotechnol. 2007;75(1):125–32.
153. Boles BR, Horswill AR. Agr-mediated dispersal of *Staphylococcus aureus* biofilms. PLoS Pathog. 2008;4(4):e1000052.
154. Mehlin C, Headley CM, Klebanoff SJ. An inflammatory polypeptide complex from *Staphylococcus epidermidis*: isolation and characterization. J Exp Med. 1999;189(6):907–18.
155. Novick RP. Staphylococcal pathogenesis and pathogenicity factors: genetics and regulation. In: Fischetti VA, Ferretti JJ, Portnoy DA, Rood JI, Novick RP, editors. Gram-positive pathogens. 2nd ed. Washington: ASM Press; 2006. p. 496–516.
156. Vuong C, Saenz HL, Götz F, Otto M. Impact of the *agr* quorum-sensing system on adherence to polystyrene in *Staphylococcus aureus*. J Infect Dis. 2000;182(6):1688–93.
157. Otto M. Staphylococcal biofilms. Curr Top Microbiol Immunol. 2008;322:207–28.
158. Otto M, O'Mahoney DS, Guina T, Klebanoff SJ. Activity of *Staphylococcus epidermidis* phenol-soluble modulin peptides expressed in *Staphylococcus carnosus*. J Infect Dis. 2004;190(4):748–55.
159. Yao Y, Sturdevant DE, Otto M. Genomewide analysis of gene expression in *Staphylococcus epidermidis* biofilms: insights into the pathophysiology of *S. epidermidis* biofilms and the role of phenol-soluble modulins in formation of biofilms. J Infect Dis. 2005;191(2):289–98.
160. Patti JM, Jonsson H, Guss B, Switalski LM, Wiberg K, Lindberg M, et al. Molecular characterization and expression of a gene encoding a *Staphylococcus aureus* collagen adhesin. J Biol Chem. 1992;267(7):4766–72.

Chapter 4
Propionibacterium acnes: An Emerging Pathogen in Biomaterial-Associated Infection

Sheila Patrick and Andrew McDowell

Abstract While it is generally accepted that *Staphylococcus* spp., including coagulase-negative staphylococci (CoNS), are associated with biomaterial-associated infection, it has become increasingly clear that *Propionibacterium acnes* is also a significant cause of such infections, especially in relation to prosthetic joint failure. *P. acnes* outnumbers CoNS in sebaceous gland-rich areas of the skin and has considerable pathogenic potential. Molecular phylogeny studies have revealed that *P. acnes* comprises major evolutionary lineages with distinct differences in the production of putative virulence determinants. Strains may, therefore, be benign skin commensals or have pathogenic and pro-inflammatory potential. The role of *P. acnes* in biofilm infections is often overlooked as, although *P. acnes* is microaerophilic-to-aerotolerant, for optimal isolation from clinical material, samples must be processed as for obligately anaerobic bacteria; biomaterials must be maintained in an anaerobic atmosphere immediately upon removal from the patient and adherent biofilm dislodged by mild ultrasound treatment. The application of non-culture methods does, however, overcome this problem and provides the potential to improve detection rates.

4.1 Introduction

The majority of biomaterial-associated infections arise from bacteria that are either permanent or transient members of the human microbiota. Molecular analyses have confirmed the overall diversity of the human microbiome and identified the predominant groups of bacteria present [1]. Given the truly phenomenal range

S. Patrick (✉) • A. McDowell
Centre for Infection and Immunity, School of Medicine, Dentistry and Biomedical Sciences, Queen's University Belfast, Belfast BT9 7BL, UK
e-mail: s.patrick@qub.ac.uk; a.mcdowell@qub.ac.uk

T.F. Moriarty et al. (eds.), *Biomaterials Associated Infection: Immunological Aspects and Antimicrobial Strategies*, DOI 10.1007/978-1-4614-1031-7_4,

of microbial groups that colonise humans, it is interesting that only a select few are frequently associated with biomaterial-associated infections. When other organisms not normally associated with such infections are occasionally isolated, it usually relates to elderly or immuno-compromised patients or individuals with another underlying condition, e.g. [2]. Implanted materials associated with any site where there is a normal microbiota (skin, lower gastrointestinal tract, oral/upper respiratory tract and female genital tract) may become colonised by opportunistic pathogens that reside within these microbial communities. Equally, if microbes "escape" from these regions into normally uncolonised body sites, including the bloodstream, as a result for example of trauma, colonisation of implanted materials and the development of infection may result. The link between trauma from dental interventions and cardiac infection deriving from the oral microbiota is perhaps one of the most generally known examples [3]. In addition to native heart valve infection, cardiac device-related biofilm infection associated with prosthetic heart valves, permanent pacemakers and implantable cardioverter defibrillators are becoming increasingly common. These infections require prolonged antibiotic therapy and frequently also device removal for cure [4, 5].

Haematogenous spread from infection at other body sites, caused by either primary or opportunistic pathogens, is another potential source of implant-colonising bacteria, but for the most part these are sporadic events. Characteristics of haematogenous infection include sudden local joint pain due to inflammation or increasing pain due to loosening [6]. An exemplar is the obligate anaerobe, *Bacteroides fragilis*. It is a member of the normal intestinal microbiota and opportunistic pathogen known to cause potentially life-threatening infection and soft tissue abscesses if it escapes from this site [7], classically as the result of the rupture of an inflamed appendix or equally as a result of abdominal surgery or bowel ischemia [8]. Biofilm infection may be the result of haematogenous spread of *B. fragilis* from abscess infection or indeed the normal microbiota of the gastrointestinal tract and female genital tract [8]. These biofilm infections include prosthetic joint and heart valve infections [4, 9]. While it is clear that implanted biomaterials may provide enhanced colonisation potential for professional pathogens and opportunistic pathogens, with a wide variety of different types of microbe capable of causing serious implant-associated infection [10], the selection of microbes most frequently associated with implant infection remains somewhat limited. This then raises the question: What defines an opportunistic pathogen in the context of biofilm infection? Access to the biomaterial surface and an ability to form biofilms are self-evidently crucial, but biofilm as a mode of growth is the rule rather than the exception for most microbes that inhabit our planet, never mind those that inhabit our bodies [11, 12]; complete genome sequence comparisons have begun to dissect what constitutes the putative virulent genotype and as a result the reasons why some microbes emerge more frequently in biofilm-related infections are becoming clearer. Genomic comparison has contributed substantively to the resolution of our understanding in the case of *Staphylococcus* spp. which can be permanent or transient members of the skin microbiota [13]. The importance of acute *Staphylococcus aureus* infection is well recognised due to documented virulence determinants; however, the acceptance of the importance of chronic infection caused by microbes such as coagulase-negative *Staphylococcus* spp. (CoNS), in particular *S. epidermidis*, has been more problematic. This is largely because these bacteria are perceived to be of low virulence and benign members of the normal human skin microbiota. Given the opportunity, however, CoNS are highly capable of causing biomaterial-associated infection. Complete genome sequencing indicates that *S. epidermidis* generally lacks the well-recognised virulence determinants which enable *S. aureus* to cause acute infection but, in addition to biofilm formation, strains of *S. epidermidis* may carry a range of other putative virulence determinant genes including lipases which are considered to contribute to chronic infection. *Staphylococcus lugdunensis*, which was first described in 1988 [14], can cause acute infection similar to *S. aureus*, although it is a CoNS. It is also associated with, for example, native and prosthetic heart valve infection, catheter-related biofilm infection and prosthetic joint infection [15, 16]. The comparison of the complete genome sequence of *S. lugdunensis* [17] with other *Staphylococcus* spp. and sequencing of other *S. lugdunensis* strains will undoubtedly aid our understanding of the spectrum of staphylococcal virulence. The role of *Staphylococcus* spp. in biomaterial infection is discussed in detail in Chaps. 2 and 3.

If *Staphylococcus* spp. are regarded as the paradigm of the spectrum of biofilm infection, the importance of *Propionibacterium acnes* is less generally accepted. When cultured, it is frequently dismissed as being of no significance, yet within this single species there are many parallels with the genus *Staphylococcus*. It is the dominant member of the human skin microbiota in many skin areas (Table 4.1), is associated consistently with implanted biomaterial infections [18–20] as well as native heart valve infection [4] and is closely linked with the inflammatory condition acne [21]. The pro-inflammatory credentials of some strains of *P. acnes* are well known to immunologists as these are commercially available for use as adjuvant. The general lack of recognition of the importance of *P. acnes* likely relates to a number of issues. In particular, the dogma is that as it is a relatively benign member of the human microbiome its presence in clinical samples reflects contamination in most contexts. Furthermore, if clinical samples are not handled and processed optimally (as for obligately anaerobic bacteria) *P. acnes* the pathogen, rather than *P. acnes* the contaminant, will be missed entirely.

This chapter focuses on the emerging recognition of *Propionibacterium acnes* as a significant cause of biofilm-related infection and addresses the issues surrounding the lack of optimal methodology for its isolation from clinical samples. Consideration is given to how this has led to an underestimation of its role in infection. In addition, the intraspecies genotypic and phenotypic diversity is discussed in relation to the potential for variability in pathogenesis and virulence amongst and within strains.

4.1.1 *Propionibacterium acnes* and Biofilm Infection

P. acnes is a Gram-positive nonspore-forming pleomorphic rod that primarily colonises sebaceous glands and hair follicles of human skin, but it is also found in the mouth, as well as the genito-urinary tract and large intestine; a common feature of

Table 4.1 Relative abundance of *Propionibacterium* spp, compared to other bacterial types, present in different human skin areas as determined by 16S ribosomal RNA gene sequence analyses of ten individuals (compiled from Grice et al. 2011)

Skin area	No. of individuals with relative abundance in range: >50%[a]	5–50%[a]
Sebaceous gland rich areas		
Glabella (between eye brows)	3	6
Alar Crease (beside nostril)	5	5
External auditory canal (inside ear)	1	9
Manubrium (upper chest)	3	7
Retroauricular crease (behind ear)[b]	4	2
Occiput (back of scalp)	2	7
Back	6	4
Moist areas		
Nare (inside nostril)	0	8
Axillary vault (armpit)[b]	0	2
Antecubital fossa (inner elbow)	1	3
Interdigital web space[c] (between middle and ring fingers)	0	4
Inguinal crease (side of groin)	0	1
Umbilicus (navel)	1	0
Toe web space	0	0
Gluteal crease (between upper buttocks)	1	2
Popliteal fossa (behind knee)	0	4
Plantar heel (bottom of heel)	0	2
Dry areas		
Volar forearm (inside mid forearm)	1	2
Hypothenar palm (palm of hand)[b]	0	4
Buttock	0	3

[a]Estimated from Grice et al. 2011 Figure S2
[b]Data available from only nine individuals
[c]Data available from only eight individuals

these environments is a lack of oxygen which reflects the anaerobic-to-microaerophilic growth characteristics of *P. acnes* [22]. Cells are 0.5–1.5 μm in diameter and 1–5 μm in length, although some strains can form filamentous-like structures [22]. Levels of *P. acnes* colonisation on the skin vary from person-to-person and from the area of the body sampled [23]. By culture, numbers can be as high as 10^6 colony-forming units/cm^2, with the neck, forehead and shoulder showing some of the highest concentrations of the bacterium compared to other sites, such as the abdomen, hip and knee where levels are lower [24]. Molecular analyses of the human skin microbiome using skin swabs or scrapes concur with this view; *Propionibacterium* spp. clearly predominate in the sebaceous gland-rich areas sampled, namely, the chest, back, face and scalp. They may also be present in abundance at dry and moist skin sites in some individuals, for example, the crease between the buttocks, forearm, inner elbow and umbilicus [1] (Table 4.1). Sampling directly from hair follicles clearly demonstrated exclusive colonisation by *P. acnes* in three healthy individuals [25]. Similarly, microscopic visualisation using specific antibodies revealed exclusive colonisation of individual hair follicles with *P. acnes* in both control and acne patients; follicles from acne patients contained predominantly large macro-colonies (>1,000 bacterial cells), whereas follicles from control patients contained predominantly micro-colonies (<100 bacterial cells) [21]. *P. acnes* is clearly well placed to contaminate surgical incision sites and it is well recognised as a cause of chronic biofilm-associated infections of indwelling medical devices, with varying proportions attributable to *P. acnes* depending on the methods used to detect the infection and the site of the implant. These include prosthetic hip, knee and shoulder joints, spinal implants, prosthetic heart valves, intravascular catheters and central nervous system shunts [18, 26–30]. The bacterium is also responsible for endophthalmitis post-eye surgery and is the most common cause of chronic post-cataract ophthalmitis [31]. *P. acnes*, however, may also be responsible for severe post-surgical infection including endocarditis, postcraniotomy infections, joint and spine infections, endophthalmitis and pansinusitis [32]. In particular, post-operative infection of prosthetic shoulder joints due to *P. acnes* appears to be an emerging clinical entity [33, 34] and the importance of *P. acnes* infection in post-neurosurgical surgery (e.g. craniotomy) is becoming increasingly recognised, even in the absence of an implanted device, which is in keeping with its predominance on the scalp [35]. While *P. acnes* can cause acute orthopaedic implant infection associated with severe tissue damage and sinus tract formation [36], its involvement in late-onset prosthetic joint infection, where the distinction between chronic low-grade infection and "aseptic loosening" of prosthetic joints is not clear, is less well recognised. Loosening in many cases is likely to be driven by subclinical bacterial biofilm infection, which may have largely resolved by the time of revision surgery. Infection can be insidious with delayed presentation of symptoms; indeed, many of the traditional signs of underlying infection are absent, e.g. [37, 38]. In addition, the inflammatory response characteristic of osteolysis could be driven by the release of bacterial components, in particular cell envelope components, from biofilms over long time periods. The inflammatory potential of these bacterial components, coupled with the microscopic biomaterial wear particles, further compounds the difficulty of diagnosis of failed joint prostheses; the bacteria that caused the inflammatory response may be in a dormant state or no longer viable at the time of revision surgery.

Another major issue in obtaining a true picture of the involvement of *P. acnes* in prosthetic joint or indeed any infection is that although in pure culture isolates may be microaerophilic-to-aerotolerant [22], for optimal isolation samples must be maintained, processed and cultured in an anaerobic environment [39]. By sampling biofilm dislodged by ultrasonication from prostheses retrieved at the time of revision operation, and adhering to strict anaerobic culture techniques, *P. acnes* is isolated as frequently as *Staphylococcus* spp. from failed devices. Of 120 retrieved hip prostheses from patients who had an initial diagnosis of "aseptic" loosening, *P. acnes* was isolated from 13% and *Staphylococcus* spp. from 11%, with both spp. isolated from three of these patients [40].

A solution to the problem of difficult detection by culture is the application of non-culture methods to detect biofilm dislodged from retrieved materials by ultrasound treatment, as discussed in detail below. While molecular analyses based on classical polymerase chain reaction (PCR) amplification can suffer from contamination problems, direct microscopical visualisation of biofilm aggregates dislodged by ultrasound from prostheses and labelled with specific antibodies provides a robust diagnostic tool that can easily discriminate skin microbiota contaminants [29, 41]. The question of non-specific contamination from whatever source is simply addressed; for example some workers have experienced problems with leakage of bags in which the prostheses are placed for ultrasound treatment [42]. In addition, the difficulties of culture-based detection in relation to anaerobic conditions, bacterial dormancy and bacterial killing as a result of the administration of prophylactic antibiotics at the time of surgery are no longer an issue. In addition these techniques, when applied to bone samples, indicate that infection is not restricted to bacteria growing in biofilm on the prosthesis surface, but may include biofilm colonisation of the adjacent bone [43]. This has important implications for the potential role of bacteria in the osteolysis observed in these patients. This returns us to the key question of the significance of *P. acnes* as both a pathogen and a member of the normal microbiota and our growing understanding of within-species diversity.

4.1.2 Population Genetic Structure of *Propionibacterium acnes*

Over the last number of years significant inroads have been made into our understanding of the population genetic structure of *P. acnes*. Initial studies based on nucleotide sequencing of the *recA* housekeeping gene and putative virulence genes clustered *P. acnes* isolates into three major clades or divisions known as type I, II and III [41, 44, 45]. Isolates within the type I cluster were also sub-divided into type IA and IB. Analysis of *P. acnes* by Multilocus Sequence Typing (MLST) (http://pubmlst.org/pacnes/), which generates a sequence type (ST) for each isolate based on partial nucleotide sequencing of core housekeeping genes (normally seven), has confirmed these major lineages (Fig. 4.1) and also revealed that the population as a whole is clonal [46, 47]. More recently, subdivision of type IA strains into the groups IA_1 and IA_2 has been proposed based on whole genome sequence and phylogenetic analyses [48] (Fig 4.1). Despite the overall clonal nature of the population, recombination events have also been shown to play a role in shaping the genetic diversity within this organism. While the major phylotypes of *P. acnes* can be resolved based on the analysis of core housekeeping genes, they share at least 99.8–99.9% identity based on 16S rRNA sequences [41, 45]. The phylotypes also display defining phenotypic and biochemical differences including sero-reactivity, cellular morphology and differential production of specific virulence determinants. *In silico* comparisons of type IA, IB and II genome sequences have found differences with respect to imported genomic islands [46, 49]. Furthermore, type IA strains have lost

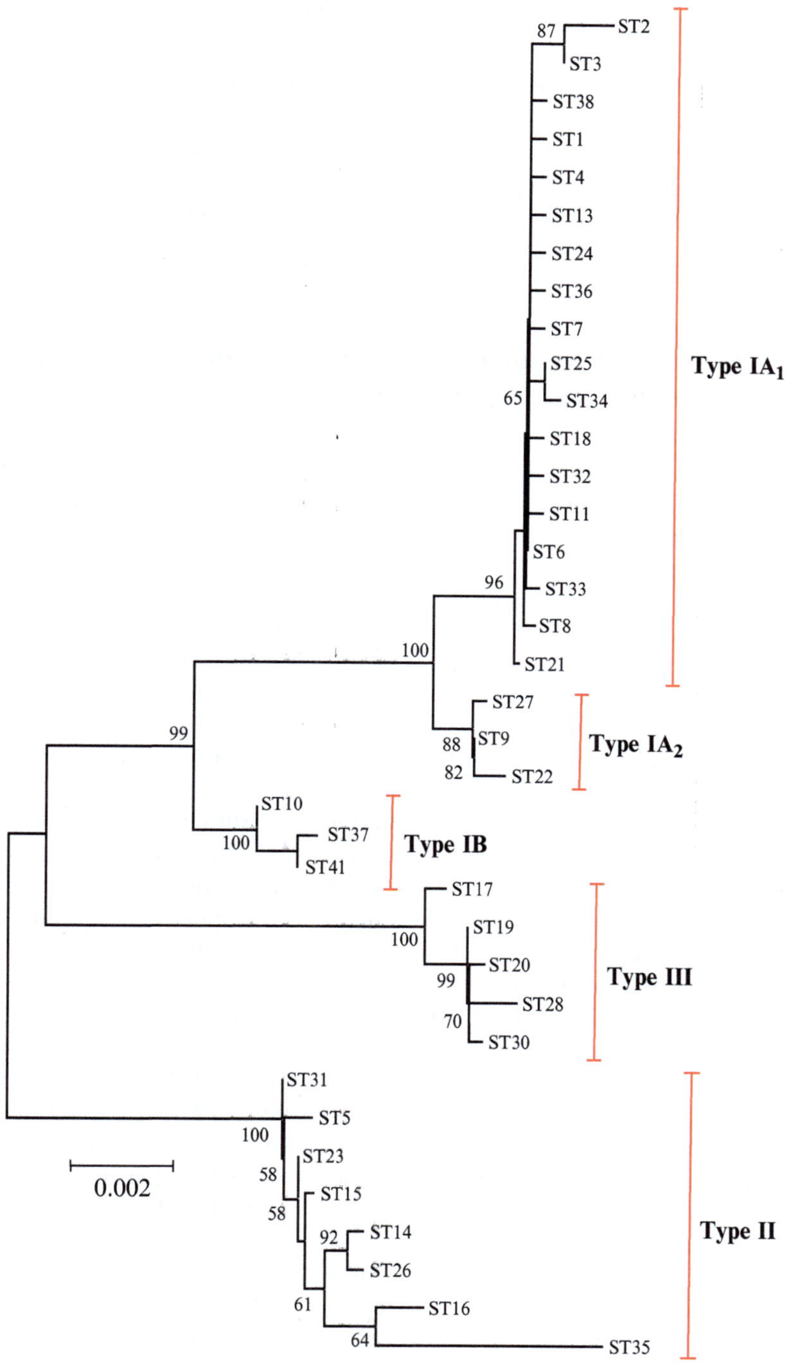

Fig. 4.1 Neighbour-joining phylogenetic tree of concatenated gene sequences (3,135 bp) from the *P. acnes* MLST scheme based on 7 loci. The tree was constructed using the Jukes–Cantor-based distance matrices on the concatenated sequences from each novel ST. The sequence input order was randomised, and bootstrapping resampling statistics were performed using 100 data sets. Bootstrap values (≥50%) are shown on the arms of the tree and the *bar* represents genetic distance. The lineage types IA_1, IA_2, IB, II and III are *highlighted*. Modified from McDowell et al. (2011) and based on STs resolved from 123 isolates covering a wide range of clinical conditions (http://pubmlst.org/pacnesi). An expanded 8 locus-based scheme has been described, McDowell et al. PLoS One in press

ten genes involved in carbohydrate metabolism but do contain genes encoding chitinase which may be important in the context of acne pathogenesis [46, 49]. Despite the presence of these different genomic regions and some insertions and deletions, a high degree of sequence conservation is observed. On this basis, most differences within and between phylotypes are likely to reflect variations in the transcriptome, with regulatory events at the transcriptional and translational levels.

4.1.3 *Propionibacterium acnes: Member of the Normal Resident Microbiota and Opportunistic Pathogen*

In addition to the biofilm infections discussed above, *P. acnes* has also been linked to synovitis–acne–pustulosis–hyperostosis–osteitis (SAPHO) syndrome [50], sarcoidosis [51] and, more recently, chronic infection of the prostate gland potentially leading to prostate cancer [52]. Although the role of *P. acnes* in the inflammatory condition acne has been debated, its involvement is now generally accepted since acne which is refractory to antibiotic treatment is associated with strains resistant to the same antibiotic [53]. The intraspecies genetic diversity of *P. acnes* and the implications of this in relation to benign versus pathogenic *lineages* are now only being unravelled and may in the future explain some of the apparent contradictions in the literature, in particular where conclusions have been drawn from the study of a single ST representing one of the phylogenetic divisions.

The immunostimulatory activity and adjuvant properties of some isolates previously, and currently by some authors, referred to as "*Corynebacterium parvum*" are well recognised and are exploited in studies of the immune system [54]. *P. acnes* has also been used as a non-specific immune stimulant in the equine industry (EqStim®) [55]. *P. acnes* is reported to trigger the expression of antimicrobial peptides and pro-inflammatory cytokines/chemokines in sebocytes [56].

P. acnes forms biofilms on a variety of different implant biomaterials including orthopaedic bone cement and titanium alloys, which, as with other bacteria, leads to enhanced antibiotic resistance [57–59]. The molecular detail of the biofilm forming process, however, has yet to be elucidated. Some strains of *P. acnes* secrete a range of components that may provide a selective advantage during colonisation as part of the normal microbiota and during opportunistic infection. Beta-haemolytic activity, which results in the complete lysis of red cells surrounding colonies growing on a blood agar plate, is associated with phylotype IA and IB STs and may relate to the *tly* gene [46, 60]. A gene family, consisting of five homologues of the *Streptococcus agalactiae* Christie–Atkins–Munch–Petersen (CAMP) factor [61], has been identified within the genome of *P. acnes* strains from all phylotypes. The *Strep. agalactiae* CAMP factor forms membrane-damaging pores [62]. Variable expression of the *P. acnes* CAMP factor homologues amongst phylotype groups has been noted [44]. In particular, strains of type II and the type IB MLST genotype ST-10 produce large quantities of CAMP factor 1 compared to type IA organisms [44]. In addition to acting as pore-forming toxins, CAMP factors may bind both IgG and

IgM immunoglobulins [62, 63]. The five CAMP factor homologues of *P. acnes* are likely to have arisen by divergence of a common ancestral gene and may now have distinct biological functions. Hyaluronate lyase may also contribute to opportunistic infection via degradation of hyaluronic acid in the extracellular matrix [64], as may neuraminidase activity [65]. Lipase production is considered highly relevant to colonisation of the lipid-rich sebaceous gland, in addition to the lipase described by Miskin and colleagues [66]. A second putative triacyl glycerol lipase and more than ten putative esterase lipases have been identified in the complete genome sequence of KPA 171202 [67]. The production of lipase and indeed neuraminidase appears especially associated with strains from the type IA division [68] and lipase production by *P. acnes* appears upregulated during biofilm growth [59].

Variation in expression of immune-reactive surface and secreted proteins has also been described [69] including variation of dermatan sulphate adhesins (DsA) which bind to the glycosaminoglycan dermatan sulphate, which is the predominant glycosaminoglycan in skin and a biological response modifier. These proteins have carboxy-terminal repeats and the amino acid motif—LPXTG—characteristic of MSCRAMMs (microbial surface components recognising adhesive molecules). Sequence variation is generated by a variable number of C residues in a C_nTC_n motif upstream of the putative signal peptide for the proteins. In addition, there are variable numbers of repeats towards the mid-region and the carboxy-terminus that alter molecular mass and expression of the proteins. This variation is probably generated by slipped strand mispairing during replication leading to phenotypic variation. Expression of DsA is associated with type IA *P. acnes* but not type IB, II or III [47]. Clearly this impacts on the properties of different isolates. The opportunity for slipped strand mispairing to cause antigenic or phase variants would provide the means for *P. acnes* to avoid host immune defences, as well as contribute to dissemination by altering adhesion specificity.

Analysis of *P. acnes* isolates from foreign body, surgical and septicemia infections in Europe has shown resistance to erythromycin (17%), clindamycin (15%) and lower levels of resistance to tetracycline (3%) [70]. Cross-resistant propionibacteria have also been detected across Europe and Asia due to the use of topical formulations of erythromycin and clindamycin in the treatment of acne [71–73]. Some strains of *P. acnes* have been reported to be resistant to metronidazole, sulphonamides and aminoglycosides [74–76]. In addition, as is well documented for other bacterial genera, the growth of *P. acnes* in biofilm on implanted materials significantly enhances antibiotic resistance [57, 58].

4.1.4 Association of *Propionibacterium acnes* Phylotypes with Biomaterial-Related Infection

Our improved understanding of the intraspecies diversity of *P. acnes* at the genetic, phenotypic and biochemical levels has stimulated research into the possibility that the different phylotypes of *P. acnes* may vary in their capacity to cause opportunistic infections and that particular "pathogenic" lineages exist alongside those that could be described as truly commensal. Evidence to support this view has come from MLST studies of *P. acnes* isolated from patients with acne [46, 47]. These studies have shown that acne appears to be associated with lineages within the type IA division, but rarely IB, II or III. Furthermore, the association of type IA strains with acne appears to reflect, at least in part, the expression of DsA proteins, thus facilitating binding of dermatan sulphate. Also, analysis of 12 *P. acnes* isolates from cancerous prostate tissue found that most were strains of phylotype IB and II providing further tentative evidence of differences between the major divisions in their capacity to cause specific types of infection [49]. Of more relevance to this discussion are studies which have investigated the association of the different *P. acnes* phylotypes with biomaterial-related infection. Investigations of 61 failed orthopaedic implants (hip, knee, shoulder and spine) by Sampdero et al. [28] found that type IB and type II strains were more frequently isolated from sonicate prepared from infected implants compared with type IA, with no type III isolates being recovered. Similar numbers of type IB and II, and increased numbers of type IA, were however also isolated from implants that did not meet the criteria for implant infection and were, therefore, defined as aseptic failure. The relationship between isolates removed from aseptically failed implants is unclear, but could reflect colonisation of the implants prior to the development of infection or alternatively contamination, or a combination of both.

Studies by Holmberg et al. [77] found that type IA isolates were more frequently recovered than types IB or II from tissue adjacent to infections of prosthetic hip and knee joints, and from post-operative mediastinitis where sternal stainless steel wires were present ($n = 45$ patients). They did not, however, attempt to isolate any organisms directly from the prosthetic devices. A similar phylotype distribution was also found amongst isolates recovered from the forehead of healthy volunteers ($n = 48$ volunteers). Biofilm formation amongst isolates associated with deep infections related to foreign material was found to be greater *in vitro* when compared to isolates from the skin of healthy volunteers [77]. Analysis of *P. acnes* isolated from endodontic lesions and associated with gutta percha (filling material) in the root canal were predominantly type II and III, while types IA and IB were primarily isolated from peri-oral skin [68]. This led the authors to speculate that types II and III may exhibit a tropism for inert foreign surfaces. All of these studies identified type IB isolates based on defining polymorphisms in *recA* and *tly* gene sequences [41]. More detailed phylogenetic analysis based on recently published MLST schemes have, however, revealed that type IB *recA* and *tly* alleles may also be found in certain type IA STs, including all type IA_2 isolates. While these alleles appear highly sensitive for type IB strains, their presence in some type IA strains may confuse identification and lead to incorrect classification in some cases. The number of strains identified as type IB in these studies may, therefore, be overestimated. The collective presence of these alleles in some type IA strains is believed to reflect conjugal replacement of a large genome segment in which these alleles are linked [46]. Despite the considerable progress that has been made in relation to molecular phylogeny, a definitive relationship between *P. acnes* phylotype and the various conditions with which it is associated remains to be determined.

4.1.5 Non-culture-Based Detection of *Propionibacterium acnes* Biofilms and Development of an Intraoperative Test

The detection of biomaterial-related biofilm infection via non-culture-based methodologies has a number of notable advantages over cultivation-based diagnostics [78]. In particular, such approaches can significantly reduce the time taken from receipt of a sample to confirmation of an underlying infection, thus facilitating earlier antimicrobial intervention. This may be especially relevant to the detection of anaerobic sp., such as *P. acnes*, which can require extended incubation times for successful growth. As highlighted earlier, the incorporation of non-culture-based methods into a diagnostic algorithm will also negate problems associated with strict adherence to anaerobic conditions during sample collection and processing, and false-negative culture results that can arise due to prior administration of prophylactic antibiotics. Removal of adherent biofilm from an implanted medical device for further analysis is probably best performed by ultrasonication. Sonication has been shown to be superior to alternate "scraping" methods for successful removal of bacteria growing as adherent biofilm on titanium and steel surfaces *in vitro* [79]. Non-culture-based detection of *P. acnes* and other pathogens in sonicate and clinical samples can be achieved using a number of different methodologies, but here we will only consider methods based on PCR amplification and microscopy. PCR-based methods for the detection of *P. acnes* normally focus on conventional or real-time amplification of 16S rRNA sequences using either species-specific or broad-range primers [20, 25, 29, 80]. While broad-range 16S rRNA-based PCR will facilitate increased diagnostic sensitivity and enable a greater range of pathogenic species to be identified, the approach can also be challenging due to the constant threat of contamination occurring at different steps in the procedure, thus leading to false-positive results. Furthermore, to identify the organism(s) present, mixed amplicons must be separated, either by the generation of clone libraries or by some form of polyacrylamide gel electrophoresis, prior to nucleotide sequencing and interrogation of online databases such as GenBank or the Ribosomal Database project [81, 82]. In the future, metagenomic analyses via routine next-generation sequencing platforms may obviate the need for such mixed amplicon separation and will facilitate the analysis of large sample numbers. As an alternative, a second "nested" PCR with "internal" species-specific primers can be carried out if particular organisms are being sought [83]. A nested PCR methodology will normally generate excellent detection limits but can suffer from contamination due to "carry-over" amplicons from previous PCR experiments. Contamination can arise from nonsterile sites surrounding the implant, from the surgical or laboratory staff or from the reagents, vials and pipettes used for DNA extraction and PCR set-up. Preventive measures include the use of

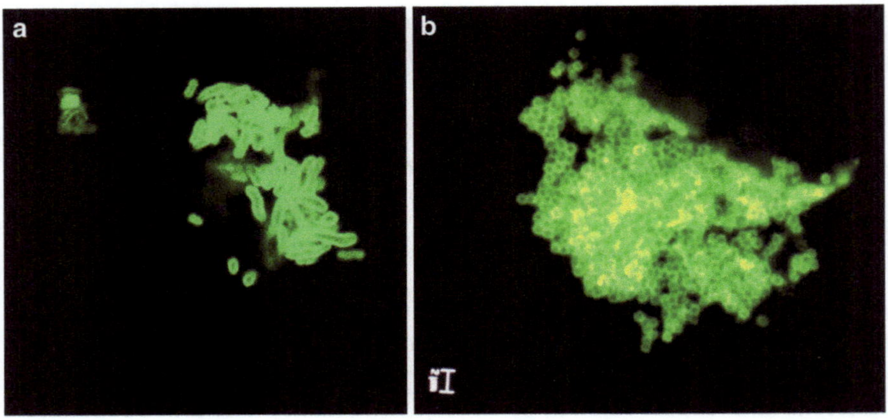

Fig. 4.2 Confocal laser scanning micrographs of *P. acnes* and *Staphylococcal* biofilms dislodged from retrieved hip prostheses by mild untrasonication. *P. acnes* biofilm was labelled with anti-*P. acnes* monoclonal antibody hybridoma culture supernatant (QUBPa3) (**a**) and Staphylococcal biofilm was labelled with anti-Staphylococcus spp. antiserum (1/400 dilution) (**b**) before incubation with a secondary antibody conjugated to the fluorescent dye FITC. Reprinted with permission from Tunney et al. (1999) [18] and McDowell et al. (2005) [60]

designated rooms for DNA extraction and PCR set-up and the use of sterile gloves, molecular grade reagents, filtered tips and HEPA-filtered safety cabinets, and equipment that stays on site. Routine decontamination of work surfaces by UV light or chemical degradation, incorporation of dUTP into the PCR reactions along with Uracil-DNA-glycosylase treatment of new samples to prevent contamination with amplicons, and appropriate use of positive and negative control samples can also help eradicate contamination [41]. The use of species-specific primers will also reduce false-positive results arising from contamination but will certainly not eliminate them, especially when attempting to detect infecting organisms that also form part of the normal skin microbiota. Furthermore, under these conditions certain species may be missed although the use of multiple primer sets in, for example, a multiplex assay may go some way to reducing this particular problem. Further issues can arise when attempting to extract DNA from *P. acnes* biofilm due to the very robust nature of the bacterium's cell wall which may result in poor yields and a reduced detection limit [60]. The use of enzymes such as lysostaphin or achromopeptidase may, however, improve cell wall lysis [84].

Alternative non-culture methods that centre on microscopy-based diagnosis include in situ hybridisation (ISH) and immunofluorescence- or immunohistochemistry (IHC)-based methods. These methods provide the opportunity to visualise, in situ, bacterial cells in a sonicate sample which has been fixed to a glass slide and can provide information on the morphology and spatial arrangement of the infecting organisms. As already mentioned, visual detection methods have a significant advantage over PCR approaches when it comes to differentiating contaminating organisms from those that form part of an adherent biofilm. The latter appear as large aggregates consisting of several layers after sonication [18, 41]

(Fig 4.2). In contrast, contaminants occur either as single cells or small aggregates of only a few cells [18, 41]. ISH utilises a single oligonucleotide probe (15–30 nucleotides), which is linked to a reporter molecule, and targets rRNA sequences within the bacterial ribosome. As with 16S rRNA-based PCR, species-specific or broad-range probes can be used [85]. These reporter groups are now normally nonradioactive and include digoxigenin (DIG), biotin and the fluorophores fluorescein isothiocyanate (FITC) and rhodamine. While the fluorophores enable direct detection (FISH), DIG and biotin reporter groups require a further secondary reaction with anti-DIG antibodies and streptavidin, respectively, which are themselves covalently linked to either horseradish peroxidase (HRP) or alkaline phosphatase enzymes, and produce a coloured precipitate upon the addition of substrate. A variant on these methods is the highly sensitive catalysed reporter deposition (CARD)-FISH which uses oligonucleotide probes directly labelled with HRP to activate fluorescein tyramine [86]. Previously, CARD-FISH and FISH have been used successfully to detect *P. acnes* in sarcoid and prostate tissue [84, 87]. Some problems can arise with ISH-based methods, however, including insufficient penetration of the oligonucleotide probe into the cell and inaccessibility of target sites on the 16S rRNA molecule. The use of peptide nucleic acid probes and unlabelled helper probes can, however, serve to reduce these problems [88, 89]. When using a fluorescent reporter group (FISH), autofluorescence of the tissue material surrounding the bacteria or indeed the bacteria themselves can also decrease the signal-to-noise ratio hindering detection [90].

Immunofluorescence- or IHC-based methods provide an alternative approach to ISH and are based on the labelling of bacterial cell surface antigens with either monoclonal or polyclonal antibodies followed by a secondary reaction with an antibody linked to a reporter molecule, such as FITC (immunofluorescence), HRP or alkaline phosphatase (IHC). This indirect approach is a more sensitive technique than direct labelling of the antibodies due to signal amplification resulting from multiple secondary antibody reactions with different antigenic sites on the primary antibody. Immunofluorescence-microscopy (IFM) methods have previously been used with some success to investigate the presence of *P. acnes* and *Staphylococcus* spp. in sonicate from retrieved prosthetic hip and shoulder joints using *P. acnes*-specific monoclonal antibodies and staphylococcal-specific monoclonal and polyclonal antibodies [18, 29]. Compared to ISH with a broad-range 16S rRNA probe, IFM will only facilitate the identification of those organisms reactive with the antibodies used. While this may be a limitation of the technique, the introduction of additional antibodies to cover those spp. most frequently found may improve this situation. For biofilm related to orthopaedic infections, antibodies to *P. acnes* and staphylococci should collectively detect the majority of cases where infection is indeed present. Of course, IFM/IHC methods will not suffer from the various problems that can arise with ISH, such as inaccessibility of the probe into the cell, as well as target sites on the 16S rRNA molecule. Combining both methods into a single IFM/FISH assay may also prove a better approach under certain circumstances with a broad-range eubacterial probe to detect all bacteria and specific antibodies targeting the main pathogens [85].

To date, no sensitive and specific intraoperative test is routinely used for the diagnosis of prosthetic joint infection [91]. The development of such a point-of-care test is desirable and overdue and would facilitate the detection of occult low-grade prosthetic joint infection at the time of revision surgery, and thus prevent a misdiagnosis of aseptic loosening and implantation of the second device without debridement and antimicrobial intervention [92]. While intraoperative histology has high specificity and sensitivity when septic loosening of a prosthesis is suspected, it has low sensitivity for the prediction of micro-organisms in samples from patients with suspected aseptic loosening and where a low-grade inflammatory response may be present due to *P. acnes* or CoNS [93, 94]. Currently, an IFM or indeed combined IFM/FISH approach for the analysis of sonicate appears the most attractive methodology for rapid intraoperative detection of biomaterial-associated biofilm versus bacterial contamination. With intraoperative frozen section analysis, turnaround times are approximately 15 min, but a turnaround time of 45–60 min would still be considered acceptable by surgeons [94]. Streamlining of the IFM method to within this time window should be possible and within our laboratory we can currently identify pure isolates of *P. acnes* or staphylococci in less than 30 min using a newly validated IFM protocol (Patrick and McDowell, unpublished). The goal will be to package the method into a simple easy-to-follow test that can be performed within a busy surgical theatre. The other major logistical issue to address will be sonication of the explanted prosthesis. It should, however, be possible to accommodate sonication in the theatre, or possibly a side room without too much problem since other procedures, such as centrifugation and vacuum mixing of polymethylmethacrylate, are already performed in operating suites by orthopaedic surgeons [94]. Furthermore, we envisage that once the slide is prepared and developed, detection of infecting bacteria could be achieved in approximately 10 min by using a microscope with a motorised stage and image analysis software for automated enumeration of labelled bacterial cells [95]. The development, validation and routine application of such a rapid test would provide significant improvements in the management and treatment of patients undergoing revision surgeries, and the ability to identify and successfully treat those patients presenting with underlying *P. acnes* or CoNS infections masquerading as aseptic loosening would also reduce the financial burden on healthcare systems.

4.1.6 Conclusions

Although *P. acnes* has traditionally been viewed as a relatively harmless member of the human microbiome, there is now a growing awareness of its role in a wide range of infections and conditions, especially infections related to indwelling medical devices where it will persist as a biofilm [18, 29]. Its role in prosthetic joint infections has been underestimated for a number of key reasons. These include subtle signs and symptoms of infection, an inappropriate lack of anaerobic processing and culture leading to poor diagnostic sensitivity and dismissal as a contaminant when isolated. To maximise detection of *P. acnes*, it is imperative that clinical samples are processed under strict anaerobic conditions from the moment the sample is retrieved, and that medical devices undergo sonication to ensure that adherent biofilm is removed. Cultures should be incubated for at least 7 days and ideally up to 14 days to prevent false-negative results. The application of non-culture-based methods will also facilitate improved detection rates for this pathogen and will likely prove the initial method of choice for rapid and accurate detection of *P. acnes*-related biofilm in the future. Detailed methods describing culture- and molecular-based detection of *P. acnes* from various clinical samples, including indwelling medical devices, have been described [47].

References

1. Grice EA, Kong HH, Conlan S, Deming CB, Davis J, Young AC, et al. Topographical and temporal diversity of the human skin microbiome. Science. 2009;324(5931):1190–2.
2. Zaninetti-Schaerer A, Van Delden C, Genevay S, Gabay C. Total hip prosthetic joint infection due to *Veillonella* species. Joint Bone Spine. 2004;71(2):161–3.
3. Shanson D. New British and American guidelines for the antibiotic prophylaxis of infective endocarditis: do the changes make sense? A critical review. Curr Opin Infect Dis. 2008;21(2):191–9.
4. Brook I. Infective endocarditis caused by anaerobic bacteria. Arch Cardiovasc Dis. 2008;101(10):665–76.
5. Fong IW. New perspectives of infections in cardiovascular disease. Curr Cardiol Rev. 2009;5(2): 87–104.
6. Zimmerli W. Infection and musculoskeletal conditions: prosthetic-joint-associated infections. Best Pract Res Clin Rheumatol. 2006;20(6):1045–63.
7. Veillon A, Zuber A. Sur quelques microbes strictement anaerobies et leur role en pathologie. Arch Med Exp Anat Pathol. 1898;10:517–45.
8. Patrick S, Duerden BI. Gram-negative non-spore forming obligate anaerobes. In: Gillespie SH, Hawkey P, editors. Principles and practice of clinical bacteriology. 2nd ed. London: Wiley; 2006. p. 541–56.
9. Lortholary O, Buu-Hoi A, Podglajen I, Boucherie JC, Pipien I, Gutmann L, et al. Endocarditis caused by multiply resistant *Bacteroides fragilis*: case report and review. Clin Microbiol Infect. 1995;1(1):44–7.
10. Marculescu CE, Berbari EF, Cockerill III FR, Osmon DR. Unusual aerobic and anaerobic bacteria associated with prosthetic joint infections. Clin Orthop Relat Res. 2006;451:55–63.
11. Allison DG, Sutherland IW. The role of exopolysaccharides in adhesion of freshwater bacteria. J Gen Microbiol. 1987;133:1319–27.
12. Fletcher M, Floodgate GD. An electron-microscopic demonstration of an acidic polysaccharide involved in the adhesion of a marine bacterium to solid surfaces. Microbiology. 1973;74:325–34.
13. Gill SR, Fouts DE, Archer GL, Mongodin EF, DeBoy RT, Ravel J, et al. Insights on evolution of virulence and resistance from the complete genome analysis of an early methicillin-resistant *Staphylococcus aureus* strain and a biofilm-producing methicillin-resistant *Staphylococcus epidermidis* strain. J Bacteriol. 2005;187(7):2426–38.
14. Freney J, Brun Y, Bes M, Meugnier H, Grimont F, Grimont PAD, et al. *Staphylococcus lugdunensis* sp. nov. and *Staphylococcus schleiferi* sp. nov., two species from human clinical specimens. Int J Syst Bacteriol. 1988;38(2):168–72.
15. Sampathkumar P, Osmon DR, Cockerill FR. Prosthetic joint infection due to *Staphylococcus lugdunensis*. Mayo Clin Proc. 2000;75(5):511–2.
16. Shah NB, Osmon DR, Fadel H, Patel R, Kohner PC, Steckelberg JM, et al. Laboratory and clinical characteristics of *Staphylococcus lugdunensis* prosthetic joint infections. J Clin Microbiol. 2010;48(5):1600–3.

17. Tse H, Tsoi HW, Leung SP, Lau SKP, Woo PCY, Yuen KY. Complete genome sequence of *Staphylococcus lugdunensis* strain HKU09-01. J Bacteriol. 2010;192(5):1471–2.
18. Tunney MM, Patrick S, Curran MD, Ramage G, Hanna D, Nixon JR, et al. Detection of prosthetic hip infection at revision arthroplasty by immunofluorescence microscopy and PCR amplification of the bacterial 16S rRNA gene. J Clin Microbiol. 1999;37(10):3281–90.
19. Zeller V, Ghorbani A, Strady C, Leonard P, Mamoudy P, Desplaces N. *Propionibacterium acnes*: an agent of prosthetic joint infection and colonization. J Infect. 2007;55(2):119–24.
20. Sampedro MF, Huddleston PM, Piper KE, Karau MJ, Dekutoski MB, Yaszemski MJ, et al. A biofilm approach to detect bacteria on removed spinal implants. Spine (Phila Pa 1976). 2010;35(12):1218–24.
21. Jahns AC, Lundskog B, Ganceviciene R, Palmer RH, Golovleva I, Zouboulis C, et al. An increased incidence of *Propionibacterium acnes* biofilms in acne vulgaris: a case-control study. Br J Dermatol 2012;167(1):50–8.
22. Patrick S, McDowell A. Propionibacterium. In: Goodfellow M, Kämpfer P, Busse HJ, Trujillo M, Suzuki K, Ludwig W, et al., editors. *The Actinobacteria* Bergey's manual of systematic bacteriology, vol. 5. 2nd ed. New York: Springer; 2012. p. 1.
23. McLorinan GC, Glenn JV, McMullan MG, Patrick S. *Propionibacterium acnes* wound contamination at the time of spinal surgery. Clin Orthop Relat Res. 2005;437:67–73.
24. Patel A, Calfee RP, Plante M, Fischer SA, Green A. *Propionibacterium acnes* colonization of the human shoulder. J Shoulder Elbow Surg. 2009;18(6):897–902.
25. Bek-Thomsen M, Lomholt HB, Kilian M. Acne is not associated with yet-uncultured bacteria. J Clin Microbiol. 2008;46(10):3355–60.
26. Brook I, Frazier EH. Infections caused by *Propionibacterium* species. Rev Infect Dis. 1991;13(5):819–22.
27. Clayton JJ, Baig W, Reynolds GW, Sandoe JA. Endocarditis caused by *Propionibacterium* species: a report of three cases and a review of clinical features and diagnostic difficulties. J Med Microbiol. 2006;55(Pt 8):981–7.
28. Sampedro MF, Piper KE, McDowell A, Patrick S, Mandrekar JN, Rouse MS, et al. Species of *Propionibacterium* and *Propionibacterium acnes* phylotypes associated with orthopedic implants. Diagn Microbiol Infect Dis. 2009;64(2):138–45.
29. Piper KE, Jacobson MJ, Cofield RH, Sperling JW, Sanchez-Sotelo J, Osmon DR, et al. Microbiologic diagnosis of prosthetic shoulder infection by use of implant sonication. J Clin Microbiol. 2009;47(6):1878–84.
30. Martin-Rabadan P, Gijon P, Alcala L, Rodriguez-Creixems M, Alvarado N, Bouza E. *Propionibacterium acnes* is a common colonizer of intravascular catheters. J Infect. 2008;56(4):257–60.
31. Aldave AJ, Stein JD, Deramo VA, Shah GK, Fischer DH, Maguire JI. Treatment strategies for postoperative *Propionibacterium acnes* endophthalmitis. Ophthalmology. 1999;106(12): 2395–401.
32. Jakab E, Zbinden R, Gubler J, Ruef C, von Graevenitz A, Krause M. Severe infections caused by *Propionibacterium acnes*: an underestimated pathogen in late postoperative infections. Yale J Biol Med. 1996;69(6):477–82.
33. Levy PY, Fenollar F, Stein A, Borrione F, Cohen E, Lebail B, et al. *Propionibacterium acnes* postoperative shoulder arthritis: an emerging clinical entity. Clin Infect Dis. 2008;46(12):1884–6.
34. Kanafani ZA, Sexton DJ, Pien BC, Varkey J, Basmania C, Kaye KS. Postoperative joint infections due to *Propionibacterium* species: a case-control study. Clin Infect Dis. 2009;49(7):1083–5.
35. Nisbet M, Briggs S, Ellis-Pegler R, Thomas M, Holland D. *Propionibacterium acnes*: an under-appreciated cause of post-neurosurgical infection. J Antimicrob Chemother. 2007;60(5):1097–103.
36. Zappe B, Graf S, Ochsner PE, Zimmerli W, Sendi P. *Propionibacterium* spp. in prosthetic joint infections: a diagnostic challenge. Arch Orthop Trauma Surg. 2008;128(10):1039–46.
37. Dodson CC, Craig EV, Cordasco FA, Dines DM, Dines JS, DiCarlo E, et al. *Propionibacterium acnes* infection after shoulder arthroplasty: a diagnostic challenge. J Shoulder Elbow Surg. 2010;19(2):303–7.
38. Söderquist B, Holmberg A, Unemo M. *Propionibacterium acnes* as an etiological agent of arthroplastic and osteosynthetic infections – two cases with specific clinical presentation including formation of draining fistulae. Anaerobe. 2010;16(3):304–6.

39. Jousimies-Somer H, Summanen P, Citron D, Baron E, Wexler HM, Finegold S. Wadsworth-KTL anaerobic bacteriology manual. 6th ed. Belmont, CA: Star Publishing; 2002.
40. Tunney MM, Patrick S, Gorman SP, Nixon JR, Anderson N, Davis RI, et al. Improved detection of infection in hip replacements. A currently underestimated problem. J Bone Joint Surg Br. 1998;80(4):568–72.
41. McDowell A, Patrick S. Evaluation of nonculture methods for the detection of prosthetic hip biofilms. Clin Orthop Relat Res. 2005;437:74–82.
42. Trampuz A, Piper KE, Hanssen AD, Osmon DR, Cockerill FR, Steckelberg JM, et al. Sonication of explanted prosthetic components in bags for diagnosis of prosthetic joint infection is associated with risk of contamination. J Clin Microbiol. 2006;44(2):628–31.
43. Patrick S, McDowell A, Glenn JV, Tunney MM. Improved detection and treatment of prosthetic joint infection. Eur Cell Mater. 2008;16 Suppl 2:30.
44. Valanne S, McDowell A, Ramage G, Tunney MM, Einarsson GG, O'Hagan S, et al. CAMP factor homologues in *Propionibacterium acnes*: a new protein family differentially expressed by types I and II. Microbiology. 2005;151(Pt 5):1369–79.
45. McDowell A, Perry AL, Lambert PA, Patrick S. A new phylogenetic group of *Propionibacterium acnes*. J Med Microbiol. 2008;57(Pt 2):218–24.
46. Lomholt HB, Kilian M. Population genetic analysis of *Propionibacterium acnes* identifies a subpopulation and epidemic clones associated with acne. PLoS One. 2010;5(8):e12277.
47. McDowell A, Gao A, Barnard E, Fink C, Murray PI, Dowson CG, et al. A novel multilocus sequence typing scheme for the opportunistic pathogen *Propionibacterium acnes* and characterisation of type I cell surface-associated antigens. Microbiology. 2011;157(Pt 7):1990–2003.
48. Hunyadkurti J, Feltoti Z, Horvath B, Nagymihaly M, Voros A, McDowell A, et al. Complete genome sequence of *Propionibacterium acnes* type IB strain 6609. J Bacteriol. 2011;193(17): 4561–2.
49. Shannon BA, Cohen RJ, Garrett KL. Polymerase chain reaction-based identification of *Propionibacterium acnes* types isolated from the male urinary tract: evaluation of adolescents, normal adults and men with prostatic pathology. BJU Int. 2006;98(2):388–92.
50. Kotilainen P, Merilahti-Palo R, Lehtonen OP, Manner I, Helander I, Mottonen T, et al. *Propionibacterium acnes* isolated from sternal osteitis in a patient with SAPHO syndrome. J Rheumatol. 1996;23(7):1302–4.
51. Eishi Y, Suga M, Ishige I, Kobayashi D, Yamada T, Takemura T, et al. Quantitative analysis of mycobacterial and propionibacterial DNA in lymph nodes of Japanese and European patients with sarcoidosis. J Clin Microbiol. 2002;40(1):198–204.
52. Cohen RJ, Shannon BA, McNeal JE, Shannon T, Garrett KL. *Propionibacterium acnes* associated with inflammation in radical prostatectomy specimens: a possible link to cancer evolution? J Urol. 2005;173(6):1969–74.
53. Eady EA, Cove JH, Holland KT, Cunliffe WJ. Erythromycin resistant propionibacteria in antibiotic treated acne patients: association with therapeutic failure. Br J Dermatol. 1989;121(1):51–7.
54. Fink H, Helming M, Unterbuchner C, Lenz A, Neff F, Martyn JA, et al. Systemic inflammatory response syndrome increases immobility-induced neuromuscular weakness. Crit Care Med. 2008;36(3):910–6.
55. Flaminio MJ, Rush BR, Shuman W. Immunologic function in horses after non-specific immunostimulant administration. Vet Immunol Immunopathol. 1998;63(4):303–15.
56. Nagy I, Pivarcsi A, Kis K, Koreck A, Bodai L, McDowell A, et al. *Propionibacterium acnes* and lipopolysaccharide induce the expression of antimicrobial peptides and proinflammatory cytokines/chemokines in human sebocytes. Microbes Infect. 2006;8(8):2195–205.
57. Tunney MM, Dunne N, Einarsson G, McDowell A, Kerr A, Patrick S. Biofilm formation by bacteria isolated from retrieved failed prosthetic hip implants in an in vitro model of hip arthroplasty antibiotic prophylaxis. J Orthop Res. 2007;25(1):2–10.
58. Ramage G, Tunney MM, Patrick S, Gorman SP, Nixon JR. Formation of *Propionibacterium acnes* biofilms on orthopaedic biomaterials and their susceptibility to antimicrobials. Biomaterials. 2003;24(19):3221–7.

59. Coenye T, Peeters E, Nelis HJ. Biofilm formation by *Propionibacterium acnes* is associated with increased resistance to antimicrobial agents and increased production of putative virulence factors. Res Microbiol. 2007;158(4):386–92.
60. McDowell A, Valanne S, Ramage G, Tunney MM, Glenn JV, McLorinan GC, et al. *Propionibacterium acnes* types I and II represent phylogenetically distinct groups. J Clin Microbiol. 2005;43(1):326–34.
61. Christie R, Atkins NE, Munch-Petersen E. A note on a lytic phenomenon shown by group B streptococci. Aust J Exp Biol Med Sci. 1944;22:197–200.
62. Lang S, Palmer M. Characterization of Streptococcus agalactiae CAMP factor as a pore-forming toxin. J Biol Chem. 2003;278(40):38167–73.
63. Brüggemann H. Insights in the pathogenic potential of *Propionibacterium acnes* from its complete genome. Semin Cutan Med Surg. 2005;24(2):67–72.
64. Steiner B, Romero-Steiner S, Cruce D, George R. Cloning and sequencing of the hyaluronate lyase gene from *Propionibacterium acnes*. Can J Microbiol. 1997;43(4):315–21.
65. von Nicolai H, Hoffler U, Zilliken F. Isolation, purification, and properties of neuraminidase from *Propionibacterium acnes*. Zentralbl Bakteriol A. 1980;247(1):84–94.
66. Miskin JE, Farrell AM, Cunliffe WJ, Holland KT. *Propionibacterium acnes*, a resident of lipid-rich human skin, produces a 33 kDa extracellular lipase encoded by gehA. Microbiology. 1997;143(5):1745–55.
67. Brüggemann H, Henne A, Hoster F, Liesegang H, Wiezer A, Strittmatter A, et al. The complete genome sequence of *Propionibacterium acnes*, a commensal of human skin. Science. 2004;305(5684):671–3.
68. Niazi SA, Clarke D, Do T, Gilbert SC, Mannocci F, Beighton D. *Propionibacterium acnes* and *Staphylococcus epidermidis* isolated from refractory endodontic lesions are opportunistic pathogens. J Clin Microbiol. 2010;48(11):3859–69.
69. Lodes MJ, Secrist H, Benson DR, Jen S, Shanebeck KD, Guderian J, et al. Variable expression of immunoreactive surface proteins of *Propionibacterium acnes*. Microbiology. 2006;152 (Pt 12):3667–81.
70. Oprica C, Nord CE. European surveillance study on the antibiotic susceptibility of *Propionibacterium acnes*. Clin Microbiol Infect. 2005;11(3):204–13.
71. Tan HH, Goh CL, Yeo MG, Tan ML. Antibiotic sensitivity of *Propionibacterium acnes* isolates from patients with acne vulgaris in a tertiary dermatological referral centre in Singapore. Ann Acad Med Singapore. 2001;30(1):22–5.
72. Ross JI, Snelling AM, Carnegie E, Coates P, Cunliffe WJ, Bettoli V, et al. Antibiotic-resistant acne: lessons from Europe. Br J Dermatol. 2003;148(3):467–78.
73. Ishida N, Nakaminami H, Noguchi N, Kurokawa I, Nishijima S, Sasatsu M. Antimicrobial susceptibilities of *Propionibacterium acnes* isolated from patients with acne vulgaris. Microbiol Immunol. 2008;52(12):621–4.
74. Pochi PE, Strauss JS. Antibiotic sensitivity of *Corynebacterium acnes* (*Propionibacterium acnes*). J Invest Dermatol. 1961;36:423–9.
75. Tung WL, Chow KC. A modified medium for efficient electrotransformation of *E. coli*. Trends Genet. 1995;11(4):128–9.
76. Wang WL, Everett ED, Johnson M, Dean E. Susceptibility of *Propionibacterium acnes* to seventeen antibiotics. Antimicrob Agents Chemother. 1977;11(1):171–3.
77. Holmberg A, Lood R, Morgelin M, Soderquist B, Holst E, Collin M, et al. Biofilm formation by *Propionibacterium acnes* is a characteristic of invasive isolates. Clin Microbiol Infect. 2009;15(8):787–95.
78. McDowell A, Patrick S. Evaluation of nonculture methods for the detection of prosthetic hip biofilms. Clin Orthop Relat Res. 2005;437:74–82.
79. Bjerkan G, Witso E, Bergh K. Sonication is superior to scraping for retrieval of bacteria in biofilm on titanium and steel surfaces in vitro. Acta Orthop. 2009;80(2):245–50.
80. Alexeyev O, Bergh J, Marklund I, Thellenberg-Karlsson C, Wiklund F, Gronberg H, et al. Association between the presence of bacterial 16S RNA in prostate specimens taken during transurethral resection of prostate and subsequent risk of prostate cancer (Sweden). Cancer Causes Control. 2006;17(9):1127–33.

81. Kolbert CP, Persing DH. Ribosomal DNA sequencing as a tool for identification of bacterial pathogens. Curr Opin Microbiol. 1999;2(3):299–305.
82. Clarridge III JE. Impact of 16S rRNA gene sequence analysis for identification of bacteria on clinical microbiology and infectious diseases. Clin Microbiol Rev. 2004;17(4):840–62. table of contents.
83. Hykin PG, Tobal K, McIntyre G, Matheson MM, Towler HM, Lightman SL. The diagnosis of delayed post-operative endophthalmitis by polymerase chain reaction of bacterial DNA in vitreous samples. J Med Microbiol. 1994;40(6):408–15.
84. Alexeyev OA, Marklund I, Shannon B, Golovleva I, Olsson J, Andersson C, et al. Direct visualization of *Propionibacterium acnes* in prostate tissue by multicolor fluorescent in situ hybridization assay. J Clin Microbiol. 2007;45(11):3721–8.
85. Ramage G, Patrick S, Houston S. Combined fluorescent in situ hybridisation and immunolabelling of *Bacteroides fragilis*. J Immunol Methods. 1998;212(2):139–47.
86. Hoshino T, Yilmaz LS, Noguera DR, Daims H, Wagner M. Quantification of target molecules needed to detect microorganisms by fluorescence in situ hybridization (FISH) and catalyzed reporter deposition-FISH. Appl Environ Microbiol. 2008;74(16):5068–77.
87. Yamada T, Eishi Y, Ikeda S, Ishige I, Suzuki T, Takemura T, et al. In situ localization of *Propionibacterium acnes* DNA in lymph nodes from sarcoidosis patients by signal amplification with catalysed reporter deposition. J Pathol. 2002;198(4):541–7.
88. Ray A, Norden B. Peptide nucleic acid (PNA): its medical and biotechnical applications and promise for the future. FASEB J. 2000;14(9):1041–60.
89. Fuchs BM, Glockner FO, Wulf J, Amann R. Unlabeled helper oligonucleotides increase the in situ accessibility to 16S rRNA of fluorescently labeled oligonucleotide probes. Appl Environ Microbiol. 2000;66(8):3603–7.
90. Moter A, Gobel UB. Fluorescence in situ hybridization (FISH) for direct visualization of microorganisms. J Microbiol Methods. 2000;41(2):85–112.
91. Patel R, Osmon DR, Hanssen AD. The diagnosis of prosthetic joint infection: current techniques and emerging technologies. Clin Orthop Relat Res. 2005;437:55–8.
92. Trampuz A, Osmon DR, Hanssen AD, Steckelberg JM, Patel R. Molecular and antibiofilm approaches to prosthetic joint infection. Clin Orthop Relat Res. 2003;414:69–88.
93. Bori G, Soriano A, Garcia S, Gallart X, Casanova L, Mallofre C, et al. Low sensitivity of histology to predict the presence of microorganisms in suspected aseptic loosening of a joint prosthesis. Mod Pathol. 2006;19(6):874–7.
94. Patel R. Biofilms and antimicrobial resistance. Clin Orthop Relat Res. 2005;437:41–7.
95. Selinummi J, Seppala J, Yli-Harja O, Puhakka JA. Software for quantification of labeled bacteria from digital microscope images by automated image analysis. Biotechniques. 2005;39(6):859–63.

Chapter 5
Bridging the Gap Between In Vitro and In Vivo Evaluation of Biomaterial-Associated Infections

Guruprakash Subbiahdoss, Joana F. da Silva Domingues, Roel Kuijer, Henny C. van der Mei, and Henk J. Busscher

Abstract Biomaterial-associated infections constitute a major clinical problem that is difficult to treat and often necessitates implant replacement. Pathogens can be introduced on an implant surface during surgery or postoperative and compete with host cells attempting to integrate the implant. The fate of a biomaterial implant has been depicted as a race between bacterial adhesion and biofilm growth on an implant surface versus tissue integration. Until today, *in vitro* studies on infection risks of biomaterials or functional coatings for orthopedic and dental implants were performed either for their ability to resist bacterial adhesion or for their ability to support mammalian cell adhesion and proliferation. Even though the concept of the race for the surface in biomaterial-associated infections has been intensively studied before *in vivo*, until recently no *in vitro* methodology existed for this purpose. Just very recently various groups have proposed coculture experiments to evaluate the simultaneous response of bacteria and mammalian cells on a surface. As an initial step towards bridging the gap between *in vitro* and *in vivo* evaluations of biomaterials, we here describe bi- and tri-culture experiments that allow better evaluation of multifunctional coatings *in vitro* and therewith bridge the gap between *in vitro* and *in vivo* studies.

5.1 Biomaterial-Associated Infections

Biomaterials play an important role in modern medicine in the restoration and maintenance of tissue, organ, or body function. A current estimate of the number of total hip replacements in the world is approximately one million a year, while the number

G. Subbiahdoss • J.F. da Silva Domingues • R. Kuijer • H.C. van der Mei • H.J. Busscher (✉)
Department of Biomedical Engineering, University Medical Center Groningen and University of Groningen, P.O. Box 196, Groningen 9700 AD, The Netherlands
e-mail: h.j.busscher@umcg.nl

T.F. Moriarty et al. (eds.), *Biomaterials Associated Infection: Immunological Aspects and Antimicrobial Strategies*, DOI 10.1007/978-1-4614-1031-7_5,

of knee replacements is more than 250,000 [1]. One of the major problems emerging from the use of biomaterial implants and medical devices is biomaterial-associated infection (BAI). On average, BAI occurs in approximately 0.5–6 % of all cases [2, 3], strongly depending on the implant site, and more often in cases of trauma or revision surgery [4–6]. Upon adhesion to a biomaterial surface, bacteria start to synthesize a hydrated matrix of extracellular polymeric substances (EPS) to form a biofilm [7]. Biofilm formation on an implant surface is a multiphase process [8] starting with reversible adhesion of single bacteria, transition to an irreversible state through physicochemical processes, and EPS production followed by bacterial growth. BAI is difficult to treat, as the biofilm mode of growth protects the infecting organisms against the host immune system and antibiotic treatment [9]. Bacteria within biofilms generally require 500–5,000 times higher doses of antibiotics than the planktonic ones suspended in body fluids [8]. In the majority of cases, the final outcome of a BAI is removal of the implant. There are various routes along which microorganisms can gain access to the implant site and cause BAI [10]. The best-documented route is direct contamination of an implant during surgery (perioperative contamination) [10]. BAI can also develop, however, due to bacteria gaining access to the implant site immediately post surgery during hospitalization (postoperative contamination) or through microbial spreading via the blood circulation from infections elsewhere in the human body [10–12]. Irrespective of the route of infection, the fate of a biomaterial implant depends mainly on the outcome of the so-called race for the surface between successful tissue integration of the biomaterial implant and bacterial colonization [10].

5.2 Routes of Infection and Infecting Organisms

Perioperative contamination implies that an implant becomes contaminated with bacteria before or during implantation into the human body. It is known that during a surgical procedure of 1 h, the total number of bacteria carrying particles falling on a wound is about 270 bacteria cm^{-2} [13]. Bacterial counts in the operating theatre are generally higher during periods of operating room personnel activity and when more people are present [13]. More recently, through the use of modern, better ventilated operation theatres (20 changes of air per hour) and impermeable patient and personnel clothing, perioperative bacterial contamination may well be less than 270 bacteria cm^{-2} falling into the wound area [14].

The second route of infection is postoperative contamination which may occur during the period of hospitalization immediately post surgery, as caused by direct contamination of open wounds or by the use of invasive devices like infusion tubes, catheters, or drains. Both perioperative and postoperative contamination can cause BAI many years after implantation, because many bacterial strains are known to be able to stay on an implant surface in a low metabolic state for several years post surgery [10]. These bacteria will favor the development of BAI, especially in immunocompromised host [10].

As a third route, BAI can result from hematogenous spreading of bacteria from infections elsewhere in the body to an implanted biomaterial. Hematogenous spreading of bacteria may result from skin infections, surgical or dental interventions, pneumonia, abscesses, or bacteremia, which can all cause temporal or chronic bacteremia resulting in infections [11]. Immune cells such as macrophages may also play a role in transporting bacteria to the implant site, as some strains are capable to survive within macrophages [15, 16]. Importantly, BAI due to hematogenous spreading of bacteria to an implant site may occur any time after implantation. Notorious in this respect are abscesses underneath the skin, developing for instance after minor injuries. Dental treatment is also known to be a cause, as even routine inspection of the dentition by the dentist or oral hygienist may give rise to bacteremia [17, 18].

In general, *Staphylococcus epidermidis* and *Staphylococcus aureus* are the most frequently isolated pathogens from infected biomaterial implants or devices, but also Gram-negative organisms as *Escherichia coli* and *Pseudomonas aeruginosa* are isolated [8, 9, 19]. Almost 50 % of infections associated with catheters, artificial joints, and heart valves are caused by *S. epidermidis* [20]. *S. aureus* is the cause of around 23 % of infections associated with prosthetic joints, especially their metal parts [20]. *P. aeruginosa* is the causative organism in around 12 % of hospital-acquired urinary tract infections, 10 % of bloodstream infections, and 7 % of all hip- and knee-joint infections [21]. Another factor that plays an important role in the pathogenesis of BAI is bacterial virulence [20]. *S. aureus* and *P. aeruginosa* infections usually progress much more aggressively than BAI caused by *S. epidermidis* [20, 21]. *S. aureus* appears more frequently in acute infections, within 4 weeks after surgery of complicated total joint arthroplasty compared to *S. epidermidis*. *S. epidermidis* strains are a common cause of "late" infections [20] and are most commonly implicated in delayed septic loosening of total joint prostheses [22] or even in presumed aseptic loosening [23], indicating their low virulence with only minor clinical symptoms of infection. Pseudomonas is also more virulent than *S. epidermidis,* which is ascribed to the more aggressive endotoxins in its EPS. The low virulence of *S. epidermidis* strains compared to *S. aureus* or *P. aeruginosa* is due to the lack of additional genes responsible for producing tissue-damaging toxins [22–24]. In *S. epidermidis* infections, biofilm formation is considered the only virulence factor as the organism lacks the genes to produce toxins and tissue-damaging exoenzymes that are produced by for instance *S. aureus* and therefore infections are usually subacute or chronic [24–27].

5.3 Bacterial and Mammalian Cell Adhesion to Biomaterials: A Race for the Surface

The introduction of bacteria by any of the above routes is the initial step in the development of BAI, although often preceded by adsorption of macromolecular components from body fluids, such as tear fluid, saliva, serum, or plasma to yield the so-called conditioning film on the biomaterial surface [11]. Bacterial adhesion can be mediated by high affinity, localized interactions between cell surface structures and specific molecular groups on a substratum surface, or overall interaction forces, such as Lifshitz–Van der Waals, electrostatic, and acid–base interactions. Conditioning films can greatly affect the interaction forces in bacterial adhesion. Many adsorbed proteins reduce bacterial adhesion, but others, like fibronectin and fibrinogen, have been shown to promote adhesion of certain *S. epidermidis* and *S. aureus* strains [8].

The best protection against BAI is rapid and complete tissue integration of an implant surface, although not all implants and devices require or allow tissue integration. Mammalian cells can adhere to both bare implant surfaces as well as to implant surfaces with adsorbed conditioning films through integrins and other receptors present upon the cell surface. Therefore, it is important to control protein adsorption to biomaterial surfaces in order to stimulate tissue integration [28], especially since mammalian cells have high affinity to the arginine–glycine–aspartic acid (RGD) motive, which is present in a number of different extracellular matrix proteins, such as fibronectin, laminin, collagen, and vitronectin [29]. Unfortunately, surface chemistries that promote tissue integration very often also mediate bacterial adhesion, like conditioning films comprising fibronectin or fibrinogen [8].

The realization that it may well be impossible to design a surface modification that discourages bacterial adhesion and at the same time encourages tissue integration is catching on. Polymer brush coatings for instance, consisting of a layer of adsorbed, highly hydrophilic poly(ethylene glycol) (PEG) in a brush configuration, resist not only protein adsorption and bacterial adhesion but also adhesion and spreading of mammalian cells [30, 31], which are imperative for successful tissue integration. This realization has led to the development of multifunctional coatings. Multifunctional coatings can be based on a nonadhesive PEG base, decorated with RGD-sequences to promote mammalian cell adhesion and spreading, but due to a low decoration density do not increase bacterial adhesion [30, 32]. The use of such bifunctional coatings enables direct control over the outcome of the race for the surface, an expression coined in 1987 by the orthopedic surgeon Anthony Gristina to describe the fate of a biomaterial implant [9].

The fate of a biomaterial implant was depicted as a race between bacterial adhesion and biofilm growth on an implant surface versus tissue integration (Fig. 5.1). If the race is won by tissue cells, then the surface is covered by tissue and less vulnerable to bacterial colonization. On the other hand, if the race is won by bacteria, the implant surface will become colonized by bacteria.

Although the concept of the race for the surface stems from 1987, it was only recently that various groups have proposed coculture experiments to evaluate the simultaneous response of bacteria and mammalian cells on a surface. Coculture experiments allow the *in vitro* evaluation of multifunctional coatings and therewith bridge the gap between *in vitro* and *in vivo* studies.

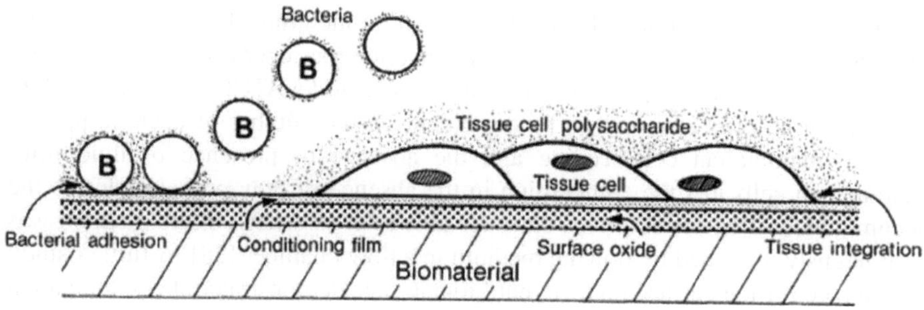

Fig. 5.1 The race between bacteria and tissue cells for the biomaterial surface [9]

5.4 Bridging the Gap Between In Vitro and In Vivo Studies

In the path towards the reduction of infection risk, biomaterials research has been focused on the development of biomaterials or coatings that can prevent bacterial adhesion and stimulate mammalian cell growth. Until today, *in vitro* studies on infection risks of biomaterials or functional coatings for orthopedic and dental implants were performed either for their ability to resist bacterial adhesion or for their ability to support mammalian cell adhesion and proliferation [32–37]. Shi et al. [32] showed that a surface composed of chitosans and RGD-containing peptides discouraged bacterial adhesion but enhanced osteoblast attachment and subsequent differentiation as indicated by increased alkaline phosphatase activity. Dexter et al. in a study suggested that an optimal concentration of seeded 3T3 fibroblasts and conditions to stimulate cell adhesion without stimulating bacterial adhesion could probably reduce infection [33]. Ploux et al. [37] showed an opposite adhesion behavior of bacteria compared to human osteoprogenitor cells on the nano-patterned surfaces prepared by pulsed plasma polymerization and UV-irradiation. In none of these *in vitro* studies, the effects of the bacterial presence on mammalian cell attachment to a biomaterial surface, cell spreading, and growth were studied, which could completely change the fate of the biomaterial implant according to the concept of the "race for the biomaterial surface." Even though the concept of the race for the surface in biomaterial-associated infections has been intensively studied before *in vivo*, until recently no *in vitro* methodology existed for this purpose.

5.4.1 *Bacterial–Mammalian Cell Interactions*

As a first step towards bridging the gap between *in vitro* and *in vivo* studies, an *in vitro* model was developed to investigate bacterial biofilm formation and mammalian cell growth in a single experiment based on the perioperative infection route [38]. In this study, bacteria were allowed to adhere to a biomaterial surface prior to mammalian cell adhesion and spreading to mimic the clinical situation where an implant becomes contaminated prior to implantation (perioperative contamination). The outcome of the competition between *S. epidermidis* ATCC 35983 and U2OS osteoblast-like cells appeared to be dependent on the number of bacteria present prior to mammalian cell seeding and the absence or presence of fluid flow. Mammalian cells lost the competition in the absence of flow probably due to the accumulation of bacterial toxins, but were able to grow when culture medium was continuously refreshed by flowing medium in a flow-chamber [38]. A further study on the race for the surface on different biomaterials demonstrated that mammalian cell interactions with biomaterials were hampered by bacterial biofilm formation on some of the most commonly used biomaterial surfaces [39]. Yet, PMMA showed better mammalian cell adhesion and spreading in the presence of adhering *S. epidermidis* ATCC 35983 than other biomaterials [39]. A study on comparison of different bacterial species involved in BAI on the competition between bacteria and mammalian cells showed that mammalian cells are bound to lose the competition for the surface in the presence of adhering, highly virulent *S. aureus* and *P. aeruginosa* [40]. Buchholz and coworkers [41] showed that in a group of 64 BAI patients (84 %) with an infection by a low-virulent organism 54 were free from infection 2 years later. Alternatively, when *S. aureus* was the causative organism, recurrent infection occurred in 28 % of all patients [41], and almost 50 % of the patients with a *Pseudomonas* infection experienced a relapse. The low virulence of *S. epidermidis* strains compared to *S. aureus* or *P. aeruginosa* is supposed to be due to the lack of additional genes responsible for producing tissue-damaging toxins [21, 22, 24]. In *S. epidermidis* infections, biofilm formation is considered the only virulence factor and therefore infections are usually subacute or chronic [25–27]. Bennion et al. [42] experimentally showed that with bacterial strains such as *S. aureus*, contamination of an implant is inoculum dependent. When the critical inoculum size is reached, local host defenses are overwhelmed and an infection is established [43].

In the concept of the race for the surface, tissue integration is an important protective factor against bacterial contamination of an implant surface [6]. Based on postoperative contamination, the effects of different degrees of mammalian cell coverage on the balance between bacterial biofilm formation and mammalian cell growth were investigated [44]. Mammalian cell growth was severely impaired when bacteria were introduced on surfaces with a low initial mammalian cell density (2.5×10^4 cells cm^{-2}) but in the presence of higher initial cell densities (17×10^4 cells cm^{-2}) contaminating staphylococci did not affect cell growth (Fig. 5.2) [44].

5.4.2 Bacterial–Macrophages Interactions

In a healthy host, the host immune system comes to the aid of tissue cells [45]. Macrophages are one of the most predominant immune cells that arrive within minutes to hours at an implant site and can remain at a biomaterial surface for several

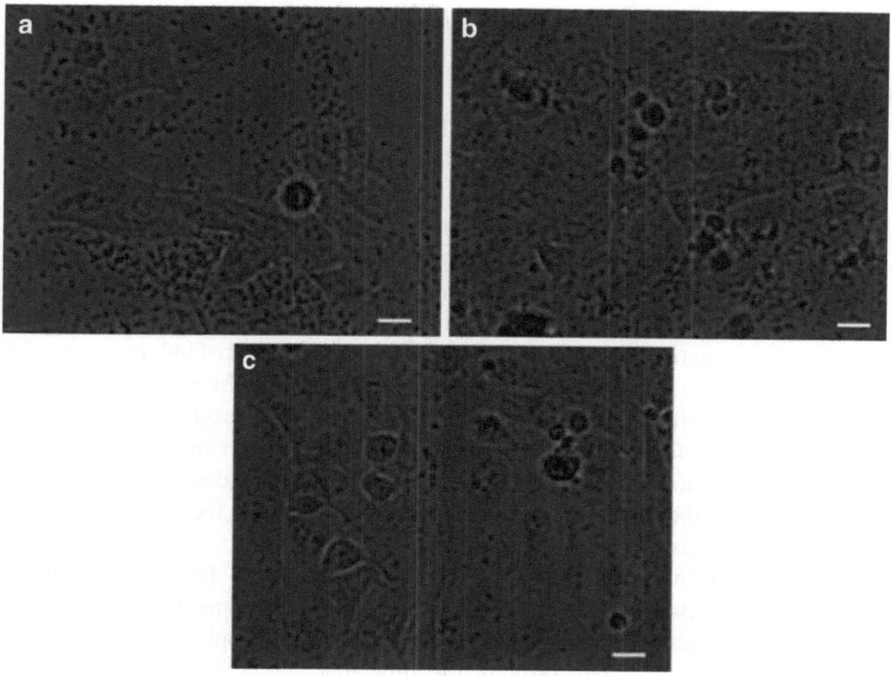

Fig. 5.2 Phase-contrast images of U2OS cells seeded to a density of (**a**) 2.5×10^4 cells/cm^2, (**b**) 8.2×10^4 cells/cm^2 and (**c**) 17×10^4 cells/cm^2 after 24 h of simultaneous growth in the presence of adhering *S. epidermidis* ATCC 35983 on PMMA. Scale bars 10 um [44]

weeks to orchestrate the inflammatory process and foreign body reactions [45]. During infection, macrophages detect bacteria via cell surface receptors that bind to bacterial ligands and opsonins [46–48]. Subsequently, macrophages ingest pathogens and activate cellular functions such as proliferation, secretion of proteins and cytokines, and respiratory burst to destroy phagocytozed microorganisms and recruit other cells from the adaptive immune system [48]. However, it has been shown that the presence of a foreign body may impair the host immune system and consequently low number of adhering bacteria can already be sufficient to cause a BAI [49]. In general, immune cells migrate, engulf, and kill invading microorganisms [50–52]. A previous study on the simultaneous interaction between macrophages and colonizing *S. epidermidis* showed that macrophage behavior is surface dependent [53]. Macrophage migration towards bacteria and phagocytosis was enhanced on highly hydrated, cross-linked PEG-based polymer coatings compared to uncoated substrata due to the weak adhesion of macrophages and bacteria to the PEG coating [53].

Kubica et al. [54] showed that intracellular *S. aureus* can survive within human macrophages for several days until bacteria escaped the intracellular confinement, proliferated in the conditioned medium, and killed the cells. Garzoni et al. [55] indicated that some coagulase-negative staphylococci could promote infection by intracellular colonization.

5.4.3 Bacterial–Mammalian Cells–Macrophages Interactions

The pathogenesis of BAI is complex and depends on factors such as bacterial virulence, physicochemical properties of the biomaterial surface, and alterations in the host defense [49]. Several studies have demonstrated that immune cells lose their ability to kill bacteria in the presence of a biomaterial [52, 56–58]. Neutrophils exhibited decreased bactericidal activity and reduced superoxide production in the presence of extracellular slime producing *S. epidermidis* [59–63], while phagocytozed *S. aureus* suppressed the production of superoxide inside macrophages [58]. In a murine model, high numbers of *S. epidermidis* could not only persist within macrophages in peri-catheter tissue without showing any signs of inflammation [49] but were also able to proliferate. Macrophages in the periphery of a biomaterial *in vivo* showed deficient intracellular killing of pathogens, resulting in a compromised local host defense [49]. In vivo, bacteria may well survive inside macrophages for prolonged periods of time. These bacteria will favor the development of BAI, especially when the physical condition of a patient disturbs the balance between bacteria and the host response [49].

In addition to bi-culture models, a tri-culture model was developed to further bridge the gap between *in vitro* and *in vivo* evaluation of biomaterials [40]. In tri-cultures, bi-cultures of mammalian cells and bacteria are supplemented with macrophages. Macrophages can prolong the time that U2OS cells can survive an attack by adhering *S. aureus* and *P. aeruginosa* on PMMA surface. However, adhering *S. aureus* and *P. aeruginosa* still cause death of all adhering U2OS cells and macrophages within 10–14 h of growth.

5.5 Conclusions

In summary, the outcome of the competition between bacteria and mammalian cells appears to depend on factors such as the physicochemical properties of the biomaterial surface, bacterial inoculum size, bacterial virulence, and mammalian cell number, attachment, density, and ability to grow. Simultaneous growth of bacteria and mammalian cells in *in vitro* bi-culture models of the competition between bacteria and mammalian cells in a flow system is useful to study biomaterial-associated infections. Bi-culture models can be extended with immune cells to further bridge the gap between *in vitro* and *in vivo* models.

It is also important to be aware of conditions distinguishing *in vitro* and *in vivo* methods. In vitro models can be used to study the short-term mechanisms by which bacteria, mammalian, and immune cells and the biomaterial interact, whereas *in vivo* models include long-term treatment effects and biological integration. Despite these limitations, the *in vitro* models described open possibilities to address many of these processes step by step and to unravel underlying mechanisms in the field of biomaterial-associated infections.

References

1. Hebert CK, Williams RE, Levy RS, Barrack RL. Cost of treating an infected total knee replacement. Clin Orthop Relat Res. 1996;331:140–5.
2. Campoccia D, Montanaro L, Arciola CR. The significance of infection related to orthopedic devices and issues of antibiotic resistance. Biomaterials. 2006;27:2331–9.
3. Trampuz A, Zimmerli W. New strategies for the treatment of infections associated with prosthetic joints. Curr Opin Investig Drugs. 2005;6:185–90.
4. Calhoun JH, Klemm K, Anger DM, Mader JT. Use of antibiotic-PMMA beads in the ischemic foot. Orthopedics. 1994;17:453–7.
5. Darouiche RO. Treatment of infections associated with surgical implants. N Engl J Med. 2004;350:1422–9.
6. Mohr VD, Eickhoff U, Haaker R, Klammer HL. External fixation of open femoral shaft fractures. J Trauma. 1995;38:648–52.
7. Costerton JW. Bacterial biofilms: a common cause of persistent infections. Science. 1999;284:1318–22.
8. Donlan RM, Costerton JW. Biofilms: survival mechanisms of clinically relevant microorganisms. Clin Microbiol Rev. 2002;15:167–93.
9. Gristina AG, Naylor PT, Myrvik QN. Musculoskeletal infection, microbial adhesion, and antibiotic resistance. Infect Dis Clin North Am. 1990;4:391–408.
10. Gristina AG. Biomaterial-centered infection: microbial adhesion versus tissue integration. Science. 1987;237:1588–95.
11. Ahlberg A, Carlsson AS, Lindberg L. Hematogenous infection in total joint replacement. Clin Orthop Relat Res. 1978;137:69–75.
12. Lidwell OM, Lowbury EJ, Whyte W, Blowers R, Stanley SJ, Lowe D. Airborne contamination of wounds in joint replacement operations: the relationship to sepsis rates. J Hosp Infect. 1983;4:111–31.
13. Fitzgerald RH. Microbiologic environment of the conventional operating-room. Arch Surg. 1979;114:772–5.
14. Verkkala K, Eklund A, Ojajarvi J, Tiittanen L, Hoborn J, Makela P. The conventionally ventilated operating theatre and air contamination control during cardiac surgery—bacteriological and particulate matter control garment options for low level contamination. Eur J Cardiothorac Surg. 1998;14:206–10.
15. Wells CL, Maddaus MA, Simmons RL. Role of the macrophage in the translocation of intestinal bacteria. Arch Surg. 1987;122:48–53.
16. Guo W, Andersson R, Ljungh A, Wang XD, Bengmark S. Enteric bacterial translocation after intraperitoneal implantation of rubber drain pieces. Scand J Gastroenterol. 1993;28:393–400.
17. Okell CC, Elliott CD. Bacteriaemia and oral sepsis with special reference to the etiology of subacute endocarditis. Lancet. 1935;2:869–75.
18. Ohara-Nemoto Y, Haraga H, Kimura S, Nemoto TK. Occurrence of staphylococci in the oral cavities of healthy adults and nasal-oral trafficking of the bacteria. J Med Microbiol. 2008;57:95–9.
19. Gristina AG. Implant failure and the immune-incompetent fibro-inflammatory zone. Clin Orthop Relat Res. 1994;298:106–18.
20. Khalil H, Williams RJ, Stenbeck G, Henderson B, Meghji S, Nair SP. Invasion of bone cells by *Staphylococcus epidermidis*. Microbes Infect. 2007;9:460–5.
21. Van Delden C, Iglewski BH. Cell-to-cell signalling and *Pseudomonas aeruginosa* infections. Emerg Infect Dis. 1998;4:551–60.
22. Robinson DA, Enright MC. Multilocus sequence typing and the evolution of methicillin-resistant *Staphylococcus aureus*. Clin Microbiol Infect. 2004;10:92–7.
23. Zimmerli W, Trampuz A, Ochsner PE. Current concepts: prosthetic-joint infections. N Engl J Med. 2004;351:1645–54.

24. Massey RC, Horsburgh MJ, Lina G, Hook M, Recker M. Opinion - the evolution and maintenance of virulence in *Staphylococcus aureus*: a role for host-to-host transmission? Nat Rev Microbiol. 2006;4:953–8.
25. Mckevitt AI, Bjornson GL, Mauracher CA, Scheifele DW. Amino-acid-sequence of a deltalike toxin from *Staphylococcus epidermidis*. Infect Immun. 1990;58:1473–5.
26. Raad I, Alrahwan A, Rolston K. *Staphylococcus epidermidis*: emerging resistance and need for alternative agents. Clin Infect Dis. 1998;26:1182–7.
27. Vuong C, Otto M. *Staphylococcus epidermidis* infections. Microbes Infect. 2002;4:481–9.
28. Ratner BD, Bryant SJ. Biomaterials: where we have been and where we are going. Annu Rev Biomed Eng. 2004;6:41–75.
29. Lebaron RG, Athanasiou KA. Extracellular matrix cell adhesion peptides: functional applications in orthopedic materials. Tissue Eng. 2000;6:85–103.
30. Schuler M, Owen GR, Hamilton DW, De Wilde M, Textor M, Brunette DM, Tosatti SGP. Biomimetic modification of titanium dental implant model surfaces using the RGDSP-peptide sequence: a cell morphology study. Biomaterials. 2006;27:4003–15.
31. VandeVondele S, Voros J, Hubbell JA. RGD-Grafted poly-L-lysine-graft-(polyethylene glycol) copolymers block non-specific protein adsorption while promoting cell adhesion. Biotechnol Bioeng. 2003;82:784–90.
32. Shi ZL, Neoh KG, Kang ET, Poh C, Wang W. Bacterial adhesion and osteoblast function on titanium with surface-grafted chitosan and immobilized RGD peptide. J Biomed Mater Res A. 2008;86A:865–72.
33. Shi Z, Neoh KG, Kang ET, Poh C, Wang W. Titanium with surface-grafted dextran and immobilized bone morphogenetic protein-2 for inhibition of bacterial adhesion and enhancement of osteoblast functions. Tissue Eng Part A. 2009;15:417–26.
34. Dexter SJ, Pearson RG, Davies MC, Camara M, Shakesheff KM. A comparison of the adhesion of mammalian cells and *Staphylococcus epidermidis* on fibronectin-modified polymer surfaces. J Biomed Mater Res. 2001;56:222–7.
35. Harris LG, Tosatti S, Wieland M, Textor M, Richards RG. *Staphylococcus aureus* adhesion to titanium oxide surfaces coated with non-functionalized and peptide-functionalized poly(L-lysine)-grafted-poly(ethylene glycol) copolymers. Biomaterials. 2004;25:4135–48.
36. Maddikeri RR, Tosatti S, Schuler M, Chessari S, Textor M, Richards RG, Harris LG. Reduced medical infection related bacterial strains adhesion on bioactive RGD modified titanium surfaces: a first step toward cell selective surfaces. J Biomed Mater Res A. 2008;84A:425–35.
37. Ploux L, Anselme K, Dirani A, Ponche A, Soppera O, Roucoules V. Opposite responses of cells and bacteria to micro/nanopatterned surfaces prepared by pulsed plasma polymerization and UV-irradiation. Langmuir. 2009;25:8161–9.
38. Subbiahdoss G, Kuijer R, Grijpma DW, Van der Mei HC, Busscher HJ. Microbial biofilm growth vs. tissue integration: "the race for the surface" experimentally studied. Acta Biomater. 2009;5:1399–404.
39. Subbiahdoss G, Grijpma DW, Van der Mei HC, Busscher HJ, Kuijer R. Microbial biofilm growth vs. tissue integration on biomaterials with different wettabilities and a polymer-brush coating. J Biomed Mater Res A. 2010;94A:533–8.
40. Subbiahdoss G, Saldarriaga Fernández IC, Da Silva Domingues JF, Kuijer R, Van der Mei HC, Busscher HJ. In vitro interactions between bacteria, osteoblast-like cells and macrophages in the pathogenesis of biomaterial-associated infections. PLoS ONE. 2011;6:e0024827.
41. Buchholz HW, Elson RA, Engelbrecht E, Lodenkamper H, Rottger J, Siegel A. Management of deep infection of total hip replacement. J Bone Joint Surg Br. 1981;63B:342–53.
42. Bennion RS, Williams RA, Wilson SE. Comparison of infectibility of vascular prosthetic materials by quantitation of median infective dose. Surgery. 1984;95:22–6.
43. Fleer A, Verhoef J. New aspects of staphylococcal infections: emergence of coagulase-negative staphylococci as pathogens. Antonie Van Leeuwenhoek. 1984;50:729–44.
44. Subbiahdoss G, Kuijer R, Busscher HJ, Van der Mei HC. Mammalian cell growth versus biofilm formation on biomaterial surfaces in an in vitro post-operative contamination model. Microbiology. 2010;156:3073–8.

45. Anderson JM. Inflammation, wound healing, and the foreign-body response. In: Ratner BD, Hoffman AS, Schoen FJ, Lemons JE, editors. Biomaterials science: an Introduction to materials in medicine. San Diego, CA: Elsevier; 2004. p. 296–304.
46. Aderem A, Underhill DM. Mechanisms of phagocytosis in macrophages. Annu Rev Immunol. 1999;17:593–623.
47. Underhill DM, Ozinsky A. Phagocytosis of microbes: complexity in action. Annu Rev Immunol. 2002;20:825–52.
48. Stuart LM, Ezekowitz RAB. Phagocytosis: elegant complexity. Immunity. 2005;22:539–50.
49. Boelens JJ, Dankert J, Murk JL, Weening JJ, Van der Poll T, Dingemans KP, Koole L, Laman JD, Zaat SAJ. Biomaterial-associated persistence of *Staphylococcus epidermidis* in pericatheter macrophages. J Infect Dis. 2000;181:1337–49.
50. Babior BM. Oxidants from phagocytes: agents of defense and destruction. Blood. 1984;64: 959–66.
51. Guenther F, Wabnitz GH, Stroh P, Prior B, Obst U, Samstag Y, Wagner C, Haensch GM. Host defence against *Staphylococcus aureus* biofilms infection: phagocytosis of biofilms by polymorphonuclear neutrophils (PMN). Mol Immunol. 2009;46:1805–13.
52. Kaplan SS, Heine RP, Simmons RL. Defensins impair phagocytic killing by neutrophils in biomaterial-related infection. Infect Immun. 1999;67:1640–5.
53. Saldarriaga Fernández IC, Da Silva Domingues JF, Van Kooten TG, Metzger S, Grainger DW, Busscher HJ, Van der Mei HC. Macrophage response to staphylococcal biofilm on cross-linked poly(ethylene) glycol polymer coatings in vitro. Eur Cell Mater. 2011;21:73–9.
54. Kubica M, Guzik K, Koziel J, Zarebski M, Richter W, Gajkowska B, Golda A, Maciag-Gudowska A, Brix K, Shaw L, Foster T, Potempa J. A potential new pathway for *Staphylococcus aureus* dissemination: the silent survival of *S. aureus* phagocytosed by human monocyte-derived macrophages. PLoS ONE. 2008;3(1):e1409.
55. Garzoni C, Kelley WL. *Staphylococcus aureus*: new evidence for intracellular persistence. Trends Microbiol. 2009;17:59–65.
56. Bonventre PF, Imhoff JG. Uptake of ^{3}H-dihydrostreptomycin by macrophages in culture. Infect Immun. 1970;2:89–95.
57. Tofte RW, Petersoson PK, Kim Y, Quie PG. Influence of serum concentration on opsonization by the classical and alternative complement pathways. Infect Immun. 1980;27:693–6.
58. Watanabe I, Ichiki M, Shiratsuchi A, Nakanishi Y. TLR2-mediated survival of *Staphylococcus aureus* in macrophages: a novel bacterial strategy against host innate immunity. J Immunol. 2007;178:4917–25.
59. Leid JG, Shirtliff HG, Costerton JW, Stoodley P. Human leukocytes adhere to, penetrate, and to respond to *Staphylococcus aureus* biofilms. Infect Immun. 2002;70:6339–45.
60. Johnson GM, Lee DA, Regelmann WE. Interference with granulocyte function by *Staphylococcus epidermidis* slime. Infect Immun. 1986;54:13–20.
61. Myrvit QN, Wagner W, Barth E, Wood P, Gristina AG. Effects of extracellular slime produced by *Staphylococcus epidermidis* on oxidative responses of rabbit alveolar macrophages. J Invest Surg. 1989;2:381–9.
62. Shanbhag A, Yang J, Lilien J, Black J. Decreased neutrophil respiratory burst on exposure to cobaltchrome alloy and polystyrene in vitro. J Biomed Mater Res. 1992;26:185–95.
63. Zimmerli W, Waldvogel F. Pathogenesis of foreign body infection. J Clin Invest. 1984;73: 1191–200.

Chapter 6
Biomaterial-Dependent Characteristics of the Foreign Body Response and *S. epidermidis* Biofilm Interactions

James M. Anderson and Jasmine D. Patel

Abstract This chapter presents our efforts to develop a better mechanistic understanding of how biomaterial interactions with blood components lead to alteration of the basic pathophysiologic mechanisms, in particular, inflammation and the foreign body response, which increase the probability of bacterial interactions, colonization, biofilm formation, and infection. In particular, we present perspectives on mechanisms of *S. epidermidis* biofilm formation, the role of surface chemistry on biofilm formation, the role of bacterial slime production in device infections, the apoptosis of adherent polymorphonuclear leukocytes in the acute inflammatory response, neutrophil mobility and phagocytosis of bacteria on biomaterials, generation of reactive oxygen and nitrogen species by biomaterial-adherent neutrophils, and quorum sensing in *S. epidermidis* biofilm formation. Our work has focused on infection mechanisms of cardiovascular prostheses and devices where blood hemodynamics and shear stress play important roles in inflammatory cell interactions with biomaterial surfaces. However, we believe that much of our results also are applicable to static implant situations found in orthopedic, cosmetic (plastic), and other surgical areas.

Abbreviations

AAP	Accumulation-associated protein
agr	Accessory gene regulator
atlE	Staphylococcal adhesin

J.M. Anderson, M.D., Ph.D. (✉)
Department of Pathology, Case Western Reserve University,
2103 Cornell Road, WRB Rm. 5105, Cleveland, OH 44106, USA
e-mail: jma6@case.edu

J.D. Patel, Ph.D.
Exponent, 3401 Market Street, Suite 300, Philadelphia, PA 19146, USA
e-mail: jpatel@exponent.com

T.F. Moriarty et al. (eds.), *Biomaterials Associated Infection: Immunological Aspects and Antimicrobial Strategies*, DOI 10.1007/978-1-4614-1031-7_6,

BMA	*n*-Butyl methacrylate
C3bi	Inactive complement component C3b, iC3b
cNOS	Constitutive nitric oxide synthetase
CXCL4	Platelet factor, PF-4
DNA	Deoxyribonucleic acid
eNOS	Endothelial nitric oxide synthetase
ERK	Extracellular-signal-related kinase
FAK	Focal adhesion kinase
Fbe	Fibrinogen-binding protein
fMLP	*N*-Formylmethionyl-leucyl-phenylalanine
IL-1	Interleukin-1
IL-4	Interleukin-4
IL-6	Interleukin-6
IL-10	Interleukin-10
IL-12	Interleukin-12
IL-13	Interleukin-13
IL-18	Interleukin-18
INFγ	Interferon gamma
iNOS	Inducible nitric oxide synthetase
L-NMMA	N^G-methyl-L-arginine, acetate salt
LTB4	Leukotriene
luxS	Accessory gene regulator
MAPC	ω-Methylacryloyloxyalkyl phosphorylcholine
MCP-1	Monocyte chemotactic protein-1
MIP-1α	Macrophage inflammatory protein-1
NADPH	Nicotinamide adenine dinucleotide phosphate
NO	Nitric oxide
NOS	Nitric oxide synthetase
OTS	Octadecyltrichlorosilane
O_2^-	Superoxide anion
PDGF	Platelet-derived growth factor
PEO	Polyethylene oxide
PIA	Polysaccharide intercellular adhesin
PMA	Phorbal myristate acetate
PMN	Polymorphonuclear leukocyte
PS/A	Polysaccharide/adhesin
RANTES	Regulated upon activation, normal T-cell cytokine
ROI	Reactive oxygen intermediate
SAM	Self-assembled monolayer
SdrG	Fibrinogen-binding protein
SME	Surface modifying end-groups
TGF-β	Transforming growth factor-beta
TNF-α	Tumor necrosis factor-alpha

6.1 Introduction

Medical device infections remain a major challenge and barrier to the successful use of medical devices, prostheses, and biomaterials. While the incidence of device-related infections is generally considered to be relatively low (1–8%), the associated morbidity and mortality of these infections is significant, making them a serious clinical concern. With the increasing use of medical implants and the projected concomitant increase in numbers of infected implants, infection represents a major impediment to the long-term clinical success of medical devices. Successful prevention and treatment strategies for dealing with medical device infections require a comprehensive understanding of the interfacial mechanisms that govern bacterial adhesion leading to infection of medical devices. Unfortunately, only a partial understanding of these interfacial mechanisms is available.

This chapter presents our efforts to develop a better mechanistic understanding of how biomaterial interactions with blood components lead to alteration of the basic pathophysiologic mechanisms, in particular, inflammation and the foreign body response, which increase the probability of bacterial interactions, colonization, biofilm formation, and infection. In particular, we present perspectives on mechanisms of *S. epidermidis* biofilm formation, the role of surface chemistry on biofilm formation, the role of bacterial slime production in device infections, the apoptosis of adherent polymorphonuclear leukocytes in the acute inflammatory response, neutrophil mobility and phagocytosis of bacteria on biomaterials, generation of reactive oxygen and nitrogen species by biomaterial-adherent neutrophils, and quorum sensing in *S. epidermidis* biofilm formation. Our work has focused on infection mechanisms of cardiovascular prostheses and devices where blood hemodynamics and shear stress play important roles in inflammatory cell interactions with biomaterial surfaces. However, we believe that much of our results also are applicable to static implant situations found in orthopedic, cosmetic (plastic), and other surgical areas.

6.1.1 The Foreign Body Reaction

6.1.1.1 The Provisional Matrix

Host reactions following implantation of biomaterials include injury, blood/material interactions, provisional matrix formation, acute inflammation, chronic inflammation, granulation tissue development, foreign body reaction, and fibrosis/fibrous capsule development (Fig. 6.1) [1–4]. In the very early process of implantation, blood/material interactions occur with protein adsorption to the biomaterial surface and development of a blood-based transient provisional matrix that forms on and around the biomaterial. The provisional matrix is the initial thrombus/blood clot at the tissue/material interface. Obviously, protein adsorption and fibrin-predominant provisional matrix formation are intimately linked in their mechanistic responses. The injury to

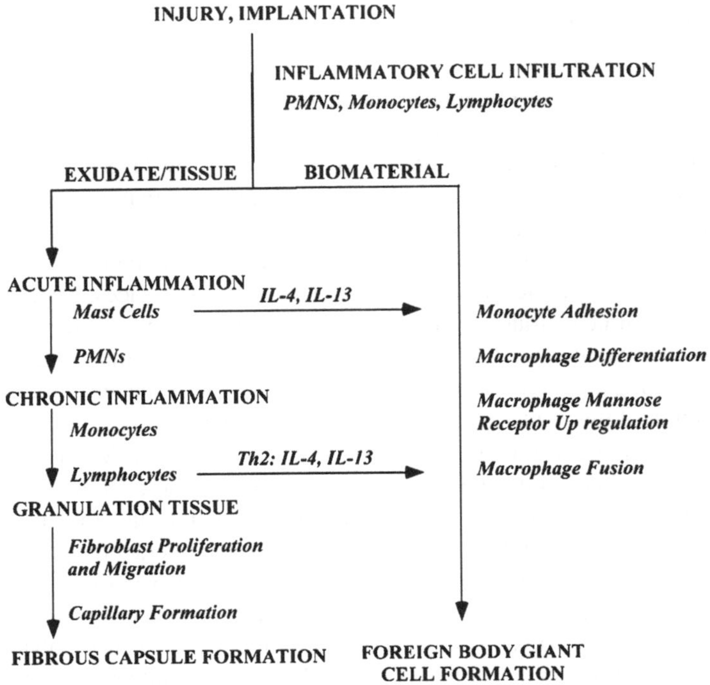

Fig. 6.1 Response to biomaterial implantation

vascularized connective tissue not only initiates the inflammatory responses (innate immunity) but also leads to thrombus formation involving activation of the extrinsic and intrinsic coagulation systems, the complement system, the fibrinolytic system, the kinin-generating system, and platelets. These protein cascades may be intimately involved in the dynamic phenomenon of protein adsorption and desorption that is known as the Vroman Effect [5]. From a wound healing perspective, blood protein deposition on a biomaterial surface is described as provisional matrix formation. The provisional matrix furnishes structural, biochemical, and cellular components to the processes of wound healing and foreign body reaction.

6.1.1.2 Acute and Chronic Inflammation

Following the initial blood/material interactions and provisional matrix formation, acute and chronic inflammation occurs in a sequential fashion as expected. The extent or degree of these responses is controlled by the extent of injury in the implantation procedure, the tissue or organ into which the device is implanted, and the extent of provisional matrix formation. Neutrophils (polymorphonuclear leukocytes, PMNs) characterize the acute inflammatory response. Mast cell degranulation with histamine release and fibrinogen adsorption is known to mediate acute inflammatory responses to implanted biomaterials [6, 7]. Interleukin-4 (IL-4) and interleukin-13 (IL-13) also are released from mast cells in a degranulation process and can play significant roles in determining the extent and degree of the subsequent development of the foreign body reaction [8, 9].

The acute inflammatory response with biomaterials usually resolves quickly, usually less than 1 week, depending on the extent of injury at the implant site. Following acute inflammation, chronic inflammation is identified by the presence of mononuclear cells, i.e., monocytes and lymphocytes, at the implant site. The presence of mononuclear cells, including lymphocytes and plasma cells, is considered chronic inflammation. This chronic inflammatory response to biomaterials is usually of short duration and is confined to the implant site. Chronic inflammation also has been used to describe the foreign body reaction where monocytes, macrophages, and foreign body giant cells are present at the biomaterial interface. With biocompatible materials, early resolution of the acute and chronic inflammatory responses occurs with the chronic inflammatory response composed of mononuclear cells usually lasting no longer than 2 weeks. The persistence of the acute and/or inflammatory responses beyond a 3 week period usually indicates an infection. Following resolution of the acute and chronic inflammatory responses, granulation tissue identified by the presence of macrophages, the infiltration of fibroblasts, and neovascularization in the new healing tissue is identified. Granulation tissue is the precursor to fibrous capsule formation and granulation tissue is separated from the implant or biomaterial by the cellular components of the foreign body reaction; a one- to two-cell layer of monocytes, macrophages, and foreign body giant cells.

6.1.1.3 Adsorption of Proteins

Biomaterials and medical devices immediately and spontaneously acquire a layer of host proteins prior to interacting with host cells. Thus it is highly probable that the types, levels, and surface conformations of the adsorbed proteins are critical determinants of the tissue reaction to such implants [10]. Conversely, the types, concentrations, and conformations of these surface-adsorbed proteins are dependent on biomaterial surface properties that dictate the adhesion and survival of cells, especially monocytes, macrophages, and FBGCs, on protein-coated surfaces. The interaction of adsorbed proteins with adhesion receptors present on inflammatory cell populations constitutes the major cellular recognition system for implantable synthetic materials and medical devices. The presence of adsorbed proteins such as albumin, fibrinogen, complement, fibronectin, vitronectin, gamma globulin, and others modulate host inflammatory cell interactions and adhesion and thus are linked to subsequent inflammatory and wound healing responses [11–14]. As previously indicated, these adsorbed proteins may desorb rapidly, i.e., the Vroman Effect, and thus present time-dependent variations in the type and level of proteins that cells encounter in the *in vivo* environment [15].

The complement system has long been recognized as a major host defense system for the interaction and removal of foreign substances *in vivo*. Complement activation and its subsequent reactions have been identified as causing adverse side effects when blood/material interactions occur with devices such as hemodialyzers, oxygenators, catheters, prostheses, stents, vascular grafts, and other devices and materials. In blood/material interactions, there is tight cross talk between the different cascade systems and platelets and leukocytes in the induction of clotting and inflammation. As complement and leukocytes are involved in biomaterial-associated thrombosis, as well as the more commonly identified coagulation factors and platelets, the true effect of a biomaterial surface on blood can be identified only when whole blood is used for interaction studies [16, 17].

6.1.1.4 Interaction with Immune Cell Integrins

The progression of events in inflammation and the foreign body response requires the extravasation and migration of monocytes/macrophages to the implant site. The guided movement of monocytes/macrophages occurs in response to chemokines and other chemoattractants.

Following blood–material interactions, platelets and the clot release chemoattractants, such as transforming growth factor (TGF-β), platelet-derived growth factor (PDGF), CXCL4 (Platelet Factor, PF4), leukotriene (LTB4), and interleukin (IL-1), that can direct macrophages to the wound site [18].

The adsorbed blood protein-modified material surface is the substrate with which the recruited monocytes/macrophages encounter and interact. The various plasma and extracellular matrix proteins deposited onto the surface give cells a means of attaching via surface receptors. Integrins are a large family of cell surface receptors that mediate cell-extracellular matrix as well as intercellular interactions [19, 20]. These adhesion molecules allow cells to migrate through the extracellular matrix and mediate signal transduction between the cell and its environments so that the cell can respond to its environment [19]. Integrins are heterodimers composed of different α and β subunits. The partnering of different subunit chains confers variability to the specificity and function of the receptor [21]. Monocytes/macrophages express integrins with three different types of β chains, β1, β2, β3. In monocytes/macrophages there are three β1 integrins, four β2 integrins, and one β3 integrin. β1 integrins include α4/β1 and α5/β1 which bind fibronectin and α6/β1 which binds laminin. Of the β2 integrins, there are αL/β2, αM/β2, and αD/β2 which are specific for intercellular adhesion molecules (ICAMs) and αX/β2 which binds complement fragment C3bi and fibrinogen. αM/β2 also interacts with a variety of other ligands such as fibrinogen, C3bi, and Factor X. Finally, monocytes/macrophages express αV/β3 of the β3 integrins which bind to vitronectin along with other RGD-containing extracellular proteins [21].

To determine monocyte, macrophage, and foreign body giant cell adhesion mechanisms, we have utilized human blood-derived monocytes in culture in a time-dependent fashion and on materials displaying different chemistries. To mimic the *in vivo* condition, we have added an additional 20% human serum to our culture media where adsorbed proteins provide ligands for integrin receptor binding and cellular adhesion. Initial monocyte adhesion is achieved through β2 integrins, in particular αM/β2(Mac-1, CD11b/CD18), by binding to various adsorbed protein ligands including fibrinogen, fibronectin, IgG, and complement fragment iC3b [12, 22]. While our efforts have focused on complement components providing the initial adhesion ligands, others have focused on fibrinogen as being the principal adsorbed protein ligand. Recent studies have suggested that multiple protein ligands may participate in the receptor–ligand binding and monocyte adhesion. In particular, complement activation on fibrinogen-adsorbed surfaces has been suggested as the primary adhesion event [16]. In our more recent studies, β1 integrins have been determined to play a role in the subsequent adhesion and IL-4-induced macrophage fusion to form foreign body giant cells. This phenomenon is time-dependent as β1 integrins are not initially detected on adherent monocytes but begin to appear during macrophage development and are strongly expressed on fusing macrophages and foreign body giant cells with increased culture time [23]. Our studies have used only human blood-derived monocytes for several reasons. Others have used murine blood and peritoneal-derived monocytes/macrophages, knockout murine models, and tumor-derived cell lines. We believe that the respective phenotypes of these models and systems are not adequately characterized to model human leukocyte behavior.

We have recently demonstrated expression and colocalization of α3, α5, or αV with β1 on fusing macrophages/FBGC at day 7 of culture as well as the strong colocalization of αM and αX with β2 in FBGC and at macrophage fusion interfaces. Therefore, IL-4-induced FBGC are characterized by the expression of αM/β2, αX/β2, α5/β1, α5/β1, α2/β1, and α3/β1, which indicates the potential interactions of complement C3b fragments, fibrin, fibrinogen, fibronectin, factor X, and vitronectin at sites of biomaterial implantations [24].

Cell adhesion to a biomaterial is dependent on the proteins that adsorb onto the surface. Surfaces with varying properties promote differential cellular integrin binding to the absorbed protein layer [25–28]. Subsequent material-dependent differences in cell adhesion and cytoskeletal rearrangement correlate to differences in phosphorylation of Focal Adhesion Kinase (FAK) and Extracellular-Signal-Regulated Kinase (ERK) [29, 30] and differences in recruitment of talin, α-actinin, and paxillin [31]. This indicates that differences in downstream integrin signaling account for morphological differences in adherent cells on material surfaces.

6.1.1.5 Control of Cell Cycle/Apoptosis

Integrin signaling is also responsible for controlling the cell cycle. Integrins are important in regulating cell death which is necessary for cell detachment and tissue remodeling [32]. Anoikis is a term for apoptosis induced by cell detachment from its supportive matrix [33]. When a cell is properly adhered to a surface, FAK mediates survival signaling. Disruption of adhesion signals promotes anoikis [34]. Monocytes initially adhere to most surfaces quite well, but fail to maintain adhesion over time [23]. Biomaterial surface chemistry influences apoptosis of adherent macrophages both *in vitro* and *in vivo* [12, 35–37]. Caspase-3, a protein involved in apoptosis signaling, is activated by neutrophils under shear stress, which leads to detachment from biomaterials [38]. Caspases are involved in cleaving gelsolin, a protein involved in regulating actin polymerization, and this in turn disrupts adhesion [34]. The disruption of adhesion leads to anoikis. Material surface chemistry can influence both adhesion and apoptosis. Materials that do not promote adhesion lead to cell detachment and anoikis. The induction of apoptosis is inversely related to fusion and we have hypothesized that macrophage fusion to form foreign body giant cells is an escape mechanism to avoid apoptosis [35].

The activation of macrophages with the subsequent release of a myriad of bioreactive agents has been extensively studied in a wide variety of *in vitro* and *in vivo* models. The adhesion of macrophages and foreign body giant cells at the surfaces of biomaterials produces a privileged microenvironment that exists between the cell membrane and the surface of the biomaterial. In a process described by Hensen as frustrated phagocytosis, macrophages and foreign body giant cells can release mediators of degradation such as reactive oxygen intermediates (ROIs, oxygen free radicals), degradative enzymes, and acid into this privileged zone between the cell membrane and biomaterial surface such that immediate buffering or inhibition of these mediators is delayed or reduced. It is known that phagolysosomes in macrophages can have acidity as low as pH 4. Biomaterial surfaces in this privileged microenvironment are thus susceptible to high concentrations of these degradative agents and to a low pH. Therefore, the chemistry of the biomaterial surface will dictate its susceptibility to biodegradation.

6.1.1.6 Impact on Macrophage Bactericidal Capacity and Activation

The adhesion of macrophages and foreign body giant cells as well as other inflammatory cells has been shown to exhibit a reduced bacteriocidal capability following adhesion to biomaterial surfaces. In part, this is related to a respiratory burst that occurs upon adhesion and leaves the adherent cell exhausted and incapable of further producing bacteriocidal molecules. In addition, the surface chemistry of the biomaterial may facilitate apoptosis (programmed cell death) that, in turn, leaves the macrophage incapable of attacking foreign organisms that may be adherent to the biomaterial [35]. We have demonstrated that biomaterial-adherent cells can undergo material-dependent apoptosis both *in vitro* and *in vivo*, thus rendering macrophages nonfunctional while the surrounding environment of the implant or biomaterial remains unaffected. Adherent inflammatory cell apoptosis has been described as an important mechanism for the persistence of cardiovascular device infection where apoptosis is induced by shear stress in cardiovascular devices [36].

Macrophages secrete an array of inflammatory mediators following activation. A resting macrophage becomes activated in response to microbial products, immune complexes, chemical mediators, certain extracellular matrix proteins, and T lymphocyte-derived cytokines. Activated macrophages are capable of secreting a wide range of cytokines such as IL-1, IL-6, IL-10, IL-12, IL-18, TNF-α, TGF-β, IL-8, MCP-1, and MIP-1α/β [39]. Adherent macrophages on biomaterials become activated in an attempt to phagocytose the biomaterial. Subsequent cytokine secretion directs the inflammatory and wound healing response to the biomaterial.

In our laboratory, we began investigating cytokine release from human adherent macrophages on biomedical polymers in the late 1980s as an indicator of the extent of macrophage activation as determined by biomaterial surface chemistry. Differential IL-1 production from macrophages cultured on biomaterials allowed us to assess the reactivity of the polymer as a measure of biocompatibility [40, 41]. IL-1 secretion was found to be dependent on the biomedical polymer and the protein pre-adsorbed on the surface [42]. IL-6 and TNF-α *in vitro* macrophage secretion was also found to be differential on variant biomedical polymers [43]. Secretion of the pro-inflammatory cytokines, IL-1β [44], IL-6, and TNF-α [45, 46], from *in vitro* adherent macrophages on a wide range of materials has been studied extensively [47–55]. These studies have shown that macrophage activation can be modulated by the surface properties, such as material surface chemistry and surface topography [56], of the biomedical polymer.

Activated macrophages have several heterogeneous phenotypes. Classical macrophage activation occurs after induction by IFN-γ and exposure to microbial products such as LPS [57]. Classically activated macrophages, whose main function is the killing of intracellular pathogens, upregulate pro-inflammatory cytokines, inhibit anti-inflammatory cytokines, and produce nitric oxide. Alternatively activated macrophages are stimulated by IL-4, IL-13, and glucocorticoids, inhibit pro-inflammatory cytokines, promote anti-inflammatory cytokines, and upregulate mannose receptors [58, 59]. Alternatively activated macrophages play a role in allergic responses, the elimination of parasites, and matrix remodeling [60].

In our culture system, a phenotypic switch in biomaterial adherent macrophages to a more alternatively activated macrophage phenotype over time is expected because in order to induce fusion IL-4 is added on days 3 and 7 [48]. IL-4 is known to induce an alternative activation phenotype in macrophages inclusive of the upregulation of mannose receptors [58] which has been found to be necessary for the formation of foreign body giant cells [61]. The biomaterial-adherent macrophage phenotype does not become entirely alternative because of the continued secretion of the chemokines RANTES and MCP-1. Using our monocyte *in vitro* cell culture, adherent macrophages/FBGCs have a differential cytokine profile than either classically or alternatively activated macrophages, indicating that biomaterial activation is unique [62]. However, the utilization of a classification system to identify macrophage activation states should be used as a framework considering the myriad of macrophage phenotypes in response to variable inflammatory mediators and pathogenic signals [63].

6.1.2 Mechanisms of *S. epidermidis* Biofilm Formation

Biofilm formation occurs in a two-step process: (1) initial adhesion to an implant surface and (2) bacterial accumulation which involves proliferation, intercellular adhesion, and exopolysaccharide slime formation, giving rise to a multilayered biofilm structure.

Initial adhesion of *S. epidermidis* to an implant surface is dictated by reversible, nonspecific physicochemical forces and irreversible, specific ligand/receptor interactions involving the implant surface, host (blood) components, and the bacterium. Adhesion directly to the surface of the device, which occurs before or at the time of implantation, is influenced by the physicochemical properties of the biomaterial surface. Bacterial adhesion occurring subsequent to implantation is modulated by host-derived proteins. Due to rapid kinetics, a layer of plasma and extracellular matrix proteins (e.g., fibronectin, albumin, fibrinogen, and vitronectin) adsorbs within seconds onto the surface of a material exposed to blood. The distribution of proteins and their native conformation on an implant surface are influenced by the surface properties of the biomaterial [64].

For primary bacterial adhesion, nonspecific interactions from physicochemical contact can occur between the bacteria and either the bare polymer surface or protein modified surfaces. Non-covalent physical forces modulate the nonspecific interactions between the bacteria and the implant surface including the long-range forces such as van der Waals and electrostatic forces. Hydrophobicity effects govern bacteria–biomaterial interactions near the material interface which can lead to bacterial attachment on the material itself or to a layer of adsorbed proteins [65, 66]. At these proximal ranges to the surface, intermolecular forces, influenced by material surface properties, including surface energy, chemical composition, charge, local media components, surface topography, and hydrodynamic forces, can enhance the adhesive capacity of the bacteria.

Bacteria are transported toward the material surface from bulk fluid by mass transport. Following transport to close proximity of the biomaterial surface, long-range interactions (>10 nm) mediated by van der Waals and electrostatic forces, and steric repulsion may direct the bacteria closer or farther from the surface. Near the surface, short-range forces allow the bacteria to contact the biomaterial surface via attractive hydrophobic forces or by interactions with the bacterial adhesins protruding from the cell wall through nonspecific hydrophobic and electrostatic forces, or specific ligand/receptor binding. The production of slime may further anchor the bacteria to the surface once attachment has taken place.

Specific interactions result in ligand/receptor binding between the bacterium and the adsorbed proteins on the biomaterial surface. Cell wall components of *S. epidermidis* such as proteins and cell surface polysaccharides may be potential receptors for extracellular matrix proteins. *S. epidermidis* has been shown to exhibit weak interactions with host proteins including fibrinogen, laminin, and albumin [67, 68], with albumin being the primary component of serum that inhibits initial *S. epidermidis* adhesion [69, 70]. These results have been confirmed by our studies which demonstrate inhibition of initial bacterial adhesion in the presence of serum proteins [71–76].

Despite the weak interactions between *S. epidermidis* and specific host proteins, various factors that govern ligand/receptor interactions have been isolated and identified. Vitronectin-mediated adhesion of *S. epidermidis* on polystyrene occurs via bacterial autolysin atlE [77, 78] and fibronectin-binding protein (Fbn) bridges *S. epidermidis* to fibronectin [79–81]. *S. epidermidis* adhesion to fibronectin is also mediated in a dose-dependent manner by bacterial cell wall teichoic acid which is a major polysaccharide component of slime in *S. epidermidis* [82, 83]. *S. epidermidis* has also been shown to bind weakly to von Willebrand factor [84]. Lastly, fibrinogen-binding proteins Fbe and SdrG are adhesins on the surface of *S epidermidis* [78, 85–87]. Adherence of *S. epidermidis* to fibrinogen is blocked by recombinant Fbe in a concentration-dependent manner and under saturating conditions [87, 88].

A study which examined the role of serum proteins in the formation of *S. epidermidis* biofilm on a biomaterial demonstrated that initial bacterial adhesion was suppressed significantly when compared to adhesion in saline or bacterial growth media [89]. Despite the initial suppression, bacterial adhesion, intercellular aggregation, and proliferation increased dramatically from 12 to 24 h of incubation, resulting in an extensive network of biofilm. Nonspecific adhesion may have occurred from competition between the host proteins for adsorption upon exposure to a biomaterial. Simultaneously, specific ligand/receptor contact may have resulted in interactions between the bacteria and host proteins. These results emphasize the importance of adsorbed proteins in the development of robust, mature biofilm.

Several molecular moieties associated with *S. epidermidis* have been identified as being involved in the primary attachment phase. AtlE autolysin, encoded by *atlE* gene, has been shown to mediate initial attachment to polystyrene. A mutant strain of *S. epidermidis* lacking atlE demonstrated reduced polystyrene attachment and vitronectin binding capabilities, and a reduced bacterial cell surface hydrophobicity [90]. AtlE is a 120 kD protein composed of a 60 kDa amidase and a 52 kDa glucosamidase that have hydrolytic activities for cell wall metabolism. In vivo studies have shown *S. epidermidis* strains lacking atlE autolysin to be much less virulent in a catheter infection model in rats, indicating the significance of this autolysin in the pathogenesis of *S. epidermidis* infection [90, 91]. Studies on gene expression of *S. epidermidis* during biofilm formation on biomaterials have shown that expression of the *atle* gene decreases slightly up to 12 h from initial adhesion, followed by a tenfold increase within 48 h [92]. This data suggests that autolysin atlE plays a role in the maturation of the *S. epidermidis* biofilm. It is possible that the upregulation of *atlE* gene also may be due to cell wall metabolic activity.

Intercellular adhesins such as PS/A and PIA have been identified in *S. epidermidis* biofilm-forming strains. The expression of PS/A, a high molecular weight N-succinylated β-1-6-linked polyglucosamine encoded by the *ica* locus, has been correlated to primary bacterial adhesion on catheters [93]. PIA, also encoded by the *ica* locus, has been shown to mediate intercellular adhesion and bacterial aggregation during the accumulation phase of biofilm formation [94–97]. In vitro studies have demonstrated mutant strains lacking PIA cannot accumulate into multilayered clusters for biofilm [96]. Because the *ica* locus encodes both PS/A and PIA, their structures are similar with each composed of a β-1-6-linked polyglucosamine backbone [98, 99].

The *ica* locus contains an operon *icaADBC* for synthesis of the polysaccharide adhesins. IcaA has been identified as a transmembrane protein with

N-acetylglucosaminyl transferase activity and the *icaD* gene product may optimize its activity as a chaperone protein that directs folding and membrane insertion of IcaA [100]. IcaC is a membrane protein which may aid in translocation of the polysaccharide adhesin to the cell surface and possibly anchor PIA [101]. IcaB, a cell surface protein, is responsible for deacetylation of the poly-*N*-acetylglucosamine adhesion [101]. A study which examined *S. epidermidis* gene expression during biofilm development on a polyurethane biomaterial showed approximately a 100-fold increase in the expression of *icaA, icaD, icaB* and a 10-fold increase in the expression of *icaC* up to 48 h of biofilm formation and maturation [92]. Interestingly, the trends showed that the largest increase in expression occurred between 12 and 24 h for each gene. Another study showed that despite the initial suppression of bacterial adhesion in the presence of serum proteins, intercellular aggregation, and proliferation increased dramatically between 12 and 24 h, resulting in an extensive network of accumulated bacteria [89]. This timeframe corresponds well to the exponential increase in the expression of the *ica* locus on polyurethane biomaterials, further supporting the role of the polysaccharide adhesin in *S. epidermidis* biofilm formation [90].

Another virulence factor that plays a role in *S. epidermidis* biofilm formation is the accumulation-associated protein (AAP), a 140 kDa protein isolated from *S. epidermidis* strains [102, 103, 104]. Strains with AAP produced more biofilm indicating its importance in *S. epidermidis* biofilm formation [103]. A study that examined the expression of the *aap* gene during biofilm formation on a biomaterial showed approximately a tenfold increase up to 48 h of biofilm formation [92]. Similar to the expression of the *ica* locus, the largest increase in *aap* expression was observed to be between 12 and 24 h of bacterial incubation [92]. This time frame corresponds well with an increase in *S. epidermidis* intercellular aggregation and proliferation between 12 and 24 h, as observed by scanning electron microscopy in another study [89]. AAP is likely involved in the accumulation phase of *S. epidermidis* biofilm formation, with increased levels of expression continuing beyond 24 h.

The establishment of infection in implanted medical devices is a complex, multifactorial process, orchestrated by upregulation/downregulation of bacterial virulence genes at different stages of adhesion, aggregation, slime formation, and finally detachment.

6.1.3 Role of Surface Chemistry on Biofilm Formation

Many indwelling implants which become infected are made of various polymeric materials such as polyvinylchloride, polyethylene, silicone rubber, and polyurethanes. Polyurethanes, in particular polyurethane elastomers, are commonly used in various biomedical devices, due to their better blood compatibility and desirable mechanical properties such as high tensile strength and high elastic modulus.

Surface properties of the implanted biomaterial can modulate the ability of the bacteria to adhere to material surfaces and subsequently initiate infection.

One approach to prevent infection associated with implants is to utilize surface modifying techniques to alter physicochemical properties which favor inhibition of bacterial adhesion, reducing the likelihood of infection and biofilm formation. Examples of surface modification are attachment of hydrophilic polymers such as polyethylene oxide (PEO) to polyurethanes which have been shown to reduce protein adsorption and bacterial adhesion [105, 106].

To govern bacterial adhesion, polymer surface chemistries can be tailored with various surface modifying endgroups (SMEs). Simultaneously, these surface properties may also dictate leukocyte adhesion on the surface. SMEs are surface active oligomers, such as PEO, fluorocarbon, and dimethylsiloxane, covalently attached to a polyurethane backbone. A monolayer forms as SMEs migrate to the surface of a material, altering the surface properties [107]. Such changes allow minimization of surface interfacial energy which may be an important factor in creating biocompatible devices. The advantages of using materials modified with SMEs allow the design of biocompatible materials, while preventing any significant changes in or loss of bulk properties of the material. These properties also pose an advantage over antibiotic—eluting biomaterials which have short-term release profiles due to depletion of the drug from its reservoirs.

The role of surface chemistry was examined by investigating bacterial adhesion and biofilm formation on a series of polyurethanes with well-defined variations in surface chemistry [72, 89]. This included a hydrophobic polyurethane as well as the polyurethane modified with fluorocarbon (very hydrophobic) or polyethylene oxide (hydrophilic) endgroups [89]. The study showed that bacterial adhesion, aggregation, and proliferation increased in the presence of serum proteins substantially over time on hydrophobic polyurethanes, resulting in extensively accumulated bacteria and thick biofilm formation. A contrasting trend was observed on the hydrophilic E80A-6PEO surface, with bacterial adhesion remaining minimal over 24 h. Additionally, an increasingly thick (~20 μm) biofilm formed on both hydrophobic surfaces while biofilm formation was impeded on the polyethylene oxide modified surface with few viable bacteria. As the biofilm matured on the hydrophobic surfaces, dead bacteria accumulated near the material surface. Furthermore, slime was detected during the initial phase of bacterial adhesion (2 h) and increased over time with the formation of biofilm. These results have shown that biomaterial surface chemistry affects phenotypic changes to biofilm formation (bacterial adhesion, aggregation, viability, and biofilm thickness) in the presence of proteins. Hydrophobic surfaces promote biofilm formation, whereas, surface modification with PEO end groups significantly reduces biofilm formation over 48 h.

Another category of materials modified to potentially minimize device-associated infections are phospholipid polymer surfaces. Polymer surfaces can be altered with poly[ω-methacryloyloxyalkyl phosphorylcholine (MAPC)-co-*n*-butyl methacrylate (BMA)] to possibly reduce the number of adherent bacteria. MAPC copolymers were created to mimic biological membranes which do not cause cell and blood protein adsorption or activation upon contact [108]. The MAPC copolymers have been shown to reduce platelet adhesion and activation [109, 110] as well as suppress an inflammatory response to adherent cells [110, 111]. The excellent compatibility of these materials has been correlated to a strong affinity for phospholipids and low protein adsorption [112–116]. One study showed that modification of polyethylene terephthalate surfaces with these phospholipid polymers resulted in reduced bacterial adhesion [71]. Furthermore, modification of the methylene chain length of the MAPC copolymer was more effective in reducing bacterial adhesion [71]. The proposed mechanism for inhibition of cell adhesion is the assembly of an organized layer of phospholipids on the surface of the material, which serves to reduce protein adsorption, and consequently creates a biocompatible, infection-resistant implant.

Biomaterial surface chemistry can not only influence initial adhesion but also has long-term effects in the process of biofilm formation (adhesion, aggregation, viability, etc.) on biomaterials.

6.1.4 Role of Slime Production in Device Infection

Some strains of *S. epidermidis* produce extracellular polysaccharides (slime) that bind bacteria to device surfaces [117]. The virulence of *S. epidermidis* can be attributed in part to its ability to colonize a biomaterial surface and produce slime (exopolysaccharide matrix) which forms a biofilm that protects against the host defense mechanisms. Slime may act as a protective biofilm, shielding bacteria from antibiotics, physiologic shear, and possibly from host cell-mediated defenses [118–120]. Bacterial proliferation, intercellular adhesion and exopolysaccharide slime formation result in a multilayered biofilm [78, 99, 121]. Intercellular adhesion and subsequent accumulation of *S. epidermidis* necessary for biofilm formation, is, in part directed by polysaccharide intercellular adhesin (PIA) [96, 97, 99, 121, 122]. A mature biofilm is composed of bacterial cell clusters embedded in an extracellular slime made of bacterial cell wall components, teichoic acid, bacterial polysaccharides, and host proteins [103, 122]. When a biofilm forms, it releases antigens that stimulate the host defense system and production of antibodies [123]. Antibiotics and the host immune response are not successful in resolving the biofilm infection, making device-associated infections persistent and recurrent. The biofilm covering the bacteria creates a diffusion barrier to antibiotics and the antibiotics are not effective, owing to the slow metabolism of the sessile bacteria from possible nutrient limitations within the biofilm [119, 120, 124–127]. Without eradication of the biofilm, patients can suffer from recurrent device infections.

6.1.5 Apoptosis of Biomaterial Adherent Leukocytes

The mechanisms underlying problematic cardiovascular device-associated infections are not well understood. The outcome of the acute response to infection is largely dependent on the function of neutrophils, so that the persistence of infection suggests that neutrophil function may be compromised because of cellular responses to shear stress. The characterization of the phenotypic and functional capacity of adherent leukocytes under shear stress has shown that adhesion and/or shear stress may play a role in the induction of apoptosis in adherent leukocytes [128]. Apoptosis is characterized by several morphological and biochemical properties, beginning with permanent, irreversible cell shrinkage, with the maintenance of an intact membrane, as well as increased membrane exposure of phosphatidylserine with loss of normal membrane asymmetry. Shive et al. used a rotating disc system to generate physiologically relevant shear stress levels (0–18 dynes/cm^2) at the surface of a poly(etherurethane urea) biomaterial which caused morphological and biochemical alterations in adherent neutrophils that are consistent with those described for apoptosis [128]. Complete neutrophil apoptosis was shown to occur at shear stress levels ≥5.6 dynes/cm^2 after only 1 h. Shear stress was also found to diminish the phagocytic ability of neutrophils adherent to this biomaterial.

Adherent neutrophils displayed irreversible cytoplasmic and nuclear condensation while maintaining intact membranes [128]. Neutrophil area and filamentous actin content demonstrated concomitant decreases with increasing levels of shear stress, and neutrophil phagocytosis of adherent bacteria decreased with increasing shear stress. Biochemical alterations included membrane phosphatidylserine exposure and DNA fragmentation. Scanning electron microscopy showed adherent neutrophils exposed to low shear were well spread, but exhibited condensed and irregular morphologies under high shear. Neutrophils exhibited characteristic multilobed nuclei, with an occasional condensed nucleus under low shear, but nuclear condensation was present in all cells exposed to high shear and these cells were incapable of interacting with bacteria. Confocal microscopy illustrated spread neutrophils under low shear localized actin in pseudopodia, whereas compact actin distribution and low-intensity staining was observed under high shear. Areas of low shear contained few annexin V-positive neutrophils, whereas high-shear areas contained sparse but exclusively annexin V-positive neutrophils. The potency of the shear–stress effect was emphasized by comparative inductive studies with adherent neutrophils under static conditions. These findings suggested a mechanism through which shear stress plays an important role in the development of cardiovascular device-associated infections.

Similarly, apoptosis of polyurethane-adherent monocytes under shear stress increased to levels of over 70% with increasing shear in a near-linear fashion ($r^2=0.713$) [129]. It was confirmed qualitatively using confocal microscopy that filamentous (F)-actin distribution was altered, DNA fragmentation occurred, and that activated caspases were involved in shear-induced apoptosis. Static studies determined that spontaneous apoptosis was material-dependent over 72 h by demonstrating marked differences between apoptosis of monocytes adherent to a polyurethane compared to an alkyl-modified glass. Treatment with TNF-α, which induces apoptosis, augmented this material dependency in a dose-dependent fashion over time. F-actin content of TNF-α-treated cells decreased to below 62% of untreated cells. We conclude that concomitant effects from both material surfaces and dynamic flow mediate human monocyte apoptosis, and may have significant implications in the context of implanted device infection.

Shear stress-induced apoptosis of adherent phagocytic cells is a mechanism through which an implanted material under flow conditions compromises host bactericidal abilities. The morbidity and mortality associated with device infections may be attributed to, in part, the failure of apoptotic cells to fulfill their bactericidal role.

6.1.6 Neutrophil Mobility and Phagocytosis of Bacteria on Biomaterials

Reduced neutrophil mobility and phagocytosis of bacteria may also be a mechanism for the increased incidence of implant-associated infection [130]. Effective neutrophil mobility should facilitate access to bacteria, prevent bacteria from developing a protective biofilm, and decrease the probability for an implant-related infection. Conversely, evidence for decreased neutrophil mobility may represent an important factor that diminishes effective recognition and phagocytosis of bacteria. Neutrophil locomotion is a spatially and a temporally coordinated process that includes extension of protrusions, formation of protrusion-substratum attachments, cytoskeletal contraction, and detachment of the cell from substratum at the tail end [131, 132]. Cell mobility can be rate-limited by either one or a combination of these mechanistic steps, depending on the environmental conditions and the nature of cell–substratum interactions [133, 134]. Neutrophil mobility and phagocytic ability was characterized on three materials, a poly(carbonate urethane), a hydrophobic surface consisting of an octadecyltrichlorosilane (OTS) self-assembled monolayer (SAM), and a glass reference. Time-lapse optical microscopy demonstrated that in the absence of adsorbed plasma proteins, the mobility of stimulated neutrophils increased with greater material hydrophobicity from glass, to polyurethane to OTS. The opposite trend was observed in the presence of adsorbed plasma proteins, such that neutrophil mobility increased with decreasing material hydrophobicity. Material surface properties are known to affect the composition, conformation, and bioactivity of adsorbed proteins [135]. This, in turn, will modulate the binding of neutrophil receptors and the subsequent signals that promote changes in the cell mobility and morphology [136]. It is the biological properties of the adsorbed protein layer that accounts for the relative cell speed of glass > polyurethane > OTS, and for the reversed order compared with materials in the absence of adsorbed proteins. In general, hydrophobic biomaterials lead to greater denaturation of adsorbed proteins, and disruption of cell–surface interactions. Consequently, the more hydrophilic materials (glass, polyurethane) should allow for least disruption of protein ligand–neutrophil receptor interactions and the subsequent signal transduction events.

Further analysis of the results showed that the mobility of *N*-formylmethionyl leucyl-phenylalanine (fMLP) stimulated neutrophils cells was inversely related to the extent of cell spreading on the materials. In the absence of plasma proteins, neutrophil phagocytosis rates increased from hydrophilic glass (3.9 ± 0.7 bacteria/h/cell) to hydrophobic OTS (7.0 ± 0.6 bacteria/h/cell). The presence of adsorbed plasma proteins increased the average bacteria phagocytosis rate on all three materials, but still displayed statistically significant material-dependent variation. The highest engulfment rate (13.5 ± 1.0 bacteria /h/cell) was observed on the polyurethane biomaterial, with lower rates on glass (9.8 ± 0.9 bacteria/h/cell) and OTS (10.6 ± 0.9 bacteria/h/cell). These data demonstrate that neutrophil mobility and capability for engulfment/ingestion of *S. epidermidis* is affected significantly by surface properties and the presence/absence of adsorbed proteins.

6.1.7 Generation of Reactive Oxygen and Nitrogen Species by Biomaterial-Adherent Neutrophils

The release of reactive oxygen species by neutrophils, particularly superoxide anion, is a well-known bactericidal mechanism, and recently nitric oxide (NO) also has been identified as an important cytotoxic mediator with enhanced NO production by upregulation of inducible nitric oxide synthase (iNOS) [137–141]. Nicotinamide adenine dinucleotide phosphate (NADPH) oxidase catalyzes a reaction converting oxygen to superoxide anion, which in turn is converted to hydrogen peroxide by spontaneous dismutation or catalyzed by superoxide dismutase. Superoxide anion is also generated by nitric oxide synthases (NOS) which typically produce nitric oxide (NO) [142].

Clinically, nitric oxide has also been identified as an important cytotoxic mediator in acute and chronic inflammatory responses with enhanced NO production in patients with sepsis, endotoxemia, and host versus graft reactions [136, 137, 142, 143, 144]. Infection has been correlated with increased NO production as measured by nitrate and nitrite levels in patients' serum and urine [137, 138, 143]. When NO production has been inhibited, microbial infection ensues [139, 140]. NO is produced by the enzymatic oxidation of L-arginine by NOS. Three isoforms of this enzyme exist—inducible NOS (iNOS), constitutive NOS (cNOS), which categorically includes endothelial NOs (eNOS), and neuronal NOS (nNOS). eNOS and nNOS are constitutively expressed and regulated by the elevation of intracellular Ca^{2+} levels and calmodulin binding [65]. iNOS, on the other hand, is expressed only when cells are stimulated with proinflammatory cytokines and/or microbial toxins, produces abundant NO for long periods of time, and its activity is dependent on calmodulin binding which occurs at basal levels of Ca^{2+} and the presence of L-arginine [145, 146]. In addition to generating NO, iNOS has also been shown to produce superoxide anion [142].

The interaction of NO and superoxide anion will result in the formation of peroxynitrite, a highly reactive species. Peroxynitrite is a potent cytotoxic oxidant that has many biological targets: inactivating ion channels, damaging DNA, and nitrating tyrosine residues that can potentially inactivate enzymes and disrupt signal transduction [147].

A diminished respiratory burst in response to stimuli has been observed with prolonged exposure to biomaterials [148–150]. This compromised response to an infectious challenge may be due to a premature biomaterial-induced activation resulting in metabolic exhaustion and depletion of its oxidative resources by a continuous release of superoxide anion.

It has been demonstrated that biomaterial-induced neutrophil activation does not cause upregulation of iNOS and activation of iNOS-mediated pathways [151]. However, upon stimulation with phorbal myristate acetate (PMA), which activates release of NO from neutrophils, iNOS was upregulated in $11.9 \pm 2.3\%$ of the biomaterial-adherent neutrophil population. These results confirm that biomaterial-induced neutrophil activation does not cause upregulation of iNOS and activation of iNOS-mediated pathways. Using confocal microscopy, intracellular nitric oxide (NO) was detected in 18.5 ± 4.4 and $50.4 \pm 4.7\%$ of adherent neutrophils when stimulated with PMA and preseeded *S. epidermidis*, respectively. A cytochrome c reduction assay measured levels of iNOS-mediated superoxide anion release in the presence/absence of a NOS inhibitor N^G-methyl-L-arginine, acetate salt (L-NMMA). Material-induced activation of adherent neutrophils generated low levels of O_2^- that elevated significantly under stimulated conditions. Moreover, the amounts released were comparable in the presence and absence of L-NMMA indicating that iNOS is not the source of O_2^- production in biomaterial adherent neutrophils, but that it may be produced primarily by other mechanisms (i.e., NAPDH oxidase). Peroxynitrite formation did occur in neutrophils challenged with bacteria. Fluid shear stress also influenced NO and O_2^- production by biomaterial adherent neutrophils with fewer cells generating NO. Superoxide anion levels were slightly elevated when exposed to shear stress.

NO and superoxide anion production by adherent neutrophils is also modulated by biomaterial surface chemistry [151]. Material-induced activation doubled NO generation in neutrophils adherent to a poly(ether urethane) with PEO endgroups in comparison to the poly(ether urethane). Under stimulated conditions, a decrease in NO production was observed on the PEO modified material and the amounts produced were lower than in unstimulated cells. Superoxide anion release doubled over time on the PEO modified biomaterial in comparison to the control biomaterial. Similar to NO generation by neutrophils, superoxide anion production decreased over time in stimulated cells on surface-modified biomaterial, whereas it increased on poly(ether urethane).

This suggests that bactericidal mechanisms in neutrophils involving NO generation (NOS pathway) are compromised more than O_2^- producing pathways (NADPH oxidase) upon exposure to biomaterials resulting in a diminished microbial killing capacity, which can increase the probability of device centered infections.

6.1.8 Neutrophil Interactions with S. epidermidis Biofilm

Effective host clearing of foreign microbes requires capable leukocyte function in locating bacteria (chemotaxis), phagocytosis, and microbial killing (reactive oxygen and nitrogen species production). Biomaterial-induced depression of leukocyte function has been shown in the presence of suspended bacteria or initially adherent

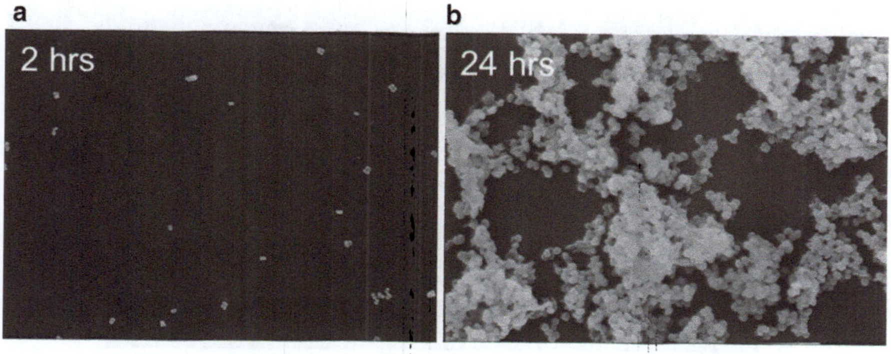

Fig. 6.2 *S. epidermidis* adhesion/biofilm formation grown for 24 h in the presence of serum proteins on a poly(ether urethane) E80A

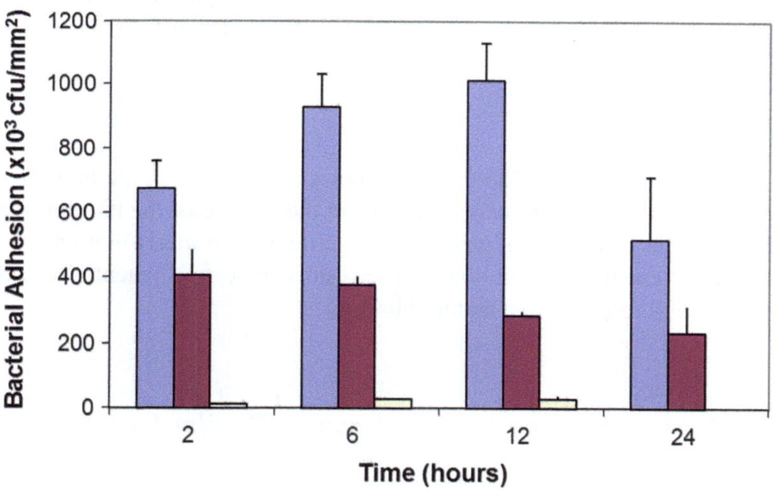

Fig. 6.3 *S. epidermidis* adhesion/biofilm formation grown for 24 h in the presence of serum proteins on a poly(ether urethane). ◻ TSB, ◼ PBS, ◻ Serum. *Not quantified due to dense layers of bacteria/biofilm

bacteria; however, little is known of leukocyte/biofilm interactions (Figs. 6.2, 6.3, 6.4 and 6.5).

In an unpublished preliminary study, the viability and neutrophil interactions with a mature biofilm were investigated. A mature biofilm was grown *in vitro* on a polyurethane biomaterial over 48 h, followed by the addition of neutrophils isolated from donors. Confocal microscopy imaging showed that after only a 2 h incubation with a mature biofilm, all the neutrophils within the biofilm were not viable. Furthermore, the results showed that neutrophils encountering thick biofilms were unable to effectively penetrate the biofilm. Figure 6.6 indicates that virtually all the bacteria within the biofilm are viable with no regions of dead bacteria observed

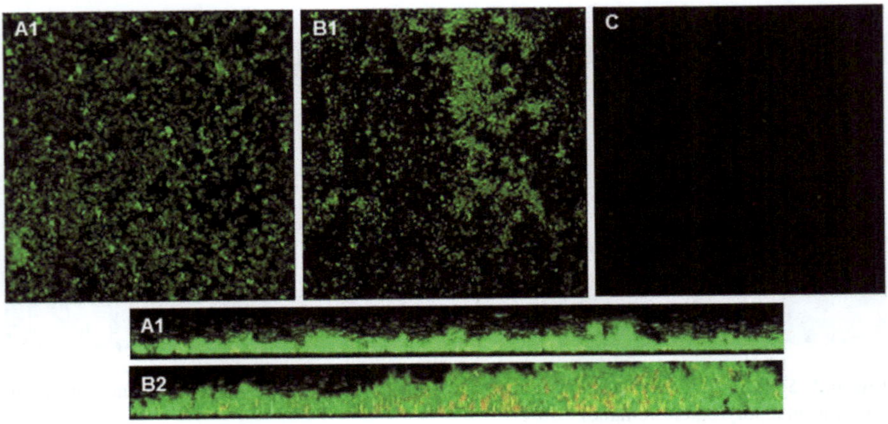

Fig. 6.4 Viability of *S. epidermidis* in a biofilm grown for 24 h in the presence of serum proteins on (**a**) poly(ether urethane), (**b**) polyurethane modified with fluorocarbon, and (**c**) polyurethane modified with PEO. *1* = Top view, *2* = Side view; *green*: viable bacteria, *red*: dead bacteria, *yellow*: overlap of viable/dead bacteria

adjacent to the neutrophils. The top view shows dark circles in the biofilm where neutrophils have managed to penetrate beyond the surface of the biofilm. It is not clear if this "hole" is a result of neutrophil engulfment of bacteria/biofilm or simple penetration by chemotactic motion. These results indicate a potentially compromised neutrophil response to a mature biofilm.

6.1.9 *Quorum Sensing in S. epidermidis Biofilm Formation*

S. epidermidis biofilm formation is regulated by the two quorum sensing systems, accessory gene regulator (agr) and luxS [152]. A *quorum sensing system* is a population-dependent intercellular signaling mechanism, which causes bacteria to change to a biofilm-like state. Because *S. epidermidis* releases a limited number of extracellular toxins, its primary virulence factor is its ability to form a biofilm on indwelling devices [153].

The *S. epidermidis* accessory gene regulator system *(agr)* activates in response to a shift from low bacterial density planktonic phase to a sessile, high density phase [153]. Enhanced *agr* activity causes upregulation of many exoproteins such as the δ-toxin, promoting biofilm detachment and downregulation of surface-associated proteins [154]. Enhanced *agr* activity causes upregulation of many exoproteins and downregulation of surface-associated proteins [154]. The *agr* system is composed of a two component regulatory system (AgrA and AgrC), a pheromone precursor protein (AgrD) and a cytoplasmic membrane protein (AgrB) believed to be involved in pheromone maturation and export [98, 99]. The pheromone promotes *agr* activity

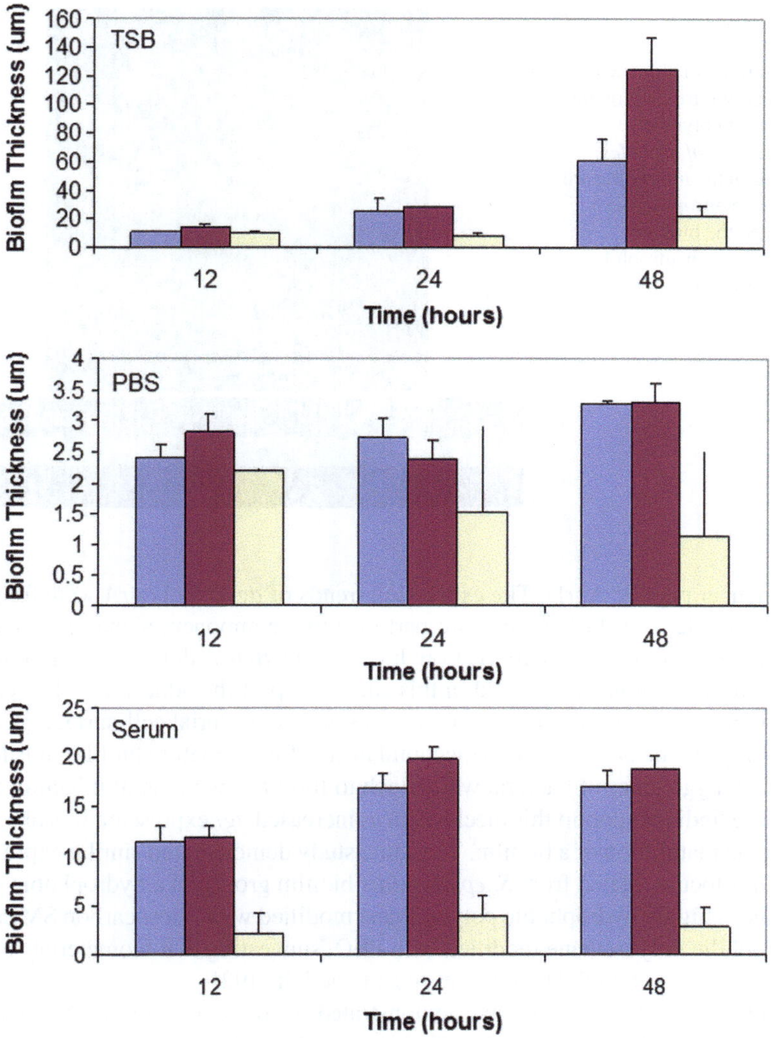

Fig. 6.5 *S. epidermidis* adhesion over 24 h on three poly(ether urethanes), in PBS, TBS, and in 20% serum proteins. ■ Elasthane 80A, ■ Elasthane 80A-6F, ■ Elasthane 80A-6PEO. *Not quantified due to dense layers of bacteria/biofilm

resulting in the synthesis of a regulatory RNA III, which in turn, causes upregulation of exoproteins and downregulation of surface proteins [153, 155, 156].

In a study on genetic mechanism of biofilm formation on a biomaterial, the expression of the genes *agrB* and *agrD* increased 10× in *S. epidermidis* biofilm adhered to a polyurethane biomaterial [92]. Similarly, expression of *agrC* and *agrA* increased ≤10×. As expected, *agrB* and *agrD* followed the same trends of expression since the gene product AgrB is responsible for the maturation and export of

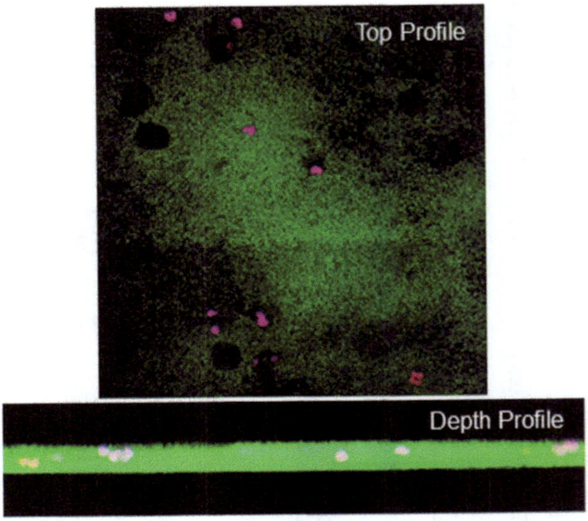

Fig. 6.6 Confocal microscopy images of neutrophil interactions with a biofilm grown for 24 h on the surface of a polyether urethane. The *black holes* observed in the top profile are areas that neutrophils have penetrated the biofilm surface. These neutrophils can be observed in the depth profile. *Green*: live bacteria; *Pink*: neutrophils

auto-inducer peptide AgrD. The expression trends of *agrC* and *agrA* were found to be comparable over 48 h as they are part of a two-component signal transduction system that promotes *agr* activity. Low levels of *agrBDCA* during the initial phase of biofilm formation, as observed in this study, support the adhesion of bacteria to the polymer surfaces, possibly due to the presence of bacterial cell-surface proteins [157, 158]. As the bacteria start to accumulate and form a mature biofilm, it is likely that small aggregates of bacteria will detach to form a new nidus of infection [159, 160]. The findings support this mechanism as increased *agr* expression was observed during the maturation of a biofilm. The same study demonstrated similar expression of the *agr* locus isolated from *S. epidermidis* biofilm grown on a hydrophobic polyurethane, a highly hydrophobic polyurethane modified with fluorocarbon SMEs and a hydrophilic polyurethane modified with PEO, suggesting that biomaterial surface chemistry does not modulate quorum sensing activity [92].

Staphylococcal *agr* system has demonstrated pheromone cross-inhibition properties between different strains of staphylococcus. *S. epidermidis* pheromone peptide is a potent inhibitor of the *S. aureus* quorum sensing system. It has been suggested that the potent cross inhibition properties of *S. epidermidis* may contribute to the predominance of this species in infection associated with medical devices [161, 162]. Similarly, the luxS system may impact biofilm formation by regulating the transcriptional activity of the *ica* locus that is responsible for PIA production [152, 163].

Surprisingly, loss of *agr* activity in *S. epidermidis* results in increased biofilm thickness and formation on polystyrene, increased attachment of bacteria, and increased expression of atlE autolysin which mediates primary bacterial attachment to polymer surfaces. However, *agr* was not found to regulate PIA production [164, 165]. Instead, the luxS system was found to downregulate PIA [152, 163]. In addition, a higher frequency of nonfunctional *agr* systems were found in clinical *S. epidermidis* strains isolated from infected joint prostheses [166]. The *agr* system may play a role in bacterial evasion of the host defense by providing resistance to reactive oxygen species activity [167].

These data show that suppression of the *agr* system can potentially enhance the pathogenesis of implant centered infection [166].

6.1.10 Conclusions

Successful prevention and treatment strategies for dealing with medical device infections require a comprehensive understanding of the interfacial mechanisms that govern bacterial adhesion leading to infection of medical devices as well as the host responses including inflammation, wound healing, and the foreign body reaction. While progress has been made in some areas such as the biomaterial-dependent surface characteristics of the foreign body response, other areas, some more than others, have only been addressed superficially and more detailed experimental efforts are necessary to achieve a better understanding of the interfacial mechanisms. In this chapter, we have presented our perspectives on mechanisms of *S. epidermidis* biofilm formation, the role of surface chemistry on biofilm formation, the role of bacterial slime production in device infections, the apoptosis of adherent polymorphonuclear leukocytes in the acute inflammatory response, neutrophil mobility and phagocytosis of bacteria on biomaterials, the generation of reactive oxygen and nitrogen species by biomaterial-adherent neutrophils, and quorum sensing in *S. epidermidis* biofilm formation. While these studies clearly identify the importance of biomaterial chemical and physical surface properties in modulating the various mechanisms by which medical device/biomaterial infections are produced, they must be considered preliminary and significantly increased efforts must be carried out to provide a better mechanistic understanding of these processes.

In considering cardiovascular devices where flow conditions are operative, the shear stress-induced apoptosis of adherent neutrophils has been identified as a mechanism by which the biomaterial may compromise host bactericidal abilities. In addition, the respiratory burst and production of bactericidal nitric oxide, NO, has been also identified as a compromised mechanism when these cells become adherent to biomaterials. Neutrophils interacting with thick biofilms are unable to effectively penetrate the biofilm to produce microbial killing. The importance of biomaterial surface chemistry in modulating *S. epidermidis* biofilm formation was identified in two quorum sensing systems where downregulation of the accessory gene regulator (agr) and luxS were identified.

Taken together, the studies presented in this chapter clearly identify the importance of biomaterial surface chemistry in modulating the foreign body response, *S. epidermidis* adhesion and subsequent biofilm formation, and the impact on efficacy of neutrophil-mediated defense. Further, the results point towards identifying and/or producing surfaces that are not conducive to *S. epidermidis* adhesion. These efforts must be expanded if we are to achieve a reduction in the incidence of medical device infections.

Acknowledgment This review was supported in part by Award Number EB000279 from the National Institute of Biomedical Imaging and Bioengineering.

References

1. Anderson JM. Multinucleated giant cells. Curr Opin Hematol. 2000;7(1):40–7.
2. Anderson JM. Biological responses to materials. Ann Rev Mater Res. 2001;31:81–110.
3. Gretzer C, Emanuelsson L, Liljensten E, Thomsen P. The inflammatory cell influx and cytokines changes during transition from acute inflammation to fibrous repair around implanted materials. J Biomater Sci Polym Ed. 2006;17(6):669–87.
4. Luttikhuizen DT, Harmsen MC, Van Luyn MJ. Cellular and molecular dynamics in the foreign body reaction. Tissue Eng. 2006;12(7):1955–70.
5. Horbett T. The role of adsorbed proteins in tissue response to biomaterials. In: Ratner B et al., editors. Biomaterials science: an introduction to biomaterials in medicine. San Diego: Elsevier Academic Press; 2004. p. 237–46.
6. Zdolsek J, Eaton JW, Tang L. Histamine release and fibrinogen adsorption mediate acute inflammatory responses to biomaterial implants in humans. J Transl Med. 2007;5:31.
7. Tang L, Jennings TA, Eaton JW. Mast cells mediate acute inflammatory responses to implanted biomaterials. Proc Natl Acad Sci USA. 1998;95(15):8841–6.
8. Keegan AD. IL-4. In: Oppenheim JJ, Feldman M, editors. Cytokine reference, vol. 1. San Diego: Academic; 2001.
9. McKenzie ANJ, Matthews DJ. IL-13. In: Oppenheim JJ, Feldman M, editors. Cytokine reference, vol. 1. San Diego: Academic; 2001.
10. Wilson CJ, Clegg RE, Leavesley DI, Pearcy MJ. Mediation of biomaterial-cell interactions by adsorbed proteins: a review. Tissue Eng. 2005;11(1–2):1–18.
11. Jenney CR, Anderson JM. Adsorbed serum proteins responsible for surface dependent human macrophage behavior. J Biomed Mater Res. 2000;49(4):435–47.
12. Brodbeck WG, Colton E, Anderson JM. Effects of adsorbed heat labile serum proteins and fibrinogen on adhesion and apoptosis of monocytes/macrophages on biomaterials. J Mater Sci Mater Med. 2003;14(8):671–5.
13. Jenney CR, Anderson JM. Adsorbed IgG: a potent adhesive substrate for human macrophages. J Biomed Mater Res. 2000;50(3):281–90.
14. Hu WJ, Eaton JW, Ugarova TP, Tang L. Molecular basis of biomaterial-mediated foreign body reactions. Blood. 2001;98(4):1231–8.
15. Xu LC, Siedlecki CA. Effects of surface wettability and contact time on protein adhesion to biomaterial surfaces. Biomaterials. 2007;28(22):3273–83.
16. Nilsson B, Ekdahl KN, Mollnes TE, Lambris JD. The role of complement in biomaterial-induced inflammation. Mol Immunol. 2007;44(1–3):82–94.
17. Gorbet MB, Sefton MV. Biomaterial-associated thrombosis: roles of coagulation factors, complement, platelets and leukocytes. Biomaterials. 2004;25(26):5681–703.
18. Broughton 2nd G, Janis JE, Attinger CE. The basic science of wound healing. Plast Reconstr Surg. 2006;117(7 Suppl):12S–34S.
19. Delon I, Brown NH. Integrins and the actin cytoskeleton. Curr Opin Cell Biol. 2007;19(1):43–50.
20. Giancotti FG, Ruoslahti E. Integrin signaling. Science. 1999;285(5430):1028–32.
21. Berton G, Lowell CA. Integrin signalling in neutrophils and macrophages. Cell Signal. 1999;11(9):621–35.
22. McNally AK, Anderson JM. Complement C3 participation in monocyte adhesion to different surfaces. Proc Natl Acad Sci USA. 1994;91(21):10119–23.
23. McNally AK, Anderson JM. Beta1 and beta2 integrins mediate adhesion during macrophage fusion and multinucleated foreign body giant cell formation. Am J Pathol. 2002;160(2):621–30.

24. McNally AK, Macewan SR, Anderson JM. Alpha subunit partners to beta1 and beta2 integrins during IL-4-induced foreign body giant cell formation. J Biomed Mater Res A. 2007;82(3):568–74.
25. Lee MH, Ducheyne P, Lynch L, Boettiger D, Composto RJ. Effect of biomaterial surface properties on fibronectin-alpha5beta1 integrin interaction and cellular attachment. Biomaterials. 2006;27(9):1907–16.
26. Lan MA, Gersbach CA, Michael KE, Keselowsky BG, Garcia AJ. Myoblast proliferation and differentiation on fibronectin-coated self assembled monolayers presenting different surface chemistries. Biomaterials. 2005;26(22):4523–31.
27. Keselowsky BG, Collard DM, Garcia AJ. Integrin binding specificity regulates biomaterial surface chemistry effects on cell differentiation. Proc Natl Acad Sci USA. 2005;102(17):5953–7.
28. Keselowsky BG, Collard DM, Garcia AJ. Surface chemistry modulates fibronectin conformation and directs integrin binding and specificity to control cell adhesion. J Biomed Mater Res A. 2003;66(2):247–59.
29. Allen LT, Tosetto M, Miller IS, O'Connor DP, Penney SC, Lynch I, Keenan AK, Pennington SR, Dawson KA, Gallagher WM. Surface-induced changes in protein adsorption and implications for cellular phenotypic responses to surface interaction. Biomaterials. 2006;27(16):3096–108.
30. Garcia AJ, Boettiger D. Integrin-fibronectin interactions at the cell-material interface: initial integrin binding and signaling. Biomaterials. 1999;20(23–24):2427–33.
31. Keselowsky BG, Collard DM, Garcia AJ. Surface chemistry modulates focal adhesion composition and signaling through changes in integrin binding. Biomaterials. 2004;25(28):5947–54.
32. Damsky CH, Ilic D. Integrin signaling: it's where the action is. Curr Opin Cell Biol. 2002;14(5):594–602.
33. Reddig PJ, Juliano RL. Clinging to life: cell to matrix adhesion and cell survival. Cancer Metastasis Rev. 2005;24(3):425–39.
34. Frisch SM, Screaton RA. Anoikis mechanisms. Curr Opin Cell Biol. 2001;13(5):555–62.
35. Brodbeck WG, Shive MS, Colton E, Nakayama Y, Matsuda T, Anderson JM. Influence of biomaterial surface chemistry on the apoptosis of adherent cells. J Biomed Mater Res. 2001;55(4):661–8.
36. Brodbeck WG, Patel J, Voskerician G, Christenson E, Shive MS, Nakayama Y, Matsuda T, Ziats NP, Anderson JM. Biomaterial adherent macrophage apoptosis is increased by hydrophilic and anionic substrates in vivo. Proc Natl Acad Sci USA. 2002;99(16):10287–92.
37. Jones JA, Dadsetan M, Collier TO, Ebert M, Stokes KS, Ward RS, Hiltner PA, Anderson JM. Macrophage behavior on surface-modified polyurethanes. J Biomater Sci Polym Ed. 2004;15(5):567–84.
38. Shive MS, Brodbeck WG, Anderson JM. Activation of caspase 3 during shear stress-induced neutrophil apoptosis on biomaterials. J Biomed Mater Res. 2002;62(2):163–8.
39. Fujiwara N, Kobayashi K. Macrophages in inflammation. Curr Drug Targets Inflamm Allergy. 2005;4(3):281–6.
40. Miller KM, Anderson JM. Human monocyte/macrophage activation and interleukin 1 generation by biomedical polymers. J Biomed Mater Res. 1988;22(8):713–31.
41. Miller KM, Huskey RA, Bigby LF, Anderson JM. Characterization of biomedical polymer-adherent macrophages: interleukin 1 generation and scanning electron microscopy studies. Biomaterials. 1989;10(3):187–96.
42. Bonfield TL, Colton E, Anderson JM. Plasma protein adsorbed biomedical polymers: activation of human monocytes and induction of interleukin 1. J Biomed Mater Res. 1989;23(6):535–48.
43. Bonfield TL, Colton E, Marchant RE, Anderson JM. Cytokine and growth factor production by monocytes/macrophages on protein preadsorbed polymers. J Biomed Mater Res. 1992;26(7):837–50.
44. Krause TJ, Robertson FM, Liesch JB, Wasserman AJ, Greco RS. Differential production of interleukin 1 on the surface of biomaterials. Arch Surg. 1990;125(9):1158–60.

45. Hwang JJ, Jelacic S, Samuel NT, Maier RV, Campbell CT, Castner DG, Hoffman AS, Stayton PS. Monocyte activation on polyelectrolyte multilayers. J Biomater Sci Polym Ed. 2005;16(2):237–51.
46. Ma N, Petit A, Yahia L, Huk OL, Tabrizian M. Cytotoxic reaction and TNF-alpha response of macrophages to polyurethane particles. J Biomater Sci Polym Ed. 2002;13(3):257–72.
47. Bonfield TL, Anderson JM. Functional versus quantitative comparison of IL-1 beta from monocytes/macrophages on biomedical polymers. J Biomed Mater Res. 1993;27(9):1195–9.
48. McNally AK, Anderson JM. Interleukin-4 induces foreign body giant cells from human monocytes/macrophages. Differential lymphokine regulation of macrophage fusion leads to morphological variants of multinucleated giant cells. Am J Pathol. 1995;147(5):1487–99.
49. Wagner VE, Bryers JD. Monocyte/macrophage interactions with base and linear- and star-like PEG-modified PEG-poly(acrylic acid) co-polymers. J Biomed Mater Res A. 2003;66(1):62–78.
50. Yun JK, DeFife K, Colton E, Stack S, Azeez A, Cahalan L, Verhoeven M, Cahalan P, Anderson JM. Human monocyte/macrophage adhesion and cytokine production on surface-modified poly(tetrafluoroethylene/hexafluoropropylene) polymers with and without protein preadsorption. J Biomed Mater Res. 1995;29(2):257–68.
51. DeFife KM, Yun JK, Azeez A, Stack S, Ishihara K, Nakabayashi N, Colton E, Anderson JM. Adhesion and cytokine production by monocytes on poly(2-methacryloyloxyethyl phosphorylcholine-co-alkyl methacrylate)-coated polymers. J Biomed Mater Res. 1995;29(4):431–9.
52. Marques AP, Reis RL, Hunt JA. Cytokine secretion from mononuclear cells cultured in vitro with starch-based polymers and poly-L-lactide. J Biomed Mater Res A. 2004;71(3):419–29.
53. Xing S, Santerre J, Labow RS, Boynton EL. Differential response to chemically altered polyethylene by activated mature human monocyte-derived macrophages. Biomaterials. 2002;23(17):3595–602.
54. Sethi RK, Neavyn MJ, Rubash HE, Shanbhag AS. Macrophage response to cross-linked and conventional UHMWPE. Biomaterials. 2003;24(15):2561–73.
55. Gretzer C, Gisselfalt K, Liljensten E, Ryden L, Thomsen P. Adhesion, apoptosis and cytokine release of human mononuclear cells cultured on degradable poly(urethane urea), polystyrene and titanium in vitro. Biomaterials. 2003;24(17):2843–52.
56. Refai AK, Textor M, Brunette DM, Waterfield JD. Effect of titanium surface topography on macrophage activation and secretion of proinflammatory cytokines and chemokines. J Biomed Mater Res A. 2004;70(2):194–205.
57. Mosser DM. The many faces of macrophage activation. J Leukoc Biol. 2003;73(2):209–12.
58. Stein M, Keshav S, Harris N, Gordon S. Interleukin 4 potently enhances murine macrophage mannose receptor activity: a marker of alternative immunologic macrophage activation. J Exp Med. 1992;176(1):287–92.
59. Gordon S. Alternative activation of macrophages. Nat Rev Immunol. 2003;3(1):23–35.
60. Mantovani A, Sica A, Sozzani S, Allavena P, Vecchi A, Locati M. The chemokine system in diverse forms of macrophage activation and polarization. Trends Immunol. 2004;25(12):677–86.
61. McNally AK, DeFife KM, Anderson JM. Interleukin-4-induced macrophage fusion is prevented by inhibitors of mannose receptor activity. Am J Pathol. 1996;149(3):975–85.
62. Jones JA, Chang DT, Meyerson H, Colton E, Kwon IK, Matsuda T, Anderson JM. Proteomic analysis and quantification of cytokines and chemokines from biomaterial surface-adherent macrophages and foreign body giant cells. J Biomed Mater Res A. 2007;83(3):585–96.
63. Mantovani A, Sozzani S, Locati M, Allavena P, Sica A. Macrophage polarization: tumor-associated macrophages as a paradigm for polarized M2 mononuclear phagocytes. Trends Immunol. 2002;23(11):549–55.
64. Gristina AGG, Myrvik QN. Bacteria and biomaterials. In: Greco RS, editor. Implant biology. New Brunswick: CRC Press; 1994. p. 131–48.
65. Bos R, van der Mei HC, Busscher HJ. Physico-chemistry of initial microbial adhesive interactions–its mechanisms and methods for study. FEMS Microbiol Rev. 1999;23(2):179–230.
66. Dickinson GM, Bisno AL. Infections associated with indwelling devices: concepts of pathogenesis; infections associated with intravascular devices. Antimicrob Agents Chemother. 1989;33(5):597–601.

67. Herrmann M, Vaudaux PE, Pittet D, Auckenthaler R, Lew PD, Schumacher-Perdreau F, Peters G, Waldvogel FA. Fibronectin, fibrinogen, and laminin act as mediators of adherence of clinical staphylococcal isolates to foreign material. J Infect Dis. 1988;158(4):693–701.
68. Vaudaux P, Pittet D, Haeberli A, Huggler E, Nydegger UE, Lew DP, Waldvogel FA. Host factors selectively increase staphylococcal adherence on inserted catheters: a role for fibronectin and fibrinogen or fibrin. J Infect Dis. 1989;160(5):865–75.
69. Galliani S, Viot M, Cremieux A, Van der Auwera P. Early adhesion of bacteremic strains of Staphylococcus epidermidis to polystyrene: influence of hydrophobicity, slime production, plasma, albumin, fibrinogen, and fibronectin. J Lab Clin Med. 1994;123(5):685–92.
70. Pascual A, Fleer A, Westerdaal NA, Verhoef J. Modulation of adherence of coagulase-negative staphylococci to Teflon catheters in vitro. Eur J Clin Microbiol. 1986;5(5):518–22.
71. Patel JD, Iwasaki Y, Ishihara K, Anderson JM. Phospholipid polymer surfaces reduce bacteria and leukocyte adhesion under dynamic flow conditions. J Biomed Mater Res A. 2005;73(3):359–66.
72. Patel JD, Ebert M, Stokes K, Ward R, Anderson JM. Inhibition of bacterial and leukocyte adhesion under shear stress conditions by material surface chemistry. J Biomater Sci Polym Ed. 2003;14(3):279–95.
73. MacKintosh EE, Patel JD, Marchant RE, Anderson JM. Effects of biomaterial surface chemistry on the adhesion and biofilm formation of Staphylococcus epidermidis in vitro. J Biomed Mater Res A. 2006;78(4):836–42.
74. Wang IW, Anderson JM, Marchant RE. Staphylococcus epidermidis adhesion to hydrophobic biomedical polymer is mediated by platelets. J Infect Dis. 1993;167(2):329–36.
75. Sapatnekar S, Kieswetter KM, Merritt K, Anderson JM, Cahalan L, Verhoeven M, Hendriks M, Fouache B, Cahalan P. Blood-biomaterial interactions in a flow system in the presence of bacteria: effect of protein adsorption. J Biomed Mater Res. 1995;29(2):247–56.
76. Shive MS, Hasan SM, Anderson JM. Shear stress effects on bacterial adhesion, leukocyte adhesion, and leukocyte oxidative capacity on a polyetherurethane. J Biomed Mater Res. 1999;46(4):511–9.
77. Li DQ, Lundberg F, Ljungh A. Characterization of vitronectin-binding proteins of Staphylococcus epidermidis. Curr Microbiol. 2001;42(5):361–7.
78. Heilmann C, Hussain M, Peters G, Gotz F. Evidence for autolysin-mediated primary attachment of Staphylococcus epidermidis to a polystyrene surface. Mol Microbiol. 1997;24(5):1013–24.
79. Shinji H, Sakurada J, Seki K, Murai M, Masuda S. Different effects of fibronectin on the phagocytosis of Staphylococcus aureus and coagulase-negative staphylococci by murine peritoneal macrophages. Microbiol Immunol. 1998;42(12):851–61.
80. Minhas T, Ludlam HA, Wilks M, Tabaqchali S. Detection by PCR and analysis of the distribution of a fibronectin-binding protein gene (fbn) among staphylococcal isolates. J Med Microbiol. 1995;42(2):96–101.
81. Williams RJ, Henderson B, Sharp LJ, Nair SP. Identification of a fibronectin-binding protein from Staphylococcus epidermidis. Infect Immun. 2002;70(12):6805–10.
82. Hussain M, Wilcox MH, White PJ. The slime of coagulase-negative staphylococci: biochemistry and relation to adherence. FEMS Microbiol Rev. 1993;10(3–4):191–207.
83. Hussain M, Heilmann C, Peters G, Herrmann M. Teichoic acid enhances adhesion of Staphylococcus epidermidis to immobilized fibronectin. Microb Pathog. 2001;31(6):261–70.
84. Li DQ, Lundberg F, Ljungh A. Binding of von Willebrand factor by coagulase-negative staphylococci. J Med Microbiol. 2000;49(3):217–25.
85. Nilsson M, Frykberg L, Flock JI, Pei L, Lindberg M, Guss B. A fibrinogen-binding protein of Staphylococcus epidermidis. Infect Immun. 1998;66(6):2666–73.
86. Davis SL, Gurusiddappa S, McCrea KW, Perkins S, Hook M. SdrG, a fibrinogen-binding bacterial adhesin of the microbial surface components recognizing adhesive matrix molecules subfamily from Staphylococcus epidermidis, targets the thrombin cleavage site in the Bbeta chain. J Biol Chem. 2001;276(30):27799–805.

87. Hartford O, O'Brien L, Schofield K, Wells J, Foster TJ. The Fbe (SdrG) protein of Staphylococcus epidermidis HB promotes bacterial adherence to fibrinogen. Microbiology. 2001;147(Pt 9):2545–52.
88. Pei L, Palma M, Nilsson M, Guss B, Flock JI. Functional studies of a fibrinogen binding protein from Staphylococcus epidermidis. Infect Immun. 1999;67(9):4525–30.
89. Patel JD, Ebert M, Ward R, Anderson JM. S. epidermidis biofilm formation: effects of biomaterial surface chemistry and serum proteins. J Biomed Mater Res A. 2007;80(3):742–51.
90. Rupp ME, Fey PD, Heilmann C, Gotz F. Characterization of the importance of Staphylococcus epidermidis autolysin and polysaccharide intercellular adhesin in the pathogenesis of intravascular catheter-associated infection in a rat model. J Infect Dis. 2001;183(7):1038–42.
91. Veenstra GJ, Cremers FF, van Dijk H, Fleer A. Ultrastructural organization and regulation of a biomaterial adhesin of Staphylococcus epidermidis. J Bacteriol. 1996;178(2):537–41.
92. Patel JD, Colton E, Ebert M, Anderson JM. Gene expression during S. epidermidis biofilm formation on biomaterials. J Biomed Mater Res A. 2012. doi: 10.1002/jbm.w.34221.
93. Muller E, Takeda S, Shiro H, Goldmann D, Pier GB. Occurrence of capsular polysaccharide/adhesin among clinical isolates of coagulase-negative staphylococci. J Infect Dis. 1993;168(5):1211–8.
94. Heilmann C, Gerke C, Perdreau-Remington F, Gotz F. Characterization of Tn917 insertion mutants of Staphylococcus epidermidis affected in biofilm formation. Infect Immun. 1996;64(1):277–82.
95. Mack D, Haeder M, Siemssen N, Laufs R. Association of biofilm production of coagulase-negative staphylococci with expression of a specific polysaccharide intercellular adhesin. J Infect Dis. 1996;174(4):881–4.
96. Mack D, Nedelmann M, Krokotsch A, Schwarzkopf A, Heesemann J, Laufs R. Characterization of transposon mutants of biofilm-producing Staphylococcus epidermidis impaired in the accumulative phase of biofilm production: genetic identification of a hexosamine-containing polysaccharide intercellular adhesin. Infect Immun. 1994;62(8):3244–53.
97. Mack D, Siemssen N, Laufs R. Parallel induction by glucose of adherence and a polysaccharide antigen specific for plastic-adherent Staphylococcus epidermidis: evidence for functional relation to intercellular adhesion. Infect Immun. 1992;60(5):2048–57.
98. McCrea KW, Hartford O, Davis S, Eidhin DN, Lina G, Speziale P, Foster TJ, Hook M. The serine-aspartate repeat (Sdr) protein family in Staphylococcus epidermidis. Microbiology. 2000;146(Pt 7):1535–46.
99. Mack D, Fischer W, Krokotsch A, Leopold K, Hartmann R, Egge H, Laufs R. The intercellular adhesin involved in biofilm accumulation of Staphylococcus epidermidis is a linear beta-1, 6-linked glucosaminoglycan: purification and structural analysis. J Bacteriol. 1996;178(1):175–83.
100. Gerke C, Kraft A, Sussmuth R, Schweitzer O, Gotz F. Characterization of the N-acetylglucosaminyltransferase activity involved in the biosynthesis of the Staphylococcus epidermidis polysaccharide intercellular adhesin. J Biol Chem. 1998;273(29):18586–93.
101. Heilmann C, Schweitzer O, Gerke C, Vanittanakom N, Mack D, Gotz F. Molecular basis of intercellular adhesion in the biofilm-forming Staphylococcus epidermidis. Mol Microbiol. 1996;20(5):1083–91.
102. Vuong C, Kocianova S, Voyich JM, Yao Y, Fischer ER, DeLeo FR, Otto M. A crucial role for exopolysaccharide modification in bacterial biofilm formation, immune evasion, and virulence. J Biol Chem. 2004;279(52):54881–6.
103. Hussain M, Herrmann M, von Eiff C, Perdreau-Remington F, Peters G. A 140-kilodalton extracellular protein is essential for the accumulation of Staphylococcus epidermidis strains on surfaces. Infect Immun. 1997;65(2):519–24.
104. Rohde H, Burdelski C, Bartscht K, Hussain M, Buck F, Horstkotte MA, Knobloch JK, Heilmann C, Herrmann M, Mack D. Induction of Staphylococcus epidermidis biofilm formation via proteolytic processing of the accumulation-associated protein by staphylococcal and host proteases. Mol Microbiol. 2005;55(6):1883–95.
105. Desai NP, Hossainy SF, Hubbell JA. Surface-immobilized polyethylene oxide for bacterial repellence. Biomaterials. 1992;13(7):417–20.

106. Desai NP, Hubbell JA. Biological responses to polyethylene oxide modified polyethylene terephthalate surfaces. J Biomed Mater Res. 1991;25(7):829–43.
107. White KA, Gill RS, Lim F, Coviello S. (1996). In: Ratner B, et al. (ed) Surface Modification of Polymeric Biomaterials. Plenum Press, New York, 155.
108. Hayward JA, Chapman D. Biomembrane surfaces as models for polymer design: the potential for haemocompatibility. Biomaterials. 1984;5(3):135–42.
109. Iwasaki Y, Kurita K, Ishihara K, Nakabayashi N. Effect of reduced protein adsorption on platelet adhesion at the phospholipid polymer surfaces. J Biomater Sci Polym Ed. 1996;8(2):151–63.
110. Iwasaki Y, Mikami A, Kurita K, Yui N, Ishihara K, Nakabayashi N. Reduction of surface-induced platelet activation on phospholipid polymer. J Biomed Mater Res. 1997;36(4):508–15.
111. Sawada S, Sakaki S, Iwasaki Y, Nakabayashi N, Ishihara K. Suppression of the inflammatory response from adherent cells on phospholipid polymers. J Biomed Mater Res A. 2003;64(3):411–6.
112. Costerton JW, Lewandowski Z, Caldwell DE, Korber DR, Lappin-Scott HM. Microbial biofilms. Annu Rev Microbiol. 1995;49:711–45.
113. Kojima M, Ishihara K, Watanabe A, Nakabayashi N. Interaction between phospholipids and biocompatible polymers containing a phosphorylcholine moiety. Biomaterials. 1991;12(2):121–4.
114. Ishihara K, Oshida H, Endo Y, Ueda T, Watanabe A, Nakabayashi N. Hemocompatibility of human whole blood on polymers with a phospholipid polar group and its mechanism. J Biomed Mater Res. 1992;26(12):1543–52.
115. Ueda T, Watanabe A, Ishihara K, Nakabayashi N. Protein adsorption on biomedical polymers with a phosphorylcholine moiety adsorbed with phospholipid. J Biomater Sci Polym Ed. 1991;3(2):185–94.
116. Ishihara K, Ziats NP, Tierney BP, Nakabayashi N, Anderson JM. Protein adsorption from human plasma is reduced on phospholipid polymers. J Biomed Mater Res. 1991;25(11): 1397–407.
117. Rozalska B, Ljungh A. Biomaterial-associated staphylococcal peritoneal infections in a neutropaenic mouse model. FEMS Immunol Med Microbiol. 1995;11(4):307–19.
118. Hermann MPG. Catheter-associated infections caused by coagulase-negative staphylococci: clinical and biological aspects. In: Seifert H, Jansen B, Farr BM, editors. Catheter-related infections. New York: Marcel Dekker; 1997. p. 79–109.
119. Vuong C, Voyich JM, Fischer ER, Braughton KR, Whitney AR, DeLeo FR, Otto M. Polysaccharide intercellular adhesin (PIA) protects Staphylococcus epidermidis against major components of the human innate immune system. Cell Microbiol. 2004;6(3): 269–75.
120. Johnson GM, Lee DA, Regelmann WE, Gray ED, Peters G, Quie PG. Interference with granulocyte function by Staphylococcus epidermidis slime. Infect Immun. 1986;54(1): 13–20.
121. Dunne Jr WM. Bacterial adhesion: seen any good biofilms lately? Clin Microbiol Rev. 2002;15(2):155–66.
122. Sadovskaya I, Vinogradov E, Flahaut S, Kogan G, Jabbouri S. Extracellular carbohydrate-containing polymers of a model biofilm-producing strain, Staphylococcus epidermidis RP62A. Infect Immun. 2005;73(5):3007–17.
123. von Eiff C, Heilmann C, Peters G. New aspects in the molecular basis of polymer-associated infections due to staphylococci. Eur J Clin Microbiol Infect Dis. 1999;18(12):843–6.
124. Costerton JW, Stewart PS, Greenberg EP. Bacterial biofilms: a common cause of persistent infections. Science. 1999;284(5418):1318–22.
125. Khardori N, Yassien M, Wilson K. Tolerance of Staphylococcus epidermidis grown from indwelling vascular catheters to antimicrobial agents. J Ind Microbiol. 1995;15(3):148–51.
126. Stewart PS. Mechanisms of antibiotic resistance in bacterial biofilms. Int J Med Microbiol. 2002;292(2):107–13.

127. Henriques M, Cerca N, Azeredo J, Oliveira R. Influence of sub-inhibitory concentrations of antimicrobial agents on biofilm formation in indwelling medical devices. Int J Artif Organs. 2005;28(11):1181–5.
128. Shive MS, Brodbeck WG, Colton E, Anderson JM. Shear stress and material surface effects on adherent human monocyte apoptosis. J Biomed Mater Res. 2002;60(1):148–58.
129. Shive MS, Salloum ML, Anderson JM. Shear stress-induced apoptosis of adherent neutrophils: a mechanism for persistence of cardiovascular device infections. Proc Natl Acad Sci USA. 2000;97(12):6710–5.
130. Zhou Y, Doerschuk CM, Anderson JM, Marchant RE. Biomaterial surface-dependent neutrophil mobility. J Biomed Mater Res A. 2004;69(4):611–20.
131. Hallett MB. Controlling the molecular motor of neutrophil chemotaxis. Bioessays. 1997;19(7):615–21.
132. Lauffenburger DA, Horwitz AF. Cell migration: a physically integrated molecular process. Cell. 1996;84(3):359–69.
133. Palecek SP, Loftus JC, Ginsberg MH, Lauffenburger DA, Horwitz AF. Integrin-ligand binding properties govern cell migration speed through cell-substratum adhesiveness. Nature. 1997;385(6616):537–40.
134. Wessels D, Vawter-Hugart H, Murray J, Soll DR. Three-dimensional dynamics of pseudopod formation and the regulation of turning during the motility cycle of Dictyostelium. Cell Motil Cytoskeleton. 1994;27(1):1–12.
135. Slack SM, Horbett TA. The Vroman effect: a critical review. In: Horbett TA, Brash J, editors. Proteins at interfaces II: fundamentals and applications. Washington, DC: American Chemical Society; 1995. p. 112–28.
136. Edwards S. Biochemistry and physiology of the neutrophils. New York: Cambridge University Press; 1994.
137. Evans TG, Thai L, Granger DL, Hibbs Jr JB. Effect of in vivo inhibition of nitric oxide production in murine leishmaniasis. J Immunol. 1993;151(2):907–15.
138. Ochoa JB, Udekwu AO, Billiar TR, Curran RD, Cerra FB, Simmons RL, Peitzman AB. Nitrogen oxide levels in patients after trauma and during sepsis. Ann Surg. 1991;214(5): 621–6.
139. Stenger S, Donhauser N, Thuring H, Rollinghoff M, Bogdan C. Reactivation of latent leishmaniasis by inhibition of inducible nitric oxide synthase. J Exp Med. 1996;183(4):1501–14.
140. MacMicking JD, North RJ, LaCourse R, Mudgett JS, Shah SK, Nathan CF. Identification of nitric oxide synthase as a protective locus against tuberculosis. Proc Natl Acad Sci USA. 1997;94(10):5243–8.
141. Johnston Jr RB, Lehmeyer JE. Elaboration of toxic oxygen by-products by neutrophils in a model of immune complex disease. J Clin Invest. 1976;57(4):836–41.
142. Xia Y, Zweier JL. Superoxide and peroxynitrite generation from inducible nitric oxide synthase in macrophages. Proc Natl Acad Sci USA. 1997;94(13):6954–8.
143. Anstey NM, Weinberg JB, Hassanali MY, Mwaikambo ED, Manyenga D, Misukonis MA, Arnelle DR, Hollis D, McDonald MI, Granger DL. Nitric oxide in Tanzanian children with malaria: inverse relationship between malaria severity and nitric oxide production/nitric oxide synthase type 2 expression. J Exp Med. 1996;184(2):557–67.
144. Langrehr JM, Hoffman RA, Billiar TR, Lee KW, Schraut WH, Simmons RL. Nitric oxide production regulates alloactivation in rat splenocyte mixed lymphocyte cultures. Transplant Proc. 1991;23(1 Pt 1):183–4.
145. Stuehr DJ. Mammalian nitric oxide synthases. Biochim Biophys Acta. 1999;1411(2–3): 217–30.
146. Ou J, Carlos TM, Watkins SC, Saavedra JE, Keefer LK, Kim YM, Harbrecht BG, Billiar TR. Differential effects of nonselective nitric oxide synthase (NOS) and selective inducible NOS inhibition on hepatic necrosis, apoptosis, ICAM-1 expression, and neutrophil accumulation during endotoxemia. Nitric Oxide. 1997;1(5):404–16.
147. Ye YZ, Strong M, Huang ZQ, Beckman JS. Antibodies that recognize nitrotyrosine. Methods Enzymol. 1996;269:201–9.

148. Giridhar G, Myrvik QN, Gristina AG. Biomaterial-induced dysfunction in the capacity of rabbit alveolar macrophages to kill Staphylococcus epidermidis RP12. J Biomed Mater Res. 1995;29(10):1179–83.
149. Kaplan SS, Basford RE, Mora E, Jeong MH, Simmons RL. Biomaterial-induced alterations of neutrophil superoxide production. J Biomed Mater Res. 1992;26(8):1039–51.
150. Hoffstein ST, Gennaro DE, Manzi RM. Surface contact inhibits neutrophil superoxide generation induced by soluble stimuli. Lab Invest. 1985;52(5):515–22.
151. Patel JD, Krupka T, Anderson JM. iNOS-mediated generation of reactive oxygen and nitrogen species by biomaterial-adherent neutrophils. J Biomed Mater Res A. 2007;80(2):381–90.
152. Kong KF, Vuong C, Otto M. Staphylococcus quorum sensing in biofilm formation and infection. Int J Med Microbiol. 2006;296(2–3):133–9.
153. Otto M, Sussmuth R, Jung G, Gotz F. Structure of the pheromone peptide of the Staphylococcus epidermidis agr system. FEBS Lett. 1998;424(1–2):89–94.
154. Vuong C, Gotz F, Otto M. Construction and characterization of an agr deletion mutant of Staphylococcus epidermidis. Infect Immun. 2000;68(3):1048–53.
155. Otto M. Staphylococcus aureus and Staphylococcus epidermidis peptide pheromones produced by the accessory gene regulator agr system. Peptides. 2001;22(10):1603–8.
156. Otto M. Quorum-sensing control in Staphylococci – a target for antimicrobial drug therapy? FEMS Microbiol Lett. 2004;241(2):135–41.
157. Novick RP. Autoinduction and signal transduction in the regulation of staphylococcal virulence. Mol Microbiol. 2003;48(6):1429–49.
158. Vuong C, Saenz HL, Gotz F, Otto M. Impact of the agr quorum-sensing system on adherence to polystyrene in Staphylococcus aureus. J Infect Dis. 2000;182(6):1688–93.
159. Yarwood JM, Bartels DJ, Volper EM, Greenberg EP. Quorum sensing in Staphylococcus aureus biofilms. J Bacteriol. 2004;186(6):1838–50.
160. Yao Y, Sturdevant DE, Otto M. Genomewide analysis of gene expression in Staphylococcus epidermidis biofilms: insights into the pathophysiology of S. epidermidis biofilms and the role of phenol-soluble modulins in formation of biofilms. J Infect Dis. 2005;191(2):289–98.
161. Otto M, Echner H, Voelter W, Gotz F. Pheromone cross-inhibition between Staphylococcus aureus and Staphylococcus epidermidis. Infect Immun. 2001;69(3):1957–60.
162. Otto M, Sussmuth R, Vuong C, Jung G, Gotz F. Inhibition of virulence factor expression in Staphylococcus aureus by the Staphylococcus epidermidis agr pheromone and derivatives. FEBS Lett. 1999;450(3):257–62.
163. Xu L, Li H, Vuong C, Vadyvaloo V, Wang J, Yao Y, Otto M, Gao Q. Role of the luxS quorum-sensing system in biofilm formation and virulence of Staphylococcus epidermidis. Infect Immun. 2006;74(1):488–96.
164. Vuong C, Gerke C, Somerville GA, Fischer ER, Otto M. Quorum-sensing control of biofilm factors in Staphylococcus epidermidis. J Infect Dis. 2003;188(5):706–18.
165. Vuong C, Kocianova S, Yao Y, Carmody AB, Otto M. Increased colonization of indwelling medical devices by quorum-sensing mutants of Staphylococcus epidermidis in vivo. J Infect Dis. 2004;190(8):1498–505.
166. Yarwood JM, Schlievert PM. Quorum sensing in Staphylococcus infections. J Clin Invest. 2003;112(11):1620–5.
167. Yao Y, Vuong C, Kocianova S, Villaruz AE, Lai Y, Sturdevant DE, Otto M. Characterization of the Staphylococcus epidermidis accessory-gene regulator response: quorum-sensing regulation of resistance to human innate host defense. J Infect Dis. 2006;193(6):841–8.

Chapter 7
Dendritic Cell–Biomaterial Interactions: Implications for the Onset and Development of the Foreign Body Response

Antonio S. Sechi and Behnaz Shokouhi

Abstract Biomaterials are used in several clinical applications. Yet they often induce a strong immune response that can lead to implant malfunction and replacement. Thus, it is of crucial importance to deeply understand the biological response to biomaterials. Here, we focus on the molecular mechanisms underlying biomaterial–dendritic cell (DC) interactions. Biomaterials regulate DC adhesion via podosomes in a β2 integrin-dependent manner. Moreover, they primarily affect DC phenotype and function by impinging on multiple Toll-like receptor signaling pathways. By putting biomaterial–DC interactions (and their consequences) in the context of the foreign body response (FBR), we propose that DCs, whose function has been altered by biomaterials, could be engaged in multiple juxtacrine and paracrine interactions with other immune cells including macrophages and neutrophils. Through this complex intercellular network, DCs could affect the immune response at the implantation site initiating (or sustaining) the series of events leading to the FBR. The detailed knowledge of biomaterial–DC interactions could be exploited to design more inert biopolymers, thus minimizing the FBR or biomaterials that elicit controlled and specific immune reactions.

7.1 Introduction

The use of biomedical polymers (i.e., biomaterials) embraces several medical applications ranging from tissue regeneration and joint replacements to delivery systems for medicines and antigens. Regardless of the application, implanted

A.S. Sechi (✉) • B. Shokouhi
Department of Cell Biology, Institute of Biomedical Engineering, Universitätsklinikum Aachen, Rheinisch-Westfälische Technische Hochschule (RWTH), Pauwelsstrasse, 30, 52074 Aachen, Germany
e-mail: Antonio.Sechi@rwth-aachen.de; bshokouhi@ukaachen.de

T.F. Moriarty et al. (eds.), *Biomaterials Associated Infection: Immunological Aspects and Antimicrobial Strategies*, DOI 10.1007/978-1-4614-1031-7_7,

biomaterials induce in the host a complex biological reaction called foreign body response (FBR), which is characterized by local inflammation (acute and chronic) and the formation of a fibrotic capsule around the implant [1, 2]. This absolutely normal biological response, which is aimed at tissue regeneration and healing, may hinder implant's functionality causing, for instance, a glucose sensor to provide inexact measurements. Hence, in the context of some medical applications, there is a major need to develop less biologically reactive biopolymers.

Classical *in vitro* tests for assessing the biocompatibility of a novel biomaterial rely primarily on the use of standard cell lines (for example, L-929 mouse fibroblasts) and routinely determine the degree of cytotoxicity of a biomaterial or, in other words, the percentage of cells that die upon exposure to it. Although simple and highly reproducible, these tests do not take into consideration the different sensitivity of cells to biomaterials and that nontoxic biomaterials can nevertheless influence key cellular functions.

Even though biomaterials are not rejected in the strict immunological sense, the strong inflammatory and fibrotic reactions at the implantation site suggest that they severely alter the homeostasis of key components of the immune system. In this context, *in vitro* and *in vivo* studies have analyzed the influence of biomaterials on various immune cells. For instance, murine or human macrophages exposed to various biomedical polymers including polydimethylsiloxane (PDMS), expanded polytetrafluoroethylene (ePTFE), and alginate react by secreting pro-inflammatory cytokines such as interleukin (IL)-1β, IL-6, tumor necrosis factor (TNF)-α, and IL-12 [3–6]. These cytokines can also be induced by the stimulation of macrophages with titanium or cobalt–chrome surfaces [7–9]. Moreover, hydrophobic and anionic surfaces promote an anti-inflammatory response and induce apoptosis in macrophages [10, 11]. The correlation between biomaterial-induced macrophage reaction and FBR has not been addressed in these investigations, although in one of them it was shown that the formation of foreign body giant cells (FBGCs), a key feature of FBR, *in vivo* is promoted by cationic surfaces [11]. Finally, chemically and physically diverse biopolymers (for instance, PTFE, polyurethane, and Dacron) typically alter the oxidative function of neutrophils inducing in these cells the release of large amounts of superoxide [12–16].

These studies clearly show that nontoxic biomedical polymers can have a profound influence on the function of several immune cells. Thus, it is of crucial importance to understand how biomedical polymers affect their function. With this premise in mind, in this chapter we discuss how biomaterials alter the function of a key immune cell type, the dendritic cells. We then also discuss the implications of dendritic cell–biomaterial interactions for the onset and development of the FBR. Finally, we provide a perspective for how a detailed knowledge of dendritic cell–biomaterial interactions could be exploited to design either biopolymers that should not evoke any significant reaction from the host immune system or biomaterials that elicit controlled and specific immune reactions (immune or therapeutic responses).

7.2 Dendritic Cells: Initiators and Organizers of the Immune Response

7.2.1 Origin and Development of DCs

Originally discovered by Steinman and Cohn in the early 1970s [17], dendritic cells (DCs) play a fundamental role in the regulation of pathogen-specific adaptive immune responses, immunologic memory, and tolerance [18, 19].

DCs are uniquely located at the interface between the host and its environment. For instance, DCs are found on body surfaces such as skin, pharynx, and vagina as well as at some internal mucosal locations including the respiratory and gastrointestinal apparatuses. At all these locations, DCs extend numerous typical membrane processes, i.e., the dendrites (hence the origin of their name; Fig. 7.1a), which confer on these cells the ability to efficiently detect and capture a wide variety of antigens ranging from pathogenic components to self-antigens.

Current models for DC development propose that most of these cells originate from hematopoietic stem cell-derived progenitors with a restricted myeloid differentiation potential. These progenitors will differentiate into common DC precursors (CDPs), which will directly generate plasmacytoid DCs (pDCs) and common or conventional DCs (cDCs) through the intermediate precursor pre-cDCs [20, 21]. A notable exception to this developmental pathway is represented by another major DC type, the Langerhans cells (LCs). These cells develop from an embryonic precursor that, after colonizing the epidermis before birth, proliferates and differentiates at this anatomical location to generate the final LC network. Adult LCs have self-renewing capability and proliferate in response to inflammatory stimuli [22, 23]. It should be pointed out that at very discrete locations or under certain physiological conditions other less frequent DC subtypes have been described such as lamina propria DCs (lpDCs) [24] and TNF- and inducible nitric oxide synthase (iNOS)-producing DCs (TipDCs) [25].

The main DC subtypes differ not only in their development but also in some physiological properties. For instance, cDCs are generally short-lived and must be continuously replaced by blood-borne precursors, whereas pDCs are relatively long-lived. pDCs, which share cytological features with plasma cells including a round shape and a prominent endoplasmic reticulum, are specialized to counteract viral pathogens by producing massive amounts of type I interferons (INFs), such as INF-α. cDCs, which also produce cytokines following their encounter with pathogens, are highly migratory cells that rapidly move from peripheral tissues to T and B zones of lymph nodes [18, 19].

7.2.2 Phenotype and Function of DCs

Let us start by defining the two major phenotypic and functional DC stages, i.e., the immature (or inactive) and the mature (or active) stage. In the immature status, DCs

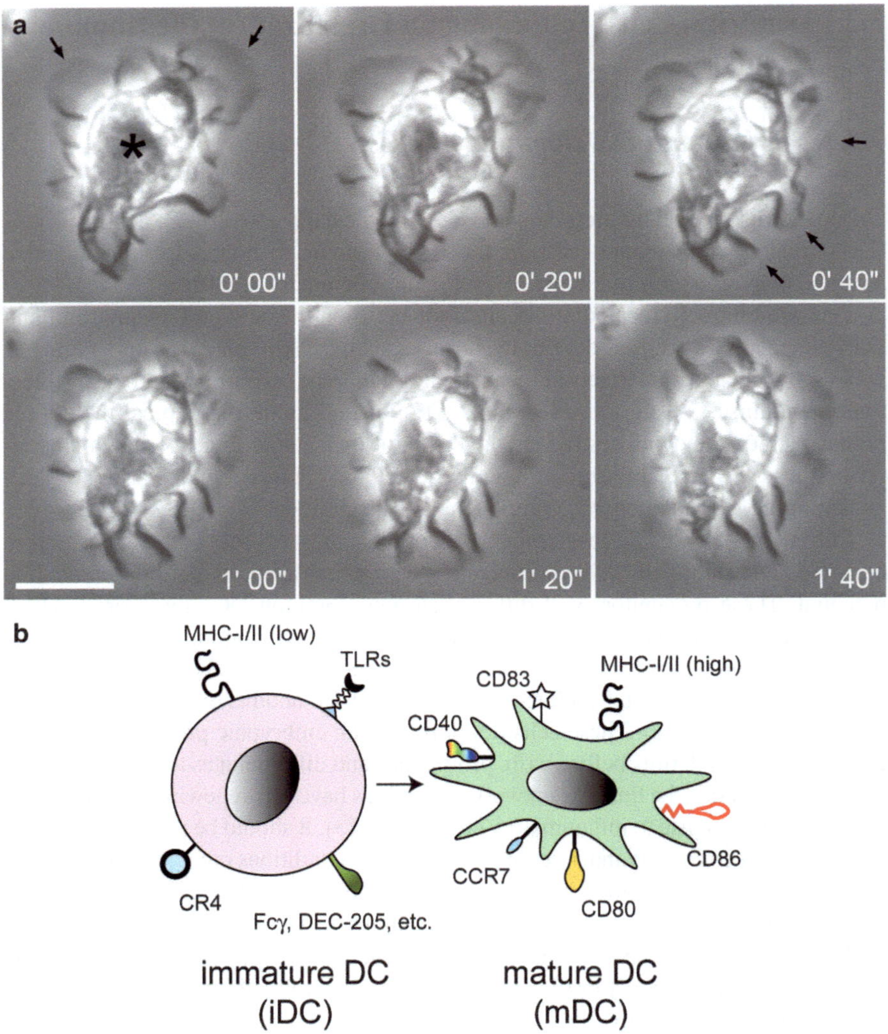

Fig. 7.1 DC morphological and phenotypical features. (**a**) Time-lapse imaging of a mouse DC plated on a glass coverslip. Numerous and active sheetlike membrane extensions formed around the cell periphery (*black arrows*). The *star* in the *upper left panel* indicates the dendritic cell body. *Scale bar* represents 5 μm. (**b**) Schematic representation of main features of immature and mature DC. Immature DCs (iDCs) express low levels of MHC-I and MHC-II and a large array of receptors for antigen detection and capture (TLRs, Fcγ, DEC-205). Following their interaction with pathogens, DCs acquire a mature phenotype (mDCs) characterized by high levels of MHC-I and MHC-II and the expression of co-stimulatory molecules such as CD40, CD80, CD86, and CD83 and the chemokine receptor CCR7

are optimally suited for antigen detection and capture through the expression of a large array of receptors including C-type lectin endocytic receptors (for instance, DEC-205), Fcγ receptors (bind to immune complexes and opsonized particles), and integrins (recognize apoptotic and necrotic cell fragments) [26]. Additional receptors required for antigen detection include proteins of the Toll-like receptor (TLR) family (see below). Immature DCs (iDCs) are also characterized by low levels of the major histocompatibility complex I and II (MHC-I, -II) and the expression of the complement receptor 4 (Fig. 7.1b). Moreover, iDCs may also express very little amounts of the molecules that they need to activate naïve T cells, namely, CD40, CD80, CD83, and CD86 (Fig. 7.1b).

Following their encounter with pathogens, DCs acquire a mature phenotype, which denotes the response of these cells to environmental stimuli allowing them to initiate an immune reaction tailored to the challenge at hand. Distinctive features of mature DCs (mDCs) are the loss or down regulation of the receptors required for antigen detection and capture and the up regulation of MHC-I and MCH-II molecules (Fig. 7.1b). Furthermore, mDCs express high levels of CD40, CD80, CD83, and CD86 and secret cytokines such as IL-12, IL-10, and IL-6. mDCs also express the receptor CCR7, which together with its ligand CCL21 robustly stimulates the migration of these cells [18, 19]. As it is discussed below, these phenotypic and functional features are fundamental for the ability of these cells to initiate strong $CD4^+$ or $CD8^+$ T-cell responses.

Two major facets of DC function are common to all DC subtypes: the recognition of pathogens and the regulation of T-cell response. DCs usually recognize invading pathogens by expressing a large array of receptors among which the TLRs play a major role. Most of the TLRs are located at the plasma membrane with just a few of them (TLR3, TLR7, TLR8, and TLR9) present on endosomal membranes. All TLRs sense a myriad of pathogen-associated molecular patterns (PAMPs). For instance, TLR1–TLR2 and TLR2–TLR6 heterodimers recognize lipoproteins and lipopeptides, whereas TLR4 recognizes another microbial wall component, the lipopolysaccharide (LPS). Bacterial flagellins are sensed only by TLR5. Endogenous and exogenous RNAs are sensed through TLR3, TLR7, and TLR8. Microbial DNAs are recognized by TLR9 [27, 28]. Regardless of these differences, the activation of TLR signaling pathways usually leads to NF-κB-dependent generation of regulatory and pro-inflammatory cytokines [27, 28].

Following pathogen detection, DCs rapidly mature and migrate to central immune structures such as the lymph nodes, where they search for, and interact with, antigen-specific $CD4^+$ or $CD8^+$ naïve T cells. As to the molecular mechanisms involved in DC-mediated T-cell activation, several studies have shown that optimal T-cell activation requires direct DC–T cell interactions and the concomitant presence of DC soluble factors (Fig. 7.2). DC–T cell association is usually achieved through a major molecular axis, which involves on the DC side MHC-I or MHC-II and on the T-cell side the T-cell receptor [29, 30]. This interaction provides the strongest stimulus (signal 1) to naïve T cells and its specificity is provided by the presence of the antigen associated with MHC-I or MHC-II molecules

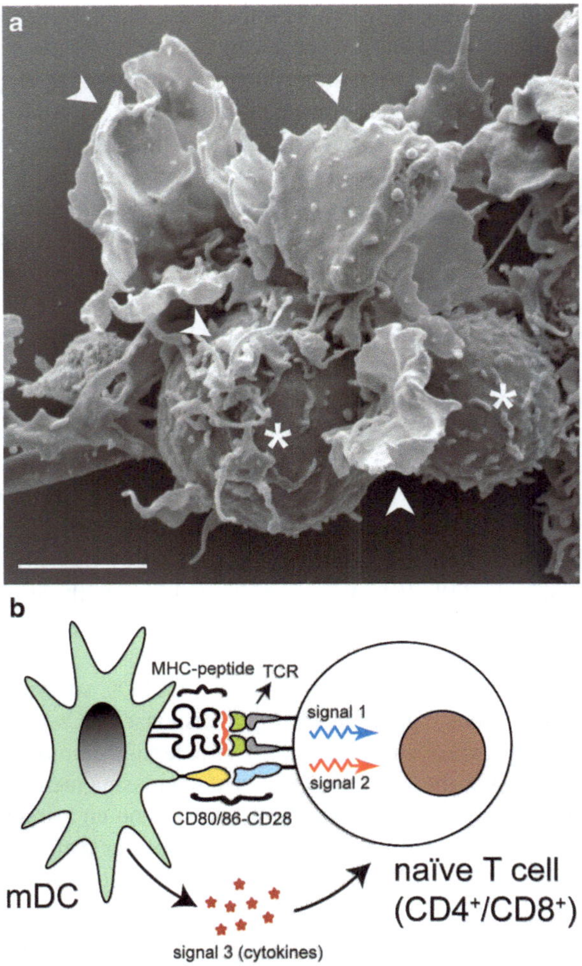

Fig. 7.2 DC-mediated T-cell activation. (**a**) Scanning electron microscopy image showing the interaction of a mouse DC with two CD4+ T cells (*white stars*). *Arrowheads* indicate the typical membrane processes generated by DCs to intimately interact with T cells. *Scale bar* represents 5 μm. (**b**) Schematic representation of the molecular events underlying DC-driven T-cell activation. DC–T cell association takes place at multiple levels involving main receptors (MHC and TCR, originating signal 1) and co-receptors (CD80/86 and CD28, originating signal 2) on both cell types. Optimal T-cell activation and differentiation also require the generation of cytokines (signal 3) by DCs

(Fig. 7.2b). The second axis of interaction involves ancillary molecules on both cell types: for instance, the co-receptors CD80 and CD86 on DCs and the co-receptor CD28 on T cells. These interactions provide a second signal that in combination with signal 1 robustly activates T cells (Fig. 7.2b). To be the most effective, signals 1 and 2 must be transduced by converging actin-dependent signaling pathways [31, 32].

Although signal 1 and signal 2 are sufficient to induce T-cell proliferation, T cells require a third DC-derived soluble signal (signal 3, Fig. 7.2b) to differentiate and acquire specific functional properties. DCs secreting IL-12, interferon (IFN)-γ, IL-23, and IL-27 can polarize T-cell response towards a T helper (Th) 1 phenotype. Th1 cells, which secrete IFN-γ and TNF-β, activate macrophages, thus inducing resistance against facultative and obligate intracellular pathogens, some mycobacteria, and protozoa [33, 34]. DCs bring about a Th2 phenotype, which is important for B cell activation and the clearance of helminth infections, through the secretion of IL-4 and monocyte chemoattractant protein (MCP)-1 and the expression of the transmembrane protein OX40L [33, 34]. Moreover, DCs can stimulate the differentiation of a recently described T-cell phenotype, the Th17 cells, which secrete IL-17 and are known to confer immune protection against some extracellular pathogens [35, 36]. Although the mechanisms by which DCs induce the differentiation of Th17 are still incompletely understood, it has been clearly demonstrated that transforming growth factor (TGF)-β and IL-6 are required to generate Th17 cells [35, 36].

DCs can guide T cells to become immunosuppressive (T regulatory) via IL-10 and TGF-β [33, 34]. DCs also induce in the T-cell clones an immunological memory that will persist for longer periods of time and will allow the immune system to rapidly respond to secondary antigen exposures. Finally, a functional class of DCs, the tolerogenic DCs, is crucial for the generation of immune tolerance through the deletion of self-reactive T-cell clones and the induction of a nonreactive state (i.e., anergy) in mature T cells [37, 38].

The crucial role of DCs in the regulation of the immune response is further highlighted by studies showing that these cells can activate natural killer (NK) cells [39, 40], a class of immune cells that have the ability to kill certain tumor cell lines. Moreover, it has been found that DCs can also regulate both the proliferation and differentiation of B cells, thus influencing the humoral response of the immune system [41, 42].

Overall, these studies show that DCs play a fundamental role in the regulation of the function of other immune cells and, as a consequence, in the development of an effective immune response.

7.3 Interaction of Dendritic Cells with Biomaterials

As already noted, biomaterials can affect the function of macrophages and neutrophils. Not surprisingly, given the crucial role of DCs in the regulation of the immune response, recent studies have begun to provide evidence for the influence of biomaterials on the function of these immune cells.

Much of the initial work has been focused on poly(lactic-*co*-glycolic acid) (PLGA), a polymer that is frequently used in antigen-delivery applications [43, 44]. Human and mouse DCs react similarly to PLGA by increasing the expression of MHC-II and the co-receptors CD40, CD80, and CD86. Similar to macrophages, a typical response of DCs to PLGA involved the secretion of the pro-inflammatory cytokines IL-6 and TNF-α [45–48]. Interestingly, mouse DCs show a comparable reaction to PLGA films and microparticles [48], whereas monocyte-derived human DCs seem to discriminate between these two forms of PLGA. In particular, human DCs are more reactive towards PLGA microparticles than films as judged by the levels of maturation markers and the magnitude of their T-cell stimulatory capability [45]. The highest reactivity of human DCs towards microparticles may be explained by the observation that these cells can uptake PLGA microparticles [49]. However, another study has shown that PLGA microparticles have no influence on human DCs in spite of being phagocytozed by these cells [50]. Moreover, mouse DCs can uptake PLGA microparticle equally well [51], but do not discriminate between the two forms of PLGA. Thus, regardless of the molecular mechanisms underlying the different DC responses to PLGA films or microparticles, it is clear that biomaterial geometry plays an important role in the interaction with DCs and could be exploited to control the biological response of these cells at implantation sites. The above studies have been complemented by findings showing that also alginate, agarose, chitosan, and titanium oxide can promote an active phenotype in human DCs [46, 47, 52].

The interaction of DCs with biomaterials not always leads to the activation of these cells. Mouse DCs exposed to silica, a component of scaffolds for bone tissue regeneration [53], respond less efficiently to the stimulation with LPS and other TLR ligands as indicated by the reduced expression of activation markers and inflammatory cytokines [54]. Put in the context of the silicosis, an occupational disease caused by the prolonged inhalation of silica particles, the impairment of DC function provides one possible explanation for the recurrent infections and deficient immune response in individuals affected by this disease. Silica is not the only biomaterial that has the ability of making DC hypo-responsive. We have recently found that mouse DCs exposed to poly-(D, L-lactide) (PDLLA), a component of drug delivery devices, do not up regulate MHC-II, CD40, CD80, and CD86 in response to LPS (our unpublished observations). Consistent with this finding, we found that PDLLA causes the least DC reaction among several biomaterials tested [55]. It would certainly be very important to determine how the ability of PDLLA to impair DC response correlates with the *in vivo* reaction to PDLLA-based implants.

The above studies were focused on just a few polymers making it difficult to comprehensively evaluate the influence of biomedical polymers on DC biology. Moreover, they did not provide any insight into the molecular mechanisms underlying DC–biomaterial interactions. To extend this knowledge, we have selected for our studies chemically and physically different biomedical polymers (degradable and nondegradable) and analyzed their effect on DCs in the context of wild-type and $TLR^{-/-}$ cellular systems. Our biomaterial library included Resomers (homopolymers of lactide or copolymers of lactide and glycolide), common materials like polytetrafluoroethylene (PTFE), and natural components such as alginate. The clinical applications of these biomaterials ranged from vascular grafts to tissue engineering [55].

We have found that some biomaterials (PDLLA, poly-dimethylsilosane (PDMS), and poly-ε-caprolactone (PCL) induce only a mild or no response in DCs as judged from the minimal (or absent) up regulation of both activation markers and pro-inflammatory cytokines [55]. On the other hand, the Resomers LT706 and RG503 and alginate behaved as potent stimulators of these cells inducing strong up regulation of activation markers and massive secretion of pro-inflammatory cytokines [55]. Interestingly, alginate, LT706, and RG503 also induced the secretion of the anti-inflammatory cytokine IL-10 [55]. To further determine to which extent DCs exposed to biomaterials were functionally active, we examined their capability to induce the proliferation of naïve $CD4^+$ T cells. Consistent with the above results, DCs exposed to alginate, LT706, and RG503 were capable to stimulate T-cell proliferation to levels similar to those induced by LPS-activated DCs. By contrast, DCs treated with PDLLA, PDMS, and PCL, which were less active in inducing the expression of surface markers and cytokines, had a lower T-cell stimulation potency [55].

Our attempts to define the biomaterial physicochemical properties responsible for the effects on DC phenotype and function have been, so far, fruitless. It should be mentioned that using simple self-assembled monolayers (SAM) of alkanethiols functionalized with COOH, NH_2, OH, or CH_3 groups, it was possible to determine that CH_3 groups (in spite of inducing high levels of IL-6 and TNF-α) were the least potent in causing human DC activation [56]. Interestingly, SAM of alkanethiols functionalized with CH_3 groups were found to induce a stronger fibrotic reaction and a more pronounced recruitment of inflammatory $CD11b^+$ (an integrin expressed on macrophages, dendritic cells, monocytes, and other immune cells) cells than SAM carrying OH or COOH functionalities [57]. By contrast, polypropylene microspheres functionalized with OH groups induce the strongest fibrotic reaction and recruitment of inflammatory $CD11b^+$ cells in mice, whereas microspheres with COOH groups cause the weakest effects [58]. The discrepancies between these two studies may be due to the different types and geometries of the biomaterials used. Nevertheless, it is clear that there is a correlation between biomaterial chemistry and both DC response and FBR magnitude. Future studies should better define this correlation by putting it in the context of, for instance, in situ changes of DC phenotype and function and more chemically complex biomaterials.

The actin cytoskeleton is important for the function of several immune cells [32, 59, 60]. Not surprisingly, DC activation is characterized by changes in actin cytoskeleton dynamics and cell adhesion [61, 62]. Particularly, DCs activated with LPS lose their typical adhesive structures, the podosomes, and acquire high motile properties [63, 64]. Consistent with these studies, we have found that biomaterials can also influence the morphology, number, and cellular distribution of podosomes in DCs (Fig. 7.3; [55]). For instance, podosomes in DCs plated on LT706 or RG503 (strong DC activators) were larger than control podosomes (non-stimulated DCs) [55]. Our findings are supported by recent observations showing that podosome-dependent DC adhesion to PLGA and the activation of these cells correlate with the expression of various types of integrins [65].

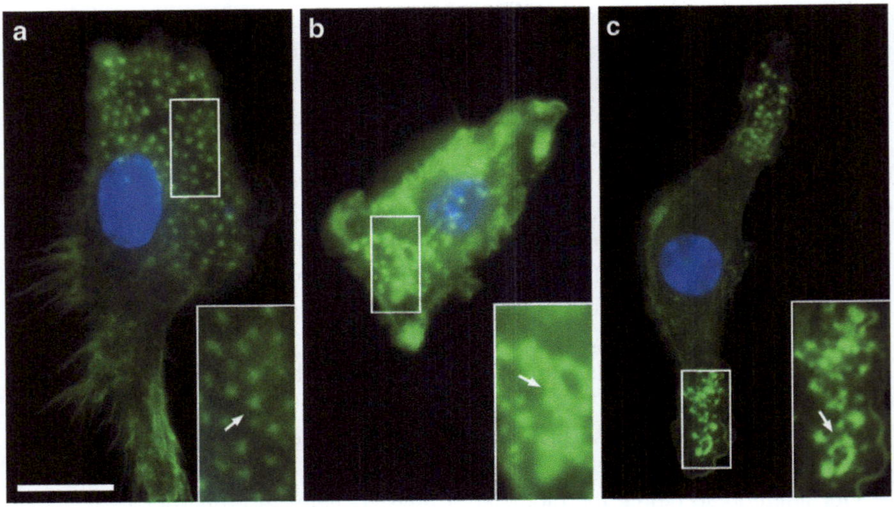

Fig. 7.3 Biomaterials alter podosome formation in DCs. Note the different size and morphology of podosomes formed on PLLA (**b**) and texin (**c**) as compared to the control (**a**). *Green*: actin labeling; *blue*: DNA labeling. *Scale bar*: 10 μm (5 μm for *insets*)

Although the above investigations have not directly addressed the molecular basis of DC–biomaterial interactions, the fact that biomaterial-induced DC activation resembled that induced by pathogens suggests that TLRs may mediate the interaction of these cells with biopolymers. We initially tested this hypothesis by examining the response of DCs lacking an essential component of TLR signaling pathways, the adapter protein MyD88 (Myeloid differentiation primary response gene 88). Regardless of the biomaterial at hand, MyD88$^{-/-}$ DCs did not respond to biomaterial exposure as judged by the inhibition of surface marker up regulation and cytokine secretion [55]. A partial exception to this behavior was represented by alginate, which still induced the secretion of RANTES (regulated upon activation, normal T cell expressed and secreted) and IL-10 in MyD88$^{-/-}$ DCs [55], suggesting that it acted also via MyD88-independent signaling pathways.

To determine which TLRs are involved in DC–biomaterial interactions, we used DCs lacking single TLRs. Precisely, deletion of TLR2, TLR4, or TLR6 impaired cytokine secretion and surface marker up regulation in DCs exposed to alginate, LT706, PDLLA, PET, and RG503 [55] (Fig. 7.4). Our findings are consistent with the observations that TLRs are involved in the reaction of macrophages to different biomaterials [66–68] and suggest that immune cells use TLR/MyD88-dependent signaling pathways as a general mechanism to sense and react to biomaterials. In this context, very recent data suggest that DCs may interact with biomaterials, primarily PLGA, also through integrin-dependent signaling pathways [65]. Finally, we found that alginate, LT706, PDLLA, and RG503 influence DC function via a combination of direct cell–biomaterial interactions and release

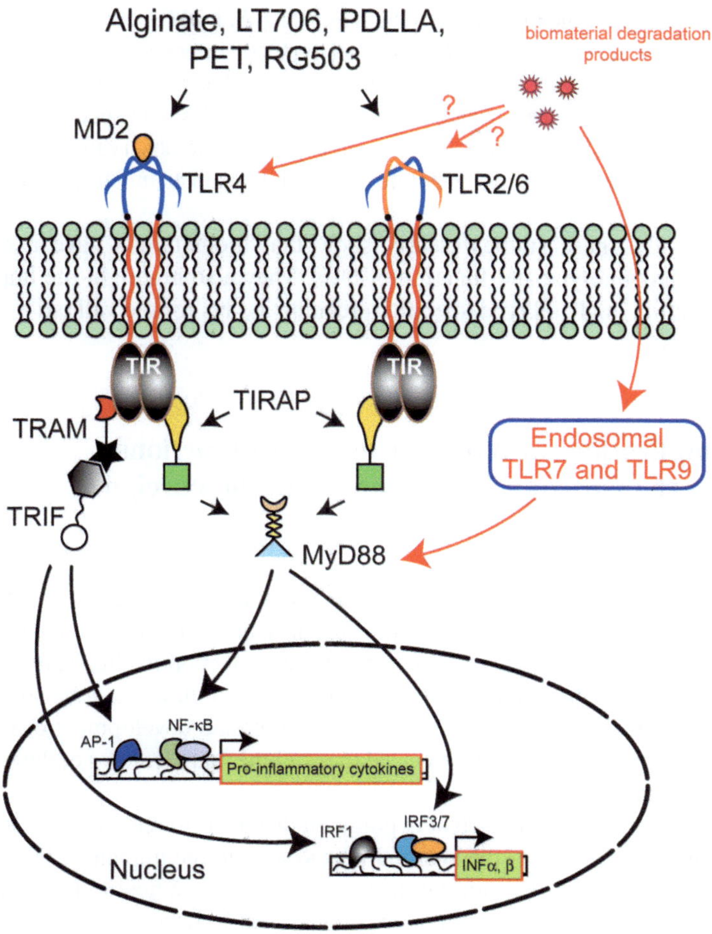

Fig. 7.4 Molecular mechanisms underlying the alteration of DC function by biomaterials. Alginate, LT706, PDLLA, PET, and RG503 activate some TLR-dependent signaling pathways that induce the generation of pro-inflammatory cytokines (via NF-κB) and interferon-α and -β (via IRF3 and IRF7). In addition, in the case of degradable biomaterials, their degradation products can also trigger DC response via endosomal TLR7 and TLR9. Degradation products may also be recognized by plasma membrane TLRs (*red arrows labeled with question marks*). *TIRAP* toll-interleukin 1 receptor (TIR) domain containing adaptor protein, *TRIF* TIR-domain-containing adapter-inducing interferon-β, *TRAM* TRIF-related adaptor molecule, *TIR Toll* interleukin-1 receptor-like domain, *IRF* interferon regulatory factor, *NF-κB* nuclear factor κB

of soluble polymer molecules. Accordingly, we have found that the endosomal TLR7 and TLR9 are involved in the recognition of these biopolymers ([55] and our unpublished data).

As to the nature of biomaterial–TLR interactions, it is known that TLRs can sense the hydrophobic moiety of biological molecules as alarm signals [69]. Thus, at least for hydrophobic biopolymers such as PDLLA, PDMS, and PET, this may be the mechanism at hand. We think that other types of interactions are also possible including the binding of TLRs to the layer of denatured host proteins that is adsorbed on the biomaterial surface or the direct interaction of TLRs with specific biomaterial chemical groups. This last possibility is supported by the observation that the host immune system is capable to generate antibodies against implanted biomaterials [70–73].

Collectively, these studies clearly show that biomaterials have a profound influence on phenotype and function of DCs. They also highlight the importance of cell function- and cell type-specific screening assays for the validation of novel biopolymers.

7.4 Implications of Biomaterial–DC Interactions for the Onset and Development of the Foreign Body Response

In the previous sections, we have discussed about the remarkable role that DCs play in the regulation of the immune response. We have also seen that biomaterials often cause inflammatory and fibrotic reactions at the implantation site. Thus, given the profound influence that biomaterials exert on DC function, would it be possible that biomaterial–DC interactions contribute to the onset and development of the FBR? Before elaborating on this issue, we need to recall the main features of the host reaction to implanted biomaterials.

At the beginning, it is important to point out that the host reaction to biomaterials is not a specific reaction of the host immune system to the implanted biopolymers, but it simply represents the normal healing reaction of the host to tissue injury. In this context, biomaterials can act as an impediment to normal tissue regeneration and healing. The first stage of the FBR consists of an acute inflammatory reaction usually of short duration. During this phase, neutrophils and macrophages infiltrate the implantation site, where they try to uptake and degrade the biomaterial [1, 2]. This acute phase is followed by a chronic inflammation characterized by the presence of many cell types (macrophages, monocytes, lymphocytes, etc.) and the initial formation of blood vessels and connective tissue. If no infectious event occurs, chronic inflammation is replaced by the formation of a granulation tissue in which fibroblasts and endothelial cells proliferate to regenerate the injured tissue and form new blood vessels, respectively [1, 2]. The granulation tissue usually develops only when tissue architecture is grossly altered and cannot be restored to its normal function. Examples include the loss of large tissue portions and the presence of foreign bodies.

In the absence of a foreign body, the normal healing process completely restores the damaged tissue. However, the presence of the implanted biomaterials triggers the formation of FBGCs and the encapsulation of the biomaterials

[1, 2]. FBGCs, the hallmark of the FBR, are formed by the fusion of macrophages in an attempt to phagocytose the much bigger (as compared to the macrophages) biomaterials. The function of FBGCs is still incompletely understood, although from the observation that biomaterial surface cracking appears just underneath FBGCs it was suggested that these cells mediate biomaterial degradation [74]. Conversely, various mechanisms have been proposed to explain FBGC formation. For instance, it has been established that both IL-4 and IL-13 play a role in macrophage fusion [75–77]. Moreover, they up regulate mannose receptors, which localize at fusion interfaces [77] and are essential for macrophage fusion [78]. Additional studies have also implicated CD44 and CD47 [79, 80] and the dendritic cell-specific transmembrane protein (DC-STAMP) in this process [81]. In the latter case, FBGC formation by macrophage fusion was completely inhibited in DC-STAMP$^{-/-}$ mice [81].

Knowledge of the involvement of DCs in the above events is limited and fragmentary. In a sheep model, it has been shown that S-100^{+} cells localize at perigraft regions suggesting the presence of DCs at these sites [82]. However, because the S-100 protein is not exclusively expressed in DCs (it can be found in melanocytes, adipocytes, and macrophages), it cannot be concluded with absolute certainty that DCs localize at implantation sites. More compelling evidence was obtained through the analysis of fibrous capsules surrounding silicone breast implants. Precisely, cells positive for CD208 (dendritic cell-lysosomal associated membrane protein; DC-LAMP) and CD209 (dendritic cell-specific intercellular adhesion molecule-3-grabbing non-integrin; DC-SIGN) were predominantly found in the zone adjacent to the implant. At this location, these cells co-localized with CD4^{+} T cells [83]. Notably, DC-LAMP is expressed only in active DCs suggesting that these cells can receive maturation stimuli (perhaps from the biomaterial) at implantation sites. Furthermore, in a rat model for granulomatous FBR (based on the subcutaneous implantation of polyvinyl sponges), cells displaying DC morphological and phenotypic features were found in a fraction of the cellular population isolated from inflammatory exudates. Interestingly, some of these cells were positive for the rat pDC markers His24 and His48 [84].

What could be, then, the role of DCs in the host reaction to biomaterials? Based on the known functions of the cytokines and chemokines secreted by these cells, we hypothesize that biomaterial-stimulated DCs attract various immune cells including T cells through the secretion of RANTES. Moreover, by generating TNF-α and IL-6, DCs could recruit macrophages and, in combination with IL-12, also enhance their phagocytic potential (Fig. 7.5). The secretion of TNF-α alone may be sufficient for the recruitment and activation of neutrophils at the implantation site. Since DCs can reduce apoptosis in neutrophils [85, 86], it is also conceivable that DCs could promote the survival of these cells at the implantation site, thus favoring or prolonging the action of neutrophils at this location.

Another possible scenario involves TNF-α and IL-1β. Through these two cytokines, which are known to induce the expression of adhesion molecules (VCAM-1, ICAM-1) on endothelial cells [87, 88], DCs could indirectly facilitate

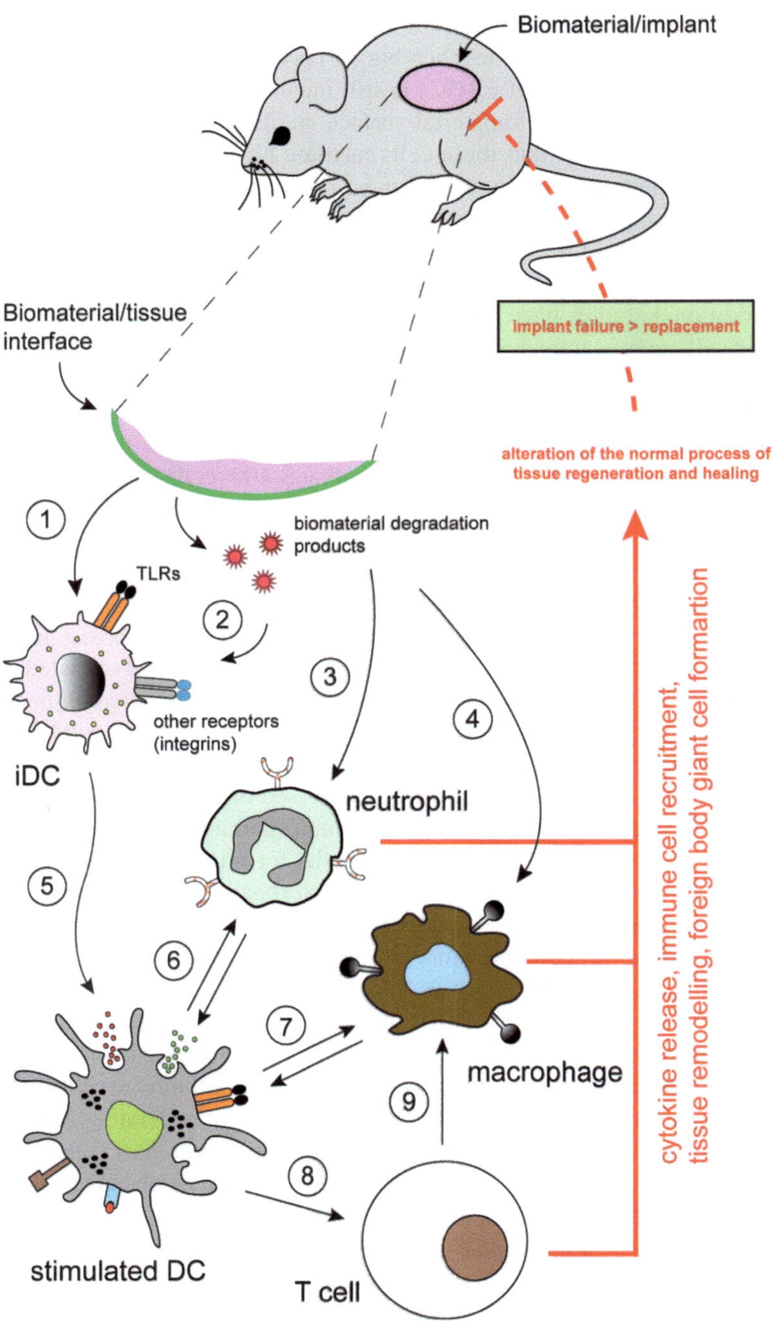

Fig. 7.5 Biomaterial–DC interactions and their potential role in the development of the foreign body reaction. The direct interaction of DCs with biomaterials (*1*) or biomaterial degradation products (*2*) mostly involves TLR signaling pathways and leads to the alteration of DC phenotype, function, and podosome formation and distribution (*5*; *green circles* in iDC: normal podosomes; *black ellipses*

the extravasation and recruitment of leukocytes. From the wound healing point of view, the growth factor-like activity of IL-1β [89] could support the proliferation of cells (i.e., fibroblasts and vascular endothelial cells) that characterizes the formation of the granulation tissue around the implant. As to the FBGC formation, DCs may contribute to this process by expressing DC-STAMP, a protein that is essential for macrophage fusion and FBGC formation (see above). Whether DCs fuse with macrophages to form FBGCs and the molecular basis of this process remain to be investigated. Furthermore, DCs could activate T cells, which in turn express IL-4 and IL-13, two cytokines that mediate macrophage fusion and FBGC formation (see above).

Final mention should be made of the possibility that immune cells at the implantation site affect DC function (Fig. 7.5). For instance, neutrophils can induce maturation of DCs via direct contact and secretion of TNF-α [85, 86], whereas macrophages can modulate DC function by producing IL-10 [90]. Since the functions of neutrophils and macrophages are also affected by biomaterials (see above), we speculate that the biomaterial-induced functional alterations of these immune cells also contribute to the changes caused by biomaterials on DCs at the implantation site.

7.5 Conclusions and Perspectives

Here, we have seen that biomaterials have a profound influence on DC function inducing these cells to secrete pro- and anti-inflammatory cytokines and to activate naïve T cells. We have also put TLR-dependent biomaterial–DC interactions (and their consequences) in the context of the FBR providing possible scenarios for the role of DCs in this process. We hypothesized that DCs, whose function has been altered by biomaterials, might be engaged in multiple juxtacrine (i.e., direct) and paracrine (i.e., indirect) interactions with other immune cells including macrophages, T cells, and neutrophils. Through this complex intercellular network, DCs might affect the functions of other immune cells (and vice versa) at the implantation

Fig. 7.5 (continued) in stimulated DC: abnormal podosomes). Alternatively, direct DC–biomaterial interactions may be mediated by integrin receptors. These "altered" DCs could then directly or indirectly interact with macrophages and neutrophils influencing their functions (*6, 7*). They could also affect the function of T cells (*8*), which then could influence macrophage function (*9*). The functions of macrophages and neutrophils could also be directly affected by the biomaterials (*3, 4*) and biomaterial-exposed macrophages and neutrophils could then, in turn, influence DC function. The combination of these intercellular interactions may determine a profound alteration of the immune system's homeostasis at the site of implantation leading to the alteration of the normal process of tissue regeneration and healing. The formation of both foreign body giant cells and a fibrous encapsulation (not depicted) may cause implant malfunctioning, thus requiring its replacement

site via a series of events (cytokine production, cell recruitment, etc.) finally leading to the FBR.

Ideal biomaterials should support or replace specific physiological or biological functions possibly causing minimal changes in the homeostasis of the host immune system. Yet, they often induce adverse host reactions calling for a deeper understanding of the biological response to biomaterials. As the discovery of a true "stealth" biomaterial still remains a remote, perhaps unreachable, goal, efforts should unquestionably be directed towards minimizing the above negative effects. Given our current knowledge of the molecular basis underlying biomaterial–DC interactions, it is conceivable that reducing the stimulation of TLR/MyD88 signaling pathways at the implantation site could result in a minimal FBR. Moreover, as also macrophages react to biomaterials through TLRs, this experimental approach could effectively reduce the FBR. To achieve this, one possible way of intervention could be the identification of physical and chemical features that are responsible for altering DC function, thus allowing the design of biomaterials in which these features are absent or modified to obtain a more inert behavior towards immune cells. This idea is supported by recent studies showing that DCs respond differently to changes in the geometry or chemistry of a polymer [56, 91]. Along this line of reasoning, it could be possible to fine-tune or program DC function to reduce, for instance, the reactivity of other immune cells at the implantation site. For instance, by inducing DC to secrete less soluble factors for macrophage and neutrophil recruitment, it may be possible to minimize the immune reaction against biomaterials. It should also be mentioned that, from a different perspective, biomaterial–DC interactions could be exploited to elicit controlled and specific reactions (immune or therapeutic responses). In this case, biomaterials could be engineered to trigger individual or multiple signaling pathways aiming at activating specific DC functional properties [92].

In order to achieve these feats, future investigations should be centered on the analysis of DC recruitment and functional status at the implantation site and on a more detailed analysis of the molecular mechanisms that underlie biomaterial–DC interactions. In vivo experiments in which DCs are selectively depleted should help to precisely define the role of these cells in the onset and development of the FBR.

Acknowledgments This work was supported by the Interdisciplinary Center for Clinical Research (IZKF) of the Medical Faculty of RWTH Aachen University.

Glossary

Adaptive (or acquired) immune response The process by which the immune system responds to an infection or vaccination by producing specific antibodies towards foreign antigens.

Antigen presentation The process by which dendritic cells (and, to a minor extent, macrophages and B cells) capture antigens and display them on their surface, thus enabling recognition by T cells.

Apoptosis A type of specialized cell death by which cells kill themselves. It plays an essential role in many processes such as tissue remodeling during embryonic development.

Biocompatibility The property of a liquid or a solid substance of being nontoxic and non-dangerous to a living cell or a tissue.

Chemokines Proteins that activate and stimulate the directional migration of immune cells.

Cytokines Soluble proteins that are secreted by immune cells and that serve to regulate the immune response.

Cytotoxicity The attribute of a substance of being toxic to living cells.

Foreign body giant cell A multinucleated cell that is generated by fusion of macrophages in response to the presence of a large foreign body.

Immunological memory The ability of the immune system to remember a specific antigen and respond rapidly and vigorously to subsequent encounters with the same antigen.

Immunological tolerance The non-reactivity of the immune system to specific antigens, primarily self-antigens.

Inflammatory response One of the earliest reactions of the immune system against tissue damage caused by pathogens, trauma, or toxins. It is characterized by pain, localized heat, skin redness, and swelling.

Interleukins A class of proteins that act to stimulate and regulate the function of several immune cell types.

Major histocompatibility complex Cell membrane antigens that are the key determinant of tissue type and transplant compatibility.

Naïve T cells A T cell that has not yet encountered its cognate antigen.

Pathogen-associated molecular patterns Molecules associated with pathogens that are recognized by TLRs or other pattern recognition receptors.

Phenotype In a cell, it represents the morphological and biochemical features as determined by its gene expression profile and interactions with the environment.

Podosome A highly dynamic, actin-rich, ringlike cellular structure frequently formed by dendritic cells and macrophages that is essential for the adhesion and motility of these cells.

Toll-like receptors A class of surface or endosomal proteins expressed by immune cells that recognize pathogen-associated molecules.

7.6 Appendix

See Table 7.1.

Table 7.1 Summary of the effect of biomaterials on human and mouse dendritic cells

Biomaterial	Biomaterial shape	DC species	CD40	CD80	CD86	MHC-II	IL-1β	IL-6	IL-10	IL-12	TNF-α	TLRs	References
Alginate	Film	Mouse	n.c.	Up	Up	Up	Up	Up	Up	Up	n.c.	2, 4, 6	[55]
Alginate	Film	Human	Down	n.d.	Down	Down	n.d.	n.d.	n.d.	n.d.	n.d.	n.d.	[46]
Agarose	Film	Human	Up	n.d.	Up	Up	n.d.	n.d.	n.d.	n.d.	n.d.	n.d.	[48]
Chitosan	Film	Human	Up	n.d.	Up	Up	n.d.	n.d.	n.d.	n.d.	n.d.	n.d.	[47]
Hyalur. acid	Film	Human	Down	n.d.	Down	Down	n.d.	n.d.	n.d.	n.d.	n.d.	n.d.	[48]
LR704	Film	Mouse	n.c.	n.c.	n.c.	n.c.	n.d.	n.d.	n.d.	n.d.	n.d.	n.d.	[55]
LT706	Film	Mouse	n.c.	up	up	up	up	up	up	up	up	2, 4, 6, 7, 9	[55]
PCL	Film	Mouse	n.c.	Up	Up	n.c.	Up	n.c.	n.c.	Up	Up	n.d.	[55]
PDLLA	Film	Mouse	n.c	n.c.	n.c.	n.c.	Up	Up	n.c.	n.c.	Up	4, 7, 9	[55]
PDMS	Film	Mouse	n.c.	n.c.	n.c.	n.c.	Up	Up	n.c.	Up	n.c.	n.d.	[55]
PET	Film	Mouse	n.c.	n.c.	n.c.	n.c.	Up	n.c.	n.c.	n.c.	n.c.	4	[55]
PLLA	Film	Mouse	n.c.	Up	n.c.	n.c.	Up	Up	n.c.	Up	Up	n.d.	[55]
PLGA	Particle	Mouse	n.d.	Up	Up	n.d.	n.d.	Up	n.d.	n.d.	Up	n.d.	[48]
PLGA	Film/particle	Human	Up	Up	Up	Up	n.d.	n.d.	n.d.	n.d.	n.d.	n.d.	[45]
PTFE	Film	Mouse	n.c.	Up	Up	n.c.	Up	Up	n.c.	Up	Up	n.d.	[55]
PVDF	Film	Mouse	n.c.	n.c.	n.c.	n.c.	Up	Up	n.c.	n.c.	n.c.	n.d.	[55]
RG503	Film	Mouse	n.c.	Up	Up	n.c.	Up	Up	Up	Up	Up	2, 4, 6, 7, 9	[55]
Texin	Film	Mouse	n.c.	Up	Up	Up	Up	n.c.	n.c.	Up	n.c.	2, 4, 6	[55]

n.c. no significant change, *n.d.* not determined, *Up* up regulated, *Down* down regulated

References

1. Anderson JM, Rodriguez A, Chang DT. Foreign body reaction to biomaterials. Semin Immunol. 2008;20(2):86–100.
2. Kenneth WW. A review of the foreign-body response to subcutaneously-implanted devices: the role of macrophages and cytokines in biofouling and fibrosis. J Diabetes Sci Technol. 2008; 2(5):768–77.
3. Bonfield TL, Colton E, Marchant RE, Anderson JM. Cytokine and growth factor production by monocytes/macrophages on protein preadsorbed polymers. J Biomed Mater Res. 1992;26(7): 837–50.
4. Anderson JM, Ziats NP, Azeez A, Brunstedt MR, Stack S, Bonfield TL. Protein adsorption and macrophage activation on polydimethylsiloxane and silicone rubber. J Biomater Sci Polym Ed. 1995;7(2):159–69.
5. Jones JA, Dadsetan M, Collier TO, Ebert M, Stokes KS, Ward RS, et al. Macrophage behavior on surface-modified polyurethanes. J Biomater Sci Polym Ed. 2004;15(5):567–84.
6. Yang D, Jones KS. Effect of alginate on innate immune activation of macrophages. J Biomed Mater Res A. 2009;90(2):411–8.
7. Sethi RK, Neavyn MJ, Rubash HE, Shanbhag AS. Macrophage response to cross-linked and conventional UHMWPE. Biomaterials. 2003;24(15):2561–73.
8. Refai AK, Textor M, Brunette DM, Waterfield JD. Effect of titanium surface topography on macrophage activation and secretion of proinflammatory cytokines and chemokines. J Biomed Mater Res A. 2004;70(2):194–205.
9. Li Y, Schutte RJ, Abu-Shakra A, Reichert WM. Protein array method for assessing *in vitro* biomaterial-induced cytokine expression. Biomaterials. 2005;26(10):1081–5.
10. Brodbeck WG, Nakayama Y, Matsuda T, Colton E, Ziats NP, Anderson JM. Biomaterial surface chemistry dictates adherent monocyte/macrophage cytokine expression *in vitro*. Cytokine. 2002;18(6):311–9.
11. Brodbeck WG, Patel J, Voskerician G, Christenson E, Shive MS, Nakayama Y, et al. Biomaterial adherent macrophage apoptosis is increased by hydrophilic and anionic substrates *in vivo*. Proc Natl Acad Sci U S A. 2002;99(16):10287–92.
12. Kaplan SS, Basford RE, Jeong MH, Simmons RL. Mechanisms of biomaterial-induced superoxide release by neutrophils. J Biomed Mater Res. 1994;28(3):377–86.
13. Kaplan SS, Basford RE, Jeong MH, Simmons RL. Biomaterial-neutrophil interactions: dysregulation of oxidative functions of fresh neutrophils induced by prior neutrophil-biomaterial interaction. J Biomed Mater Res. 1996;30(1):67–75.
14. Moore MA, Kaplan DS, Picciolo GL, Wallis RR, Kowolik MJ. Effect of cellulose acetate materials on the oxidative burst of human neutrophils. J Biomed Mater Res. 2001;55(3):257–65.
15. Patel JD, Krupka T, Anderson JM. iNOS-mediated generation of reactive oxygen and nitrogen species by biomaterial-adherent neutrophils. J Biomed Mater Res A. 2007;80(2):381–90.
16. Santos TC, Marques AP, Silva SS, Oliveira JM, Mano JF, Castro AG, et al. *In vitro* evaluation of the behaviour of human polymorphonuclear neutrophils in direct contact with chitosan-based membranes. J Biotechnol. 2007;132(2):218–26.
17. Steinman RM, Cohn ZA. Identification of a novel cell type in peripheral lymphoid organs of mice. I. Morphology, quantitation, tissue distribution. J Exp Med. 1973;137(5):1142–62.
18. Banchereau J, Steinman RM. Dendritic cells and the control of immunity. Nature. 1998;392(6673):245–52.
19. Mellman I, Steinman RM. Dendritic cells: specialized and regulated antigen processing machines. Cell. 2001;106(3):255–8.
20. Auffray C, Sieweke MH, Geissmann F. Blood monocytes: development, heterogeneity, and relationship with dendritic cells. Annu Rev Immunol. 2009;27:669–92.
21. Naik SH, Sathe P, Park HY, Metcalf D, Proietto AI, Dakic A, et al. Development of plasmacytoid and conventional dendritic cell subtypes from single precursor cells derived *in vitro* and *in vivo*. Nat Immunol. 2007;8(11):1217–26.

22. Chorro L, Sarde A, Li M, Woollard KJ, Chambon P, Malissen B, et al. Langerhans cell (LC) proliferation mediates neonatal development, homeostasis, and inflammation-associated expansion of the epidermal LC network. J Exp Med. 2009;206(13):3089–100.
23. Merad M, Ginhoux F, Collin M. Origin, homeostasis and function of Langerhans cells and other langerin-expressing dendritic cells. Nat Rev Immunol. 2008;8(12):935–47.
24. Bogunovic M, Ginhoux F, Helft J, Shang L, Hashimoto D, Greter M, et al. Origin of the lamina propria dendritic cell network. Immunity. 2009;31(3):513–25.
25. Serbina NV, Salazar-Mather TP, Biron CA, Kuziel WA, Pamer EG. TNF/iNOS-producing dendritic cells mediate innate immune defense against bacterial infection. Immunity. 2003; 19(1):59–70.
26. Banchereau J, Briere F, Caux C, Davoust J, Lebecque S, Liu YJ, et al. Immunobiology of dendritic cells. Annu Rev Immunol. 2000;18:767–811.
27. Takeuchi O, Akira S. Signaling pathways activated by microorganisms. Curr Opin Cell Biol. 2007;19(2):185–91.
28. Kawai T, Akira S. The roles of TLRs, RLRs and NLRs in pathogen recognition. Int Immunol. 2009;21(4):317–37.
29. Huppa JB, Davis MM. T-cell-antigen recognition and the immunological synapse. Nat Rev Immunol. 2003;3(12):973–83.
30. Dustin ML. Insights into function of the immunological synapse from studies with supported planar bilayers. Curr Top Microbiol Immunol. 2010;340:1–24.
31. Sechi AS, Buer J, Wehland J, Probst-Kepper M. Changes in actin dynamics at the T-cell/APC interface: implications for T-cell anergy? Immunol Rev. 2002;189:98–110.
32. Sechi AS, Wehland J. Interplay between TCR signalling and actin cytoskeleton dynamics. Trends Immunol. 2004;25(5):257–65.
33. Kapsenberg ML. Dendritic-cell control of pathogen-driven T-cell polarization. Nat Rev Immunol. 2003;3(12):984–93.
34. de Jong EC, Smits HH, Kapsenberg ML. Dendritic cell-mediated T cell polarization. Springer Semin Immunopathol. 2005;26(3):289–307.
35. Ghilardi N, Ouyang W. Targeting the development and effector functions of TH17 cells. Semin Immunol. 2007;19(6):383–93.
36. Stockinger B, Veldhoen M, Martin B. Th17 T cells: linking innate and adaptive immunity. Semin Immunol. 2007;19(6):353–61.
37. Steinman RM, Hawiger D, Nussenzweig MC. Tolerogenic dendritic cells. Annu Rev Immunol. 2003;21:685–711.
38. Morelli AE, Thomson AW. Tolerogenic dendritic cells and the quest for transplant tolerance. Nat Rev Immunol. 2007;7(8):610–21.
39. Andrews DM, Andoniou CE, Scalzo AA, van Dommelen SL, Wallace ME, Smyth MJ, et al. Cross-talk between dendritic cells and natural killer cells in viral infection. Mol Immunol. 2005;42(4):547–55.
40. Ferlazzo G, Tsang ML, Moretta L, Melioli G, Steinman RM, Munz C. Human dendritic cells activate resting natural killer (NK) cells and are recognized via the NKp30 receptor by activated NK cells. J Exp Med. 2002;195(3):343–51.
41. Dubois B, Bridon JM, Fayette J, Barthelemy C, Banchereau J, Caux C, et al. Dendritic cells directly modulate B cell growth and differentiation. J Leukoc Biol. 1999;66(2):224–30.
42. Fayette J, Dubois B, Vandenabeele S, Bridon JM, Vanbervliet B, Durand I, et al. Human dendritic cells skew isotype switching of CD40-activated naive B cells towards IgA1 and IgA2. J Exp Med. 1997;185(11):1909–18.
43. Elamanchili P, Diwan M, Cao M, Samuel J. Characterization of poly(D, L-lactic-co-glycolic acid) based nanoparticulate system for enhanced delivery of antigens to dendritic cells. Vaccine. 2004;22(19):2406–12.
44. Elamanchili P, Lutsiak CM, Hamdy S, Diwan M, Samuel J. "Pathogen-mimicking" nanoparticles for vaccine delivery to dendritic cells. J Immunother. 2007;30(4):378–95.
45. Yoshida M, Babensee JE. Poly(lactic-co-glycolic acid) enhances maturation of human monocyte-derived dendritic cells. J Biomed Mater Res A. 2004;71(1):45–54.

46. Babensee JE, Paranjpe A. Differential levels of dendritic cell maturation on different biomaterials used in combination products. J Biomed Mater Res A. 2005;74(4):503–10.
47. Yoshida M, Babensee JE. Differential effects of agarose and poly(lactic-co-glycolic acid) on dendritic cell maturation. J Biomed Mater Res A. 2006;79(2):393–408.
48. Yoshida M, Mata J, Babensee JE. Effect of poly(lactic-co-glycolic acid) contact on maturation of murine bone marrow-derived dendritic cells. J Biomed Mater Res A. 2007;80(1):7–12.
49. Lutsiak ME, Robinson DR, Coester C, Kwon GS, Samuel J. Analysis of poly(D, L-lactic-co-glycolic acid) nanosphere uptake by human dendritic cells and macrophages in vitro. Pharm Res. 2002;19(10):1480–7.
50. Waeckerle-Men Y, Scandella E, Uetz-Von Allmen E, Ludewig B, Gillessen S, Merkle HP, et al. Phenotype and functional analysis of human monocyte-derived dendritic cells loaded with biodegradable poly(lactide-co-glycolide) microspheres for immunotherapy. J Immunol Methods. 2004;287(1–2):109–24.
51. Newman KD, Elamanchili P, Kwon GS, Samuel J. Uptake of poly(D, L-lactic-co-glycolic acid) microspheres by antigen-presenting cells in vivo. J Biomed Mater Res. 2002;60(3):480–6.
52. Schanen BC, Karakoti AS, Seal S, Drake 3rd DR, Warren WL, Self WT. Exposure to titanium dioxide nanomaterials provokes inflammation of an *in vitro* human immune construct. ACS Nano. 2009;3(9):2523–32.
53. Arcos D, Vallet-Regi M. Sol–gel silica-based biomaterials and bone tissue regeneration. Acta Biomater. 2010;6(8):2874–88.
54. Beamer CA, Holian A. Silica suppresses Toll-like receptor ligand-induced dendritic cell activation. FASEB J. 2008;22(6):2053–63.
55. Shokouhi B, Coban C, Hasirci V, Aydin E, Dhanasingh A, Shi N, et al. The role of multiple toll-like receptor signalling cascades on interactions between biomedical polymers and dendritic cells. Biomaterials. 2010;31(22):5759–71.
56. Shankar SP, Petrie TA, Garcia AJ, Babensee JE. Dendritic cell responses to self-assembled monolayers of defined chemistries. J Biomed Mater Res A. 2010;92(4):1487–99.
57. Barbosa JN, Madureira P, Barbosa MA, Aguas AP. The influence of functional groups of self-assembled monolayers on fibrous capsule formation and cell recruitment. J Biomed Mater Res A. 2006;76(4):737–43.
58. Kamath S, Bhattacharyya D, Padukudru C, Timmons RB, Tang L. Surface chemistry influences implant-mediated host tissue responses. J Biomed Mater Res A. 2008;86(3):617–26.
59. Krause M, Sechi AS, Konradt M, Monner D, Gertler FB, Wehland J. Fyn-binding protein (Fyb)/SLP-76-associated protein (SLAP), Ena/vasodilator-stimulated phosphoprotein (VASP) proteins and the Arp2/3 complex link T cell receptor (TCR) signaling to the actin cytoskeleton. J Cell Biol. 2000;149(1):181–94.
60. Coppolino MG, Krause M, Hagendorff P, Monner DA, Trimble W, Grinstein S, et al. Evidence for a molecular complex consisting of Fyb/SLAP, SLP-76, Nck, VASP and WASP that links the actin cytoskeleton to Fcgamma receptor signalling during phagocytosis. J Cell Sci. 2001;114(Pt 23):4307–18.
61. Ross R, Jonuleit H, Bros M, Ross XL, Yamashiro S, Matsumura F, et al. Expression of the actin-bundling protein fascin in cultured human dendritic cells correlates with dendritic morphology and cell differentiation. J Invest Dermatol. 2000;115(4):658–63.
62. Shutt DC, Daniels KJ, Carolan EJ, Hill AC, Soll DR. Changes in the motility, morphology, and F-actin architecture of human dendritic cells in an *in vitro* model of dendritic cell development. Cell Motil Cytoskeleton. 2000;46(3):200–21.
63. West MA, Wallin RP, Matthews SP, Svensson HG, Zaru R, Ljunggren HG, et al. Enhanced dendritic cell antigen capture via toll-like receptor-induced actin remodeling. Science. 2004;305(5687):1153–7.
64. van Helden SF, Krooshoop DJ, Broers KC, Raymakers RA, Figdor CG, van Leeuwen FN. A critical role for prostaglandin E2 in podosome dissolution and induction of high-speed migration during dendritic cell maturation. J Immunol. 2006;177(3):1567–74.
65. Rogers TH, Babensee JE. The role of integrins in the recognition and response of dendritic cells to biomaterials. Biomaterials. 2011;32(5):1270–9.

66. Maitra R, Clement CC, Crisi GM, Cobelli N, Santambrogio L. Immunogenecity of modified alkane polymers is mediated through TLR1/2 activation. PLoS One. 2008;3(6):e2438.
67. Flo TH, Ryan L, Latz E, Takeuchi O, Monks BG, Lien E, et al. Involvement of toll-like receptor (TLR) 2 and TLR4 in cell activation by mannuronic acid polymers. J Biol Chem. 2002;277(38):35489–95.
68. Grandjean-Laquerriere A, Tabary O, Jacquot J, Richard D, Frayssinet P, Guenounou M, et al. Involvement of toll-like receptor 4 in the inflammatory reaction induced by hydroxyapatite particles. Biomaterials. 2007;28(3):400–4.
69. Seong SY, Matzinger P. Hydrophobicity: an ancient damage-associated molecular pattern that initiates innate immune responses. Nat Rev Immunol. 2004;4(6):469–78.
70. Pastor JC, Puente B, Telleria J, Carrasco B, Sanchez H, Nocito M. Antisilicone antibodies in patients with silicone implants for retinal detachment surgery. Ophthalmic Res. 2001;33(2):87–90.
71. Zippel R, Wilhelm L, Marusch F, Koch A, Urban G, Schlosser M. Antigenicity of polyester (Dacron) vascular prostheses in an animal model. Eur J Vasc Endovasc Surg. 2001;21(3):202–7.
72. Schlosser M, Wilhelm L, Urban G, Ziegler B, Ziegler M, Zippel R. Immunogenicity of polymeric implants: long-term antibody response against polyester (Dacron) following the implantation of vascular prostheses into LEW.1A rats. J Biomed Mater Res. 2002;61(3):450–7.
73. Wilhelm L, Zippel R, von Woedtke T, Kenk H, Hoene A, Patrzyk M, et al. Immune response against polyester implants is influenced by the coating substances. J Biomed Mater Res A. 2007;83(1):104–13.
74. Zhao Q, Topham N, Anderson JM, Hiltner A, Lodoen G, Payet CR. Foreign-body giant cells and polyurethane biostability: *in vivo* correlation of cell adhesion and surface cracking. J Biomed Mater Res. 1991;25(2):177–83.
75. McNally AK, Anderson JM. Interleukin-4 induces foreign body giant cells from human monocytes/macrophages. Differential lymphokine regulation of macrophage fusion leads to morphological variants of multinucleated giant cells. Am J Pathol. 1995;147(5):1487–99.
76. Kao WJ, McNally AK, Hiltner A, Anderson JM. Role for interleukin-4 in foreign-body giant cell formation on a poly(etherurethane urea) *in vivo*. J Biomed Mater Res. 1995;29(10):1267–75.
77. DeFife KM, Jenney CR, McNally AK, Colton E, Anderson JM. Interleukin-13 induces human monocyte/macrophage fusion and macrophage mannose receptor expression. J Immunol. 1997;158(7):3385–90.
78. McNally AK, DeFife KM, Anderson JM. Interleukin-4-induced macrophage fusion is prevented by inhibitors of mannose receptor activity. Am J Pathol. 1996;149(3):975–85.
79. Cui W, Ke JZ, Zhang Q, Ke HZ, Chalouni C, Vignery A. The intracellular domain of CD44 promotes the fusion of macrophages. Blood. 2006;107(2):796–805.
80. Han X, Sterling H, Chen Y, Saginario C, Brown EJ, Frazier WA, et al. CD47, a ligand for the macrophage fusion receptor, participates in macrophage multinucleation. J Biol Chem. 2000;275(48):37984–92.
81. Yagi M, Miyamoto T, Sawatani Y, Iwamoto K, Hosogane N, Fujita N, et al. DC-STAMP is essential for cell-cell fusion in osteoclasts and foreign body giant cells. J Exp Med. 2005;202(3):345–51.
82. Bobryshev YV, Inder SJ, Cherian SM, Lord RS, Ao PY, Hawthorne WJ, et al. Colonisation of prosthetic grafts by immunocompetent cells in a sheep model. Cardiovasc Surg. 2001;9(2):166–76.
83. Wolfram D, Rainer C, Niederegger H, Piza H, Wick G. Cellular and molecular composition of fibrous capsules formed around silicone breast implants with special focus on local immune reactions. J Autoimmun. 2004;23(1):81–91.
84. Vasilijic S, Savic D, Vasilev S, Vucevic D, Gasic S, Majstorovic I, et al. Dendritic cells acquire tolerogenic properties at the site of sterile granulomatous inflammation. Cell Immunol. 2005;233(2):148–57.
85. Megiovanni AM, Sanchez F, Robledo-Sarmiento M, Morel C, Gluckman JC, Boudaly S. Polymorphonuclear neutrophils deliver activation signals and antigenic molecules to dendritic cells: a new link between leukocytes upstream of T lymphocytes. J Leukoc Biol. 2006;79(5):977–88.

86. van Gisbergen KP, Sanchez-Hernandez M, Geijtenbeek TB, van Kooyk Y. Neutrophils mediate immune modulation of dendritic cells through glycosylation-dependent interactions between Mac-1 and DC-SIGN. J Exp Med. 2005;201(8):1281–92.
87. Myers CL, Wertheimer SJ, Schembri-King J, Parks T, Wallace RW. Induction of ICAM-1 by TNF-alpha, IL-1 beta, and LPS in human endothelial cells after downregulation of PKC. Am J Physiol. 1992;263(4 Pt 1):C767–72.
88. Iwasawa K, Kameyama T, Ishikawa H, Sawa Y. Induction of ICAM-1 and VCAM-1 on the mouse lingual lymphatic endothelium with TNF-alpha. Acta Histochem Cytochem. 2008;41(5):115–20.
89. Dinarello CA. IL-1: discoveries, controversies and future directions. Eur J Immunol. 2010; 40(3):599–606.
90. Denning TL, Wang YC, Patel SR, Williams IR, Pulendran B. Lamina propria macrophages and dendritic cells differentially induce regulatory and interleukin 17-producing T cell responses. Nat Immunol. 2007;8(10):1086–94.
91. Petersen LK, Xue L, Wannemuehler MJ, Rajan K, Narasimhan B. The simultaneous effect of polymer chemistry and device geometry on the *in vitro* activation of murine dendritic cells. Biomaterials. 2009;30(28):5131–42.
92. Chen H, Li P, Yin Y, Cai X, Huang Z, Chen J, et al. The promotion of type 1 T helper cell responses to cationic polymers *in vivo* via toll-like receptor-4 mediated IL-12 secretion. Biomaterials. 2010;31(32):8172–80.

Chapter 8
Tissue Colonization in Biomaterial-Associated Infection

Sebastian A.J. Zaat

Abstract Biomedical devices made of biomaterials predispose to infection as they provide surfaces for biofilm formation by microorganisms. Moreover, their presence in host tissue also compromises the local host immune response, allowing bacteria to persist in the vicinity of medical devices to cause infection. Biofilm formation, particularly by staphylococci, has been described in depth in Chaps. 2 and 6. This chapter therefore focuses on the colonization of peri-biomaterial tissue and host cells by bacteria, particularly staphylococci, on the characteristics of staphylococci residing intracellularly, the efficacy of antibiotics against intracellular staphylococci, and the pathogenic process leading to peri-implant tissue colonization and how immune modulation can contribute to prevent this.

8.1 Biomaterials Enhance Susceptibility to Infection

It has been recognized for more than half a century that the presence of a foreign body predisposes for infection. In 1957, Elek and Conen performed a classic study in which they assessed the infection-enhancing effect of stitches in human volunteers. Stitches were contaminated with graded numbers of *Staphylococcus aureus* (then referred to as *Staphylococcus pyogenes*) and either pulled through the skin, or left in place with the bacteria. The dose required to cause infection with the stitch in place was at least 10,000-fold lower than when the stitch was pulled through. The volunteers had lesions progressing "to the size of an orange," which led the authors to conclude that "this experiment convincingly demonstrated the enhancing effect

S.A.J. Zaat (✉)
Department of Medical Microbiology, Center for Infection and Immunity Amsterdam (CINIMA), University of Amsterdam, Amsterdam, The Netherlands
e-mail: s.a.zaat@amc.uva.nl

T.F. Moriarty et al. (eds.), *Biomaterials Associated Infection: Immunological Aspects and Antimicrobial Strategies*, DOI 10.1007/978-1-4614-1031-7_8,

of the stitch, but led to great difficulty in finding further volunteers" [44]. Studies in laboratory animals repeatedly confirmed the infection-enhancing effect of a foreign body [64, 110, 136, 158].

Those earlier studies were focused on *S. aureus*, a major pathogen involved in wound and foreign body material infection. *Staphylococcus epidermidis*, the harmless skin commensal, was later also shown to be a major pathogen in the presence of inserted or implanted medical devices [101, 119]. Other coagulase-negative staphylococci, yeasts, enterococci, streptococci, *Propionibacterium*, and different gram-negative species can cause these infections as well, depending on the device and site of application [60, 151]. Scanning electron microscopy studies of catheters retrieved from patients showed that bacteria were present as biofilms, encased in elaborated extracellular polysaccharide, also referred to as "slime" [99–102, 115, 116, 121]. These studies led to the still expanding research field of biofilms in medical device infection [32, 39, 52, 111, 132].

In the early 1980s, studies by Zimmerli et al. showed that biofilm formation was not the only process responsible for the infection-enhancing effect of foreign bodies [156, 157]. The use of subcutaneous tissue cages in which biomaterials are placed allowed studying the influence of different materials on the efficacy of the host immune system. Over time the tissue fluid including host cells is sampled for analysis of cellular content and function. Neutrophils harvested from these cages at 14 days after implantation appeared particularly reduced in their phagocytic and bactericidal properties [156, 157, 159]. Apparently, the environment of the tissue cage with the biomaterial inactivated neutrophils after a certain period.

Also in bone the presence of a foreign body strongly predisposes for infection. In a rabbit model, only 50 CFU were sufficient to cause infection after a cemented hemiarthroplasty had been implanted, whereas in the absence of a foreign body an inoculum of 10,000 CFU was required for infection [133].

8.2 Bacteria in Peri-Implant Tissue and Inside Host Macrophages

So, bacteria causing medical device infection have at least two major ways to evade the host defense; they can form biofilms and they can take advantage of the reduced efficacy of the host immune response. To study the pathogenesis of infection of subcutaneous implants, Christensen et al. devised a mouse model [36]. In this model segments of materials are implanted subcutaneously and infected with defined doses of bacteria, often *S. epidermidis* or other coagulase-negative staphylococci. Traditionally, implant sites were inspected for purulence, and the implants, but not the surrounding tissue, were quantitatively cultured. *S. epidermidis*, *Staphylococcus schleiferi*, and *Staphylococcus lugdunensis* were shown to be more virulent than *Staphylococcus warneri* and *Staphylococcus hominis* [78], and *S. epidermidis* strains producing "slime" (extracellular polysaccharide) were more virulent than non-slime producing strains [35, 117].

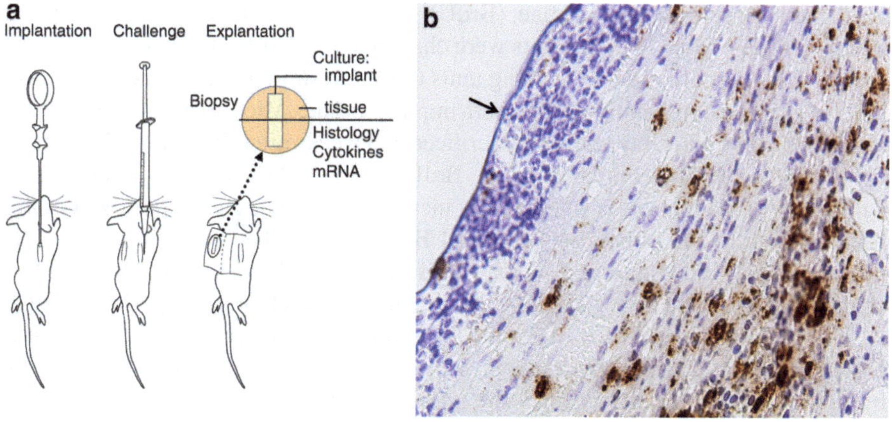

Fig. 8.1 Mouse model of biomaterial-associated infection of subcutaneous implants. (**a**) Materials are implanted subcutaneously with minimal tissue damage, the incision is closed with a single stitch, a bacterial inoculum is delivered by injection along the implant. At selected time points mice are terminated and biopsies are collected for evaluation of (1) implant colonization, (2) surrounding tissue colonization, (3) cytokines in the tissue, (4) transcriptome analysis (microarrays/deep sequencing), and (5) (immune) histology. (**b**) *S. epidermidis* bacteria are immunostained *brown* using antibodies against lipoteichoic acid, a major cell wall component of Staphylococci and other gram-positive bacteria. The bacteria are predominantly localized within macrophages in the peri-implant tissue at a certain distance from the implant–tissue interface. *Arrow*, implant–tissue interface. Figure modified from [31] with permission of the publisher

We have used this mouse model extensively to study the pathogenesis of *S. epidermidis* biomaterial-associated infection (BAI). In our studies we routinely analyzed both the implanted materials and the surrounding tissue (Fig. 8.1a). To our surprise, it was generally not so much the biomaterial that yielded high numbers of bacteria, but rather the surrounding tissue. Bacteria were retrieved from the tissue at lower challenge doses, and for longer periods than from the implanted material [18, 21]. Histology of tissue surrounding polyvinylpyrrolidone (PVP)-coated polyamide showed that the *S. epidermidis* bacteria were present within the tissue at a certain distance from the tissue–biomaterial interface rather than on the material itself, and predominantly within host cells (Fig. 8.1b) [17, 27]. Immunohistochemistry showed that large numbers of bacteria were localized inside macrophages, and transmission electron microscopy showed cocci with the morphology of viable and dividing bacteria inside host cells [17]. Moreover, numbers of infected macrophages, and numbers of bacteria per macrophage increased over time, suggesting possible intracellular growth [17]. This phenomenon also occurred at slightly higher bacterial challenge doses in tissue surrounding segments of non-coated polyamide, silicone elastomer coated with PVP (SEpvp), and non-coated silicone elastomer (SE) [17, 27, 28]. Tissue colonization was observed with different strains of *S. epidermidis* and in C57BL/6 as well as in BALB/c mice [17, 18, 27]. In a study by Vuong et al. peri-implant tissue colonization by *S. epidermidis* was also seen in rabbits [150].

To assess whether peri-implant tissue truly is a niche for *S. epidermidis* persistence and *in vivo* replication, we performed a mouse study using the nucleotide analogue 5-bromo-2-deoxyuridine (BrdU), which can be detected with specific antibodies [29]. Mice with implants were challenged with a relatively low inoculum of *S. epidermidis* (10^6 colony forming units (CFU)) resulting in low-grade infection at 2 weeks, predominantly of the peri-implant tissue. Mice then received BrdU injections every other day and were sacrificed at 3 weeks after the start of the experiment, so 1 week after initiation of the BrdU regimen. At 3 weeks the number of culture-positive tissues was significantly increased, indicating *in vivo* growth of the bacteria. Immunohistochemistry revealed BrdU-positive bacteria within the peri-implant tissue, showing that *S. epidermidis* present in the tissue at 2 weeks after challenge had been actively dividing during the third week of the experiment [29].

The peri-implant tissue of the mice also appeared to offer protection against a regimen of rifampicin and vancomycin. *S. epidermidis* was cleared from the implants by the antibiotics but persisted in the tissue [30]. Confirming and extending our results to a different pathogen, Engelsman et al. studied the role of tissue colonization by *S. aureus* in a reinfection model [46]. They implanted silicone rubber disks with pregrown *S. aureus* biofilms subcutaneously in mice. After removal of the infected implant and treatment with rifampicin and vancomycin novel, sterile implants were implanted in the pockets. Despite the antibiotic therapy the original challenge strain was recultured from these novel implants and their surrounding tissues [46], confirming that tissue can act as a reservoir for infecting bacteria.

8.3 Is Peri-Implant Tissue a Niche for Infecting Bacteria in Humans?

The difficulty to eradicate tissue-colonizing bacteria is a well-known problem in particular in orthopedic implant infection. Despite removal of the infected device and extensive debridement to remove any biofilm remnants, patients need antibiotic treatment for prolonged periods in order to render the implantation area culture-negative (Chaps. 1 and 9). In these patients bacteria apparently survive in the peri-implant tissue. This colonization may in part derive from an initial colonization of the implant–tissue interface, involving the presence of a biofilm. Although often inferred from *in vitro* experiments, positive cultures and the identification of aggregates of bacteria in sonicates of infected explanted prostheses [138, 141], the presence of an actual biofilm of bacteria directly attached to the implant surface in situ has only sparsely been documented [134]. Direct evidence was shown in studies on explanted infected implants [54, 106, 141]. A combination of extended culture and in situ hybridization using 16S rRNA-specific probes showed presence of *S. epidermidis* in connective tissue from a patient with prosthetic hip loosening [75]. The most elaborate approach to study whether viable bacteria are actually present on or around implants in situ, has been the combined application of confocal microscopy using Live-Dead stains for bacteria, fluorescent in situ hybridization, and reverse transcriptase polymerase chain reaction (RT-PCR) [134]. In aspirate, cement and tissue obtained during revision surgery of a chronically infected total elbow arthroplasty, viable bacteria were seen in microscopy using Live-Dead staining, and the RT-PCR confirmed the presence of viable *S. aureus* bacteria. The bacteria were seen in clusters, either associated with cement or with fibrous threads within the retrieved tissue, proving the presence of viable bacteria in biofilms on the material and also within the peri-implant tissue.

Similar to bone, also for soft-tissue implants limited data are available on the actual presence of biofilms in situ. In a series of revisions of failed silicone breast implants and tissue expanders, the tissue surrounding the biomaterials that had capsular contractions yielded positive cultures in 50% of the cases where the implant was infected [147]. In one case the tissue yielded bacterial growth whereas the implant itself was culture-negative. Injected biomaterial fillers used in cosmetic surgery can cause dramatic adverse reactions. In a small series of patients, 15 of 28 of biopsies from their adverse reactions appeared to contain bacteria, whereas none of five control biopsies did [37]. In a large prospective study bacteria were detected in an even higher percentage of the samples from adverse reactions, whereas those of control subjects were all negative for bacteria (L.H. Christensen, personal communication), clearly showing a relationship between bacterial presence and adverse effects.

In order to study the possible colonization of human peri-biomaterial soft tissue systematically, we performed a study on catheters and surrounding tissue of deceased human intensive care patients [31]. These patients had not died because of infection, but had been elected for autopsy for other reasons. After removal of the skin, biopsies were obtained including catheter and surrounding subcutaneous tissue. The biopsies were segmented and alternating segments were studied by quantitative culture and by histology. For culture, the segments were further divided into small blocs, and bacteria adherent to the catheter segments and present in homogenates of the tissue blocs were quantitatively cultured. Figure 8.2a shows a characteristic result. Bacteria were retrieved from the catheter, but often more bacteria were cultured from the tissue surrounding the catheter, usually with numbers decreasing from the skin-contacting to the deeper catheter parts. Interestingly, bacteria were even cultured from parts of the biopsies which had not been in contact with the catheter. The major species cultured were *S. epidermidis* and *Enterococcus faecalis*. Immunohistochemistry confirmed the presence of bacteria in the tissue of each of the culture-positive biopsies (Fig. 8.2b). The bacteria in the explanted patient tissue were capable of incorporation BrdU within a 4 h incubation time, showing they were capable of replicating in situ [29].

Not only staphylococci have been reported to reside within peri-implant tissue. *Propionibacterium acnes* causing prosthesis infections in bone, did not only grow as a biofilm on the prosthesis, but was also cultured from adjacent bone [118]. Moreover *P. acnes* was isolated from tissue surrounding infected prosthetic hip and knee joints and also from tissue of postoperative mediastinitis with sternal stainless steel wires present, where the prosthetic devices themselves were culture-negative [60]. These studies indicate that peri-implant tissue colonization may be an important aspect of BAI.

The relative importance of biofilm formation and tissue colonization in the pathogenesis of BAI in patients will depend on the type of application. In case the medical device is residing in the bloodstream, it is highly likely that biofilm formation will be

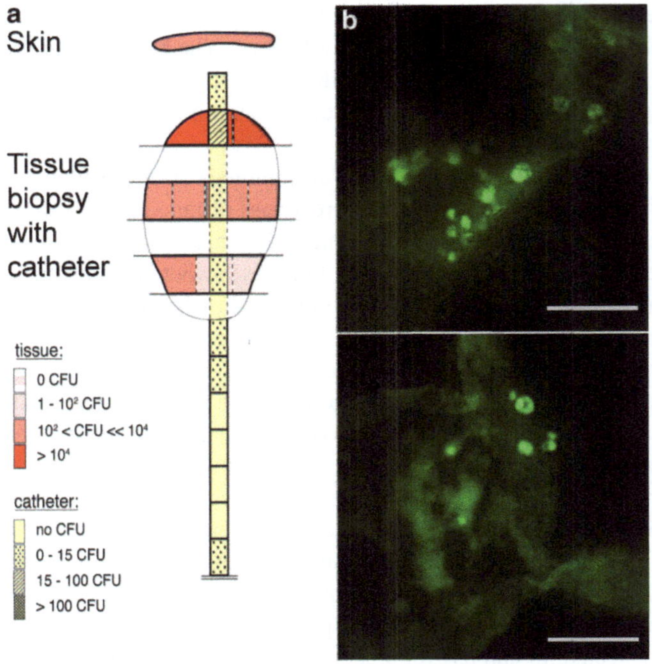

Fig. 8.2 Catheter and peri-catheter tissue colonization in a biopsy of human peri-catheter tissue. (**a**) Processing scheme and culture results of a biopsy representative for culture-positive biopsies obtained from deceased ICU patients, who had died from noninfectious causes [31]. In a procedure carefully avoiding contamination, skin was removed first, and a biopsy was collected using new sterile instrumentation. The biopsy was sectioned, and alternating sections were used for histology and culture. For culture, segments were divided into smaller blocs, which were homogenized and quantitatively cultured. (**b**) Fluorescence immunohistochemistry with Alexa-488-labeled antibody against lipoteichoic acid, a major outer cell surface constituent of gram-positive bacteria. Bacteria were detected in the peri-catheter tissue with this technique in all biopsies which had been positive in culture

essential for bacterial persistence. In case the bacteria colonize the tissue surrounding an inserted or implanted device, there is less strict need for strong adherence and accumulation capacity of the bacteria. Such bacterial isolates might therefore even lack particular genes encoding proteins involved in biofilm formation. For *S. epidermidis* extensive epidemiology has been performed to link biofilm genes to pathogenic potential for infection of a wide range of devices in different patient populations. This is extensively reviewed in Chap. 2. The frequency of presence of genes involved in the production of biofilm factors such as polysaccharide intercellular adhesin (PIA) produced by gene products of the *ica* (intercellular adhesin) gene cluster, Adherence/accumulation-associated protein (Aap), and Extracellular matrix binding protein (Embp) is often increased in patient isolates as compared to commensal isolates from the same patient groups (see Chap. 2). Interestingly, in many studies there always is a proportion of clinical isolates that lack biofilm-forming capacity and/or biofilm-associated genes [2, 3, 33, 41, 48, 70, 72]. These isolates are often considered contaminants, but it is also possible that they would be tissue/intracellular survivors with as yet unidentified virulence characteristics contributing to survival in this niche, rather than to biofilm formation.

8.4 Propensity of Staphylococci to Survive Inside Host Cells

There is growing evidence that *S. aureus* can be regarded as a facultative intracellular pathogen [139]. Microscopic evidence for intracellular localization of *S. aureus* in clinical samples has mainly been provided for *S. aureus* osteomyelitis [38, 106, 160]. In live bone tissue excised from a patient with recurrent, long-term osteomyelitis, bacteria were observed inside osteoblasts and osteoclasts [38]. In orthopedic joint revision patients *S. aureus* was detected in peri-implant fibroblasts in biopsies from joint aspiration and/or intraoperative tissue [24].

In vitro, *Staphylococcus aureus* is known to be able to survive inside a variety of cell types [91], either in professional phagocytes [4, 12, 16, 38, 53, 76, 126, 131] or in nonprofessional phagocytic cells such as epithelial [69, 96, 155], endothelial [96, 113], and adipocyte-like cells [58], fibroblasts and osteoblasts [38, 45, 65, 137]. Studies on this topic have recently extensively been reviewed by Garzoni and Kelley [51]. In vivo, in experimental mouse peritonitis, *S. aureus* bacteria are present intracellularly within peritoneal cells, predominantly macrophages, retrieved from peritoneal washings [123–126].

Although less frequently reported, *S. epidermidis* also has the capacity to survive inside host cells *in vitro* [112], although survival is lower than of *S. aureus* strains when tested in parallel [113]. In vivo, *S. epidermidis* resides intracellularly in peri-implant tissue in mice in the presence of implants [17, 19, 27–31, 153, 154]. When large inocula are applied, *S. epidermidis* has also been observed in association with host macrophages in the absence of implants [42].

8.4.1 Small Colony Variant Phenotype

The most deeply investigated characteristic contributing to intracellular survival of staphylococci is the "small colony variant (SCV)" phenotype. SCVs are slow growing variants which form small colonies on agar plates have a reduced susceptibility to aminoglycoside antibiotics and are difficult to diagnose and treat [51, 120, 139]. SCVs from clinical specimens consistently are either deficient in electron transport because of deficiencies in the synthesis of menadione or haemin, or in the synthesis of thiamine, which is required for the synthesis of menadione [51, 120, 139]. Clinical *S. aureus* SCVs have been isolated from patients with pacemaker-related bloodstream infections [128], SCVs of *S. epidermidis* and *Staphylococcus* capitis were retrieved from infected pacemakers [8, 148], and *S. epidermidis* from prosthetic heart valves [7]. *S. aureus* SCVs are internalized by various nonprofessional phagocytes, such as endothelial [9, 57] and epithelial cells [15], keratinocytes [149], and osteoblasts [61]. Wild type *S. aureus* strains readily cause lysis or apoptosis of host cells, predominantly through their alpha-toxin [10, 59, 145], and for neutrophils also by the Penton-Valentine leucocidin (PVL) and the phenol-soluble modulins (PMS) [97], a group of secreted staphylococcal cytolytic peptides [152]. SCVs have lower levels of alpha-toxin expression than wild type *S. aureus* [67, 71] and therefore do not destruct the host cells. The relative expression of PVL and of PSM of SCVs has not yet been reported, but down-regulation of these toxins might also contribute to intracellular survival.

The SCV status has been considered to be a specialized adaptive phenotype for intracellular persistence of *S. aureus* [139]. Whereas wild type *S. aureus* induce extensive alterations in the gene and protein expression of endothelial cells [55, 103], SCVs hardly induce detectable changes, even though they are present inside the cells in higher numbers [139]. In comparative studies with wild type and SCV *S. aureus*, wild type *S. aureus* more readily entered the cytoplasm and induced apoptosis of the cells [10, 59], thus eliminating their protective niche and inducing more local immune activation. So, the SCV phenotype allows *S. aureus* to maintain the host cell as a safe shelter.

The evolutionary driving force for the development of the capacity to form SCV is not likely to be antibiotic selective pressure, since antibiotics have only been applied in humans for a very brief period in evolutionary terms. Since SCV tend to have a reduced susceptibility to cationic antimicrobial proteins [122], an important element of the host cell intracellular antimicrobial armamentum, reduced susceptibility to these proteins may have been the selective advantage for the capacity of SCV development. Interestingly, the intracellular milieu is considered a stimulus for SCV formation [146], and intracellular SCV have even been shown to even survive within lysosomes of endothelial cells [127].

The mechanism of switching between normal and SCV phenotype has been partly resolved in studies on the genetic make-up of clinical *S. aureus* menadione auxothophic SCV isolates. These SCVs carried mutations in the *menB* gene, encoding naphtoate synthase involved in the synthesis of menadione. These mutations either were a deletion, or point mutations resulting in a premature stopcodon or in an amino-acid substitution rendering naphtoate synthase nonfunctional [79]. Upon reculturing, the mutants produced compensatory mutations at a relatively high frequency of 2×10^{-8}, which grew as wild type colonies, were not menadione-auxotrophic, had wild type levels of alpha-toxin production, and were susceptible to aminoglycoside antibiotics. Thus, *S. aureus* can switch between wild type and SCV phenotypes by mutation and compensatory mutations, allowing rapid adaptation of the population to the environmental conditions. This poses a realistic problem for treatment, necessitating the identification and use of antibiotics active against SCV.

S. epidermidis is far less virulent than *S. aureus* and does not produce alpha-toxin. Wild type *S. epidermidis* strains will therefore not easily kill host cells when they are internalized. Reduction of virulence by entering the SCV status therefore may not be essential for *S. epidermidis* to be able to survive in host cells without destroying them. However, just as *S. aureus*, *S. epidermidis* will be exposed to the intracellular defense mechanisms such as cationic antimicrobial proteins when taken up by host phagocytes. The reduced susceptibility to these defense systems then provides the SCVs with a selective advantage over the wild type bacteria. SCVs of *S. epidermidis* have indeed been retrieved from infected medical devices [7, 8, 148]. The reduced susceptibility to antibiotics because of the SCV phenotype likely will have contributed to their survival in the vicinity of the implanted devices. Of note, an *S. epidermidis hemB* mutant strain with an SCV phenotype had increased synthesis of the PIA, one of the major polymers of the extracellular "slime" [1]. As this polymer reduces phagocyte bactericidal activity [66, 131] and susceptibility to antimicrobial proteins [151] this may additionally contribute to enhanced capacity of intracellular survival for the SCVs.

8.4.2 Other Intracellular Survival Characteristics of Staphylococci

Besides the SCV phenotype, staphylococci may have other characteristics contributing to survival in host tissue or intracellularly. Vuong et al. [150] reported that an *S. epidermidis* strain mutated in the *agr* system ("a general regulator"; the quorum-sensing system involved in regulation of virulence-related genes) was more capable of biofilm formation *in vivo* in a rabbit subcutaneous implant infection model than the wild type strain. This led the authors to conclude that the *agr* mutation increased virulence, in the sense of implant colonization. However, the *agr* mutation reduced the capacity of the bacteria to survive in the peri-implant tissue, implying that Agr-controlled genes may contribute to *S. epidermidis* survival in host tissue. Among others, Agr controls the expression of phenol-soluble modulins, which are strong agonists for the neutrophil formyl peptide receptor 2 (FPR2/ALX) [73]. Inactivation of Agr in *S. aureus* or *S. epidermidis* strongly reduces neutrophil responses [74], possibly increasing survival of the *S. epidermidis agr* mutant strains in tissue [151]. As Agr influences activity of many other genes, further research on the influence of *agr*-regulated genes on survival in tissue and possibly intracellular survival is warranted.

For *S. aureus*, the *sar* global regulator of virulence factors is involved in intracellular survival in neutrophils in a mouse peritonitis model [53]. The wild type *S. aureus* strain DB was present in neutrophils in peritoneal lavage of infected mice and transfer of these infected neutrophils to naïve mice induced infection. Intracellular survival after uptake by mouse or human neutrophils stimulated with the chemokines IL-8 or MIP-2, respectively, was lower for the *sarA* mutant than for the wild type *S. aureus*. Only the wild type bacteria partially degraded phagosome membranes and entered the cytoplasm. This suggests that in professional phagocytes such as PMN, full virulence of *S. aureus* is required for maximal bacterial survival. It has not yet been assessed which of the SarA-regulated virulence factor(s) in particular were involved in the intracellular persistence/replication.

8.5 Antibiotic Activity Against Intracellular Staphylococci

Antibiotic regimes for BAI generally are optimized to treat bacteria in biofilms, i.e., the antibiotics should penetrate biofilms and eliminate bacteria even in dormant state, as often encountered in biofilms [63, 120, 158]. Treatment of intracellular bacteria additionally requires that antibiotics penetrate host tissue, accumulate within cells, and reach the cellular compartment where the bacteria reside, such as phagosomes, phagolysosomes, or the cytoplasm. Moreover, the susceptibility of bacteria to antibiotics is often changed intracellularly, reducing their efficacy [40, 109, 114, 140, 142]. To characterize antimicrobial activity within host cells, it is therefore not sufficient to measure antibiotic uptake per se, but the actual activity against internalized bacteria should be assessed. In particular, antibiotics should exert a bactericidal effect against intracellular bacteria, since a mere bacteriostatic effect will allow the bacteria to persist intracellularly. Such bacteria can cause infection at later time points due to their regrowth after treatment has ended.

8.5.1 In Vitro Test System for Antibiotic Activity Against Intracellular Staphylococci

Various cell types have been used to study intracellular activity of antibiotics against staphylococci, most often against *S. aureus* strains. The most often used type of cells are the human THP-1 monocyte/macrophages [5], which are considered to be largely devoid of intracellular bactericidal activity, thus allowing evaluation of the bactericidal effect of externally applied antibiotics [114]. Cells with internalized bacteria are exposed to the antibiotics of interest, numbers of CFU of bacteria residing within the cells are recorded at different time points, and the decrease or the fold reduction relative to the inoculum is reported. More complex measures for intracellular activity, such as C_s and EC_{50} values are calculated based on the effects observed with graded concentrations of antibiotics [12, 87]. C_s is the concentration which causes a net bacteriostatic effect, i.e. where no change in bacterial numbers relative to the inoculum is observed. The EC_{50} value is calculated from the E_{min} and E_{max}. E_{min} is the fold change in numbers of CFU (usually a positive value due to intracellular growth of the bacteria) relative to the inoculum in the absence of antibiotic (a so-called "infinitely low concentration"), and E_{max} is the maximal reduction of the bacterial population at an infinitely large antibiotic concentration, where all binding sites of a bacterial target are considered to be saturated [12, 87]. EC_{50} is defined as the antibiotic concentration resulting in bacterial numbers exactly in the middle between E_{min} and E_{max}. It should be noted that an EC_{50} value does not provide information on the level of inhibition or killing; an antibiotic that only marginally reduces intracellular growth can have the same EC50 as an antibiotic with strong bactericidal activity. The most relevant parameter for intracellular activity obviously is actual reduction of numbers of CFU relative to the inoculum, with a true bactericidal effect generally defined as a 3-log reduction in numbers of CFU within 24 h. Table 8.1 summarizes the extensive data on antibiotics studied for their intracellular activity in THP-1 cells, J774 mouse macrophages, and in a mouse peritonitis intracellular activity model, focusing on bactericidal capacity for internalized staphylococci. The data from studies, which have reported the activity in terms of E_{min}, E_{max}, and CE_{50} values, have been used to derive the levels of bactericidal activity as they are listed in Table 8.1.

8.5.2 *In Vivo Model for Antibiotic Activity Against Intracellular Staphylococci*

Although elegant, the *in vitro* systems of staphylococci internalized by host cells cannot fully mimic the complexity of an entire living organism. Sandberg et al. therefore developed a mouse peritonitis model to study intracellular antibiotic activity *in vivo* [123]. Their model is a mouse peritonitis model, where the pathogen, often *S. aureus*, is injected intraperitoneally, and the mice are treated by subcutaneous injection of the antibiotic of choice. At the desired time point after initiation of treatment (typically 4 h), the mice are sacrificed and a peritoneal lavage is performed. Cells are collected by centrifugation, to separate the intracellular from the extracellular bacteria. The effect of the antibiotic is assessed by comparison of treated mice with infected but non-treated controls. In these studies the activities of antibiotics in the mouse model have been compared with the efficacy of the same antibiotics in the *in vitro* macrophage model [26, 123–126] (Table 8.1).

8.5.3 *Intracellular Staphylocidal Capacity of Antibiotics In Vitro and In Vivo*

The studies in Table 8.1 have compared a number of different parameters and antibiotics. Although the cell type used often is the same (THP-1 cells), the bacterial strains differ, which may have caused differences in the outcome of the studies. It is clear, however, that there are vast differences in intracellular bactericidal capacity among the tested antibiotics and that many antibiotics have low activity intracellularly. Antibiotics that do show good activity (+++ or ++++ in Table 8.1) are the fluoroquinolones moxifloxacin, gemifloxacin, delafloxacin (particularly at pH 5.5), levofloxacin, and garenoxacin, the lipoglycopeptides oritavancin and telavancin, and to a lesser degree quinupristin/dalfopristin (Table 8.1). Rifampicin (or rifampin), an antibiotic known to readily penetrate biofilms and therefore advised for regimens to treat orthopedic infections [158, 159], appears to have only moderate intracellular activity on its own, but increased activity against *S. aureus* SCV was observed when rifampicin was combined with moxifloxacin and oritavancin [108]. Vancomycin had a very poor intracellular activity.

Table 8.1 Intracellular activity of antibiotics against *S. aureus* and *S. epidermidis* in human THP-1 macrophages, mouse J774 macrophages and other cell types, and in the mouse peritonitis model

Study subject	Cell type	Pathogen	Antibiotics	Intracellular antibacterial effect[a]	Additional remarks	Reference
Activity of moxifloxacin against intracellular community-acquired methicillin-resistant *Staphylococcus aureus*: comparison with clindamycin, linezolid and co-trimoxazole and attempt at defining an intracellular susceptibility breakpoint	THP-1 macrophages Keratinocytes	*S. aureus*	Moxifloxacin Clindamycin Linezolid Co-trimoxazole	++ to +++ ++ + to ++ N	Activities similar in THP-1 and keratinocytes	[84]
Activity of fusidic acid against extracellular and intracellular *Staphylococcus aureus*: influence of pH and comparison with linezolid and clindamycin	THP-1 macrophages	*S. aureus*	Fusidic acid Linezolid Clindamycin	+ + +	Fusidic acid active against *S. aureus* in broth as well as intracellularly, with no cross-resistance to other antibiotics. Accumulation in THP-1 cells of fusidic acid, clindamycin and linezolid was increased, reduced and unaffected, respectively, by lowering the pH from 7.4 to 5.5	[87]
Cellular accumulation of fluoroquinolones is not predictive of their intracellular activity: studies with gemifloxacin, moxifloxacin, and ciprofloxacin in a pharmacokinetic/pharmacodynamic model of uninfected and infected macrophages	Murine J774 macrophages	*S. aureus*	Gemifloxacin Moxifloxacin Ciprofloxacin	+++ ++ ++	Accumulation decreased in the following order: gemifloxacin, moxifloxacin and then ciprofloxacin, but larger cellular concentrations of gemifloxacin than of moxifloxacin or ciprofloxacin were needed to obtain a similar target effect	[142]

Activity of finafloxacin, a novel fluoroquinolone with increased activity at acid pH, towards extracellular and intracellular *Staphylococcus aureus*, *Listeria monocytogenes* and *Legionella pneumophila*	THP-1 macrophages	*S. aureus*	Finafloxacin pH 7.4 pH 5.5 Ciprofloxacin pH 7.4 pH 5.5	 ++ ++ ++ +		[89]
Contrasting effects of acidic pH on the extracellular and intracellular activities of the anti-gram-positive fluoroquinolones moxifloxacin and delafloxacin against *Staphylococcus aureus*	THP-1 macrophages	*S. aureus*	Moxifloxacin pH 7.4 pH 5.5 Delafloxacin pH 7.4 pH 5.5	 +++ +++ +++ ++++	At pH 5.5 a tenfold higher intracellular accumulation of delafloxacin was measured than at pH 7.4	[86]
Intra- and extracellular activities of dicloxacillin and linezolid against a clinical *Staphylococcus aureus* strain with a small-colony-variant phenotype in an in vitro model of THP-1 macrophages and an in vivo mouse peritonitis model	THP-1 macrophages In vivo mouse peritonitis model	*S. aureus*, SCV	Dicloxacillin Linezolid	+ to ++ + to ++	Activity against normal and SCV *S. aureus* was similar, SCVs were slightly more susceptible In in vivo mouse peritonitis model with one dose of antibiotic immediately after challenge, an initial suppression of *S. aureus* by both antibiotics was seen, but outgrowth occurred in later stage of infection, particularly of SCV	[126]
Cellular pharmacodynamics of the novel biaryloxazolidinone radezolid: studies with infected phagocytic and nonphagocytic cells, using *Staphylococcus aureus*, *Staphylococcus epidermidis*, *Listeria monocytogenes*, and *Legionella pneumophila*	Human keratinocytes, endothelial cells, bronchial epithelial cells, osteoblasts, macrophages, and rat embryo fibroblasts	*S. aureus* *S. epidermidis*	Radezolid	*S. aureus* + S. *epidermidis* ++	Radezolid was approximately tenfold more active than linezolid in all cell types, and had more rapid bactericidal activity	[85]

(continued)

Table 8.1 (continued)

Study subject	Cell type	Pathogen	Antibiotics	Intracellular antibacterial effect[a]	Additional remarks	Reference
Activity of quinupristin/ dalfopristin against extracellular and intracellular *Staphylococcus aureus* with various resistance phenotypes	THP-1 macrophages	*S. aureus*	Quinupristin/ dalfopristin	++ to +++ (S*)		[14]
Intra- and extracellular activities of dicloxacillin against *Staphylococcus aureus* in vivo and in vitro	THP-1 macrophages In vivo mouse peritonitis model	*S. aureus*	Dicloxacillin	THP-1: + Mouse peritonitis: + to ++	In the in vivo peritonitis model a similar tenfold reduction was observed as in the in vitro THP-1 macrophage model Multiple dosing resulted in a 2 log reduction (++) in the mouse peritonitis model	[124]
Intra- and extracellular activity of linezolid against *Staphylococcus aureus* in vivo and in vitro	THP-1 macrophages In vivo mouse peritonitis model	*S. aureus*	Linezolid	THP-1: + Mouse peritonitis: N	Linezolid failed to prevent increase in intracellular CFU in the mouse peritonitis model	[125]
Intracellular activity of the peptide antibiotic NZ2114: studies with *Staphylococcus aureus* and human THP-1 monocytes, and comparison with daptomycin and vancomycin	THP-1 macrophages	*S. aureus*	NZ2114 Daptomycin Vancomycin	+ to ++ + +		[25]
Cellular accumulation and pharmacodynamic evaluation of the intracellular activity of CEM-101, a novel fluoroketolide, against *Staphylococcus aureus*, *Listeria monocytogenes*, and *Legionella pneumophila* in human THP-1 macrophages	THP-1 macrophages	*S. aureus*	CEM-101 Telithromycin, Azithromycin, Clarithromycin	+ + N +	CEM-101 had slightly better intracellular activity than telithromycin and clarithromycin. Azithromycin did not prevent intracellular growth	[90]

Cellular pharmacokinetics and intracellular activity of torezolid (TR-700): studies with human macrophage (THP-1) and endothelial (HUVEC) cell lines	THP-1 macrophages HUVEC (human umbilical vein endothelial cells)	*S. aureus*	Torezolid Linezolid	THP-1 cells: + HUVEC: ++	Five to tenfold more potent intracellularly (lower concentration needed to achieve a bacteriostatic effect) than linezolid against phagocytosed *S. aureus*	[81]
Activities of ceftobiprole and other cephalosporins against extracellular and intracellular (THP-1 macrophages and keratinocytes) forms of methicillin-susceptible and methicillin-resistant *Staphylococcus aureus*	THP-1 macrophages Keratinocytes	*S. aureus*	Ceftobiprole Cephalexin, Cefuroxime, Cefoxitin, Ceftriaxone	+ to ++ N to + N to + N to + N to ++	The conventional cephalosporins were less active than ceftobiprole in THP-1 cells, and inactive in keratinocytes	[83]
Intracellular activity of antibiotics against *Staphylococcus aureus* in a mouse peritonitis model	In vivo mouse peritonitis model	*S. aureus*	Dicloxacillin Cefuroxime Gentamicin Azithromycin Rifampin	++++ ++++ + + ++	Treatment was initiated at 2 h postinfection, and mice were terminated for CFU enumeration 4 h later, so at 6 h postinfection	[123]
Intracellular activity of antibiotics in a model of human THP-1 macrophages infected by a *Staphylococcus aureus* small-colony variant strain isolated from a cystic fibrosis patient: pharmacodynamic evaluation and comparison with isogenic normal-phenotype and revertant strains	THP-1 macrophages	*S. aureus* SCV, SCV revertants and normal phenotypes	Oxacillin Fusidic acid Clindamycin Gentamicin Rifampin Vancomycin, Linezolid, Quinupristin–dalfopristin Daptomycin Tigecycline, Moxifloxacin Telavancin Oritavancin	+ + + + ++ + + + + + + ++ + +++	Data in this table to the left are for SCV. All drugs except tigecycline and oritavancin showed approximately one order of magnitude higher reduction of numbers of CFU intracellular activity against normal or revertant phenotypes than against SCV	[107]

(continued)

Table 8.1 (continued)

Study subject	Cell type	Pathogen	Antibiotics	Intracellular antibacterial effect[a]	Additional remarks	Reference
Intracellular activity of antibiotics in a model of human THP-1 macrophages infected by a *Staphylococcus aureus* small-colony variant strain isolated from a cystic fibrosis patient: study of antibiotic combinations	THP-1 macrophages	*S. aureus* SCV	Rifampin Oritavancin Rifampin with oritavancin Quinupristin-dalfopristin Quinupristin–dalfopristin with rifampin Quinupristin–dalfopristin with oritavancin Moxifloxacin Moxifloxacin with rifampin Moxifloxacin with oritavancin	++ +++ ++++ + ++ +++ ++ ++ ++++ ++ ++ ++++	Rifampin and oritavancin were synergistic at all concentration ratios investigated. Oritavancin and moxifloxacin were also synergistic but only at high concentrations of oritavancin. Rifampin and moxifloxacin were additive	[108]
Plectasin shows intracellular activity against *Staphylococcus aureus* in human THP-1 monocytes and in a mouse peritonitis model	THP-1 macrophages In vivo mouse peritonitis model	*S. aureus*	Plectasin	THP-1 cells: ++ Mouse peritonitis: ++		[26]
Activities of antistaphylococcal antibiotics towards the extracellular and intraphagocytic forms of *Staphylococcus aureus* isolates from a patient with persistent bacteraemia and endocarditis	THP-1 macrophages	*S. aureus*	Rifampicin Linezolid Fusidic acid Vancomycin, Daptomycin, Quinupristin–Dalfopristin Oritavancin	N to ++ + to ++ + + + to ++ ++ ++ to +++		[80]

Modulation of the cellular accumulation and intracellular activity of daptomycin towards phagocytized *Staphylococcus aureus* by the P-glycoprotein (MDR1) efflux transporter in human THP-1 macrophages and Madin-Darby canine kidney cells	THP-1 macrophages Madin-Darby canine kidney (MDCK) cells	*S. aureus*	Daptomycin	THP-1 cells: ++ MDCK cells: ++		[82]
Pharmacodynamic evaluation of the intracellular activities of antibiotics against *Staphylococcus aureus* in a model of THP-1 macrophages	THP-1 macrophages	*S. aureus*	Azithromycin	+	Only oxacillin, levofloxacin, garenoxacin, moxifloxacin, oritavancin had true intracellular antimicrobial effect (>2 log reduction at Cmax for humans)	[12]
			Telithromycin	+		
			Gentamicin	+		
			Linezolid	++		
			Penicillin V	++		
			Nafcillin	++		
			Ampicillin	++		
			Oxacillin	+++		
			Teicoplanin	++		
			Vancomycin	++		
			Oritavancin	+++		
			Rifampin	++		
			Ciprofloxacin	++		
			Levofloxacin	+++		
			Garenoxacin	+++		
			Moxifloxacin	+++		
Evaluation of the extracellular and intracellular activities (human THP-1 macrophages) of telavancin versus vancomycin against methicillin-susceptible, methicillin-resistant, vancomycin-intermediate and vancomycin-resistant *Staphylococcus aureus*	THP-1 macrophages	*S. aureus*	Telavancin	+++		[13]
			Vancomycin	++		

(continued)

Table 8.1 (continued)

Study subject	Cell type	Pathogen	Antibiotics	Intracellular antibacterial effect[a]	Additional remarks	Reference
Factors influencing the intracellular activity of fluoroquinolones: a study using levofloxacin in a *Staphylococcus aureus* THP-1 monocyte model	THP-1 macrophages	*S. aureus*	Levofloxacin	++		[109]
Activity of three beta-lactams (ertapenem, meropenem and ampicillin) against intraphagocytic *Listeria monocytogenes* and *Staphylococcus aureus*	THP-1 macrophages	*S. aureus*	Ertapenem Meropenem Ampicillin	+ + +		[88]
Comparative activity of quinolones (ciprofloxacin, levofloxacin, moxifloxacin and garenoxacin) against extracellular and intracellular infection by *Listeria monocytogenes* and *Staphylococcus aureus* in J774 macrophages	Mouse J774 macrophages	*S. aureus*	Ciprofloxacin Levofloxacin, Moxifloxacin Garenoxacin	+ + ++ ++	Antibiotics were tested at a maximum concentration of 4 mg/L	[129]
Cellular pharmacokinetics and pharmacodynamics of the glycopeptide antibiotic oritavancin (LY333328) in a model of J774 mouse macrophages	Mouse J774 macrophages	*S. aureus*	Oritavancin	++++		[144]

Quantitative analysis of gentamicin, azithromycin, telithromycin, ciprofloxacin, moxifloxacin, and oritavancin (LY333328) activities against intracellular *Staphylococcus aureus* in mouse J774 macrophages	Mouse J774 macrophages	*S. aureus*	Gentamicin, Azithromycin, Telithromycin, Ciprofloxacin, Moxifloxacin Oritavancin	++ N + + ++ ++++	[130]
Discrepancy between uptake and intracellular activity of moxifloxacin in a *Staphylococcus aureus*-human THP-1 monocytic cell model	THP-1 macrophages	*S. aureus*	Moxifloxacin	+++	[114]

[a]Maximal antibacterial effects at any concentration upto the maximal concentration in humans upon treatment, reported for the listed antibiotics are based on fold changes relative to the inoculum after 24 h of incubation. The antibacterial effects are summarized as follows: *N* no bacteriostatic or bactericidal effect, i.e. increase of numbers of CFU; *S* static effect, no change in numbers of CFU, S* bacteriostatic for one test strain, + upto tenfold reduction of numbers of CFU (low activity), ++ between 10- and 100-fold reduction (moderate activity), +++ between 100- and 1,000-fold reduction (good activity), ++++, over 1,000-fold reduction of numbers of CFU (true bactericidal activity). The *S. aureus* strains used in the studied differed in antibiotic resistance and clonal relationship. Only the small colony phenotype is indicated. In several studies other bacterial species are tested in parallel; these data have not been included

The activity of antibiotics in the mouse peritonitis model were largely similar to their activities in the *in vitro* cell systems. In case of differences, antibiotics tended to be less capable of killing intracellular staphylococci *in vivo* than *in vitro* (Table 8.1). Results with dicloxacillin were not consistent between studies. This antibiotic was reported to show true bactericidal activity (++++ in Table 8.1) *in vivo* in one study [123], but required multiple doses to reach a moderate intracellular activity (++) in the peritonitis model in a follow-up study [124], where it also had low *in vitro* activity against *S. aureus* in THP-1 macrophages. The reason for these differences is not clear.

One of the most potent antibiotics against intracellular staphylococci appears to be the investigational lipoglycopeptide oritavancin (Table 8.1) which strongly accumulates intracellularly [143]. There has been a concern for oritavancin-associated intracellular deposition of concentric lamellar structures and finely granular material and other material in giant vesicles within phagocytic (J774 mouse macrophages) and nonphagocytic (rat embryo fibroblasts) cells, which is reminiscent of a mixed-lipid storage disorder [144]. The accumulation of oritavancin did not, however, interfere with the capacity of either mouse RAW264.7 or human HL60 phagocyte cell lines to kill *Candida albicans* and enhanced killing of *S. aureus*. Killing of *Acinetobacter baumannii* in HL60 cells was modestly reduced but not affected in RAW264.7 cells [11].

8.5.4 Antibiotics Acting Against Intracellular Small Colony Variants

Intracellular *S. aureus* SCVs are more difficult to kill than their corresponding normal colony strains. In THP-1 macrophages, oxacillin, fusidic acid, clindamycin, gentamicin, vancomycin, linezolid, and quinupristin–dalfopristin all caused a maximal 1-log reduction in CFU of intracellular *S. aureus* SCV after 24 h, whereas rifampin was slightly more active. The corresponding wild type and SCV revertant strains were more readily killed, with tenfold less CFU surviving intracellularly than of the SCV strains [107] (Table 8.1). The use of combinations of antibiotic showed promise to eradicate intracellular *S. aureus* SCV [108]. Oritavancin alone was already capable to cause a 2- to 3-log reduction, but together with rifampin true bactericidal activity was reached (>3-log reduction). Rifampin alone had moderate activity. Rifampin and oritavancin were synergistic at all concentration ratios investigated. Oritavancin and moxifloxacin were also synergistic but at high oritavancin concentrations only. Rifampin and moxifloxacin were additive. Quinupristine/dalfopristine had low activity and did not enhance activity of the other antibiotics [108]. Intracellular *S. aureus* wild type and SCV were almost equally susceptible to dicloxacillin and linezolid, causing a 1- to 2-log reduction inside THP-1 macrophages. In the *in vivo* mouse peritonitis model with one dose of antibiotic immediately after challenge, an initial suppression of *S. aureus* by both antibiotics was seen, but outgrowth occurred in later stages of infection, particularly of SCV [126]. These studies indicate that careful selection may provide antibiotic regimens capable of killing intracellular *S. aureus* SCV, one of the most difficult tasks to achieve. It would be very interesting to test antibiotic combinations, particularly oritavancin with rifampin, in the mouse peritonitis model.

8.5.5 Novel Antimicrobial Agents to Kill Intracellular Staphylococci

In view of the ever increasing antimicrobial resistance development, novel antimicrobials are urgently needed. A novel fluoroquinolone, JNJ-Q2, has not yet specifically been tested for intracellular activity, but had lower MIC values for *S. aureus* isolates, including MRSA (methicillin-resistant *S. aureus*), than contemporary fluoroquinones [104]. JNJ-Q2 was as active as moxifloxacin and more active than vancomycin in mouse MSSA (methicillin-suseceptible *S. aureus*) and MRSA septicemia, respectively. JNJ-Q2 reduced numbers of MRSA CFU in murine skin infection models to a larger extent than ciprofloxacin, moxifloxacin, linezolid, and vancomycin and did not select for ciprofloxacin resistance, whereas ciprofloxacin and moxifloxacin did select for such resistance [47].

Studies on entirely novel classes of antimicrobials are of paramount importance to complement the conventional antibiotic classes. Cationic antimicrobial peptides (AMP) are such a category of potential novel agents. Plectasin, a defensin-type AMP from the fungus *Pseudoplectania nigrella* targets the bacterial cell wall precursor lipid II, a mode of action not displayed by conventional antibiotics. Plectasin has been shown to have moderate capacity to kill intracellular *S. aureus* in THP-1 cells as well as in the mouse peritonitis model [26]. NZ2114, a variant of plectasin, has similar capacity to kill *S. aureus* inside THP-1 cells [25]. There is a large diversity of AMPs, which are produced by all forms of life. A number of AMPs have already reached clinical phase 2 or 3 testing [56], and application of AMPs in antimicrobial coatings for biomaterials is a subject of increasing interest [50, 68, 94, 98]. Thus, the capacity of AMP to eradicate intracellular bacteria in addition to their capacity to kill biofilm bacteria, deserves further investigation.

8.5.6 Tissue Cage Model for Intracellular Killing of S. aureus in Biomaterial-Associated Infection

The tissue cage model has been used to assess antibiotic activity in BAI *in vivo* in a number of studies reviewed by Zimmerli and Moser [158]. The model is also suited for detailed analysis of intracellular antibiotic activity [105]. Upon treatment, host cells residing in the body fluid in the tissue cages are harvested and intracellular bacteria quantified in a way similar to the procedure in the mouse peritonitis model

[123], to assess efficacy of intracellular antibiotic activity. In this model, the vast majority of infecting *S. aureus* bacteria were found to be localized intracellularly. In a comparison with oxacillin, levofloxacin was more effective in reducing the numbers of CFU, causing a 100-fold reduction of the intracellular staphylococci over an 8-day treatment period [105], but still approximately 3×10^4 CFU of *S. aureus* per ml survived. This study shows that the tissue cage model is useful to assess activity against bacteria within tissue fluid in tissue cages. To study intracellular activity against bacteria in tissue, which might be less accessible for antibiotics than bacteria in the tissue cage fluid, is a next challenge. Such models are truly necessary in order to develop antimicrobial strategies to eradicate bacteria causing BAI not only in the biofilms but also in their intracellular niches.

8.5.7 *Even the Best Antibiotics Fail to Eradicate Intracellular Staphylococci*

A point of major concern arising from the collective data on intracellular activity is, that even the antibiotics with the highest intracellular activity only partially eradicate the intracellular bacterial populations, even when applied at the concentrations maximally reached in treated humans (Table 8.1). With an intracellular bacterial population of approximately 10^7 CFU in the *in vitro* cell-based assays, the maximal reduction in numbers of CFU are in the order of 3-logs, and in the *in vivo* peritonitis model *S. aureus* shows regrowth after 24 h, even at optimized dosage and timing of the antibiotics [123]. There is some room for increasing antibiotic concentrations locally, e.g. when released from biomaterial surfaces, and this may improve the level of intracellular activity slightly [12], but we have to face the reality that a significant intracellular reservoir of bacteria will persist. This implies that in the patient, even the antibiotics with the best intracellular antibacterial effect likely will only partially suppress bacterial colonization of host cells. Thus, the final clearance of an infection depends to a large extent on the bactericidal capacity of the host immune cells that have internalized these bacteria. In BAI this is a major problem, since the presence of the biomaterial causes reduction of the efficacy of the host immune response and of intracellular bactericidal activity, as discussed earlier. It is therefore vital to address the immune-deranging effects of biomaterials, particularly in the presence of bacteria.

8.6 What Causes Survival of Bacteria in Tissue and Inside Host Cells and Can It Be Corrected?

The presence of a biomaterial causes a foreign body response, characterized by an initial acute inflammatory reaction followed by a decline of pro-inflammatory condition and development of a chronic inflammation, with tissue repair and final fibrosis/encapsulation (Chap. 2). These events are orchestrated by a number of specific inflammatory mediators, such as interleukins and chemokines. Biomaterial physicochemical and mechanical properties can influence the level of inflammation and the foreign body response cascade of events. A new field of research is to actually take benefit of these biomaterial characteristics, by designing materials with optimal adjuvant properties to enhance immune responses to antigens in vaccine development [49].

8.6.1 Presence of Bacteria Influences Immune Response to Biomaterials

Similar to the situation in the field of immunomodulatory biomaterials, there is also a combined presence of foreign body material and bacteria in the case of BAI, provoking a combined response. The nature of this inflammatory response is depending on the physicochemical characteristics of the biomaterial and of the inflammatory capacity of the bacteria. Unfortunately, there are no paradigms to predict the nature of the inflammatory response in relation to biomaterial characteristics yet, and this knowledge needs to be developed.

Enhancement of the inflammatory response to biomaterials by bacterial components has probably for the first time been shown in studies assessing differences in responses of monocytes to biomaterials. Human monocytes or macrophages exposed to biomaterials induced production of cytokines [22, 23, 34, 135]. However, detectable levels of the pro-inflammatory cytokines interleukin-1, TNF-alpha, and interleukin-6 were only found after addition of LPS, a major outer membrane component of Gram-negative bacteria. Moreover, the addition of LPS was also required to reveal differences in cytokine-inducing properties between the biomaterials [22, 23, 34, 135]. Thus, the inflammatory reactions in the tissue in response to a foreign body are modulated by bacterial components present in case of infection. Moreover, since implantation or insertion of a medical device inevitably will involve low level contamination by bacteria from the environment [62, 95] and/or the skin of the patient [6, 43], small numbers of bacteria will in general be present and may influence the host response to the biomaterial. The essential role of this costimulation is of pivotal importance in order to understand the infection-potentiating propensity of different materials.

8.6.2 Tailored Immune Modulation to Prevent Biomaterial-Associated Infection

In our studies in the mouse subcutaneous implant infection model, we observed strongly different types of immune responses with different materials infected with the standard experimental BAI strain of *S. epidermidis*, strain RP62a [36]. Silicone elastomer (SE) coated with polyvinylpyrrolide hydrogel (pvp) had a strongly reduced adherence of *S. epidermidis in vitro* compared to non-modified SE [18]. However, both in rabbits and mice, subcutaneously implanted SEpvp infected with *S. epidermidis* caused a strong and protracted pro-inflammatory response, which delayed the normal sequence of events of the foreign body response and was associated with high susceptibility to bacterial persistence [18, 27–30, 77, 153, 154]. The persistent pro-inflammatory status was correlated with a high and protracted level of the pro-inflammatory cytokine IL-1 beta in the peri-implant tissue [21]. In IL-1 receptor knock-out mice, this aberrant high pro-inflammatory response did not occur and susceptibility to infection was reduced to levels observed with non-coated SE [20]. This proved a causal relationship between the high IL-1 beta levels and the deranged immune response and ensuing susceptibility to infection and showed that it was possible to reduce susceptibility to infection by appropriate immune modulation.

The experimental material polyvinylpyrrolidone-coated polyamide behaved quite differently from SEpvp, yet also was very susceptible to protracted infection by *S. epidermidis*. Although the tissue of the mice at 14 days after challenge macroscopically did not show any signs of inflammation, large numbers of bacteria were cultured from the implants as well as the surrounding tissue, apparently not giving rise to an appropriate immune response. This was most likely due to the fact that the bacteria mostly were "hidden" inside macrophages in the peri-implant tissue, at a distance of 10–20 cells from the biomaterial-tissue interface [17]. Numbers of macrophages with bacteria and numbers of bacteria per macrophage increased over time, suggesting bacterial growth in the tissue and/or inside the macrophages. In line with this, a group a mice to be analyzed after 6 weeks had to be terminated at 3 weeks because they developed serious illness. These mice had disseminated infection as judged by positive cultures of blood and organs, implying that the tissue had been the reservoir of bacteria that caused this systemic implant-associated infection [17].

The cytokine profiles of the mice showed a remarkable lack of interferon-gamma at 5 days postimplantation/infection, which is normally increased at that time point in these experiments. As interferon-gamma is an important activator of macrophage intracellular killing mechanisms, we treated mice with interferon-gamma. This treatment restored susceptibility to infection to normal levels, and no macrophages with high numbers of bacteria were observed in the tissue anymore [19]. This again shows that immune dysregulation due to the combined presence of biomaterials and bacteria can be corrected by appropriate immune modulation, provided that the precise nature of the immune response to the particular biomaterial is known.

A good recent example of a controlled immune modulatory system is a nanoscale polypeptide IL-12-containing coating [92]. IL-12 helps to polarize naïve T-cells into the pro-inflammatory Th1 type T-cells and stimulates those cells to produce interferon-gamma which will upregulate the bactericidal mechanisms of the macrophages. Thus, the biomaterial coating will increase the macrophage bactericidal capacity. Indeed, in a rat femur fracture fixation model, the coating reduced *S. aureus* infection by 50 and 70 % at 6 and 21 days postinfection, respectively [92]. Additionally adding macrophage chemotactic protein 1 (MCP-1) to the coating had an adverse effect on the efficacy [93]. This illustrates that local immune modulation certainly has potential, but that careful tailoring of the combinations of agents and controlled release devices is needed to develop biomaterials which will prevent infection through optimization of the host's own immune defense.

8.7 Conclusions and Outlook

The work summarized in this chapter shows that in BAI, the causative bacteria not only form biofilms but also colonize the tissue surrounding medical devices, in animal models as well as in patients. Staphylococcal SCVs can be considered an adapted form of these bacteria for intracellular survival and are even harder to kill with antibiotics than their wild type forms. A wide range of antibiotics fail to kill intracellular staphylococci. Antibiotics that do have a true bactericidal effect, such as several fluoroquinolones and lipoglycopeptides alone or in combination with rifampin, do not fully eradicate the intracellular staphylococci. Thus, even the best antibiotics tested are unable to fully clear an intracellular infection, either *in vitro* or in mouse models. There are encouraging results, though, showing that immune modulation may be a way to assist the host's immune system to eradicate the invading bacteria, even in presence of the medical device which causes immune dysfunction. The type of modulation will need to be tailored to the nature of the immune derangement caused by the specific material, particularly when bacteria are also influencing this immune response. Combinations of immune modulators and antibiotics may provide better solutions than either of these classes of agents alone.

Thus, research on medical devices in view of infection prevention or control should focus on (1) prevention of immune dysregulation by the materials, (2) identify immune modulation strategies to correct dysregulation, (3) develop test systems to assess the immune modulating capacity of novel biomaterials, such as dendritic cells (Chap. 7) or whole animal models, (4) test antibiotics for intracellular bactericidal activity and translate this information to the clinic, and (5) develop novel antimicrobials with low risk of resistance development, with activity against bacteria both in biofilms and inside host cells.

Acknowledgments The research on biomaterial-associated infection at our Department builds on the work of Professor Jacob Dankert, who passed away much too early. Much of our own research described in this chapter has been performed by Corine Broekhuizen and Jaap Jan Boelens during their PhD periods, with excellent support of Leonie de Boer, Kim Schipper, Jean Luc Murk, and Jan Meeldijk.

References

1. Al Laham N, Rohde H, Sander G, et al. Augmented expression of polysaccharide intercellular adhesin in a defined *Staphylococcus epidermidis* mutant with the small-colony-variant phenotype. J Bacteriol. 2007;189:4494–501.
2. Arciola CR, Baldassarri L, Montanaro L. Presence of *icaA* and *icaD* genes and slime production in a collection of staphylococcal strains from catheter-associated infections. J Clin Microbiol. 2001;39:2152–6.
3. Arciola CR, Campoccia D, Gamberini S, et al. Search for the insertion element IS256 within the *ica* locus of *Staphylococcus epidermidis* clinical isolates collected from biomaterial-associated infections. Biomaterials. 2004;25:4117–25.
4. Arciola CR, Visai L, Testoni F, et al. Concise survey of *Staphylococcus aureus* virulence factors that promote adhesion and damage to peri-implant tissues. Int J Artif Organs. 2011;34:771–80.
5. Auwerx J. The human leukemia cell line, THP-1—a multifaceted model for the study of monocyte-macrophage differentiation. Experientia. 1991;47:22–31.
6. Bach A, Eberhardt H, Frick A, Schmidt H, Bottiger BW, Martin E. Efficacy of silver-coating central venous catheters in reducing bacterial colonization. Crit Care Med. 1999;27:515–21.
7. Baddour LM, Christensen GD. Prosthetic valve endocarditis due to small-colony staphylococcal variants. Rev Infect Dis. 1987;9:1168–74.
8. Baddour LM, Barker LP, Christensen GD, Parisi JT, Simpson W. Phenotypic variation of *Staphylococcus epidermidis* in infection of transvenous endocardial pacemaker electrodes. J Clin Microbiol. 1990;28:676–9.
9. Balwit JM, van Langevelde P, Vann JM, Proctor RA. Gentamicin-resistant menadione and hemin auxotrophic *Staphylococcus aureus* persist within cultured endothelial cells. J Infect Dis. 1994;170:1033–7.
10. Bantel H, Sinha B, Domschke W, Peters G, Schulze-Osthoff K, Jänicke RU. Alpha-toxin is a mediator of *Staphylococcus aureus*-induced cell death and activates caspases via the intrinsic death pathway independently of death receptor signaling. J Cell Biol. 2001;155:637–48.
11. Baquir B, Lemaire S, Van Bambeke F, Tulkens PM, Lin L, Spellberg B. Macrophage killing of bacterial and fungal pathogens is not inhibited by intense intracellular accumulation of the lipoglycopeptide antibiotic oritavancin. Clin Infect Dis. 2012;54:S229–32.
12. Barcia-Macay M, Seral C, Mingeot-Leclerq M, Tulkens P, Van Bambeke F. Pharmacodynamic evaluation of the intracellular activities of antibiotics against *Staphylococcus aureus* in a model of THP-1 macrophages. Antimicrob Agents Chemother. 2006;50:841–51.
13. Barcia-Macay M, Lemaire S, Mingeot-Leclercq M-P, Tulkens PM, Van Bambeke F. Evaluation of the extracellular and intracellular activities (human THP-1 macrophages) of telavancin versus vancomycin against methicillin-susceptible, methicillin-resistant, vancomycin-intermediate and vancomycin-resistant *Staphylococcus aureus*. J Antimicrob Chemother. 2006;58:1177–84.
14. Baudoux P, Lemaire S, Denis O, Tulkens PM, Van Bambeke F, Glupczynski Y. Activity of quinupristin/dalfopristin against extracellular and intracellular *Staphylococcus aureus* with various resistance phenotypes. J Antimicrob Chemother. 2010;65:1228–36.
15. Bayles KW, Wesson CA, Liou LE, et al. Intracellular *Staphylococcus aureus* escapes the endosome and induces apoptosis in epithelial cells intracellular. Infect Immun. 1998;66:336–42.
16. Beam TR. Sequestration of staphylococci at an inaccessible focus. Lancet. 1979;2(8136):227–8.
17. Boelens JJ, Dankert J, Murk JL, et al. Biomaterial-associated persistence of *Staphylococcus epidermidis* in pericatheter macrophages. J Infect Dis. 2000;181:1337–49.
18. Boelens JJ, Zaat SAJ, Meeldijk J, Dankert J. Subcutaneous abscess formation around catheters induced by viable and nonviable *Staphylococcus epidermidis* as well as by small amounts of bacterial cell wall components. J Biomed Mater Res. 2000;50:546–56.
19. Boelens JJ, van der Poll T, Dankert J, Zaat SAJ. Interferon-gamma protects against biomaterial-associated *Staphylococcus epidermidis* infection in mice. J Infect Dis. 2000;181:1167–71.

20. Boelens JJ, van der Poll T, Zaat SAJ, Murk JL, Weening JJ, Dankert J. Interleukin-1 receptor type I gene-deficient mice are less susceptible to *Staphylococcus epidermidis* biomaterial-associated infection than are wild-type mice. Infect Immun. 2000;68:6924–31.
21. Boelens JJ, Zaat SAJ, Murk JL, Weening JJ, Van der Poll T, Dankert J. Enhanced susceptibility to subcutaneous abscess formation and persistent infection around catheters is associated with sustained interleukin-1 beta levels. Infect Immun. 2000;68:1692–5.
22. Bonfield TL, Colton E, Anderson JM. Plasma protein adsorbed biomedical polymers: activation of human monocytes and induction of interleukin-1. J Biomed Mater Res. 1989;23:535–48.
23. Bonfield TL, Colton E, Marchant RE, Anderson JM. Cytokine and growth factor production by monocytes/macrophages on protein preadsorbed polymers. J Biomed Mater Res. 1992;26: 837–50.
24. Bosse MJ, Gruber HE, Ramp W. Internalization of bacteria by osteoblasts in a patient with recurrent, long-term osteomyelitis. J Bone Joint Surg Am. 2005;87A:1343–7.
25. Brinch KS, Tulkens PM, Van BF, Frimodt-Moller N, Hoiby N, Kristensen HH. Intracellular activity of the peptide antibiotic NZ2114: studies with *Staphylococcus aureus* and human THP-1 monocytes, and comparison with daptomycin and vancomycin. J Antimicrob Chemother. 2010;65:1720–4.
26. Brinch KS, Sandberg A, Baudoux P, et al. Plectasin shows intracellular activity against *Staphylococcus aureus* in human THP-1 monocytes and in a mouse peritonitis model. Antimicrob Agents Chemother. 2009;53:4801–8.
27. Broekhuizen CAN, De Boer L, Schipper K, et al. Peri-implant tissue is an important niche for *Staphylococcus epidermidis* in experimental biomaterial-associated infection in mice. Infect Immun. 2007;75:1129–36.
28. Broekhuizen CAN, De Boer L, Schipper K, et al. The influence of antibodies on *Staphylococcus epidermidis* adherence to polyvinylpyrrolidone-coated silicone elastomer in experimental biomaterial-associated infection in mice. Biomaterials. 2009;30:6444–50.
29. Broekhuizen CAN, Sta M, Vandenbroucke-Grauls CM, Zaat SAJ. Microscopic detection of viable *Staphylococcus epidermidis* in peri-implant tissue in experimental biomaterial-associated infection, identified by bromodeoxyuridine incorporation. Infect Immun. 2010;78:954–62.
30. Broekhuizen CA, De Boer L, Schipper K, et al. *Staphylococcus epidermidis* is cleared from biomaterial implants but persists in peri-implant tissue in mice despite rifampicin/vancomycin treatment. J Biomed Mater Res A. 2008;85:498–505.
31. Broekhuizen CA, Schultz MJ, van der Wal AC, et al. Tissue around catheters is a niche for bacteria associated with medical device infection. Crit Care Med. 2008;36:2395–402.
32. Busscher HJ, van der Mei HC. How do bacteria know they are on a surface and regulate their response to an adhering state? PLoS Pathog. 2012;8:e1002440.
33. Cafiso V, Bertuccio T, Santagati M, et al. Presence of the *ica* operon in clinical isolates of *Staphylococcus epidermidis* and its role in biofilm production. Clin Microbiol Infect. 2004;10:1081–8.
34. Cardona MA, Simmons RL, Kaplan SS. TNF and IL-1 generation by human monocytes in response to biomaterials. J Biomed Mater Res. 1992;26:851–9.
35. Christensen GD, Baddour LM, Madison BM, et al. Colonial morphology of staphylococci on Memphis agar: phase variation of slime production, resistance to beta-lactam antibiotics, and virulence. J Infect Dis. 1990;161:1153–69.
36. Christensen GD, Simpson WA, Bisno AL, Beachey EH. Experimental foreign body infections in mice challenged with slime-producing *Staphylococcus epidermidis*. Infect Immun. 1983; 40:407–10.
37. Christensen L, Breiting V, Janssen M, Vuust J, Hogdall E. Adverse reactions to injectable soft tissue permanent fillers. Aesthetic Plast Surg. 2005;29:34–48.
38. Ciampolini J, Harding KG. Pathophysiology of chronic bacterial osteomyelitis. Why do antibiotics fail so often? Postgrad Med J. 2000;76:479–83.
39. Costerton JW, Stewart PS, Greenberg EP. Bacterial biofilms: a common cause of persistent infections. Science. 1999;284:1318–22.

40. Darouiche R, Hamill RJ. Antibiotic penetration of and bactericidal activity within endothelial cells. Antimicrob Agents Chemother. 1994;38:1059–64.
41. De Silva GDI, Kantzanou M, Justice A, et al. The *ica* operon and biofilm production in coagulase-negative staphylococci associated with carriage and disease in a neonatal intensive care unit. J Clin Microbiol. 2002;40:382–8.
42. Deighton MA, Borland R, Capstick JA. Virulence of *Staphylococcus epidermidis* in a mouse model: significance of extracellular slime. Epidemiol Infect. 1996;117:267–80.
43. Dobbins BM, Kite P, Kindon A, McMahon MJ, Wilcox MH. DNA fingerprinting analysis of coagulase negative staphylococci implicated in catheter related bloodstream infections. J Clin Pathol. 2002;55:824–8.
44. Elek SD, Conen PE. The virulence of *Staphylococcus* pyogenes for man: a study of the problems of wound infection. Br J Exp Pathol. 1957;38(6):573.
45. Ellington JK, Harris M, Webb L, et al. A mechanism for the indolence of osteomyelitis. J Bone Joint Surg Br. 2003;85-B:918–21.
46. Engelsman AF, Saldarriaga-Fernandez IC, Nejadnik MR, et al. The risk of biomaterial-associated infection after revision surgery due to an experimental primary implant infection. Biofouling. 2010;26:761–7.
47. Fernandez J, Hilliard JJ, Morrow BJ, et al. Efficacy of a new fluoroquinolone, JNJ-Q2, in murine models of *Staphylococcus aureus* and *Streptococcus pneumoniae* skin, respiratory, and systemic infections. Antimicrob Agents Chemother. 2011;55:5522–8.
48. Frank KL, Hanssen AD, Patel R. *icaA* is not a useful diagnostic marker for prosthetic joint infection. J Clin Microbiol. 2004;42:4846–9.
49. Franz S, Rammelt S, Scharnweber D, Simon JC. Immune responses to implants—a review of the implications for the design of immunomodulatory biomaterials. Biomaterials. 2011;32:6692–709.
50. Gao G, Lange D, Hilpert K, et al. The biocompatibility and biofilm resistance of implant coatings based on hydrophilic polymer brushes conjugated with antimicrobial peptides. Biomaterials. 2011;32:3899–909.
51. Garzoni C, Kelley WL. *Staphylococcus aureus*: new evidence for intracellular persistence. Trends Microbiol. 2009;17:59–65.
52. Gotz F, Heilmann C, Cramton SE. Molecular basis of catheter associated infections by staphylococci. Adv Exp Med Biol. 2000;485:103–11.
53. Gresham HD, Lowrance JH, Caver TE, et al. Survival of *Staphylococcus aureus* inside neutrophils contributes to infection. J Immunol. 2000;164:3713–22.
54. Gristina A, Costerton J. Bacterial adherence to biomaterials and tissue; the significance of its role in clinical sepsis. J Bone Joint Surg Am. 1985;67:264–73.
55. Grundmeier M, Tuchscherr L, Brück M, et al. Staphylococcal strains vary greatly in their ability to induce an inflammatory response in endothelial cells. J Infect Dis. 2010;201:871–80.
56. Guaní-Guerra E, Santos-Mendoza T, Lugo-Reyes SO, Terán LM. Antimicrobial peptides: general overview and clinical implications in human health and disease. Clin Immunol. 2010;135:1–11.
57. Hamill RJ, Vann JM, Proctor RA. Phagocytosis of *Staphylococcus aureus* by cultured bovine aortic endothelial cells: model for postadherence events in endovascular infections. Infect Immun. 1986;54:833–6.
58. Hanses F, Kopp A, Bala M, et al. Intracellular survival of *Staphylococcus aureus* in adipocyte-like differentiated 3T3-L1 cells is glucose dependent and alters cytokine, chemokine, and adipokine secretion. Endocrinology. 2011;152:4148–57.
59. Haslinger B, Strangfeld K, Peters G, Schulze-Osthoff K, Sinha B. *Staphylococcus aureus* alphatoxin induces apoptosis in peripheral blood mononuclear cells: role of endogenous tumour necrosis factor-alpha and the mitochondrial death pathway. Cell Microbiol. 2003;5:729–41.
60. Holmberg A, Lood R, Mörgelin M, et al. Biofilm formation by *Propionibacterium acnes* is a characteristic of invasive isolates. Clin Microbiol Infect. 2009;15:787–95.
61. Hudson MC, Ramp WK, Nicholson NC, Williams AS, Nousiainen MT. Internalization of *Staphylococcus aureus* by cultured osteoblasts. Microb Pathog. 1995;19:409–19.
62. Hughes SP, Anderson FM. Infection in the operating room. J Bone Joint Surg Br. 1999;81:754–5.

63. Høiby N, Bjarnsholt T, Givskov M, Molin S, Ciofu O. Antibiotic resistance of bacterial biofilms. Int J Antimicrob Agents. 2010;35:322–32.
64. James RC, MacLeod CM. Induction of staphylococcal infections in mice with small inocula introduced on sutures. Br J Exp Pathol. 1961;42:266–77.
65. Jevon M, Guo C, Ma B, et al. Mechanisms of internalization of *Staphylococcus aureus* by cultured human osteoblasts. Infect Immun. 1999;67:2677–81.
66. Johnson G, Lee D, Regelmann W, Gray E, Peters G, Quie PG. Interference with granulocyte function by *Staphylococcus epidermidis* slime. Infect Immun. 1986;54:13–20.
67. Kahl BC, Belling G, Becker P, et al. Thymidine-dependent *Staphylococcus aureus* small-colony variants are associated with extensive alterations in regulator and virulence gene expression profiles. Infect Immun. 2005;73:4119–26.
68. Kazemzadeh-Narbat M, Kindrachuk J, Duan K, Jenssen H, Hancock REW, Wang R. Antimicrobial peptides on calcium phosphate-coated titanium for the prevention of implant-associated infections. Biomaterials. 2010;31:9519–26.
69. Kim KW, Im J, Jeon JH, Lee H-G, Yun C-H, Han SH. *Staphylococcus aureus* induces IL-1β expression through the activation of MAP kinases and AP-1, CRE and NF-κB transcription factors in the bovine mammary gland epithelial cells. Comp Immunol Microbiol Infect Dis. 2011;34:347–54.
70. Klug D, Wallet F, Kacet S, Courcol RJ. Involvement of adherence and adhesion *Staphylococcus epidermidis* genes in pacemaker lead-associated infections. J Clin Microbiol. 2003;41:3348–50.
71. Kohler C, Eiff CV, Peters G, Proctor RA, Hecker M, Engelmann S. Physiological characterization of a heme-deficient mutant of *Staphylococcus aureus* by a proteomic approach. J Bacteriol. 2003;185:6928–37.
72. Koskela A, Nilsdotter-Augustinsson A, Persson L, Söderquist B. Prevalence of the ica operon and insertion sequence IS256 among *Staphylococcus epidermidis* prosthetic joint infection isolates. Eur J Clin Microbiol Infect Dis. 2009;28:655–60.
73. Kretschmer D, Gleske A-K, Rautenberg M, et al. Human formyl peptide receptor 2 senses highly pathogenic *Staphylococcus aureus*. Cell Host Microbe. 2010;7:463–73.
74. Kretschmer D, Nikola N, Dürr M, Otto M, Peschel A. The virulence regulator agr controls the staphylococcal capacity to activate human neutrophils via the formyl peptide receptor 2. J Innate Immun. 2012;4:201–12.
75. Krimmer V, Merkert H, Eiff CV, et al. Detection of *Staphylococcus aureus* and *Staphylococcus epidermidis* in clinical samples by 16S rRNA-directed in situ hybridization. J Clin Microbiol. 1999;37:2667–73.
76. Kubica M, Guzik K, Koziel J, et al. A potential new pathway for *Staphylococcus aureus* dissemination: the silent survival of *S aureus* phagocytosed by human monocyte-derived macrophages. PLoS One. 2008;3(1):e1409.
77. Kwakman PHS, te Velde AA, Vandenbroucke-Grauls CMJE, van Deventer SJH, Zaat SAJ. Treatment and prevention of *Staphylococcus epidermidis* experimental biomaterial-associated infection by bactericidal peptide 2. Antimicrob Agents Chemother. 2006;50:3977–83.
78. Lambe Jr DW, Ferguson KP, Keplinger JL, Gemmell CG, Kalbfleisch J. Pathogenicity of *Staphylococcus* lugdunensis, *Staphylococcus* schleiferi, and 3 other coagulase-negative staphylococci in a mouse model and possible virulence factors. Can J Microbiol. 1990;36:455–63.
79. Lannergard J, von Eiff C, Sander G, et al. Identification of the genetic basis for clinical menadione-auxotrophic small-colony variant isolates of *Staphylococcus aureus*. Antimicrob Agents Chemother. 2008;52:4017–22.
80. Lemaire S, Kosowska-Shick K, Julian K, Tulkens PM, Van Bambeke F, Appelbaum PC. Activities of antistaphylococcal antibiotics towards the extracellular and intraphagocytic forms of *Staphylococcus aureus* isolates from a patient with persistent bacteraemia and endocarditis. Clin Microbiol Infect. 2008;14:766–77.
81. Lemaire S, Van Bambeke F, Appelbaum PC, Tulkens PM. Cellular pharmacokinetics and intracellular activity of torezolid (TR-700): studies with human macrophage (THP-1) and endothelial (HUVEC) cell lines. J Antimicrob Chemother. 2009;64:1035–43.
82. Lemaire S, Van Bambeke F, Mingeot-Leclercq MP, Tulkens PM. Modulation of the cellular accumulation and intracellular activity of daptomycin towards phagocytized *Staphylococcus*

aureus by the P-glycoprotein (MDR1) efflux transporter in human THP-1 macrophages and madin-darby canine kidney cells. Antimicrob Agents Chemother. 2007;51:2748–57.
83. Lemaire S, Glupczynski Y, Duval V, Joris B, Tulkens PM, Van Bambeke F. Activities of ceftobiprole and other cephalosporins against extracellular and intracellular (THP-1 macrophages and keratinocytes) forms of methicillin-susceptible and methicillin-resistant *Staphylococcus aureus*. Antimicrob Agents Chemother. 2009;53:2289–97.
84. Lemaire S, Kosowska-Shick K, Appelbaum PC, Glupczynski Y, Van Bambeke F, Tulkens PM. Activity of moxifloxacin against intracellular community-acquired methicillin-resistant *Staphylococcus aureus*: comparison with clindamycin, linezolid and co-trimoxazole and attempt at defining an intracellular susceptibility breakpoint. J Antimicrob Chemother. 2011;66:596–607.
85. Lemaire S, Kosowska-Shick K, Appelbaum PC, Verween G, Tulkens PM, Van Bambeke F. Cellular pharmacodynamics of the novel biaryloxazolidinone radezolid: studies with infected phagocytic and nonphagocytic cells, using *Staphylococcus aureus*, *Staphylococcus epidermidis*, Listeria monocytogenes, and Legionella pneumophila. Antimicrob Agents Chemother. 2010;54:2549–59.
86. Lemaire S, Tulkens PM, Van Bambeke F. Contrasting effects of acidic pH on the extracellular and intracellular activities of the anti-gram-positive fluoroquinolones moxifloxacin and delafloxacin against *Staphylococcus aureus*. Antimicrob Agents Chemother. 2011;55:649–58.
87. Lemaire S, Van Bambeke F, Pierard D, Appelbaum PC, Tulkens PM. Activity of fusidic acid against extracellular and intracellular *Staphylococcus aureus*: influence of pH and comparison with linezolid and clindamycin. Clin Infect Dis. 2011;52:S493–503.
88. Lemaire S, Van Bambeke F, Mingeot-Leclercq M-P, Tulkens PM. Activity of three beta-lactams (ertapenem, meropenem and ampicillin) against intraphagocytic Listeria monocytogenes and *Staphylococcus aureus*. J Antimicrob Chemother. 2005;55:897–904.
89. Lemaire S, Van Bambeke F, Tulkens PM. Activity of finafloxacin, a novel fluoroquinolone with increased activity at acid pH, towards extracellular and intracellular *Staphylococcus aureus*, Listeria monocytogenes and Legionella pneumophila. Int J Antimicrob Agents. 2011;38:52–9.
90. Lemaire S, Van Bambeke F, Tulkens PM. Cellular accumulation and pharmacodynamic evaluation of the intracellular activity of CEM-101, a novel fluoroketolide, against *Staphylococcus aureus*, Listeria monocytogenes, and Legionella pneumophila in human THP-1 macrophages. Antimicrob Agents Chemother. 2009;53:3734–43.
91. Lewis K. Persister cells, dormancy and infectious disease. Nat Rev Microbiol. 2007;5:48–56.
92. Li B, Jiang B, Boyce BM, Lindsey BA. Multilayer polypeptide nanoscale coatings incorporating IL-12 for the prevention of biomedical device-associated infections. Biomaterials. 2009;30:2552–8.
93. Li B, Jiang B, Dietz MJ, Smith ES, Clovis NB, Rao KMK. Evaluation of local MCP-1 and IL-12 nanocoatings for infection prevention in open fractures. J Orthop Res. 2010;28:48–54.
94. Liu L, Xu K, Wang H, et al. Self-assembled cationic peptide nanoparticles as an efficient antimicrobial agent. Nat Nanotechnol. 2009;4:457–63.
95. Loftus RW, Koff MD, Burchman CC, et al. Transmission of pathogenic bacterial organisms in the anesthesia work area. Anesthesiology. 2008;109:399–407.
96. Lâm T-T, Giese B, Chikkaballi D, et al. Phagolysosomal integrity is generally maintained after *Staphylococcus aureus* invasion of nonprofessional phagocytes but is modulated by strain 6850. Infect Immun. 2010;78:3392–403.
97. Löffler B, Hussain M, Grundmeier M, et al. *Staphylococcus aureus* panton-valentine leukocidin is a very potent cytotoxic factor for human neutrophils. PLoS Pathog. 2010;6:e1000715.
98. Ma M, Kazemzadeh-Narbat M, Hui Y, et al. Local delivery of antimicrobial peptides using self-organized TiO(2) nanotube arrays for peri-implant infections. J Biomed Mater Res A. 2011;100A:278–85.
99. Marrie TJ, Costerton JW. Morphology of bacterial attachment to cardiac pacemaker leads and power packs. J Clin Microbiol. 1984;19:911–4.
100. Marrie TJ, Costerton JW. Scanning and transmission electron microscopy of in situ bacterial colonization of intravenous and intra-arterial catheters. J Clin Microbiol. 1984;19:687–93.
101. Marrie TJ, Nelligan J, Costerton JW. A scanning and transmission electron microscopic study of an infected endocardial pacemaker lead. Circulation. 1982;66:1339–41.

102. Marrie TJ, Noble MA, Costerton JW. Examination of the morphology of bacteria adhering to peritoneal dialysis catheters by scanning and transmission electron microscopy. J Clin Microbiol. 1983;18:1388–98.
103. Matussek A, Strindhall J, Stark L, et al. Infection of human endothelial cells with *Staphylococcus aureus* induces transcription of genes encoding an innate immunity response. Scand J Immunol. 2005;61:536–44.
104. Morrow BJ, He W, Amsler KM, et al. In vitro antibacterial activities of JNJ-Q2, a new broad-spectrum fluoroquinolone. Antimicrob Agents Chemother. 2010;54:1955–64.
105. Murillo O, Pachón ME, Euba G, et al. Intracellular antimicrobial activity appearing as a relevant factor in antibiotic efficacy against an experimental foreign-body infection caused by *Staphylococcus aureus*. J Antimicrob Chemother. 2009;64:1062–6.
106. Neut D, Van Horn J, Van Kooten T, Van der Mei H, Busscher H. Detection of biomaterial-associated infections in orthopaedic joint implants. Clin Orthop Relat Res. 2003;413:261–8.
107. Nguyen HA, Denis O, Vergison A, et al. Intracellular activity of antibiotics in a model of human THP-1 macrophages infected by a *Staphylococcus aureus* small-colony variant strain isolated from a cystic fibrosis patient: pharmacodynamic evaluation and comparison with isogenic normal-phenotype. Antimicrob Agents Chemother. 2009;53:1434–42.
108. Nguyen HA, Denis O, Vergison A, Tulkens PM, Struelens MJ, Van Bambeke F. Intracellular activity of antibiotics in a model of human THP-1 macrophages infected by a *Staphylococcus aureus* small-colony variant strain isolated from a cystic fibrosis patient: study of antibiotic combinations. Antimicrob Agents Chemother. 2009;53:1443–9.
109. Nguyen HA, Grellet J, Paillard D, Dubois V, Quentin C, Saux M-C. Factors influencing the intracellular activity of fluoroquinolones: a study using levofloxacin in a *Staphylococcus aureus* THP-1 monocyte model. J Antimicrob Chemother. 2006;57:883–90.
110. Noble WC. Production of subcutaneous staphylococcal skin lesions in mice. Br J Exp Pathol. 1965;46:254–62.
111. Otto M. Staphylococcus epidermidis—the "accidental" pathogen. Nat Rev Microbiol. 2009; 7:555–67.
112. Oviedo-Boyso J, Bravo-Patiño A, Cajero-Juárez M, Valdez-Alarcón JJ, Baizabal-Aguirre VM. TNF-alpha reduces the level of *Staphylococcus epidermidis* internalization by bovine endothelial cells. FEMS Microbiol Lett. 2009;292:92–9.
113. Oviedo-Boyso J, Cardoso-Correa BI, Cajero-Juárez M, Bravo-Patiño A, Valdez-Alarcón JJ, Baizabal-Aguirre VM. The capacity of bovine endothelial cells to eliminate intracellular *Staphylococcus aureus* and *Staphylococcus epidermidis* is increased by the proinflammatory cytokines TNF-alpha and IL-1beta. FEMS Immunol Med Microbiol. 2008;54:53–9.
114. Paillard D, Grellet J, Dubois V, Saux M-C, Quentin C. Discrepancy between uptake and intracellular activity of moxifloxacin in a *Staphylococcus aureus*—human THP-1 monocytic cell model. Antimicrob Agents Chemother. 2002;46:288–93.
115. Passerini L, Lam K, Costerton JW, King EG. Biofilms on indwelling vascular catheters. Crit Care Med. 1992;20:665–73.
116. Passerini L, Phang PT, Jackson FL, Lam K, Costerton JW, King EG. Biofilms on right heart flow-directed catheters. Chest. 1987;92:440–6.
117. Patrick CC, Hetherington SV, Roberson PK, Henwick S, Sloas MM. Comparative virulence of *Staphylococcus epidermidis* isolates in a murine catheter model. Pediatr Res. 1995;37:70–4.
118. Patrick S, Mcdowell A, Glenn JV, Tunney MM. Improved detection and treatment of prosthetic joint infection. Eur Cell Mater. 2008;16:30.
119. Peters G, Locci R, Pulverer G. Microbial colonization of prosthetic devices. II. Scanning electron microscopy of naturally infected intravenous catheters. Zentralbl Bakteriol Mikrobiol Hyg B. 1981;173:293–9.
120. Proctor RA, von Eiff C, Kahl BC, et al. Small colony variants: a pathogenic form of bacteria that facilitates persistent and recurrent infections. Nat Rev Microbiol. 2006;4:295–305.
121. Raad I, Costerton W, Sabharwal U, Sacilowski M, Anaissie E, Bodey GP. Ultrastructural analysis of indwelling vascular catheters: a quantitative relationship between luminal colonization and duration of placement. J Infect Dis. 1993;168:400–7.

122. Sadowska B, Bonar A, von Eiff C, et al. Characteristics of *Staphylococcus aureus*, isolated from airways of cystic fibrosis patients, and their small colony variants. FEMS Immunol Med Microbiol. 2002;32:191–7.
123. Sandberg A, Hessler JHR, Skov RL, Blom J, Frimodt-Møller N. Intracellular activity of antibiotics against *Staphylococcus aureus* in a mouse peritonitis model. Antimicrob Agents Chemother. 2009;53:1874–83.
124. Sandberg A, Jensen KS, Baudoux P, Van Bambeke F, Tulkens PM, Frimodt-Møller N. Intra- and extracellular activities of dicloxacillin against *Staphylococcus aureus* in vivo and in vitro. Antimicrob Agents Chemother. 2010;54:2391–400.
125. Sandberg A, Jensen KS, Baudoux P, Van Bambeke F, Tulkens PM, Frimodt-Møller N. Intra- and extracellular activity of linezolid against *Staphylococcus aureus* in vivo and in vitro. J Antimicrob Chemother. 2010;65:962–73.
126. Sandberg A, Lemaire S, Van Bambeke F, et al. Intra- and extracellular activities of dicloxacillin and linezolid against a clinical *Staphylococcus aureus* strain with a small-colony-variant phenotype in an in vitro model of THP-1 macrophages and an in vivo mouse peritonitis model. Antimicrob Agents Chemother. 2011;55:1443–52.
127. Schröder A, Kland R, Peschel A, von Eiff C, Aepfelbacher M. Live cell imaging of phagosome maturation in *Staphylococcus aureus* infected human endothelial cells: small colony variants are able to survive in lysosomes. Med Microbiol Immunol. 2006;195: 185–94.
128. Seifert H, Wisplinghof H, Schnabel P, Von Eiff C. Small colony variants of *Staphylococcus aureus* and pacemaker-related infection. Emerg Infect Dis. 2003;9:1316–8.
129. Seral C, Barcia-Macay M, Mingeot-Leclercq MP, Tulkens PM, Van Bambeke F. Comparative activity of quinolones (ciprofloxacin, levofloxacin, moxifloxacin and garenoxacin) against extracellular and intracellular infection by Listeria monocytogenes and *Staphylococcus aureus* in J774 macrophages. J Antimicrob Chemother. 2005;55:511–7.
130. Seral C, Bambeke FV, Tulkens PM, Bambeke V. Quantitative analysis of gentamicin, azithromycin, telithromycin, ciprofloxacin, moxifloxacin, and oritavancin (LY333328) activities against intracellular Staphylococcus aureus in mouse J774 macrophages. Antimicrob Agents Chemother. 2003;47:2283–92.
131. Shiau AL, Wu CL. The inhibitory effect of *Staphylococcus epidermidis* slime on the phagocytosis of murine peritoneal macrophages is interferon-independent. Microbiol Immunol. 1998;42:33–40.
132. Sjollema J, Sharma PK, Dijkstra RJB, et al. The potential for bio-optical imaging of biomaterial-associated infection in vivo. Biomaterials. 2010;31:1984–95.
133. Southwood RT, Rice JL, McDonald PJ, Hakendorf PH, Rozenbilds M. Infection in experimental hip arthroplasties. J Bone Joint Surg Br. 1985;67:229–31.
134. Stoodley P, Nistico L, Johnson S, et al. Direct demonstration of viable *Staphylococcus aureus* biofilms in an infected total joint arthroplasty. A case report. J Bone Joint Surg Am. 2008;90:1751–8.
135. Swartbol P, Truedsson L, Pärsson H, Norgren L. Tumor necrosis factor-alpha and interleukin-6 release from white blood cells induced by different graft materials in vitro are affected by pentoxifylline and iloprost. J Biomed Mater Res. 1997;36:400–6.
136. Taubler JH, Kapral FA. Staphylococcal population changes in experimentally infected mice: infection with suture-adsorbed and unadsorbed organisms grown in vitro and in vivo. J Infect Dis. 1966;116:257.
137. Testoni F, Montanaro L, Poggi A, Visai L, Campoccia D, Arciola CR. Internalization by osteoblasts of two *Staphylococcus aureus* clinical isolates differing in their adhesin gene pattern. Int J Artif Organs. 2011;34:789–98.
138. Trampuz A, Piper KE, Hanssen AD, et al. Sonication of explanted prosthetic components in bags for diagnosis of prosthetic joint infection is associated with risk of contamination. J Clin Microbiol. 2006;44:628–31.
139. Tuchscherr L, Heitmann V, Hussain M, et al. *Staphylococcus aureus* small-colony variants are adapted phenotypes for intracellular persistence. J Infect Dis. 2010;202:1031–40.
140. Tulkens PM. Intracellular distribution and activity of antibiotics. Eur J Clin Microbiol Infect Dis. 1991;10:100–6.

141. Tunney MM, Patrick S, Curran MD, et al. Detection of prosthetic hip infection at revision arthroplasty by immunofluorescence microscopy and PCR amplification of the bacterial 16S rRNA gene. J Clin Microbiol. 1999;37:3281–90.
142. Vallet CM, Marquez B, Ngabirano E, et al. Cellular accumulation of fluoroquinolones is not predictive of their intracellular activity: studies with gemifloxacin, moxifloxacin and ciprofloxacin in a pharmacokinetic/pharmacodynamic model of uninfected and infected macrophages. Int J Antimicrob Agents. 2011;38:249–56.
143. Van Bambeke F, Carryn S, Seral C, et al. Cellular pharmacokinetics and pharmacodynamics of the glycopeptide antibiotic oritavancin (LY333328) in a model of J774 mouse macrophages. Antimicrob Agents Chemother. 2004;48:2853–60.
144. Van Bambeke F, Saffran J, Mingeot-Leclercq M-P, Tulkens PM. Mixed-lipid storage disorder induced in macrophages and fibroblasts by oritavancin (LY333328), a new glycopeptide antibiotic with exceptional cellular accumulation. Antimicrob Agents Chemother. 2005;49:1695–700.
145. Vann J, Proctor R. Cytotoxic effects of ingested *Staphylococcus aureus* on bovine endothelial cells: role of *S. aureus* alpha-hemolysin. Microb Pathog. 1988;4:443–53.
146. Vesga O, Groeschel MC, Otten MF, Brar DW, Vann JM, Proctor RA. *Staphylococcus aureus* small colony variants are induced by the endothelial cell intracellular milieu. J Infect Dis. 1996;173:739–42.
147. Virden CP, Dobke MK, Stein P, Parsons CL, Frank D. Subclinical infection of the silicone breast implant surface as a possible cause of capsular contracture. Aesthetic Plast Surg. 1992;16:173–9.
148. Von Eiff C, Vaudaux P, Kahl BC, et al. Bloodstream infections caused by small-colony variants of coagulase-negative staphylococci following pacemaker implantation. Clin Infect Dis. 1999;29:932–4.
149. Von Eiff C, Becker K, Metze D, Lubritz G. Intracellular persistence of Staphylococcus aureus small-colony variants within keratinocytes: a cause for antibiotic treatment failure in a patient with Darier' s disease. Clin Infect Dis. 2001;32:1643–7.
150. Vuong C, Kocianova S, Yao Y, Carmody AB, Otto M. Increased colonization of indwelling medical devices by quorum-sensing mutants of *Staphylococcus epidermidis* in vivo. J Infect Dis. 2004;190:1498–505.
151. Waldvogel F, Bisno A, editors. Infections associated with indwelling medical devices. Washington, DC: ASM Press; 2000.
152. Wang R, Braughton KR, Kretschmer D, et al. Identification of novel cytolytic peptides as key virulence determinants for community-associated MRSA. Nat Med. 2007;13:1510–4.
153. Zaat S, Broekhuizen C, De Boer L, et al. Biomaterial-associated infection: breaking out of the biofilm. Eur Cell Mater. 2008;16:10.
154. Zaat S, Broekhuizen C, Riool M. Host tissue as a niche for biomaterial-associated infection. Future Microbiol. 2010;5:1149–51.
155. Zautner AE, Krause M, Stropahl G, et al. Intracellular persisting *Staphylococcus aureus* is the major pathogen in recurrent tonsillitis. PLoS One. 2010;5:e9452.
156. Zimmerli W, Lew PD, Waldvogel FA. Pathogenesis of foreign body infection Evidence for a local granulocyte defect. J Clin Invest. 1984;73:1191–200.
157. Zimmerli W, Waldvogel FA, Vaudaux P, Nydegger HE. Pathogenesis of foreign body infection: description and characteristics of an animal model. J Infect Dis. 1982;146:487.
158. Zimmerli W, Moser C. Pathogenesis and treatment concepts of orthopaedic biofilm infections. FEMS Immunol Med Microbiol. 2012;1:1–11.
159. Zimmerli W, Sendi P. Pathogenesis of implant-associated infection: the role of the host. Semin Immunopathol. 2011;33:295–306.

Chapter 9
Systemic Treatment Options for Medical Device-Associated Infection

Oscar Murillo, Jaime Lora-Tamayo, and Javier Ariza

Abstract Difficulties in eradicating medical device-related infections are primarily related to the presence of bacterial biofilms. The foreign body can often facilitate such infections, which may be usually caused by non-aggressive microorganisms with the ability to form biofilms even at low inoculum size.

The biofilm is responsible for several phenotypic changes in the bacteria including increased minimal bactericidal concentrations (tolerance to antibiotics). Other factors that have also been related to difficulties in the treatment of medical device-related infections are functional abnormalities in the activity of phagocytic cells in contact with the foreign body and the presence of intracellular bacteria.

While the anatomical location of the medical device can determine certain aspects of the treatment of these infections, this therapy must include an appropriate and lengthy antibiotic treatment combined with adequate surgical intervention. Antimicrobial therapy needs to be carefully designed, and the antibiotics to use against device-related infections can be chosen according both to their activity against bacterial biofilms and nongrowing microorganisms, and to their intracellular efficacy.

The specific characteristics of medical device-related infections, as well as the difficulties involved in their treatment, mean that multidisciplinary medical teams are required to ensure the optimal approach to and management of this pathology.

O. Murillo (✉) • J. Lora-Tamayo • J. Ariza
Department of Infectious Diseases, IDIBELL, Hospital Universitari Bellvitge, Barcelona, Spain
e-mail: omurillo@bellvitgehospital.cat

T.F. Moriarty et al. (eds.), *Biomaterials Associated Infection: Immunological Aspects and Antimicrobial Strategies*, DOI 10.1007/978-1-4614-1031-7_9,

9.1 Introduction

Medical devices are widely used in current clinical practice and can be placed in several anatomical locations and serve various functions (e.g., prosthetic joint, cardiac prosthetic valve, ventriculoperitoneal shunt, etc.). Overall, they are improving the quality of human lives, although serious concerns arise when they become infected, since such infections are difficult to treat [1, 2].

Difficulties in eradicating foreign-body infections are primarily related to the presence of bacterial biofilms [3]. The biofilm is responsible for several phenotypic changes in the bacteria, including increased minimal bactericidal concentrations (tolerance to antibiotics) [3–5]. Other factors that have been related to difficulties in the treatment of medical device-related infections are functional abnormalities in the activity of phagocytic cells in contact with the foreign body and the presence of intracellular bacteria, as well as the lack of an effective humoral immune response [6–10].

The treatment of device-related infections must include an appropriate and lengthy antibiotic therapy combined with adequate surgical intervention [1, 11]. In general, antibiotics alone are only able to suppress these infections, which are unlikely to be completely eradicated without surgery [12]. The antibiotics used in this setting are usually administered at high doses and must be chosen for their efficacy against biofilm cells, which may differ from their minimal bactericidal concentration against planktonic cells in standard *in vitro* studies [13, 14]. Given the above, it is essential to identify the microorganisms responsible for the infection so as to determine the most appropriate antibiotic regime.

The specific characteristics of medical device-related infections, as well as the difficulties involved in their treatment, mean that multidisciplinary medical teams are invariably required to ensure the optimal approach to and management of this pathology. While the anatomical location of the medical device can determine certain aspects of the treatment of these infections, this is beyond the scope of this chapter, and we instead review a number of pathologies that are representative of foreign-body infections of this kind. Specifically, we focus on infections related to implanted intravascular devices (prosthetic valve endocarditis and vascular grafts, excluding catheter infections), to extravascular devices (prosthetic joint infections), and to the central nervous system (ventricular shunt infections). Other device-related infections will be briefly discussed.

9.2 General Principles

9.2.1 Systemic Antibiotic Therapy

Antibiotic therapy must be carefully selected by taking into account the particular features of foreign-body infections, especially (1) the presence of biofilm and bacterial phenotypic changes (tolerance to antibiotics) and (2) the functional abnormalities of phagocytic cells and the existence of intracellular bacteria [3, 4, 6, 9, 10].

The bacterial biofilm constitutes a community of interconnected microorganisms, within a self-produced polysaccharide matrix (named glycocalyx), that are adhered to a surface [15]. Currently, biofilm bacteria are involved in the pathogenesis of as many as 60 % of human bacterial infections [3, 15]. The microorganisms within the biofilm exhibit several phenotypic changes, including the expression of tolerance to antibiotics (very high values of minimal bactericidal concentrations) [16, 17]. This phenomenon was previously noted with *in vitro* studies [18, 19] and also *in vivo* by Chuard et al. using a tissue-cage model of staphylococcal infection [4]. In the latter study the authors showed that several antimicrobials had MBC values more than 100-fold greater for ex vivo bacteria from the tissue cage fluid than for *in vitro* log-phase. This increase in MBC values is reported for all kinds of antibiotics, although they are not all affected in the same way [18, 19]. Among *staphylococci*, their susceptibilities to beta-lactams, glycopeptides, and aminoglycosides are significantly altered, whereas fluoroquinolones and rifampin remain more active and bactericidal against bacterial biofilms.

In addition to the difficulties in eradicating foreign-body infections caused by these phenotypic changes, the presence of the biofilm itself may interfere with the diffusion of some antibiotics in the way to reach their bacterial targets. Thus, the antimicrobials can be trapped within the biofilm in various ways depending on their physical and chemical properties and, so, they can only achieve lower concentrations (lower than the MBC) or they can be inactivated [20, 21].

Because of this low susceptibility to antibiotics within biofilms, the use of standard *in vitro* studies (the minimal inhibitory concentration (MIC) and the MBC in the log-phase) fails to correlate with the *in vivo* efficacy against foreign-body infections. Moreover, microorganisms in the *in vitro* stationary phase enter a nongrowing state and are tolerant to antibiotics in the same way as the bacteria within the biofilm do. Research has shown that the bactericidal action in the stationary phase of a drug correlates better with its efficacy against *in vivo* foreign-body infections [13, 18, 19, 22].

The usual pharmacodynamic parameters of antibiotics are MIC related, i.e., determined from log-phase cultures, and as such they may be less than adequate for predicting the outcome of device-related infections. Previous experimental results suggest that the use of these pharmacodynamic parameters is more appropriate in relation to the stationary-phase MBC (MBC_{stat}), i.e., the time the drug remains above the MBC_{stat} (T>MBC_{stat}) or the area under the curve/MBC_{stat} (AUC/MBC_{stat}) [13, 19]. However, to the best of our knowledge no studies have investigated this in a clinical setting and more information is also needed to determine whether pharmacokinetic/pharmacodynamic parameters must be related to levels achieved in serum or at the site of infection (i.e., the surface of prosthetic joint). Thus, given that standard *in vitro* studies (MIC and MBC in the log-phase) and their MIC-related pharmacodynamic parameters fail to predict the outcome of foreign-body infections, *in vitro* studies in the stationary phase provide readily available and useful information about drug efficacy against these infections.

Finally, phagocytic cells in contact with foreign bodies present functional abnormalities that affect their ability to phagocytize and kill bacteria. This fact plays an important role in the pathogenesis of device-related infections, as previously described by Zimmerli et al. using a tissue-cage model with infection by *Staphylococcus aureus* [6]. In our experience, these functional abnormalities of phagocytic cells result in the presence of a high number of intracellular bacteria [23]. The role of intracellular bacteria in the maintenance and recurrence of foreign-body infections (i.e., prosthetic joint infections) is not clearly established [24–26]. However, such bacteria are also phenotypically adapted (they are more resistant to antibiotics) and are better protected against antimicrobial activity in this location [27]. By way of example, recent experimental reports have shown the presence of *S. aureus* inside osteoblast cells in a variant form that presents greater cell wall thickness [24]. Moreover, the mechanism used to reach the intracellular environment differs for several kinds of antibiotics and, so, the concentrations and the degree of intracellular killing allowed are both drug dependent [28, 29].

In our opinion, the antibiotics to use against device-related infections can be chosen according both to their activity against nongrowing microorganisms (stationary-phase bacteria) and bacterial biofilms and their intracellular efficacy. Obviously, it is desirable to optimize the pharmacokinetic/pharmacodynamic parameters of antibiotics in the context of device-related infections (i.e., the use of high doses of antimicrobials, the continuous infusion of beta-lactams, etc.). Moreover, therapeutic strategies to diminish the risk of resistance developing and to achieve greater efficacy (i.e., the use of combination therapy) are also advisable. These are further reasons why it is essential to identify the microorganisms responsible for the device-associated infection, thereby enabling the physician to determine the most appropriate antibiotic regimen.

9.2.2 Biofilm-Guided Therapies

Our enhanced understanding of biofilms has opened up new lines of research to improve the treatment of device-related infections. The quorum-sensing system (cell-to-cell signaling within the biofilm) has provided the basis for the use of synthetic molecules that can mimic these signals and so block the formation and maintenance of biofilms. One of the quorum-sensing systems involved in staphylococcal virulence use the inducer RNAIII-activating protein (RAP) to induce the phosphorylation of the target of RAP (TRAP) [30]. Several earlier studies by Balaban et al. have shown the *in vitro* and *in vivo* efficacy of the RNAIII-inhibiting peptide (RIP, inhibitor of the RAP signal) in the prevention of staphylococcal biofilm infections [31, 32].

Results on the activity of RIP inhibitor in the treatment of staphylococcal foreign-body infections are more contradictory; while Balaban et al. reported good *in vivo* efficacy against a 2-day-old biofilm infection using a graft rat model [33], our group showed no activity against a 3-day-old staphylococcal foreign-body infection using the tissue-cage rat model [34].

More recent and promising studies have shown the efficacy of low concentration of clarithromycin in the eradication of polysaccharide glycocalix around bacterial colonies although the mechanisms through which it eradicates this glycocalix are not clear [35]. Moreover, these studies have reported that clarithromycin enhances the efficacy of other antibiotics against biofilm infection [36].

9.3 Treatment of Prosthetic Joint Infection

Prosthetic joint infections are difficult to treat. In contrast to other more frequent infections, the number of controlled studies in this setting is scarce. Consequently, current concepts about antimicrobial treatment in this setting are based on data from animal models, expert opinions, and a limited number of well-designed randomized studies and a greater number of observational series [1]. The complete eradication of prosthetic joint infections requires a combination of adequate antimicrobial therapy and surgical interventions [1, 2]; currently, there are no global consensus to recommend the same clinical approach and management of these infections. Probably, the most useful guidelines in this setting are the classification of Tsukayama [37] and the classification of Zimmerli [1]; they have some disagreements that are still under discussion. Tsukayama classifies prosthetic joint infections into the following: (1) early postoperative infection (developed less than 1 month after the surgery), (2) late chronic infection (developed 1 month after the surgery and presented with an insidious clinical evolution), (3) acute hematogenous infection (associated with a documented or suspected bacteremia and an acute onset of symptoms), and (4) positive intraoperative cultures (those cases previously considered as aseptic loosening that presented positive microbiological cultures at the time of surgery without obvious signs of infection). According to classification of Tsukayama, acute hematogenous and early postoperative infections can be treated with debridement and prosthesis retention, whereas others are managed with prosthesis exchange.

For its part, classification of Zimmerli includes the following: (1) early infections (developed less than 3 months after surgery), (2) delayed infections (3–24 months after surgery), and (3) late infections (more than 24 months after surgery). According to the classification of Zimmerli early and delayed infections are usually acquired during implantation of the prosthesis, whereas late infections are acquired by hematogenous seeding. When the duration of symptoms is less than 3 weeks, early and hematogenous infections can be managed with debridement and retention of prosthesis.

As presented above, several therapies can be used, including the complete removal of the prosthesis (in one- or two-stage surgical exchange) or debridement with implant retention. In all cases, however, long-term antimicrobial therapy is needed, although the optimal duration of antibiotic treatment has yet to be well defined. Table 9.1 summarizes the choice of antibiotics according to the microorganism. Specific aspects in the systemic antibiotic treatment of prosthetic joint infections are commented below and are based on previous guidelines and personal opinion.

Table 9.1 Treatment of prosthetic joint infection caused by common microorganisms[a]

Microorganism	First-line treatment	Alternative treatment
S. aureus or coagulase-negative staphylococci methicillin-susceptible	Levofloxacin 750–1,000 mg/day p.o. plus Rifampin 600–900 mg/day p.o. or Cloxacillin 2 g/6 h i.v. plus Rifampin i.v./p.o. for 1–2 weeks, followed by Levofloxacin p.o. plus Rifampin p.o.	Rifampin p.o. plus [b]Trimethoprim-sulfamethoxazole 1 DS/8 h p.o. or Clindamycin 600 mg/8 h p.o.
S. aureus or coagulase-negative staphylococci methicillin-resistant (rifampin-susceptible)	Vancomycin 1 g/12 h i.v. plus Rifampin i.v./p.o. for 1–2 weeks, followed by Rifampin p.o. plus Levofloxacin p.o. or Trimethoprim–sulfamethoxazole p.o. or Linezolid 600 mg/12 h p.o. or Clindamycin p.o.	Daptomycin 10 mg/kg/day i.v. plus Rifampin i.v./p.o. for 1–2 weeks, followed by Rifampin p.o. plus Levofloxacin p.o. or Trimethoprim–sulfamethoxazole p.o. or Linezolid 600 mg/12 h p.o. or Clindamycin p.o.
S. aureus or coagulase-negative staphylococci methicillin-resistant (rifampin-resistant)	Vancomycin i.v. for 4–6 weeks, followed by Linezolid or Trimethoprim–sulfamethoxazole p.o.	Daptomycin 10 mg/kg/day i.v. for 4–6 weeks, followed by Linezolid or Trimethoprim–sulfamethoxazole p.o. or Daptomycin for 8 weeks Linezolid for 8 weeks
Streptococcus sp.	Ceftriaxone 2 g/day i.v. for 2–4 weeks, followed by Amoxicillin 1 g/8 h p.o.	
Enterococcus faecalis	Ampicillin 2 g/6 h i.v. plus Ceftriaxone 2 g/day i.v. for 2–4 weeks, followed by Amoxicillin 1 g/8 h p.o.	Ampicillin i.v. plus Gentamicin 2–3 mg/kg/day (single daily dose) for 2–4 weeks, followed by Amoxicillin 1 g/8 h p.o.
Enterobacteriaceae	Ciprofloxacin 750 mg/12 h p.o.	[c]Ceftriaxone 1 g/day i.v. followed by Cefuroxime 500 mg/8 h p.o. Ertapenem 1 g/day i.v. Trimethoprim–sulfamethoxazole p.o.
Pseudomonas aeruginosa	Ciprofloxacin 1 g/12 h p.o. plus Ceftazidime 2 g/8 h i.v. for 2–4 weeks, followed by Ciprofloxacin p.o.	Piperacillin–Tazobactam 4 g/8 h i.v. or Imipenem 500 mg/8 h i.v. ± Ciprofloxacin p.o. [d]Colistin 2×10^6 U/8 h i.v.
Anaerobes	Metronidazole 500 mg/8 h i.v. or p.o.	Clindamycin 600 mg/8 h i.v. or p.o.

[a] References [1, 57] modified according to authors' opinion. For more details see the text (Sect. 3)
[b] DS double strength (trimethoprim, 160 mg, and sulfamethoxazole, 800 mg)
[c] Alternative treatments for enterobacteriaceae must be selected according to antimicrobial susceptibility patterns in each case
[d] Treatment of multidrug-resistant Gram-negative bacteria (i.e., *P. aeruginosa* or *A. baumannii*) must be individualized

9.3.1 *Staphylococcal Infections*

As regards the most appropriate antibiotics against prosthetic joint infections, experimental and clinical studies have led to rifampin being currently considered the first-line therapy against staphylococcal infections [14, 38, 39]. However, it should never be used in isolation because of the rapid emergence of resistance. The efficacy of rifampin in this context is classically related to its bactericidal action against nongrowing bacteria, activity against intracellular bacteria, and an ability to diffuse into the biofilm. In terms of the *in vitro* bactericidal action of rifampin against nongrowing *S. aureus*, contradictory results have been reported [40–42]. Our own experience with stationary-phase studies (where both the nutrient-restricted media and high bacterial inocula were used) suggests that the activity of rifampin is greatly affected by bacterial inoculum size and the emergence of resistant strains during the *in vitro* study (data not published).

Current guidelines suggest that the combination of fluoroquinolones and rifampin is the most desirable treatment against staphylococcal prosthetic joint infections [1], this being based on previous experimental [13] and, somewhat limited, clinical studies using ciprofloxacin or ofloxacin [14, 39]. This combined therapy strongly reduces the risk of emergence of rifampin or quinolone resistance while allowing the benefits of rifampin in the setting of foreign-body infection. However, an *in vitro* antagonistic effect has been described for this combination, although it has yet to be corroborated by clinical experience of foreign-body infections [40, 42].

In recent years, it has been recommended that instead of ciprofloxacin or ofloxacin, rifampin should be combined with high doses (750–1,000 mg/day) of the newer fluoroquinolone levofloxacin, although clinical results are still scarce [1]. Levofloxacin presents an improved *in vitro* bactericidal action against growing and nongrowing bacteria and its efficacy has been evaluated by our own group and others against experimental staphylococcal foreign-body infection, both alone and in combination at usual and high doses (500 mg and 750–1,000 mg/day, respectively) [22, 42, 43]. Although we noted good *in vivo* efficacy of levofloxacin at high doses (according to the increase in its pharmacodynamic parameters) with respect to that for usual doses, we also observed an antagonistic effect of rifampin on the efficacy of levofloxacin at high doses (but not at usual doses). Furthermore, the *in vivo* efficacy of the combination levofloxacin–rifampin was not improved by increasing the dosage of levofloxacin [42]. We believe, therefore, that the most effective dose of levofloxacin in combination with rifampin, together with the potentially antagonistic effects of rifampin on the efficacy of levofloxacin, must be carefully evaluated in clinical studies. In this regard, the safety and efficacy of newer fluoroquinolone in monotherapy (levofloxacin and moxifloxacin) should be also tested for the treatment of orthopedic implant-related staphylococcal infections due to their potential advantages in this setting (i.e., bactericidal activity against nongrowing bacteria and good intracellular activity). To our knowledge, only a recent clinical study has shown the safety and efficacy of moxifloxacin in this regard [44].

Although the majority of methicillin-susceptible *S. aureus* (MSSA) infections can be treated with this quinolone–rifampin combination, this is not the case for a great number of methicillin-resistant *S. aureus* (MRSA) and coagulase-negative *Staphylococcus* (CNS) since methicillin resistance is commonly accompanied in some geographical regions with quinolone resistance [45]. In this context, other rifampin combinations should be considered (whenever possible): vancomycin, clindamycin, trimethoprim/sulfamethoxazole, fusidic acid, and the more recently developed drugs linezolid, tigecycline, and daptomycin.

Linezolid (at a dosage of 600 mg/12 h) is the first member of the oxazolidinone class of antibiotics. It is effective for the treatment of serious gram-positive infections, including those caused by multidrug-resistant *S. aureus* and *Enterococcus faecium*. Linezolid inhibits bacterial protein synthesis by binding to 23S rRNA in the site of the 50S ribosome. It is an alternative therapy for prosthetic joint infections caused by methicillin-resistant *Staphylococcus* strains because of its antimicrobial efficacy, good oral bioavailability, and good diffusion into biofilms. However, linezolid only presents *in vitro* bacteriostatic activity against growing and nongrowing bacteria, and alone it shows low activity against *in vivo* foreign-body infection, although this can be improved in combination with rifampin [46, 47]. Well-designed, randomized clinical studies using linezolid in prosthetic joint infections are needed, although a limited number of studies have already reported satisfactory results [48–50].

Tigecycline is the first member of the glycylcyclines class of antimicrobial agents that are derivative of the tetracycline antibiotics. It exhibits bacteriostatic activity by binding to the 30S ribosomal subunit and inhibiting protein translation. The clinical experience of tigecycline against staphylococcal prosthetic joint infections is anecdotal. However, according to *in vitro* and experimental studies its efficacy would appear to be limited by bacteriostatic activity and the low concentrations achieved locally in bone and joints at standard dosages [51, 52]. The efficacy of tigecycline was previously shown in the tissue-cage infection model in comparison with that for vancomycin [53]; recently, our own results (data not published) show a low *in vivo* efficacy in this model, although its combination with rifampin improves the activity and ensures protection against the emergence of resistant strains.

Daptomycin is the first drug of a new class of antibiotics, the lipopeptides, for which the mechanism of action is not completely known. It seems to be a more promising drug in the setting of prosthetic joint infections because of its bactericidal activity against growing and nongrowing bacteria and its proven efficacy against experimental foreign-body infections, especially when used at high doses and in combination with rifampin [47, 54–56]. However, the most appropriate dosage for avoiding clinical failures and the emergence of resistant strains in the context of prosthetic joint infections needs to be better defined.

9.3.2 Non-staphylococcal Infections

Fluoroquinolones (e.g., ciprofloxacin at 750–1,000 mg/12 h) also play a leading role in the treatment of foreign-body infections caused by enterobacteria and *Pseudomonas aeruginosa*. They present good bactericidal action against

Gram-negative rods in the stationary phase, achieve high intracellular concentrations, and have good, long-term oral availability. The use of aminoglycosides in foreign-body infections, on the other hand, is limited by their low activity against biofilms [1, 57]. The efficacy of beta-lactams can be severely affected by reduced susceptibility to antimicrobials of the bacteria in biofilms, although the presence of fluoroquinolone-resistant enterobacteria or *P. aeruginosa* makes their use necessary. In this context, the optimization of pharmacodynamic efficacy parameters using continuous infusion of beta-lactams can be useful [58].

The most appropriate treatment of device-related infections by multidrug-resistant Gram-negative bacteria (i.e., *P. aeruginosa* or *Acinetobacter baumannii*) remains a matter of great concern. The efficacy of less common antibiotics (e.g., colistin or fosfomycin) in this setting is not well known, and their potential combination with other drugs has yet to be well evaluated [59–61].

The treatment of *Streptococcus* spp. and *Enterococcus faecalis* also requires further research. The use of beta-lactams is recommended, despite their low efficacy against biofilms, while the combined activity of beta-lactams and rifampin in this context has not been widely studied [1, 62]. In the particular case of infections by *E. faecalis*, our group reported the efficacy of the combination ampicillin–ceftriaxone in the treatment of prosthetic joint infections [63].

9.4 Treatment of Prosthetic Valve Endocarditis and Pacemaker Infections

9.4.1 Prosthetic Valve Endocarditis

The frequency of prosthetic valve endocarditis is most common during the first 3 months following the cardiac surgery, decreases during the next 6–12 months, and thereafter remains constant over time with a frequency ranging between 0.5 and 0.6 % [64, 65]. There are two types of prosthetic valve endocarditis, early infections (appearing in the first 12 months after device implantation) and late infections (occurring subsequently). The pathogenesis, etiology, and clinical presentation of these two types of endocarditis are different [66]. Early infections are mainly caused by staphylococci (both *S. aureus* and CNS in almost 50 % of cases), followed by *enterococci*, gram-negative rods, and others [67]. In the late infections although *staphylococci* remain frequent, the microorganisms responsible are more similar to those observed on native valve endocarditis, with a predominance of viridans group *streptococci*, *Streptococcus bovis*, *Enterococci* (mainly *E. faecalis*), and HACEK microorganisms (*Haemophilus* spp., *Actinobacillus actinomycetemcomitans*, *Cardiobacterium hominis*, *Eikenella* spp., and *Kingella* spp.) [66–68].

The complex pathogenesis of infective endocarditis involves the initial formation of vegetations (fibrin and platelets bind to endothelial surface) on which the microorganisms may adhere and grow; the presence of prosthetic valves provides additional complexity because of the formation of biofilms [12, 66]. In patients with prosthetic valve endocarditis the vegetations are located on the line between the valvular ring and prosthesis and can cause its dehiscence and after, the myocardial invasion with periprosthetic abscess formation. In the early infections, prosthetic valves are usually colonized during surgery or immediately after, therefore most cases are nosocomial infections coming from the skin of patients or clinical staff. In cases of late infections, the prosthesis is colonized by hematogenous seeding coming from distant foci (i.e., urinary or gastrointestinal tract, …).

Antimicrobial therapy of prosthetic valve endocarditis commonly consists of at least 6 weeks of intravenous treatment with bactericidal drugs and the definitive cure may require surgical valve replacement. In contrast to the choice of antibiotics for prosthetic joint infection, with efficacy against biofilm and nongrowing bacteria (i.e., fluoroquinolones), the treatment of intravascular device infections (endocarditis, vascular grafts) is based on the use of antibiotics with bactericidal *in vitro* activity against planktonic microorganisms (i.e., beta lactams, aminoglycosides). These discrepancies are not easy to explain since both infections involve bacterial biofilms; under our opinion may have other differences. Thus, the presence of a high bacterial burden (greater than 10^8–10^{10} CFU/ml) and the existence of bacteria in the log-phase of growth in the intravascular setting may both contribute in prosthetic valve endocarditis. In contrast, prosthetic joint infections involve moderate inoculums (10^6 CFU/ml) and stationary-phase bacteria.

Surgical intervention in combination with antibiotics is indicated in some situations, such as cardiac insufficiency with severe prosthetic valve dysfunction, presence of perivalvular extension (abscess) or valve dehiscence, multiple peripheral septic emboli, persistence of bacteremia despite adequate antimicrobial treatment, or recurrence of endocarditis.

In terms of choosing the specific antibiotic therapy, Table 9.2 summarizes the antibiotic treatment according to microorganisms. For a more extent detail, we recommend the current guidelines produced by American [69] and European Societies [70].

9.4.2 Staphylococcal Infections

In prosthetic valve endocarditis caused by *staphylococci*, rifampin plays a key role based on *in vitro* and *in vivo* experiments. Infections by CNS are treated with a vancomycin–rifampin combination for at least 6 weeks, with the addition of gentamicin during the first 2 weeks [71, 72].

While MSSA infections can be treated with the combination oxacillin–rifampin–gentamicin (in a similar way to the treatment of CNS) [69], the best treatment against MRSA remains under evaluation. Currently, prosthetic valve endocarditis by MRSA is a serious infection with high derived mortality and the need for surgical intervention in most cases. The combination vancomycin–rifampin–gentamicin (when this can be used because of rifampin susceptibility and the absence of clinical contraindications) is the standard reference therapy [69]. In a randomized study of

Table 9.2 Treatment of prosthetic valve infection caused by common microorganisms[a]

Microorganism	First-line treatment	Alternative treatment
Viridans group streptococci or *S. bovis* penicillin-susceptible (CMI < 0.12 μg/ml)	Penicillin G sodium 12–18 million U/day i.v. (continuously or divided in 4–6 doses) for 6 weeks or Ceftriaxone 2 g/24 h i.v. for 6 weeks plus Gentamicin 3 mg/kg/day i.v. for 2 weeks	Vancomycin 30 mg/kg/day i.v. (divided in two doses) for 6 weeks
Viridans group streptococci or *S. bovis* penicillin relatively or fully resistant (CMI > 0.12 μg/ml)	Penicillin G sodium 24 million U/day i.v. (continuously or divided in 4–6 doses) for 6 weeks or Ceftriaxone 2 g/24 h i.v. for 6 weeks plus Gentamicin 3 mg/kg/day i.v. for 6 weeks	Vancomycin for 6 weeks
Staphylococci (*S. aureus* or coagulase-negative *Staphylococcus*)	*Oxacillin-susceptible* Cloxacillin 12 g/day (divided in six doses) for 6 weeks plus Rifampin 900 mg/day (divided in three doses) for 6 weeks plus Gentamicin 3 mg/kg/day (divided in three doses) for 2 weeks *Oxacillin-resistant* Vancomycin plus Rifampin plus Gentamicin	Vancomycin plus Rifampin plus Gentamicin for 6 weeks
Enterococci	Penicillin G sodium 24 million U/day i.v. for 6 weeks or Ampicillin 12 g/day (divided in six doses) for 6 weeks plus Gentamicin for 6 weeks	Vancomycin plus Gentamicin for 6 weeks or Ampicillin plus Ceftriaxone 2 g/12 h i.v. for 6 weeks
HACEK microorganisms	Ceftriaxone for 6 weeks or Ampicillin for 6 weeks	Ciprofloxacin 800–1,000 mg/day p.o. (divided in two doses)

[a]References [69, 70] modified according to authors' opinion. For more details see the text (Sect. 4)

S. aureus bacteremia, daptomycin (at a dosage of 6 mg/kg/day) in association with gentamicin was no better than comparators, although it was approved for use in right native valve endocarditis [73]. Daptomycin at high doses (greater than 6 mg/kg/day) has shown efficacy against *in vitro* simulated endocarditis and in an animal model of endocarditis by MRSA, but its bactericidal effect is clearly diminished against high inoculum infections [74, 75]. Nowadays, the use of daptomycin in left prosthetic valve infections by MRSA can be considered as an alternative therapy, although both its efficacy (alone or in combination) and dosage (i.e., doses higher than 8–10 mg/kg/day) need to be well evaluated [76].

The use of other anti-MRSA drugs, such as linezolid and tigecycline, has yet to be studied, although their *in vitro* bacteriostatic activities would seem to predict a limited efficacy in a difficult-to-treat infection where bactericidal killing is desirable [51, 77]. In this regard, the use of fosfomycin plus a beta-lactam (i.e., fosfomycin–imipenem) may offer an alternative treatment against prosthetic valve endocarditis by MRSA, this being based on the documented synergistic and bactericidal *in vitro* activity of this combination and the limited clinical experience [78, 79].

As in the case of MRSA infections, more data are needed to define the optimal treatment of prosthetic valve endocarditis by other multidrug-resistant microorganisms (e.g., *P. aeruginosa*, vancomycin-resistant *Enterococci*, etc.).

9.4.3 Pacemaker Infections

These infections can be classified into two groups: (1) infections in the pacemaker generator pocket and (2) intravascular infection of electrodes. The latter includes pacemaker infective endocarditis; in these cases vegetation formation is not only in the tricuspid valve but also anywhere along the course of the electrode [66, 80].

Microorganisms more often involved in pacemaker infections are staphylococci (*S. aureus* and CNS); among pacemaker infective endocarditis occurring in the first month after the device implantation, *S. aureus* is involved in almost 80 % of cases, whereas in late endocarditis CNS are the predominant microorganisms [66].

The treatment of pacemaker infections usually requires antibiotic therapy and pacemaker system (generator and electrode) removal with device implantation at a new site when bacteremia is controlled. However, there are no controlled studies comparing cure rates for antibiotic treatment alone versus the combination of antimicrobials and pacemaker removal. As a particular situation, in those cases of pacemaker generator pocket infection without concomitant bacteremia, the only removal of the generator in association with antibiotic therapy is a valid therapeutic option. A long length of time that pacemaker is in place is associated with technical difficulties and complications at lead removal; for these cases several surgical techniques for electrode extraction have been described [81].

Adequate antibiotic therapy for pacemaker infection or pacemaker infective endocarditis can be guided according to comments described above for prosthetic valve endocarditis (see also Table 9.2).

9.5 Treatment of Vascular Graft Infections

Although vascular grafts (i.e., femoropopliteal, aortic) only become infected in a small percentage of cases, such infections are very severe and associated with high rates of mortality. The combination of antibiotic therapy and surgical intervention is the most appropriate strategy to eradicate this infection [82, 83]. Excision of the infected vascular graft and its replacement by another prosthetic device or autologous venous system are the surgical options usually considered. In the particular case of aortic graft infections, an axillofemoral bypass may be performed prior to excision of the infected graft [84]. Since these infections are associated with septicemia, the antibiotic treatment can be managed in accordance with the above comments regarding prosthetic valve endocarditis. The definitive duration of antibiotic therapy may be longer if the presence of an abscess is identified.

9.6 Treatment of Ventricular Shunt Infections

These infections are managed through the combination of antibiotic therapy and shunt removal [85–87]. Surgical intervention is usually performed in two steps; the infected shunt can be externalized for a few days (in selected cases), or it may be removed at the outset of the diagnosis when implanting an external ventricular or a lumbar catheter. In all cases, systemic antibiotics are administered (for a minimum of 3–7 days) and repeated analyses of cerebrospinal fluid must be performed in order to ensure sterility prior to insertion of the new ventricular shunt. When this new shunt is inserted, the antibiotic therapy should be prolonged for a further 7 days, or for 14 days in the case of Gram-negative rods or *E. faecalis* infections.

In this setting, antibiotics have to cross the hematoencephalic barrier and thus, they are selected according to their penetration through this barrier or are often used at high doses. In this regard, the intrathecal administration of antibiotics is an alternative option: vancomycin for staphylococcal infections, aminoglycosides for Gram-negative rods, or colistin for multidrug microorganisms (i.e., *P. aeruginosa* or *A. baumannii*). The role of antibiotics with antibiofilm efficacy (i.e., the addition of rifampin) has not been well established.

The treatment of ventricular shunt infections with antibiotics alone (i.e., without removing the shunt) has been associated with a poor curative ratio and high mortality. Thus, this conservative strategy must only be used against noncomplicated infections by *Streptococcus pneumoniae*, *Neisseria meningitidis*, or *Haemophilus influenzae*, or in those cases in which surgical intervention is contraindicated.

9.7 Treatment of Other Medical Device-Associated Infections

9.7.1 Breast Implants

The incidence of breast implant infections is considered low, although the complexities of establishing a consensus on diagnostic definitions make accurate estimates difficult. The management of these infections usually involves a two-stage replacement procedure [88, 89]. Similarly to what has been discussed above, the infected implant is removed in the first stage, and excision of the capsule can also be considered. A 10–14-day course of systemic antibiotics should then be administered, although the most appropriate duration for this therapy is not standardized. A new breast implant is subsequently inserted in second-stage surgery. Whether or not the contralateral implant should also be removed remains a matter of controversy.

9.7.2 *Penile Implants*

The treatment of penile implant infections is also based on surgical removal of the infected implant and antibiotic therapy. In this setting, there are diverging views about the stages of surgical management [90, 91]. Although a single-stage replacement has been used in selected patients, the preferred approach is a two-stage replacement. In the latter case, the procedure is similar to that described above, with removal of the infected implant in the first stage, followed by a 10–14-day course of systemic antibiotics. A new penile implant will then be inserted some months later.

9.8 Summary

Medical device-related infections are difficult to treat owing primarily to the presence of bacterial biofilms. Their slow development, the phenotypic changes undergone by the bacteria that make them tolerant to antibiotics, and the ineffective cellular and humoral immune responses are all characteristic features of these biofilm-related infections.

Standard *in vitro* studies evaluating the bactericidal effect of antibiotics in the log-phase and their MIC-related pharmacodynamic parameters fail to predict the outcome of device-related infections. In this way, the use of parameters reflecting drug bactericidal activity against biofilm and nongrowing (stationary phase) cells has proved more appropriate for correlating with antibiotic efficacy in the treatment of these infections.

The curative treatment of medical device-related infections involves a combination of several surgical interventions (with the removal of the foreign body) and the long-term use of antimicrobials. The antibiotics must be carefully chosen for their bactericidal activity against biofilms and intracellular bacteria and their ability to diffuse into biofilms. Currently, rifampicin and fluoroquinolones play a key role in the treatment of some device-related infections caused by several staphylococcal strains and Gram-negative rods but the need for new drugs is advisable. Our progressive understanding of biofilms should open up new lines of research to improve the treatment of device-related infections. In this regard, promising results with the use of molecules that block the quorum-sensing system or others that eradicate the glycocalix must be confirmed. Finally, the knowledge of genetic basis for the biofilm formation could delineate new potential targets for future therapies.

References

1. Zimmerli W, Trampuz A, Ochsner PE. Prosthetic-joint infections. N Engl J Med. 2004;351(16):1645–54.
2. Del Pozo JL, Patel R. Clinical practice Infection associated with prosthetic joints. N Engl J Med. 2009;361(8):787–94.
3. Stewart PS, Costerton JW. Antibiotic resistance of bacteria in biofilms. Lancet. 2001;358 (9276):135–8.
4. Chuard C, Lucet JC, Rohner P, Herrmann M, Auckenthaler R, Waldvogel FA, Lew DP. Resistance of *Staphylococcus aureus* recovered from infected foreign body *in vivo* to killing by antimicrobials. J Infect Dis. 1991;163(6):1369–73.
5. Donlan RM, Costerton JW. Biofilms: survival mechanisms of clinically relevant microorganisms. Clin Microbiol Rev. 2002;15(2):167–93.
6. Zimmerli W, Lew PD, Waldvogel FA. Pathogenesis of foreign body infection. Evidence for a local granulocyte defect. J Clin Invest. 1984;73(4):1191–200.
7. Proctor RA, Balwit JM, Vesga O. Variant subpopulations of *Staphylococcus aureus* as cause of persistent and recurrent infections. Infect Agents Dis. 1994;3(6):302–12.
8. Proctor RA, von Eiff C, Kahl BC, Becker K, McNamara P, Herrmann M, Peters G. Small colony variants: a pathogenic form of bacteria that facilitates persistent and recurrent infections. Nat Rev Microbiol. 2006;4(4):295–305.
9. Boelens JJ, Dankert J, Murk JL, Weening JJ, van der Poll T, Dingemans KP, Koole L, Laman JD, Zaat SA. Biomaterial-associated persistence of *Staphylococcus epidermidis* in pericatheter macrophages. J Infect Dis. 2000;181(4):1337–49.
10. Broekhuizen CA, de Boer L, Schipper K, Jones CD, Quadir S, Feldman RG, Dankert J, Vandenbroucke-Grauls CM, Weening JJ, Zaat SA. Peri-implant tissue is an important niche for *Staphylococcus epidermidis* in experimental biomaterial-associated infection in mice. Infect Immun. 2007;75(3):1129–36.
11. Darouiche RO. Treatment of infections associated with surgical implants. N Engl J Med. 2004;350(14):1422–9.
12. Costerton JW. Biofilm theory can guide the treatment of device-related orthopaedic infections. Clin Orthop Relat Res. 2005;437:7–11.
13. Widmer AF, Frei R, Rajacic Z, Zimmerli W. Correlation between *in vivo* and *in vitro* efficacy of antimicrobial agents against foreign body infections. J Infect Dis. 1990;162(1):96–102.
14. Zimmerli W, Widmer AF, Blatter M, Frei R, Ochsner PE. Role of rifampin for treatment of orthopedic implant-related staphylococcal infections: a randomized controlled trial. Foreign-Body Infection (FBI) Study Group. JAMA. 1998;279(19):1537–41.
15. Costerton JW, Stewart PS, Greenberg EP. Bacterial biofilms: a common cause of persistent infections. Science. 1999;284(5418):1318–22.
16. Gilbert P, Collier PJ, Brown MR. Influence of growth rate on susceptibility to antimicrobial agents: biofilms, cell cycle, dormancy, and stringent response. Antimicrob Agents Chemother. 1990;34(10):1865–8.
17. Anwar H, Dasgupta MK, Costerton JW. Testing the susceptibility of bacteria in biofilms to antibacterial agents. Antimicrob Agents Chemother. 1990;34(11):2043–6.
18. Blaser J, Vergeres P, Widmer AF, Zimmerli W. In vivo verification of in vitro model of antibiotic treatment of device-related infection. Antimicrob Agents Chemother. 1995;39(5):1134–9.
19. Zimmerli W, Frei R, Widmer AF, Rajacic Z. Microbiological tests to predict treatment outcome in experimental device-related infections due to *Staphylococcus aureus*. J Antimicrob Chemother. 1994;33(5):959–67.
20. Farber BF, Kaplan MH, Clogston AG. *Staphylococcus epidermidis* extracted slime inhibits the antimicrobial action of glycopeptide antibiotics. J Infect Dis. 1990;161(1):37–40.
21. Singh R, Ray P, Das A, Sharma M. Penetration of antibiotics through *Staphylococcus aureus* and *Staphylococcus epidermidis* biofilms. J Antimicrob Chemother. 2010;65(9):1955–8.

22. Murillo O, Domenech A, Garcia A, Tubau F, Cabellos C, Gudiol F, Ariza J. Efficacy of high doses of levofloxacin in experimental foreign-body infection by methicillin-susceptible *Staphylococcus aureus*. Antimicrob Agents Chemother. 2006;50(12):4011–7.
23. Murillo O, Pachon ME, Euba G, Verdaguer R, Carreras M, Cabellos C, Cabo J, Gudiol F, Ariza J. Intracellular antimicrobial activity appearing as a relevant factor in antibiotic efficacy against an experimental foreign-body infection caused by *Staphylococcus aureus*. J Antimicrob Chemother. 2009;64(5):1062–6.
24. Ellington JK, Harris M, Hudson MC, Vishin S, Webb LX, Sherertz R. Intracellular *Staphylococcus aureus* and antibiotic resistance: implications for treatment of staphylococcal osteomyelitis. J Orthop Res. 2006;24(1):87–93.
25. Gresham HD, Lowrance JH, Caver TE, Wilson BS, Cheung AL, Lindberg FP. Survival of *Staphylococcus aureus* inside neutrophils contributes to infection. J Immunol. 2000;164(7):3713–22.
26. Sendi P, Rohrbach M, Graber P, Frei R, Ochsner PE, Zimmerli W. *Staphylococcus aureus* small colony variants in prosthetic joint infection. Clin Infect Dis. 2006;43(8):961–7.
27. Maurin M, Raoult D. Intracellular organisms. Int J Antimicrob Agents. 1997;9(1):61–70.
28. Barcia-Macay M, Seral C, Mingeot-Leclercq MP, Tulkens PM, Van Bambeke F. Pharmacodynamic evaluation of the intracellular activities of antibiotics against *Staphylococcus aureus* in a model of THP-1 macrophages. Antimicrob Agents Chemother. 2006;50(3):841–51.
29. Carryn S, Chanteux H, Seral C, Mingeot-Leclercq MP, Van Bambeke F, Tulkens PM. Intracellular pharmacodynamics of antibiotics. Infect Dis Clin North Am. 2003;17(3):615–34.
30. Gov Y, Bitler A, Dell'Acqua G, Torres JV, Balaban N. RNAIII inhibiting peptide (RIP), a global inhibitor of *Staphylococcus aureus* pathogenesis: structure and function analysis. Peptides. 2001;22(10):1609–20.
31. Balaban N, Giacometti A, Cirioni O, Gov Y, Ghiselli R, Mocchegiani F, Viticchi C, Del Prete MS, Saba V, Scalise G, Dell'Acqua G. Use of the quorum-sensing inhibitor RNAIII-inhibiting peptide to prevent biofilm formation in vivo by drug-resistant *Staphylococcus epidermidis*. J Infect Dis. 2003;187(4):625–30.
32. Balaban N, Goldkorn T, Gov Y, Hirshberg M, Koyfman N, Matthews HR, Nhan RT, Singh B, Uziel O. Regulation of *Staphylococcus aureus* pathogenesis via target of RNAIII-activating protein (TRAP). J Biol Chem. 2001;276(4):2658–67.
33. Balaban N, Cirioni O, Giacometti A, Ghiselli R, Braunstein JB, Silvestri C, Mocchegiani F, Saba V, Scalise G. Treatment of *Staphylococcus aureus* biofilm infection by the quorum-sensing inhibitor RIP. Antimicrob Agents Chemother. 2007;51(6):2226–9.
34. Garrigos C, Murillo O, Euba G, et al. Study of the efficacy of RNA III inhibiting peptide (RIP) alone and in combination with vancomycin in an experimental foreign-body infection by methicillin-resistan *Staphylococcus aureus*. 50th International Conference on Antimicrobial Agents and Chemotherapy (ICAAC 2010); 2010 Sep 12–15; Boston: B-710
35. Yasuda H, Ajiki Y, Koga T, Yokota T. Interaction between clarithromycin and biofilms formed by *Staphylococcus epidermidis*. Antimicrob Agents Chemother. 1994;38(1):138–41.
36. Parra-Ruiz J, Vidaillac C, Rose WE, Rybak MJ. Activities of high-dose daptomycin, vancomycin, and moxifloxacin alone or in combination with clarithromycin or rifampin in a novel *in vitro* model of *Staphylococcus aureus* biofilm. Antimicrob Agents Chemother. 2010;54(10):4329–34.
37. Tsukayama DT, Estrada R, Gustilo RB. Infection after total hip arthroplasty. A study of the treatment of one hundred and six infections. J Bone Joint Surg Am. 1996;78(4):512–23.
38. Widmer AF, Gaechter A, Ochsner PE, Zimmerli W. Antimicrobial treatment of orthopedic implant-related infections with rifampin combinations. Clin Infect Dis. 1992;14(6):1251–3.
39. Drancourt M, Stein A, Argenson JN, Zannier A, Curvale G, Raoult D. Oral rifampin plus ofloxacin for treatment of *Staphylococcus*-infected orthopedic implants. Antimicrob Agents Chemother. 1993;37(6):1214–8.
40. Bahl D, Miller DA, Leviton I, Gialanella P, Wolin MJ, Liu W, Perkins R, Miller MH. In vitro activities of ciprofloxacin and rifampin alone and in combination against growing and non-growing strains of methicillin-susceptible and methicillin-resistant *Staphylococcus aureus*. Antimicrob Agents Chemother. 1997;41(6):1293–7.

41. Eng RH, Padberg FT, Smith SM, Tan EN, Cherubin CE. Bactericidal effects of antibiotics on slowly growing and nongrowing bacteria. Antimicrob Agents Chemother. 1991;35(9):1824–8.
42. Murillo O, Pachon ME, Euba G, Verdaguer R, Tubau F, Cabellos C, Cabo J, Gudiol F, Ariza J. Antagonistic effect of rifampin on the efficacy of high-dose levofloxacin in staphylococcal experimental foreign-body infection. Antimicrob Agents Chemother. 2008;52(10):3681–6.
43. Trampuz A, Murphy CK, Rothstein DM, Widmer AF, Landmann R, Zimmerli W. Efficacy of a novel rifamycin derivative, ABI-0043, against *Staphylococcus aureus* in an experimental model of foreign-body infection. Antimicrob Agents Chemother. 2007;51(7):2540–5.
44. San Juan R, Garcia-Reyne A, Caba P, Chaves F, Resines C, Llanos F, Lopez-Medrano F, Lizasoain M, Aguado JM. Safety and efficacy of moxifloxacin monotherapy for treatment of orthopedic implant-related staphylococcal infections. Antimicrob Agents Chemother. 2010;54(12):5161–6.
45. Vindel A, Cuevas O, Cercenado E, Marcos C, Bautista V, Castellares C, Trincado P, Boquete T, Perez-Vazquez M, Marin M, Bouza E. Methicillin-resistant *Staphylococcus aureus* in Spain: molecular epidemiology and utility of different typing methods. J Clin Microbiol. 2009;47(6):1620–7.
46. Baldoni D, Haschke M, Rajacic Z, Zimmerli W, Trampuz A. Linezolid alone or combined with rifampin against methicillin-resistant *Staphylococcus aureus* in experimental foreign-body infection. Antimicrob Agents Chemother. 2009;53(3):1142–8.
47. Murillo O, Garrigos C, Pachon ME, Euba G, Verdaguer R, Cabellos C, Cabo J, Gudiol F, Ariza J. Efficacy of high doses of daptomycin versus alternative therapies against experimental foreign-body infection by methicillin-resistant *Staphylococcus aureus*. Antimicrob Agents Chemother. 2009;53(10):4252–7.
48. Bassetti M, Di Biagio A, Cenderello G, Del Bono V, Palermo A, Cruciani M, Bassetti D. Linezolid treatment of prosthetic hip Infections due to methicillin-resistant *Staphylococcus aureus* (MRSA). J Infect. 2001;43(2):148–9.
49. Bassetti M, Vitale F, Melica G, Righi E, Di Biagio A, Molfetta L, Pipino F, Cruciani M, Bassetti D. Linezolid in the treatment of Gram-positive prosthetic joint infections. J Antimicrob Chemother. 2005;55(3):387–90.
50. Jover-Saenz A, Gaite FB, Ribelles AG, Porcel-Perez JM, Garrido-Calvo S. Linezolid treatment of total prosthetic knee infection due to methicillin-resistant *Staphylococcus epidermidis*. J Infect. 2003;47(1):87–8.
51. Entenza JM, Moreillon P. Tigecycline in combination with other antimicrobials: a review of *in vitro*, animal and case report studies. Int J Antimicrob Agents. 2009;34(1):8.e1–9.
52. Rodvold KA, Gotfried MH, Cwik M, Korth-Bradley JM, Dukart G, Ellis-Grosse EJ. Serum, tissue and body fluid concentrations of tigecycline after a single 100 mg dose. J Antimicrob Chemother. 2006;58(6):1221–9.
53. Vaudaux P, Fleury B, Gjinovci A, Huggler E, Tangomo-Bento M, Lew DP. Comparison of tigecycline and vancomycin for treatment of experimental foreign-body infection due to methicillin-resistant *Staphylococcus aureus*. Antimicrob Agents Chemother. 2009;53(7):3150–2.
54. Mascio CT, Alder JD, Silverman JA. Bactericidal action of daptomycin against stationary-phase and nondividing *Staphylococcus aureus* cells. Antimicrob Agents Chemother. 2007;51(12):4255–60.
55. John AK, Baldoni D, Haschke M, Rentsch K, Schaerli P, Zimmerli W, Trampuz A. Efficacy of daptomycin in implant-associated infection due to methicillin-resistant *Staphylococcus aureus*: importance of combination with rifampin. Antimicrob Agents Chemother. 2009; 53(7) :2719–24.
56. Garrigos C, Murillo O, Euba G, Verdaguer R, Tubau F, Cabellos C, Cabo J, Ariza J. Efficacy of usual and high doses of daptomycin in combination with rifampin versus alternative therapies in experimental foreign-body infection by methicillin-resistant *Staphylococcus aureus*. Antimicrob Agents Chemother. 2010;54(12):5251–6.
57. Ariza J, Euba G, Murillo O. Orthopedic device-related infections. Enferm Infecc Microbiol Clin. 2008;26(6):380–90.
58. Roberts JA, Paratz J, Paratz E, Krueger WA, Lipman J. Continuous infusion of beta-lactam antibiotics in severe infections: a review of its role. Int J Antimicrob Agents. 2007;30(1):11–8.

59. Falagas ME, Kasiakou SK. Colistin: the revival of polymyxins for the management of multidrug-resistant gram-negative bacterial infections. Clin Infect Dis. 2005;40(9):1333–41.
60. Falagas ME, Kastoris AC, Karageorgopoulos DE, Rafailidis PI. Fosfomycin for the treatment of infections caused by multidrug-resistant non-fermenting Gram-negative bacilli: a systematic review of microbiological, animal and clinical studies. Int J Antimicrob Agents. 2009;34(2):111–20.
61. Li J, Nation RL, Turnidge JD, Milne RW, Coulthard K, Rayner CR, Paterson DL. Colistin: the re-emerging antibiotic for multidrug-resistant Gram-negative bacterial infections. Lancet Infect Dis. 2006;6(9):589–601.
62. Meehan AM, Osmon DR, Duffy MC, Hanssen AD, Keating MR. Outcome of penicillin-susceptible streptococcal prosthetic joint infection treated with debridement and retention of the prosthesis. Clin Infect Dis. 2003;36(7):845–9.
63. Euba G, Lora-Tamayo J, Murillo O, Pedrero S, Cabo J, Verdaguer R, Ariza J. Pilot study of ampicillin-ceftriaxone combination for treatment of orthopedic infections due to *Enterococcus faecalis*. Antimicrob Agents Chemother. 2009;53(10):4305–10.
64. Ivert TS, Dismukes WE, Cobbs CG, Blackstone EH, Kirklin JW, Bergdahl LA. Prosthetic valve endocarditis. Circulation. 1984;69(2):223–32.
65. Calderwood SB, Swinski LA, Waternaux CM, Karchmer AW, Buckley MJ. Risk factors for the development of prosthetic valve endocarditis. Circulation. 1985;72(1):31–7.
66. Karchmer AW, Longworth DL. Infections of intracardiac devices. Infect Dis Clin North Am. 2002;16(2):477–505. Xii.
67. Murdoch DR, Corey GR, Hoen B, Miro JM, Fowler Jr VG, Bayer AS, Karchmer AW, Olaison L, Pappas PA, Moreillon P, Chambers ST, Chu VH, Falco V, Holland DJ, Jones P, Klein JL, Raymond NJ, Read KM, Tripodi MF, Utili R, Wang A, Woods CW, Cabell CH. Clinical presentation, etiology, and outcome of infective endocarditis in the 21st century: the International Collaboration on Endocarditis-Prospective Cohort Study. Arch Intern Med. 2009;169(5):463–73.
68. Gandelman G, Frishman WH, Wiese C, Green-Gastwirth V, Hong S, Aronow WS, Horowitz HW. Intravascular device infections: epidemiology, diagnosis, and management. Cardiol Rev. 2007;15(1):13–23.
69. Baddour LM, Wilson WR, Bayer AS, Fowler Jr VG, Bolger AF, Levison ME, Ferrieri P, Gerber MA, Tani LY, Gewitz MH, Tong DC, Steckelberg JM, Baltimore RS, Shulman ST, Burns JC, Falace DA, Newburger JW, Pallasch TJ, Takahashi M, Taubert KA. Infective endocarditis: diagnosis, antimicrobial therapy, and management of complications: a statement for healthcare professionals from the Committee on Rheumatic Fever, Endocarditis, and Kawasaki Disease, Council on Cardiovascular Disease in the Young, and the Councils on Clinical Cardiology, Stroke, and Cardiovascular Surgery and Anesthesia, American Heart Association: endorsed by the Infectious Diseases Society of America. Circulation. 2005;111(23):e394–434.
70. Horstkotte D, Follath F, Gutschik E, Lengyel M, Oto A, Pavie A, Soler-Soler J, Thiene G, von Graevenitz A, Priori SG, Garcia MA, Blanc JJ, Budaj A, Cowie M, Dean V, Deckers J, Fernandez Burgos E, Lekakis J, Lindahl B, Mazzotta G, Morais J, Smiseth OA, Vahanian A, Delahaye F, Parkhomenko A, Filipatos G, Aldershvile J, Vardas P. Guidelines on prevention, diagnosis and treatment of infective endocarditis executive summary; the task force on infective endocarditis of the European society of cardiology. Eur Heart J. 2004;25(3):267–76.
71. Archer GL, Johnston JL, Vazquez GJ, Haywood 3rd HB. Efficacy of antibiotic combinations including rifampin against methicillin-resistant *Staphylococcus epidermidis*: *in vitro* and *in vivo* studies. Rev Infect Dis. 1983;5 Suppl 3:S538–42.
72. Kobasa WD, Kaye KL, Shapiro T, Kaye D. Therapy for experimental endocarditis due to *Staphylococcus epidermidis*. Rev Infect Dis. 1983;5 Suppl 3:S533–7.
73. Fowler Jr VG, Boucher HW, Corey GR, Abrutyn E, Karchmer AW, Rupp ME, Levine DP, Chambers HF, Tally FP, Vigliani GA, Cabell CH, Link AS, DeMeyer I, Filler SG, Zervos M, Cook P, Parsonnet J, Bernstein JM, Price CS, Forrest GN, Fatkenheuer G, Gareca M, Rehm SJ, Brodt HR, Tice A, Cosgrove SE. Daptomycin versus standard therapy for bacteremia and endocarditis caused by *Staphylococcus aureus*. N Engl J Med. 2006;355(7):653–65.

74. Marco F, de la Maria CG, Armero Y, Amat E, Soy D, Moreno A, del Rio A, Almela M, Mestres CA, Gatell JM, Jimenez de Anta MT, Miro JM. Daptomycin is effective in treatment of experimental endocarditis due to methicillin-resistant and glycopeptide-intermediate *Staphylococcus aureus*. Antimicrob Agents Chemother. 2008;52(7):2538–43.
75. Sakoulas G, Eliopoulos GM, Alder J, Eliopoulos CT. Efficacy of daptomycin in experimental endocarditis due to methicillin-resistant *Staphylococcus aureus*. Antimicrob Agents Chemother. 2003;47(5):1714–8.
76. Falagas ME, Giannopoulou KP, Ntziora F, Vardakas KZ. Daptomycin for endocarditis and/or bacteraemia: a systematic review of the experimental and clinical evidence. J Antimicrob Chemother. 2007;60(1):7–19.
77. Falagas ME, Manta KG, Ntziora F, Vardakas KZ. Linezolid for the treatment of patients with endocarditis: a systematic review of the published evidence. J Antimicrob Chemother. 2006;58(2):273–80.
78. Miro JM, Del Rio A, Moreno A, et al. Efficacy and safety of fosfomycin plus imipenem for the treatment of methicillin-resistant *Staphylococcus aureus* (MRSA) native valve endocarditis (NVE): preliminary results of a clinical trial. 48th International Conference on Antimicrobial Agents and Chemotherapy (ICAAC 2008); 2008 Oct 25–28; Washington, DC: L-1523
79. Sieradzki K, Tomasz A. Suppression of beta-lactam antibiotic resistance in a methicillin-resistant *Staphylococcus aureus* through synergic action of early cell wall inhibitors and some other antibiotics. J Antimicrob Chemother. 1997;39(Suppl A):47–51.
80. Baddour LM, Bettmann MA, Bolger AF, Epstein AE, Ferrieri P, Gerber MA, Gewitz MH, Jacobs AK, Levison ME, Newburger JW, Pallasch TJ, Wilson WR, Baltimore RS, Falace DA, Shulman ST, Tani LY, Taubert KA. Nonvalvular cardiovascular device-related infections. Circulation. 2003;108(16):2015–31.
81. Bracke FA, Meijer A, van Gelder LM. Pacemaker lead complications: when is extraction appropriate and what can we learn from published data? Heart. 2001;85(3):254–9.
82. Reilly L. Aortic graft infection: evolution in management. Cardiovasc Surg. 2002;10(4):372–7.
83. Kudva A, Hye RJ. Management of infectious and cutaneous complications in vascular access. Semin Vasc Surg. 1997;10(3):184–90.
84. Yeager RA, Taylor Jr LM, Moneta GL, Edwards JM, Nicoloff AD, McConnell DB, Porter JM. Improved results with conventional management of infrarenal aortic infection. J Vasc Surg. 1999;30(1):76–83.
85. Martinez E, Rello J, Coll P. Clinical diagnosis of ventriculostomy-related infections. Lancet. 1994;344(8928):1015–6.
86. Yogev R, Bisno A. Infections of central nervous system shunts. In: Waldvogel F, Bisno AL, editors. Infections associated with indwelling medical devices. 3rd ed. Washington, DC: ASM Press, 2000; 231–46.
87. Schreffler RT, Schreffler AJ, Wittler RR. Treatment of cerebrospinal fluid shunt infections: a decision analysis. Pediatr Infect Dis J. 2002;21(7):632–6.
88. Pittet B, Montandon D, Pittet D. Infection in breast implants. Lancet Infect Dis. 2005;5(2): 94–106.
89. Gabriel SE, Woods JE, O'Fallon WM, Beard CM, Kurland LT, Melton 3rd LJ. Complications leading to surgery after breast implantation. N Engl J Med. 1997;336(10):677–82.
90. Mulcahy JJ. Surgical management of penile prosthesis complications. Int J Impot Res. 2000;12 Suppl 4:S108–11.
91. Wilson SK, Delk 2nd JR. Inflatable penile implant infection: predisposing factors and treatment suggestions. J Urol. 1995;153(3 Pt 1):659–61.

Chapter 10
Preventive Measures Against Transcutaneous Device Infections

Paul H.S. Kwakman and Sebastian A.J. Zaat

Abstract Transcutaneous medical devices are indispensible in medicine. Infection is the most frequently reported complication of indwelling devices and is associated with substantial costs, morbidity, and even mortality. Since antibiotics have limited efficacy in the treatment of such infections, removal of the device is required to eradicate the infection in a considerable number of cases. Therefore, measures to prevent contamination of devices during and after insertion are of crucial importance to minimize the incidence of device-related infection. Since the patient's own skin microflora is considered a major source of infection of transcutaneous devices, reduction of skin colonization at the insertion site of devices has high priority as a means to reduce the incidence of infection. Strategies to reduce the risk for contamination of transcutaneous devices with skin bacteria include (1) hygiene measures during surgery, (2) promoting integration of implanted devices with host tissues, (3) surface modification of the device to prevent adherence of bacteria, and (4) topical antimicrobial prophylaxis. Use of antibiotics for topical antimicrobial prophylaxis is strongly discouraged in view of the risk for resistance development. Antiseptics can be effective to reduce the incidence of infection of transcutaneous devices, but application of these compounds is mainly restricted to superficial skin disinfection. In addition, there are increasing concerns regarding antiseptic resistance development related to the widespread use of these agents. Therefore, alternative antimicrobial strategies are urgently needed. The potential of antimicrobial peptides and of honey as novel antimicrobial agents to prevent infection of transcutaneous devices is discussed.

P.H.S. Kwakman (✉) • S.A.J. Zaat
Department of Medical Microbiology, Center for Infection and Immunity Amsterdam (CINIMA), Academic Medical Center, University of Amsterdam, Meibergdreef 15, Amsterdam 1105, The Netherlands
e-mail: p.h.kwakman@amc.uva.nl

T.F. Moriarty et al. (eds.), *Biomaterials Associated Infection: Immunological Aspects and Antimicrobial Strategies*, DOI 10.1007/978-1-4614-1031-7_10,

10.1 Introduction

Transcutaneous devices are increasingly used in various medical and surgical disciplines and are indispensible in medicine. Transcutaneous devices can be categorized into two classes: (1) non-implanted devices including catheters and drains and (2) implanted devices including soft-tissue anchored implants such as ventricular assist devices, transcutaneous port systems (e.g., small bowel stomas) and catheters intended for long-term use such as continuous ambulatory peritoneal dialysis, hemodialyisis and chemotherapy, and bone-anchored implants like external fixators, intra-osseous transcutaneous amputation prostheses, auricular, craniofacial and dental implants.

Infection is the most frequently reported complication of indwelling devices and is associated with substantial costs, morbidity, and even mortality. It is well-known that the presence of a foreign body predisposes the host for infection [1, 2]. The major factors for increased susceptibility of indwelling devices to infection are local impairment of host immune defenses and the formation of biofilms by bacteria adhering to these devices [1]. Coagulase-negative staphylococci (CoNS) and *Staphylococcus aureus* are the most common causative agents of biomaterial-associated infection (BAI) [3]. Bacteria can cause infection either by adhering to foreign bodies and subsequent biofilm formation or by colonizing the tissue surrounding these materials [1, 4]. Established BAI are very difficult to treat due to limited efficacy of antibiotics in such situations [3]. Reduced susceptibility of sessile bacteria present in biofilms and poor penetration of antibiotics through the biofilm matrix are considered the predominant causes of this lack of efficacy of antibiotics [4, 5]. In a considerable number of cases, removal of the foreign body is required to eradicate the infection. Therefore, measures to prevent contamination of biomaterials during and after insertion are of crucial importance to minimize the incidence of BAI. In this chapter, we summarize available information on the etiology of infections of transcutaneous medical devices. Since the data indicate the patient's skin microflora as one of the main sources of infection, we subsequently discuss current and future strategies to prevent infection of devices through this route.

10.2 Catheter-Related Infections

10.2.1 Magnitude of the Problem

Catheters are by far the most frequently used transcutaneous devices. Central venous catheters (CVCs) are the predominant type of catheters that are at high risk for infection [6]. Since many patients with CVCs have severe underlying illnesses, catheter-related infections are a major problem associated with high costs, morbidity, and mortality. In the USA alone, three million CVCs are used yearly [7]. Reported rates of catheter-related blood stream infections (CRBSI) for CVCs range from 1.4 to 7.7 episodes per 1,000 catheter days [8–10]. A total of 250,000 episodes of CRBSI are reported annually in the USA [6], of which 80,000 episodes occur in critically ill patients [11]. It has been estimated that CRBSIs are responsible for 28,000 deaths per year, an average of 9–12 extra days of hospital stay per patient [12, 13] with an estimated additional cost of care ranging from $3,000 to $56,000 per case [11, 14, 15]. Generally, catheters that are suspected for infection are removed and when necessary, a new catheter is placed at a different location. Reimplantation at the original location over a guide-wire is strongly advised against because of the high risk of reinfection [16]. Only under circumstances where removal of a catheter is impossible, it is attempted to salvage infected catheters by prolonged administration of antibiotics, but the success rates are very disappointing [17].

10.2.2 Etiology of Catheter-Related Bloodstream Infection

A bloodstream infection is defined as CRBSI when the same organism is cultured from the catheter and from a peripherally drawn blood culture, combined with the absence of other apparent primary sources of infection. Coagulase-negative staphylococci and *S. aureus* are responsible for the majority of CRBSI cases, while enterococci, various Gram-negative bacteria and *Candida* species are responsible for most of the remaining cases [14]. Microbial colonization of the catheter insertion site is a major risk factor for bloodstream infections related to the use of short-term CVCs (<5–7 days in place) [18–20], while colonization of the catheter hub is the major risk factor associated with the infection of long-term catheters [16, 21, 22]. This suggests that bacteria originating from the skin are the major source for infection of short-term catheters and that contamination of the catheter hub is the major cause of infections related to long-term use of CVCs.

Since staphylococci are a prominent part of the normal human skin microflora, it is essential to be able to discriminate between isolates causing CRBSI and commensal bacteria in order to reliably assess an etiologic relationship. Pulsed field gel-electrophoresis (PFGE) is the gold standard for molecular typing of staphylococci, and also random amplification of amplified DNA (RAPD) is described as an appropriate technique with high discriminative power [23]. Only few studies have however been performed to actually study the source of bacteria causing CRBSI.

Dobbins et al. [24] investigated the etiology of 21 cases of CRBSI caused by CoNS. These cases were defined on the basis of the same organism being isolated from the catheter and from peripheral blood cultures as identified by standard phenotypic methods. Isolates from blood, catheter surface, skin at the catheter insertion site and from the catheter hub were selected based on distinct morphological features and were analyzed by PFGE. For five cases defined as CRBSI using standard phenotypic methods, molecular typing revealed that different isolates were actually recovered from the catheter and from peripheral blood [24]. These cases were classified as false positive and not included in further analysis.

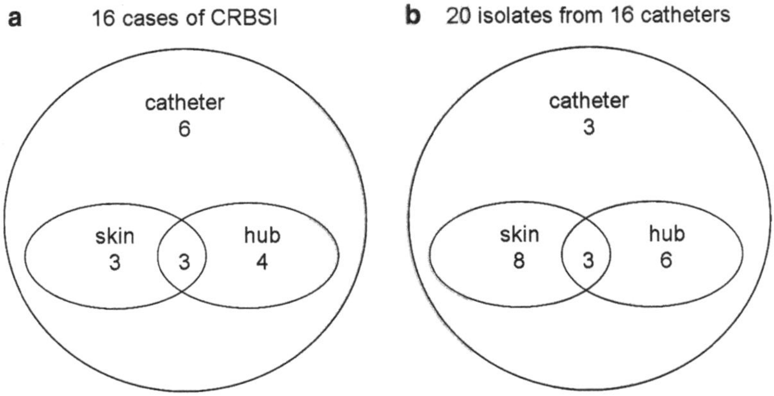

Fig. 10.1 Assessment of the etiology of catheter infections using pulsed-field gel electrophoresis. The etiology of 16 cases of CRBSI due to coagulase-negative Staphylococci (**a**) and of 20 isolates cultured from 16 infected central venous catheters (**b**) as studied by Dobbins et al. [24] and Bach et al. [25], respectively. Both studies performed molecular typing of bacteria cultured from the skin at the catheter insertion site and from the catheter hub to assess the potential contribution of these sites as the source of catheter infection. The data presented indicate the number of isolates from the infected catheters, and the numbers of identical isolates cultured from skin, hub, or both

Of the 16 remaining cases, in three cases the causative isolate was only isolated from the skin at the catheter insertion site, in four cases from only the hub and in three cases from the skin as well as from the hub (Fig. 10.1a). In six cases, the causative isolate was not cultured from the skin or the hub. These data clearly indicate that both the skin at the catheter insertion site and the hub are suspected sources of CRBSI. In the cases where both the skin and the hub were colonized by the CRBSI-causing isolate, it is not possible to determine the initial source of infection but a possible route of infection is colonization of the hub by bacteria from the patient's skin, with subsequent catheter colonization and CRBSI.

Bach et al. used a similar molecular approach to determine the relative contribution of skin and catheter hub colonization as source for catheter tip colonization [25]. Bacterial isolates from the internal and external catheter surfaces, skin, catheter hub, and blood were compared by molecular typing using PFGE. Twenty different isolates were identified on 16 colonized catheters. Eight of these isolates were also cultured from skin at the respective catheter insertion sites but not from the hub, six isolates were retrieved only from the hub and in three cases, an isolate cultured from the catheter was found on both the skin and hub (Fig 10.1b). Only three isolates cultured from catheter tips were not found on the skin or on the hub.

In cases where the causative isolate was not identified at the catheter insertion site or the hub, infection may have been caused by exogenous contamination, e.g., via healthcare workers, contaminated infusate or hematogenous seeding of the catheter. Alternatively, the isolate causing infection may actually have originated from the patient's endogenous flora, but may have been missed in the analysis for several reasons. Firstly, bacterial isolates present in the deeper layers of skin may not be collected by the skin swab. Secondly, for each colony phenotype only one colony was picked for typing. Since many types may have similar colony morphology, some types might thus have been omitted from molecular analysis when only one colony was selected for each colony phenotype. Thirdly, there is considerable temporal variation in the composition of the skin microflora [26, 27]. In the case of skin as the primary source of infection, temporal changes in composition of the microflora are possibly enhanced by frequent disinfection of the catheter insertion site. Consequently, at the time of sampling when CRBSI has become manifest, a causative isolate may have disappeared from the skin. As a result, the contribution of skin bacteria as source for CRBSI may be underestimated.

To summarize, molecular techniques are very useful to sensitively assess similarity between individual bacterial isolates and are an excellent tool to identify potential sources of infection, but it is not possible to precisely quantify the contribution of specific sources as origin of infection. Since isolates responsible for CRBSI are very frequently also retrieved from the skin at the catheter insertion site, this does however indicate the importance of skin microflora as source of CRBSI and thus supports the high priority for reduction of skin colonization as a means to reduce the incidence of CRBSI [16].

10.3 Transcutaneous Implant-Related Infections

10.3.1 Soft Tissue-Anchored Implants: Magnitude of the Problem

Soft tissue-anchored implants include complex implantable devices such as ventricular assist devices (VADs), transcutaneous port systems such as small bowel stomas, and catheters intended for long-term use in continuous ambulatory peritoneal dialysis, hemodialysis and chemotherapy. Among soft-tissue anchored implants the highest incidence of implant-related infection is reported for implantable VADs, with infections occurring in 19–56% of all implanted devices [5, 28]. VADs have a pump located inside the body and a power source located outside. A driveline connects the pump to the power source through a small hole in the abdomen. Implantable VADs are mainly used to span the time period to a heart transplantation (mostly a period of weeks to several months), but these devices are also increasingly used as a permanent solution, e.g., for patients not eligible for a heart transplantation. Infectious complications result in about $80,000 additional costs for every infected VAD [28] and also result in attributable mortality in the case of VAD-related endocarditis [29]. In 2009, approximately 1,700 and 430 VADs have been implanted in patients in the USA and in Europe, respectively, and the numbers of patients receiving an implantable VAD are increasing dramatically [30].

Typically frequencies of infection for other soft tissue-anchored devices are much lower, generally between 2 and 6% [5]. Despite this lower incidence, these implant-related infections are also a major problem due to the large and increasing numbers of implants used and because of the difficulties to treat such infections.

10.3.2 Bone-Anchored Devices: Magnitude of the Problem

Bone-anchored transcutaneous devices include external fixators, intra-osseous transcutaneous amputation prostheses, and auricular, craniofacial and dental implants. The frequencies of infection of bone-anchored devices are much higher than those of soft-tissue implanted devices and catheters and often exceed 10% [31–33] For external fixator-related infections, the reported frequencies of infection range from 0.5% to even as high as 50% ([34–36]). Similar to other transcutaneous devices, these infections are mainly caused by *S. epidermidis* and *S. aureus* [4].

External fixation is an essential component of damage control orthopedics used for stabilization of bone fractures. In this procedure an exterior scaffold-like frame is fixed to metal pins or wires which are inserted into bone fragments through small incisions in the skin. A major advantage is that the use of external fixators allows gradual realignment with frequent small adjustments to the external apparatus making it extremely useful in the management of deformity correction and bone loss.

The large variation in pin track infection frequencies may in part be explained by a lack of a clear definition of pin tract infection [35, 36]. In addition to infection of the pin tract, about 4% of patients develop chronic osteomyelitis [37]. Pins are mainly colonized by *S. epidermidis* and *S. aureus* [38] and treatment of infection often requires removal of the pin and tissue debridement [34]. It is thought that care of pin sites will reduce the frequency of pin site infections, loosening of fixation, and osteomyelitis, but there currently is no generally accepted protocol for pin site care [35]. Care may involve disinfection of the skin at the pin insertion site with hydrogen peroxide, povidone iodine, or chlorhexidine, but no obvious advantage for any protocol has been established [34, 35]. This may be related to the poor quality and lack of sufficient power of most clinical trials performed [35].

10.3.3 Etiology of Transcutaneous Implant-Related Infections

As with CVCs, staphylococci are the major pathogens involved in transcutaneous implant-related infections, and Gram-negative bacilli, streptococci and *Candida* spp. are responsible for most of the remaining infections [28, 39–42]. The high frequencies of staphylococci as causative agents of these device-related infections again suggest skin microflora as the major source of infection. Contamination from the patient's own skin or from the physician during surgery is indeed considered the most prominent cause of implant infection [42]. For transcutaneous implants, postoperative contamination is another potential cause of infection due to a lack of barrier function of the skin. Only few studies however assessed the etiology of such implant infections. In patients with VAD, mobility and frequent manipulation of the drivelines easily cause driveline trauma impeding the healing of the exit-site wound. This results in an open tract from the skin via the driveline to other parts of the device, allowing bacteria to enter, grow, and cause infection of various components of the device. Driveline infections are the most common cause of VAD infections, but more than half of all VAD infections include multiple infected sites [28]. In a study by Herrmann et al. the etiology of VAD infections was studied by typing the isolates by PFGE (43). In 3 of 12 cases identical isolates were cultured from the infected device and from the skin at the driveline exit site [43]. As already discussed for the etiology of CVC infections, for several reasons a causative isolate may be missed in molecular typing of skin microflora which could thus result in underestimation of the role of skin microflora in implant infection.

Most other implanted transcutaneous devices have a much longer dwell time than VADs. This may further complicate identification of the source of late-onset infections. In fact, to the best of our knowledge there are no reports on the etiology of device-related infections at the level of molecular typing other than those described above. In summary, the few studies that assessed the etiology of transcutaneous device-related infection in molecular detail reinforce the presumed importance of the skin microflora as the major cause of transcutaneous device-related infections.

10.4 Preventive Anti-infective Strategies

Effective treatment of implant infections often requires removal of the device. In the case of infected catheters or external fixator pins, a new catheter or pin is inserted at a distant site when a new device is required [8, 44]. In situations where the devices must be replaced at the site of infection, and replacement can be postponed, a two-stage approach is frequently used since immediate replacement of infected devices is prone to rapid reinfection [5]. In the first stage the implant is removed and antibiotic therapy is initiated. In the second stage, a new device is placed but only when the infection is fully eradicated. This is very debilitating for the patient and in some situations even impossible because some devices are life-sustaining. In these cases, surgical debridement and/or prolonged treatment with antibiotics with the implant in place is attempted. This may control clinical manifestations but rarely eradicates the infection [3, 5]. In view of the limited efficacy of antibiotics to clear implant infections and the risk for resistance development with prolonged application of antibiotics, a strong emphasis on strategies to prevent device-related infections is warranted.

Strategies to prevent transcutaneous device-related infections are primarily aimed at reducing the chance of skin bacteria reaching and colonizing these devices. The main strategies are (1) hygiene measures during surgery to reduce the risk of contamination of the device during insertion or implantation, (2) promote integration of implanted devices with host tissues, (3) surface modification of the device to prevent adherence of bacteria, and (4) topical antimicrobial prophylaxis to reduce skin colonization at the entry site of the device.

10.4.1 Hygiene Measures

While traditionally most approaches to reduce the incidence of CRBSI were based on technology, several more recent studies have emphasized the importance of hygiene measures for catheter insertion and catheter care as an important strategy to reduce the incidence of infection. Eggimann et al. used a comprehensive strategy to implement best practice strategies on all levels of catheter insertion and catheter care for ICU patients with various types of venous catheters [45]. Compared to a control period, implementation of these measures reduced the incidence of CRBSI from 2.4 to 0.8 episodes per 1,000 catheter days for all catheter types combined and from 6.6 to 2.3 episodes per 1,000 catheter days for CVCs. Pronovost et al. modified this elaborate protocol into a simplified strategy composed of the five most important elements: good hand hygiene, maximal sterile barriers, skin disinfection with chlorhexidine, use of the subclavia as the preferred site of insertion, and removal of the catheter as soon as possible [10]. These so-called "care bundles" reduced the incidence of CRBSI in ICU patients from 7.7 to 1.4 cases per 1,000 catheter days. In recent years, the introduction of these practice-based catheter care protocols has resulted in a substantial reduction in incidence of CRBSI [7, 46, 47].

Full implementation of care bundles requires intensive education and a very strong commitment of the medical and nursing staff. As a consequence, many healthcare facilities have incompletely adopted these evidence-based practices [48]. Strategies to improve compliance to catheter care protocols are therefore strongly required.

It has been estimated that approximately one-third of all infections of implantable devices could be prevented with strict hygiene measures [49]. In the case of transcutaneous implants, strict hygiene measures are essential until implants are fully integrated in host tissue. Despite the strict adherence to hygiene guidelines, many wounds are contaminated during surgical procedures [42]. Due to the huge impact of device-related infections, perioperative systemic antimicrobial prophylaxis has become standard practice for all implantable devices [50]. Quality standards for antimicrobial prophylaxis have been published 18 years ago [51], but there still is insufficient data to establish evidence-based guidelines [52]. It is recommended to administer a single dose or maximally apply 1 day intravenous administration of an antibiotic with an antimicrobial spectrum as narrow as possible to limit the risk for resistance development [3]. Since hygiene measures do not reduce the incidence of infection of transcutaneous devices to zero, additional anti-infective measures are required [8].

10.4.2 Integration of Implanted Devices with Host Tissues

For prevention of transcutaneous implant failure non-antimicrobial strategies like firm immobilization and proper design of implants are generally regarded the most crucial aspects. These measures are intended to reduce the risk for mechanical implant loosening, for inward growth of epithelial cells along the implant resulting in pocket formation, and for infection, which are the three most common causes of transcutaneous implant failure [31, 53–57]. Firm fixation of devices is essential to prevent mechanical loosening. Since firm fixation also enhances wound healing, it might reduce the risk for implant infection as well. There are several reports on device modifications that physically prevent inward growth of epithelial cells or which enhance epithelial cells and keratinocytes to attach to the biomaterial in order to reduce this inward growth [56, 58, 59]. Good integration between the implant and the surrounding tissue will restore the skin barrier and might thus reduce the risk for infection. This approach is however still in an experimental phase, so additional anti-infective strategies are required.

10.4.3 Antiseptics

Antiseptics are among the agents utilized to modify catheters and implantable devices to render them antimicrobial, as is extensively discussed in several chapters of this book. In the remainder of the current paragraph we will focus on topical antimicrobial applications of antiseptics.

Despite disinfection of the insertion site with chlorhexidine or povidone iodine prior to insertion of a catheter and regular disinfection during catheter use, approximately 30% of skin swabs of catheter insertion sites are culture-positive [18, 19, 60]. Povidone iodine antiseptic ointment and bacitracin/gramicidin/polymyxin B ointment are recommended to apply at exit sites of hemodialysis catheters, since these catheters are at high risk for infection [11, 61]. Antimicrobial prophylaxis using antibiotics is however strongly discouraged for other applications of catheters because of increased risk for resistance development and the potential risk for fungal overgrowth [8].

When the incidence of CRBSI does not decrease upon adherence to basic preventive measures the use of chlorhexidine-impregnated sponge dressings is recommended for catheters with an expected stay of >5 days [8]. In the largest trial to date, chlorhexidine sponges were even effective to reduce the rate of CRBSI in ICU patients when baseline level of infection was already low; the incidence of CRBSI was reduced from 1.3 episodes/1,000 catheter days in the control group to 0.4 episodes/1,000 catheter days in the chlorhexidine sponge group ($P=0.005$) [62]. Also for cancer patients receiving chemotherapy, efficacy of chlorhexidine sponges to reduce the incidence of CRBSI was reported [63].

Chlorhexidine is a biguanide compound that is used as an antiseptic, disinfectant and preservative in the pharmaceutical and cosmetic industry. It is a cationic membrane-active molecule that induces leakage of microbial membranes resulting in collapse of the membrane potential [64]. Although microbial resistance for antiseptics like chlorhexidine does not develop as rapidly as for antibiotics, there is increasing concern for antiseptic resistance. Various multidrug efflux pumps, e.g., AcrAB-TolC in *Escherichia coli*, MexCD-OmpJ in *Pseudomonas aeruginosa*, and QacA, QacB, and SMR in *S. aureus*, confer upto fourfold reduced susceptibility to chlorhexidine and other antiseptic as well as to antibiotic compounds [65–67]. Since the *in vitro* minimal bactericidal concentrations of chlorhexidine for bacteria expressing such efflux pumps are still well below the concentration used in clinical practice, the clinical relevance of these resistance mechanisms have been questioned [68]. A recent study with ICU patients in the UK however showed that the *in vivo* efficacy of chlorhexidine is severely impaired in qacA/B-positive methicillin-resistant *S. aureus* (MRSA) isolates despite only a threefold increase in *in vitro* MBC values compared to *qacA/B*-negative MRSA isolates [69]. In this study, chlorhexidine effectively eradicated *qacA/B*-negative isolates from the patient's body surfaces, but *qacA/B*-positive isolates were not affected by chlorhexidine antisepsis, and transmission of these isolates was also not prevented [69]. Although these results should be verified with randomized controlled trials, they clearly indicate that antiseptic-resistance mechanisms are highly relevant for clinical practice. With the recent expansion in applications of chlorhexidine, including chlorhexidine-impregnated catheters, the risk for selection and rapid spread of antiseptic-resistant bacteria is of great concern [65, 70, 71]. This highlights the importance of surveillance for resistance to chlorhexidine.

Preoperative disinfection of the skin with antiseptic agents is widely accepted as standard care prior to implantation of transcutaneous devices, but there are no general guidelines for postoperative topical antimicrobial prophylaxis. Exit site care may involve skin antisepsis with chlorhexidine or hydrogen peroxide during dressing change [72], but in most situations no topical postoperative antimicrobial prophylaxis is given. Antiseptic agents generally lack efficacy to prevent postoperative wound infection and may even impair wound healing [73]. Although topical application of antibiotics may be effective to prevent postoperative infections [73], their use for this purpose is discouraged due to the high risk for development of antibiotic resistance [3]. Therefore alternative antimicrobial strategies are urgently needed for situations where hygiene measures alone are not sufficient to prevent CRBSI.

10.4.4 Novel Topical Antimicrobial Agents

Infection is the major cause for failure of indwelling devices, but the currently available agents for antimicrobial prophylaxis—antiseptics and antibiotics—are of limited value to prevent device-related infections. Therefore alternative antimicrobial agents are required for prevention of implant infections. In recent years there has been an increasing interest in topical application of antimicrobial peptides (AMPs) and honey.

10.4.4.1 Antimicrobial Peptides

AMPs are a crucial component of the innate immune system of virtually all higher organisms [74, 75]. In general, AMPs are small, cationic peptides with highly distinct positively charged regions and hydrophobic regions [76, 77]. Most AMPs disturb the integrity of negatively charged microbial membranes [78], while some appear to interact with intracellular biomolecules such as DNA or proteins [79, 80]. In contrast to antibiotics, AMPs mostly do not have specific targets but rather interact with general structures of microorganisms. Due to this mode of action the risk for development of microbial resistance is regarded low [75] and AMPs thus are an interesting class of antimicrobial compounds for prophylactic applications.

Pexiganan, an analogue of the natural AMP magainin, has been tested in a model for transcutaneous osseointegrative implant infection [81]. In this model, titanium alloy pins were secured into the tibia of rabbits. A pexiganan acetate ointment was applied topically at the pin site 1 day after surgery, followed by daily application starting at 5 days postsurgery. From 2 weeks postsurgery the pin insertion sites were challenged weekly with 10^8 colony-forming units of *S. aureus*. Animals were followed for 24 weeks or were sacrificed upon signs of infection. In the treatment group only one of four rabbits had proven pin tract infection, while all eight rabbits of the control group were infected. Thus, pexiganan acetate reduced the incidence of pin tract infection in rabbits by 75%, indicating that this agent has promising activity to prevent transcutaneous implant infections.

Several other AMPs have been tested in human or animal studies for their efficacy to reduce colonization of skin or biomaterials. Of all AMPs studied, omiganan has advanced farthest towards potential therapeutic application. In a phase III clinical trial to prevent catheter-related bloodstream infections, this peptide reduced catheter colonization and local catheter infections compared to povidone iodine, but did not reach a statistically significant reduction in incidence of CRBSI, probably due to a small sample size [82].

Omiganan has also been assessed in a gel formulation for topical antimicrobial activity in an ex vivo pig skin model and in a whole-animal guinea pig skin model. After 24 h of application a 1% omiganan gel reduced the level of colonization by *S. epidermidis* by approximately 5- and 3-log in the ex vivo pig skin and the guinea pig models, respectively [83]. This gel also substantially decreased skin colonization with other microorganisms including *S. aureus* and *C. albicans*. Moreover, in a comparative analysis, the efficacy of the 1% omiganan gel was similar to that of a commercially available mupirocin ointment in the ex vivo pig skin model, indicating the potential of this peptide for topical applications.

The efficacy of coating of urethral stents with the bovine antimicrobial peptide BMAP-28 to treat *S. aureus* and *Enterococcus faecalis* infection has been tested in a rat model [84]. At 5 days postinfection stents coated with this peptide showed approximately 100-fold reduction in numbers of *S. aureus* or *E. faecalis* adhering to the stent and in numbers of bacteria cultured from urine compared to uncoated controls.

Giacometti et al. studied the efficacy of various antimicrobial peptides in a rat model for graft-associated infection. Subcutaneously placed dacron grafts were presoaked in a solution of antimicrobial peptides prior to implantation and directly after implantation, a bacterial inoculum was injected along the implanted graft. At 7 days postinfection the level of microbial colonization of grafts was quantified. In this model, buforin II and ranalexin reduced the level of graft colonization with methicillin-resistant *S. epidermidis* (MRSE) by 6 logs and 4 logs, respectively [85], and a dermaseptin derivative reduced the level of graft colonization with MRSA or MRSE by 4 logs compared to untreated controls [86].

The *in vivo* efficacy of the synthetic antimicrobial peptide BP2 has been tested in a murine model for *S. epidermidis* biomaterial-associated infection [87]. Segments of silicon elastomer catheters were implanted subcutaneously and a bacterial inoculum was injected along the implant. In a treatment model, BP2 was injected at 1 h after challenge with *S. epidermidis* and in a prophylactic model at 3 h prior to bacterial challenge. In the treatment model, an 80% reduction in the number of culture-positive implants and a 100-fold reduction in the survival of *S. epidermidis* in peri-implant tissue was observed at 24 h post-challenge. In the prophylactic model, the numbers of CFU cultured from biomaterial implants and peri-implant tissue were reduced by 85% and 90%, respectively.

AMPs injected along, or noncovalently bound to biomaterials might rapidly lose their activity due to dilution. Recent data from the group of Hancock indicate that the synthetic antimicrobial peptide Tet-20 retains potent antimicrobial activity *in vitro* and *in vivo* when covalently bound to polymer brushes on titanium surfaces. Such activity opens new possibilities for the coating of biomaterials with antimicrobials to prevent device-related infections. Clearly, AMPs have potential as novel agents to prevent device-related infection, either using such peptides as topical antimicrobial agents or as a coating of devices.

10.4.4.2 Honey

Honey has been used as a topical antimicrobial agent for thousands of years for prevention and treatment of wound infections [88]. With its long history of use honey is regarded as safe and several honeys and honey-based products are now registered as medical devices for use in wound dressing. Currently, manuka-based honeys and Revamil® are the major honeys approved for clinical application [89]. These medical-grade honeys are sterilized by gamma irradiation to destroy potentially present bacterial spores [90, 91]. The composition in antimicrobial compounds varies considerably among honeys, resulting in highly distinct antimicrobial properties.

Manuka honey is produced from the manuka bush (*Leptospermum scoparium*) indigenous to New Zealand and Australia. Because of the large batch-to-batch variation in antibacterial activity [92], batches of manuka honey are individually tested for activity. Revamil® is produced in greenhouses by a proprietary, standardized process. We demonstrated that 11 batches of this honey had less than a factor two difference in bactericidal activity against *Bacillus subtilis* [93].

Honey has broad spectrum antimicrobial activity, including activity against antibiotic-resistant microorganisms, due to the presence of a combination of different antimicrobial components. Well-known antimicrobial factors in honey are the high sugar content [approximately 80% (w/w)], production of hydrogen peroxide by the enzyme glucose oxidase upon moderate dilution of honey and the low pH. More recently, the antimicrobial compound methylglyoxal (MGO) and the antimicrobial peptide bee defensin-1 have been identified in honey [94–96]. Since honey exerts its antimicrobial activity via multiple mechanisms the risk for resistance development is negligible and resistance has so far never been described [97, 98].

Revamil® currently is the only honey of which all factors contributing to the bactericidal activity have been identified [96]. The activity of this honey is based on the high sugar concentration, hydrogen peroxide, bee defensin-1, MGO, and the low pH [96]. The antibacterial activity of manuka honeys depends largely on the presence of high levels of methylglyoxal, and also the high sugar content, low pH and several as yet unidentified compounds [89]. The bee defensin-1 concentration is negligible, at least in the honey preparation tested [89]. Interestingly, both Revamil® and manuka honey have potent broad spectrum antimicrobial activity despite the major differences in their modes of action [89]. There are clear indications for the presence of several additional, as yet unidentified antimicrobial compounds in yet other honeys [89, 99]. Clearly, not all honeys have similar antimicrobial properties. This illustrates the importance to characterize the antimicrobial activity honeys intended for clinical applications and to define the compounds contributing to the activity of these honeys.

In a study with healthy human volunteers we determined that Revamil® honey reduced the frequency of positive forearm skin cultures by 76% and reduced the level of colonization 100-fold (Fig. 10.2) [93]. This indicates that this honey has interesting properties in view of topical antimicrobial prophylactic applications. We are currently performing a randomized controlled trial to assess the efficacy of daily topical application of this medical-grade honey to reduce the frequency of positive skin cultures at the catheter insertion sites of ICU patients.

Manuka honey has been compared with mupirocin ointment in a randomized controlled trial for infection-free survival of tunneled, cuffed, internal jugular venous hemodialysis catheters [98]. Earlier studies had already shown that application of mupirocin at the catheter insertion site significantly reduced the incidence of catheter-related bacteremias and increased the time to first bacteremia for hemodialysis patients [100, 101]. A major drawback of mupirocin however is the rapid development of microbial resistance to this antibiotic [102, 103]. In the follow-up study with manuka honey, similar rates of catheter-related bacteremia and infection-free survival time compared to mupirocin ointment were achieved [98]. Although the latter study was not adequately powered to demonstrate equivalence between honey and mupirocin, the results indicate that thrice-weekly application of manuka honey to tunneled, cuffed hemodialysis catheters reduces the risk of catheter-related infections. Thus, honey could be an interesting alternative to antibiotics for topical antimicrobial prophylaxis to prevent infection of transcutaneous devices.

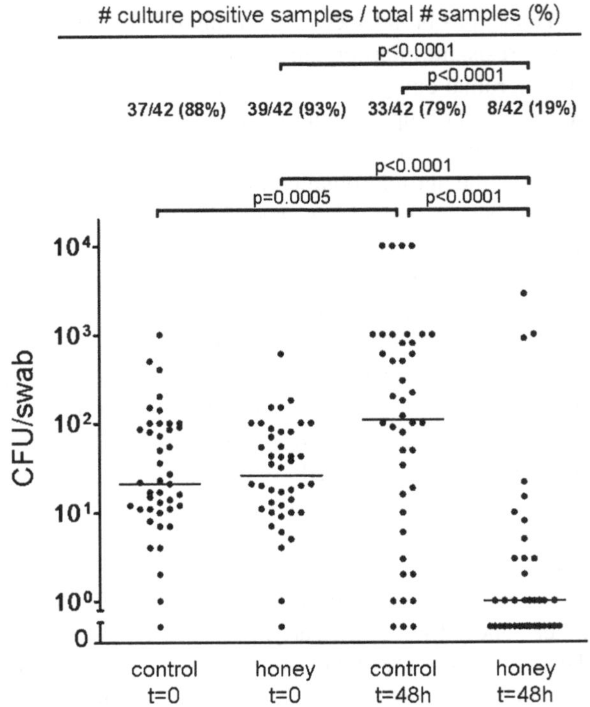

Fig. 10.2 Efficacy of honey to reduce forearm skin colonization. Number of culture-positive samples and numbers of CFU cultured from control and honey-treated skin patches. An aliquot of 0.5 ml Revamil® honey was applied on one forearm skin patch, and a second patch on the same arm served as a no-treatment control. Both patches were covered with polyurethane dressing. Prior to, and after 2 days of application, the skin patches were sampled to quantitatively assess the level of microbial skin colonization. Figure reprinted with permission from Clinical Infectious Diseases

10.5 Concluding Remarks

Bacteria from the patient's own skin are considered a major source for device-related infections, but limited data are available since the etiology of such infections has been assessed using adequate molecular typing of the isolates in only a few relatively small studies. Proper hygiene measures are of crucial importance to minimize the risk for implant-related infection. Antibiotics can be effective to prevent transcutaneous device-related infections, but due to risk for resistance development their use is strongly discouraged for most prophylactic applications. Antiseptics are effective for instantaneous disinfection of intact skin but not for long-term prevention of infections of open wounds or insertion sites. In addition, also for antiseptics reduced microbial susceptibility has been reported, and there are concerns for spreading of resistance genes related to the widespread use of these agents. Therefore novel antimicrobial agents are required for prevention of transcutaneous device-related infections. Antimicrobial peptides and honey have potential for such applications.

Glossary

Antimicrobial peptides (AMPs) Evolutionarily conserved component of the innate immune system found among all classes of life. The combined presence of cationic and hydrophobic domains AMPs results in specificity for interaction with negatively charged microbial membranes.

External fixator Surgical treatment to set bone fractures in cases where a cast would not allow proper alignment of the fracture. Holes are drilled into uninjured areas of bones around the fracture and pins or wires are screwed into the holes. Outside the body, a scaffold-like frame is fixed to the pins or wires to make a rigid support.

Molecular typing DNA-based methods to discriminate microbial isolates at a subspecies level to determine diversity and epidemiology of infections.

Ventricular assist device (VAD) Implanted mechanical device to partially or completely replace the function of the left and/or right heart ventricle. Mostly used as a bridge to heart transplantation.

References

1. Zimmerli W, Sendi P. Pathogenesis of implant-associated infection: the role of the host. Semin Immunopathol. 2011;33:295–306.
2. Elek SD, Conan PE. The virulence of Staphylococcus pyogenes for man; a study of the problems of wound infection. Br J Exp Pathol. 1957;38:573–86.
3. Trampuz A, Zimmerli W. Antimicrobial agents in orthopaedic surgery: prophylaxis and treatment. Drugs. 2006;66:1089–105.
4. Esteban J, Cordero-Ampuero J. Treatment of prosthetic osteoarticular infections. Expert Opin Pharmacother. 2011;12:899–912.
5. Darouiche RO. Treatment of infections associated with surgical implants. N Engl J Med. 2004;350:1422–9.
6. Maki DG, Kluger DM, Crnich CJ. The risk of bloodstream infection in adults with different intravascular devices: a systematic review of 200 published prospective studies. Mayo Clin Proc. 2006;81:1159–71.
7. Edgeworth J. Intravascular catheter infections. J Hosp Infect. 2009;73:323–30.
8. O'Grady NP, Alexander M, Burns LA, Dellinger EP, Garland J, Heard SO, Lipsett PA, Masur H, Mermel LA, Pearson ML, Raad II, Randolph AG, Rupp ME, Saint S. Guidelines for the prevention of intravascular catheter-related infections. Clin Infect Dis. 2011;52:e162–93.
9. National Nosocomial Infections Surveillance (NNIS) System Report, data summary from January 1992 through June 2004, issued October 2004. Am J Infect Control 32(8):470–85.
10. Pronovost P, Needham D, Berenholtz S, Sinopoli D, Chu H, Cosgrove S, Sexton B, Hyzy R, Welsh R, Roth G, Bander J, Kepros J, Goeschel C. An intervention to decrease catheter-related bloodstream infections in the ICU. N Engl J Med. 2006;355:2725–32.
11. Mermel LA. Prevention of intravascular catheter-related infections. Ann Intern Med. 2000; 132:391–402.
12. Soufir L, Timsit JF, Mahe C, Carlet J, Regnier B, Chevret S. Attributable morbidity and mortality of catheter-related septicemia in critically ill patients: a matched, risk-adjusted, cohort study. Infect Control Hosp Epidemiol. 1999;20:396–401.

13. Renaud B, Brun-Buisson C. Outcomes of primary and catheter-related bacteremia. A cohort and case-control study in critically ill patients. Am J Respir Crit Care Med. 2001;163:1584–90.
14. O'Grady NP, Alexander M, Dellinger EP, Gerberding JL, Heard SO, Maki DG, Masur H, McCormick RD, Mermel LA, Pearson ML, Raad II, Randolph A, Weinstein RA. Guidelines for the prevention of intravascular catheter-related infections. Clin Infect Dis. 2002;35:1281–307.
15. Burke JP. Infection control - a problem for patient safety. N Engl J Med. 2003;348:651–6.
16. Safdar N, Kluger DM, Maki DG. A review of risk factors for catheter-related bloodstream infection caused by percutaneously inserted, noncuffed central venous catheters: implications for preventive strategies. Medicine (Baltimore). 2002;81:466–79.
17. Raad I, Hanna H, Maki D. Intravascular catheter-related infections: advances in diagnosis, prevention, and management. Lancet Infect Dis. 2007;7:645–57.
18. Bouza E, Munoz P, Burillo A, Lopez-Rodriguez J, Fernandez-Perez C, Perez MJ, Rincon C. The challenge of anticipating catheter tip colonization in major heart surgery patients in the intensive care unit: are surface cultures useful? Crit Care Med. 2005;33:1953–60.
19. Safdar N, Maki D. The pathogenesis of catheter-related bloodstream infection with noncuffed short-term central venous catheters. Intensive Care Med. 2004;30:62–7.
20. Bisno AL, Waldvogel FA. Infections associated with indwelling medical devices. Washington DC: ASM Press; 1994.
21. Moro ML, Vigano EF, Cozzi LA. Risk factors for central venous catheter-related infections in surgical and intensive care units. The Central Venous Catheter-Related Infections Study Group. Infect Control Hosp Epidemiol. 1994;15:253–64.
22. Conly JM, Grieves K, Peters B. A prospective, randomized study comparing transparent and dry gauze dressings for central venous catheters. J Infect Dis. 1989;159:310–9.
23. Casey AL, Worthington T, Caddick JM, Hilton AC, Lambert PA, Elliott TSJ. RAPD for the typing of coagulase-negative staphylococci implicated in catheter-related bloodstream infection. J Infect. 2006;52:282–9.
24. Dobbins BM, Kite P, Kindon A, McMahon MJ, Wilcox MH. DNA fingerprinting analysis of coagulase negative staphylococci implicated in catheter related bloodstream infections. J Clin Pathol. 2002;55:824–8.
25. Bach A, Eberhardt H, Frick A, Schmidt H, Bottiger BW, Martin E. Efficacy of silver-coating central venous catheters in reducing bacterial colonization. Crit Care Med. 1999;27:515–21.
26. Gao Z, Tseng CH, Pei Z, Blaser MJ. Molecular analysis of human forearm superficial skin bacterial biota. Proc Natl Acad Sci U S A. 2007;104:2927–32.
27. Grice EA, Segre JA. The skin microbiome. Nat Rev Microbiol. 2011;9:244–53.
28. Gordon RJ, Quagliarello B, Lowy FD. Ventricular assist device-related infections. Lancet Infect Dis. 2006;6:426–37.
29. Monkowski DH, Axelrod P, Fekete T, Hollander T, Furukawa S, Samuel R. Infections associated with ventricular assist devices: epidemiology and effect on prognosis after transplantation. Transpl Infect Dis. 2007;9:114–20.
30. Birks EJ. The comparative use of ventricular assist devices: differences between Europe and the United States. Tex Heart Inst J. 2010;37:565–7.
31. von Recum AF. Applications and failure modes of percutaneous devices: a review. J Biomed Mater Res. 1984;18:323–36.
32. Pendegrass CJ, Gordon D, Middleton CA, Sun SN, Blunn GW. Sealing the skin barrier around transcutaneous implants: in vitro study of keratinocyte proliferation and adhesion in response to surface modifications of titanium alloy. J Bone Joint Surg Br. 2008;90:114–21.
33. Abu-Serriah MM, McGowan DA, Moos KF, Bagg J. Extra-oral endosseous craniofacial implants: current status and future developments. Int J Oral Maxillofac Surg. 2003;32:452–8.
34. Parameswaran AD, Roberts CS, Seligson D, Voor M. Pin tract infection with contemporary external fixation: how much of a problem? J Orthop Trauma. 2003;17:503–7.
35. Lethaby A, Temple J, Santy J. Pin site care for preventing infections associated with external bone fixators and pins. Cochrane Database Syst Rev. 2008;4:CD004551.
36. Lee-Smith J, Santy J, Davis P, Jester R, Kneale J. Pin site management. Towards a consensus: part 1. J Orthop Nurs. 2001;5:37–42.

37. Green SA, Ripley MJ. Chronic osteomyelitis in pin tracks. J Bone Joint Surg Am. 1984; 66A:1092–8.
38. Collinge CA, Goll G, Seligson D, Easley KJ. Pin tract infections: silver vs uncoated pins. Orthopedics. 1994;17:445–8.
39. Steckelberg JM, Osmon DR. Prosthetic joint infection. In: Bisno AL, Waldvogel FA, editors. Infections associated with indwelling medical devices. Washington, DC: American Society for Microbiology; 2000. p. 173–209.
40. Hugonnet S, Sax H, Eggimann P, Chevrolet JC, Pittet D. Nosocomial bloodstream infection and clinical sepsis. Emerg Infect Dis. 2004;10:76–81.
41. Von EC, Peters G, Heilmann C. Pathogenesis of infections due to coagulase-negative staphylococci. Lancet Infect Dis. 2002;2:677–85.
42. Schierholz JM, Beuth J. Implant infections: a haven for opportunistic bacteria. J Hosp Infect. 2001;49:87–93.
43. Herrmann M, Weyand M, Greshake B, Von EC, Proctor RA, Scheld HH, Peters G. Left ventricular assist device infection is associated with increased mortality but is not a contraindication to transplantation. Circulation. 1997;95:814–7.
44. Trampuz A, Zimmerli W. Diagnosis and treatment of infections associated with fracture-fixation devices. Injury. 2006;37:59–66.
45. Eggimann P, Harbarth S, Constantin MN, Touveneau S, Chevrolet JC, Pittet D. Impact of a prevention strategy targeted at vascular-access care on incidence of infections acquired in intensive care. Lancet. 2000;355:1864–8.
46. Gastmeier P, Geffers C. Prevention of catheter-related bloodstream infections: analysis of studies published between 2002 and 2005. J Hosp Infect. 2006;64:326–35.
47. Zingg W, Walder B, Pittet D. Prevention of catheter-related infection: toward zero risk? Curr Opin Infect Dis. 2011;24:377–84.
48. Warren DK, Yokoe DS, Climo MW, Herwaldt LA, Noskin GA, Zuccotti G, Tokars JI, Perl TM, Fraser VJ. Preventing catheter-associated bloodstream infections: a survey of policies for insertion and care of central venous catheters from hospitals in the prevention epicenter program. Infect Control Hosp Epidemiol. 2006;27:8–13.
49. Jarvis WR. Selected aspects of the socioeconomic impact of nosocomial infections: morbidity, mortality, cost, and prevention. Infect Control Hosp Epidemiol. 1996;17:552–7.
50. Haas DW, Kaiser AB. Antimicrobial prophylaxis of infections associated with foreign bodies. In: Bisno AL, Waldvogel FA, editors. Infections associated with indwelling medical devices. Washington, DC: American Society for Microbiology; 2000. p. 395–406.
51. Dellinger EP, Gross PA, Barrett TL, Krause PJ, Martone WJ, McGowan Jr JE, Sweet RL, Wenzel RP. Quality standard for antimicrobial prophylaxis in surgical procedures. The Infectious Diseases Society of America. Infect Control Hosp Epidemiol. 1994;15:182–8.
52. Jaeger M, Maier D, Kern WV, Sudkamp NP. Antibiotics in trauma and orthopedic surgery – a primer of evidence-based recommendations. Injury. 2006;37 Suppl 2:S74–80.
53. Pukstad BS, Ryan L, Flo TH, Stenvik J, Moseley R, Harding K, Thomas DW, Espevik T. Non-healing is associated with persistent stimulation of the innate immune response in chronic venous leg ulcers. J Dermatol Sci. 2010;59:115–22.
54. Shunmugaperumal T. Microbial colonization of medical devices and novel preventive strategies. Recent Pat Drug Deliv Formul. 2010;4:153–73.
55. Lyko F, Foret S, Kucharski R, Wolf S, Falckenhayn C, Maleszka R. The honey bee epigenomes: differential methylation of brain DNA in queens and workers. PLoS Biol. 2010;8: e1000506.
56. Branemark PI, Albrektsson T. Titanium implants permanently penetrating human-skin. Scand J Plast Reconstr Surg. 1982;16:17–21.
57. Johansson ML, Thomsen P, Hulten L, Halvorsen PS, Fosse E, Edwin B. Integration between a percutaneous implant and the porcine small bowel. J Biomed Mater Res B Appl Biomater. 2011;98:101–9.
58. Pendegrass CJ, Goodship AE, Blunn GW. Development of a soft tissue seal around bone-anchored transcutaneous amputation prostheses. Biomaterials. 2006;27:4183–91.

59. Isackson D, McGill LD, Bachus KN. Percutaneous implants with porous titanium dermal barriers: an in vivo evaluation of infection risk. Med Eng Phys. 2011;33:418–26.
60. Guidet B, Nicola I, Barakett V, Gabillet JM, Snoey E, Petit JC, Offenstadt G. Skin versus hub cultures to predict colonization and infection of central venous catheter in intensive care patients. Infection. 1994;22:43–8.
61. Sesso R, Barbosa D, Leme IL, Sader H, Canziani ME, Manfredi S, Draibe S, Pignatari AC. Staphylococcus aureus prophylaxis in hemodialysis patients using central venous catheter: effect of mupirocin ointment. J Am Soc Nephrol. 1998;9:1085–92.
62. Timsit JF, Schwebel C, Bouadma L, Geffroy A, Garrouste-Org M, Pease S, Herault MC, Haouache H, Calvino-Gunther S, Gestin B, Armand-Lefevre L, Leflon V, Chaplain C, Benali A, Francais A, Adrie C, Zahar JR, Thuong M, Arrault X, Croize J, Lucet JC. Chlorhexidine-impregnated sponges and less frequent dressing changes for prevention of catheter-related infections in critically ill adults: a randomized controlled trial. JAMA. 2009;301:1231–41.
63. Ruschulte H, Franke M, Gastmeier P, Zenz S, Mahr KH, Buchholz S, Hertenstein B, Hecker H, Piepenbrock S. Prevention of central venous catheter related infections with chlorhexidine gluconate impregnated wound dressings: a randomized controlled trial. Ann Hematol. 2009;88:267–72.
64. Kuyyakanond T, Quesnel LB. The mechanism of action of chlorhexidine. FEMS Microbiol Lett. 1992;100:211–5.
65. Levy SB. Active efflux, a common mechanism for biocide and antibiotic resistance. J Appl Microbiol. 2002;92(Suppl):65S–71.
66. Longtin J, Seah C, Siebert K, McGeer A, Simor A, Longtin Y, Low DE, Melano RG. Distribution of antiseptic resistance genes qacA, qacB, and smr in methicillin-resistant Staphylococcus aureus isolated in Toronto, Canada, from 2005 to 2009. Antimicrob Agents Chemother. 2011;55:2999–3001.
67. Fraud S, Campigotto AJ, Chen Z, Poole K. MexCD-OprJ multidrug efflux system of Pseudomonas aeruginosa: involvement in chlorhexidine resistance and induction by membrane-damaging agents dependent upon the AlgU stress response sigma factor. Antimicrob Agents Chemother. 2008;52:4478–82.
68. Meyer B, Cookson B. Does microbial resistance or adaptation to biocides create a hazard in infection prevention and control? J Hosp Infect. 2010;76:200–5.
69. Batra R, Cooper BS, Whiteley C, Patel AK, Wyncoll D, Edgeworth JD. Efficacy and limitation of a chlorhexidine-based decolonization strategy in preventing transmission of methicillin-resistant Staphylococcus aureus in an intensive care unit. Clin Infect Dis. 2010;50:210–7.
70. Edgeworth JD. Has decolonization played a central role in the decline in UK methicillin-resistant Staphylococcus aureus transmission? A focus on evidence from intensive care. J Antimicrob Chemother. 2011;66:II41–7.
71. Russell AD. Bacterial adaptation and resistance to antiseptics, disinfectants and preservatives is not a new phenomenon. J Hosp Infect. 2004;57:97–104.
72. Holman WL, Pamboukian SV, McGiffin DC, Tallaj JA, Cadeiras M, Kirklin JK. Device related infections: are we making progress? J Card Surg. 2010;25:478–83.
73. Lio PA, Kaye ET. Topical antibacterial agents. Infect Dis Clin North Am. 2009;23:945–63. ix.
74. Zasloff M. Antimicrobial peptides of multicellular organisms. Nature. 2002;415:389–95.
75. Devine DA, Hancock RE. Cationic peptides: distribution and mechanisms of resistance. Curr Pharm Des. 2002;8:703–14.
76. Epand RM, Vogel HJ. Diversity of antimicrobial peptides and their mechanisms of action. Biochim Biophys Acta. 1999;1462:11–28.
77. Matsuzaki K. Why and how are peptide-lipid interactions utilized for self-defense? Magainins and tachyplesins as archetypes. Biochim Biophys Acta. 1999;1462:1–10.
78. Nguyen LT, Haney EF, Vogel HJ. The expanding scope of antimicrobial peptide structures and their modes of action. Trends Biotechnol. 2011;29:464–72.
79. Shai Y. Mode of action of membrane active antimicrobial peptides. Biopolymers. 2002; 66:236–48.

80. Kobayashi S, Takeshima K, Park CB, Kim SC, Matsuzaki K. Interactions of the novel antimicrobial peptide buforin 2 with lipid bilayers: proline as a translocation promoting factor. Biochemistry. 2000;39:8648–54.
81. Chou TGR, Petti CA, Szakacs J, Bloebaum RD. Evaluating antimicrobials and implant materials for infection prevention around transcutaneous osseointegrated implants in a rabbit model. J Biomed Mater Res A. 2010;92A:942–52.
82. Hancock REW, Sahl HG. Antimicrobial and host-defense peptides as new anti-infective therapeutic strategies. Nat Biotechnol. 2006;24:1551–7.
83. Rubinchik E, Dugourd D, Algara T, Pasetka C, Friedland HD. Antimicrobial and antifungal activities of a novel cationic antimicrobial peptide, omiganan, in experimental skin colonisation models. Int J Antimicrob Agents. 2009;34:457–61.
84. Orlando F, Ghiselli R, Cirioni O, Minardi D, Tomasinsig L, Mocchegiani F, Silvestri C, Skerlavaj B, Riva A, Muzzonigro G, Saba V, Scalise G, Zanetti M, Giacometti A. BMAP-28 improves the efficacy of vancomycin in rat models of gram-positive cocci ureteral stent infection. Peptides. 2008;29:1118–23.
85. Giacometti A, Cirioni O, Barchiesi F, Del Prete MS, Fortuna M, Caselli F, Scalise G. In vitro susceptibility tests for cationic peptides: comparison of broth microdilution for bacteria that grow aerobically. Antimicrob Agents Chemother. 2000;44:1694–6.
86. Balaban N, Gov Y, Giacometti A, Cirioni O, Ghiselli R, Mocchegiani F, Orlando F, D'Amato G, Saba V, Scalise G, Bernes S, Mor A. A chimeric peptide composed of a dermaseptin derivative and an RNA III-inhibiting peptide prevents graft-associated infections by antibiotic-resistant staphylococci. Antimicrob Agents Chemother. 2004;48:2544–50.
87. Kwakman PHS, Te Velde AA, Vandenbroucke-Grauls CMJE, van Deventer SJH, Zaat SAJ. Treatment and prevention of Staphylococcus epidermidis experimental biomaterial-associated infection by bactericidal peptide 2. Antimicrob Agents Chemother. 2006;50:3977–83.
88. Majno G. Man and wound in the ancient world. Cambridge, MA: Harvard University Press; 1975.
89. Kwakman PHS, Te Velde AA, de Boer L, Vandenbroucke-grauls CMJE, Zaat SAJ. Two major medicinal honeys have different mechanisms of bactericidal activity. PLoS ONE. 2011;6:e17709.
90. Postmes T, van den Bogaard AE, Hazen M. Honey for wounds, ulcers, and skin graft preservation. Lancet. 1993;341:756–7.
91. Snowdon JA, Cliver DO. Microorganisms in honey. Int J Food Microbiol. 1996;31:1–26.
92. Allen KL, Molan PC, Reid GM. A survey of the antibacterial activity of some New Zealand honeys. J Pharm Pharmacol. 1991;43:817–22.
93. Kwakman PHS, Van den Akker JPC, Guclu A, Aslami H, Binnekade JM, de Boer L, Boszhard L, Paulus F, Middelhoek P, Te Velde AA, Vandenbroucke-Grauls CMJE, Schultz MJ, Zaat SAJ. Medical-grade honey kills antibiotic-resistant bacteria in vitro and eradicates skin colonization. Clin Infect Dis. 2008;46:1677–82.
94. Adams CJ, Boult CH, Deadman BJ, Farr JM, Grainger MN, Manley-Harris M, Snow MJ. Isolation by HPLC and characterisation of the bioactive fraction of New Zealand manuka (Leptospermum scoparium) honey. Carbohydr Res. 2008;343:651–9.
95. Mavric E, Wittmann S, Barth G, Henle T. Identification and quantification of methylglyoxal as the dominant antibacterial constituent of Manuka (Leptospermum scoparium) honeys from New Zealand. Mol Nutr Food Res. 2008;52:483–9.
96. Kwakman PHS, Te Velde AA, de Boer L, Speijer D, Vandenbroucke-Grauls CMJE, Zaat SAJ. How honey kills bacteria. FASEB J. 2010;24:2576–82.
97. Cooper RA, Jenkins L, Henriques AFM, Duggan RS, Burton NF. Absence of bacterial resistance to medical-grade manuka honey. Eur J Clin Microbiol Infect Dis. 2010;29: 1237–41.
98. Johnson DW, van Eps C, Mudge DW, Wiggins KJ, Armstrong K, Hawley CM, Campbell SB, Isbel NM, Nimmo GR, Gibbs H. Randomized, controlled trial of topical exit-site application of honey (Medihoney) versus mupirocin for the prevention of catheter-associated infections in hemodialysis patients. J Am Soc Nephrol. 2005;16:1456–62.

99. Kwakman PH, Zaat SA. Antibacterial components of honey. IUBMB Life. 2012;64:48–55.
100. Johnson DW, MacGinley R, Kay TD, Hawley CM, Campbell SB, Isbel NM, Hollett P. A randomized controlled trial of topical exit site mupirocin application in patients with tunnelled, cuffed haemodialysis catheters. Nephrol Dial Transplant. 2002;17:1802–7.
101. Mccann M, Moore ZE. Interventions for preventing infectious complications in haemodialysis patients with central venous catheters. Cochrane Database Syst Rev. 2010;1:CD006894.
102. Zakrzewska-Bode A, Muytjens HL, Liem KD, Hoogkamp-Korstanje JA. Mupirocin resistance in coagulase-negative staphylococci, after topical prophylaxis for the reduction of colonization of central venous catheters. J Hosp Infect. 1995;31:189–93.
103. Miller MA, Dascal A, Portnoy J, Mendelson J. Development of mupirocin resistance among methicillin-resistant Staphylococcus aureus after widespread use of nasal mupirocin ointment. Infect Control Hosp Epidemiol. 1996;17:811–3.

Chapter 11
Infections Associated with Implanted Dental Devices

Georgios N. Belibasakis, Thomas Thurnheer, and Nagihan Bostanci

Abstract Contemporary Restorative and Regenerative Dentistry mandates the use of implantable devices, as part of the overall treatment plan. The ultimate aim is to restore missing teeth or regenerate defective tissues. This can be achieved by the implementation of devices such osseointegrated dental implants or tissue regeneration materials, respectively. The oral cavity is rich in microbiota, which have the capacity to form polymicrobial biofilm communities on natural or artificial surfaces. It is therefore inevitable that implanted dental devices are also prone to microbial colonisation, and associated oral infections, such as peri-implantitis. Treatment of these infections involves the elimination of the causative factor (biofilms) and restoration of the structure and function of the affected tissues. The present chapter is discussing the aetiology, pathogenesis, diagnosis and therapeutic challenges of these newly emerged infections of the oral cavity.

11.1 Introduction

The most prevalent oral diseases, such as dental caries and periodontal diseases, are of infectious aetiology. They are attributed to the formation of complex polymicrobial biofilm communities on tooth surfaces. The treatment of these infections requires elimination of the causative biofilms and restoration of the structure and

G.N. Belibasakis (✉) • T. Thurnheer
Oral Microbiology and Immunology, Institute of Oral Biology, Center of Dental Medicine, Faculty of Medicine, University of Zürich, Plattenstrasse 11, Zürich 8032, Switzerland
e-mail: george.belibasakis@zzm.uzh.ch; thomas.thurnheer@zzm.uzh.ch

N. Bostanci
Oral Translational Research, Institute of Oral Biology, Center of Dental Medicine, Faculty of Medicine, University of Zürich, Plattenstrasse 11, Zürich 8032, Switzerland
e-mail: nagihan.bostanci@zzm.uzh.ch

T.F. Moriarty et al. (eds.), *Biomaterials Associated Infection: Immunological Aspects and Antimicrobial Strategies*, DOI 10.1007/978-1-4614-1031-7_11,

function of the affected tissues. The progress of Restorative and Regenerative Dentistry has mandated the use of implantable devices, such as osseointegrated dental implants or tissue regeneration materials, as part of the overall treatment plan. Inevitably, these devices are as well prone to colonisation by oral biofilms and subsequently oral infections. Consequently, an uncharted territory of oral infections has emerged, posing new therapeutic challenges. These issues are detailed further in the present chapter.

11.2 The Oral Cavity as an Ecological Niche for Microbiota

The oral cavity is a unique niche of the human body permitting communication with the external environment. As it is the beginning of the digestive system and forms a common pathway with the respiratory system, it serves key biological functions. Ingestion of food, initiation of the digestive process, sensory functions (i.e. taste and smell), formation of speech as well as aesthetics are of vital importance for human life and wellbeing. Structurally, the oral cavity is special in that it consists of both soft mucosal surfaces and hard dental tissues all bathed in saliva, the secretory product of the salivary glands. It is worth noting that teeth are the only hard tissue of the human body directly exposed to the external environment, and their non-shedding surfaces are uniquely linked to the oral (gingival) mucosa, via a soft–hard tissue interface known as junctional epithelium. For a more elaborative review see [1].

As an ecological niche, the oral cavity is continually colonised by microorganisms known as the resident oral microbiota or microflora. At present, the diversity of oral microbiota is estimated to comprise more than 700 different species [2]. Within these species, numerous different strains may also be present. The oral microbiota has co-evolved with the host over thousands of years, adapting to the local conditions of their habitat. This means that they naturally attach to the various soft and hard surfaces of the oral cavity and thereafter grow in the form of microbial communities, depending on the availability of nutrients, ability to colonise surfaces, as well as withstand the local host defences. Clinically, these communities are visible as "dental plaque". Dental plaque possesses all the characteristics of a biofilm [3]. A biofilm is a highly structured and spatially organised matrix-embedded diverse microbial population, adherent to each other and/or to surfaces or interfaces [4]. Within biofilms, bacteria communicate with each other by secretion of diffusible signalling molecules. Thus, as constituents of a biofilm, bacteria exhibit differential gene expression compared to their planktonic (single cell) state [5, 6]. Importantly, biofilms are less penetrable by the components of the host immune system and less susceptible to antimicrobial agents [7, 8], rendering the individual bacteria in the biofilm more virulent than their free planktonic state counterparts. In health, there is a tight balance between the resident biofilm microbiota and the host's response. Changes in the conditions of the local microenvironment and the host response may cause a swift in the composition of the biofilm microflora. Under the newly established circumstances, the predominant bacterial species may enhance their virulence with the potential to cause disease. This is the basis of the "ecological plaque hypothesis", which describes the relationship between the host and bacteria in health and disease [9].

In the case of oral bacteria, there are no true pathogens as such. Instead, species that may be found at low numbers in health, can act as pathogens when certain switches occur within their ecological niche. These are best characterised as "opportunistic pathogens". Koch's classical postulates do not apply to opportunistic or putative pathogens of the oral cavity. Hence, the following modifications of these postulates were proposed in the case of oral microbiota, in order to better reflect oral infectious diseases, such as dental caries and periodontal disease [10]: (a) the putative pathogens should be associated with the disease (detected at higher levels or proportions compared to health), (b) elimination or suppression of these species should be associated with inhibition of disease progression or clinical improvement, (c) the suspected species should be able to affect the immune and/or inflammatory responses of the host, (d) application of the putative pathogen in an experimental animal model should be able to reproduce aspects of the disease and (e) the putative pathogen should possess virulence factors able to induce pathogenic mechanisms of action.

Since the major oral infectious diseases are of polymicrobial nature, the putative oral pathogens exert their virulence potential as constituents of oral biofilm communities. Hence, oral biofilm models are discussed in more detail in the following section.

11.3 Oral Biofilms

The ability of microorganisms to grow attached to solid surfaces was first described by Zobell and Anderson in 1936 [11]. However, it took more than 40 years until it was recognised that planktonic microorganisms are rarely found in nature and that most bacteria grow in biofilms attached to a surface [4, 12]. Dental plaque is a particularly conspicuous manifestation of a biofilm that is temporarily disrupted by the mechanical forces of tooth brushing, but can invariably reconstitute itself [13]. Ever since W. D. Miller's fundamental studies on the oral microbiota more than a century ago [14], the chemoprophylaxis of oral infections has become an important issue, with many substances being studied for their control, such as antibiotics or antiseptic mouthwashes [15]. However, the morphological and physiological status of the microorganisms within a biofilm, as well as their location, can render them more resistant to antimicrobials. Bacterial cells in a biofilm can be ≥500 times more resistant to antimicrobial agents than their planktonic counterparts [4]. Therefore, an issue of crucial importance in the evaluation of new antimicrobials for the control or eradication of oral biofilms is the assay system used.

11.3.1 Supragingival Biofilm Model

The classical *in vitro* measures of antimicrobial potency utilising planktonic monocultures, along with the prolonged contact times of these assays are poor predictors of the clinical efficacy of antiseptic mouthrinses. This is because bacterial colonisation occurs in the form of biofilms, and mouthrinses are practically applied for only short periods of time [3, 16]. To overcome such limitations, chemostat flow cells [17, 18], modified Robbins devices [19], and other apparatuses [20–22] were developed to form and study oral biofilms *in vitro*. While these devices have contributed to our understanding of microbial adhesion and biofilm formation, they can be cumbersome to construct and/or difficult to maintain over long periods of time. An *in vitro* system has been devised for modelling supragingival oral biofilms, which are typically formed on the tooth surface above the gingival (gum) margin. The discussed model is based on a batch culture approach: biofilms composed of *Actinomyces naeslundii, Streptococcus oralis, Fusobacterium nucleatum, Streptococcus sobrinus, Veillonella dispar* and *Candida albicans* are formed on a salivary pellicle-coated hydroxyapatite or bovine enamel discs in 24-well cell culture dishes and incubated anaerobically at 37°C [23, 24]. The reproducibility, composition, 3D structure and properties of these *in vitro* biofilms, such as mass transport of macromolecules, have been thoroughly characterised (Fig. 11.1) [25–27]. For instance, through this work it became evident that biofilms formed in the presence of oral streptococci and sucrose consist of microbial colonies embedded in a compact extracellular polysaccharide hydrogel that lacks channels. Due to this property, it is not possible for high molecular weight dextrans to diffuse through the biofilm mass, as opposed to low molecular weight dextrans. This finding could explain the low cariogenic potential of starch, as compared to low molecular weight saccharides.

A particular strength of this "Zürich Biofilm Model" is that it is open to modifications and can be adjusted to suit particular research purposes, such as the testing of antimicrobial efficacy [28], the assessment of compounds resulting in de- or remineralisation of enamel [24, 29], the evaluation of the microbial colonisation of dental material surfaces, or the study of strains genetically engineered for the expression of their putative virulence factors [30–32]. In yet another application, it was demonstrated that results obtained from this *in vitro* biofilm model correspond well with those obtained from an *in vivo* dental caries model in rats [33]. More specifically, when the same microbiota used in the *in vitro* biofilm model were applied to this *in vivo* model, followed by a carbohydrate-containing diet, the resulting enamel demineralization levels were comparable between the two experimental models. Thus, the *in vitro* biofilm model is an important contribution towards the elimination of animal experimentation, implementing the 3R principles (refinement, replacement, reduction).

11.3.2 Subgingival Biofilm Model

The combination of bacterial culture analyses [34] and investigations on the diversity of characteristic 16S rRNA motifs [35, 36] has revealed a very complex

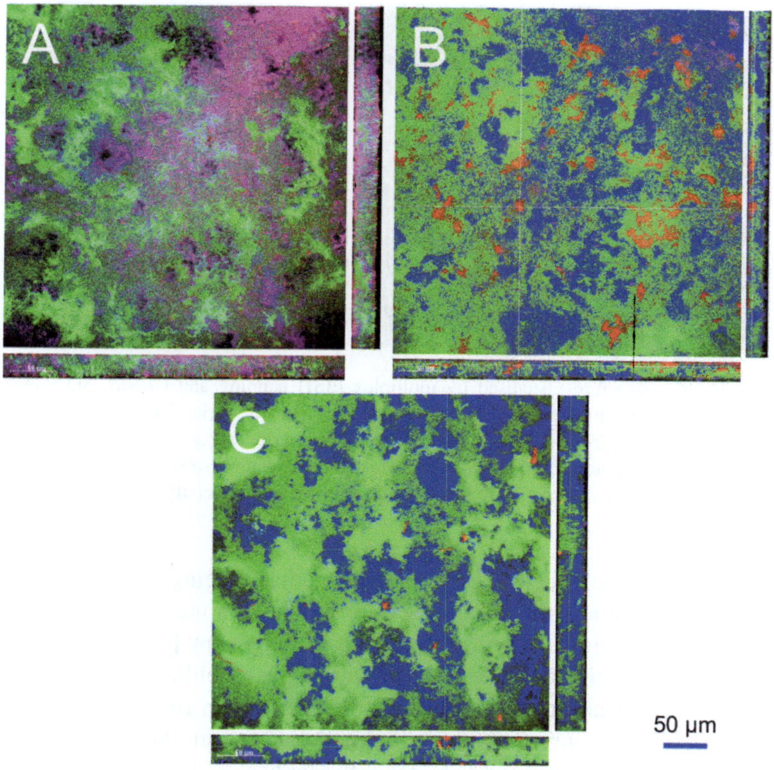

Fig. 11.1 Size dependent dextran diffusion into biofilms. Biofilms were exposed for 30 min to Texas-Red-conjugated dextrans of 3 kDa (**a**), 10 kDa (**b**) and 40 kDa (**c**). The penetration of the dextrans into the biofilms is recognisable in the CLSM images from the distribution of *red* colour. Bacterial cells were labelled by the DNA stain Syto13 (*green*), the EPS matrix was visualised with calcofluor (*blue*). Note that 40 kDa dextran can no longer penetrate the EPS matrix

microbial ecology of the biofilms formed in the periodontal pocket, the pathological gingival tissue pouch developed in periodontitis [37]. The biofilm formed within this periodontal pocket is clinically deemed as subgingival plaque (below the gingival margin). It is important to emphasise that subgingival biofilms are formed *in vivo* under substantially different nutritional conditions, compared to supragingival biofilms. Subgingivally, the most significant source of nutrients is the inflammation-dependent flow of gingival crevicular fluid (GCF), whereas supragingivally the major source is the influx of food-derived nutrients. A particularly influential component of food-derived nutrients is sucrose, which is hardly present in subgingival biofilms. Sucrose is quickly and efficiently transformed into a variety of extracellular polysaccharides. These will in turn become deposited within the extracellular matrix of supragingival biofilms and define to a large extent their diffusional properties [25, 33].

Subgingival biofilms are difficult to study as a whole, due to technical limitations. There are only very few structural investigations in this respect, as they rely

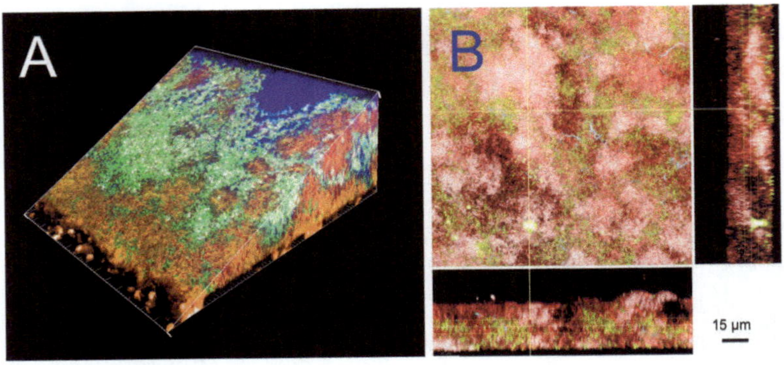

Fig. 11.2 Biofilm structure visualised by multiplex FISH staining and CLSM. 3D image of an optically sectioned 64.5 h 10-species biofilm stained for total bacteria, *Fusobacterium nucleatum* (*orange*), Veillonella dispar (*white*) and extracellular polysaccharides (*blue*) (**a**). Transverse and vertical cross-sections through a biofilm stained for Fusobacterium nucleatum (reddish), Treponema denticola (*light blue*) and total bacteria (**b**). Courtesy of Thomas Ammann, University of Zürich, CH

on extractions of whole teeth together with the adjacent gingival tissues, followed by extensive and meticulous processing (embedding, demineralization, sectioning) before the final examination by light or electron microscopy [38, 39]. To overcome such technical limitations, a new *in vitro* "subgingival biofilm model" was considered. This subgingival biofilm model uses the same batch culture approach as the previously described supragingival biofilm model. However, the subgingival model is formed in the absence of sucrose by ten different bacterial species that in addition to *F. nucleatum, A. naeslundii, S. oralis* and *V. dispar* (also part of the supragingival model), includes *Streptococcus intermedius* and the putative periodontal pathogens *Campylobacter rectus, Porphyromonas gingivalis, Prevotella intermedia, Tannerella forsythia* and *Treponema denticola*. These species were selected based on evidence from the literature that they are frequently, and in high numbers, detected in progressive periodontal pockets, and most commonly utilised in virulence experiments [40]. Thus, this short list of organisms reflect a widely considered "periodontopathogenic flora" [34, 41, 42]. In addition, these species are feasible to culture and enumerate, and can be analysed by fluorescence in situ hybridization (FISH) and/or immuno-fluorescence as additional species-specific detection assays. Confocal laser scanning microscopy (CLSM) images of this model are shown in Fig. 11.2.

11.3.3 Co-culture of Oral Biofilm Models with Eukaryotic Cells

The Zürich *in vitro* biofilm can also serve as an appropriate model for the study of host–bacteria interactions, in order to study mechanisms of inflammation and infection. Recently, supragingival and subgingival biofilm growth and survival has been optimised under various nutritional and atmospheric conditions. The biofilm coated discs are ultimately brought into co-culture with eukaryotic cells, with the biofilm surface facing towards the cell layer, separated only by a 1 mm thick ring. This conformation allows for ample interactions between the biofilm and the challenged cells, in a manner that resembles the *in vivo* tooth–biofilm–tissue interface. For the first time it was made possible to challenge human gingival epithelial cell cultures with intact viable subgingival biofilms *in vitro*, in order to investigate apoptotic and cytokine-stimulating responses in these cells [43]. In a further approach, other cells of dental and periodontal relevance, such as human dental pulp and periodontal ligament cells, were challenged by the supragingival biofilm, and the effects on osteolytic cytokine gene expression were investigated [44]. These models are amenable to further adjustments in order to address scientific questions, pertinent to the aetiology and pathogenesis of periodontal diseases, and may also aid the validation of interventional (i.e. pharmaceutical) approaches for the treatment of this cluster of diseases.

11.4 Biofilm-Related Oral Infections

Oral diseases are highly prevalent among the global population and they constitute a considerable economic burden for national health systems and economies. The main oral diseases, dental caries and periodontal diseases, are of infectious aetiology. They are essentially biofilm-related diseases, but their pathogenesis is largely attributed to the susceptibility and responses of the host.

11.4.1 Dental Caries

Dental caries is a biofilm-related disease that affects most of the population in industrialised countries. By definition, dental caries is the localised destruction of the hard dental tissues as a result of demineralisation induced by acid-producing bacteria of the oral biofilm. The acidic products are derived from the fermentation of dietary carbohydrates by acid-producing and acid-tolerant bacterial species of the biofilms. The demineralisation process starts from the enamel surface and can progress into the underlying less hard dentinal tissue. This can eventually allow for the invasion of bacteria into the soft pulp tissue causing inflammation (pulpitis) and potentially tooth necrosis. Similar effects can be initiated not only on the enamel but also on the cementum of the root, a situation more likely to occur in the older population, where the cementum is exposed as a result of gingival recession due to periodontal disease. Although no single bacterial species can be solely implicated in the aetiology of dental caries, enamel caries is highly associated with mutans streptococci and lactobacilli, whereas root caries is also associated with *Actinomyces* species. The growth of mutans streptococci and lactobacilli is favoured in acidic environments, which can in turn produce more acid at a faster rate, enhancing demineralisation. Thus, factors that cause drop of the local pH, such as increased intake of dietary carbohydrates or consumption of acid-containing drinks, may provide an adequate ecological shift in the biofilm niche that would result in caries progression.

11.4.2 Periodontal Diseases

11.4.2.1 Definition of Periodontal Diseases

Periodontal diseases are a major group of biofilm-associated oral infections, affecting the periodontal (tooth-supporting) tissues. Biofilms accumulated on the dental surface trigger and propagate an inflammatory response by the surrounding gingival tissue. If the established inflammation remains within this tissue, the condition is "gingivitis". This condition affects virtually the whole global population and largely manifests as swelling and bleeding of the gingiva, a trait which is easily identified by the patient. Gingivitis is reversible, provided that the biofilm presence is eradicated from the tooth surface by proper oral hygiene. However, if the biofilm presence persists over longer periods of time, it may lead to an exacerbated local inflammatory response, resulting in lysis of the junctional epithelial barrier. The local inflammation will then spread deeper, affecting the tooth-supporting alveolar bone and interconnecting periodontal ligament. Once the alveolar bone is resorbed and the periodontal ligament attachment to the tooth is lost as a result of this inflammatory process, the condition is then termed "periodontitis". Of note, gingivitis always precedes periodontitis, but not all gingivitis cases will develop into periodontitis. The progression of periodontitis is also characterised by the formation of pathological "periodontal pockets" and potentially gingival recession (apical retraction of the gingival margin). The disease can culminate in tooth loss if left untreated, and therefore impair the function of mastication and compromise aesthetics. Periodontitis is the main cause of tooth loss in the industrialised world, with a prevalence of 1/3 of the adult population, rendering it perhaps the most common infectious disease in humans [45]. Based on clinical features, the latest classification of periodontal diseases recognises Chronic Periodontitis and Aggressive Periodontitis, as the two most common forms of periodontitis [46]. The former has a slow progression rate and affects mainly the adult population, whereas the latter affects adolescents or younger adults, and exhibits a rapid progression rate, which is usually not commensurate with the levels of oral hygiene.

11.4.2.2 Microbiology of Periodontal Diseases

The healthy periodontal environment of the gingival crevice is characterised by Gram positive aerobic, or facultative aerobic bacterial species, such as *Streptococcus* and *Actinomyces*, whereas Gram negative and obligate anaerobic species are relatively sparse. The model of "ecological plaque hypothesis" fits well the aetiology of periodontitis. Changes in the local microenvironment of the periodontal tissues may create conditions of higher pH, temperature, anaerobiosis and inflammation, favouring the colonisation and growth of proteolytic, Gram negative obligate anaerobic species. The resident microbiota of gingivitis is not unique to this form of periodontal disease, but represents a transitional stage between periodontal health and periodontitis. This is characterised by a switch from *Streptococci*, to higher proportions of *Actinomyces* and *Capnocytophaga* species, as well as some Gram negative anaerobic species. The progression to periodontitis is associated with a switch of this biofilm microflora to one consisting of more Gram negative, obligate anaerobic and motile species, which are mainly proteolytic and asaccharolytic. The most well accepted of such species are *P. gingivalis*, *T. forsythia* and *T. denticola*, followed by *P. intermedia*, *F. nucleatum*, *C. rectus*, *Treponema socranskii*, *Dialister pneumosintes* and others [42]. It is also well established that *Aggregatibacter actinomycetemcomitans* is particularly associated with (localised) Aggressive Periodontitis. Notably, the employment of molecular detection methodologies has led to the identification of novel, as-yet uncultivable microorganisms within oral biofilms. An interesting finding is the presence of methanogenic *Archaea* in severe periodontitis, but their absence in health [47].

11.4.2.3 Pathogenesis of Periodontal Diseases

On the histopathological level, early classical studies performed on the inflammatory nature of periodontal disease are still up-to-date [48]. The initial host response starts as classic acute exudative vasculitis of the gingiva, followed by an early lesion of T- and B-lymphocyte infiltrates and further loss of interstitial connective tissue structure. The established lesion (2–3 weeks) is characterised by the predominance of antibody producing B-lymphocytes, but no bone loss at this stage and the histopathological events are clinically similar to gingivitis. The condition may remain at this stage for years, or develop into an advanced lesion. Whether this occurs or not, is largely dependent on susceptibility differences between individuals, or even between periodontal sites of the same individual [49]. Apart from the predominance of lymphocytes, the advanced lesion stage is also accompanied by destruction of the alveolar bone and periodontal ligament tissue, hence clinically manifesting as periodontitis.

On the molecular level, the periodontal tissues respond to the bacterial challenge by a marked production of inflammatory mediators, such as cytokines of the interleukin (IL)-1 family, tumour necrosis factor (TNF)-α and prostaglandin E_2 [50, 51]. Their exacerbated production may collectively lead to excessive production of matrix metalloproteinases, enzymes responsible for degrading the connective tissue extracellular matrix in periodontitis [52]. However, alveolar bone destruction is regulated by members of the TNF ligand and receptor super-families, namely, receptor activator of nuclear factor-κB ligand (RANKL), its cellular receptor RANK and its soluble decoy receptor osteoprotegerin (OPG) [53]. When binding to RANK on the surface of osteoclast precursors, RANKL triggers their differentiation into osteoclasts, thus activating bone resorption. On the contrary, when OPG binds to RANKL it inhibits this action. The expression of the RANKL–RANK–OPG triad is tightly regulated by the inflammatory events in periodontal disease. A locally elevated RANKL/OPG ratio has been demonstrated in periodontitis [54, 55]. This indication could potentially serve as a molecular diagnostic marker for the disease, particularly in GCF, the inflammatory exudate of gingiva. The composition of GCF greatly varies between health and periodontal disease.

11.5 Infections of Implanted Dental Devices

Having described the major oral diseases associated with biofilms, this section deals with the biofilm-related infections of implanted dental devices. In the dental clinical view, the term "implanted device" directly alludes to osseointegrated dental implants. Dental implants constitute an integral treatment option of Restorative Dentistry, when considering the replacement of missing teeth due to disease processes or congenital or physical reasons. In the broader meaning of the term, one could also consider surgically installed pre-manufactured devices that facilitate tissue regeneration as implanted devices. Examples of which are guided tissue regeneration (GTR) barrier membranes used in Periodontics and Oral Surgery practice. These are artificial scaffolds used to prevent a healing soft tissue from invading a bony defect, thus allowing the slower growing underlying bone to fill this space.

11.5.1 Peri-implant Infections

11.5.1.1 Dental Implants

Dental implants are primarily titanium-based machined devices that are surgically implanted within the jaw bone ("endosseous") via the oral mucosa, in order to replace a missing tooth or teeth. Their shape is typically, cylindrical with a threaded body of varying sizes (3–5 mm diameter and 7–20 mm length), customised for the available area and volume of the jaw bone. Following surgical installation, a direct functional and structural biological connection will be developed between the living bone and the metallic surface of the implant, known as osseointegration [56]. The mucosa lining the bone will also heal around the implant's cervical region, establishing a transmucosal attachment that protects the underlying bone from the stimuli of the oral cavity (Fig. 11.3). A prosthodontic restoration structurally and aesthetically resembling the crown of the tooth is finally firmly adapted to the abutment of the implant.

Although the clinical and histological features of the peri-implant mucosa resemble those of healthy gingiva surrounding the teeth, a major difference is that

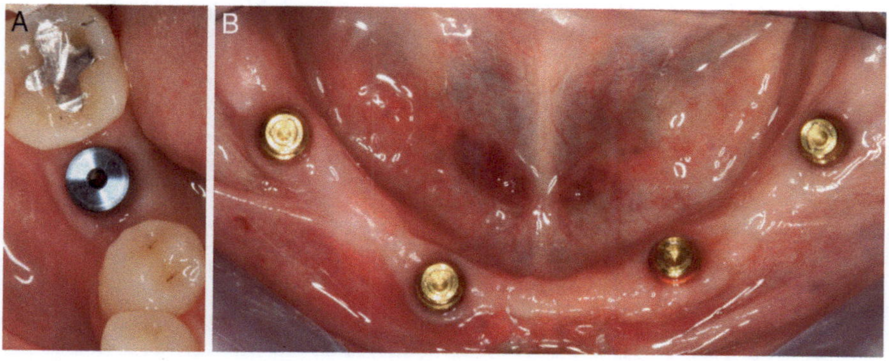

Fig. 11.3 Replacement of a single missing tooth by an osseointegrated dental implant (**a**). Placement of four dental implants on an edentulous mandible for restoration with an overdenture (**b**). The implants have been surgically installed and once the healing is completed the crown or the overdenture are adapted on the already osseointegrated implants. Courtesy of Dr. Evangelos Patellis, Willisau, CH

the underlying connective tissue, which in teeth is perpendicularly inserted into the cementum (Sharpey's fibres), is missing in the case of implants. Instead, the collagen fibres of the respective connective tissue run parallel to the implant and lack the capacity to penetrate into its surface [57, 58]. As a result, the peri-implant sulcus between the implant surface and the mucosa is deeper than the gingival sulcus of natural teeth [59]. Importantly, dental implants are directly fixed to the bone by osseointegration, whereas natural teeth are interconnected to the alveolar bone via the mediating soft periodontal ligament tissue that allows for slight tooth movement and provides blood supply. Consequently, dental implants lack the capacity for minute movements and the blood supply from the ligament. Blood supply to the peri-implant tissues is provided solely by the supra-periosteal blood vessels on the periphery of the supporting bone [60]. As a result, reduced numbers of immune cells may be recruited in the peri-implant tissues, which may therefore be more susceptible to biofilm infection, compared to periodontal tissues.

11.5.1.2 Definition of Peri-implant Infections

Dental implants can also be affected by oral biofilm-induced inflammation, resulting in failure of their function, or even their complete loss. When the inflammation is restricted to the soft peri-implant mucosal tissue, without any signs of supporting bone loss, the condition is termed peri-implant mucositis. However, when the inflammation causes loss of the supporting bone, the condition is then termed peri-implantitis (Fig. 11.4). It is considered as a "late failure" of implant function, meaning that the failing implant had previously been successfully osseointegrated. These pathological entities are analogous to gingivitis and periodontitis occurring in natural teeth and in fact they exhibit closely related histopathological and microbiological

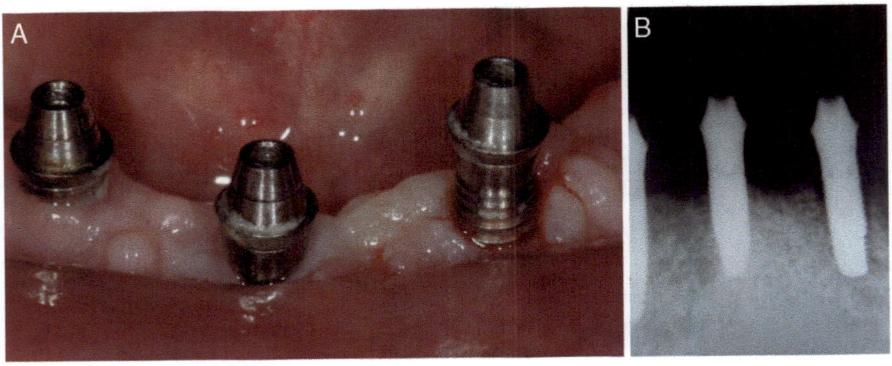

Fig. 11.4 A case of peri-implantitis in the mandible. These three implants were supporting an overdenture, which had to be removed (**a**). Note the inflammation of the mucosa, particularly around the right implant where bleeding is also evident. The threads of this implant are now exposed to the oral environment. Also note the accumulation of a white biofilm layer on the neck of the implant abutments. The radiographic evaluation shows considerable bone loss around the implants, indicating peri-implantitis (**b**)

characteristics [61]. Peri-implant mucositis occurs in approximately 80% of subjects with dental implants, affecting 50% of the implant sites, whereas peri-implantitis has been variably identified in 28–56% of subjects and 12–43% of implant sites, respectively [62].

11.5.1.3 Microbiology of Peri-implant Infections

The exposed implant surfaces can be colonised by oral bacteria, forming biofilms. The colonisation starts as early as 30 min after implant installation, and the same species can be found on the implant surfaces several months later [63, 64]. The bacterial composition of the biofilms formed around healthy implants closely resembles that of their neighbouring periodontally healthy teeth [65]. Therefore, the resident oral microbiota of natural teeth may act as a reservoir for biofilm formation on implant surfaces [61]. The surface characteristics of the implant are likely to affect the capacity for biofilm formation, although it is not clear if microbial composition is also affected. Higher surface roughness and higher free energy are predisposing factors for biofilm formation, resulting in faster bacterial deposition rate and greater thickness of the biofilm [66]. However, surface chemistry and design characteristics of the implant-abutment interface are also important determinants of biofilm formation [67].

In peri-implant health, the biofilms are characterised by Gram positive aerobic cocci and rods, although limited Gram negative species can also be found [63, 68], but at lower levels and slower growth rates, compared to natural teeth [64, 69, 70]. When peri-implant mucositis develops, the implant-associated biofilms are characterised by cocci, motile rods and spirochetes, at proportions comparable to gingivitis in natural teeth [71]. Further on, the switch to peri-implantitis is associated with biofilms comprising predominantly of Gram negative, motile and anaerobic species, such as *P. gingivalis*, *T. forsythia*, *T. denticola*, *F. nucleatum*, *P. intermedia* and occasionally *A. actinomycetemcomitans*, resembling the microbiota characteristic of periodontitis [72–74]. Nevertheless, certain bacterial species that are not commonly detected in periodontitis, such as *Staphylococcus aureus, Staphylococcus epidermidis, Enterobacter aerogenes, Enterobacter cloacae, Escherichia coli* and *C. albicans*, could frequently be found at peri-implantitis sites [65, 75–79]. Within the same individual with periodontitis, the bacterial flora detected on the implant surfaces closely resembles that of the infected periodontal pockets [80]. Hence, the remaining natural teeth can act as a reservoir for contamination of implant surfaces and peri-implant tissues. This could indicate that patients with a prior history of periodontitis may be at a higher risk for the development of peri-implantitis [81–83]. However, stringent oral hygiene measures can reduce this risk, thus resulting in higher implant success rate, irrespective of prior history of periodontitis [84].

11.5.1.4 Pathogenesis of Peri-implant Infections

The biofilm formed on the surface of dental implants is likely to cause a host inflammatory and immune response by the surrounding peri-implant tissues. Indeed, in human experimental models it has been demonstrated that biofilm formation on implants favours the development of peri-implant mucositis, in a manner that resembles gingivitis [61]. Studies analysing human biopsies of peri-implant diseases demonstrate that the inflammatory lesion in mucositis is dominated by T-lymphocyte infiltrates, restricted to the epithelial attachment with the implant [62]. The sulcular epithelium becomes acanthotic and the underlying connective tissue exhibits interstitial collagen loss and vascular changes, while dense inflammatory cell infiltrates now dominate both tissues [85]. However, when compared to gingivitis, the established inflammatory lesion in peri-implant mucositis is more extensive [86–88].

Peri-implantitis will occur if the inflammation is expanded beyond the epithelial barrier and connective tissue, to affect the underlying implant-supporting bone. This lesion is characterised by increased proportions of T- and B-lymphocytes, polymorphonuclear leukocytes and macrophages [89, 90]. Therefore, the immunological features of peri-implantitis are sequentially similar to those of periodontitis, albeit of greater magnitude [61]. A considerable difference from periodontitis is that peri-implant inflammation can potentially expand into the bone marrow tissue [87], which may account for the rapid progression of the disease. This pathological feature is arguably attributed to the lack of periodontal ligament (Sharpey's) fibres inserting into the implant surface, which would have prevented the spread of inflammation and bacterial invasion into the bone tissue [91].

11.5.1.5 Diagnosis of Peri-implant Infections

Similar to periodontal diseases, the diagnostic criteria for peri-implant infections are mainly clinical and radiographic [92]. The clinical features of peri-implant mucositis are largely inflammatory and include swelling and redness of the surrounding mucosa. On clinical examination, evaluation of bleeding on probing is the major sign of this condition. In the case of peri-implantitis, the consensus clinical features are: (a) vertical destruction of the bone crest, traditionally characterised by a "saucer-shaped" crater around the exposed portion of the implant, (b) a peri-implant pocket greater than 4 mm, (c) bleeding or suppuration during pocket probing, (d) visual tissue redness and swelling and (e) implant mobility at more progressed stages (Fig. 11.4). It is interesting that the crater-like bony defect around the implant is rather symmetrical, as opposed to natural teeth where usually one site is more affected than another. Radiographic evaluation is necessary 1 year after placement, followed by biannual checks.

As clinical signs of early peri-implantitis are not always evident, there is a need for additional diagnostic aids that would reveal subclinical disease activity. These could include markers of active host tissue destruction in peri-implant crevicular fluid (PICF), which is the GCF analogue in dental implants. Therefore, both GCF and PICF are non-invasive molecular "windows" to study the state and progression of periodontal and peri-implant diseases, respectively [93]. The RANKL–OPG system could be a candidate diagnostic marker for peri-implantitis, as clinical studies have demonstrated that an elevated RANKL/OPG ratio in GCF can be indicative of the occurrence of periodontitis [54, 55]. Specifically, in sampled healthy or gingivitis sites where bone loss is not clinically evident, this ratio is considerably lower than in periodontitis, where bone loss is a major clinical diagnostic feature.

It is important to note that several factors can confer an increased risk for the development of peri-implantitis [94]. These include (a) poor oral hygiene, (b) smoking (c) systemic conditions, such as diabetes mellitus, (d) genetic susceptibility, (e) alcohol consumption and (f) prior history of periodontitis. The first four are common risk factors with periodontitis, whereas the last may indicate further susceptibility to local infections.

11.5.1.6 Treatment of Peri-implant Infections

Similarly to the case of periodontitis, the ultimate aim of peri-implantitis treatment is to remove the causative microbial factor, resolve the associated inflammation and restore the functionality of the implant. Accordingly, the following considerations should be taken into account for the treatment of peri-implantitis [95]: (a) the disturbance or removal of the local biofilm from the implant surface or peri-implant tissues, (b) the decontamination and smoothening of the implant surface, (c) surgical correction of the architecture of the created peri-implant defect, (d) optimization of oral hygiene procedures, and (e) re-osseointegration. The mechanical debridement of the contaminated implant surfaces can be achieved by either surgical or non-surgical intervention. The surgical approach is recommended in deeper defects, aiming to provide a better access to the contaminated implant surface for the removal of the subgingival biofilms. Moreover, surgery is necessitated when the treatment plan incorporates regenerative techniques for augmenting the lost bone tissue around the implant, such as guided bone regeneration (GBR). This initial treatment phase should be followed by a maintenance phase of repeated visits for professional removal of further biofilm accumulation from the exposed implant surfaces. Despite the professional intervention, for maintaining a long-term successful therapeutic outcome it is imperative to ensure a high level of oral hygiene by the patient.

Given the infectious nature of peri-implant mucositis and peri-implantitis, supportive therapy using antimicrobial agents, such as local or systemic antibiotics and antiseptic mouthrinses, may also be useful. The use of antibiotics as adjunctive treatment modalities for peri-implantitis can be beneficial at sites with pocket depths greater than 6 mm [95–98].

11.5.2 Infections of Implantable Tissue Regeneration Devices

This section aims to discuss infections of implanted devices that promote regeneration of oral tissues lost as a result of disease. Such examples are encountered in the fields of Periodontics and Oral Surgery, where synthetic barrier membranes are surgically applied as devices used to facilitate tissue regeneration.

11.5.2.1 The Use of Guided Tissue Regeneration Membrane Devices

Periodontal therapy has evolved from solely eliminating the causative factors (i.e. biofilms or calculus, which is the mineralised variant of biofilms) and resolution of the associated inflammation, to the use of a wide array of advanced plastic and regenerative surgical procedures. GTR is a surgical approach that utilises the implantation of an impermeable barrier membrane between the cleaned exposed root surface and the gingival tissue, aiming to regain the lost periodontal apparatus [99–101]. If such a device is not placed as a hindrance over the surgically exposed defect, then the rapidly proliferating cells of the gingival tissues will occupy the space during the healing process [102]. As a result, the slower growing alveolar bone and periodontal ligament would be prevented from repopulating the defect, thus compromising the successful healing outcome that mandates the structural reconstitution particularly of these two tissues. Therefore, the GTR technique employs a thin manufactured barrier membrane to cover the defect (Fig. 11.5c), allowing for the selective and uninterrupted growth of alveolar bone and periodontal ligament. Barrier membranes are also protective of the blood clot during the early phases of healing, although they confer no superior biological advantages with regards to osteoinductive (bone stimulatory) properties.

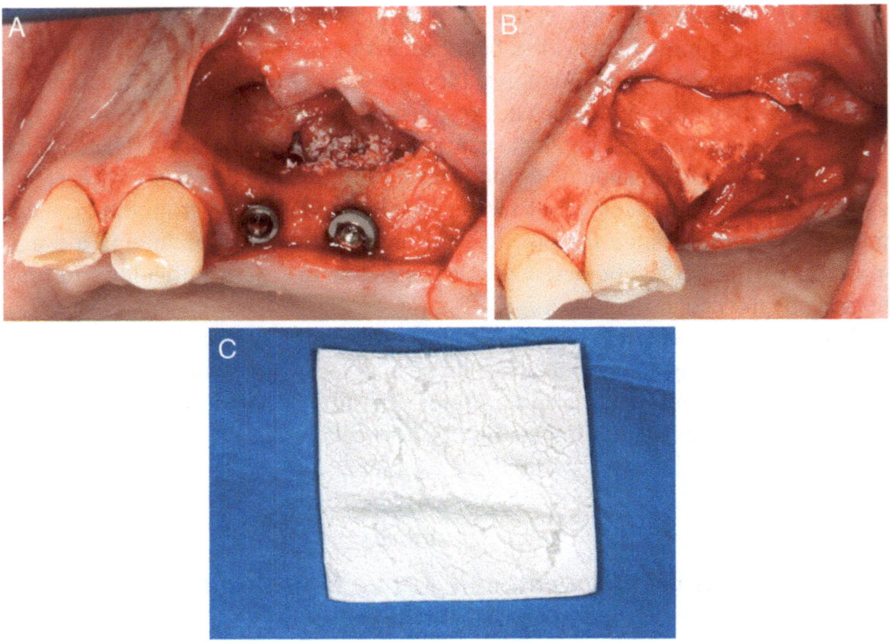

Fig. 11.5 Use of GBR technique for sinus lifting in combination with implant placement. After the floor of the maxillary sinus was surgically lifted, two implants were installed on the residual bone crest and the bony defect was then packed with a bovine bone graft (**a**). Thereafter, the filled defect was covered with a membrane made of native collagen (**b**), and the flap was repositioned and sutured. A collagen-based GTR membrane (**c**). Courtesy of Dr. Evangelos Patellis, Willisau, CH

Physical barrier membranes are also routinely used in Oral Surgery, with the aim to regenerate lost bone due to pathological conditions. In these instances, the procedure is rather known as GBR. Compared to periodontal GTR, it is technically and biologically less demanding, as the principal aim is to regenerate only bone tissue, without concomitant regeneration of the periodontal ligament. GBR is commonly used to augment bone tissue volume for the subsequent placement of dental implants in residual alveolar bone crests, or in combination with lifting the floor of a low maxillary sinus (Fig. 11.5). GBR is also particularly successful when implants are placed immediately after tooth extraction [103, 104]. The implant survival rate in sites with regenerated bone using barrier membranes was reported between 79 and 100% after 5 years of function [105].

A wide range of materials, natural or synthetic, have been used for manufacturing GTR membranes, providing them with bio-resorbable or non-resorbable properties. Non-resorbable membranes made of methylcellulose acetate were successfully used in the first clinical GTR case ever [101]. However, such membranes were fragile and susceptible to tearing forces, so their clinical use proved to be limited. They were later replaced by non-resorbable expanded polytetrafluoroethylene (ePTFE) membranes. Despite their technical advantages and encouraging clinical outcomes, the fact that a second surgical procedure is required for their removal has been restrictive for their routine use. For this purpose, bio-resorbable membranes have been developed from various biocompatible material, including polyglycoside synthetic polymers, collagen (Fig. 11.5c) and calcium sulphate, which do not require a second surgical stage for their removal [106]. These are slowly degraded within their host bed, while providing in the meantime healing stability for the in-growth of alveolar bone and periodontal ligament.

11.5.2.2 Infections of Guided Tissue Regeneration Membrane Devices

Despite the progress and optimization of regenerative procedures in oral and periodontal surgery, the treatment outcome can often be compromised by factors such as smoking, poor oral hygiene and residual periodontal disease [107–109]. Hence, a local infection of the surgical site and the regenerative device is possible, albeit not surprising as the procedure takes place in an environment with high bacterial load. Therefore, prior infection control is a crucial factor for success [110].

Membrane exposure is a major complication of GTR or GBR, which can permit the colonisation of oral bacteria establishing biofilms and causing a local infection [111]. Early reports have indicated high infection rates (between 50 and 100%), when membranes were exposed to the oral environment post-surgically [112, 113]. The bacterial colonisation can start as early as 3 min after placement of the membrane [114]. The potential sources of membrane contamination are likely to be saliva, the tongue or areas of active residual periodontal infection. Not surprisingly, the bacterial species associated with a failing GTR therapy are putative periodontal pathogens, such as *P. gingivalis, A. actinomycetemcomitans, T. forsythensis, P. intermedia, P. micros*, *C. rectus* and *Capnocytophaga*, indicating that residual periodontal infection or poor post-operative oral hygiene could compromise the treatment outcome [110, 115, 116]. The material of the membrane can be of importance for the type of bacteria colonised. In vitro studies demonstrated a high attachment specificity of *S. mutans* on ePTFE, polylactic acid and collagen membranes, and a less specific attachment of *A. actinomycetemcomitans* and *F. nucleatum* on most types of membranes [117]. *P. gingivalis* was shown to preferentially attach on polyglactin and collagen membranes, but not to polyglactic acid, synthetic glycolide and lactide copolymer ones, and in fact could penetrate through the colonised membrane [118].

Clinically, an association has been demonstrated between poor treatment outcomes and increasing microbial colonisation of the membranes [119–121]. A successful tissue regeneration outcome is largely dependent on reducing the load of periodontal pathogens on the surgical site, and the totality of the oral cavity, prior to membrane insertion, as they can act as reservoirs for infection. This can be less problematic in GBR, where a periodontally diseased tooth does not constitute part of the surgical site. Initial full mouth periodontal treatment will reduce the bacterial biofilm load from the diseased site and the oral cavity in general. During the surgery, local application of antimicrobial agents to the diseased root surfaces, such as iodophore, tetracycline or metronidazole gel, may reduce further the local bacterial load [114], while careful suturing of the surgical flaps for complete coverage of the membrane can reduce further the risk of infection. Systemic antibiotic therapy prior to membrane insertion is a preventive measure, whereas appropriate local antimicrobial agents, such as chlorohexidine mouthwash, are suggested afterwards. Frequent professional biofilm removal is also recommended during the healing period, although subgingival scaling or probing of the sites should be avoided, in order not to disturb the maturing healing tissue. During this time, compliance of the patient to effective oral hygiene measures is imperative. In the case that post-operative complications (membrane exposure and infection) would still occur, earlier removal of the membrane may be necessitated.

11.6 Conclusion

Technological advancements have led to the manufacturing of implanted dental devices that are used as part of contemporary Restorative Dentistry. Dental implants can replace missing teeth, fulfilling the functional and aesthetic restoration of the dental arches. GTR membranes facilitate the regain of oral tissue that was lost as a result of disease process. However, both types of devices are susceptible to colonisation and biofilm formation by oral bacteria, potentially resulting in infection of the supporting oral tissues that can compromise any therapeutic benefit. These "contemporary" infections are now incorporated into the list of oral diseases that the dental practitioner has to deal with. As these are still an uncharted territory, several therapeutic challenges remain to be solved. Hence, it is imperative that the clinician is well aware of these infections and should take under consideration the related risks prior to treatment planning.

References

1. Schroeder HE, Listgarten MA. The junctional epithelium: from strength to defense. J Dent Res. 2003;82(3):158–61.
2. Aas JA, Paster BJ, Stokes LN, Olsen I, Dewhirst FE. Defining the normal bacterial flora of the oral cavity. J Clin Microbiol. 2005;43(11):5721–32.
3. Marsh PD. Dental plaque: biological significance of a biofilm and community life-style. J Clin Periodontol. 2005;32 Suppl 6:7–15.
4. Costerton JW, Lewandowski Z, Caldwell DE, Korber DR, Lappin-Scott HM. Microbial biofilms. Annu Rev Microbiol. 1995;49:711–45.
5. Sauer K, Camper AK, Ehrlich GD, Costerton JW, Davies DG. Pseudomonas aeruginosa displays multiple phenotypes during development as a biofilm. J Bacteriol. 2002;184(4):1140–54.
6. Costerton W, Veeh R, Shirtliff M, Pasmore M, Post C, Ehrlich G. The application of biofilm science to the study and control of chronic bacterial infections. J Clin Invest. 2003;112(10):1466–77.
7. Gilbert P, Das J, Foley I. Biofilm susceptibility to antimicrobials. Adv Dent Res. 1997; 11(1): 160–7.
8. Marsh PD. Plaque as a biofilm: pharmacological principles of drug delivery and action in the sub- and supragingival environment. Oral Dis. 2003;9 Suppl 1:16–22.
9. Marsh PD. Are dental diseases examples of ecological catastrophes? Microbiology. 2003;149(Pt 2):279–94.

10. Socransky SS. Criteria for the infectious agents in dental caries and periodontal disease. J Clin Periodontol. 1979;6(7):16–21.
11. Zobell CE, Anderson DQ. Observations on the multiplication of bacteria in different volumes of stored sea water and the influence of oxygen tension and solid surfaces. Biol Bull. 1936;71:324.
12. Costerton JW, Geesey GG, Cheng KJ. How bacteria stick. Sci Am. 1978;238(1):86–95.
13. Marsh PD, Bradshaw DJ. Dental plaque as a biofilm. J Ind Microbiol. 1995;15(3):169–75.
14. Miller WD. The micro-organisms of the human mouth: the local and general diseases which are caused by them (Unaltered reprint of the original work by Willoughby D. Miller (1853–1907) published in, 1890 in Philadelphia). Basel: Karger; 1973.
15. Allaker RP, Douglas CW. Novel anti-microbial therapies for dental plaque-related diseases. Int J Antimicrob Agents. 2009;33(1):8–13.
16. Shapiro S, Guggenheim B. Chemoprophylaxis in the oral cavity: 'plus on change les choses, plus elles devraient rester les mêmes'. In: Shapiro S, Guggenheim B, editors. Oral biology at the turn of the century. Misconceptions, challenges and prospects. Basel: Karger; 1998. p. 226–38.
17. Stoodley P, Lewandowski Z, Boyle JD, Lappin-Scott HM. The formation of migratory ripples in a mixed species bacterial biofilm growing in turbulent flow. Environ Microbiol. 1999;1(5):447–55.
18. Foster JS, Kolenbrander PE. Development of a multispecies oral bacterial community in a saliva-conditioned flow cell. Appl Environ Microbiol. 2004;70(7):4340–8.
19. Larsen T, Fiehn NE. Development of a flow method for susceptibility testing of oral biofilms in vitro. APMIS. 1995;103(5):339–44.
20. Bradshaw DJ, McKee AS, Marsh PD. Effects of carbohydrate pulses and pH on population shifts within oral microbial communities in vitro. J Dent Res. 1989;68(9):1298–302.
21. Kinniment SL, Wimpenny JW, Adams D, Marsh PD. Development of a steady-state oral microbial biofilm community using the constant-depth film fermenter. Microbiology. 1996;142(Pt 3):631–8.
22. Ceri H, Olson ME, Stremick C, Read RR, Morck D, Buret A. The Calgary Biofilm Device: new technology for rapid determination of antibiotic susceptibilities of bacterial biofilms. J Clin Microbiol. 1999;37(6):1771–6.
23. Guggenheim B, Giertsen E, Schupbach P, Shapiro S. Validation of an in vitro biofilm model of supragingival plaque. J Dent Res. 2001;80(1):363–70.
24. Guggenheim B, Guggenheim M, Gmur R, Giertsen E, Thurnheer T. Application of the Zurich biofilm model to problems of cariology. Caries Res. 2004;38(3):212–22.
25. Thurnheer T, Gmur R, Shapiro S, Guggenheim B. Mass transport of macromolecules within an in vitro model of supragingival plaque. Appl Environ Microbiol. 2003;69(3):1702–9.
26. Thurnheer T, Gmur R, Guggenheim B. Multiplex FISH analysis of a six-species bacterial biofilm. J Microbiol Methods. 2004;56(1):37–47.
27. Reese S, Guggenheim B. A novel TEM contrasting technique for extracellular polysaccharides in in vitro biofilms. Microsc Res Tech. 2007;70(9):816–22.
28. Shapiro S, Giertsen E, Guggenheim B. An in vitro oral biofilm model for comparing the efficacy of antimicrobial mouthrinses. Caries Res. 2002;36(2):93–100.
29. Gmur R, Giertsen E, Van der Veen MH, Josselin D, de Jong E, Ten Cate JM, Guggenheim B. In vitro quantitative light-induced fluorescence to measure changes in enamel mineralization. Clin Oral Investig. 2006;10(3):187–95.
30. Van Der Ploeg JR, Guggenheim B. Deletion of gtfC of Streptococcus mutans has no influence on the composition of a mixed-species in vitro biofilm model of supragingival plaque. Eur J Oral Sci. 2004;112(5):433–8.
31. Thurnheer T, van der Ploeg JR, Giertsen E, Guggenheim B. Effects of Streptococcus mutans gtfC deficiency on mixed oral biofilms in vitro. Caries Res. 2006;40(2):163–71.
32. Senadheera MD, Guggenheim B, Spatafora GA, Huang YC, Choi J, Hung DC, et al. A VicRK signal transduction system in Streptococcus mutans affects gtfBCD, gbpB, and ftf expression, biofilm formation, and genetic competence development. J Bacteriol. 2005;187(12):4064–76.
33. Thurnheer T, Giertsen E, Gmur R, Guggenheim B. Cariogenicity of soluble starch in oral in vitro biofilm and experimental rat caries studies: a comparison. J Appl Microbiol. 2008;105(3):829–36.

34. Moore WE, Moore LV. The bacteria of periodontal diseases. Periodontol 2000. 1994;5:66–77.
35. Paster BJ, Boches SK, Galvin JL, Ericson RE, Lau CN, Levanos VA, et al. Bacterial diversity in human subgingival plaque. J Bacteriol. 2001;183(12):3770–83.
36. Wade WG. Non-culturable bacteria in complex commensal populations. Adv Appl Microbiol. 2004;54:93–106.
37. Bartold PM, Walsh LJ, Narayanan AS. Molecular and cell biology of the gingiva. Periodontol 2000. 2000;24:28–55.
38. Listgarten MA. Structure of the microbial flora associated with periodontal health and disease in man. A light and electron microscopic study. J Periodontol. 1976;47(1):1–18.
39. Zijnge V, van Leeuwen MB, Degener JE, Abbas F, Thurnheer T, Gmur R, et al. Oral biofilm architecture on natural teeth. PLoS One. 2010;5(2):e9321.
40. Darveau RP. Periodontitis: a polymicrobial disruption of host homeostasis. Nat Rev Microbiol. 2010;8(7):481–90.
41. Tanner A, Maiden MF, Macuch PJ, Murray LL, Kent Jr RL. Microbiota of health, gingivitis, and initial periodontitis. J Clin Periodontol. 1998;25(2):85–98.
42. Socransky SS, Haffajee AD. Periodontal microbial ecology. Periodontol 2000. 2005;38:135–87.
43. Guggenheim B, Gmur R, Galicia JC, Stathopoulou PG, Benakanakere MR, Meier A, et al. In vitro modeling of host-parasite interactions: the 'subgingival' biofilm challenge of primary human epithelial cells. BMC Microbiol. 2009;9:280.
44. Belibasakis GN, Meier A, Guggenheim B, Bostanci N. Oral biofilm challenge regulates the RANKL-OPG system in periodontal ligament and dental pulp cells. Microb Pathog. 2011; 50(1):6–11.
45. Eke PI, Genco RJ. CDC Periodontal Disease Surveillance Project: background, objectives, and progress report. J Periodontol. 2007;78(7 Suppl):1366–71.
46. Armitage GC. Development of a classification system for periodontal diseases and conditions. Ann Periodontol. 1999;4(1):1–6.
47. Lepp PW, Brinig MM, Ouverney CC, Palm K, Armitage GC, Relman DA. Methanogenic Archaea and human periodontal disease. Proc Natl Acad Sci U S A. 2004;101(16):6176–81.
48. Page RC, Schroeder HE. Pathogenesis of inflammatory periodontal disease. A summary of current work. Lab Invest. 1976;34(3):235–49.
49. Schenkein HA. Host responses in maintaining periodontal health and determining periodontal disease. Periodontol 2000. 2006;40:77–93.
50. Offenbacher S. Periodontal diseases: pathogenesis. Ann Periodontol. 1996;1(1):821–78.
51. Page RC. The role of inflammatory mediators in the pathogenesis of periodontal disease. J Periodontal Res. 1991;26(3 Pt 2):230–42.
52. Birkedal-Hansen H. Role of matrix metalloproteinases in human periodontal diseases. J Periodontol. 1993;64(5 Suppl):474–84.
53. Liu YC, Lerner UH, Teng YT. Cytokine responses against periodontal infection: protective and destructive roles. Periodontol 2000. 2010;52(1):163–206.
54. Bostanci N, Ilgenli T, Emingil G, Afacan B, Han B, Toz H, et al. Differential expression of receptor activator of nuclear factor-kappaB ligand and osteoprotegerin mRNA in periodontal diseases. J Periodontal Res. 2007;42(4):287–93.
55. Bostanci N, Ilgenli T, Emingil G, Afacan B, Han B, Toz H, et al. Gingival crevicular fluid levels of RANKL and OPG in periodontal diseases: implications of their relative ratio. J Clin Periodontol. 2007;34(5):370–6.
56. Carlsson L, Rostlund T, Albrektsson B, Albrektsson T, Branemark PI. Osseointegration of titanium implants. Acta Orthop Scand. 1986;57(4):285–9.
57. Berglundh T, Lindhe J, Ericsson I, Marinello CP, Liljenberg B, Thomsen P. The soft tissue barrier at implants and teeth. Clin Oral Implants Res. 1991;2(2):81–90.
58. Buser D, Weber HP, Donath K, Fiorellini JP, Paquette DW, Williams RC. Soft tissue reactions to non-submerged unloaded titanium implants in beagle dogs. J Periodontol. 1992; 63(3):225–35.
59. Berglundh T, Lindhe J. Dimension of the periimplant mucosa. Biological width revisited. J Clin Periodontol. 1996;23(10):971–3.

60. Berglundh T, Lindhe J, Jonsson K, Ericsson I. The topography of the vascular systems in the periodontal and peri-implant tissues in the dog. J Clin Periodontol. 1994;21(3):189–93.
61. Heitz-Mayfield LJ, Lang NP. Comparative biology of chronic and aggressive periodontitis vs. peri-implantitis. Periodontol 2000. 2010;53:167–81.
62. Zitzmann NU, Berglundh T. Definition and prevalence of peri-implant diseases. J Clin Periodontol. 2008;35(8 Suppl):286–91.
63. Furst MM, Salvi GE, Lang NP, Persson GR. Bacterial colonization immediately after installation on oral titanium implants. Clin Oral Implants Res. 2007;18(4):501–8.
64. Quirynen M, Vogels R, Peeters W, van Steenberghe D, Naert I, Haffajee A. Dynamics of initial subgingival colonization of 'pristine' peri-implant pockets. Clin Oral Implants Res. 2006; 17(1):25–37.
65. Salvi GE, Furst MM, Lang NP, Persson GR. One-year bacterial colonization patterns of *Staphylococcus aureus* and other bacteria at implants and adjacent teeth. Clin Oral Implants Res. 2008;19(3):242–8.
66. Teughels W, Van Assche N, Sliepen I, Quirynen M. Effect of material characteristics and/or surface topography on biofilm development. Clin Oral Implants Res. 2006;17 Suppl 2:68–81.
67. Subramani K, Jung RE, Molenberg A, Hammerle CH. Biofilm on dental implants: a review of the literature. Int J Oral Maxillofac Implants. 2009;24(4):616–26.
68. Koka S, Razzoog ME, Bloem TJ, Syed S. Microbial colonization of dental implants in partially edentulous subjects. J Prosthet Dent. 1993;70(2):141–4.
69. Quirynen M, Alsaadi G, Pauwels M, Haffajee A, van Steenberghe D, Naert I. Microbiological and clinical outcomes and patient satisfaction for two treatment options in the edentulous lower jaw after 10 years of function. Clin Oral Implants Res. 2005;16(3):277–87.
70. Quirynen M, Vogels R, Pauwels M, Haffajee AD, Socransky SS, Uzel NG, et al. Initial subgingival colonization of 'pristine' pockets. J Dent Res. 2005;84(4):340–4.
71. Pontoriero R, Tonelli MP, Carnevale G, Mombelli A, Nyman SR, Lang NP. Experimentally induced peri-implant mucositis. A clinical study in humans. Clin Oral Implants Res. 1994;5(4):254–9.
72. Mombelli A, van Oosten MA, Schurch Jr E, Land NP. The microbiota associated with successful or failing osseointegrated titanium implants. Oral Microbiol Immunol. 1987;2(4):145–51.
73. Shibli JA, Melo L, Ferrari DS, Figueiredo LC, Faveri M, Feres M. Composition of supra- and subgingival biofilm of subjects with healthy and diseased implants. Clin Oral Implants Res. 2008;19(10):975–82.
74. Hultin M, Gustafsson A, Hallstrom H, Johansson LA, Ekfeldt A, Klinge B. Microbiological findings and host response in patients with peri-implantitis. Clin Oral Implants Res. 2002; 13(4):349–58.
75. Leonhardt A, Dahlen G, Renvert S. Five-year clinical, microbiological, and radiological outcome following treatment of peri-implantitis in man. J Periodontol. 2003;74(10):1415–22.
76. Leonhardt A, Grondahl K, Bergstrom C, Lekholm U. Long-term follow-up of osseointegrated titanium implants using clinical, radiographic and microbiological parameters. Clin Oral Implants Res. 2002;13(2):127–32.
77. Leonhardt A, Renvert S, Dahlen G. Microbial findings at failing implants. Clin Oral Implants Res. 1999;10(5):339–45.
78. Rams TE, Roberts TW, Feik D, Molzan AK, Slots J. Clinical and microbiological findings on newly inserted hydroxyapatite-coated and pure titanium human dental implants. Clin Oral Implants Res. 1991;2(3):121–7.
79. Rosenberg ES, Torosian JP, Slots J. Microbial differences in 2 clinically distinct types of failures of osseointegrated implants. Clin Oral Implants Res. 1991;2(3):135–44.
80. Agerbaek MR, Lang NP, Persson GR. Comparisons of bacterial patterns present at implant and tooth sites in subjects on supportive periodontal therapy, I. Impact of clinical variables, gender and smoking. Clin Oral Implants Res. 2006;17(1):18–24.
81. Karoussis IK, Muller S, Salvi GE, Heitz-Mayfield LJ, Bragger U, Lang NP. Association between periodontal and peri-implant conditions: a 10-year prospective study. Clin Oral Implants Res. 2004;15(1):1–7.

82. Karoussis IK, Salvi GE, Heitz-Mayfield LJ, Bragger U, Hammerle CH, Lang NP. Long-term implant prognosis in patients with and without a history of chronic periodontitis: a 10-year prospective cohort study of the ITI Dental Implant System. Clin Oral Implants Res. 2003;14(3):329–39.
83. Heitz-Mayfield LJ, Huynh-Ba G. History of treated periodontitis and smoking as risks for implant therapy. Int J Oral Maxillofac Implants. 2009;24(Suppl):39–68.
84. Ellegaard B, Baelum V, Karring T. Implant therapy in periodontally compromised patients. Clin Oral Implants Res. 1997;8(3):180–8.
85. Sanz M, Alandez J, Lazaro P, Calvo JL, Quirynen M, van Steenberghe D. Histo-pathologic characteristics of peri-implant soft tissues in Branemark implants with 2 distinct clinical and radiological patterns. Clin Oral Implants Res. 1991;2(3):128–34.
86. Ericsson I, Berglundh T, Marinello C, Liljenberg B, Lindhe J. Long-standing plaque and gingivitis at implants and teeth in the dog. Clin Oral Implants Res. 1992;3(3):99–103.
87. Lindhe J, Berglundh T, Ericsson I, Liljenberg B, Marinello C. Experimental breakdown of peri-implant and periodontal tissues. A study in the beagle dog. Clin Oral Implants Res. 1992;3(1):9–16.
88. Berglundh T, Lindhe J, Marinello C, Ericsson I, Liljenberg B. Soft tissue reaction to de novo plaque formation on implants and teeth. An experimental study in the dog. Clin Oral Implants Res. 1992;3(1):1–8.
89. Gualini F, Berglundh T. Immunohistochemical characteristics of inflammatory lesions at implants. J Clin Periodontol. 2003;30(1):14–8.
90. Berglundh T, Gislason O, Lekholm U, Sennerby L, Lindhe J. Histopathological observations of human periimplantitis lesions. J Clin Periodontol. 2004;31(5):341–7.
91. Schou S, Holmstrup P, Reibel J, Juhl M, Hjorting-Hansen E, Kornman KS. Ligature-induced marginal inflammation around osseointegrated implants and ankylosed teeth: stereologic and histologic observations in cynomolgus monkeys (*Macaca fascicularis*). J Periodontol. 1993;64(6):529–37.
92. Mombelli A. Aging and the periodontal and peri-implant microbiota. Periodontol 2000. 1998;16:44–52.
93. Uitto VJ. Gingival crevice fluid—an introduction. Periodontol 2000. 2003;31:9–11.
94. Heitz-Mayfield LJ. Peri-implant diseases: diagnosis and risk indicators. J Clin Periodontol. 2008;35(8 Suppl):292–304.
95. Mombelli A. Microbiology and antimicrobial therapy of peri-implantitis. Periodontol 2000. 2002;28:177–89.
96. Salvi GE, Persson GR, Heitz-Mayfield LJ, Frei M, Lang NP. Adjunctive local antibiotic therapy in the treatment of peri-implantitis II: clinical and radiographic outcomes. Clin Oral Implants Res. 2007;18(3):281–5.
97. Mombelli A, Feloutzis A, Bragger U, Lang NP. Treatment of peri-implantitis by local delivery of tetracycline. Clinical, microbiological and radiological results. Clin Oral Implants Res. 2001;12(4):287–94.
98. Mombelli A, Lang NP. Antimicrobial treatment of peri-implant infections. Clin Oral Implants Res. 1992;3(4):162–8.
99. Gottlow J, Nyman S, Karring T, Lindhe J. New attachment formation as the result of controlled tissue regeneration. J Clin Periodontol. 1984;11(8):494–503.
100. Nyman S, Gottlow J, Karring T, Lindhe J. The regenerative potential of the periodontal ligament. An experimental study in the monkey. J Clin Periodontol. 1982;9(3):257–65.
101. Nyman S, Lindhe J, Karring T, Rylander H. New attachment following surgical treatment of human periodontal disease. J Clin Periodontol. 1982;9(4):290–6.
102. Listgarten MA, Rosenberg MM. Histological study of repair following new attachment procedures in human periodontal lesions. J Periodontol. 1979;50(7):333–44.
103. Chen ST, Wilson Jr TG, Hammerle CH. Immediate or early placement of implants following tooth extraction: review of biologic basis, clinical procedures, and outcomes. Int J Oral Maxillofac Implants. 2004;19(Suppl):12–25.
104. Hammerle CH, Chen ST, Wilson Jr TG. Consensus statements and recommended clinical procedures regarding the placement of implants in extraction sockets. Int J Oral Maxillofac Implants. 2004;19(Suppl):26–8.

105. Hammerle CH, Jung RE, Feloutzis A. A systematic review of the survival of implants in bone sites augmented with barrier membranes (guided bone regeneration) in partially edentulous patients. J Clin Periodontol. 2002;29 Suppl 3:226–31. discussion 32–3.
106. Wang HL, MacNeil RL. Guided tissue regeneration. Absorbable barriers. Dent Clin North Am. 1998;42(3):505–22.
107. Cortellini P, Pini Prato G, Tonetti MS. Periodontal regeneration of human intrabony defects with bioresorbable membranes A controlled clinical trial. J Periodontol. 1996;67(3):217–23.
108. Tonetti MS, Prato GP, Cortellini P. Factors affecting the healing response of intrabony defects following guided tissue regeneration and access flap surgery. J Clin Periodontol. 1996;23(6):548–56.
109. Nowzari H, Matian F, Slots J. Periodontal pathogens on polytetrafluoroethylene membrane for guided tissue regeneration inhibit healing. J Clin Periodontol. 1995;22(6):469–74.
110. Slots J, MacDonald ES, Nowzari H. Infectious aspects of periodontal regeneration. Periodontol 2000. 1999;19:164–72.
111. Mombelli A, Lang NP, Nyman S. Isolation of periodontal species after guided tissue regeneration. J Periodontol. 1993;64(11 Suppl):1171–5.
112. Yoshinari N, Tohya T, Kawase H, Matsuoka M, Nakane M, Kawachi M, et al. Effect of repeated local minocycline administration on periodontal healing following guided tissue regeneration. J Periodontol. 2001;72(3):284–95.
113. Urbani G, Graziani A, Lombardo G, Caton JG. Clinical results with exposed polyglactin 910 resorbable membranes for guided tissue regeneration. Int J Periodontics Restorative Dent. 1997;17(1):41–51.
114. Nowzari H, MacDonald ES, Flynn J, London RM, Morrison JL, Slots J. The dynamics of microbial colonization of barrier membranes for guided tissue regeneration. J Periodontol. 1996;67(7):694–702.
115. Machtei EE, Cho MI, Dunford R, Norderyd J, Zambon JJ, Genco RJ. Clinical, microbiological, and histological factors which influence the success of regenerative periodontal therapy. J Periodontol. 1994;65(2):154–61.
116. Nowzari H, Slots J. Microorganisms in polytetrafluoroethylene barrier membranes for guided tissue regeneration. J Clin Periodontol. 1994;21(3):203–10.
117. Wang HL, Yuan K, Burgett F, Shyr Y, Syed S. Adherence of oral microorganisms to guided tissue membranes: an in vitro study. J Periodontol. 1994;65(3):211–8.
118. Ricci G, Rasperini G, Silvestri M, Cocconcelli PS. In vitro permeability evaluation and colonization of membranes for periodontal regeneration by *Porphyromonas gingivalis*. J Periodontol. 1996;67(5):490–6.
119. Smith MacDonald E, Nowzari H, Contreras A, Flynn J, Morrison J, Slots J. Clinical and microbiological evaluation of a bioabsorbable and a nonresorbable barrier membrane in the treatment of periodontal intraosseous lesions. J Periodontol. 1998;69(4):445–53.
120. Machtei EE. The effect of membrane exposure on the outcome of regenerative procedures in humans: a meta-analysis. J Periodontol. 2001;72(4):512–6.
121. Selvig KA, Kersten BG, Chamberlain AD, Wikesjo UM, Nilveus RE. Regenerative surgery of intrabony periodontal defects using ePTFE barrier membranes: scanning electron microscopic evaluation of retrieved membranes versus clinical healing. J Periodontol. 1992;63(12):974–8.

Chapter 12
Animal Models of Orthopedic Implant-Related Infection

Lorenzo Calabro, Cameron Lutton, Ahmed Fouad Seif El Din, R. Geoff Richards, and T. Fintan Moriarty

Abstract Musculoskeletal infection remains a great challenge in orthopedic and trauma surgery. Despite best medical and surgical practice and significant advances in research and development, bone and implant associated infections are still difficult to diagnose, impossible to prevent in all cases and require invasive and debilitating treatment. The development and safe clinical implementation of novel preventative, therapeutic or diagnostic strategies requires the use of animal models of infection, which provide crucial evidence regarding performance, cytocompatibility, biocompatibility, and safety prior to clinical implementation.

Many animal models of musculoskeletal infection have been described in the literature; however, there remains a dearth of fully standardized or universally accepted reference models hindering advancement in the field. The following chapter provides an overview of the animal models available for the study of musculoskeletal infection, the latest advances that are expected to improve them, and some of the most important scientific output achieved using these models.

L. Calabro (✉)
AO Research Institute Davos, Clavadelerstrasse 9, CH7270 Davos, Switzerland

Queensland University of Technology, Brisbane, Australia
e-mail: l_calabro@hotmail.com

C. Lutton • R.G. Richards • T.F. Moriarty
AO Research Institute Davos, Clavadelerstrasse 9, CH7270 Davos, Switzerland

A.F. Seif El Din
AO Research Institute Davos, Clavadelerstrasse 9, CH7270 Davos, Switzerland

Orthopaedic Surgery and Trauma Department, Kasr Al-Ainy Cairo University Hospitals, Cairo, Egypt

T.F. Moriarty et al. (eds.), *Biomaterials Associated Infection: Immunological Aspects and Antimicrobial Strategies*, DOI 10.1007/978-1-4614-1031-7_12,

12.1 Introduction

Infection of the bone and surrounding tissues is one of the most common and costly complications encountered in orthopedic surgery. Despite best modern surgical practices, neither prophylaxis nor treatment of these infections is 100% effective and considering that infected cases can result in a doubling of operative costs and a quadrupling of ward costs [1], improvements in both prophylaxis and treatment options are urgently required. Local antibiotic delivery vehicles and vaccines are two of the most promising approaches aimed at reducing the incidence of these infections, the development of which will require the use of animal models. Animal models of bone infection provide crucial evidence of performance, biocompatibility and safety of such novel interventional strategies prior to clinical implementation in addition to providing answers to many basic research questions.

A large number of animal models of musculoskeletal infection have been described in the literature, reflecting a lack of standardization in these animal models, and the disparity of infections that fall under the umbrella term "musculoskeletal infection." The varied forms of musculoskeletal infection include, for example; osteomyelitis, implant-related osteomyelitis, superficial cellulitis, myositis, septic arthritis, or soft-tissue abscesses. All are distinct types of musculoskeletal infection and it is therefore of no surprise that numerous models have been created to investigate the pathology or potential interventions for each. When one then also considers the diversity in animal species, anatomical location, or pathogenic organisms that may be involved, the reason for such a large variety of animal models becomes apparent.

The selection of the most appropriate animal model of musculoskeletal infection may be a difficult task despite the availability of many potential models. The salient clinical features of the particular infection type should be replicated to the best possible extent in the chosen animal model. This may include factors that significantly contribute to the classification of the infection and the resultant treatment protocols that are used, such as: the presence or absence of an implant; the type of implant used; the means by which the bacteria arrived at the eventual site of infection (hematogenous, iatrogenic, environmental or from a contiguous source of infection); the identity and antibiotic resistance of the causative organism; whether the infection is acute or chronic, and finally, whether the model is aimed at testing prophylaxis, treatment, diagnosis or pathology.

The focus of our research has been on the fixation of bones fractured after trauma and the infections that may develop surrounding these devices. Animal models are indispensible in this field [2]. Infections associated with fracture fixation devices involve the presence of a foreign body in the host, which adds an additional dimension not present in non-implant-related infections such as septic arthritis or non-implant-related osteomyelitis. The mere presence of a foreign body in a surgical site increases susceptibility to infection [3] and many infection models actually utilize foreign bodies to specifically potentiate the infection. Studies into fracture fixation in particular also add additional, unique challenges for the development of infection models: fractures typically occur after a traumatic event such as a motor vehicle accident with extensive soft tissue damage and wounds exposed to environmental contamination, which are normally not observed in elective prosthetic joint replacement surgery for example. Depending on anatomic location and injury type, the implant system used to fix a fracture vary with the most common options being an external fixator, a plate, or an intramedullary (IM) nail. The resulting biomechanical stability of the fixed bone construct will have a significant impact on the healing of the fracture and to faithfully replicate a contaminated or infected fracture fixation device, the chosen model would ideally need to incorporate not only the fracture, but also the correct post-fixation biomechanics to allow the host defenses to fight an infection in clinically relevant circumstances.

The present chapter aims to provide an overview of existing animal models employed in research into musculoskeletal infection including those investigating basic surgical principles, implant design factors, bacterial pathogenicity, and antibiotic therapy. It is beyond the scope of any single chapter to attend to all the diverse models of musculoskeletal infection, and so we have focused on models of osteomyelitis associated with trauma. A representative sample of the most recent or relevant models are presented, as well as some of the main outcomes these models have provided for the advancement of patient care over the last few decades. Finally, we will present a brief discussion of where these models may be somehow improved, again with particular focus upon infection in fracture fixation.

12.2 Species Selection

The ideal species for an orthopedic implant-related infection model would have molecular, cellular, structural and mechanical features akin to human bone, a size and temperament allowing easy, low-cost maintenance and handling, a well-documented genetic and immunologic profile, and be sufficiently robust to endure medical and surgical interventions that reflect clinical practice. Unsurprisingly, no such species exists and the selection of any one model involves compromise and prioritization amongst potentially numerous research ideals. This conclusion is known as the "Krogh principle" after the Nobel prize winning Danish physiologist who wrote in 1929 that "For a large number of problems there will be some animal of choice or a few such animals on which it can be most conveniently studied" [4]. Thorough evaluation of any particular hypothesis may necessitate preclinical testing in more than one species as is typical of pharmaceutical trials where both a small and large animal model is recommended [5].

Interspecies comparison of bony architecture and fracture susceptibility, important parameters in the clinical practice of fracture fixation and implant-related osteomyelitis studies, reveal significant variation and do not conform to a direct proportional comparison [5–9]. The relationship between bone mineral density and fracture resistance varies between species and the degree of this variation varies with anatomical location [6]. In general, ruminants do not have a well-organized Haversian system [7, 9, 10] whilst rabbit bone has a primary lamellar structure distinct from that of humans, as well as a comparatively higher ratio of bone mineral to collagen [9]. Aerssens et al. compared femoral cortical and lumbar trabecular bone samples from four animal species with those from human cadavers and found that canine and porcine femora had very similar ash and collagen phase proportions to humans, but the respective trabecular bone samples were vastly different [6]. Another finding from the same study was that while ovine and bovine bone mineral content was similar, the fracture stress in ovine samples was approximately twice that of bovine. Wang et al. also compared femoral cortical samples from canine, bovine and human cadavers and concurrently found that the mineral and organic phase proportions of canine and human bone were similar [9]. Subsequent microscopic analysis showed that canine and human bone shared a similar secondary osteonal structure, but despite these findings, canine bone had significantly higher fracture toughness, which is a measure of the energy required to propagate an existing microfracture in bone. The same study also analyzed rabbit and baboon samples and concluded that the baboon bone had the microstructure and fracture resistance that was most similar to human bone, with canine bone placing second. Fracture toughness in rabbits was not significantly different to that of human bone but it must be noted that this study used a single layer sandwich sample for mechanical testing and the results may not take species specific functional loading or fracture patterns into account [9].

In general, larger animals such as nonhuman primates, sheep, goats, and dogs are better able to tolerate surgical interventions and their bony geometry can accommodate human scale prostheses under known loading conditions [7]. This is extremely important, as mechanobiological variables including fixation stability are known to influence both fracture healing and infection susceptibility [11, 12]. Smaller animals such as rats, mice, and guinea pigs perform less well against these criteria due to the perceived difficulty in achieving mechanically defined or controlled conditions; however, such small animals may be housed and fed more easily allowing larger numbers and more statistically powerful conclusions at an acceptable financial cost [7, 13]. Rabbits have been used frequently for osteomyelitis models due to their relatively easy maintenance, reproducible infection profile, and apparent tolerance to implantation of readily available prostheses [14, 15]. Technological advances in small scale fixation devices and the availability of transgenic rodents has recently challenged the generalization that small animals are not suitable for mechanobiological studies with fracture fixation implants, with custom designed implants allowing small animal fracture models to better emulate clinical conditions [16, 17]. Matthys and Perren developed an internal fixator, analogous to a locked plate for use in mouse femoral osteotomy models[17] and subsequent variations on this implant allow the investigator to choose between stable or flexible fixation, and hence choose between a model with a model with primary or secondary fracture healing [18]. Garcia et al. similarly developed an interlocking nail for mouse femoral models with stable and less stable fixation options [16]. Studies into implant-related osteomyelitis in rodent models of infection is clearly within reach and opens up the possibility of rapid progression in the field.

Historically the success of animal osteomyelitis models has been determined by the degree to which radiographic, histological and microbiological outcomes mirror those found in human disease [14]. Inoculation protocols and surgical approaches are typically manipulated to meet this end though it has been noted that some animals such as rats and dogs are more difficult to infect than others such as rabbits [14]. Furthermore, many human viral pathogens may not cause predictable disease in a particular animal or any disease at all [19], though this has not been established for bacterial pathogens in osteomyelitis. Interspecies differences in immune response are also important to acknowledge in infection trials. Traditionally immunobiological research has used nonhuman primate models in preclinical testing for their similar genetic profile but it is noted that even chimpanzees have differences in major histocompatibility complex (MHC) and human leukocyte antigen (HLA) genes and that these differences can influence disease process [19]. Transgenic mice with humanized immune systems may provide a novel solution. In this technique human hematopoietic stem cells are introduced into immune-deficient mice allowing characterization of a human immune response in a preclinical model [20]. These animals create the potential for detailed molecular characterization in preclinical models of human infection [7, 20], though again, such animals have not, to our knowledge, been used to date in the field of osteomyelitis or implant-related osteomyelitis.

Intra and interspecies heterogeneity confounds any simple rule in selecting animal models for orthopedic implant-related infection. The researcher must be aware of the physiologic and mechanical qualities of commonly used animal models and consider these in the context of the available implant technologies and the experimental hypothesis.

12.3 Overview of Animal Models of Musculoskeletal Infection

Numerous animal models of musculoskeletal infection exist (Tables 12.1 and 12.2, Fig. 12.1). The primary outcome measurements most often determined from these studies include bacterial pathogenicity or virulence factors, impact of treatment strategies, or the performance of a novel biomaterial such as bone cement, implant coating or the implant itself. In the course of investigating these issues, more or less complex models are required depending on the exact research question. The majority of these studies create a local infection in bone courtesy of a bacterial inoculum and the infection is potentiated by either an implanted device or a sclerosing agent. Secondary traumatic factors such as a fracture or soft tissue trauma may be present in the clinical cases modeled in these animals; however, they are usually not involved. In cases where, for example, the impact of fracture type, soft tissue injury or hematoma is investigated to determine influence on the development or treatment of infection, a more complex model including trauma may be required. In the following sections both minimally traumatic models and those including the influence of tissue trauma shall be described.

Table 12.1 Established infection models

Primary author	Year	Model	Pathogen	Question/goal/intervention
Implant-related infection models				
Scheman [21]	1941	Rabbit	*S. aureus*	Repeatable OM model
Norden [22]	1970	Rabbit	*S. aureus*	Repeatable chronic OM model
Norden [77]	1975	Rabbit	*S. aureus*	Comparing antibiotic regimens
Rissing [23]	1985	Rat	*S. aureus*	Repeatable OM model
Andriole [27]	1973	Rabbit	*S. aureus*	Repeatable OM model
Petty [28]	1985	Dog	*S. aureus*; *S. epidermidis*; *E coli*	Effect of implant material on infection resistance
Cordero[29]	1994	Rabbit	*S. aureus*	Effect of metal type and surface on infection resistance
Melcher [33]	1994	Rabbit	*S. aureus*	Hollow vs. solid IM nail on infection resistance
Melcher [34]	1995	Rabbit	*S. aureus*	Reamed vs. unreamed IM nail on infection resistance
Hendricks [39]	2001	Rat	*S. aureus*; *P. aeruginosa*	Bacterial synergy in spinal OM
Horn [36]	2005	Rabbit	*S. aureus*	Cannulated vs. Hollow vs. solid nail on infection resistance
Salgado [10]	2005	Goat	*S. aureus*	Repeatable OM model
Li [88]	2008	Mouse	*S. aureus*	In vivo infection imaging
Hematogenous osteomyelitis models				
Deysine [40]	1976	Dog	*S. aureus*	Repeatable OM model
Emslie [41]	1986	Chicken	*S. aureus*	Repeatable OM model
Whalen [44]	1988	Rabbit	*S. aureus*	Physeal OM in immature bone
Hienz [43]	1995	Rat	*S. aureus*	Repeatable OM model
Poultsides [42]	2008	Rabbit	MRSA	Repeatable OM model

Fracture + Implant models				
Rittman [64]	1974	Sheep	*S. aureus*	Primary vs. secondary fracture healing
Passl [63]	1984	Guinea pig	*S. aureus*; *E. coli*	Repeatable OM model
Worlock [65]	1988	Rabbit	*S. aureus*	Repeatable OM model
Worlock [66]	1994	Rabbit	*S. aureus*	Effect of fixation stability on infection susceptibility
Curtis [68]	1995	Goat	*S. aureus*	Ex fix vs. IM nail, reamed or not for infection susceptibility
Hill [69]	2002	Sheep	*S. aureus*	Infect susceptibility of primary IM nailing contaminated fractures
Lindsey [70]	2010	Rat	*S. aureus*	Immunological analysis
Bone defect Osteomyelitis models				
Chen [73]	2002	Rat	*S. aureus*	Osteoinduction effect of OP-1 in infected bone defect
Southwood [24]	2004	Rat	*S. aureus*	Osteoinduction effect of BMP-2 in infected bone defect
Stewart [74]	2010	Rat	*S. aureus*; *E. coli*	Gentamicin paste on a degradable scaffold
Smeltzer [26]	1997	Rabbit	*S. aureus*	Repeatable OM model

OM osteomyelitis, *IM* intramedullary, *OP*-1 osteogenic protein 1, *BMP*-2 bone morphogenic protein 2

Table 12.2 Infection prophylaxis and treatment models

Primary author	Year	Model	Pathogen	Question/goal/intervention
Prophylaxis models				
Rodeheaver [79]	1983	Rabbit	*S. aureus*; *E. coli*	ALBC
Lucke [90]	2003	Rat	*S. aureus*	PDLLA/gentamicin coated nail
Darouiche [94]	1998	Rabbit	*S. aureus*	Antiseptic coated nail with fracture
Antoci [89]	2007	Rat	*S. aureus*	Titanium + covalently bonded vancomycin nail
Huneault [84]	2004	Dog	*S. aureus*	High amylose starch + ciprofloxacin
Treatment models				
Fitzgerald [104]	1983	Dog	*S. aureus*	ALBC nail
Evans [80]	1993	Rabbit	*S. aureus*	ALBC
Giavaresi [102]	2008	Rabbit	MRSA	ALBC coated nail + gent/vanc
Isiklar [108]	1996	Rabbit	*S. epidermidis*	IV antibiotics
Orhan [106]	2006	Rat	MRSA	Chitosan spheres + cipro
Moskowitz [97]	2010	Rabbit	*S. aureus*	Polyelectrolyte coating + gent
Monzon [109]	2002	Rat	*S. aureus*	IV antibiotics
Garvin [105]	1994	Dog	*S. aureus*	PLA/PGA + gent
Cornell [101]	1993	Rabbit	*S. aureus*	HA + gentamicin
Shirtliff [103]	2002	Rabbit	MRSA	HA + vancomycin
Huneault [84]	2004	Dog	*S. aureus*	High amylose starch + cipro
Del Pozo [107]	2009	Rabbit	*S. epidermidis*	Electrical current to implant

ALBC antibiotic loaded bone cement, *HA* hydroxyapatite, *MR* methicillin resistant, *PDLLA* poly D, L lactic acid, *PLA/PGA* polylactide/polyglycolide, *cipro* ciprofloxacin

12.3.1 Minimally Traumatic Models of Infection

In most animal models of infection the trauma is limited to that required for the surgical approach, implant placement and inoculation of bacteria. These models usually involve the injection of a bacterial suspension and the insertion of a biomaterial/foreign body into the metaphyseal bone or intramedullary canal, minimizing damage to the bone and the surrounding tissue. Such models are typically used to investigate biomaterial specific effects on infection susceptibility as well as providing a useful starting point for the evaluation of therapeutic agents. They are also the most clinically relevant models to examine various aspects of prosthetic joint infection, which are elective surgeries and usually do not occur subsequent to a traumatic event.

Early attempts at developing minimally traumatic models of osteomyelitis used sclerosing agents such as sodium morrhuate to initiate tissue damage to the bone. Sodium morrhuate is a salt of morrhuic acid, which is derived from a mixture of fatty acids obtained from cod-liver oil. Sodium morrhuate causes local vascular thrombosis, leading to necrotic tissue formation and potentiates an infection in bone by limiting the ability of the immune system to respond to the bacteria. Scheman and colleagues first developed a model of osteomyelitis resembling human disease in rabbit tibiae using sodium morrhuate as a sclerosing agent [21]. They had previously found that efforts to create osteomyelitis with intravenous inoculation caused

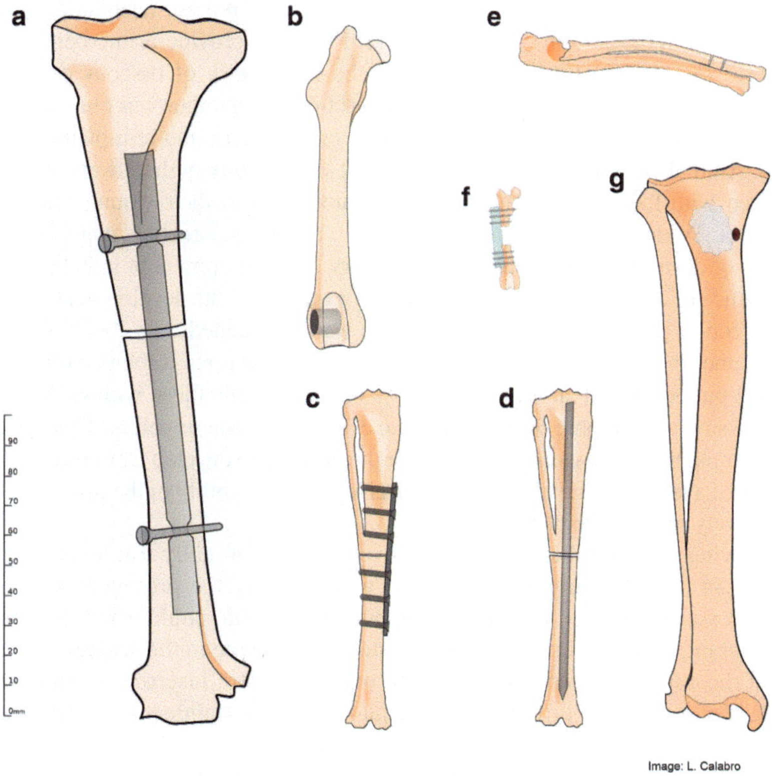

Fig. 12.1 Scale diagram of selected animal osteomyelitis models. (**a**) Sheep tibia as per Hill et al. [69]. (**b**) Rabbit femur as per An et al. [153]. (**c**) and (**d**) Rabbit tibiae fixation as per Worlock et al. [66]. (**e**) Rabbit radius as per Smeltzer et al. [26]. (**f**) Rat femoral fixation as per Chen et al. [73]. (**g**) Dog tibia with bone cement foreign body as per Fitzgerald et al. [104]

inconsistent results, and that direct inoculation of *Staphylococcus aureus* alone into bone failed to create pathology mimicking chronic osteomyelitis [21]. Norden and Kennedy built on the Scheman model, demonstrating that a chronic, progressive lesion could consistently be created with a lower concentration of sodium morrhuate and in some cases be sustained for up to 6 months [22]. In this study, all rabbits in the group that received both a sclerosing agent and *S. aureus* developed progressive osteomyelitis, while no rabbits in the group inoculated with *S. aureus* alone developed infections [22]. It was also noted that not all strains of *S. aureus* were capable of causing osteomyelitis in this model though the precise reasons for this were not established. Minor variations on this technique of Scheman et al. have subsequently been used to create similar models in the rat [23] and goat [10]. Concerns have been raised with the use of sodium morrhuate in that it reduces the clinical relevance of the model due to impaired local vascularization and circulation and may have other undesired side effects [24]. Additionally, the radiographic appearance of aseptic necrosis which can be caused by sclerosing agents may resemble infection making it difficult to determine the separate impact of both radiographically [25].

Smeltzer et al. developed a model of infection that did not use sclerosing agents but instead relied upon portions of devascularized bone to provide a microenvironment conducive to infection [26], demonstrating the importance of osseous perfusion in osteomyelitis. In the Smeltzer model, a small section of the radius was removed, inoculated with 2×10^6 colony forming units (CFU) of *S. aureus*, and reimplanted into the wound. Animals with the reimplanted bone had significantly higher levels of infection compared with the animals who were inoculated but had their wound filled with a hematoma. Andriole et al. demonstrated that neither sclerosing agents nor large amounts of devascularized tissue were necessary for development of infection but that the presence of a foreign body was all that was required for the development of a stable bone infection [27]. Rabbit medullary canals were inoculated with 2×10^8 CFU of *S. aureus* either in the presence or absence of a stainless steel pin. Rabbits with the stainless steel pin developed an infection in 88% of cases, while those without the pin did not show any radiographic or microbiologic evidence of osteomyelitis. This suggested that the implant was providing a surface for the bacteria to develop a biofilm and/or the presence of a large foreign object was interfering with the ability of the immune system to effectively remove the bacteria.

Following on from this Petty et al. developed a minimally traumatic model of infection in dogs that highlighted that it was not only the presence of a foreign body that was important, but that the type of biomaterial could affect the chances of developing an infection [28]. They created a small hole in the greater trochanter, reamed the canal, and removed the bone marrow before inserting cylindrical test cylinders of a range of orthopedic materials including stainless steel alloy, cobalt chromium alloy, high molecular weight polyethylene, prepolymerized polymethylmethacrylate (PMMA) bone cement, and PMMA bone cement at the dough stage injected into the canal. Three species of bacteria associated with orthopedic infections (*S. aureus*, *S. epidermidis*, and *Escherichia coli*) were used to create the infection. Bacterial inocula varied between 1×10^2 and 1×10^8 CFU. In this model the authors specifically introduced a lag phase into the inoculum by storing the bacteria on ice for 15 min prior to injection in an attempt to simulate the conditions in which the bacteria may contaminate the wound during a surgery. They demonstrated that the presence of a biomaterial was enough to reduce the inoculum required for infection to develop by 50% when compared with controls without implants, and that there was variability in the infection rate between the materials, with the PMMA (dough) being most susceptible to infection. Cordero et al. demonstrated that not only can the material or composition of the implant play a role but also the surface texture [29]. Cylinders of either cobalt–chromium–molybdenum alloy or titanium–aluminum–vanadium alloy were implanted into rabbit femora. Each implant had either a smooth polished surface or a porous coating of the metal in question. In both cases the porous surfaces required significantly less bacteria to cause an infection. In particular, porous cobalt chrome needed 40 times less bacteria to create osteomyelitis compared to the smooth polished surface, possibly because the pores provide a site for bacteria to multiply while shielding them from host immune defenses. In contrast, another study showed that the microrough surface of commercially available titanium implants confers greater infection resistance than an equivalent polished stainless steel surface in a rabbit model using a miniature fracture fixation implant [30]. This difference was attributed to the different biocompatibility and tissue adherence between the two surfaces but would appear to be a less decisive factor than implant architecture [30, 31]. In a similar model, low contact PC-fix plates were compared to the dynamic compression plates used in the above study and were shown to have much lower infection susceptibility, presumably due to preservation of the underlying periosteum [11]. More recent examples using similar models of infection have elucidated the role of more subtle effects including surface topography, variations in surface chemistry and surface energy [32]. It would appear that of these factors, implant architecture is the most decisive factor [31].

The effects of implant design have also been investigated using minimally traumatic models of infection. Melcher et al. used a rabbit tibia model in an attempt to define the relative infection resistance of a variety of different IM nail designs. They initially demonstrated that hollow slotted nails were more susceptible to infection than a solid design, a finding attributed to the creation of an avascular dead space within the hollow nail [33]. In subsequent studies reamed nails were found to be significantly more susceptible to infection than unreamed, and stainless steel nails more susceptible than titanium [34, 35]. These findings appeared to confirm two hypotheses: that preservation of the intramedullary blood supply in unreamed nailing; and the enhanced osseointegration seen with microrough titanium implants are factors in infection resistance. These findings were repeated in the same animal model in a subsequent study which also evaluated cannulated nails [36]. The results for the newer cannulated nail design were similar to those for hollow implants. Interestingly these findings have not been corroborated in prospective randomized clinical trials [37, 38], highlighting the difficulty in extrapolating from preclinical animal models to clinical reality.

The previously described models have focused on long bones such as tibia and femur as a site for the development of infection, although other bone sites are also clinically relevant for models of infection. Hendricks et al. developed a model of infection using the spinous process of a rat [39] wherein they implanted a stainless steel wire through the process and initiated infection using either *S. aureus*, *Pseudomonas aeruginosa* or a combination of both. They showed that low concentrations of both bacteria ($<10^3$ CFUs) significantly increased the infection rates when compared to similar concentrations of individual bacteria. On increasing the inoculation dose ($\sim 10^6$ CFUs) they found the converse to be true: individual species bacteria were able to cause high infection rates whereas high concentrations of both bacterial species caused a drop in infection rates. Interestingly, only *S. aureus* was then cultured from the polymicrobial inoculated infections.

12.3.1.1 Hematogenous Osteomyelitis Models

Osteomyelitis caused by hematogenous seeding following bacteremia is acknowledged as a distinct sub classification of disease. This is a particular issue in pediatric medicine where septic arthritis and osteomyelitis in the adjacent metaphyses of long bones is relatively common. It is also a common cause for late infection in previously well-functioning prosthetic joints. Acknowledging this distinct etiology, a number of authors have attempted to create osteomyelitis with a hematogenous seeding of bacteria. Similar to direct inoculation methods, *S. aureus* is the most commonly used bacterial species and a focal bone lesion is typically created either concurrently or prior to inoculation. An early model created by Deysine and colleagues involved injecting a combination of barium and 5×10^5 CFU of *S. aureus* simultaneously into the tibial nutrient artery of dogs [40]. The authors reliably created acute osteomyelitis, but there was an unacceptably high mortality rate from sepsis. Other similar models in chickens [41] and rabbits [42] unsurprisingly encountered the same problem, describing narrow safety margins in inoculum dose when administered systemically. Hienz et al. however were able to create a model using rats with no mortality during the 14 day study period [43]. They injected sodium morrhuate locally into the mandible and tibia of rats and inoculated varying doses of *S. aureus* intravenously into the tail vein to determine the dose required to infect 50% (ID_{50}) and 100% (ID_{100}) of animals. Animals inoculated with bacteria but spared the focal sodium morrhuate injection did not develop osteomyelitis, again demonstrating the importance of local perturbation of bone physiology. An alternative approach was taken by Whalen et al. who designed a model to mimic pediatric hematogenous osteomyelitis [44]. Using immature rabbits, they created a partial growth plate fracture at the proximal tibial metaphysis by application of an external force. A *S. aureus* inoculum between 7×10^7 and 1×10^8 CFU was then administered intravenously via an ear vein, reliably causing focal acute osteomyelitis without metastatic infection. This model has been adapted for application in mice in order to study immune response by taking advantage of the greater availability of molecular tools for the mouse [45].

12.3.2 Traumatic Models of Infection

Infection rates associated with fracture fixation are consistently higher than those in elective orthopedic implant surgery such as arthroplasty [46, 47]. Osteomyelitis occurs in 5–22% of open fractures and this rate can double for tibial shaft fractures [46]. Surgical fixation of closed fractures is associated with lower risk at 1–2% [48]. A number of factors contribute to the higher incidence of implant-related infection in trauma including contaminated and devitalized soft tissue, systemic inflammatory response, procedural, implant and host issues[2, 46, 49]. A variety of animal models have been developed to better understand the etiology and natural history of osteomyelitis associated with trauma and fracture fixation. Typically an anesthetized animal will be subjected to a standardized trauma and fixation, with a perioperative bacterial inoculation, usually of *S. aureus*. These models can be categorized according to the nature of the trauma inflicted; the subsequent management of this trauma;

the animal used; and the anatomic location within the animal. Given this kaleidoscope of options, animal models in the existing literature are unsurprisingly highly varied.

12.3.2.1 Standardizing Trauma

Clinical trauma studies are confounded by the heterogeneity of injury mechanisms and fracture patterns, and indeed the first methodological challenge in an animal model is the need for creation of a repeatable injury. The most controlled way of achieving this is to create an osteotomy. Osteotomy healing has been shown to grossly resemble low-energy random fracture healing in sheep [50]. An obvious drawback, however, is the need for surgical exposure which in the case of intramedullary fixation would otherwise be unnecessary and not observed clinically. Furthermore, osteotomy does not re-create the typical forces involved in clinical trauma, nor the more complex fracture patterns and soft tissue trauma.

The most common alternative to osteotomy is a three point bend method, with a controlled force delivered by various means. Ashurst et al. [51] devised a method of creating a reproducible fracture in the mid shaft of rabbit tibiae. In this model only the medial cortex was osteotomized, a plate was then placed in situ and a three point bend device subsequently used to fracture the remaining cortex. Jackson et al. [52] had previously created a blunt impaction guillotine device to create a closed fracture over rat femora which had already been implanted with an intramedullary pin. The force was delivered with a pneumatic impulse from a compressed air cylinder. Bonnarens and Einhorn [53] simplified the Jackson model by creating a similar guillotine device relying only on the force of gravity. They prepinned the femora of 40 Sprague-Dawley rats and were able to produce a transverse mid shaft femoral fracture in 38 of 40 animals and a short oblique fracture in the other 2. Marturano et al. [54] more recently refined this device for use in mice.

In order to generate consistent fracture patterns, relatively low energy mechanisms are used. Clinically however, high energy injuries are associated with worse outcomes in terms of fracture healing and infection, at least partly due to concomitant soft tissue injury [46, 55, 56]. Isolated studies have sought to reproduce high energy mechanisms quite literally such as by gunshot, but the varied results confound comparison [57]. An alternative approach is to create a low energy fracture or osteotomy and then a subsequent experimental soft tissue injury. Soft tissue injury has been variously inflicted by a direct muscular crush [58], resection of part or all the muscle overlying a fracture [56, 58] or with a guillotine device in the manner of Bonnarens and Einhorn [59, 60]. Kalicke et al. [61] demonstrated in a rat model that a closed soft tissue injury can increase the risk of infection independent of bacterial contamination but an animal model combining a fracture, infection and an additional standardized soft tissue injury does not, to our knowledge, exist.

The means of pathogen inoculation have also been the subject of variation depending on specific research questions. In an early study, Matsumoto et al. sought to examine the efficacy of topical antibiotic spray in combat wounds sustained in

American troops during the Vietnam War [62]. Pathogens isolated from infected war wounds were enriched in broth and mixed with soil samples from the field to model environmental contamination in rabbits. More typically a predetermined dose of a strain of *S. aureus* isolated from a clinical implant-associated infection is inoculated to the fracture site or intramedullary canal [32]. The optimal timing of inoculation relative to fracture and fixation is the subject of speculation as detailed in the following section.

12.3.2.2 Combining Trauma and Infection

Many studies, including those above, have characterized the nature of fracture healing with regards to experimental variables such as pharmaceutical interventions, soft tissue injury or fixation devices and techniques. Few authors however have investigated fracture associated osteomyelitis and the existing literature is highly heterogeneous. The potential ethical issues and complications inherent in these models are exemplified in one early study where efforts were made to reliably induce either *E. coli* or *S. aureus* osteomyelitis in a guinea pig femoral osteotomy [63]. Subject animals were divided into small groups ($n=7$) depending on which pathogen was used and whether or not the fracture was fixed. The authors reported an attrition rate in some groups of greater than 50%, and microbiologic outcomes which were potentially confounded by a high rate of fracture nonunion and contaminant bacteria. Rittman and Perren established an early experimental model in sheep to evaluate the impact of fixation stability on healing in an infected fracture [64]. An osteotomy was created in the tibial midshaft of 25 mountain sheep under general anesthetic. This was subsequently fixed with two strain-gauge plates in compression mode, a broad plate in compression mode, or a broad plate modified to be somewhat flexible over the osteotomy site. A pathogenic strain of *S. aureus* was inoculated percutaneously to the osteotomy site 1 week postoperatively and then repeatedly at intervals until local infection was confirmed by culture of an aspirate. All animals were eventually infected and radiologic, histologic, and mechanical characterization of fracture healing ensued. The authors found that a large implant offering greater stability was more likely to result in fracture union in this setting.

Worlock et al. developed a model to reliably cause fracture fixation associated osteomyelitis in rabbits without causing systemic illness [65]. Using the fracture and plate fixation method of Ashurst and Perren in rabbit tibiae, an *S. aureus* strain isolated from a patient with chronic osteomyelitis was inoculated immediately following wound closure. With an inoculum of 1×10^7 CFU, osteomyelitis developed in four of five animals without causing sepsis. The same group adapted this model for a later study evaluating the impact of fixation stability on infection risk in fractures fixed with a stable construct (dynamic compression plate) or an unstable construct (nonlocked intramedullary pin) [66]. Fractures fixed with stable fixation were significantly less likely to develop osteomyelitis than those with unstable fixation when inoculated with 10^7 CFU of bacteria [91% vs. 40% ($p \leq 0.01$)]. The authors noted a high early fixation failure rate in the stable group (21.4%) which was attributed to technical difficulties with the procedure. A rabbit tibia osteomyelitis model was also created by Hamel and colleagues using a bifocal diaphyseal osteotomy and therefore devascularizing a segment of tibia [67]. Relatively unstable fixation was used in the form of either an intramedullary pin or external fixator but the study was only designed to evaluate acute response to antibiotics and animals were sacrificed prior to fracture healing [67].

Curtis et al. [68] compared different treatment options in heavily contaminated open tibial fractures in goats. A saw osteotomy was created and then fixed with an external fixator, a reamed nail, or an unreamed nail. Following fixation, *S. aureus* was inoculated at the osteotomy site on an absorbable gelatin sponge. Clinical, radiographic, and microbiological evaluation demonstrated significantly fewer and less severe infections in fractures fixed with external fixation than in those fixed with either IM nail. Hill et al. created an infected osteotomy in sheep tibia fixed with an IM nail [69]. This model differed from other osteotomy models in that the inoculum was introduced 6 h prior to fixation in order to more closely mirror a typical clinical scenario. The authors reasoned that the reaming process often used in IM nailing technique disseminated pathogens within the IM canal, a crucial step in the infection process which is omitted in models where inoculation follows fixation. Two groups of six sheep were used: a control group where osteotomy, inoculation and contamination occurred alone and a treatment group where debridement, lavage, and systemic antibiotics were instituted. The authors introduced a high dose of *S. aureus* (3×10^8 CFU) on a bovine collagen sponge, and demonstrated that despite local debridement and lavage prior to nailing, contamination immediately following reaming had spread both proximally and distally along the tibia. By completion of the study all animals in both groups had developed osteomyelitis and the authors concluded that primary intramedullary nailing in severely contaminated open long bone fractures was unsafe regardless of debridement.

Lindsey et al. [70] created a rat femur fracture model using a blunt force trauma mechanism modified from Bonnarens and Einhorn. In this model a closed fracture was inflicted prior to any fixation. This was reported to produce a standard fracture pattern. Surgery followed wherein the fracture was inoculated and left for 1 h before an intramedullary Kirschner wire was implanted. An inoculum of 10^2 CFU produced an infection rate of 90–100% in the 21 day follow-up period and this dose of inoculum substantially challenged the immune system—as evidenced by decreased serum levels of IL-12—without causing overwhelming sepsis.

Rat femur models with infected critical size cortical bone defects have also been created allowing for the evaluation of osteoinductive agents and or osteoconductive scaffold materials in the context of osteomyelitis. Chen and colleagues have used one such model to evaluate a range of interventions in the context of osteomyelitis [71–73]. This model involves surgically removing a 6 mm segment of the femoral diaphysis and bridging the defect with a 4 mm thick polyacetyl plate, and threaded Kirschner wires. The defect is fixed and inoculated with between 103 and 10^6 CFU of *S. aureus* and the authors have used a range of histologic and microbiologic outcomes to demonstrate the effect on fracture healing of agents such as recombinant human osteogenic protein-1 and bone morphogenic protein 2 (rhOP-1/rhBMP-2). Southwood and colleagues created a similar model to evaluate the effects of BMP-2 delivery by adenovirus gene transfer [24]. Stewart et al. recently designed a rat femur model attempting to simulate the key features of open long bone fracture [74]. A blunt impact guillotine was used to create a comminuted fracture, which was subsequently debrided, creating a 5 mm cortical bone defect. A cylindrical scaffold was implanted and stabilized with an IM Kirschner wire. Achieving stable fixation is perhaps unsurprisingly a major challenge in small animal bone defect models. Two of the aforementioned models employed large and perhaps unconventional internal fixation constructs [24, 71]. Chen et al. reported a high incidence of fixation failure and acknowledge that this may confound results regarding infection [73]. In addition to the descriptive models above, some authors have sought to evaluate the effect of various treatment modalities on osteomyelitis associated with fractures. Those papers are dealt in Sect. 12.4.1.

As can be seen, fracture infection models juggle a number of mechanical, temporal and microbiologic variables and are necessarily complex. The small number of existing studies is unsurprisingly heterogeneous and comparison is difficult. The mechanism of trauma must inevitably be determined with a specific research question in mind, while questions such as the optimal timing of inoculation remain unresolved. Recent development of animal specific implants with modifiable stability options should help to standardize fixation in future studies [75]. Standardized animal models for fracture infection will be integral to advancing prophylaxis and treatment technology in fracture associated osteomyelitis.

12.4 Understanding to Be Gained in Animal Models

While animal models of osteomyelitis can improve our understanding of the mechanisms of infection and the role that various factors such as implants play, their ultimate value is in the development of new therapies to prevent or treat infection in humans. There are two important and distinct aspects to this: prophylaxis to prevent infection; and the treatment of established osteomyelitis. Given the relatively low incidence of infection in closed fracture fixation, prospective, randomized clinical trials require prohibitively large numbers to reach powerful conclusions, making animal models invaluable in both scenarios.

12.4.1 Improve Treatments

12.4.1.1 Prophylaxis Models

Clinical prophylaxis and management of trauma-related osteomyelitis involves a multimodal medical and surgical approach. The details vary, but always build on the foundation principles of prompt debridement, delivery of an appropriate antibiotic regimen, and where necessary, fracture stabilization and soft tissue cover. Parenteral broad spectrum antibiotic prophylaxis in open fractures and closed fracture fixation is universally accepted and mandated widely by clinical guidelines [76]. Animal models have been used predominantly to confirm their efficacy, tailor regimens, and characterize the pharmacologic parameters involved [77]. Antibiotic loaded bone cement (ALBC), when combined with systemic antibiotic administration has also been confirmed by large scale registry data as an effective prophylactic tool for arthroplasty [78]. Like systemic antibiotic prophylaxis, its clinical use in arthroplasty grew sporadically during the 1970s and was only later characterized and analyzed in controlled animal models [79]. In trauma-related osteomyelitis, where biofilm forming bacteria are so often implicated, there is a potential benefit of local antibiotic delivery. ALBC beads, most often loaded with gentamicin have also been used since the 1970s but their role is less well accepted [80]. They play no structural role in management, and therefore involve implantation of yet more foreign material in a potentially contaminated wound. As in arthroplasty, clinical use predates published animal trials, although *in vivo* models have been created to better understand pharmacokinetics and prophylactic efficacy [80]. The limited utility of ALBC beads and a trend toward cementless arthroplasty have driven the search for alternative methods of local antibiotic delivery. Animal models have been and continue to be essential in preclinical evaluation of this technology.

Technology explored using *in vivo* models for prophylactic local delivery of antiseptic or antibiotic includes collagen sheets [81], calcium phosphate pellets [82], polysaccharide (chitosan) beads [83], cross-linked high amylose starch implants [84], alginate beads, biodegradable polymer beads and coatings [85–87], cytokine nanocoating [88], and covalently bonded antibiotic [89]. In vivo models may be designed to analyze drug release profiles, biocompatibility, effect on fracture healing, and infection susceptibility and resistance. A prominent example is the rat model designed by Lucke et al. to evaluate a gentamicin impregnated poly-D,L-lactic acid (PDLLA) coated nail that has since been approved for clinical use [90]. Kirschner wires were coated with a PDLLA polymer containing gentamicin and implanted in the tibial medullary canal along with 1×10^{3} CFU of *S. aureus*. Compared with control groups (wires without antibiotics, and PDLLA-only coated wires) there was a significant reduction in clinical symptoms of osteomyelitis in the treatment group. In follow-up experiments using the same rat model the antibiotic burst release profile and osseous drug concentrations at progressive time points were described. Supporting *in vitro* trials were conducted to explore the influence of different polymer:solvent:antibiotic ratios on release profiles [91]. This study was conducted in parallel with a prospective clinical trial of the nail in open tibial fractures [92, 93] This system is only one of many currently being developed but this series of studies nonetheless demonstrates the role of the animal model in the "bench top to bedside" journey.

The model of Darouiche et al. is also noteworthy as one of the first *in vivo* models evaluating implant coatings for prophylaxis and for the fact that it incorporates a fracture into the osteomyelitis model [94]. A saw osteotomy was made in rabbit tibiae which were subsequently fixed with a chlorhexidine and chloroxylenol coated intramedullary Kirschner wire. Detailed evaluation of fracture healing was not described but the authors noted that histopathology results were similar between experimental and control groups. Fracture-infection prophylaxis models are otherwise lacking in the literature, although it is worth noting that the rat model investigating the PDLLA coating system described above has been adapted, with addition of a closed fracture [95, 96]. To date this model has only been used to evaluate the effect of local delivery of agents such as simvastatin or zoledronic acid on healing rather than an infection trial [95, 96].

More recently, alternative antimicrobial loaded coatings or implants with antimicrobial agents covalently bound to their surface have been engineered [89, 97]. Antoci et al. created titanium rods with vancomycin bound directly to the implant and showed in a rat femur *S. aureus* osteomyelitis model that these rods were more resistant to infection than those without vancomycin [89]. In vitro experiments demonstrate that the surface bound vancomycin is stable and also effectively reduces *S. epidermidis* colonization. The vancomycin coating did not foster antibiotic resistance in an 8 week *in vitro* trial but further investigation is needed to rule out this possibility [98]. Furanones, which appear to inhibit biofilm formation by interrupting bacterial signaling rather than a direct anti-microbial effect have also been bound to implants but to date *in vivo* osteomyelitis trials have not been performed [99, 100].

12.4.1.2 Treatment Models

Managing established, fixation associated osteomyelitis is a distinctly more challenging undertaking than preventing infection. Biofilm formation, poor osseous perfusion of antibiotics and abscess formation necessitate a multimodal medical and surgical approach. In vivo models for treatment are likewise more complex. At least two surgical procedures are required—one to inoculate, and after an observation period to allow the infection to develop, a second surgical procedure is performed to treat the infection. The length of the intervening period has the potential to greatly influence results as does the nature of the therapeutic intervention and these two factors vary widely in the existing literature. Crucially, some authors debride necrotic or infected bone in both treatment and control groups [101, 102] while others do so in only one [103] or neither [104]. It is clear that interventions which are effective in prophylaxis will not necessarily be effective in treatment. Fitzgerald et al. demonstrated this with Gentamicin loaded PMMA (Palacos™) [104]. Osteomyelitis was prevented in canine tibiae following *S. aureus* inoculation in 90% of cases but the same impregnated implant was ineffective in treating established *S. aureus* osteomyelitis in the same model, though the extent, if any, of debridement was not clearly described. A treatment role for ALBC as part of a multimodal approach has however been supported, if not confirmed, in subsequent models [80]. Evans et al. showed in a rabbit radius model that ALBC beads when combined with debridement were significantly more effective than debridement alone in treating osteomyelitis established in a devascularized bone segment [80]. Only when local and parenteral antibiotic therapy was combined, however, was infection eradicated in all animals. Garvin et al. modified the Fitzgerald canine model to compare the efficacy of gentamicin loaded PMMA implant and a degradable gentamicin loaded 50:50 polylactide–polyglycolide copolymer implant [105]. Both implants were inserted after surgical debridement and compared to a control group with parenterally administered gentamicin. After 4 weeks, infection was eradicated in 89% of animals with ALBC implants and 100% of those with the degradable implant. Both performed significantly better than parenteral antibiotics but the study was not powered to detect a difference between them. Giavaresi and colleagues, using a rabbit model, also combined debridement with ALBC—this time on a coated stainless steel implant—with promising microbiological outcomes in comparison to control groups [102].

The therapeutic value of degradable local delivery options has also been evaluated in *in vivo* models with mixed results. Antibiotic loaded calcium hydroxyapatite implants were used to treat established *S. aureus* osteomyelitis in rabbit tibiae in two separate trials [101, 103]. Even when combined with debridement, neither study demonstrated reliable eradication of infection, although it must be noted that the hydroxyapatite implant performed as well as ALBC when the two were compared [103]. More recently, Orhan et al. engineered ciprofloxacin-eluting pectin and chitosan microspheres and used a rat tibia model to evaluate their effectiveness in treating *S. aureus* osteomyelitis [106]. Chitosan sphere treated bone yielded significantly less bacteria than control groups but again, infection was only rarely eradicated. Low intensity electrical current, delivered directly to the infected implant has also been explored with promising results in a rabbit model [107].

Other authors have sought to refine treatment models by selecting specific strains of bacteria known to form biofilms. Isiklar et al. inoculated a biofilm forming strain of *S. epidermidis* into rabbit femora along with a stainless steel screw to evaluate the efficacy of different antibiotic regimes [108]. They found that vancomycin alone was ineffective in treating the infection, but when combined with rifampicin it appeared to eradicate infection in 90% of subjects. Monzon et al. also demonstrated the importance of IV antibiotics against biofilm infections in a novel model [109]. Instead of a inoculating the bacteria into the bone using a suspension, they initiated the infection using a biofilm. Stainless steel implants were preinoculated with a highly adherent *S. aureus* 12 h prior to implantation. The implants were then introduced into the tibia of rats and left for 42 days to develop a chronic infection. Using this approach, they showed that the infection was able to spread to the bone and all rats in the biofilm group developed osteomyelitis. A range of antibiotics were subsequently evaluated, with results suggesting that the presence of the biofilm significantly reduced the effectiveness of the drugs, notably including Vancomycin. None of five antibiotics tested were able to reliably sterilize the implant, while Cefuroxime and Vancomycin were, however, able to sterilize the surrounding bone.

It is difficult to compare any of the above models to each other for the methodological reasons detailed above. Animal models have the potential to improve treatments but the complexity of providing clinically relevant bone and soft tissue trauma, the scale of the model and the technical challenges in developing sustained release systems has limited the progression of new therapies. These significant challenges are yet to be overcome

12.5 Bacteria, Biofilm and Pathogenesis

Infections associated with fracture fixation devices are caused by a relatively limited number of bacterial species in the majority of cases, most of whom are either commensal, transient colonizers of the human skin or opportunistic pathogens. The most regularly identified bacterial species include *S. aureus*, coagulase negative staphylococci (CNS) such as *S. epidermidis*, Gram negative pathogens such as *P. aeruginosa* and *E. coli*, plus various Enterococci, Streptococci, and anaerobes such as *Propionibacterium acnes* [48]. Many other species have been isolated from comparatively fewer patients, with some of the more notable species including non-tuberculosis mycobacteria [110] and *Acinetobacter baumannii* [111] in military personnel. At 27%, polymicrobial i.e. infections involving two or more infecting organisms are second only to *S. aureus* in prevalence in infections associated with fracture fixation devices [48, 112, 113] and with the availability of ever more sophisticated molecular detection techniques capable of detecting pathogens that are difficult to grow in the laboratory, the range of microorganisms identified and the frequency of polymicrobial infections may be expected to increase with increasing uptake of these sophisticated detection systems [114]. The availability of animal models of infection involving these microorganisms mirrors clinical prevalence, with *S. aureus* and *S. epidermidis* accounting for the majority of orthopedic implant-related infection animal studies.

S. aureus is the most commonly cultured microorganism from patients with infections associated with implanted fracture fixation devices, with rates reported at 30% [48] and along with the CNS (22%) they are the causative organisms in approximately 50% of cases [48]. It is therefore of no surprise that the majority of animal models investigating infection associated with fracture fixation devices have modeled *S. aureus* infections [26, 31, 88, 115–120] and, to a lesser extent, *S. epidermidis* infections [107, 121–126]. The animal models used to investigate the potential for various microorganisms to cause an infection suffer from many of the limitations already described in this chapter, such as lack of clinical relevance; however, in addition these studies are further complicated by the use of numerous different strains of the same bacterial species. For example, the specific bacterial strains used in these studies has varied from the use of clinical strains isolated from patients with a similar infection [31, 118], to well characterized laboratory isolates [118] and more recently genetically modified strains [88]. The relative merits of each strain type are largely dependent upon the particular needs of the study in question[127], though the use of a widely available clinical isolate such as UAMS-1 for example [26] may be a particularly suitable for the study of *S. aureus* osteomyelitis isolate given the large amount of characterization data available [26, 128–130].

The CNS are also often isolated from chronic orthopedic infections [131] although animal models of CNS osteomyelitis seem to be less prevalent within the published literature [107, 121–126, 132, 133]. Those studies investigating CNS most often focus upon *S. epidermidis* with only few reports of animal models investigating other CNS species. S. lugdunensis and *S. shlieferi* are accepted bone pathogens and animal models of infection caused by these microorganisms have been described [121, 124, 125] in the literature though not in osteomyelitis models. Polymicrobial infections are also somewhat underrepresented in the literature of implant-related osteomyelitis after trauma, despite their prevalence in infection in fracture fixation. Polymicrobial infection models have been described for *S. aureus* and *P. aeruginosa* coinfections [39, 134] and also *S. epidermidis* and bacteroides coinfections [125, 126, 132]. The relative paucity of polymicrobial infection models may be related to the difficulty in setting up a polymicrobial model with complications arising in the administration of the inoculation, the determination of the relative contribution to pathogenesis and selective culture of the microorganisms post-euthanasia. Models of infection due to anaerobic species of bacteria are also somewhat scare in the literature, mainly limited to studies from the 1980s and 1990s of implant-related osteomyelitis due to *Bacteroides fragilis* [135] and *P. acnes* [136, 137], though again, the models used were not specifically mimicking post-traumatic infection of a fracture.

Most animal models of orthopedic implant-related infection go beyond the mere ability of a particular microorganism to cause infection and aim to provide supportive or confirmatory evidence of the role of specific factors possessed by the microorganism in the pathogenesis of the infection. One such factor is the ability to form a biofilm. Infections associated with implanted orthopedic devices are widely acknowledged to involve biofilm formation [138, 139], whereby the biofilm offers the constituent cells protection from host defense mechanisms and antibiotic therapy [140]. Furthermore, biofilms are commonly accepted to be the primary reason for the difficulty in successfully treating orthopedic implant-related infections and thus a greater understanding of the role of biofilm formation *in vivo* has been a focus of research and one which is also reliant on animal models for a level of evidence that is not possible to achieve *in vitro*. Biofilm formation involves complex genetic regulatory mechanisms that vary significantly from species to species and may also be significantly variable within different isolates of the same species [129, 141]. A comprehensive mechanism for biofilm formation amongst strains of even just the same species is as yet unavailable and for well-studied *S. aureus* strains it appears there are multiple mechanisms for biofilm formation, further complicating the ability to fully characterize these processes [128, 142, 143]. In one study, an artificially constructed biofilm deficient mutant (negative for a fully functional *rbf* gene that is involved in biofilm formation) has been shown to display reduced virulence in a murine foreign body infection model [118]. However, the multitude of mechanisms involved in the formation of a biofilm make generalizations difficult. This is illustrated by another *in vivo* study which showed that the virulence of biofilm forming *S. aureus* isolates in a guinea pig tissue cage model was equivalent to isogenic mutants lacking a functional *ica* locus, which codes for a polysaccharide involved in biofilm formation [123]. Fluckiger et al. also showed, again unfortunately in a nonorthopedic model, that ica-deficient *S. aureus* and *S. epidermidis* mutants displayed similar growth and adherence in guinea pigs in comparison with the isogenic parental strains but that wild type strains did significantly compromise mutant growth when inoculated together in competition *in vivo* [122]. A full description of biofilm formation as a staphylococcal virulence factor is beyond the scope of this chapter but is further described elsewhere in the present book.

Virulence factors other than biofilm formation are also known to play a role in implant-related osteomyelitis. In vitro studies have shown that virulence factors such as antibiotic resistance determinants, toxins and adhesins are widely detected in isolates causing implant-related osteomyelitis [144]. A large number of animal studies investigating the pathogenesis of osteomyelitis have studied the role of specific surface adhesins on the pathogenesis of infection. Many of these studies have focused upon the microbial surface components recognizing adhesive matrix molecules (MSCRAMMs) of *S. aureus* [145, 146], many of which have been identified though the specific ligands for each are still being continuously updated [147]. Some adhesins have been shown in animal models to be important virulence factors in staphylococcal osteomyelitis. For example, wild type *S. aureus* isolates expressing a collagen adhesin (Cna) were found to be more capable of causing septic arthritis in a hematogenous model in the mouse, without prior local bone damage, as evidenced by increased destruction of joint cartilage and underlying bone in comparison with Cna negative mutant strains [119]. In a similar study with a similar mouse model, Elasri et al. showed that Cna was not actually required to establish a joint infection, but that it was required to cause infection of the bone [117]. In yet another study using a similar model no role was found for Cna, whereas another MSCRAMM involved in fibronectin binding was found to play a role in the development of infection [148]. In another study to clarify the role of MSCRAMMS on the pathogenesis of implant-related osteomyelitis, Darouiche et al. showed in an implant-related rabbit osteomyelitis model that the ability of a mutant strain of *S. aureus* lacking MSCRAMMS for fibrinogen, fibronectin and collagen was equally capable of causing an infection in the bone compared with the wild-type strain possessing all the above mentioned MSCRAMMS [149]. There is clearly a role for animal studies in clarifying the seemingly conflicting importance and role of MSCRAMMS in the pathogenesis of implant-related osteomyelitis, though progress in this field will always be complicated due to the possession of multiple MSCRAMMS and a redundancy in ligand binding. In the absence (or loss through mutation) of one MSCRAMM, alternative MSCRAMMS may assume a greater role in the development of the infection and thus confound the results and mask the influence of the adhesin. The availability of standardized, affordable and clinically relevant mouse models of bone infection would assist in the provision of this data. This is because relatively large numbers of animals may be required to clarify the role of adhesins on the pathogenesis of infection and overcome the apparent conflicting data on the importance of MSCRAMMS in the development of implant-related osteomyelitis.

Staphylococcal toxins also play a role in pathogenesis of implant-related osteomyelitis. One such example is Panton-Valentine leukocidin (PVL) [150], which is emerging in association with the spread of PVL positive community-acquired methicillin resistant *S. aureus* (MRSA) [151]. Due to this increasing prevalence of such toxins, animal models are utilized to provide greater understanding in their role in the pathogenesis of infection. In a rabbit model of non-implant-related osteomyelitis, whereby a sclerosing agent was administered according to the model of Norden et al. [22] Crèmieux et al. showed that PVL does play a role in the development of bone infection in rabbits by increasing bacterial spread within the bone and persistence in comparison with isogenic mutants [115]. A similar role for other important toxins and virulence factors, such as alpha and beta toxin or protein A, are described in the literature for other infections [152], though not specifically in the context of implant-related osteomyelitis. Another potential attribute possessed by bacteria that may contribute to the establishment of osteomyelitis may be the natural mutation rate of the bacterium. For instance increased mutation as displayed by hypermutators may lead to more rapid induction of antibiotic resistance within the strain or provide the strain along with other possible benefits. However, Daurel et al. showed in a chronic osteomyelitis model in the rat that the mutation rate in *S. aureus* (inactivated *mutL* gene) did not play a role in the pathogenesis of infection [116].

Animal models clearly have begun to play a vital role in understanding the molecular mechanisms behind the pathogenesis of implant-related osteomyelitis. This is true not only from the perspective of bacterial virulence, but also in terms of the host immune response to the infection. In the case of infection after trauma, the concurrent presence of bacteria and host response to the bacterium and the destructive tissue damage may only be somewhat reliably replicated via the use of animal models. As already described however, improved models of infection are required to follow the pathogenesis for infection within hosts displaying the clinically salient features of the infection, this is not only true for interventional studies but also true for bacterial pathogenicity studies.

12.6 Future Directions

It is imperative that researchers strive to create animal models which will generate meaningful outcomes while doing the least harm to the lowest possible number of animals [7]. Animal models in the existing literature are highly heterogeneous. Reliable, repeatable "standardized" models would not only enable interstudy comparison but ultimately reduce the number of animals required. A number of recent technological advances have the potential to refine existing models in the field.

In vivo longitudinal quantitative monitoring of infection via the use bioluminescent bacteria offers one such opportunity. Previously, thorough quantitative microbiological or PCR evaluation of osteomyelitis required the subject animal to be euthanized. Where data at different time points is required, this necessitates cross sectional evaluation of groups and increases the animal numbers required. Li et al. characterized the longitudinal progression of tibial osteomyelitis in a murine model, using bioluminescent *S. aureus* and the Xenogen IVIS camera system (Xenogen Co.

Alameda, CA) [88]. The results were compared to radiography along with μCT, histology, and PCR after sacrifice, allowing the authors to chart temporal variation in bacterial metabolic activity as well as number. A second development, already touched on, involves the development of custom designed implant systems for small laboratory animals which open up a number of opportunities. Bioluminescent imaging is only available in small animal models such as mice due to the lack of penetration of the signal through the significantly deeper soft and hard tissues in large animals. This may be seen as a disadvantage, but as mentioned in Sect.12.2 the availability of custom designed, biomechanically correct and stable fixation options and genetically defined or engineered strains of mice and rats has vastly increased the potential utility of these animals in orthopedic trauma research.

Standardization and refinement of fixation methods in larger animals would also improve existing models. Rabbits, for example have been frequently used in the field (Table 12.1) but the nature of implant or fixation and the anatomical location used vary. Where noninfective complication rates, such as fracture post-fixation, are reported they are often higher than is optimal. Effective and repeatable fixation systems which mirror clinical practice and materials and reliably heal fractures in larger subject animals without complication would also improve current practice.

Consensus and validation of inoculation methods could benefit future animal models. As described above the dose, timing, location and vehicle of pathogen inoculation can be manipulated. The safety of osteomyelitis models relies on the ability to deliver the pathogen and hence create infection at the desired locus without creating sepsis or metastatic infection. Some researchers achieve this by cultivating biofilm on an implant preimplantation, while others implant an object such as a collagen sponge and still more simply inject a liquid inoculum to the area. The success in creating an infection in an individual animal is obviously impacted by immune and host factors individual to each animal and the confounding nature of this variation is typically overcome by increasing sample size. The corollary of this of course is that a reliable and widely applicable method of delivering the desired pathogen to the area of interest would enable a reduction in animal numbers. Finally, as detailed above, there are very few models for some important clinical aspects of orthopedic infection: late developing prosthetic joint infection caused by bacteremia and hematogenous seeding is one such area while fracture associated osteomyelitis is another. Models are described above in both of these areas but could be improved with factors such as stable clinically relevant fixation, repeatable creation of trauma, and, importantly, explicit description of safety and ethical issues such as animal exclusion due to unexpected complications.

12.7 Conclusion

Animal models are indispensible in trauma research. Antibiotic resistance within the most common pathogens along with increased demand for orthopedic implant surgery driven by an ageing demographic will ensure that musculoskeletal infection continues to pose a major challenge to clinicians. At the same time it is clear that current infection rates and treatment algorithms could be improved with new technology. Safe and expedient implementation of this technology relies on well designed and clinically relevant *in vivo* simulation using animal models. As can be seen, the design variables involved in creating a model are manifold and inevitably demand compromise. The particular flavor of compromise chosen must be catered to specific research questions but this perhaps does not justify the profound heterogeneity observed in existing models. An inherent limitation of animal models is the difficulty in extrapolating results not only to humans, but to other animals or to the same model in another laboratory. A standardized and repeatable approach to, for example, simple fracture creation, or soft tissue crush injury, or stable long bone fixation in rodents, would enable greater comparison and ultimately increase the validity of animal trauma research. As mentioned above, recent advances in some of these areas have been made and the burgeoning availability of tools for longitudinal *in vivo* monitoring of infection is another exciting development in the field. These refinements and others should enable the creation of robust, controlled, and consistent models which allow strong scientific conclusions with a minimum of harm to animals.

Glossary

Implant-related Osteomyelitis Osteomyelitis occurring associated with an implant, most commonly prosthetic joint replacements or fracture fixation devices.

Osteomyelitis Osteomyelitis is an acute or chronic infection of the bone or bone marrow.

References

1. Edwards C, Counsell A, Boulton C, Moran CG. Early infection after hip fracture surgery: risk factors, costs and outcome. J Bone Joint Surg Br. 2008;90(6):770–7.
2. Schlegel U, Perren SM. Surgical aspects of infection involving osteosynthesis implants: implant design and resistance to local infection. Injury. 2006;37 Suppl 2:S67–73.
3. Elek SD, Conen PE. The virulence of *Staphylococcus pyogenes* for man; a study of the problems of wound infection. Br J Exp Pathol. 1957;38(6):573–86.
4. Krogh A. The progress of physiology. Science. 1929;70(1809):200–4.
5. Egermann M, Goldhahn J, Schneider E. Animal models for fracture treatment in osteoporosis. Osteoporos Int. 2005;16 Suppl 2:S129–38.
6. Aerssens J, Boonen S, Lowet G, Dequeker J. Interspecies differences in bone composition, density, and quality: potential implications for in vivo bone research. Endocrinology. 1998; 139(2):663–70.
7. Auer JA, Goodship A, Arnoczky S, Pearce S, Price J, Claes L, et al. Refining animal models in fracture research: seeking consensus in optimising both animal welfare and scientific validity for appropriate biomedical use. BMC Musculoskelet Disord. 2007;8:72.

8. Patel M, Rojavin Y, Jamali AA, Wasielewski SJ, Salgado CJ. Animal models for the study of osteomyelitis. Semin Plast Surg. 2009;23(2):148–54.
9. Wang X, Mabrey JD, Agrawal CM. An interspecies comparison of bone fracture properties. Biomed Mater Eng. 1998;8(1):1–9.
10. Salgado CJ, Jamali AA, Mardini S, Buchanan K, Veit B. A model for chronic osteomyelitis using *Staphylococcus aureus* in goats. Clin Orthop Relat Res. 2005;436:246–50.
11. Arens S, Eijer H, Schlegel U, Printzen G, Perren SM, Hansis M. Influence of the design for fixation implants on local infection: experimental study of dynamic compression plates versus point contact fixators in rabbits. J Orthop Trauma. 1999;13(7):470–6.
12. Claes L, Eckert-Hubner K, Augat P. The effect of mechanical stability on local vascularization and tissue differentiation in callus healing. J Orthop Res. 2002;20:1099–105.
13. An YH, Friedman RJ. Animal models of orthopedic implant infection. J Invest Surg. 1998;11(2):139–46.
14. An YH, Kang QK, Arciola CR. Animal models of osteomyelitis. Int J Artif Organs. 2006; 29(4):407–20.
15. Norden CW. Lessons learned from animal models of osteomyelitis. Rev Infect Dis. 1988; 10(1):103–10.
16. Garcia P, Holstein J, Histing T, Burkhardt M, Culemann U, Pizanis A, et al. A new technique for internal fixation of femoral fractures in mice: impact of stability on fracture healing. J Biomech. 2008;41:1689–96.
17. Matthys R, Perren S. Internal fixator for use in the mouse. Injury. 2009;40:S103–9.
18. Histing T, Garcia P, Matthys R, Leidinger M, Holstein J, Kristen A, et al. An internal locking plate to study intramembranous bone healing in a mouse femur fracture model. J Orthop Res. 2010;28(3):397–402.
19. Legrand N, Ploss A, Balling R, Becker PD, Borsotti C, Brezillon N, et al. Humanized mice for modeling human infectious disease: challenges, progress, and outlook. Cell Host Microbe. 2009;6(1):5–9.
20. Shultz LD, Ishikawa F, Greiner DL. Humanized mice in translational biomedical research. Nat Rev Immunol. 2007;7(2):118–30.
21. Scheman L, Janota M, Lewin P. The production of experimental osteomyelitis. JAMA. 1941;117(18):1525–9.
22. Norden CW. Experimental osteomyelitis. I. A description of the model. J Infect Dis. 1970;122(5):410–8.
23. Rissing JP, Buxton TB, Weinstein RS, Shockley RK. Model of experimental chronic osteomyelitis in rats. Infect Immun. 1985;47(3):581–6.
24. Southwood LL, Frisbie DD, Kawcak CE, Ghivizzani SC, Evans CH, McIlwraith CW. Evaluation of Ad-BMP-2 for enhancing fracture healing in an infected defect fracture rabbit model. J Orthop Res. 2004;22(1):66–72.
25. Volk A, Cremieux AC, Belmatoug N, Vallois JM, Pocidalo JJ, Carbon C. Evaluation of a rabbit model for osteomyelitis by high field, high resolution imaging using the chemical-shift-specific-slice-selection technique. Magn Reson Imaging. 1994;12(7):1039–46.
26. Smeltzer MS, Thomas JR, Hickmon SG, Skinner RA, Nelson CL, Griffith D, et al. Characterization of a rabbit model of staphylococcal osteomyelitis. J Orthop Res. 1997; 15(3):414–21.
27. Andriole VT, Nagel DA, Southwick WO. A paradigm for human chronic osteomyelitis. J Bone Joint Surg Am. 1973;55(7):1511–5.
28. Petty W, Spanier S, Shuster JJ, Silverthorne C. The influence of skeletal implants on incidence of infection. Experiments in a canine model. J Bone Joint Surg Am. 1985;67(8):1236–44.
29. Cordero J, Munuera L, Folgueira MD. Influence of metal implants on infection. An experimental study in rabbits. J Bone Joint Surg Br. 1994;76(5):717–20.
30. Arens S, Schlegel U, Printzen G, Ziegler WJ, Perren SM, Hansis M. Influence of materials for fixation implants on local infection. An experimental study of steel versus titanium DCP in rabbits. J Bone Joint Surg Br. 1996;78(4):647–51.

31. Moriarty TF, Debefve L, Boure L, Campoccia D, Schlegel U, Richards RG. Influence of material and microtopography on the development of local infection in vivo: experimental investigation in rabbits. Int J Artif Organs. 2009;32(9):663–70.
32. Moriarty TF, Schlegel U, Perren S, Richards RG. Infection in fracture fixation: can we influence infection rates through implant design? J Mater Sci Mater Med. 2010;21(3):1031–5.
33. Melcher GA, Claudi B, Schlegel U, Perren SM, Printzen G, Munzinger J. Influence of type of medullary nail on the development of local infection. An experimental study of solid and slotted nails in rabbits. J Bone Joint Surg Br. 1994;76(6):955–9.
34. Melcher GA, Metzdorf A, Schlegel U, Ziegler WJ, Perren SM, Printzen G. Influence of reaming versus nonreaming in intramedullary nailing on local infection rate: experimental investigation in rabbits. J Trauma. 1995;39(6):1123–8.
35. Melcher GA, Hauke C, Metzdorf A, Perren SM, Printzen G, Schlegel U, et al. Infection after intramedullary nailing: an experimental investigation on rabbits. Injury. 1996;27 Suppl 3:SC23–6.
36. Horn J, Schlegel U, Krettek C, Ito K. Infection resistance of unreamed solid, hollow slotted and cannulated intramedullary nails: an in-vivo experimental comparison. J Orthop Res. 2005;23(4):810–5.
37. Finkemeier CG, Schmidt AH, Kyle RF, Templeman DC, Varecka TF. A prospective, randomized study of intramedullary nails inserted with and without reaming for the treatment of open and closed fractures of the tibial shaft. J Orthop Trauma. 2000;14(3):187–93.
38. Larsen LB, Madsen JE, Hoiness PR, Ovre S. Should insertion of intramedullary nails for tibial fractures be with or without reaming? A prospective, randomized study with 3.8 years' follow-up. J Orthop Trauma. 2004;18(3):144–9.
39. Hendricks KJ, Burd TA, Anglen JO, Simpson AW, Christensen GD, Gainor BJ. Synergy between *Staphylococcus aureus* and *Pseudomonas aeruginosa* in a rat model of complex orthopaedic wounds. J Bone Joint Surg Am. 2001;83-A(6):855–61.
40. Deysine M, Rosario E, Isenberg HD. Acute hematogenous osteomyelitis: an experimental model. Surgery. 1976;79(1):97–9.
41. Emslie KR, Nade S. Pathogenesis and treatment of acute hematogenous osteomyelitis: evaluation of current views with reference to an animal model. Rev Infect Dis. 1986;8(6):841–9.
42. Poultsides LA, Papatheodorou LK, Karachalios TS, Khaldi L, Maniatis A, Petinaki E, et al. Novel model for studying hematogenous infection in an experimental setting of implant-related infection by a community-acquired methicillin-resistant *S. aureus* strain. J Orthop Res. 2008;26(10):1355–62.
43. Hienz SA, Sakamoto H, Flock JI, Morner AC, Reinholt FP, Heimdahl A, et al. Development and characterization of a new model of hematogenous osteomyelitis in the rat. J Infect Dis. 1995;171(5):1230–6.
44. Whalen JL, Fitzgerald Jr RH, Morrissy RT. A histological study of acute hematogenous osteomyelitis following physeal injuries in rabbits. J Bone Joint Surg Am. 1988;70(9):1383–92.
45. Chadha HS, Fitzgerald Jr RH, Wiater P, Sud S, Nasser S, Wooley PH. Experimental acute hematogenous osteomyelitis in mice. I. Histopathological and immunological findings. J Orthop Res. 1999;17(3):376–81.
46. Patzakis MJ, Wilkins J. Factors influencing infection rate in open fracture wounds. Clin Orthop Relat Res. 1989;243:36–40.
47. Zimmerli W, Trampuz A, Ochsner PE. Prosthetic-joint infections. N Engl J Med. 2004;351 (16):1645–54.
48. Trampuz A, Zimmerli W. Diagnosis and treatment of infections associated with fracture-fixation devices. Injury. 2006;37 Suppl 2:S59–66.
49. Bowen TR, Widmaier JC. Host classification predicts infection after open fracture. Clin Orthop Relat Res. 2005;433:205–11.
50. Dumont C, Kauer F, Bohr S, Schmidtmann U, Knopp W, Engelhardt T, et al. Long-term effects of saw osteotomy versus random fracturing on bone healing and remodeling in a sheep tibia model. J Orthop Res. 2009;27(5):680–6.
51. Ashhurst DE, Hogg J, Perren SM. A method for making reproducible experimental fractures of the rabbit tibia. Injury. 1982;14(3):236–42.

52. Jackson RW, Reed CA, Israel JA, Abou-Keer FK, Garside H. Production of a standard experimental fracture. Can J Surg. 1970;13(4):415–20.
53. Bonnarens F, Einhorn TA. Production of a standard closed fracture in laboratory animal bone. J Orthop Res. 1984;2(1):97–101.
54. Marturano JE, Cleveland BC, Byrne MA, O'Connell SL, Wixted JJ, Billiar KL. An improved murine femur fracture device for bone healing studies. J Biomech. 2008;41(6):1222–8.
55. Gustilo RB, Anderson JT. Prevention of infection in the treatment of one thousand and twenty-five open fractures of long bones: retrospective and prospective analyses. J Bone Joint Surg Am. 1976;58(4):453–8.
56. Landry P, Marino A, Sadasivan K, Albright J. Effect of soft-tissue trauma on the early periosteal response of bone to injury. J Trauma. 2000;48(3):479.
57. Li BC, Zhang JJ, Xu C, Zhang LC, Kang JY, Zhao H. Treatment of rabbit femoral defect by firearm with BMP-4 gene combined with TGF-beta1. J Trauma. 2009;66(2):450–6.
58. Utvag S, Grundnes O, Rindal D, Reikeras O. Influence of extensive muscle injury on fracture healing in rat tibia. J Orthop Trauma. 2003;17(6):430–5.
59. Claes L, Maurer-Klein N, Henke T, Gerngross H, Melnyk M, Augat P. Moderate soft tissue trauma delays new bone formation only in the early phase of fracture healing. J Orthop Res. 2006;24(6):1178–85.
60. Melnyk M, Henke T, Claes L, Augat P. Revascularisation during fracture healing with soft tissue injury. Arch Orthop Trauma Surg. 2008;128:1159–65.
61. Kälicke T, Schlegel U, Printzen G, Schneider E, Muhr G, Arens S. Influence of a standardized closed soft tissue trauma on resistance to local infection. An experimental study in rats. J Orthop Res. 2003;21(2):373–8.
62. Matsumoto T, Hardaway III RM, Dobek AS, Noyes HE. Antibiotic topical spray applied in a simulated combat wound. Arch Surg. 1967;95(2):288–94.
63. Passl R, Muller C, Zielinski CC, Eibl MM. A model of experimental post-traumatic osteomyelitis in guinea pigs. J Trauma. 1984;24(4):323–6.
64. Rittman WW, Perren S. Cortical bone healing after internal fixation and fracture. New York: Springer; 1974.
65. Worlock P, Slack R, Harvey L, Mawhinney R. An experimental model of post-traumatic osteomyelitis in rabbits. Br J Exp Pathol. 1988;69(2):235–44.
66. Worlock P, Slack R, Harvey L, Mawhinney R. The prevention of infection in open fractures: an experimental study of the effect of fracture stability. Injury. 1994;25(1):31–8.
67. Hamel A, Caillon J, Jacqueline C, Rogez JM, Potel G. Internal device decreases antibiotic's efficacy on experimental osteomyelitis. J Child Orthop. 2008;2(3):239–43.
68. Curtis MJ, Brown PR, Dick JD, Jinnah RH. Contaminated fractures of the tibia: a comparison of treatment modalities in an animal model. J Orthop Res. 1995;13(2):286–95.
69. Hill PF, Clasper JC, Parker SJ, Watkins PE. Early intramedullary nailing in an animal model of a heavily contaminated fracture of the tibia. J Orthop Res. 2002;20(4):648–53.
70. Lindsey BA, Clovis NB, Smith ES, Salihu S, Hubbard DF. An animal model for open femur fracture and osteomyelitis: part I. J Orthop Res. 2010;28(1):38–42.
71. Chen X, Tsukayama DT, Kidder LS, Bourgeault CA, Schmidt AH, Lew WD. Characterization of a chronic infection in an internally-stabilized segmental defect in the rat femur. J Orthop Res. 2005;23(4):816–23.
72. Chen X, Schmidt AH, Mahjouri S, Polly Jr DW, Lew WD. Union of a chronically infected internally stabilized segmental defect in the rat femur after debridement and application of rhBMP-2 and systemic antibiotic. J Orthop Trauma. 2007;21(10):693–700.
73. Chen X, Kidder LS, Lew WD. Osteogenic protein-1 induced bone formation in an infected segmental defect in the rat femur. J Orthop Res. 2002;20(1):142–50.
74. Stewart RL, Cox JT, Volgas D, Stannard J, Duffy L, Waites KB, et al. The use of a biodegradable, load-bearing scaffold as a carrier for antibiotics in an infected open fracture model. J Orthop Trauma. 2010;24(9):587–91.
75. Holstein J, Garcia P, Histing T, Kristen A, Scheuer C, Menger M, et al. Advances in the establishment of defined mouse models for the study of fracture healing and bone regeneration. J Orthop Trauma. 2009;23:S31.

76. Frommelt L. Principles of systemic antimicrobial therapy in foreign material associated infection in bone tissue, with special focus on periprosthetic infection. Injury. 2006;37 Suppl 2:S87–94.
77. Norden CW. Experimental osteomyelitis. IV. Therapeutic trials with rifampin alone and in combination with gentamicin, sisomicin, and cephalothin. J Infect Dis. 1975;132(5):493–9.
78. Espehaug B, Engesaeter LB, Vollset SE, Havelin LI, Langeland N. Antibiotic prophylaxis in total hip arthroplasty. Review of 10,905 primary cemented total hip replacements reported to the Norwegian arthroplasty register, 1987 to 1995. J Bone Joint Surg Br. 1997;79(4):590–5.
79. Rodeheaver GT, Rukstalis D, Bono M, Bellamy W. A new model of bone infection used to evaluate the efficacy of antibiotic-impregnated polymethylmethacrylate cement. Clin Orthop Relat Res. 1983;178:303–11.
80. Evans RP, Nelson CL. Gentamicin-impregnated polymethylmethacrylate beads compared with systemic antibiotic therapy in the treatment of chronic osteomyelitis. Clin Orthop Relat Res. 1993;295:37–42.
81. Riegels-Nielsen P, Espersen F, Holmich LR, Frimodt-Moller N. Collagen with gentamicin for prophylaxis of postoperative infection. *Staphylococcus aureus* osteomyelitis studied in rabbits. Acta Orthop Scand. 1995;66(1):69–72.
82. Beardmore AA, Brooks DE, Wenke JC, Thomas DB. Effectiveness of local antibiotic delivery with an osteoinductive and osteoconductive bone-graft substitute. J Bone Joint Surg Am. 2005;87(1):107–12.
83. Aimin C, Chunlin H, Juliang B, Tinyin Z, Zhichao D. Antibiotic loaded chitosan bar. An in vitro, in vivo study of a possible treatment for osteomyelitis. Clin Orthop Relat Res. 1999;366:239–47.
84. Huneault LM, Lussier B, Dubreuil P, Chouinard L, Desevaux C. Prevention and treatment of experimental osteomyelitis in dogs with ciprofloxacin-loaded crosslinked high amylose starch implants. J Orthop Res. 2004;22(6):1351–7.
85. Diefenbeck M, Mückley T, Hofmann GO. Prophylaxis and treatment of implant-related infections by local application of antibiotics. Injury. 2006;37 Suppl 2:S95–104.
86. Ueng SW, Yuan LJ, Lee N, Lin SS, Chan EC, Weng JH. In vivo study of biodegradable alginate antibiotic beads in rabbits. J Orthop Res. 2004;22(3):592–9.
87. Ambrose CG, Clyburn TA, Louden K, Joseph J, Wright J, Gulati P, et al. Effective treatment of osteomyelitis with biodegradable microspheres in a rabbit model. Clin Orthop Relat Res. 2004;421:293–9.
88. Li D, Gromov K, Soballe K, Puzas JE, O'Keefe RJ, Awad H, et al. Quantitative mouse model of implant-associated osteomyelitis and the kinetics of microbial growth, osteolysis, and humoral immunity. J Orthop Res. 2008;26(1):96–105.
89. Antoci Jr V, Adams CS, Hickok NJ, Shapiro IM, Parvizi J. Vancomycin bound to Ti rods reduces periprosthetic infection: preliminary study. Clin Orthop Relat Res. 2007;461:88–95.
90. Lucke M, Schmidmaier G, Sadoni S, Wildemann B, Schiller R, Haas NP, et al. Gentamicin coating of metallic implants reduces implant-related osteomyelitis in rats. Bone. 2003; 32(5):521–31.
91. Strobel C, Schmidmaier G, Wildemann B. Changing the release kinetics of gentamicin from poly(D, L-lactide) implant coatings using only one polymer. Int J Artif Organs. 2011; 34(3):304–16.
92. Schmidmaier G, Lucke M, Wildemann B, Haas N, Raschke M. Prophylaxis and treatment of implant-related infections by antibiotic-coated implants: a review. Injury. 2006;37(2):S105–12.
93. Fuchs T, Stange R, Schmidmaier G, Raschke MJ. The use of gentamicin-coated nails in the tibia: preliminary results of a prospective study. Arch Orthop Trauma Surg. 2011;131(10):1419–25.
94. Darouiche RO, Farmer J, Chaput C, Mansouri M, Saleh G, Landon GC. Anti-infective efficacy of antiseptic-coated intramedullary nails. J Bone Joint Surg Am. 1998;80(9):1336–40.
95. Greiner SH, Wildemann B, Back DA, Alidoust M, Schwabe P, Haas NP, et al. Local application of zoledronic acid incorporated in a poly(D, L-lactide)-coated implant accelerates fracture healing in rats. Acta Orthop. 2008;79(5):717–25.
96. Pauly S, Luttosch F, Morawski M, Haas NP, Schmidmaier G, Wildemann B. Simvastatin locally applied from a biodegradable coating of osteosynthetic implants improves fracture healing comparable to BMP-2 application. Bone. 2009;45(3):505–11.

97. Moskowitz JS, Blaisse MR, Samuel RE, Hsu HP, Harris MB, Martin SD, et al. The effectiveness of the controlled release of gentamicin from polyelectrolyte multilayers in the treatment of *Staphylococcus aureus* infection in a rabbit bone model. Biomaterials. 2010;31(23):6019–30.
98. Antoci Jr V, Adams CS, Parvizi J, Davidson HM, Composto RJ, Freeman TA, et al. The inhibition of *Staphylococcus epidermidis* biofilm formation by vancomycin-modified titanium alloy and implications for the treatment of periprosthetic infection. Biomaterials. 2008;29(35):4684–90.
99. Baveja JK, Willcox MD, Hume EB, Kumar N, Odell R, Poole-Warren LA. Furanones as potential anti-bacterial coatings on biomaterials. Biomaterials. 2004;25(20):5003–12.
100. Lonn-Stensrud J, Landin MA, Benneche T, Petersen FC, Scheie AA. Furanones, potential agents for preventing *Staphylococcus epidermidis* biofilm infections? J Antimicrob Chemother. 2009;63(2):309–16.
101. Cornell CN, Tyndall D, Waller S, Lane JM, Brause BD. Treatment of experimental osteomyelitis with antibiotic-impregnated bone graft substitute. J Orthop Res. 1993;11(5):619–26.
102. Giavaresi G, Borsari V, Fini M, Giardino R, Sambri V, Gaibani P, et al. Preliminary investigations on a new gentamicin and vancomycin-coated PMMA nail for the treatment of bone and intramedullary infections: an experimental study in the rabbit. J Orthop Res. 2008;26(6):785–92.
103. Shirtliff ME, Calhoun JH, Mader JT. Experimental osteomyelitis treatment with antibiotic-impregnated hydroxyapatite. Clin Orthop Relat Res. 2002;401:239–47.
104. Fitzgerald Jr RH. Experimental osteomyelitis: description of a canine model and the role of depot administration of antibiotics in the prevention and treatment of sepsis. J Bone Joint Surg Am. 1983;65(3):371–80.
105. Garvin KL, Miyano JA, Robinson D, Giger D, Novak J, Radio S. Polylactide/polyglycolide antibiotic implants in the treatment of osteomyelitis. A canine model. J Bone Joint Surg Am. 1994;76(10):1500–6.
106. Orhan Z, Cevher E, Mulazimoglu L, Gurcan D, Alper M, Araman A, et al. The preparation of ciprofloxacin hydrochloride-loaded chitosan and pectin microspheres: their evaluation in an animal osteomyelitis model. J Bone Joint Surg Br. 2006;88(2):270–5.
107. Del Pozo JL, Rouse MS, Euba G, Kang CI, Mandrekar JN, Steckelberg JM, et al. The electricidal effect is active in an experimental model of *Staphylococcus epidermidis* chronic foreign body osteomyelitis. Antimicrob Agents Chemother. 2009;53(10):4064–8.
108. Isiklar ZU, Darouiche RO, Landon GC, Beck T. Efficacy of antibiotics alone for orthopaedic device related infections. Clin Orthop Relat Res. 1996;332:184–9.
109. Monzon M, Garcia-Alvarez F, Lacleriga A, Amorena B. Evaluation of four experimental osteomyelitis infection models by using precolonized implants and bacterial suspensions. Acta Orthop Scand. 2002;73(1):11–9.
110. Goodhart GL. *Mycobacterium fortuitum* osteomyelitis following trauma. J Orthop Trauma. 1993;7(2):142–5.
111. Caricato A, Montini L, Bello G, Michetti V, Maviglia R, Bocci MG, et al. Risk factors and outcome of *Acinetobacter baumanii* infection in severe trauma patients. Intensive Care Med. 2009;35(11):1964–9.
112. Glueck DA, Charoglu CP, Lawton JN. Factors associated with infection following open distal radius fractures. Hand (N Y). 2009;4(3):330–4.
113. Mader JT, Cripps MW, Calhoun JH. Adult posttraumatic osteomyelitis of the tibia. Clin Orthop Relat Res. 1999;360:14–21.
114. Costerton JW, Post JC, Ehrlich GD, Hu FZ, Kreft R, Nistico L, et al. New methods for the detection of orthopedic and other biofilm infections. FEMS Immunol Med Microbiol. 2011;61(2):133–40.
115. Cremieux AC, Dumitrescu O, Lina G, Vallee C, Cote JF, Muffat-Joly M, et al. Panton-valentine leukocidin enhances the severity of community-associated methicillin-resistant *Staphylococcus aureus* rabbit osteomyelitis. PLoS One. 2009;4(9):e7204.
116. Daurel C, Prunier AL, Chau F, Garry L, Leclercq R, Fantin B. Role of hypermutability on bacterial fitness and emergence of resistance in experimental osteomyelitis due to *Staphylococcus aureus*. FEMS Immunol Med Microbiol. 2007;51(2):344–9.

117. Elasri MO, Thomas JR, Skinner RA, Blevins JS, Beenken KE, Nelson CL, et al. *Staphylococcus aureus* collagen adhesin contributes to the pathogenesis of osteomyelitis. Bone. 2002; 30(1):275–80.
118. Luong TT, Lei MG, Lee CY. *Staphylococcus aureus* Rbf activates biofilm formation in vitro and promotes virulence in a murine foreign body infection model. Infect Immun. 2009; 77(1):335–40.
119. Patti JM, Bremell T, Krajewska-Pietrasik D, Abdelnour A, Tarkowski A, Ryden C, et al. The *Staphylococcus aureus* collagen adhesin is a virulence determinant in experimental septic arthritis. Infect Immun. 1994;62(1):152–61.
120. Weiss EC, Spencer HJ, Daily SJ, Weiss BD, Smeltzer MS. Impact of sarA on antibiotic susceptibility of *Staphylococcus aureus* in a catheter-associated in vitro model of biofilm formation. Antimicrob Agents Chemother. 2009;53(6):2475–82.
121. Ferguson KP, Lambe Jr DW, Keplinger JL, Kalbfleisch JH. Comparison of the pathogenicity of three species of coagulase-negative Staphylococcus in a mouse model with and without a foreign body. Can J Microbiol. 1991;37(9):722–4.
122. Fluckiger U, Ulrich M, Steinhuber A, Doring G, Mack D, Landmann R, et al. Biofilm formation, icaADBC transcription, and polysaccharide intercellular adhesin synthesis by staphylococci in a device-related infection model. Infect Immun. 2005;73(3):1811–9.
123. Francois P, Tu Quoc PH, Bisognano C, Kelley WL, Lew DP, Schrenzel J, et al. Lack of biofilm contribution to bacterial colonisation in an experimental model of foreign body infection by *Staphylococcus aureus* and *Staphylococcus epidermidis*. FEMS Immunol Med Microbiol. 2003;35(2):135–40.
124. Lambe Jr DW, Ferguson KP, Keplinger JL, Gemmell CG, Kalbfleisch JH. Pathogenicity of *Staphylococcus lugdunensis*, *Staphylococcus schleiferi*, and three other coagulase-negative staphylococci in a mouse model and possible virulence factors. Can J Microbiol. 1990;36(7):455–63.
125. Lambe Jr DW, Ferguson KP, Mayberry-Carson KJ, Tober-Meyer B, Costerton JW. Foreign-body-associated experimental osteomyelitis induced with *Bacteroides fragilis* and *Staphylococcus epidermidis* in rabbits. Clin Orthop Relat Res. 1991;266:285–94.
126. Mayberry-Carson KJ, Tober-Meyer B, Gill LR, Lambe Jr DW, Hossler FE. Effect of ciprofloxacin on experimental osteomyelitis in the rabbit tibia, induced with a mixed infection of *Staphylococcus epidermidis* and *Bacteroides thetaiotaomicron*. Microbios. 1990;64(258):49–66.
127. Campoccia D, Montanaro L, Moriarty TF, Richards RG, Ravaioli S, Arciola CR. The selection of appropriate bacterial strains in preclinical evaluation of infection-resistant biomaterials. Int J Artif Organs. 2008;31(9):841–7.
128. Beenken KE, Dunman PM, McAleese F, Macapagal D, Murphy E, Projan SJ, et al. Global gene expression in *Staphylococcus aureus* biofilms. J Bacteriol. 2004;186(14):4665–84.
129. Cassat J, Dunman PM, Murphy E, Projan SJ, Beenken KE, Palm KJ, et al. Transcriptional profiling of a *Staphylococcus aureus* clinical isolate and its isogenic agr and sarA mutants reveals global differences in comparison to the laboratory strain RN6390. Microbiology. 2006;152(Pt 10):3075–90.
130. Herbert S, Ziebandt AK, Ohlsen K, Schafer T, Hecker M, Albrecht D, et al. Repair of global regulators in *Staphylococcus aureus* 8325 and comparative analysis with other clinical isolates. Infect Immun. 2010;78(6):2877–89.
131. Sivadon V, Rottman M, Chaverot S, Quincampoix JC, de Avettand V, Mazancourt P, et al. Use of genotypic identification by sodA sequencing in a prospective study to examine the distribution of coagulase-negative Staphylococcus species among strains recovered during septic orthopedic surgery and evaluate their significance. J Clin Microbiol. 2005;43(6):2952–4.
132. Mayberry-Carson KJ, Tober-Meyer B, Lambe Jr DW, Costerton JW. Osteomyelitis experimentally induced with *Bacteroides thetaiotaomicron* and *Staphylococcus epidermidis*. Influence of a foreign-body implant. Clin Orthop Relat Res. 1992;280:289–99.
133. Marberry KM, Kazmier P, Simpson WA, Christensen GD, Phaup JG, Hendricks KJ, et al. Surfactant wound irrigation for the treatment of staphylococcal clinical isolates. Clin Orthop Relat Res. 2002;403:73–9.

134. Hidaka S. An experimental study on pyogenic osteomyelitis with special reference to polymicrobial infections. Nippon Seikeigeka Gakkai Zasshi. 1985;59(4):429–41.
135. Johansson A, Svensson O, Blomgren G, Eliasson G, Nord CE. Anaerobic osteomyelitis. A new experimental rabbit model. Clin Orthop Relat Res. 1991;265:297–301.
136. Blomgren G, Lundquist H, Nord CE, Lindgren U. Late anaerobic haematogenous infection of experimental total joint replacement. A study in the rabbit using *Propionibacterium acnes*. J Bone Joint Surg Br. 1981;63B(4):614–8.
137. Blomgren G. Hematogenous infection of total joint replacement. An experimental study in the rabbit. Acta Orthop Scand Suppl. 1981;187:1–64.
138. Costerton JW. Biofilm theory can guide the treatment of device-related orthopaedic infections. Clin Orthop Relat Res. 2005;437:7–11.
139. Stoodley P, Nistico L, Johnson S, Lasko LA, Baratz M, Gahlot V, et al. Direct demonstration of viable *Staphylococcus aureus* biofilms in an infected total joint arthroplasty. A case report. J Bone Joint Surg Am. 2008;90(8):1751–8.
140. Gristina AG, Costerton JW. Bacterial adherence and the glycocalyx and their role in musculoskeletal infection. Orthop Clin North Am. 1984;15(3):517–35.
141. Smith K, Perez A, Ramage G, Lappin D, Gemmell CG, Lang S. Biofilm formation by Scottish clinical isolates of *Staphylococcus aureus*. J Med Microbiol. 2008;57(Pt 8):1018–23.
142. Fitzpatrick F, Humphreys H, O'Gara JP. Evidence for icaADBC-independent biofilm development mechanism in methicillin-resistant *Staphylococcus aureus* clinical isolates. J Clin Microbiol. 2005;43(4):1973–6.
143. O'Neill E, Pozzi C, Houston P, Humphreys H, Robinson DA, Loughman A, et al. A novel *Staphylococcus aureus* biofilm phenotype mediated by the fibronectin-binding proteins, FnBPA and FnBPB. J Bacteriol. 2008;190(11):3835–50.
144. Campoccia D, Speziale P, Ravaioli S, Cangini I, Rindi S, Pirini V, et al. The presence of both bone sialoprotein-binding protein gene and collagen adhesin gene as a typical virulence trait of the major epidemic cluster in isolates from orthopedic implant infections. Biomaterials. 2009;30(34):6621–8.
145. Patti JM, Allen BL, McGavin MJ, Hook M. MSCRAMM-mediated adherence of microorganisms to host tissues. Annu Rev Microbiol. 1994;48:585–617.
146. Patti JM, Hook M. Microbial adhesins recognizing extracellular matrix macromolecules. Curr Opin Cell Biol. 1994;6(5):752–8.
147. Vazquez V, Liang X, Horndahl JK, Ganesh VK, Smeds E, Foster TJ, et al. Fibrinogen is a ligand for the *S. aureus* MSCRAMM Bbp (bone sialoprotein-binding protein). J Biol Chem. 2011;286(34):29797–805.
148. Johansson A, Flock JI, Svensson O. Collagen and fibronectin binding in experimental staphylococcal osteomyelitis. Clin Orthop Relat Res. 2001;382:241–6.
149. Darouiche RO, Landon GC, Patti JM, Nguyen LL, Fernau RC, McDevitt D, et al. Role of *Staphylococcus aureus* surface adhesins in orthopaedic device infections: are results model-dependent? J Med Microbiol. 1997;46(1):75–9.
150. Dohin B, Gillet Y, Kohler R, Lina G, Vandenesch F, Vanhems P, et al. Pediatric bone and joint infections caused by Panton-Valentine leukocidin-positive *Staphylococcus aureus*. Pediatr Infect Dis J. 2007;26(11):1042–8.
151. Tristan A, Bes M, Meugnier H, Lina G, Bozdogan B, Courvalin P, et al. Global distribution of Panton-Valentine leukocidin—positive methicillin-resistant Staphylococcus aureus, 2006. Emerg Infect Dis. 2007;13(4):594–600.
152. Patel AH, Nowlan P, Weavers ED, Foster T. Virulence of protein A-deficient and alpha-toxin-deficient mutants of *Staphylococcus aureus* isolated by allele replacement. Infect Immun. 1987;55(12):3103–10.
153. An YH, Bradley J, Powers DL, Friedman RJ. The prevention of prosthetic infection using a cross-linked albumin coating in a rabbit model. J Bone Joint Surg Br. 1997;79(5):816–9.

Part II
Biomaterial-Based Strategies to Reduce Medical Device-Associated Infection

Chapter 13
Antimicrobial Medical Devices in Preclinical Development and Clinical Use

Benjamin D. Brooks, Amanda E. Brooks, and David W. Grainger

Abstract Medical devices are increasingly used worldwide for an expanding repertoire of patient clinical needs. Biomaterials and medical device designs have become progressively more complex to accommodate diverse demands for performance and safety *in vivo*. While a majority of these implants satisfy their clinical expectations with safety and efficacy in their specific applications, a minority of implants induce serious adverse events with substantial health and economic consequences. One recognized challenge is the growing clinical problem with implant-associated infections. Increasing number and types of implants used in patients have resulted in increasing numbers of biomaterial-associated infections. Researchers and medical device manufacturers have responded to this challenge with intensified attention to innovating device designs, surgical implantation protocols, and biomaterials to minimize infection opportunities. Medical devices with claims to limit microbial adhesion and colonization using combinations of pharmacological, topological, and materials chemistry approaches have been brought into clinical use with the intent of reducing device-related infections. Many types of catheters, stents, orthopedic devices, contact lenses, surgical meshes, shunts, sutures, cardiovascular replacements, and many other device categories offer antimicrobial enhancements. Approaches include different biomaterials chemistries that intrinsically resist microbial colonization or that deter active growth on contact, surface modifications that

B.D. Brooks • A.E. Brooks
Department of Pharmaceutics and Pharmaceutical Chemistry, University of Utah, 20 South 2030 East, BPRB 194, Salt Lake City, UT 84112-5820, USA
e-mail: david.grainger@hsc.utah.edu

D.W. Grainger (✉)
Department of Pharmaceutics and Pharmaceutical Chemistry, University of Utah, 20 South 2030 East, BPRB 194, Salt Lake City, UT 84112-5820, USA

Department of Bioengineering, University of Utah, 20 South 2030 East, BPRB 194, Salt Lake City, UT 84112-5820, USA
e-mail: david.grainger@hsc.utah.edu

T.F. Moriarty et al. (eds.), *Biomaterials Associated Infection: Immunological Aspects and Antimicrobial Strategies*, DOI 10.1007/978-1-4614-1031-7_13,

produce topologies observed to limit pathogen attachment, medicinal, antiseptic or bioactive coatings, direct antimicrobial attachment to surfaces, or drug impregnation within the biomaterial, and extended release strategies that control antimicrobial agent release from the device over time after implantation.

13.1 Introduction

13.1.1 Increased Implanted Medical Device Use and Diversification

Development and use of synthetic biomaterials in medical device [1] clinical applications has expanded dramatically within the past half century, growing in both the total number of devices implanted and the breadth of new devices offered as surgical implants [2]. The implanted medical device market is estimated at over $300 billion dollars worldwide, growing at 20% annually [3–5]. Given the approximately five million people implanted with medical devices in 2010, this number is expected to balloon within the next 10 years because of device innovations, aging global population demographics, increased medical device access in developed countries, and a greater demand for higher patient quality of life offered by many medical devices [4, 6].

13.1.2 Biomaterial-Associated Infections: Pathogen-Colonized Implanted Medical Devices

Biomaterial-associated infections (BAIs) remain a major unsolved clinical challenge, despite long-standing innovation efforts across all general classes of implanted device designs, biomaterials, operating room facilities, and surgical techniques. Worldwide, BAI incidence approaches one million cases per year [7]. BAIs result from successful, persistent bacterial adherence to an implant before, during, or after surgery, allowing colonization and subsequent bacterial growth on the implant, creating a clinical grade infection on and around the device and, potentially leading to a systemic infection with severe consequences. BAIs are most commonly associated with intravascular, cardiovascular, neurosurgical, orthopedic, ophthalmological, and dental devices [8]. Despite efforts to create an aseptic surgical environment, acute infections associated with implant surgery are a primary source of BAIs, with later (i.e., chronic) and hematogenously sourced (i.e., blood-transported from another site) infections accounting for remaining BAIs [9]. Common medical consequences of these infections include tissue necrosis, systemic infections, multiple surgeries, skyrocketing medical costs, device malfunction, and possibly death [10].

Clinical treatment of BAIs generally involves immediate removal of the infected implant and weeks of parenteral antibiotic therapy to clear the infection prior to new implantation of another device as a revision. As a typical example, revision surgeries for infected total hip arthroplasties increase the costs of a prosthesis placement substantially. Total hip joint prosthetic septic loosening is treated clinically by two-stage revision surgeries. First, the infected hip prosthesis is surgically removed along with substantial amounts of bone surrounding and often cemented to the implant. Subsequently, the soft tissue adjacent to the prosthesis is surgically debrided, and the patient is infused with high antibiotic doses through both local and systemic infusions as well as local, temporarily placed drug delivery implants (i.e., plastic joint spacers and strings of drug-releasing plastic beads in the former implant site). These antimicrobial doses are prescribed to kill local pathogens in the devascularized joint space and from surrounding tissues prior to surgical placement of a new hip prosthesis. Increasingly reduced efficacy of systematically administered antibiotics against implant-associated infections due to antimicrobial resistance is a major clinical concern to this standard routine [11]. Additionally, given that the patient must endure weeks without a hip joint, remaining bed-ridden and immobile, patient discomfort and morbidity is also substantial, and associated ancillary healthcare costs are high. Alternatively, local or topical administration of antibiotics is also used to (1) minimize systemic dosing and drug toxicity, and (2) maintain antibiotic concentrations at the implant site above bacterial minimum inhibitory concentrations for extended durations [16]. This alternative antibiotic delivery can be successful, with local doses far smaller than those required systemically. For these reasons, local, targeted antimicrobial therapy is often proposed to be superior to routine systemic approaches; however, the challenge remains in engineering pharmaceutical and delivery solutions that maintain safety and efficacy for each individual device application and pathogen type [12–15]. As a second example, catheter-associated blood stream infections commonly occur, with substantial costs of $12,321 per case, and attributable mortality averaging 27%, rising to 35% for ICU patients [16]. With increasing number of implanted medical devices, the economic and medical impact of BAIs remains a critical area to address, given that 60–70% of hospital-acquired (nosocomial) infections (HAIs) are associated with implanted medical devices [17], and the Centers for Disease Control and Prevention (CDC) estimate that the annual costs of HAIs in the United States total $4.5–$11 billion [18]. Compounding the situation further for health care providers is pressure from insurance companies and from government-run health care programs to reduce or eliminate payouts for preventable infections, of which HAIs are a major contributor [19].

Preventative and prophylactic steps to minimize BAI include reduction in host bacterial load prior to and during device implantation using improved surgical suite hygiene and tactical improvements and common clinically routine antimicrobial regimens [9]. Systemic antibiotic prophylaxis remains the clinical standard of care in this regard [20, 21]. However, reduced vascularity surrounding implants post-surgery, a natural result of implant placement, tissue trauma, and abnormal healing around implants, decreases antibiotic transport to the implant site [4, 15]. These complicating factors allow biofilm formation at sites of implantation, further reducing the efficacy of systemic antibiotic treatments against biofilms given their known resistance to antibiotics. There are many reasons for this resistance including (1) poor and delayed penetration of drug into and through the biofilm's complex exopolysaccharide/proteoglycan matrix, (2) low metabolic activity and physiological changes from quiescent bacteria in biofilm that remain unaffected by antibiotics, and (3) sigma factors that allow the cell to survive in a biofilm oxygen-poor environment [22].

Bacterial biofilms represent a clinically significant form and opportunistic phenotype for implant-associated sessile bacteria. Sessile, adherent bacteria and pathogens residing within a biofilm are intrinsically more resistant to host clearance and drug therapies than planktonic or non-biofilm-forming pathogens. In the biofilm phenotype, bacteria adhere to surfaces and proliferate, aggregating eventually into communal clusters. Within such communal cell clusters, bacteria begin to communicate using quorum sensing signals that influence each other's growth state and behaviors. Two distinct features of the biofilm bacterial colony are (1) production of an extracellular polymeric substance (EPS), a biopolymeric mixture of proteins and polysaccharides serving to encapsulate the bacterial cluster on a surface [10] and (2) the production of unique soluble communication molecules (typically acyl homoserine lactone (AHL) derivatives in Gram negative species, small peptides in Gram positive species and auto-inducer (AI2) molecules in both species) called quorum sensing molecules [23]. These two features of communal biofilm bacterial behavior are distinct from their planktonic and non-biofilm-forming analogues, leading to antibiotic-resistant and more virulent pathogenic bacteria that are notoriously difficult for host phagocytes to clear from surfaces. Furthermore, these traits make the biofilm a particular threat to implanted device surfaces. Biofilm formation is a common hallmark mechanism of a BAI as biofilm phenotypes exhibit more than 1,000 times greater minimum inhibitory concentrations (MIC) of common clinical antibiotics than non-biofilm strains [9, 10, 24, 25].

13.1.3 Problems of Antibiotic Resistance

Antibiotic resistance associates with implanted medical devices in two ways. First, increasing numbers and types of microbes especially species of staphylococcus, streptococcus, and pseudomonas [26–28] now associated with BAIs are strains exhibiting antimicrobial resistance. Since the discovery of antimicrobial resistance, BAIs are increasingly associated with drug-resistant microbes: ~60% of BAIs result from resistant microbes [26–28]. Second and more importantly, government regulatory agencies, especially the US FDA, consider use of antimicrobial agents as promoting the "emergence of resistance to the antimicrobial agent" [29]. Initially, antibiotic resistance is hypothesized to occur when a microbe is exposed to an antimicrobial agent at below its MIC, allowing pathogen survival and adaptation under drug stress [28]. Thus, according to the FDA, the potential that a drug-releasing medical device may induce antimicrobial resistance is a "serious clinical problem" and that (antimicrobial) "use on a medical device should outweigh the associated risk [of an organism developing antimicrobial resistance]" [29].

13.1.4 Microbes Associated with BAIs

While types of infectious organisms cultured on implanted devices are highly diverse, a BAI is caused most often by commensal bacteria, often sourced from the host, that opportunistically form biofilms on implants [9]. Gram positive bacteria including *Staphylococcus aureus* and *Staphylococcus epidermidis* are located on or around 90% of all implanted medical devices, and as such account for the majority of infectious BAIs [30–33]. Other common gram negative organisms associated with BAIs include *Escherichia coli*, *Pseudomonas aeruginosa*, *Proteus mirabilis*, and *Proteus vulgaris* [34, 35]. Another mechanism of BAI is bacterial entry, survival, localization and even proliferation within phagolysosomes inside host macrophages in peri-implant tissues. Host macrophages naturally respond to wound trauma associated with implant placement, and routinely attempt to clear debris and microbes from the site by phagocytosis [36]. Macrophage uptake of some bacterial strains provides a niche for pathogen survival internally within these cells. This internalized pathogen survival and proliferation within host inflammatory cells offers protection against both the host immune system and antibiotics [36], as well as transport to other tissue sites as macrophages migrate throughout the tissue.

Fungi, especially Candida spp., also represent a significant cause of BAI [26]. *Candida albicans* not only forms polymicrobial biofilms *in vivo* with *S. aureus*, but both species are shown to gain a fitness advantage by altered sensitivity to antimicrobial agents acquired through their interactions. In other studies, *C. albicans* and *S. epidermidis* were shown to create a biofilm capable of inhibiting penetration of the antifungal fluconazole and capable of providing protection against the antibiotic vancomycin [37, 38].

While statistics associated with implant infections are based largely on classical microbiological culture methods, certainty in identifying causal implant-associated pathogens is confounded by evidence showing that only a minority of organisms producing such infections are culturable [22]. A large majority of biofilm bacteria are asserted not to be readily culturable using traditional microbiological techniques [38–40]. Live/dead discrimination of bacteria is also difficult: PCR or genomics-based results from implants or biopsies may support the presence of a pathogen from signatures in recovered genetic material, but this does not provide evidence for when that pathogen was alive at that site, or if it is attributed to producing the infection in question [40].

13.1.5 Material Properties of Implanted Devices

General biocompatible expectations for implanted medical devices include both mechanical/structural and interfacial topological/chemical properties that do not compromise appropriate device function and reliability within host tissues specific for each application. However, biocompatibility is an ambiguous term with only contextual meaning for each device in each application. There are no universal biocompatibility criteria except a general acceptance of properties governing medically safe and efficacious responses in a particular application. Unfortunately, biomaterial implants result in higher rates of bacterial colonization and subsequent biofilm formation. Data show that an implant significantly reduces the minimum infectious dose of *S. aureus* leading to abscess formation by up to a 100,000 fold, demonstrating the intrinsically compromising nature of any implanted material toward endogenous host infection response [9, 41, 42]. Few approved clinical devices were originally designed and approved to carry antimicrobial drugs or properties, however. Current attempts to simply add antimicrobial benefits onto existing device classes are not always simple and effective. Challenges achieving biocompatibility in many diverse clinical contexts for implantable medical devices are further compounded when antimicrobial surface properties are added to the biomaterial. Drug dosing from the implant must be carefully engineered, on-device drug modifications require different sterilization regimens than device alone, device coatings used to block microbial attachment also interfere with host cell recruitment and device integration, and topological modifications can change device mechanical properties.

13.1.6 Device Regulatory Approval Requirements

All implanted medical devices are required by the FDA and other nations' regulatory agencies to meet minimum safety and efficacy requirements. Moreover, such approval is given only for prescribed clinical indications. The safety and efficacy of the medical device is determined by "weighing any probable benefit to health from the use of the device against any probable risk of injury or illness from such use". Thus, a determination of safety and effectiveness of a device and its predicate involves an assessment of their risks and benefits" [29]. To be approved as implanted medical devices [1], extensive studies must be conducted including those supporting *in vitro*, preclinical animal, and human clinical trials [29].

Importantly, medical devices that contain an antimicrobial drug as a secondary mode of action are often considered "combination products" by the FDA [43]. Combination medical products comprise "two or more regulated components." Combination products often result in complex jurisdictional concerns for the manufacturer since not all centers within the FDA have comparable regulatory requirements and approval rates. In an attempt to address this concern, the FDA created "Inter Center Agreements" or "ICAs" between each of the Centers (CDER, CBER and CDRH), to establish the FDA general approach to review of combination devices. In the last 20 years the number of combination products has increased as well as the complexity and overlapping mechanisms of action by which products may achieve their intended medical purpose. In 2002, the FDA created the Office of Combination Products ("OCP") to help assign which Center has primary jurisdiction for review of combination products where the jurisdiction is unclear or in dispute. Further, the agency has created algorithms and draft guidance documents to help facilitate the process; however, the routes for combination product review are still evolving and will likely require additional regulations to complete and provide a clear process forward.

Recent FDA guidance documents outline the technical hurdles required to receive the label of an antimicrobial device. Notably, this does not address the critical complicating question regarding the product's primary mode of action as a device or a drug. Classification as primarily a device with secondary drug properties or primarily as a drug with secondary device properties is not always clear, and this decision significantly changes the regulatory path and regulatory hurdles for an antimicrobial device and may lead to obscurity in labeling to direct the product approval into the "easier" regulatory path. For most implanted medical device classes, antimicrobial addition in chemical (i.e., drug or bonded coating) or morphological (i.e., surface patterns) constitutes a secondary activity to the device's primary mode of action (i.e., tibial nail). The efficacy requirements, i.e., "Indications for Use," for labeling an antimicrobial device are that the device "reduces or prevents device-related infections" and "reduces or inhibits microbial colonization on a medical device" [29]. FDA approval also requires a rationale for adding the antimicrobial agent to the device and must "include an identification of the target pathogens to the genus and species level," must "include a discussion of the underlying scientific or clinical rationale (or both, as applicable) for adding the specific antimicrobial agent to the device," and must include an assessment of the potential for antimicrobial resistance due to the device. Further requirements include characterization of the device with the added antimicrobial agent and an extensive characterization of the antimicrobial agent including identity and formulation, concentration, application to the device, mechanism of action, antimicrobial spectrum of the agent, its release kinetics, toxicity, and minimum effective concentration [29].

In addition to FDA and other regulatory requirements, product demands include a surface with extended timeframes for eliciting antimicrobial activity, broad spectrum antibiotic effectiveness, low *in vivo* toxicity, biocompatibility, low cost, easily produced, and disposable with minimal environmental challenges [44].

The result of regulatory requirements is that antimicrobial enhancements to even currently approved medical devices must undergo extensive testing for safety and efficacy in order to gain approval for human use. Data considered valid for "reducing or preventing device-related infections" and "reducing or inhibiting microbial colonization on a medical device" are often convincing from *in vitro* anti-microbial assays, but less convincing, and most often unconvincing, when applied to preclinical infection models *in vivo*. Even worse, results from animal preclinical testing do not often translate to efficacy in human use [45]. This discrepancy has produced a "credibility gap" in the development, approval and marketing and anti-microbial strategies in many device categories, with resulting confusion over claims to anti-infective efficacy and obfuscation over mechanisms of action and clinical benefits to patients.

13.1.7 Medical Device Surfaces as an Infection Nidus

Pre- and perioperative systemic administration of prophylactic antibiotics to reduce the incidence of BAI is the clinical standard of care and yet mildly controversial due to the risks for antibiotic resistance [46, 47]. A more sophisticated, direct approach to reducing BAI is to reduce or inhibit the microbe's ability to adhere, colonize and/or develop into a biofilm on an implanted device. Hence, the device's biomaterial surface topological and chemical properties are important for resisting bacterial adherence, at least *in vitro*. In vivo, the device is subject to an instant and constant barrage of surface-adsorbing solutes and cells from host milieu, including 70 mg/ml of plasma proteins comprising some 20,000 distinct molecular species, and many host cell types, including blood platelets and their rich array of growth factors. The result of any implantation is the rapid formation of a proteinaceous conditioning layer on the entire exposed device surface. The composition and conformation of proteins in this adherent film *in vivo* are neither characterized much less controlled or controllable; yet, it is likely that bacteria at the implant site introduced from many sources adhere and interact with the material via this conditioning film. Bacterial receptors (i.e., MSCRAMMS, adhesins) specific for binding certain proteins in this film (e.g., collagens, fibronectins, fibrin) are well-recognized as virulence factors facilitating bacterial-implant colonization [48]. Local implant trauma from placement, blood coagulation, the acute host foreign body inflammatory response, eventual wound healing and revascularization responses, and implant material surface properties all affect the device's conditioning proteinaceous film and consequently the types of bacteria able to interact with the device and its localized environment. Causal mechanisms linking the conditioning film with each adhering pathogen, and methods to alter or prevent them are challenging to discover and to mitigate. Some surfaces such as silicone exhibit significantly higher bacteria colonization rates than other implant surfaces [49–51]. While these protein and cellular interactions have been attributed to almost every surface property, they have not been substantially mitigated by any surface modification to date. Current clinical strategies for reducing infection on implanted medical devices include anti-adhesive surface modifications, surface and bulk impregnation with antibiotics, tethered antibiotics, and antimicrobial polymer coatings [9, 10, 24]. These approaches have met with varied success from a bioengineering perspective and distinctly less-impressive performance from a clinical outcomes perspective [9, 10, 24]. Further complications arise from the variety of different implant locations, differing patient health status and immune competencies, different surgical techniques and facilities, and the diverse infectious agents possible [17, 52].

13.1.8 Types of BAIs

BAIs are an increasing problem associated with more than one million HAIs in the United States related to medical devices [7]. Cardiac device-related infections are a major contributor to this problem as the rate of infection for these devices increased 124% between 1990 and 1999. Specifically, prosthetic valve infections increased 50% over the same period [53]. With the ever-expanding number of antibiotic resistant species, this problem will continue to grow. Table 13.1 outlines the diverse types of medical devices currently implanted [54]. The remainder of this section focuses on preclinical and clinically approved medical devices with antimicrobial surfaces.

For the purpose of this chapter, implants are divided into three primary categories: (1) devices that are fully implanted within the body, such as orthopedic device, meshes, and heart valves (see Sect. 13.3), (2) devices that reside within or in intimate tissue contact the body but without penetration or invasion through tissue such as contact lenses, endotracheal (tracheal) tubes, aural tubes, and urinary catheters in contact with various epithelial membranes (see Sect. 13.4), and (3) percutaneous devices that transverse through epithelium such as percutaneous catheters and dental implants (see Sect. 13.5). All three device categories pose unique infection challenges.

Table 13.1 Clinical device categories, examples, and types of antimicrobial surfaces

Category	Examples of devices	Types of antimicrobial surfaces
Cardiovascular implants	Catheters, central venous (CVC) and arterial, stents, pacemakers, heart valves	Silver, hydrophilic, drug eluting
Neural implants	CSF shunts, neurologic stimulators	Antibiotic eluting, hydrophilic
Nephrology and urology implants	Catheters: peritoneal dialysis, urinary and hemodialysis catheters, penile prostheses	Silver, hydrophilic, drug eluting
Ocular devices	Contacts lenses, vitreous body prostheses, retinal prostheses, iris prostheses	Selenium, furanone eluting
Orthopedic implants	Joint prostheses; pins, screws, intramedullary nails, intervertebral disks; intramedullary nails	Antibiotic eluting
Aural implants	Cochlear implants, tympanostomy tubes,	Silver
Dental implants	Dental implants	Silver, hydrophilic, drug eluting
Non-dental oral devices	Endotracheal tubes, voice prosthetics	Silver

13.2 Antimicrobial Surfaces

Antimicrobial devices can be characterized in many different ways including the mechanism by which the antimicrobial is bound within or to the material such as (1) tethered or bonded antimicrobial, (2) device-impregnated antimicrobial, (3) coating-releasing antimicrobial, or (4) intrinsically antimicrobial surfaces (see Fig. 13.1). This section outlines various mechanisms to engineer an antimicrobial surface.

13.2.1 Device Surface Treatments

Early strategies for developing antimicrobial surfaces exploited various coatings or surface treatments that reduced microbial adhesion [55, 56]. This approach has been highly explored for decades. Initial research focused on making device surfaces hydrophilic to increase hydration [55, 56], reduce protein adsorption and therefore limit bacterial colonization of the conditioning film [52, 57]. Silicone implantable devices were rationalized to exhibit increased bacterial adhesion because they were

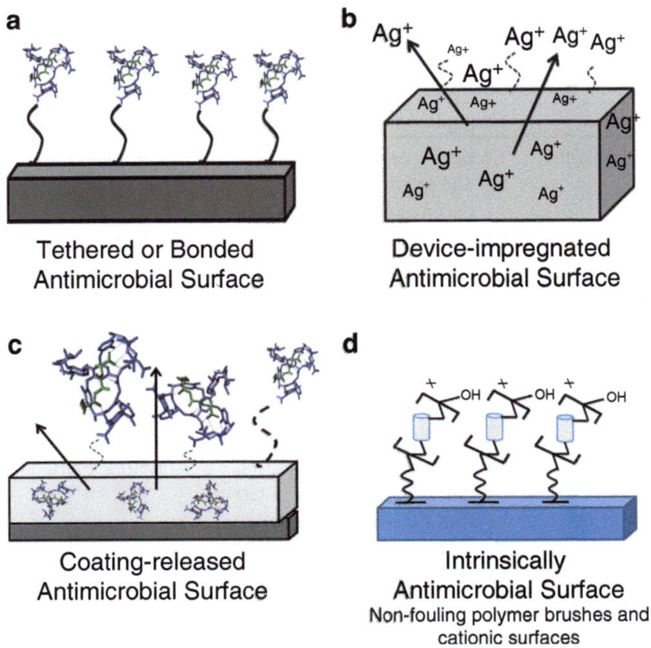

Fig. 13.1 Mechanisms by which antimicrobial agents are bound within or to the material including (**a**) tethered or bonded antimicrobial, (**b**) device-impregnated antimicrobial, (**c**) coating-releasing antimicrobial, or (**d**) intrinsically antimicrobial surfaces

hydrophobic [52, 57]. Later efforts focused on more sophisticated coatings and surface treatments that altered device surface physiochemical properties by steric hindrance, hydration, conformation, and even topologies. These "passive" treatments include poly(ethylene glycol) coatings [58], poly(ethylene oxide) brushes [59], hydrophilic poly(vinylpyrrolidone) coating [60], and hydrophilic polyurethane coatings [61], along with many other chemistry modifications [49]. Nonetheless, despite some distinctive behaviors from simplistic *in vitro* bacterial assays, these coatings and treatments demonstrate limited and controversial anti-pathogen effectiveness in animal studies [49, 52]. Further, some treatments are only effective against the initial round of pathogens, failing to perform under sustained or repeated bacterial assaults. [17] After exhaustive analysis and past innovation, device coating utility alone seems insufficient to sustain *in vivo* antimicrobial activity as a clinical remedy to device infection. While *in vitro* studies under controlled and sometimes even contrived assay conditions can demonstrate reduced microbial colonization of biomaterials surfaces, the reliability and longevity of this protection *in vivo* is rarely established (specifically see Sects. 13.5.1 and 13.6).

13.2.2 Silver-Containing Antimicrobial Surfaces

Silver exhibits antimicrobial activity in numerous formulations, has a rich history of antimicrobial use in device applications, and is generally regarded as safe (GRAS) at prescribed concentrations [62, 63]. Despite concerns, clinical development of antimicrobial resistance to silver to date is rare [44, 64]. As a result, silver has been incorporated into catheters, endotracheal tubes, prosthetic heart valves and many other devices under development [65, 66]. Silver's revival in conventional medicine as an antimicrobial resulted from silver sulfadiazine (SSD) topical cream [67] utilized in wound and burn care. Silver kills a broad spectrum of bacteria associated with BAIs including *P. aeruginosa*, *E. coli*, *S. aureus*, and *S. epidermidis* [24]. Silver also kills antibiotic-resistant strains of these microbes and exhibits activity against other pathogen classes as well (e.g., fungi and yeast). Silver has an active antimicrobial effect in the ionic form, Ag^{+}, not in elemental metallic form via three mechanisms: (1) inhibiting membrane transport processes [68], (2) blocking cell replication [69], and (3) disrupting cell metabolism [44, 70]. Silver ion (Ag^{+}, and in some exotic cases, Ag^{2+}) is proposed to bind to DNA and also key thiol groups in proteins, disrupting functions of bacterial cell membranes and crucial metabolic proteins and enzymes [70]. The ability of silver to minimize antimicrobial resistance is attributed to the combination of all three bactericidal mechanisms acting in tandem [71].

Silver has diverse forms as a device-based antimicrobial. It is generally coated in elemental form (Ag0 metal) onto the device surface, impregnated in various forms (metallic silver particles, silver salts, complexes, and chelates) within the device matrix (i.e., polymer) or applied within a polymer coating as a reservoir to control silver ion release. Nanoparticulate silver possesses unique physical properties due to high surface area to volume ratios, surface oxidation states, and ionic solubilities.

Consequently, nanosilver formulations on devices represent a newer formulation strategy to minimize BAIs [72]. While other nanoscale metals also possess promising antimicrobial properties, nanosilver has been the most promising because of low nanoscale particle or chemical toxicity exhibited to date [73], although this remains controversial and unapproved for device formulation.

Recent work developing silver-coated devices continues to prove silver's antimicrobial effects *in vitro* in many forms. However, antimicrobial testing and standardization of silver-releasing surfaces for reliable, predictive clinical efficacy is not only challenging but also costly to obtain efficacy data for regulatory purposes [74, 75]. The "elephant in the room" in any consideration of silver for antimicrobial device use is the evidence for clinical efficacy: silver's efficacy *in vivo* is an ongoing unresolved controversy in many applications [76, 77], and the freedom to operate (i.e., right to practice this method without patent infringement on a device) with any given technology on a device is complex to determine. New nanosilver technologies with antimicrobial applications include a microporous hydroxyapatites incorporating nanosilver [78], a silver-encapsulated, micro-arc oxidized, calcium phosphate coating [66], glasses containing silver oxide [79–81], silver embedded within nanostructured oxide matrices [82], and oxidized silver in polyamide coatings [83, 84]. Given the existing number of clinically approved precedent antimicrobial applications utilizing many forms of silver, newly emerging silver delivery systems are likely to see more commercial interest seeking regulatory approvals than other less developed antimicrobial device technologies. Nonetheless, properly designed and powered clinical trials will need to be conducted to provide convincing evidence for silver device-based efficacy to a skeptical user base.

13.2.3 Antibiotic Surfaces

Antibiotic surfaces on medical devices have a love-hate relationship with government regulatory agencies that apparently embrace the concept of reducing BAIs, but with increasing concerns over antibiotic resistance [17]. Current clinically approved approaches for device-based antibiotics exploit either device-drug impregnation, such as catheters on which the antimicrobial agents are claimed to be immobilized, or in drug-releasing coatings (polymer and ceramic) that provide a strong early phase drug bolus release to eliminate bacterial presence and device adhesion [24]. One prominent example of this design is antibiotic-releasing bone cements in which antibiotics including gentamicin, tobramycin, and also vancomycin [85] are loaded into bulk poly(methylmethacrylate) (PMMA) bone cement formulations intra-operatively, often placed around total joint arthroplasties. Several vendors market clinically approved antibiotic-releasing bone cements containing either gentamicin or tobramycin. Typically drug cement is mixed in a 1:40 mass/mass ratios, suspending drug solids in the liquid cement resin using the identical operating room mixing equipment used for standard bone cements. Substantial antibiotic release occurs within several days to kill pathogens, using local release to the wound site to minimize systemic toxicity and provide initial drug concentrations above the MIC at the site. However, loaded drug dose release is far from complete (estimated only 40% bioavailability with the remainder trapped within the PMMA matrix [9, 86, 87]) and drug release kinetics drop below the MIC after a week or two [88, 89]. Nonetheless, infection still occurs on these antibiotic-releasing bone cements [90]. While infection rates are reduced with antibiotic releasing bone cement, concerns with creating additional antibiotic resistant strains are increasing.

The Synthes antibiotic-releasing polymer-coated tibial nail (see Sect. 13.3.3 for details) seeks to avoid this issue by extending effective local antibiotic release kinetics to weeks; however, while this type of polymer-coated drug release device approach has been shown to reduce orthopedic infections in certain trauma situations [91, 92], not all clinical device applications benefit from coatings or longer drug release kinetics. Efficacy appears to be host tissue site and device specific and likely also involves issues of pathogen virulence, host comorbidities and health status, and polymicrobial involvement at the implant site that vary from patient to patient.

Active surfaces that release effective doses of antimicrobial agents over longer periods of time (weeks) at concentrations above the MIC may be required to both provide reliable efficacy and also appease regulatory concerns over antibiotic resistance [29]. However, as most devices utilize drug coating [43, 86, 93–96], or impregnation [97–99] drug dosing can be very limited on-device to achieve these goals with sufficient drug release. Several approaches, including applying newly emerging antimicrobials with higher potencies (equating to lower on-board doses, lower MICs) [100], new coating formulations that improve drug loading and release control, and new antibiotic developments that seek to avoid pathogen resistance mechanisms, seek to overcome this goal. One other important strategy involves covalent attachment of antimicrobial chemistry [101–107] or active antibiotic drug directly to a device surface [9, 108–110]. Antimicrobials including penicillin, vancomycin, selenium, and several other clinically common antibiotics have been chemically immobilized on biomaterials surfaces [111–115]. Non-covalent but tightly adsorbed (bound) antibiotic layers on device surfaces represent a further iteration on this concept. Clinically recognized examples of this idea include catheters adsorbed with silver/chlorhexidine or minocylcine/rifampicin (see Sect. 13.5.1 below). Similar to silver-releasing formulations, many approaches seek to extend the release of antibiotics locally from the implanted device, including bioactive ceramic coating containing hydroxyapatite (HA) [116], polylactic acid and copolymers with glycolic acid [92, 117, 118], PET (polyethyleneterephthalate, Dacron™), PCL and PEG integrated with tobramycin coatings on allograft bone [49], among many others.

13.3 Implanted Device Infection Propagation Within a Host

Implanted devices are inserted or grafted surgically within the body. Such devices can become contaminated by pathogens perioperatively, postoperatively, and hematogenously. Perioperative infection occurs during surgery and is commonly associated with commensal ubiquitous microbes from skin and hair of those present in the surgical suite [119]. Postoperative infection results during post-surgery hospital care, linked to open wounds or wound penetration. Both perioperatively and postoperatively sourced infections can yield BAIs many years after device implantation because bacteria retained on a biomaterial implant surface can retain senescent or low metabolic capacities for years before emerging to produce a new infection [12]. Hematogenously sourced implant infections, where circulating pathogens in blood from an unrelated infection nidus seed an implant site, often years after device placement, represent a minority of BAIs [120]. More recently, host macrophages responding to implant sites as a consequence of normal wound healing have been shown to transport phagocytosed bacteria to the implant surface, where certain strains capable of surviving within macrophages then lyse the macrophages and seed the implant [121, 122]. Completely internalized implanted devices in general have lower infection rates than extracorporeal or percutaneous devices because of reduced opportunities for external pathogens to contaminate, colonize, and infect the device.

13.3.1 Prosthetic Heart Valves

Replacement cardiac valves are either synthetic mechanical or xenograft tissue constructs, with each type having a distinct set of advantages and disadvantages. Among these, tissue-based heart valves are not as susceptible as synthetic valves to prosthetic valve endocarditis (PVE). PVE is divided into three types: early PVE cases occur within 2 months of implant and are usually hospital-acquired resulting from catheter insertion and maintenance. Late PVE occurs more than 12 months after surgery and are community acquired. The third category of PVEs occurs 2–12 months after surgery, distinguished by their induction by a combination of early and late PVEs [123]. All three types of PVE are caused most commonly by coagulase-negative Staphylococci, *S. aureus* and Enterococcus spp. are also causes [123]. These microbes can originate from host flora or from other implanted devices, such as central venous catheters or from dental work. Biofilms associated with PVE form on surrounding tissue and on the valve's sewing cuff used to attach the device to tissue [124, 125]. A first major clinical application of antimicrobial technology came in the form of clinical trials of an elemental silver-coated prosthetic heart valve (Silzone) produced by St. Jude Medical [126]. Silver was coated onto the sewing cuff on the heart valve to prevent bacterial colonization of the valve. Initial biocompatibility testing with these valves was promising. However, 4 years into the clinical trial, patients began to experience elevated rates of paravalvular leakage resulting from inhibition of normal fibroblast function [127]. As a result of this leakage, the device never advanced beyond clinical trials. Newer heart valves with nanocomposite carbon coated with embedded nanosilver, as well as polyhedral oligomeric silsesquioxane and poly(carbonate urea) urethane, are being investigated for next-generation implantable valves and cardiovascular devices [44, 128].

13.3.2 Intravascular Stents

Vascular stents are used to maintain arterial flow in diseased, occluded blood vessels. With heart disease representing the United States' largest cause of death, the market for stents is large and growing, totaling an estimated $1.8 billion in 2010 with projections growing at an annual rate of 14% [7]. Stents provide structure and support to address vascular stenosis at a lesion in a blood vessel. Stenting is now the preferred interventional prophylactic method in heart disease, with >1 million stents implanted annually worldwide [9] and >600,000 in the United States [10]. Vascular stents are either bare metal or drug-eluting designs, with drug-eluting stents comprising the majority of all implanted stents [129]. Drug-eluting stunts are bare metal stents with their metal struts covered in a thin polymer/drug combination that provides controlled drug release, usually an anti-proliferative drug. Infection in coronary arteries is rare; however, when it occurs, stent infection produces significant morbidity and mortality [130]. The stent infection rate is surprisingly low (1.3–6%) given the synthetic material composition of stents, and that most eluted drugs are immunosuppressive [131]. With the low incidence of stent infection, drug eluting stents designed primarily to elute antimicrobials do not exist. The primary drugs eluted from drug eluting stents are anti-mitotic and anti-stenotic drugs [132]. In addition to desired anti-proliferative effects, several of these drugs, however, are indeed also antimicrobial. J&J's Cypher™ stents utilize the drug sirolimus or rapamycin. Sirolimus is a powerful immunosuppressant, anti-proliferative, and antifungal. Labeled use of sirolimus stents does not include indications as an antimicrobial; however, a recent study indicates sirolimus has intrinsic antimicrobial properties [133] The Cypher™ stent polymer is composed of 67/33% rations of two non-erodible polymers—polyethylene-co-vinyl acetate (PEVA) and poly *n*-butyl methacrylate (PBMA). The polymer is applied to a parylene C-coated stainless steel stent followed by another top-layer of PBMA to control release of sirolimus. The Sahajanand Supralimus Core stent utilizes a 50/50% poly-L-lactic acid (PLLA) and poly-D,L-lactide (PDLLA) and polyvinyl pyrrolidone as a polymer blend to control release. Guidant suspended use of another antimicrobial stent, the actinomycin D-eluting stent, using a drug originally used as an antibiotic but later found efficacious as a chemotherapeutic. Nonetheless, due to clinical trial safety data, stent development was suspended [78] Table 13.2).

13.3.3 Orthopedic Implants

A wide variety of devices and instrumentation are used to repair, replace, or stabilize bone pieces and joints. There are numerous types of implants including hip, knee, shoulder and elbow prosthesis, plates, nails, rods, wires, pins, and screws. Current markets for orthopedic implants are growing, with projected revenues to reach $23 billion in 2012 [40]. Typical implant infection rates are 1% for primary

Table 13.2 Selected clinically used antimicrobial vascular stents

Product	Manufacturer	Status	Agents	References
Cypher	Cordis (J&J)	FDA, CE Mark	Sirolimus (rapamycin)	[78, 133–135]
Cypher Select™ Plus	Cordis (J&J)	CE Mark	Sirolimus (rapamycin)	[78, 133–135]
Supralimus-Core™	Sahajanand Medical Technology	Trials	Sirolimus (rapamycin)	[78, 133–135]
Antinomycin D stent	Guidant	Suspended	Antinomycin D	[78, 136–138]

hip implants, 4% for knee implants, and more than 15% for trauma-associated implants [40]. Types of infections associated with orthopedic implants vary widely; however, infection risk is rising as number of implants increases and bacterial attachment on the device is a lifetime risk. Revision surgeries, due to both septic and aseptic complications, represent an even more substantial risk for infection with a 5–40% infection rate [40]. Common causative microbes include commensal bacteria, but especially *S. aureus*. Increasing evidence suggests that many of these infections are polymicrobial [40] and that *S. aureus* is the predominant culturable organism [40].

Two common clinical approaches to reducing infection in orthopedic infections are prophylactic systemic antibiotic administration and polymer-based bone cements impregnated with the aminoglycoside antibiotic, gentamicin [43, 139]. These approaches, however, are unable to address increasing BAI rates. A frequent surgical complication associated with distraction-related or external fixation is percutaneous pin tract infection [140]. These infections cause the pins and/or wires used to loosen in bone anchoring positions, and often lead to additional infections in the surrounding tissues. Smith and Nephew (UK) developed a device that delivers the antibiotic gentamicin for prophylaxis and/or infection treatment in pin tract infections: its OrthoGuard AB Antimicrobial Sleeve is a polyurethane tubing coated "on the inner and outer surfaces with an antimicrobial coating of gentamicin complexed with lauryl sulfate in a matrix of nitrocellulose and polyurethane." Studies show this device used over metallic pins is safe and effective in this use [141] (Table 13.3).

Synthes has recently reported development of an antibiotic-loaded degradable polymer sleeve for bone fixation plates [142]. Triclosan and gentamicin release are shown to extend beyond 1-week while the polymer sleeve degrades slowly around the metal plate on sheep tibia. Sheep cellular reactions at sleeve-plate tibial implant sites were slightly increased over sleeve-less plated controls, with changes related primarily to macrophages and foreign body cells. Bone formation was not affected. Antimicrobial efficacy was not reported.

Synthes markets an antimicrobial polymer-coated tibial nail in Europe coated with a thin layer of degradable poly(D,L-lactic acid) (PDLLA) impregnated with gentamicin. The polymer covers the nail surface including the cannulation, allowing local delivery of the antibiotic directly to the tibial implant site. This local delivery to the intramedullary canal means significantly more antibiotic directed to the target region for significantly longer periods of time (~2 weeks) than through systemic delivery [92, 143].

Table 13.3 Selected antimicrobial orthopedic devices approved for human clinical use

Product	Manufacturer	Status	Agents	References
PROTect Tibial Nail	Synthes	CE Mark	Gentamicin in PDLLA	[92, 143]
OrthoGuard AB™ Antimicrobial Sleeve	Smith & Nephew	FDA	Gentamicin	[141]

13.3.4 Surgical Meshes and Pouches

Surgical meshes provide scaffolding and support for tissue reconstruction, organ prolapse, and/or treatment for large wounds, commonly comprising Teflon®, Gore-Tex®, polypropylene, titanium, or other polymers. Mesh infections are a problem, with the incidence of incisional hernia mesh infections occurring about 1% during endoscopic placement and more than 15% for open surgeries [144]. Perioperative mesh contamination is the primary infection mechanism as all meshes have regions where bacteria may adhere (often leading to BAI). Common infective agents for mesh infection are the commensal pathogens including *S. aureus*, Streptococcus spp., *E. coli*, Enterococcus and Mycobacteria [144] (Table 13.4).

The number of approved antimicrobial surgical meshes is small, especially considering the number of antimicrobial technologies available for this application including silver, antibiotic, and antiseptic technologies. Dualmesh Plus™ and MycroMesh Plus™ are antimicrobial ePTFE meshes coated with a silicone coating impregnated with a silver carbonate and chlorhexidine. DualMesh is used for peritoneal reconstruction, while MycroMesh is used for hernia repair. Both silver and chlorhexidine are common antimicrobials. Used together in other applications, silver and chlorhexidine have claims to be synergistic and to minimize antimicrobial resistance [148], although further published studies are needed to verify these claims. Both Dualmesh Plus and MycroMesh Plus exhibited decreased infection rates and comparable tissue integration with ePTFE surgical meshes [145]. AIGISRX™ from TyRx is a surgical pouch that incorporates a controlled release polymer impregnated with minocycline and rifampin (both at 86 $\mu g/cm^2$) to minimize infections in cardiac pacemaker applications [146]. Lastly, FEX Textiltechnik has PVDF+PAAc+gentamicin covalently attached surgical mesh surfaces [147].

13.3.5 Cerebrospinal Shunts

The most common treatment for hydrocephalus is a cerebral shunt implant that facilitates drainage of cerebrospinal fluid (CSF) to maintain near-normal intracranial pressures [149]. Between 25,000 and 40,000 CSF shunt implant operations are performed annually in the United States, with ~7,000 of these procedures replacing clogged or infected shunts [54, 150]. Shunt infection rates range from 0.3 to 12.9% depending on the surgical environment and surgeon [151]. While

Table 13.4 Selected antimicrobial surgical meshes

Product	Manufacturer	Status	Agents	References
GORE® MYCROMESH® PLUS (SOFT TISSUE PATCHES)	Gore	FDA	Silver carbonate/ chlorhexidine diacetate	[145]
GORE® DUALMESH® PLUS (DUALMESH EMERGE & EMERGE PLUS BIOMATERIAL, SOFT TISSUE PATCH PLUS)	Gore	FDA	Silver carbonate/ chlorhexidine diacetate	[145]
AIGIS$_{RX}$™ antibacterial envelope	TyRx Pharma, Inc.	FDA	Minocycline and rifampin	[146]
FEG Textiltechnik	FEX	Europe	Gentamicin	[147]

surgical rules and standard procedures to minimize shunt infections [151] have been in place for nearly two decades, shunt infections remain a challenge [54]. Direct medical costs associated with shunt infection are estimated to be $100 million annually as shunt infection occurs in up to 27% of implanted patients, with significant associated morbidity and mortality [54, 150]. *S. aureus*, *S. epidermis*, and *C. albicans* are common microbial agents found in shunt infections. Often, systemic prophylactic antibiotics are used to minimize infection risks associated with CSF shunts. Shunt fibrosis and extracranial infections secondary to shunt infections are not effectively treated by systemic anti-fibrotics and antibiotics [150].

Antibiotic-impregnated (CNS) shunts (AIS) were developed to reduce shunt infections, showing that most if not all studies conducted using AIS show significant decreases in shunt-associated infections when compared to non-AIS devices [150, 152]. The BACTISEAL® shunt from Depuy is the market leader in AIS. BACTISEAL® AIS are manufactured with silicone impregnated with both 0.15% clindamycin and 0.054% rifampicin. Both internal and external surfaces of the shunt eluting these bioactive agents are thus antimicrobial, and have been shown to be effective in reducing infection, especially against gram positive bacteria, in clinical trials for 42–56 days [153]. Even with significant AIS infection reduction benefits over this time frame, studies have shown that clinicians are reluctant to adopt AIS due to cost [154].

13.3.6 Dental Implants

Dental implants are used in tooth replacements, periodontal ligament and soft tissue repair, orofacial bone replacement applications resulting from trauma, neoplasia, infection and disease, and/or congenital defects. The popular biomaterial for dental implants is inert alloplastic (i.e., nonbiological biomaterials) including titanium and titanium alloys, ceramics including calcium hydroxyapatite phosphates and aluminum oxides, metals such as gold, and amalgams and other alloys are also used. Also very common for endossus implants are coatings of titanium or hydroxyapatite to facilitate osseo-integration [155]. Currently over one million dental implants worldwide are implanted annually from 80 different manufacturers [156]. Implant success rates are as high as 90–95% which is striking given the diversity and abundance of flora in the oral cavity and the minimal sterility of the typical "office theatre" operating environment [157]. However, alveolar and mandibular bone is highly vascularized and pro-healing, reducing propensity for infections, despite the hard–soft tissue interface with oral mucosa. Additionally, natural oral flora is likely necessarily and remarkably probiotic in preventing unwanted microbial invasion [158]. The most common infective microbes for dental implants are the same as those usually associated with other BAIs and not those associated with periodontitis or dental abscesses. These include coagulase-negative staphylococci, coliforms, and Candida spp., with *S. aureus* exhibiting a strong affinity for titanium implants [102, 103].

Interestingly, as the oral maxillofacial region generally heals well with its rich vascular perfusion of bone, dental implants often serve as a implant testing grounds for future implant materials and designs intended for orthopedic applications. However, no antimicrobial dental devices are clinically approved in the United States. Possible reasons for the lack of antimicrobial dental devices are (1) the lack of clinical need to address such infection since oral bone heals well except in cases of diabetes, age, or smoking, (2) the unique ability to deliver topical antibiotics effectively to dental implant sites without having to associate directly with the implant, and/or (3) the potential difficulty in regulatory approval of antimicrobial dental devices due to clinical trial design problems required to show statistical reductions in infection rates over already small rates (e.g., highly powered studies involving large patient number at high cost). Evidence supporting hypothesis (2) include the three FDA-approved localized topical delivery mechanisms presently existing for dental applications, including a tetracycline in an ethylene/vinyl acetate copolymer fiber, a doxycycline hyclate gel, and minocycline microspheres. All three deliver localized antibiotic in the 1 mg/mL range, effectively treating most cases of periodontitus [159, 160].

13.4 Epithelial-Contacting Devices

Epithelial-contacting medical devices reside upon or against epithelial tissue (i.e., skin, cornea, oral mucosa) but remain in contact with the external environment. Since these devices reside technically 100% external to this tissue, do not interface with blood or other tissues directly, and remain in direct contact with the external environment, they are most susceptible to infections associated with host-derived commensal microorganisms. Moreover, these devices often function in microorganism-rich environments (i.e., intestine, mouth, urethra, trachea, eye), making implant colonization virtually guaranteed. For devices contacting epithelium are therefore efficiently and continuously seeded with diverse inoculum that colonizes the implant. This then forms the nidus for continual seeding of epithelial tissues adjacent to the implant, producing problematic infections of tissues very refractory to treatment.

13.4.1 Contact Lenses

Contact lenses for corrective, cosmetic, or therapeutic purposes are fabricated from many different types and combinations of silicone, vinyl or acrylic hydrogels and placed directly on the cornea of the eye. Nearly 110 million people use contact lenses worldwide and use is increasing [161]. Microbial keratitis (MK) results from microorganisms attaching to the lens surface both on- and off-eye (i.e., in the lens case, or as a result of hand-cleaning of the lens), forming a biofilm, and/or subsequent transfer to the ocular surface [162]. While many commensal microorganisms can lead to MK and biofilm formation and/or ocular infection, *S. aureus* and other Gram negative bacteria, especially *P. aeruginosa*, are common pathogens [162]. Approximately 10% of lenses, 50% of all types of care solutions, and up to 30% of preserved products have been shown to harbor microorganisms [162]. The outcomes for MK vary widely from a painful and inconvenient episode treated with antibiotics to hospitalization with potential vision loss (11–13% of MK cases) [163–165]. More than 50% of people with MK spend more than $1000 of direct costs to treat the problem; fortunately [165], only 1 in 30,000 contact lens wearers will suffer from MK [166] (Table 13.5).

Given their ubiquitous use and long-recognized infection risk, contact lenses have long been a focus for antimicrobial technologies. While several contact lens cleaning solutions contain antimicrobial components (antiseptics, bacteriostatics), no contact lens-associated antimicrobial technologies are currently marketed. However, a few concepts are in human clinical trials. Furanones are analogues of naturally occurring AHLs—quorum sensing molecules that interfere with biofilm development by disrupting bacterial quorum sensing, making pathogens more susceptible to antimicrobial technologies and the host's natural immune defenses. Willcox et al. have covalently attached a furanone (fimbrolide) to a high Dk (i.e., oxygen permeable) contact lens [167] and showed that the furanone-coated lens reduced adhesion of microorganisms tested; moreover, no significant differences were observed in host ocular responses. Encouraging results were also seen in clinical trials of a selenium-coated lens [168]. Selenium can generate superoxide free radicals that oxidize bacterial cells, halting their growth [169], however, these radicals are nonselective and can also harm ocular epithelium. Given economic constraints for the lens market (i.e., shift to disposable use, low price point, low profit margin, mass production), and the unique microbial and drug toxicity challenges of the lens ocular environment, antimicrobial technologies in this device sector have not moved to clinical use. Lens infections and contamination have frequently been attributed to human use-contaminated contact lens storage cases [170]. This has produced a substantial amount of effort directed at eliminating lens case microbial adhesion and growth using antimicrobial-impregnated or immobilized case materials [171] and antimicrobial lens wash solutions [172].

Table 13.5 Selected antimicrobial contact lenses in testing

Product	Manufacturer	Status	Agents	References
Furanone-coated contact lens	Institute for Eye Research (Australia)	Clinical trials	Furanone	[167]
Selenium-coated contact lens	LV Prasad Eye Institute (India)	Clinical trials	Selenium	[168, 169]

13.4.2 Endotracheal Tubes

Endotracheal tubes temporarily maintain a patient's airway open during anesthesia, acute critical care, or a traumatically compromised airway. These tubes generally comprise of polymers of silicone, latex rubber, and polyvinyl chloride (PVC), with PVC the primary biomaterial and other polymers used for specialty applications. Endotracheal tubes are inserted through the mouth into the upper airway, placed against and contacting the tracheal epithelial surface. Significantly, this device placement bypasses common endothelial defense mechanisms, blocks the upward mucosal ciliated transport and clearance mechanism operative in the airway. Moreover, the device is continuously exposed to ambient air during respiration cycles, and undergoes repeated mechanical tissue irritation due to breathing cycles. Hence, patients with this device are susceptible to ventilator-associated pneumonia (VAP) [173]. The general mechanism of infection in VAP is colonization of the endotracheal tube surface with a pathogenic bacteria followed by bacterial aspiration into the lower airways [173, 174]. The pathogen, typically an antibiotic resistant, community-acquired pathogen, commonly *S. pneumoniae, H. influenzae,* or *S. aureus*, infects and overwhelms the lung's parenchyma. Additional pathogens associated with late onset VAP include *S. pneumonia*, *Haemophilis influenza*, or *Morazella catarrhalis* [174, 175]. In US hospitals VAP results in 26,000 deaths [175] and adds $40,000 per patient per hospital admission to the cost of stay [176].

To combat VAP, the US FDA approved a silver-coated endotracheal tube (Agento I.C. ™ produced by C.R. Bard). This endotracheal tube is designed for patient placement for more than 24 h. Both the interior and exterior surfaces excluding the cuff and inflation lumen are coated with silver formulations. The interior surface has a proprietary hydrophilic silver coating. Clinical trials comparing the AGENTO® I.C. to an uncoated endotracheal tube found 4.8% incidence of VAP infection with the AGENTO® I.C. versus a 7.5% infection rate with controls. AGENTO® I.C. use was also attributed to delayed VAP onset [177].

13.4.3 Voice Prostheses

The voice prosthetic device is implanted to simulate the larynx, enabling those with a larynectomy to speak. These devices are pneumatic, comprising an array of different designs and biomaterials, including the Groningen button, Provox, Tokyo and

Van Hunen prostheses, and electronic devices including the Western Electric electrolarynx, Tait oral vibrator, Cooper-Rand electrolarynx, and the Ticchioni pipe. Voice prosthetics are classified as an extracorporeal device that, as implants in the oropharyngeal tract, directly contacting numerous epithelial tissues and are continuously exposed to salivary excretions, oral flora, consumed food, drink, and inspired and expired air from the patient. These extreme environmental exposures naturally lead to high microorganism contamination burdens. Hence, the lifetime of voice prostheses is limited primarily due to biofilm formation [178]. The most common microorganism colonizing the voice prostheses is Candida [179], causing the vast majority of silicone implant infections. The main bacterial microorganism colonizing the prostheses is streptococci [178].

While promising approaches to reduce microorganism contamination on the voice prosthetic are pursued at a research level [180, 181], only one is in clinical trials presently: the Blom-Singer® Advantage [99]. This device utilizes an impregnated silver oxide in the device matrix in order to reduce microbial colonization. The average voice prosthetic lifetime increased from 36 days to over 110 days with the Blom-Singer Advantage. Many other earlier-stage technologies are being developed for application in voice prostheses and are mentioned below in the Sect. 13.6.

13.4.4 Urinary Catheters

Catheters are clinically popular, long, flexible tubes introduced into a natural body cavity, duct, or vessel to allow passage of fluid or distention of a passageway. Over 30 million urinary catheters are inserted into patient urethras each year, many used unnecessarily [182]. Catheter-associated urinary tract infection (CA-UTI) is the most common nosocomial infection worldwide; nearly 95% of all urinary tract infections (UTIs) in hospital ICUs result from urinary catheters [183]. Most CA-UTIs result from endogenous colonic bacteria introduced from the urethral maetus and patient skin flora upon catheter insertion and during subsequent placement in the urethra. Common microbes causing CA-UTI include *S. epidermidis*, *Enterococcus faecalis*, *E. coli*, and *Proteus mirabilis* but also can include *P. aeruginosa, Klebsiella pneumonia, Providencia suartia, Morganella morganii*, and *Acinetobacter calcoaceticus* [22]. The catheter as a foreign body resting on the urethral epithelia facilitates bacterial adhesion, often leading to biofilm formation on both its external and internal surfaces and eventually leading to entry into the bladder [184]. The catheter environment is rigorous, with continuous exposure and direct contact with the external environment and patient skin provides a continuous source of contamination. Furthermore, constant device contact with urine prompts catheter encrustation—a complex reaction resulting in deposition of mineralized organic deposits from urine to the catheter lumen. This deposit is a nidus for infection and the result is a substantial risk for BAI within increasing residence time [185]. Thus, very high infection rates are reported for

Table 13.6 Selected antimicrobial urinary catheters

Device name	Manufacturer	Surface mechanism	References
SilvaGard® catheter	I-Flow/Acrymed	Ionic Silver Nanoparticles	[193]
KENDALL DOVER® Inorganic Silver Coatings (DOVER SILVER TEMPERATURE SENSING FOLEY CATHETER, DOVER SILVER HYDROGEL COATED SILICONE FOLEY, CHRONIC HEMODIALYSIS CATHETER)	Tyco Healthcare	Silver infused polymer	[194]
Silvertouch™ Coating (Foley Catheter)	Medline Industries	Silver infused polymer	
BACTI-GUARD® Silver Alloy Coating (BARDEX I.C.PLUS FOLEY CATHETER)	C.R. Bard	Silver	[195–199]

urinary catheters remaining in patients beyond 1 week [184], mandating frequent changes that are irritating and costly. In fact, after 30 days of urinary catheterization, 100% of patients in the United Kingdom developed a bacterial infection, accounting for ~40% of the documented nosocomial infections [186].

Common strategies for reduction of CA-UTIs include treatment with systemic antibiotics, antimicrobial ointments and lubricants applied during insertion, bladder installation and irrigation, antimicrobial agents in the collection bags, antimicrobial catheter coatings, catheter impregnation with silver and other antimicrobial agents [22]. Many of these are not directly catheter-associated. A listing of selected antimicrobial catheters is provided in Table 13.6. Studies to determine the efficacy of silver ion-releasing antimicrobial catheters suggest that polymer-coated silver oxide had mixed efficacy, while silver hydrogel catheters experienced reduced infection rates by as much as 30% [22]. However, further studies are necessary to determine the actual clinical efficacy of silver-treated catheters.

Historically, catheters were fabricated from natural latex rubber (urinary, Foley; 1930s) or silicone (Silastic, total parenteral nutrition, Brovic; 1973; expanded capacity and applications Hickmann; 1977 [187]). While some commercially available urinary catheters still rely on latex for their bulk material they maintain a silicone coating to prevent latex allergy. Furthermore, due to increasingly prevalent latex allergies, silicone is an increasingly popular bulk and coating material for urinary catheters, possessing rigidity that allows for thinner polymer walls and a larger open lumen [186]. In addition to latex sensitivities, four out of seven commercially available urinary catheters have demonstrated cytotoxicity in patients [186]. The selection of a catheter biomaterial represents a tradeoff between rigidity, which facilitates placement (polyurethane), and increased flow due to a larger diameter (silicone) [188, 189]. Preclinical animal models suggest an increased susceptibility to infection and thrombus formation with silicone catheters [190, 191]; whereas, polyurethane-based catheters are documented to react strongly to alcohol, limiting the topical antibiotic formulations that can be applied at the insertion site [188]. Although the development of bulk materials with intrinsic anti-adhesive properties presents a distinct advantage in the "race for the surface" [192], no catheter material or coating to date prevents host protein adhesion and the consequent conditioning films that promote infection.

13.5 Percutaneous Implant Infections

Percutaneous devices have components that reside extracorporeally, specifically that traverse epithelial tissue beds, with linked device components that reside both external to and inside the body. Components outside of the body are continuously exposed to microbes from tissue and exogenous sources that contaminate the device. Significantly, because these external components are physically connected through epithelial tissue to device components inside the body, the device provides a continuous, direct biomaterials conduit for microbial migration and colonization of components within host tissues. As a consequence these devices are very susceptible to infection, making their performance life span *in vivo* relatively short.

13.5.1 Vascular Catheters

Vascular access is achieved using three primary modes: (1) subcutaneous catheters, (2) totally implanted systems, or (3) peripherally inserted percutaneous catheters [187]. Vascular access catheters are considered essential to most current medical practices [200], with 150 million catheters purchased each year. Of these, approximately 145 million are peripheral catheters and 3–5 million are central venous catheters [201]. Most vascular access catheters are fabricated from polyurethane or silicone [188]. The majority of short-term or acute care catheters, some 70–80% of the total global catheter supply, are produced by one of five vascular catheter manufacturers in the United States: Covidien, Arrow International, Bard Access Systems, AngioDynamics, and Medcomp. Four of these manufacturers—Covidien (MAHURKAR), Arrow, Bard (Niagra), and MedComp (Duo Flow)—fabricate acute care catheters from thermoplastic polyurethane. This maintains tubing rigidity when inserted and then softens when exposed to body temperature. Alternatively, AngioDynamics Schon XL and Medcomp Hemocath use silicone-based catheter systems for acute care. All of these manufacturers also produce chronic or long-term care catheters with the addition of Boston Scientific and Spire (both US). Interestingly, although the majority of chronic care catheters are made from thermoplastic polyurethane, Covidien and Medcomp, which manufacture their acute care catheters from this material, have chosen to manufacture their chronic care systems (Permcath, Covidien, and Hemo-Cath and Hemo-Flow, Medcomp) from silicone. Table 13.7 shows selected antimicrobial commercial catheters.

While generally regarded as safe, effective, and indispensable to modern medical practices, catheters are also highly susceptible to BAI. Catheter-related bloodstream infections (CRBSI) are a leading cause of HAI, numbering over 50,000 infections per year [202], with the incidence being directly correlated with the length of hospitalization. These infections not only compromise patient health, resulting in significant morbidity, but also dramatically increase medical costs [203, 204]. Although percutaneous catheters represent a much larger market (the length of time inserted is much shorter for percutaneous catheters), central venous catheters are the largest cause of infection. It is estimated that 80–90% of CRBSIs arise from central venous catheters, adding between $29,000–$56,000 to the cost of hospitalization and treatment [205] and resulting in 7,000–14,000 deaths per year or up to a 35% case-fatality rate. This infection morbidity from catheterization alone culminates in a staggering $2.3 billion dollar cost to the healthcare industry [206]. The risk of acquiring a CRBSI ranges from 0.9 to 8% with variables such as catheter care, hospital unit (e.g., surgical department, burn unit, intensive care unit, etc.), and duration of catheter placement all playing a role in CRBSI incidence. Importantly, the process of catheter insertion alone damages skin integrity, resulting in greater risks for BAI [207]. Not only is the infection risk linked to repeat catheterization, but it is also impacted by the length of catheterization.

Limiting infection risk and determining a treatment strategy must also consider the important connection between the preconditioning host protein biomaterial coating, infection, and thrombosis. Using host proteins such as fibronectin and fibrinogen adsorbed upon device insertion, opportunistic microbes produce biofilms [192, 205, 208, 209]. Furthermore, infection and thrombosis on intravascular devices are almost inextricably connected in their etiologies [210–214]; device-triggered thrombus providing a rich environment for bacterial adhesion and growth, and bacterial adhesion predisposes the surface to complement activation, platelet adhesion and thrombus formation [210, 211, 215–217]. Thus, efforts aimed to control infection must also consider relationships to anti-thrombotic strategies. Several material and procedural modifications—standard care and insertion guidelines [199, 206] (not discussed here), physical catheter modifications (e.g., hubs, antibiotic catheter locks, cuffs, dressings), catheter biomaterial modifications (e.g., topography, intrinsically antimicrobial materials)—have been adopted. Heparin is a common clinical anticoagulant commercialized as an FDA-approved device coating (Carmeda, Sweden). Heparin-coated catheters have extensive history for inhibiting thrombosis and infection in vascular catheters, and the analogous encrustation and infection of urinary catheters. However, they also are reported to provide ancillary anti-infective activity [218, 219]. Covidien markets a combination percutaneous catheter—Tal Palindrome Sapphire—to combat thrombosis with a heparin coating and infection with an antimicrobial silver ion subcutaneous sleeve.

All intravascular catheters—hemodialysis, intravenous, or central lines—provide surfaces for BAI biofilm formation on both internal and external surfaces [220]. While biofilms may form anywhere on a catheter, within 3 days of insertion, the

Table 13.7 Selected commercial antimicrobial catheters [252, 265–281]

Device name	Manufacturer	Surface mechanism	References
ARROWg+ard Blue PLUS® Surface (PRESSURE INJECTable CENTRAL VENOUS CATHETER, QUAD-LUMEN CENTRAL VENOUS CATHETER, MULTI-LUMEN CENTRAL VENOUS CATHETER TWO-LUMEN HEMODIALYSIS CATHERIZATION KIT)	Arrow Intl.	Silver	[257]
ARROWg+ard®/Chlorag+ard® Surface [AC 3 LUMEN CENTRAL VENOUS ACCESS KIT, PRESSURE INJECTable CENTRAL VENOUS CATHETER, (PERIPHERALLY INSERTED CENTRAL CATHETER (PICC))]	Arrow Intl.	Chlorhexidine	[257]
Palindrome™ H Hemodialysis Catheter with Heparin Coating	Kendall	Heparin/silver	[270–273]
ULTRASITE® Ag LAD	B. Braun Medical	Silver	[274]
Antimicrobial CLAVE®	ICU Medical	Silver	
Agion® Antimicrobial Technology (MAXGUARD ADVANCED LAD)	Medegen Medical Manufacturing	Silver	[275–277]
V-LINK with VITALSHIELD® (CLEARLINK LUER ACTIVATED DEVICE AND EXTENSION SETS)	Baxter Healthcare	Silver	[278]
LifeShield™ Antimicrobial CLAVE™ (LATEX-FREE MICROBORE EXTENSION SET)	Hospira	Silver	[279, 280]
Spi-Argent™ (PERITONEAL CATHETHER, IRRIGATION CATHETER, MEDIASTINAL SILICONE DRAIN, RETRO SILVER HEMODIALYSIS CATHETERS)	SIL-MED	Nanocrystalline silver	[281–283]
CODMAN® BACTISEAL® (EVD, ENDOSCOPIC VENTRICULAR, and BARIUM STRIPED CATHETERS)	DePuy	Antibiotic: clindamycin and rifampicin	[97, 284–286]
Cook Spectrum® (VASCULAR CATHETERS)	Cook Medical	Antibiotic: minocycline and rifampicin	[248, 249]

different environmental and catheter conditions result in biofilm formation on external surfaces forming predominantly for catheters in place <10 days; however, the probability of luminal biofilms increases with catheter duration >30 days [207], highlighting the need for effective antimicrobial approaches active on both internal and external surfaces with different antimicrobial properties, and possibly different drugs and drug release kinetics, appropriate for each case [209, 212]. Biofilm formation on intravascular catheters can also be facilitated by catheter materials as well as surface coatings on the catheter including the host's proteinaceous conditioning film, type and properties of catheter-contacting tissue during placement, and types of infective microbes [207]. Historically, catheters have been made from extruded PVC, silicone, polyurethanes, and TeflonTM. Importantly, a pannus (fibrous attached membrane) forms on the luminal surface of the catheter due to the natural extended blood contact activation response to the implanted material. Microbial adhesion to this pannus in the catheter lumen promotes infection potentially acting as a nidus for bacterial growth [212, 221].

Approaches to address this overwhelming and seemingly inescapable problem must recognize not only the pathogen source but also consider the pathogen phenotype (i.e., virulent, quiescent, proliferative, biofilm-associated, planktonic, etc.) in determining proper approaches for abatement [189], as sessile and planktonic pathogens display distinct sensitivities to antimicrobial strategies. The primary causative organisms are Staphylococcus spp. and Candida spp. adhered to the catheter surface, depending on the site of insertion [200, 222–224]. In particular, CRI are often caused by coagulase-negative staphylococci, primarily *S. epidermidis*, *S. aureus*, *P. aeruginosa*, and *E. faecalis* [224]. Candida commonly forms biofilms with other microbes, potentiating a polymicrobial infection risk[207].

Although standard of care guidelines have reduced the risk of infection, they are inadequate to completely address the scope of the BAI problem. Physical catheter design modifications must also be considered for infection resistance. Antimicrobial lock therapy is becoming an increasingly popular treatment alternative. A high concentration of antibiotic is flushed into the lumen of the infected catheter for an extended period [225]. Unfortunately, antimicrobial lock solutions are usually inadequate to treat Candida spp. infections, and the current recommendation is removal and replacement [226]. Furthermore, infusion of high concentrations of antibiotics might increase risks of antibiotic resistance, promoting skepticism in the medical community regarding the value of this treatment strategy [189].

Over the past three decades, some catheter-related anti-infective strategies have integrated mechanical design alternatives (e.g., skin cuffs) into current technologies or applying pharmaceutical innovations to better control thrombosis and infection. Cuffing and tunneling of percutaneous catheters reduces the infection rate to 77% [188]. However, such mechanical alterations have found a modicum of success in reducing infection for short-term catheterization but wider acceptance in the commercial medical device arena. A combination of physical catheter design modifications with biomaterial modifications offers opportunities for long-term mitigation of infectious complications. Covidien markets the Tal Palindrome Ruby, which has a silver ion-releasing sleeve bound between the hub and the cuff. The position of this sleeve along the length of the catheter allows this design to deliver silver ions to the catheter surface to reduce colonization in the tunnel track [188].

Catheters can also be completely modified with polymeric coatings that attempt to control bacterial adhesion and hinder colonization. PTFE coatings were introduced in the 1960s to ameliorate patient discomfort during insertion; however, in terms of impact on infection, there is no conclusive evidence to support the concept that PTFE-coated catheters hinder or reduce bacterial adhesion [227]. By contrast, hydrogel coatings have been reported to enhance bacterial migration on catheter surfaces [186]. Alternatively, polymeric hydrogel coatings on a silicone catheter, despite their ability to retain silver [228–230] or liposomal encapsulated antibiotics [231, 232] have exhibited inadequate material properties to withstand the forces necessary for function [186]. Additionally, polymer microbe-resistant surfaces, while colonization-resistant *in vitro*, are seriously compromised by the formation of the ubiquitous host protein conditioning film *in vivo*. Currently, the only clinically available polymer-modified catheter is Hydrocath® (Becton Dickinson) [60], a hydrophilic poly(vinylpyrrolidone)-coated polyurethane catheter. Importantly, some work suggests that hydrophilic catheters promote bacterial surface adhesion [233–236]. Chemically tailored polymers that exhibit certain regular topological features to direct and manipulate protein adhesion and subsequent bacterial interactions provide a passive, yet innovative, option for infection-resistant materials [237]. Channels, ridges, pillars, pits, and biomimetic ribs in specific dimensions and patterns (e.g., Sharklet™AF technology, USA) have all been fabricated on a silicone rubber elastomers using photolithography to prevent surface fouling and are presently being developed as antimicrobial surfaces [238].

These materials-only approaches are now decades old and relatively unconvincing in their performance *in vivo*: independent of additional adjunct pharmaceuticals, they are largely incapable of addressing the clinical infection problem alone. The Hydrocath catheter base coating has been useful for producing several further material innovations in its performance; its hydrophilic coating has allowed successful incorporation of antiseptic benzalkonium chloride (Hydrocath Assure, BD; Multi-Med, Edwards Life Sciences) [239] and iodine [240]. Alternatively, benzalkonium chloride has also been bound by heparin coatings (AMC Thromboshield treatment, Edwards Life Sciences) [209] and suspended in a Hydromer gel (Hydromer, Inc.) [209]. Unfortunately, although these designs have shown promising results in preclinical animal models and a limited number of clinical trials [240–244], meta-analysis revealed the inability of benzalkonium chloride coatings to prevent CRBSIs [209]. A 1991 clinical trial of antibiotic-bonded catheters pretreated with the cationic surfactant tridodecylmethylammonium chloride and subsequently bound with the anionic antibiotic, cefazolin, concluded that antibiotic-bonded catheters were safe and effective. However, as of 2011, this technology has still not entered the marketplace [222].

Antimicrobial surfaces can also be prepared by soaking the biomaterial in an antibiotic solution, either adsorbing or chemically binding the antimicrobial to device surfaces, or directly impregnating the antibiotic into the coating material

[237]. Angiotech received 510 k clearance from the FDA in 2008 for a central venous catheter with a 5-fluorouracil (5-FU) coating to mitigate concerns about acquired antibiotic and silver bacterial resistance [245]. Contrary to classic antibiotic modes of action, 5-FU makes the catheter surface inhospitable to bacterial colonization [246]. Tethering various anti-infective agents (silver-containing compounds, quaternary ammonium compounds, heparin, chlorhexidine, antibiotics) directly to catheter surfaces represents a relatively effective, well-known approach [189, 247].

While dozens of reports describe various antibiotics adsorbed to or coated on biomaterials surfaces, the only marketed antibiotic-coated vascular catheter shown to effectively control infection is a minocycline–rifampicin coated catheter (marketed as the Cook Spectrum™ catheter, Cook Critical Care) where both the internal and external surfaces are impregnated with minocycline and rifampicin. In a multicenter trial, this minocycline–rifampicin-coated catheter was found to be threefold less likely to be colonized when compared to a chlorhexidine silver sulfadiazine (CHSS)-coated catheter and 12-fold less likely to lead to a life-threatening catheter-related blood stream infection [248, 249]. While more expensive, the use of an antimicrobial-bonded catheter may offset the long-term costs of catheter-related infection treatment [189, 247]. Interestingly, the medical community remains either skeptical that antimicrobial efficacy claims are exaggerated, or wary of such strategies for prophylactic use due to the potential development drug-resistant bacteria. Thus, antibiotic-bound catheters are only recommended for high-risk patients with anticipated catheterization greater than 5 days [209].

Silver in various forms has an extensive history of antimicrobial use on catheters to inhibit adhesion of both gram-positive and gram-negative microbes [250], particularly for urinary catheters [251, 252]. Two silver-based antimicrobial urinary catheters—Bardex I.C. and Lubrisil I.C. (CR Bard) have been FDA-approved in the United States [117]. Silver ion (Ag^+), the oxidized form of metallic silver, is considered the primary antimicrobial bioactive species and is sourced in various forms: from metallic silver coatings, nanoparticles, ion implants, from silica-based zeolite complexes, as slightly soluble salts (e.g., SSD), and other solids (silver oxide, alloys, bioglasses) [253]. Bard markets two hemodialysis catheters with their BioBloc coating eluting SSD to prevent bacterial adhesion. However, cytotoxic, thrombotic, and other adverse reactions to silver, possible pathogen resistance, controversial efficacy claims, as well as material processing and its crowded, complex intellectual property space all plague its mainstream application and use [252, 254]. Combinations of silver with another antiseptic have provided new options for clinical applications. In this regard, a polyurethane catheter impregnated with small amounts of chlorhexidine and silver sulfadiazine (CHSS) (ArrowGard, Arrow International) is clinically approved [209]. This first-generation CHSS catheter, introduced over a decade ago, is coated only on its external surface. However, the second-generation design (ArrowGard Plus) is coated on its external and luminal surfaces, providing efficacy beyond the 15 days [255] seen in the first-generation design. Although both catheter designs are claimed to offer significant reductions in colonization and resulting catheter-related blood stream infection [205, 256, 257], a limited number of reports detail resistance to chlorhexidine *in vitro* [258, 259] as well as anaphylactic reactions in patients, most likely due to chlorhexidine use, in Japan and the United Kingdom [260–262]. CHSS catheters were withdrawn from the Japanese market in 1997 [148, 263] amid conflicting reports of clinical safety. A combination silver-carbon/platinum-incorporated polyurethane catheter (Oligon) facilitating Ag+ ion release to the surrounding tissue was marketed as a peripherally implanted central catheter (Olympic™, Vygon, UK). Unfortunately, this catheter suffered from a mechanical instability and was withdrawn from the market in Germany. A more recent design using a similar technology (Oligon Vantex®, Edwards Life Science) provides this same anti-infective capability and does not seem to suffer from mechanical problems [264, 265].

In summary, claims for clinical efficacy of antimicrobial-coated catheters compared to uncoated catheters in reducing CRBSIs come from several published randomized controlled trials, meta-analyses and careful performance reviews [266]. Nonetheless, factors to consider in adopting these catheters are not so simple, complicated by their increased costs relative to uncoated catheters, uncertainty in the patient risks associated with each CRBSI, and clinician uncertainty over the efficacy of various antimicrobial catheters marketed [247]. This clinical uncertainty is reflected in a recent survey of hospitals finding that only 32% of US Veterans Affairs hospitals and 38% of non-Veterans Affairs hospitals adopted antimicrobial-impregnated catheters [267]. The most recent US Centers for Disease Control and Prevention recommendations for preventing CRBSIs reflect complex decision processes for these devices: "[here at start and also at end with] antimicrobial or antiseptic-impregnated CVC [central venous catheters] should be used in adults whose catheter is expected to remain in place >5 days if, after implementing a comprehensive strategy to reduce rates of catheter-related BSI, the rate remains above the goal set by the individual institution based on benchmark rates and local factors. The comprehensive strategy should include the following three components: educating persons who insert and maintain catheters, use of maximal sterile barrier precautions, and a 2% chlorhexidine preparation for skin antisepsis during CVC insertion (category IB)" [206]. Thus, the decision to use an antimicrobial catheter is not simply a device-based decision. Medical care providers must carefully consider several factors, including baseline rates of CRBSI, costs of treating the infection, number of patients requiring central venous catheters, duration of catheterization and the use of other preventive measures that may reduce the risk of infection, such as the catheter infection-prevention bundle [268]. The next step is assessing device efficacy in this context and should compare the effectiveness of drug-impregnated catheters with other clinical actions recently shown to reduce infection risk, such as chlorhexidine-impregnated sponge dressings [269].

13.5.2 Sutures

While modern surgical techniques and facilities continue to evolve and progress, surgical site infection persists in modern surgery not only as a common complication but also a major cause of postoperative morbidity [287]. Surgical site infection rates vary worldwide—the US CDC estimates are 2.8% [288] and European rates range from 1.5 to 20% [287, 289]. Given over 50 million annual surgeries or wound closures in the United States, surgical site infections occur in over 500,000 patients every year [290]. While estimates are difficult to calculate, the costs are estimated to be 1.47–19.1 billion Euros for European infections [289]. Many of these costs might be ameliorated with appropriate wound closure.

Wound closure methods in surgery are also evolving, with new technologies in surgical adhesives, glues and staples introduced frequently [291, 292]. Nonetheless, sutures remain the primary method of choice for closing surgical sites, selected in over 80% of all wound closure cases. Sutures are classified into two major categories: absorbable and non-absorbable. US suture markets currently are approaching $1 billion/year, with non-absorbable sutures (primarily nylons) leading the market. Originally, sutures were composed of biological materials including catgut and silk. More modern sutures materials are now synthetic and include the absorbable materials (polyglycolic acid, polyactic acid, and polydioxanone) and non-absorbable materials (nylon and polypropylene). While being classified as non-absorbable or absorbable, sutures are also characterized as natural or synthetic and multifilament (e.g., braided) or monofilament. No single material is ideal for all wound closure applications, and often compromises are made in use, associated with surgical site context. Ideal performance characteristics of suture materials include strength, elasticity, hypoallergenicity, and non-wicking behavior. Non-wicking properties are an important feature in minimizing bacterial colonization of the suture. Antimicrobial properties of suture materials are increasingly important. Sutures with clinically proven antimicrobial activities are listed in Table 13.8. Since infection potential is reduced by elimination of the biomaterial nidus from the host, an intuitive conclusion would be that absorbable materials would represent attractive suture material because of their elimination within several weeks in many cases. However, this rationale is not governing current suture selection by clinicians.

In 2002, Ethicon received FDA approval for its antimicrobial suture that utilizes the antiseptic and bacteriostatic, triclosan, also known as irgosan or IRGACARE® MP. Triclosan is used in a large number of antimicrobial products including hand soaps and toothpastes—so common in household products that an FDA study found triclosan in the urine of 75% of Americans [293]. Ethicon utilizes triclosan for their complete line of antimicrobial sutures, including Vicryl Plus (polyglactin 910), Monocryl Plus (poliglecaprone 25), and PDS Plus (polydioxanone). Ethicon antimicrobial sutures utilize the base suture materials with small amounts of impregnated calcium stearate and <1.0% IRGACARE® MP.

Perceptions of clinical efficacy and performance of antimicrobial sutures are mixed. While numerous studies suggest that Ethicon's antimicrobial sutures are both safe and effective (Table 13.8, [294–303]), a small meta-analysis of clinical trials conducted by the Canadian Agency on Drugs and Technologies in Health Care shows mixed efficacy results [304]. Nevertheless, they concluded that small differences in costs between normal suture materials and antimicrobial suture materials ($21 versus $30 per surgery) versus the costs associated with reduction in surgical site infections warrant further clinical investigation. Further complicating efficacy issues is the fact that the FDA is investigating the safety and efficacy of triclosan as

Table 13.8 Clinically approved antimicrobial suture materials

Product	Manufacturer	Status	Agents	References
Vicryl, Monocryl, and PDS Plus	Ethicon, Inc.	FDA	Triclosan	[294–303]

a general antimicrobial [293]. In the words of the FDA, "We are engaged in an ongoing scientific and regulatory review of the safety of triclosan in FDA-regulated products" [293]. Concerns with antimicrobial resistance are one of the primary concerns associated with triclosan and will cloud future-of-use strategies for device makers until this is resolved.

Several emerging technologies applied to antimicrobial sutures appear on the horizon. Silver technologies, already discussed extensively here in this chapter, are also being incorporated into suture copolymers by manufacturers such as X-Static. Antibiotic-containing sutures are also being developed. Polymedix is developing its PolyCide™-containing antimicrobial suture material. Polycide is, "like PolyMedix's antibiotic PMX-30063 which is currently in a Phase 2 clinical trial for treatment of Staphylococcus infections," as an antimicrobial that disrupts cell membrane [305]. Other antibiotics that have seen some degree of interest as antimicrobial sutures include neomycin, gentamicin, penicillin, sulfonamide, quaternary ammonium complexes, chlorhexidine, and iodine [306]. Applications of these antimicrobials include either dipped coating (with pretreatments and/or polymer carriers) or chemical surface modifications to generate a covalent attachment. These technologies, however, require significant *in vivo* studies before clinical use [306].

13.6 Emerging Antimicrobial Devices

While beyond the scope of this chapter, several early stage concepts are worth brief descriptions as possibly future emerging technologies to combat BAIs. These include (1) nitric oxide-releasing surfaces, (2) photoactive TiO_2 coatings, (3) emerging biosurfactants, (4) new antimicrobial biopharmaceutical agents, (5) new polymers that innately reduce bacterial adhesion *in vivo*, and (6) polymerized antimicrobial agents that degrade to release the bioactive small molecule drug monomers. Nitric oxide (NO) is a highly reactive biologically produced, cell-sourced radical that has proven *in vitro* to be a powerful, broad spectrum antimicrobial agent [307]. Recent progress in NO-releasing surfaces includes organic–inorganic composite films [308], NO-releasing gold nanoparticles [309], NO-releasing silica nanoparticles [310, 311], NO-releasing nanoparticle impregnated hydrogel/glasspowder composite [312], among some other NO-releasing polymer coating formulations [313, 314]. Fujishima and Honda [315] developed TiO_2 surfaces that when exposed to UV radiation generate reactive oxygen and hydroxyl radicals [316]. These have widely recognized antimicrobial effects. This radical-producing technology has advantages in terms of UV-based turn-off/turn-on effects, and the obvious practical disadvantages associated with required access to UV light-based sources as triggers for these effects within human tissue or the implant site. This free radical production approach has not been reported in antimicrobial contexts for medical implants in tissue. Biosurfactants are naturally derived amphiphilic compounds that accumulate at interfaces as they possess both hydrophilic and hydrophobic chemistry that is surface active. Many biosurfactants are produced by bacteria as part of their endogenous defensive mechanisms against other bacteria. Medical applications for these molecules include catheters coatings [317], silicone rubber devices [318], contact lenses [317] and specific applications where genetically engineered or synthetic mimics of biosurfactant molecules might be optimized for structure–activity relationships against BAI pathogens and further applied [318]. Zwitterionic surfaces, incorporated onto devices as coatable copolymers bearing zwitterionic side chains of several different chemistries (e.g., phosphorylcholine, carboxybetaine, and sulfobetaine) are a continuing interest as bacterial-resistant coatings, exhibiting interesting behaviors *in vivo* [319–321]. In addition to new surfaces, the antimicrobial device landscape is broadening to consider release of different antimicrobials including (1) antimicrobial peptides both natural [322] and synthetic [323–325], (2) antimicrobial proteins and therapeutic anti-infective antibodies and antibody fragments [326], (3) antimicrobial coatings derived from aqueous polymer suspensions in contrast to organic solvent-cast coatings [327, 328], and (4) synthetic macromolecular pro-drugs [329] such as vancomycin-based polymers that degrade on-device to release their monomeric components as antibiotics [330–332]. These emerging technologies offer promising new options for studies reducing medical device infection rates.

13.7 Summary

Implanted medical devices will increase dramatically as device technologies improve and as the number of adults treated with implants increase, many in developing countries. The risk of infection for persons with medical devices (so-called BAIs) is substantial and, with antibiotic resistant pathogens increasing without effective mitigation, the probabilities for infections are increasing. Clinical approaches to this challenge are multifactorial, involving surgical techniques, operating room protocols and facilities, pre- and post-surgery hospital procedures, patient-personnel contact, device designs, patient profiling and personalization, and antimicrobial innovations. Current clinically approved methods for reducing BAI include many types of antimicrobial device coatings that either intrinsically limit pathogen colonization through topology and chemistry, or actively kill through antibiotic release or immobilization. Other contributions to the problem include improved sterile surgical techniques and methodologies, and altered device designs such as catheter impregnated cuffs and antibiotic lock therapies.

While these approaches demonstrate some metric of effectiveness in reducing infection risk, all current devices including those with antimicrobial technologies remain susceptible to BAIs. In many cases, existing clinical data for improved antimicrobial device performance are under-powered and inconclusive, or outright contradictory. In almost all cases, *in vitro* efficacy for a given strategy is produced in bacterial assay test beds that have little relevance to *in vivo* operating conditions for infections and provide little predictive reliability to *in vivo* expectations for BAIs. Creative product marketing often substitutes for actual supporting clinical evidence for effectiveness claims against BAIs. That is, while device labeling claims may not claim anti-infective efficacy, marketing claims can be distinctly different implying "antimicrobial" properties with descriptive product language such as "antibiotic-eluting" or "gentamicin-containing." Despite compelling clinical needs, this current multifactorial debate over the risk-benefits for antimicrobial device technologies has injected substantial skepticism into the field of antimicrobial medical device development and applications, as well as regulatory oversight driven by FDA concern for safety and efficacy and in possibly promoting antimicrobial resistant pathogens [29]. In this regard, antimicrobial device research should better proceed down research and development validation pathways that more readily and convincingly translate into clinical applications that address BAIs. The US Center for Disease Control has outlined several key research needs for better addressing clinical issues that link biofilms, devices and BAIs [176]. First, the ability to collect, assay and quantify reproducible features of biofilms must be advanced for each type of device and pathogen. Second, development of model biofilm systems should be better developed to closely simulate *in vivo* conditions encountered by the device. Third, improved understanding for how biofilms relate to antimicrobial drug resistance [333] and how biofilms spread to seed other types of infections (e.g., bloodstream) must be acquired. These variables must then be correlated to device designs and materials aspects to guide rational device-based design innovations to address BAIs.

Lastly, medical device innovations often get stranded in academic publications at a basic level without successful transition to further efforts impacting development, on-device testing and possible approval on medical devices. Researchers need to develop technologies to the point that businesses are enticed to finish the development of a product. Too often with these ideas, the simple, easy colonization experiments *in vitro* are performed with exaggerated claims to antimicrobial efficacy, and no proof-of-concept under relevant conditions of BAI or device infection. Often these studies have been a "red herring" to the field, reducing the credibility in delivering meaningful ideas and new technologies important and impacting to clinical use. Commercial investment and government agencies should provide incentives to help protect, select and direct the more practical promising antimicrobial device applications into key validation protocols to certify better technologies against BAIs and bridge the research-to-product gap.

Acknowledgments The authors acknowledge support from the University of Utah Technology Commercialization Office and from NIH grant EB01473 and EB00894.

Glossary

Antimicrobial A pharmaceutical agent used to mitigate, eradicate, and/or prevent microbial infections.

Biomaterial Material, either synthetic or natural, that is suitable for contact with living tissue or bodily fluid as a part of a medical device.

Biomaterial Associated Infection (BAI) A microbial infection on, around, or resulting from biomaterial contact with human tissues or fluids.

Biofilm An interface-associated colony or colonies of microbes embedded in a complex extracellular matrix (ECM) that is composed primarily of polymers and proteins. Microbes in biofilms have additional resistance to antimicrobials due to reduced metabolism and the protective features of the ECM.

Hospital acquired infection (HAI) Microbial infections without evidence of inoculation or incubation at before admission to a healthcare environment.

Implanted device Medical device that is completely inserted or grafted into the body.

Medical Device (US FDA standard definition) An instrument, apparatus, implement, machine, contrivance, implant, in vitro reagent, or other similar or related article, including a component part, or accessory which is:

Recognized in the official National Formulary, or the US Pharmacopoeia, or any supplement to them,

Intended for use in the diagnosis of disease or other conditions, or in the cure, mitigation, treatment, or prevention of disease, in man or other animals, or

Intended to affect the structure or any function of the body of man or other animals, and which does not achieve any of its primary intended purposes through chemical action within or on the body of man or other animals and which is not dependent upon being metabolized for the achievement of any of its primary intended purposes.

Microbe A microorganism, especially bacteria, fungi, viruses, and protozoa, that causes infection.

Nosocomial See Hospital-acquired infections.

Percutaneous Implants A medical device that traverses the epithelial layer.

References

1. US Food and Drug Administration. Is the product a medical device? 2002; http://www.fda.gov/medicaldevices/deviceregulationandguidance/overview/classifyyourdevice/ucm051512.htm, accessed June, 2012.
2. Simchi A, et al. Recent progress in inorganic and composite coatings with bactericidal capability for orthopaedic applications. Nanomedicine. 2011;7(1):22–39.
3. Chu PK, Liux, editors. Biomaterials fabrication and processing handbook. Boca Raton: CRC Press, Taylor & Francis Group; 2008. p. ix.
4. Zimmerli W, Trampuz A. Implant-associated infections. In Bjarnsholt T, Moser C, Jensen P, Rigshopitalet HS, Hoiby N (Eds): Biofilm infections. Springer Science; 2011. pp. 69–89.

5. Medical Devices Industry Outlook—April 2011. 2011 [cited 2012 1/5/2012]. http://www.informationweek.com/news/mobility/business/232301163
6. Popat KC, et al. Titania nanotubes: a novel platform for drug-eluting coatings for medical implants? Small. 2007;3(11):1878–81.
7. Darouiche RO. Treatment of infections associated with surgical implants. N Engl J Med. 2004;350(14):1422–9.
8. von Eiff C, et al. Infections associated with medical devices: pathogenesis, management and prophylaxis. Drugs. 2005;65(2):179–214.
9. Zilberman M, Elsner JJ. Antibiotic-eluting medical devices for various applications. J Control Release. 2008;130:202–15.
10. Hall-Stoodley L, Costerton JW, Stoodley P. Bacterial biofilms: from the natural environment to infectious diseases. Nat Rev Microbiol. 2004;2(2):95–108.
11. Gilbert P, Collier PJ, Brown MR. Influence of growth rate on susceptibility to antimicrobial agents: biofilms, cell cycle, dormancy, and stringent response. Antimicrob Agents Chemother. 1990;34(10):1865–8.
12. Gristina A. Biomaterial-centered infection: microbial adhesion versus tissue integration. Clin Orthop Relat Res. 2004;(427):4–12.
13. Gristina AG. Biomaterial-centered infection: microbial adhesion versus tissue integration. Science. 1987;237(4822):1588–95.
14. Springer BD, et al. Systemic safety of high-dose antibiotic-loaded cement spacers after resection of an infected total knee arthroplasty. Clin Orthop Relat Res. 2004;427:47–51.
15. Zalavras CG, Patzakis MJ, Holtom P. Local antibiotic therapy in the treatment of open fractures and osteomyelitis. Clin Orthop Relat Res. 2004;427:86–93.
16. Laupland KB, et al. Cost of intensive care unit-acquired bloodstream infections. J Hosp Infect. 2006;63(2):124–32.
17. Bryers JD. Medical biofilms. Biotechnol Bioeng. 2008;100(1):1–18.
18. Pollack A. Rising threat of infections unfazed by antibiotics. In: New York Times; 2010.
19. Walker EP. Medicaid to quit paying for preventable events. In: Medpage Today. Washington, DC; 2011.
20. Price JS, et al. Controlled release of antibiotics from coated orthopedic implants. J Biomed Mater Res. 1996;30(3):281–6.
21. Ruszczak Z, Friess W. Collagen as a carrier for on-site delivery of antibacterial drugs. Adv Drug Deliv Rev. 2003;55(12):1679–98.
22. Donlan RM, Costerton JW. Biofilms: survival mechanisms of clinically relevant microorganisms. Clin Microbiol Rev. 2002;15(2):167–93.
23. Parsek MR, Greenberg EP. Sociomicrobiology: the connections between quorum sensing and biofilms. Trends Microbiol. 2005;13(1):27–33.
24. Hetrick EM, Schoenfisch MH. Reducing implant-related infections: active release strategies. Chem Soc Rev. 2006;35(9):780–9.
25. Smith AW. Biofilms and antibiotic therapy: is there a role for combating bacterial resistance by the use of novel drug delivery systems? Adv Drug Deliv Rev. 2005;57(10): 1539–50.
26. Francolini I, Donelli G. Prevention and control of biofilm-based medical-device-related infections. FEMS Immunol Med Microbiol. 2010;59(3):227–38.
27. Gold HS, Moellering Jr RC. Antimicrobial-drug resistance. N Engl J Med. 1996;335(19):1445–53.
28. Gransden WR. Antibiotic resistance. Nosocomial gram-negative infection. J Med Microbiol. 1997;46(6):436–9.
29. FDA. Premarket Notification [510(k)] Submissions for Medical Devices that include antimicrobial agents. FDA; 2007.
30. Davis SC, et al. Microscopic and physiologic evidence for biofilm-associated wound colonization *in vivo*. Wound Repair Regen. 2008;16(1):23–9.
31. Otto M. Staphylococcal Biofilms, Vol 322. In: Romeo T, editor. Current Topics in Microbiology and Immunology:Bacterial Biofilms. Springer Verlag: Berlin; 2008.

32. Yao Y, et al. Factors characterizing *Staphylococcus epidermidis* invasiveness determined by comparative genomics. Infect Immun. 2005;73(3):1856–60.
33. Harris LG, Richards RG. Staphylococci and implant surfaces: a review. Injury. 2006;37 Suppl 2:S3–14.
34. An YH, Friedman RJ. Prevention of sepsis in total joint arthroplasty. J Hosp Infect. 1996;33(2):93–108.
35. Friedman RJ. Infection in total joint arthroplasty from distal intravenous lines. A case report. J Arthroplasty. 1988;3(Suppl):S69–71.
36. Broekhuizen CA, et al. *Staphylococcus epidermidis* is cleared from biomaterial implants but persists in peri-implant tissue in mice despite rifampicin/vancomycin treatment. J Biomed Mater Res A. 2008;85(2):498–505.
37. Al-Fattani MA, Douglas LJ. Penetration of Candida biofilms by antifungal agents. Antimicrob Agents Chemother. 2004;48(9):3291–7.
38. Adam B, Baillie GS, Douglas LJ. Mixed species biofilms of *Candida albicans* and *Staphylococcus epidermidis*. J Med Microbiol. 2002;51(4):344–9.
39. Rohacek M, et al. Bacterial colonization and infection of electrophysiological cardiac devices detected with sonication and swab culture. Circulation. 2010;121(15):1691–7.
40. Trampuz A, Widmer AF. Infections associated with orthopedic implants. Curr Opin Infect Dis. 2006;19(4):349–56.
41. Poelstra KA, et al. Prophylactic treatment of gram-positive and gram-negative abdominal implant infections using locally delivered polyclonal antibodies. J Biomed Mater Res. 2002;60(1):206–15.
42. Elek SD, Conen PE. The virulence of *Staphylococcus pyogenes* for man; a study of the problems of wound infection. Br J Exp Pathol. 1957;38(6):573–86.
43. Wu P, Grainger DW. Drug/device combinations for local drug therapies and infection prophylaxis. Biomaterials. 2006;27(11):2450–67.
44. Chaloupka K, Malam Y, Seifalian AM. Nanosilver as a new generation of nanoproduct in biomedical applications. Trends Biotechnol. 2010;28(11):580–8.
45. Hartung T. Thoughts on limitations of animal models. Parkinsonism Relat Disord. 2008;14 Suppl 2:S81–3.
46. Levy SB. Active efflux, a common mechanism for biocide and antibiotic resistance. J Appl Microbiol. 2002;92(Suppl):65S–71.
47. Sampath LA, Tambe SM, Modak SM. *In vitro* and *in vivo* efficacy of catheters impregnated with antiseptics or antibiotics: evaluation of the risk of bacterial resistance to the antimicrobials in the catheters. Infect Control Hosp Epidemiol. 2001;22(10):640–6.
48. Patti JM, et al. MSCRAMM-mediated adherence of microorganisms to host tissues. Annu Rev Microbiol. 1994;48:585–617.
49. Rodrigues L, et al. Inhibition of microbial adhesion to silicone rubber treated with biosurfactant from *Streptococcus thermophilus* A. FEMS Immunol Med Microbiol. 2006;46(1):107–12.
50. Rodrigues L, et al. Biosurfactant from *Lactococcus lactis* 53 inhibits microbial adhesion on silicone rubber. Appl Microbiol Biotechnol. 2004;66(3):306–11.
51. Rodrigues L, et al. Influence of biosurfactants from probiotic bacteria on formation of biofilms on voice prostheses. Appl Environ Microbiol. 2004;70(7):4408–10.
52. Habash M, Reid G. Microbial biofilms: their development and significance for medical device-related infections. J Clin Pharmacol. 1999;39(9):887–98.
53. Lee JH, et al. Prosthetic valve endocarditis: clinicopathological correlates in 122 surgical specimens from 116 patients (1985–2004). Cardiovasc Pathol. 2011;20(1):26–35.
54. Sampedro MF, Patel R. Infections associated with long-term prosthetic devices. Infect Dis Clin North Am. 2007;21(3):785–819. x.
55. Darouiche RO. Antimicrobial coating of devices for prevention of infection: principles and protection. Int J Artif Organs. 2007;30(9):820–7.
56. Darouiche RO. Antimicrobial approaches for preventing infections associated with surgical implants. Clin Infect Dis. 2003;36(10):1284–9.
57. Linke D. Bacterial adhesion: chemistry, biology and physics. London: Springer; 2011.

58. Kingshott P, Wei J, Bagge-Ravn D, Gadegaard N, Gram L. Covalent attachment of poly(ethylene glycol) to surfaces, critical for reducing bacterial adhesion. Langmuir. 2003;19:6912–21.
59. Kaper HJ, Busscher HJ, Norde W. Characterization of poly(ethylene oxide) brushes on glass surfaces and adhesion of *Staphylococcus epidermidis*. J Biomater Sci Polym Ed. 2003; 14(4):313–24.
60. Tebbs SE, Sawyer A, Elliott TS. Influence of surface morphology on *in vitro* bacterial adherence to central venous catheters. Br J Anaesth. 1994;72(5):587–91.
61. Nagel JA, Dickinson RB, Cooper SL. Bacterial adhesion to polyurethane surfaces in the presence of pre-adsorbed high molecular weight kininogen. J Biomater Sci Polym Ed. 1996;7(9):769–80.
62. Gosheger G, et al. Silver-coated megaendoprostheses in a rabbit model—an analysis of the infection rate and toxicological side effects. Biomaterials. 2004;25(24):5547–56.
63. Lee D, Cohen RE, Rubner MF. Antibacterial properties of Ag nanoparticle loaded multilayers and formation of magnetically directed antibacterial microparticles. Langmuir. 2005;21(21):9651–9.
64. Rojas IA, Slunt JB, Grainger DW. Polyurethane coatings release bioactive antibodies to reduce bacterial adhesion. J Control Release. 2000;63(1–2):175–89.
65. Blaker JJ, Nazhat SN, Boccaccini AR. Development and characterisation of silver-doped bioactive glass-coated sutures for tissue engineering and wound healing applications. Biomaterials. 2004;25(7–8):1319–29.
66. Song WH, Ryu HS, Hong SH. Antibacterial properties of Ag (or Pt)-containing calcium phosphate coatings formed by micro-arc oxidation. J Biomed Mater Res A. 2009;88(1):246–54.
67. Chang TW, Weinstein L. Prevention of herpes keratoconjunctivitis in rabbits by silver sulfadiazine. Antimicrob Agents Chemother. 1975;8(6):677–8.
68. Jung WK, et al. Antibacterial activity and mechanism of action of the silver ion in *Staphylococcus aureus* and *Escherichia coli*. Appl Environ Microbiol. 2008;74(7):2171–8.
69. Yang W, et al. Food storage material silver nanoparticles interfere with DNA replication fidelity and bind with DNA. Nanotechnology. 2009;20(8):085102.
70. Feng QL, et al. A mechanistic study of the antibacterial effect of silver ions on *Escherichia coli* and *Staphylococcus aureus*. J Biomed Mater Res. 2000;52(4):662–8.
71. Shrivastava, Siddhartha and Bera, Tanmay and Roy, Arnab and Singh, Gajendra and Ramachandrarao, P and Dash, Debabrata. Characterization of enhanced antibacterial effects of novel silver nanoparticles Nanotechnology. 2007; 18(22):5.
72. Rai M, Yadav A, Gade A. Silver nanoparticles as a new generation of antimicrobials. Biotechnol Adv. 2009;27(1):76–83.
73. Samuel U, Guggenbichler JP. Prevention of catheter-related infections: the potential of a new nano-silver impregnated catheter. Int J Antimicrob Agents. 2004;23 Suppl 1:S75–8.
74. Chopra I. The increasing use of silver-based products as antimicrobial agents: a useful development or a cause for concern? J Antimicrob Chemother. 2007;59(4):587–90.
75. Darouiche RO. Device-associated infections: a macroproblem that starts with microadherence. Clin Infect Dis. 2001;33(9):1567–72.
76. Stickler DJ. Biomaterials to prevent nosocomial infections: is silver the gold standard? Curr Opin Infect Dis. 2000;13(4):389–93.
77. Hermans MH. Silver-containing dressings and the need for evidence. Adv Skin Wound Care. 2007;20(3):166–73. quiz 174–5.
78. Zilberman M, et al. Drug-eluting medical implants. Handb Exp Pharmacol. 2010;197:299–341.
79. Kawashita M, et al. Antibacterial silver-containing silica glass prepared by sol–gel method. Biomaterials. 2000;21(4):393–8.
80. Balamurugan A, et al. An *in vitro* biological and anti-bacterial study on a sol–gel derived silver-incorporated bioglass system. Dent Mater. 2008;24(10):1343–51.
81. Bellantone M, Coleman NJ, Hench LL. Bacteriostatic action of a novel four-component bioactive glass. J Biomed Mater Res. 2000;51(3):484–90.
82. Wu C, et al. Novel sphene coatings on Ti-6Al-4V for orthopedic implants using sol–gel method. Acta Biomater. 2008;4(3):569–76.
83. Kumar R, Howdle S, Munstedt H. Polyamide/silver antimicrobials: effect of filler types on the silver ion release. J Biomed Mater Res B Appl Biomater. 2005;75(2):311–9.

84. Kumar R, Munstedt H. Silver ion release from antimicrobial polyamide/silver composites. Biomaterials. 2005;26(14):2081–8.
85. Chohfi M, et al. Pharmacokinetics, uses, and limitations of vancomycin-loaded bone cement. Int Orthop. 1998;22(3):171–7.
86. Kanellakopoulou K, Giamarellos-Bourboulis EJ. Carrier systems for the local delivery of antibiotics in bone infections. Drugs. 2000;59(6):1223–32.
87. Walenkamp GH, Vree TB, van Rens TJ. Gentamicin-PMMA beads. Pharmacokinetic and nephrotoxicological study. Clin Orthop Relat Res. 1986;205:171–83.
88. Diefenbeck M, Muckley T, Hofmann GO. Prophylaxis and treatment of implant-related infections by local application of antibiotics. Injury. 2006;37 Suppl 2:S95–104.
89. Neut D, et al. The effect of mixing on gentamicin release from polymethylmethacrylate bone cements. Acta Orthop Scand. 2003;74(6):670–6.
90. Neut D, et al. Biomaterial-associated infection of gentamicin-loaded PMMA beads in orthopaedic revision surgery. J Antimicrob Chemother. 2001;47(6):885–91.
91. Fuchs T, et al. The use of gentamicin-coated nails in the tibia: preliminary results of a prospective study. Arch Orthop Trauma Surg. 2011;131(10):1419–25.
92. Schmidmaier G, et al. Prophylaxis and treatment of implant-related infections by antibiotic-coated implants: a review. Injury. 2006;37 Suppl 2:S105–12.
93. Brohede U, et al. Multifunctional implant coatings providing possibilities for fast antibiotics loading with subsequent slow release. J Mater Sci Mater Med. 2009;20(9):1859–67.
94. Schierholz JM, et al. Controlled release of antibiotics from biomedical polyurethanes: morphological and structural features. Biomaterials. 1997;18(12):839–44.
95. Kwok CS, Horbett TA, Ratner BD. Design of infection-resistant antibiotic-releasing polymers.II. Controlled release of antibiotics through a plasma-deposited thin film barrier. J Control Release. 1999;62(3):301–11.
96. Kwok CS, et al. Design of infection-resistant antibiotic-releasing polymers: I. Fabrication and formulation. J Control Release. 1999;62(3):289–99.
97. Govender ST, Nathoo N, van Dellen JR. Evaluation of an antibiotic-impregnated shunt system for the treatment of hydrocephalus. J Neurosurg. 2003;99(5):831–9.
98. Joseph TN, Chen AL, Di Cesare PE. Use of antibiotic-impregnated cement in total joint arthroplasty. J Am Acad Orthop Surg. 2003;11(1):38–47.
99. Kress P, Schäfer P, Schwerdtfeger FP. Clinical use of a voice prosthesis with a flap valve containing silver oxide (Blom-Singer Advantage), biofilm formation, in-situ lifetime and indication. Laryngorhinootologie. 2006;85(12):893–6.
100. Hancock RE. Mechanisms of action of newer antibiotics for Gram-positive pathogens. Lancet Infect Dis. 2005;5(4):209–18.
101. Tiller JC, et al. Designing surfaces that kill bacteria on contact. Proc Natl Acad Sci USA. 2001;98(11):5981–5.
102. Endo Y, Tani T, Kodama M. Antimicrobial activity of tertiary amine covalently bonded to a polystyrene fiber. Appl Environ Microbiol. 1987;53(9):2050–5.
103. Gottenbos B, et al. *In vitro* and *in vivo* antimicrobial activity of covalently coupled quaternary ammonium silane coatings on silicone rubber. Biomaterials. 2002;23(6):1417–23.
104. Flemming RG, et al. Bacterial colonization of functionalized polyurethanes. Biomaterials. 2000;21(3):273–81.
105. Fuchs AD, Tiller JC. Contact-active antimicrobial coatings derived from aqueous suspensions. Angew Chem Int Ed Engl. 2006;45(40):6759–62.
106. Gabriel M, et al. Preparation of LL-37-grafted titanium surfaces with bactericidal activity. Bioconjug Chem. 2006;17(2):548–50.
107. Gelman MA, et al. Biocidal activity of polystyrenes that are cationic by virtue of protonation. Org Lett. 2004;6(4):557–60.
108. Ketonis C, et al. Antibacterial activity of bone allografts: comparison of a new vancomycin-tethered allograft with allograft loaded with adsorbed vancomycin. Bone. 2011;48(3):631–8.
109. Parvizi J, et al. Frank Stinchfield Award. Titanium surface with biologic activity against infection. Clin Orthop Relat Res. 2004;429:33–8.

110. Antoci Jr V, et al. Covalently attached vancomycin provides a nanoscale antibacterial surface. Clin Orthop Relat Res. 2007;461:81–7.
111. Aumsuwan N, Heinhorst S, Urban MW. The effectiveness of antibiotic activity of penicillin attached to expanded poly(tetrafluoroethylene) (ePTFE) surfaces: a quantitative assessment. Biomacromolecules. 2007;8(11):3525–30.
112. Aumsuwan N, et al. Attachment of ampicillin to expanded poly(tetrafluoroethylene): surface reactions leading to inhibition of microbial growth. Biomacromolecules. 2008;9(7):1712–8.
113. Antoci Jr V, et al. The inhibition of *Staphylococcus epidermidis* biofilm formation by vancomycin-modified titanium alloy and implications for the treatment of periprosthetic infection. Biomaterials. 2008;29(35):4684–90.
114. Yang J, Welby JL, Meyerhoff ME. Generic nitric oxide (NO) generating surface by immobilizing organoselenium species via layer-by-layer assembly. Langmuir. 2008;24(18):10265–72.
115. Cha W, Meyerhoff ME. Catalytic generation of nitric oxide from S-nitrosothiols using immobilized organoselenium species. Biomaterials. 2007;28(1):19–27.
116. Teller M, et al. Release of gentamicin from bone regenerative materials: an *in vitro* study. J Biomed Mater Res B Appl Biomater. 2007;81(1):23–9.
117. Gollwitzer H, et al. Antibacterial poly(D, L-lactic acid) coating of medical implants using a biodegradable drug delivery technology. J Antimicrob Chemother. 2003;51(3):585–91.
118. Gollwitzer H, et al. Biomechanical and allergological characteristics of a biodegradable poly(D, L-lactic acid) coating for orthopaedic implants. J Orthop Res. 2005;23(4):802–9.
119. Fitzgerald Jr RH. Microbiologic environment of the conventional operating room. Arch Surg. 1979;114(7):772–5.
120. Steckelberg J, Osmon D. Prosthetic Joint Infection. In: Waldvogel FA, Bisno AL, editors. Infections Associated with Indwelling Medical Devices. American Society of Microbiologists, Washington: D.C.; 2000.
121. Wells CL, Maddaus MA, Simmons RL. Role of the macrophage in the translocation of intestinal bacteria. Arch Surg. 1987;122(1):48–53.
122. Guo W, et al. Enteric bacterial translocation after intraperitoneal implantation of rubber drain pieces. Scand J Gastroenterol. 1993;28(5):393–400.
123. Hill EE, et al. Evolving trends in infective endocarditis. Clin Microbiol Infect. 2006;12(1):5–12.
124. Illingworth BL, et al. *In vivo* efficacy of silver-coated (Silzone) infection-resistant polyester fabric against a biofilm-producing bacteria, *Staphylococcus epidermidis*. J Heart Valve Dis. 1998;7(5):524–30.
125. Carrel T, et al. Definitive cure of recurrent prosthetic endocarditis using silver-coated St. Jude Medical heart valves: a preliminary case report. J Heart Valve Dis. 1998;7(5):531–3.
126. Grunkemeier GL, Jin R, Starr A. Prosthetic heart valves: objective performance criteria versus randomized clinical trial. Ann Thorac Surg. 2006;82(3):776–80.
127. Jamieson WR, et al. Seven-year results with the St Jude Medical Silzone mechanical prosthesis. J Thorac Cardiovasc Surg. 2009;137(5):1109–15. e2.
128. Ghanbari H, et al. Polymeric heart valves: new materials, emerging hopes. Trends Biotechnol. 2009;27(6):359–67.
129. Atiyeh BS, et al. Effect of silver on burn wound infection control and healing: review of the literature. Burns. 2007;33(2):139–48.
130. Kaufmann BA, et al. Coronary stent infection: a rare but severe complication of percutaneous coronary intervention. Swiss Med Wkly. 2005;135(33–34):483–7.
131. Deiparine MK, et al. Endovascular stent infection. J Vasc Surg. 1996;23(3):529–33.
132. Takahashi H, Letourneur D, Grainger DW. Delivery of large biopharmaceuticals from cardiovascular stents: a review. Biomacromolecules. 2007;8(11):3281–93.
133. Yan Y, et al. Rapamycin can inhibit the development of *Chlamydia pneumoniae*, which might partly contribute to the prevention of in-stent restenosis. Cardiovasc Drugs Ther. 2010; 24(3):189–95.
134. Spaulding C, et al. Four-year follow-up of TYPHOON (trial to assess the use of the CYPHer sirolimus-eluting coronary stent in acute myocardial infarction treated with BallOON angioplasty). JACC Cardiovasc Interv. 2011;4(1):14–23.

135. Zhang F, Dong L, Ge J. Meta-analysis of five randomized clinical trials comparing sirolimus- versus paclitaxel-eluting stents in patients with diabetes mellitus. Am J Cardiol. 2010;105(1):64–8.
136. Zimarino M, De Caterina R. Actinomycin D-eluting stents do not reduce the risk of restenosis. Commentary. Evid Based Cardiovasc Med. 2005;9(1):41–3.
137. Serruys PW, et al. Actinomycin-eluting stent for coronary revascularization: a randomized feasibility and safety study: the ACTION trial. J Am Coll Cardiol. 2004;44(7):1363–7.
138. Silber S. When are drug-eluting stents effective? A critical analysis of the presently available data. Z Kardiol. 2004;93(9):649–63.
139. Espehaug B, et al. Antibiotic prophylaxis in total hip arthroplasty. Review of 10,905 primary cemented total hip replacements reported to the Norwegian arthroplasty register, 1987 to 1995. J Bone Joint Surg Br. 1997;79(4):590–5.
140. Mahan J, et al. Factors in pin tract infections. Orthopedics. 1991;14(3):305–8.
141. Forster H, et al. Bactericidal activity of antimicrobial coated polyurethane sleeves for external fixation pins. J Orthop Res. 2004;22(3):671–7.
142. von Plocki SC, et al. Biodegradable sleeves for metal implants to prevent implant-associated infection: an experimental *in vivo* study in sheep. Vet Surg. 2012;41:410–21.
143. Lucke M, et al. Gentamicin coating of metallic implants reduces implant-related osteomyelitis in rats. Bone. 2003;32(5):521–31.
144. Dietz UA, Spor L, Germer CT. Management of mesh-related infections. Chirurg. 2011; 82(3):208–17.
145. Bhende S, et al. Infection potentiation study of synthetic and naturally derived surgical mesh in mice. Surg Infect (Larchmt). 2007;8(3):405–14.
146. Hansen LK, et al. *In vivo* model of human pathogen infection and demonstration of efficacy by an antimicrobial pouch for pacing devices. Pacing Clin Electrophysiol. 2009;32(7):898–907.
147. Junge K, et al. Gentamicin supplementation of polyvinylidenfluoride mesh materials for infection prophylaxis. Biomaterials. 2005;26(7):787–93.
148. Elliott TS. An update on antimicrobial central venous catheters. J Hosp Infect. 2007;65 Suppl 2:34–8.
149. Drake JM, Kestle JR, Tuli S. CSF shunts 50 years on—past, present and future. Childs Nerv Syst. 2000;16(10–11):800–4.
150. Farber SH, et al. Cost analysis of antibiotic-impregnated catheters in the treatment of hydrocephalus in adult patients. World Neurosurg. 2010;74(4–5):528–31.
151. Choux M, et al. Shunt implantation: reducing the incidence of shunt infection. J Neurosurg. 1992;77(6):875–80.
152. Eymann R, et al. Clinical and economic consequences of antibiotic-impregnated cerebrospinal fluid shunt catheters. J Neurosurg Pediatr. 2008;1(6):444–50.
153. Wong GK, et al. Antibiotics-impregnated ventricular catheter versus systemic antibiotics for prevention of nosocomial CSF and non-CSF infections: a prospective randomised clinical trial. J Neurol Neurosurg Psychiatry. 2010;81(10):1064–7.
154. Parker SL, et al. Comparison of hospital cost and resource use associated with antibiotic-impregnated versus standard shunt catheters. Clin Neurosurg. 2011;58:122–5.
155. Pye AD, et al. A review of dental implants and infection. J Hosp Infect. 2009;72(2):104–10.
156. Jokstad A, et al. Quality of dental implants. Int Dent J. 2003;53(6 Suppl 2):409–43.
157. Sones AD. Complications with osseointegrated implants. J Prosthet Dent. 1989;62(5):581–5.
158. Reid G, et al. Microbiota restoration: natural and supplemented recovery of human microbial communities. Nat Rev Microbiol. 2011;9(1):27–38.
159. Walker C, Karpinia K. Rationale for use of antibiotics in periodontics. J Periodontol. 2002;73(10):1188–96.
160. Williams RC, et al. Treatment of periodontitis by local administration of minocycline microspheres: a controlled trial. J Periodontol. 2001;72(11):1535–44.
161. Gabriel MM, Weisbarth RE. Developing antimicrobial surfaces for silicone hydrogels. CIBA Medical; 2009.
162. Szczotka-Flynn LB, Pearlman E, Ghannoum M. Microbial contamination of contact lenses, lens care solutions, and their accessories: a literature review. Eye Contact Lens. 2010; 36(2):116–29.

163. Nilsson SE, Montan PG. The annualized incidence of contact lens induced keratitis in Sweden and its relation to lens type and wear schedule: results of a 3-month prospective study. CLAO J. 1994;20(4):225–30.
164. Cheng KH, et al. Incidence of contact-lens-associated microbial keratitis and its related morbidity. Lancet. 1999;354(9174):181–5.
165. Keay L, et al. Factors affecting the morbidity of contact lens-related microbial keratitis: a population study. Invest Ophthalmol Vis Sci. 2006;47(10):4302–8.
166. Goldstein MH, Kowalski RP, Gordon YJ. Emerging fluoroquinolone resistance in bacterial keratitis: a 5-year review. Ophthalmology. 1999;106(7):1313–8.
167. Zhu H, et al. Fimbrolide-coated antimicrobial lenses: their *in vitro* and *in vivo* effects. Optom Vis Sci. 2008;85(5):292–300.
168. Willcox MD. New strategies to prevent *Pseudomonas keratitis*. Eye Contact Lens. 2007;33(6 Pt 2):401–3. discussion 410–1.
169. Mathews SM, et al. Prevention of bacterial colonization of contact lenses with covalently attached selenium and effects on the rabbit cornea. Cornea. 2006;25(7):806–14.
170. Larkin DF, Kilvington S, Easty DL. Contamination of contact lens storage cases by Acanthamoeba and bacteria. Br J Ophthalmol. 1990;74(3):133–5.
171. Dantam J, Zhu H, Stapleton F. Biocidal efficacy of silver-impregnated contact lens storage cases *in vitro*. Invest Ophthalmol Vis Sci. 2011;52(1):51–7.
172. Farber BF, et al. A novel antibiofilm technology for contact lens solutions. Ophthalmology. 1995;102(5):831–6.
173. Collard HR, Saint S, Matthay MA. Prevention of ventilator-associated pneumonia: an evidence-based systematic review. Ann Intern Med. 2003;138(6):494–501.
174. Pruitt B, Jacobs M. Best-practice interventions: how can you prevent ventilator-associated pneumonia? Nursing. 2006;36(2):36–41. quiz 41–2.
175. Ventilator-associated pneumonia (VAP) event. In: CDC, editor; 2011.
176. Tablan OC, Anderson LJ, Besser R, Bridges C, Hajjeh R. Guidelines for preventing health-care-associated pneumonia, 2003. Recommendations of CDC and the Healthcare Infection Control Practices Advisory Committee. Part 1. Issues on preventing health care associated pneumonia. In: C.f.D. Control, editor. Atlanta, GA; 2003.
177. Kollef MH, et al. Silver-coated endotracheal tubes and incidence of ventilator-associated pneumonia: the NASCENT randomized trial. JAMA. 2008;300(7):805–13.
178. Free RH, et al. Biofilm formation on voice prostheses: influence of dairy products *in vitro*. Acta Otolaryngol. 2000;120(1):92–9.
179. Mahieu HF, et al. Candida vegetations on silicone voice prostheses. Arch Otolaryngol Head Neck Surg. 1986;112(3):321–5.
180. Rodrigues L, et al. Strategies for the prevention of microbial biofilm formation on silicone rubber voice prostheses. J Biomed Mater Res B Appl Biomater. 2007;81(2):358–70.
181. De Prijck K, et al. Prevention of *Candida albicans* biofilm formation by covalently bound dimethylaminoethylmethacrylate and polyethylenimine. Mycopathologia. 2010;170(4):213–21.
182. Reddy ST, et al. Micropatterned surfaces for reducing the risk of catheter-associated urinary tract infection: an *in vitro* study on the effect of sharklet micropatterned surfaces to inhibit bacterial colonization and migration of uropathogenic *Escherichia coli*. J Endourol. 2011;25(9):1547–52.
183. Hooton TM, et al. Diagnosis, prevention, and treatment of catheter-associated urinary tract infection in adults: 2009 International Clinical Practice Guidelines from the Infectious Diseases Society of America Oxford Journals Medicine. Clin Infect Dis. 2009;50(5):625–63.
184. Warren JW. Catheter-associated urinary tract infections. Infect Dis Clin North Am. 1997; 11(3):609–22.
185. Stickler DJ, Jones GL, Russell AD. Control of encrustation and blockage of Foley catheters. Lancet. 2003;361(9367):1435–7.
186. Lawrence EL, Turner IG. Materials for urinary catheters: a review of their history and development in the UK. Med Eng Phys. 2005;27(6):443–53.
187. Horattas MC, et al. Changing concepts in long-term central venous access: catheter selection and cost savings. Am J Infect Control. 2001;29(1):32–40.

188. Tal MG, Ni N. Selecting optimal hemodialysis catheters: material, design, advanced features, and preferences. Tech Vasc Interv Radiol. 2008;11(3):186–91.
189. Sousa C, Henriques M, Oliveira R. Mini-review: antimicrobial central venous catheters—recent advances and strategies. Biofouling. 2011;27(6):609–20.
190. Solomon DD, et al. An *in vivo* method for the evaluation of catheter thrombogenicity. J Biomed Mater Res. 1987;21(1):43–57.
191. Sherertz RJ, et al. Contribution of vascular catheter material to the pathogenesis of infection: the enhanced risk of silicone *in vivo*. J Biomed Mater Res. 1995;29(5):635–45.
192. Subbiahdoss G, et al. Microbial biofilm growth versus tissue integration: "the race for the surface" experimentally studied. Acta Biomater. 2009;5(5):1399–404.
193. Roe D, et al. Antimicrobial surface functionalization of plastic catheters by silver nanoparticles. J Antimicrob Chemother. 2008;61(4):869–76.
194. Tambyah PA, Maki DG. The relationship between pyuria and infection in patients with indwelling urinary catheters: a prospective study of 761 patients. Arch Intern Med. 2000;160(5):673–7.
195. Saint S, et al. The efficacy of silver alloy-coated urinary catheters in preventing urinary tract infection: a meta-analysis. Am J Med. 1998;105(3):236–41.
196. Liedberg H. Catheter induced urethral inflammatory reaction and urinary tract infection. An experimental and clinical study. Scand J Urol Nephrol Suppl. 1989;124:1–43.
197. Liedberg H, Lundeberg T. Silver alloy coated catheters reduce catheter-associated bacteriuria. Br J Urol. 1990;65(4):379–81.
198. Rosch W, Lugauer S. Catheter-associated infections in urology: possible use of silver-impregnated catheters and the Erlanger silver catheter. Infection. 1999;27 Suppl 1:S74–7.
199. Maki DG, Tambyah PA. Engineering out the risk for infection with urinary catheters. Emerg Infect Dis. 2001;7(2):342–7.
200. Schierholz JM, et al. The antimicrobial efficacy of a new central venous catheter with long-term broad-spectrum activity. J Antimicrob Chemother. 2000;46(1):45–50.
201. Bullard KM, Dunn DL. Bloodstream and intravascular catheter infections. In: Holzheimer RG, Mannick JA, editors. Surgical treatment: evidence-based and problem-oriented. Munich: Zuckschwerdt; 2001.
202. Widmer AF. Central Venous Catheters. In: Seifert H, Jansen B, Farr BM, editors. Catheter-Related Infections. Marcel Dekker: New York; 1997.
203. McGee DC, Gould MK. Preventing complications of central venous catheterization. N Engl J Med. 2003;348(12):1123–33.
204. Frasca D, Dahyot-Fizelier C, Mimoz O. Prevention of central venous catheter-related infection in the intensive care unit. Crit Care. 2010;14(2):212.
205. Maki DG, et al. Prevention of central venous catheter-related bloodstream infection by use of an antiseptic-impregnated catheter. A randomized, controlled trial. Ann Intern Med. 1997; 127(4):257–66.
206. O'Grady NP, et al. Guidelines for the prevention of intravascular catheter-related infections. Centers for Disease Control and Prevention. MMWR Recomm Rep. 2002;51((RR-10)):1–29.
207. Donlan RM. Biofilm elimination on intravascular catheters: important considerations for the infectious disease practitioner. Clin Infect Dis. 2011;52(8):1038–45.
208. Schierholz JM, et al. Antimicrobial substances and effects on sessile bacteria. Zentralbl Bakteriol. 1999;289(2):165–77.
209. Casey AL, et al. Antimicrobial central venous catheters in adults: a systematic review and meta-analysis. Lancet Infect Dis. 2008;8(12):763–76.
210. Boersma RS, et al. Thrombotic and infectious complications of central venous catheters in patients with hematological malignancies. Ann Oncol. 2008;19(3):433–42.
211. Mohammad SF. Enhanced risk of infection with device-associated thrombi. ASAIO J. 2000;46(6):S63–8.
212. Raad II, Hanna HA. Intravascular catheter-related infections: new horizons and recent advances. Arch Intern Med. 2002;162(8):871–8.
213. Stillman RM, et al. Etiology of catheter-associated sepsis. Correlation with thrombogenicity. Arch Surg. 1977;112(12):1497–9.

214. Diskin CJ, et al. Heparin and biofilm: is this the risk factor for catheter-related sepsis? Am J Kidney Dis. 2008;52(1):197–8. author reply 198.
215. Baumgartner JN, Cooper SL. Influence of thrombus components in mediating *Staphylococcus aureus* adhesion to polyurethane surfaces. J Biomed Mater Res. 1998;40(4):660–70.
216. Bos HM, et al. Evidence that bacteria prefer to adhere to thrombus. ASAIO J. 1996; 42(5):M881–4.
217. van Rooden CJ, et al. Infectious complications of central venous catheters increase the risk of catheter-related thrombosis in hematology patients: a prospective study. J Clin Oncol. 2005;23(12):2655–60.
218. Appelgren P, et al. Surface heparinization of central venous catheters reduces microbial colonization *in vitro* and *in vivo*: results from a prospective, randomized trial. Crit Care Med. 1996;24(9):1482–9.
219. Tenke P, et al. Bacterial biofilm formation on urologic devices and heparin coating as preventive strategy. Int J Antimicrob Agents. 2004;23 Suppl 1:S67–74.
220. Davey ME, O'Toole GA. Microbial biofilms: from ecology to molecular genetics. Microbiol Mol Biol Rev. 2000;64(4):847–67.
221. Brunstedt MR, et al. Bacteria/blood/material interactions. I. Injected and preseeded slime-forming Staphylococcus epidermidis in flowing blood with biomaterials. J Biomed Mater Res. 1995;29(4):455–66.
222. Kamal GD, et al. Reduced intravascular catheter infection by antibiotic bonding. A prospective, randomized, controlled trial. JAMA. 1991;265(18):2364–8.
223. Kundu B, et al. Development of porous HAp and beta-TCP scaffolds by starch consolidation with foaming method and drug-chitosan bilayered scaffold based drug delivery system. J Mater Sci Mater Med. 2010;21(11):2955–69.
224. Costerton JW, Stewart PS, Greenberg EP. Bacterial biofilms: a common cause of persistent infections. Science. 1999;284(5418):1318–22.
225. Berrington A, Gould FK. Use of antibiotic locks to treat colonized central venous catheters. J Antimicrob Chemother. 2001;48(5):597–603.
226. Mermel LA, et al. Clinical practice guidelines for the diagnosis and management of intravascular catheter-related infection: 2009 Update by the Infectious Diseases Society of America. Clin Infect Dis. 2009;49(1):1–45.
227. Monson T, Kunin CM. Evaluation of a polymer-coated indwelling catheter in prevention of infection. J Urol. 1974;111(2):220–2.
228. Ahearn DG, et al. Effects of hydrogel/silver coatings on *in vitro* adhesion to catheters of bacteria associated with urinary tract infections. Curr Microbiol. 2000;41(2):120–5.
229. Thibon P, et al. Randomized multi-centre trial of the effects of a catheter coated with hydrogel and silver salts on the incidence of hospital-acquired urinary tract infections. J Hosp Infect. 2000;45(2):117–24.
230. Bologna RA, et al. Hydrogel/silver ion-coated urinary catheter reduces nosocomial urinary tract infection rates in intensive care unit patients: a multicenter study. Urology. 1999;54(6):982–7.
231. Cormio L, et al. Bacterial adhesion to urethral catheters: role of coating materials and immersion in antibiotic solution. Eur Urol. 2001;40(3):354–8. discussion 359.
232. Pugach JL, et al. Antibiotic hydrogel coated Foley catheters for prevention of urinary tract infection in a rabbit model. J Urol. 1999;162(3 Pt 1):883–7.
233. Boks NP, et al. Bond-strengthening in staphylococcal adhesion to hydrophilic and hydrophobic surfaces using atomic force microscopy. Langmuir. 2008;24(22):12990–4.
234. Boks NP, et al. Residence time dependent desorption of *Staphylococcus epidermidis* from hydrophobic and hydrophilic substrata. Colloids Surf B Biointerfaces. 2008;67(2):276–8.
235. Boks NP, et al. Mobile and immobile adhesion of staphylococcal strains to hydrophilic and hydrophobic surfaces. J Colloid Interface Sci. 2009;331(1):60–4.
236. Boks NP, et al. Forces involved in bacterial adhesion to hydrophilic and hydrophobic surfaces. Microbiology. 2008;154(Pt 10):3122–33.
237. Banerjee I, Pangule RC, Kane RS. Antifouling coatings: recent developments in the design of surfaces that prevent fouling by proteins, bacteria, and marine organisms. Adv Mater. 2010;23(6):690–718.

238. Chung KK, et al. Impact of engineered surface microtopography on biofilm formation of *Staphylococcus aureus*. Biointerphases. 2007;2(2):89–94.
239. Tebbs SE, Elliott TS. Modification of central venous catheter polymers to prevent *in vitro* microbial colonisation. Eur J Clin Microbiol Infect Dis. 1994;13(2):111–7.
240. Jansen B, et al. In-vitro efficacy of a central venous catheter ('Hydrocath') loaded with teicoplanin to prevent bacterial colonization. J Hosp Infect. 1992;22(2):93–107.
241. Tamilvanan S, Venkateshan N, Ludwig A. The potential of lipid- and polymer-based drug delivery carriers for eradicating biofilm consortia on device-related nosocomial infections. J Control Release. 2008;128(1):2–22.
242. Sampath LA, et al. Infection resistance of surface modified catheters with either short-lived or prolonged activity. J Hosp Infect. 1995;30(3):201–10.
243. Romano G, et al. Efficacy of a central venous catheter (Hydrocath) loaded with teicoplanin in preventing subcutaneous staphylococcal infection in the mouse. Zentralbl Bakteriol. 1993;279(3):426–33.
244. Crocker IC, et al. A novel electrical method for the prevention of microbial colonization of intravascular cannulae. J Hosp Infect. 1992;22(1):7–17.
245. Angiotech. Angiotech CVC with anti-microbial surface. 2011 [cited 22 Aug 2011]. http://www.angiotech.com/focus-markets/5fu-cvc/.
246. Walz JM, et al. Anti-infective external coating of central venous catheters: a randomized, noninferiority trial comparing 5-fluorouracil with chlorhexidine/silver sulfadiazine in preventing catheter colonization. Crit Care Med. 2010;38(11):2095–102.
247. Halton KA, et al. Cost effectiveness of antimicrobial catheters in the intensive care unit: addressing uncertainty in the decision. Crit Care. 2009;13(2):R35.
248. Raad I, et al. Central venous catheters coated with minocycline and rifampin for the prevention of catheter-related colonization and bloodstream infections. A randomized, double-blind trial. The Texas Medical Center Catheter Study Group. Ann Intern Med. 1997; 127(4):267–74.
249. Johnson JR, Delavari P, Azar M. Activities of a nitrofurazone-containing urinary catheter and a silver hydrogel catheter against multidrug-resistant bacteria characteristic of catheter-associated urinary tract infection. Antimicrob Agents Chemother. 1999;43(12):2990–5.
250. Jansen B, et al. *In vitro* evaluation of the antimicrobial efficacy and biocompatibility of a silver-coated central venous catheter. J Biomater Appl. 1994;9(1):55–70.
251. Shunmugaperumal T. Chapter 11: Polymer-based antimicrobial delivery carriers. In: Shunmugaperumal T, editor. Biofilm eradication and prevention: a pharmaceutical approach to medical device infections. Wiley: Hoboken; 2010. pp. 359–417.
252. Stevens KN, et al. The relationship between the antimicrobial effect of catheter coatings containing silver nanoparticles and the coagulation of contacting blood. Biomaterials. 2009;30(22):3682–90.
253. Lansdown A. Silver in healthcare: its antimicrobial efficacy and safety in use. Issues in toxicology. Cambridge: RSC; 2010.
254. Schierholz JM, Beuth J, Pulverer G. Silver coating of medical devices for catheter-associated infections? Am J Med. 1999;107(1):101–2.
255. Schierholz JM, Beuth J, Pulverer G. The quantity and duration of the antimicrobial efficacy of a chlorhexidine and silversulfadiazine impregnated catheter. J Hosp Infect. 1999;42(2): 163–5.
256. Brun-Buisson C, et al. Prevention of intravascular catheter-related infection with newer chlorhexidine-silver sulfadiazine-coated catheters: a randomized controlled trial. Intensive Care Med. 2004;30(5):837–43.
257. Rupp ME, et al. Effect of a second-generation venous catheter impregnated with chlorhexidine and silver sulfadiazine on central catheter-related infections: a randomized, controlled trial. Ann Intern Med. 2005;143(8):570–80.
258. Tattawasart U, et al. Development of resistance to chlorhexidine diacetate and cetylpyridinium chloride in Pseudomonas stutzeri and changes in antibiotic susceptibility. J Hosp Infect. 1999;42(3):219–29.
259. Trautner BW, Darouiche RO. Catheter-associated infections: pathogenesis affects prevention. Arch Intern Med. 2004;164(8):842–50.

260. Terazawa E, et al. Severe anaphylactic reaction due to a chlorhexidine-impregnated central venous catheter. Anesthesiology. 1998;89(5):1296–8.
261. Oda T, et al. Anaphylactic shock induced by an antiseptic-coated central venous [correction of nervous] catheter. Anesthesiology. 1997;87(5):1242–4.
262. Russ BR, Maddern PJ. Anaphylactic reaction to chlorhexidine in urinary catheter lubricant. Anaesth Intensive Care. 1994;22(5):611–2.
263. Burlington B. FDA Public Health Notice: Potential Hypersensitivity Reactions to Chlorhexidine-Impregnated Medica Devices. US Food and Drug Administration: Silver Spring; 1998.
264. Ranucci M, et al. Impact of oligon central venous catheters on catheter colonization and catheter-related bloodstream infection. Crit Care Med. 2003;31(1):52–9.
265. Darouiche RO. Anti-infective efficacy of silver-coated medical prostheses. Clin Infect Dis. 1999;29(6):1371–7. quiz 1378.
266. Veenstra DL, et al. Efficacy of antiseptic-impregnated central venous catheters in preventing catheter-related bloodstream infection: a meta-analysis. JAMA. 1999;281(3):261–7.
267. Krein SL, et al. Use of central venous catheter-related bloodstream infection prevention practices by US hospitals. Mayo Clin Proc. 2007;82(6):672–8.
268. Safdar N. Antimicrobial catheters in the ICU: is the juice worth the squeeze? Crit Care. 2009;13(3):148.
269. Timsit JF, et al. Chlorhexidine-impregnated sponges and less frequent dressing changes for prevention of catheter-related infections in critically ill adults: a randomized controlled trial. JAMA. 2009;301(12):1231–41.
270. Mojibian H, et al. Initial clinical experience with a new heparin-coated chronic hemodialysis catheter. Hemodial Int. 2009;13(3):329–34.
271. Tal MG. Comparison of recirculation percentage of the palindrome catheter and standard hemodialysis catheters in a swine model. J Vasc Interv Radiol. 2005;16(9):1237–40.
272. Kakkos SK, et al. Effectiveness of a new tunneled catheter in preventing catheter malfunction: a comparative study. J Vasc Interv Radiol. 2008;19(7):1018–26.
273. Spector M, et al. Clinical outcome of the Tal Palindrome chronic hemodialysis catheter: single institution experience. J Vasc Interv Radiol. 2008;19(10):1434–8.
274. Mendelson M, Loin-Chen B, Finkelstein-Blond L, Kogan G, Hollinger I. Study of Introcan Safety IV Catheter (IVC) (B.Braun Medical Inc.) for the prevention of percutaneous injuries (PIs) in healthcare workers (HCWs). In: The Society for Healthcare Epidemiology of America; 2003.
275. Cowan MM, et al. Antimicrobial efficacy of a silver-zeolite matrix coating on stainless steel. J Ind Microbiol Biotechnol. 2003;30(2):102–6.
276. Loertzer H, et al. Use of catheters with the AgION antimicrobial system in kidney transplant recipients to reduce infection risk. Transplant Proc. 2006;38(3):707–10.
277. Galeano B, Korff E, Nicholson WL. Inactivation of vegetative cells, but not spores, of *Bacillus anthracis*, *B. cereus*, and *B. subtilis* on stainless steel surfaces coated with an antimicrobial silver- and zinc-containing zeolite formulation. Appl Environ Microbiol. 2003;69(7):4329–31.
278. Chernecky C, Waller J. In vitro comparisons of two antimicrobial intravenous connectors. Clin Nurs Res. 2011;20(1):101–9.
279. Maragakis LL, et al. Increased catheter-related bloodstream infection rates after the introduction of a new mechanical valve intravenous access port. Infect Control Hosp Epidemiol. 2006;27(1):67–70.
280. Ryder M. Bacterial transfer through needleless connectors: a comparison of nine devices. In: SHEA, 2007. Center for Biofilm Engineering.
281. Bambauer R, et al. Large bore catheters with surface treatments versus untreated catheters for vascular access in hemodialysis. Artif Organs. 2004;28(7):604–10.
282. Bambauer R, et al. Large bore catheters with surface treatments versus untreated catheters for blood access. J Vasc Access. 2001;2(3):97–105.
283. Kampf G, et al. Microbicidal activity of a new silver-containing polymer, SPI-ARGENT II. Antimicrob Agents Chemother. 1998;42(9):2440–2.
284. Bayston R, et al. Prevention of hydrocephalus shunt catheter colonisation *in vitro* by impregnation with antimicrobials. J Neurol Neurosurg Psychiatry. 1989;52(5):605–9.
285. Bayston R, Lambert E. Duration of protective activity of cerebrospinal fluid shunt catheters impregnated with antimicrobial agents to prevent bacterial catheter-related infection. J Neurosurg. 1997;87(2):247–51.

286. Zabramski JM, et al. Efficacy of antimicrobial-impregnated external ventricular drain catheters: a prospective, randomized, controlled trial. J Neurosurg. 2003;98(4):725–30.
287. Galal I, El-Hindawy K. Impact of using triclosan-antibacterial sutures on incidence of surgical site infection. Am J Surg. 2011;202(2):133–8.
288. Barie PS. Surgical site infections: epidemiology and prevention. Surg Infect (Larchmt). 2002;3 Suppl 1:S9–21.
289. Leaper DJ, et al. Surgical site infection - a European perspective of incidence and economic burden. Int Wound J. 2004;1(4):247–73.
290. Nichols RL. Preventing surgical site infections: a surgeon's perspective. Emerg Infect Dis. 2001;7(2):220–4.
291. Tajirian AL, Goldberg DJ. A review of sutures and other skin closure materials. J Cosmet Laser Ther. 2010;12(6):296–302.
292. Hochberg J, Meyer KM, Marion MD. Suture choice and other methods of skin closure. Surg Clin North Am. 2009;89(3):627–41.
293. US Food and Drug Administration. Triclosan: what consumers should know;2010.
294. Ford HR, et al. Intraoperative handling and wound healing: controlled clinical trial comparing coated VICRYL plus antibacterial suture (coated polyglactin 910 suture with triclosan) with coated VICRYL suture (coated polyglactin 910 suture). Surg Infect (Larchmt). 2005;6(3):313–21.
295. Stadler S, Fleck T. Triclosan-coated sutures for the reduction of sternal wound infections? A retrospective observational analysis. Interact Cardiovasc Thorac Surg. 2011;13(3):296–9.
296. Fleck T, et al. Triclosan-coated sutures for the reduction of sternal wound infections: economic considerations. Ann Thorac Surg. 2007;84(1):232–6.
297. Deliaert AE, et al. The effect of triclosan-coated sutures in wound healing. A double blind randomised prospective pilot study. J Plast Reconstr Aesthet Surg. 2009;62(6):771–3.
298. Stone J, Gruber TJ, Rozzelle CJ. Healthcare savings associated with reduced infection rates using antimicrobial suture wound closure for cerebrospinal fluid shunt procedures. Pediatr Neurosurg. 2010;46(1):19–24.
299. Rozzelle CJ, Leonardo J, Li V. Antimicrobial suture wound closure for cerebrospinal fluid shunt surgery: a prospective, double-blinded, randomized controlled trial. J Neurosurg Pediatr. 2008;2(2):111–7.
300. Mangram AJ, et al. Guideline for Prevention of Surgical Site Infection, 1999. Centers for Disease Control and Prevention (CDC) Hospital Infection Control Practices Advisory Committee. Am J Infect Control. 1999;27(2):97–132. quiz 133–4; discussion 96.
301. Gomez-Alonso A, et al. Study of the efficacy of Coated VICRYL Plus Antibacterial suture (coated Polyglactin 910 suture with Triclosan) in two animal models of general surgery. J Infect. 2007;54(1):82–8.
302. Storch ML, Rothenburger SJ, Jacinto G. Experimental efficacy study of coated VICRYL plus antibacterial suture in guinea pigs challenged with *Staphylococcus aureus*. Surg Infect (Larchmt). 2004;5(3):281–8.
303. Barbolt TA. Chemistry and safety of triclosan, and its use as an antimicrobial coating on Coated VICRYL* Plus Antibacterial Suture (coated polyglactin 910 suture with triclosan). Surg Infect (Larchmt). 2002;3 Suppl 1:S45–53.
304. Ho C, Spry C. Antibacterial sutures for wound closure after surgery: a review of the clinical effectiveness and long-term adverse effects. Canadian Agency for Drugs and Technologies in Health; 2008.
305. Caperelli L. PolyMedix presents data showing antimicrobial activity of sutures containing PolyCide[05/28/201109/19/2011].http://www.reuters.com/article/2011/06/28/idUS109125+28-Jun-2011+BW20110628.
306. Chu C-C, Von Fraunhofer JA, Greilser H, editors. Wound closure biomaterials and devices. Boca Raton: CRC; 1997.
307. Moncada S, Higgs EA. Molecular mechanisms and therapeutic strategies related to nitric oxide. FASEB J. 1995;9(13):1319–30.
308. Zhang H, et al. Nitric oxide-releasing fumed silica particles: synthesis, characterization, and biomedical application. J Am Chem Soc. 2003;125(17):5015–24.

309. Rothrock AR, Donkers RL, Schoenfisch MH. Synthesis of nitric oxide-releasing gold nanoparticles. J Am Chem Soc. 2005;127(26):9362–3.
310. Shin JH, Metzger SK, Schoenfisch MH. Synthesis of nitric oxide-releasing silica nanoparticles. J Am Chem Soc. 2007;129(15):4612–9.
311. Hetrick EM, et al. Bactericidal efficacy of nitric oxide-releasing silica nanoparticles. ACS Nano. 2008;2(2):235–46.
312. Friedman AJ, et al. Sustained release nitric oxide releasing nanoparticles: characterization of a novel delivery platform based on nitrite containing hydrogel/glass composites. Nitric Oxide. 2008;19(1):12–20.
313. Frost MC, Reynolds MM, Meyerhoff ME. Polymers incorporating nitric oxide releasing/generating substances for improved biocompatibility of blood-contacting medical devices. Biomaterials. 2005;26(14):1685–93.
314. Reynolds MM, et al. Tailored synthesis of nitric oxide-releasing polyurethanes using O-protected diazeniumdiolated chain extenders. J Mater Chem. 2010;20(15):3107–14.
315. Fujishima A, Honda K. Electrochemical photolysis of water at a semiconductor electrode. Nature. 1972;238(5358):37–8.
316. Hufschmidt D, et al. Photocatalytic water treatment: fundamental knowledge required for its practical application. Water Sci Technol. 2004;49(4):135–40.
317. Mireles II JR, Toguchi A, Harshey RM. Salmonella enterica serovar typhimurium swarming mutants with altered biofilm-forming abilities: surfactin inhibits biofilm formation. J Bacteriol. 2001;183(20):5848–54.
318. Rodrigues L, et al. Biosurfactants: potential applications in medicine. J Antimicrob Chemother. 2006;57(4):609–18.
319. Hirota K, et al. Coating of a surface with 2-methacryloyloxyethyl phosphorylcholine (MPC) co-polymer significantly reduces retention of human pathogenic microorganisms. FEMS Microbiol Lett. 2005;248(1):37–45.
320. Cho WK, Kong B, Choi IS. Highly efficient non-biofouling coating of zwitterionic polymers: poly((3-(methacryloylamino)propyl)-dimethyl(3-sulfopropyl)ammonium hydroxide). Langmuir. 2007;23(10):5678–82.
321. Cheng G, et al. Zwitterionic carboxybetaine polymer surfaces and their resistance to long-term biofilm formation. Biomaterials. 2009;30:5234–40.
322. Boman HG. Antibacterial peptides: basic facts and emerging concepts. J Intern Med. 2003;254(3):197–215.
323. Thaker HD, et al. Synthetic mimics of antimicrobial peptides from triaryl scaffolds. J Med Chem. 2011;54(7):2241–54.
324. Tew GN, et al. De novo design of antimicrobial polymers, foldamers, and small molecules: from discovery to practical applications. Acc Chem Res. 2010;43(1):30–9.
325. Som A, et al. Synthetic mimics of antimicrobial peptides. Biopolymers. 2008;90(2):83–93.
326. Nagy E, Giefing C, von Gabain A. Anti-infective antibodies: a novel tool to prevent and treat nosocomial diseases. Expert Rev Anti Infect Ther. 2008;6(1):21–30.
327. Pedron S et al. Combinatorial approach for fabrication of coatings to control bacterial adhesion. J Biomater Sci Polym Ed. 2011; doi:10.1163/092050611X589329.
328. Rosenberg LE, et al. Salicylic acid-based poly(anhydride esters) for control of biofilm formation in *Salmonella enterica* serovar Typhimurium. Lett Appl Microbiol. 2008;46(5):593–9.
329. Tashiro T. Antibacterial and bacterium adsorbing macromolecules. Macromol Mater Eng. 2001;286(2):63–87.
330. Lawson MC, Bowman CN, Anseth KS. Vancomycin derivative photopolymerized to titanium kills *S. epidermidis*. Clin Orthop Relat Res. 2007;461:96–105.
331. Lawson MC, et al. Inhibition of *Staphylococcus epidermidis* biofilms using polymerizable vancomycin derivatives. Clin Orthop Relat Res. 2010;468(8):2081–91.
332. Lawson MC, et al. Polymerizable vancomycin derivatives for bactericidal biomaterial surface modification: structure-function evaluation. Biomacromolecules. 2009;10(8):2221–34.
333. Donlan RM. Biofilms and device-associated infections. Emerg Infect Dis. 2001;7(2):277–81.

Chapter 14
Silver Containing Biomaterials

Neil Poulter, Krasimir Vasilev, Stefani S. Griesser, and Hans J. Griesser

Abstract Despite considerable research and development efforts, the problem of infections related to biomedical devices and implants persists. Silver has attracted considerable interest for its ability to mitigate bacterial colonization of biomaterials surfaces *in vitro* and has been used in some commercial products such as wound bandages. Silver ion releasing biomaterials are thus considered to be promising candidates for rendering surfaces of biomedical devices and implants resistant to bacterial attachment. Here we review a number of strategies used for the design of antibacterial coatings containing silver. We also discuss the continuing controversy regarding the potential for silver ions to exert adverse effects on human cells and tissue. Finally we briefly compare the silver release approach with the alternative strategy of antibacterial coatings comprising organic antibiotics covalently coupled onto biomaterials surfaces.

14.1 Introduction

Biomaterial-associated infections (BAIs), caused by bacteria attaching and proliferating on surfaces of biomedical devices and implants, are a significant issue in implant surgery and also with many other biomedical devices [1–3]. BAIs often are

N. Poulter • S.S. Griesser
Ian Wark Research Institute, University of South Australia, Mawson Lakes, SA 5095, Australia

Mawson Institute, University of South Australia, Mawson Lakes, SA 5095, Australia

K. Vasilev
Mawson Institute, University of South Australia, Mawson Lakes, SA 5095, Australia

H.J. Griesser (✉)
Ian Wark Research Institute, University of South Australia, Mawson Lakes, SA 5095, Australia
e-mail: hans.griesser@unisa.edu.au

T.F. Moriarty et al. (eds.), *Biomaterials Associated Infection: Immunological Aspects and Antimicrobial Strategies*, DOI 10.1007/978-1-4614-1031-7_14,

not detected at an early stage; they pose a substantial health risk to patients, often requiring reoperation and replacement of the infected device, and incur considerable costs to the health care system. The severity of BAIs varies greatly among devices [1] and patients but can be serious, and even fatal in some instances. With contact lenses for example, serious lens-related infections are rare, as users remove contact lenses when the eye becomes sore upon bacterial infection. There are, however, no analogous early warning signs of bacterial infection for many implants and devices; the outbreak of an infection can be masked by the ongoing soreness of tissue inflammation after surgery, and BAI diagnosis often occurs when a full-blown infection has already caused damage to host tissue.

Health care systems try to minimize the risk of infection on short-term biomedical devices by prophylactic measures. Catheters are replaced at frequent intervals, but such preventative replacement schedules impose considerable costs to the health care system. With implants, the problem cannot be addressed as readily. Improved procedures in sterilization and operating theaters have reduced significantly the frequency of infections of devices and implants, but infections still occur. In trauma surgery, such as accident victims, the wound tissue may be contaminated with bacteria introduced during the trauma. Planktonic bacteria circulating in the vascular system may also attach on to the implant surface, multiply, and form a biofilm that eventually leads to infection.

Attached and growing bacterial colonies soon produce an extracellular polymeric substance (EPS), which protects them against antibiotics and the host body's innate and adaptive defense systems [2, 4]. Thus, bacterial biofilms on biomedical device surfaces are much more difficult to eradicate by antibiotics than circulating bacteria; hence the need for surgical removal of a substantial proportion of infected implants and surrounding affected tissues. Accordingly, a promising strategy for reducing the occurrence of BAIs is to prevent the initial attachment of bacteria to implant and device surfaces. This has spurred research efforts on the development of antibacterial surfaces and thin coatings that can be applied to biomedical devices to confer resistance to bacterial colonization. They should deter BAIs while not affecting other properties such as the visual clarity of contact lenses or the flexibility of vascular grafts. Antibacterial biomaterial surfaces also should not cause adverse effects on cells, tissues, or fluids (e.g., blood, tear fluid) of the human host.

Here we review reports on a prominent strategy for the development of antibacterial surfaces for biomedical devices that use silver as a biocide. Considerable attention has focused for several decades on the use of silver for combating BAIs. The antibacterial properties of silver have been utilized since the time of the ancient Greeks, and other ancient civilizations such as the Egyptians, Romans, and Mesopotamians, also were aware of the benefits of silver for tableware and storage containers. With the need to combat BAIs, silver has received enormous attention as a leading strategy. We provide a summary of research on this topic. Given the enormity of the literature and that many studies overlap considerably, a comprehensive overview is neither feasible nor instructive, and thus we select examples that illustrate various approaches and findings. Our intention is to provide a critical discussion rather than a collation of the huge number of publications in this area.

Synthetic materials, be they polymeric, metallic, or ceramic, offer great flexibility for the design of biomedical devices. However, bacteria are able to attach and form colonies not only on biologically derived surfaces but also on many synthetic materials surfaces, including biomaterials. Thus, the use of biomedical devices and implants entails a significant risk of infection via bacterial colonization of the biomaterial surface. A few polymers can kill bacteria [5–8] or prevent bacteria from attaching [9–12], but those polymers are often not suitable as bulk materials for the fabrication of devices due to other considerations such as strength or flexibility. A strategy to circumvent this quandary is to use a mechanically suitable biomaterial for the bulk of the device and apply a thin coating to confer protection against bacterial colonization.

Bactericidal or attachment-resistant polymers have been deployed as surface coatings onto existing devices. However, for biomedical device applications, this approach may *in vivo* be compromised by the adsorption of proteins burying the active surface layer and providing a platform for bacterial attachment onto the adsorbed protein layer [13]. Accordingly, much attention has focused on strategies comprising the local release of antibacterial agents into the wound site and adjacent tissue. Such a localized approach to combating infections, as opposed to systemic administration, offers advantages such as much more efficient delivery.

The extensive literature on the development of biomedical devices that release antibacterial agents has been the subject of several reviews, for example [14–17]. Various antimicrobial compounds have been loaded into polymers or polymer composite films, including organic antibiotics [18–20], silver [21–23], and nitric oxide [24] for various intended biomedical applications. Other metal ions have also been investigated, including zinc [25–28], copper [29–31], and bismuth [32, 33]. A large part of this body of literature, however, focuses on the release of silver, and the approach of releasing silver to combat biofilm formation and BAIs has found application for commercial products.

14.2 Release of Silver Ions from Polymers and Polymeric Coatings

The release of silver ions from biomedical devices has attracted considerable attention. Compared to systemic delivery, a key advantage of local delivery at a specific site is that high local doses can be administered without exceeding the systemic toxicity level and risking, for example, renal and liver complications. Thus, increased doses can be applied at the site of a medical implant, or a wound dressing, or a catheter for example.

An important issue is the kinetics of release. Fast release provides relatively high doses but short-term action. Slower release may not reach the required therapeutic level, and might also lead to bacterial resistance at the release site due to survival of strains that adapt. The released silver ions must either act before a protective EPS protects the colony or be able to diffuse through the EPS in sufficient concentrations. Most reported systems achieve release over periods of a few days, which may suffice to prevent bacterial colonization problems during the stages of acute inflammation and initial wound healing.

Systems comprising silver, silver compounds and silver nanoparticles have enjoyed a great deal of attention [34]. Many polymers suitable for biomedical devices, such as silicones, can be blended, or loaded by in-diffusion, with silver nanoparticles or silver salts which then release silver ions when in contact with a fluid environment. A number of commercial products containing silver have emerged, including wound dressings, bandages for burns and chronic wounds, silver and silver nanoparticles coated catheters, and other medical devices. Examples are antimicrobial wound dressings, including Acticoat™ (Smith & Nephew) and Melgisorb Ag™ (Mölnlycke). They work very effectively in many clinical situations by releasing silver ions into the wound environment, preventing infection from taking hold.

However, there is cause for concern for three principal reasons. First, many antimicrobial compounds are somewhat cytotoxic, and silver is no exception. Acticoat™ has been found to show some in-vitro toxicity with keratinocytes [35]. The goal is to use an antimicrobial at concentrations below its cytotoxic level but above its antimicrobial MIC (Minimum Inhibitory Concentration). Second, continual exposure of a bacterial population to an antimicrobial agent, especially at subcritical concentrations when not all the population is killed, is a highly effective method to accelerate the evolution of resistant species [36]. Third, the issue of shape-mediated toxicity with particulate, nanocrystalline silver in dressings is still poorly understood. Asharani et al. found silver nanoparticle toxicity in developing zebra fish and found evidence of both systemic and developmental toxicity [37]. These factors do not necessarily argue against the use of silver as an antimicrobial agent, but suggest that having silver ions continually present in a wound may not be desirable.

The efficiency of some coatings and devices has been subject to controversy, and clinical concerns have also surfaced, as discussed below. A number of factors, both from the coating and the environment, come into play in determining the efficiency of silver based coating. It was found that coatings based on metallic silver are least efficient. This is because silver needs to be in its oxidized form (Ag^+) in order to exhibit antibacterial action. Ag^+ ions can be complexed by chloride ions to produce AgCl which precipitates, and this reduces the available local concentrations of silver ions. Another important problem is that in many cases, due the high solubility of ionic silver, most of the loaded silver is released very quickly from the coating, thus limiting the time of protection. A further concern is that little is known about the subsequent fate of silver ions, in terms of excretion or accumulation at specific body sites.

Liu et al. [38] investigated various methods of controlling silver diffusion from biomaterials, ranging from accelerated release through to slowed release (Fig. 14.1). The approaches of pre-oxidation or particle size reduction (i.e., increased nanoparticle surface area) provide routes to enhanced silver release in slow-leaching systems; whereas complexing silver with sulfides, amines, or citrates may be used to slow down silver release in the case of biomaterial surfaces where the silver release is too rapid. As silver ions are released from the surface

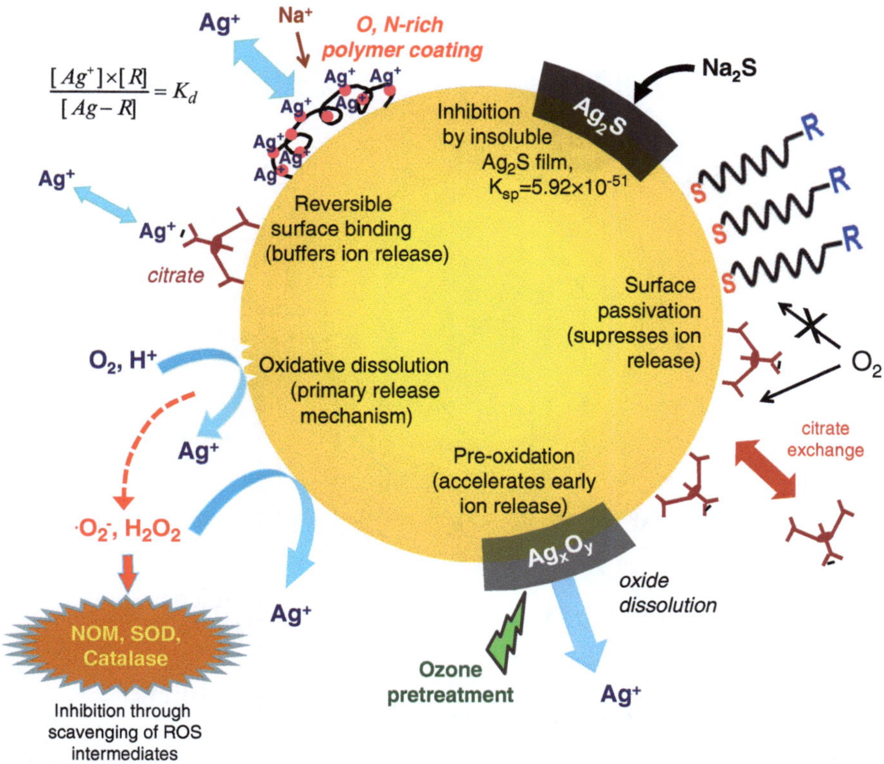

Fig. 14.1 A summary of chemical approaches which may be used to control silver ion release from antibacterial biomaterials [38]

of silver nanoparticles following oxidation of surface atoms, the extent of oxidation of silver nanoparticles and the surface-to-volume ratio are important operational parameters. The sterilization techniques of ethylene oxide and steam autoclaving both lead to enhanced surface oxidation of silver nanoparticles and hence usually faster release of silver ions compared with non-sterilized analogous samples, a factor which has been overlooked in some studies. It is thus essential to characterize silver ion release rates and antibacterial activity with samples sterilized by a method appropriate to the intended biomedical device application.

There have been substantial research efforts towards engineering thin polymer films loaded with silver and silver nanoparticles in order to achieve more prolonged release through both chemical and physical means. Many studies show effectiveness *in vitro*, but these new approaches are yet to be verified *in vivo*. While some studies have applied contiguous silver film coatings onto substrates such as fibers [39], most studies focus on silver (nano)particles incorporated within polymeric matrices. For biomedical devices it may be advantageous to use thin polymer films containing silver nanoparticles fully encapsulated within the polymer matrix, as

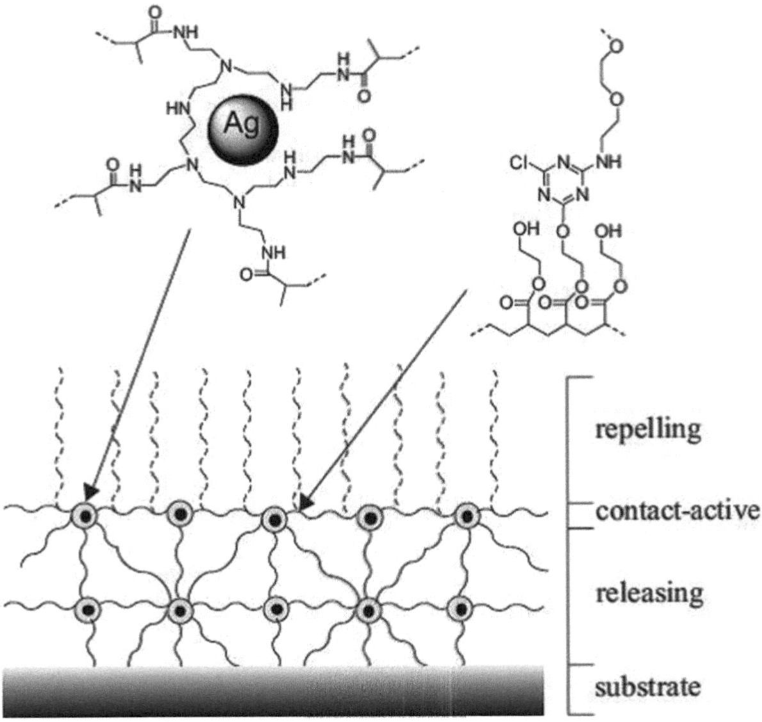

Fig. 14.2 Schematic representation of a non-fouling, contact active silver releasing polymer matrix. Imine functional groups within the polymer matrix coordinate the silver ions prior to reduction to nanoparticles, which become trapped in the bulk coating, while polyethylene glycol groups anchored at the surface serve to prevent cells from binding [41]

opposed to the situation where larger silver particles partly protrude from the polymer film layer, in which case it has been alleged that protruding asperities can "puncture" both bacterial and eukaryotic cells [40]. Ho et al. [41] presented an interesting approach (Fig. 14.2), first forming silver nanoparticles within a thin film of polyethylene imine. The polymer was copolymerized with 2-hydroxyethyl acrylate, which allowed further surface grafting of polyethylene glycol (PEG). In this manner coatings with two functions were generated, with the PEG layer preventing bacterial adhesion and the silver ions released from the coating providing biocidal activity against *Staphylococcus aureus.*

Vimala et al. [42] prepared hydrogel networks based on cross-linked poly(acrylamide) and three different carbohydrate polymers—gum acacia, carboxymethylcellulose, and starch. Silver nanoparticles were loaded into the hydrogel networks via in situ reduction of silver nitrate using sodium borohydride as reducing agent. Lee at al. [43] generated multilayer films, by layer-by-layer assembly, rich in catechol groups and used these groups to reduce silver ions to metallic silver when immersed in an aqueous metal salt solution. This resulted in the generation of silver nanoparticles within the films. Another study [44] synthesized silver nanoparticles in commercial cross-linked Amberlite resins. Sulfhydryl and $-N(CH_2COOH)_2$ groups present on these resins were used for Ag^+ chelation. Reductants were used to reduce the Ag^+ ions under alkaline pH. The coatings were bactericidal against *Escherichia coli* within a few minutes [44]. An approach by Ramstedt et al. utilized the properties of polyanionic electrolyte brushes to trap ionic silver within a film of atom transfer radical polymerized poly(3-sulfopropylmethacrylate) [45]. This method allows silver release by ion exchange with cations present in physiological media (e.g., Na^+ K^+), causing *in vitro* bactericidal action against *S. aureus*. The brushes were "rechargeable" upon soaking in aqueous silver nitrate after depletion of the loaded silver, which could be advantageous for multiple use applications such as contact lenses. However, the highly soluble nature of ionic silver in these films creates problems with excessive outdiffusion of silver and potential cytotoxicity in a mammalian environment due to an elevated local silver concentration [46].

An interesting approach comprises the simultaneous deposition of silver and the polymer matrix coating, thus embedding silver atoms, atomic clusters, and silver nanoparticles within a coating during its application rather than in a separate step. Thin layers of poly(tetrafluoroethylene) (PTFE) can be sputter-deposited with concurrent incorporation of nanosilver [47], or plasma polymerization can be used to form a polymeric thin film while simultaneously adding silver by sputtering [48]. The advantages of vapor deposition of polymeric matrices for nanosilver are that vacuum deposition often creates better interfacial adhesion and better coating uniformity at lower, controllable thicknesses, compared with solvent coating processes. However, the issue of even more limited loading and particle size, and thus limited duration of antibacterial action needs to be considered. The issue of silver loading into co-sputtered plasma polymer films has been addressed by Korner et al., demonstrating the that input power may be used to facilitate loading, but silver release from such systems is still under investigation [49].

An alternative approach comprises the creation of a polymeric coating by plasma polymerization, which is a convenient vacuum coating method for depositing thin polymeric films onto various substrates, followed by loading with silver nanoparticles. Such coated articles can be stored dry, and when brought into contact with aqueous solutions or body fluids, oxidation of silver atoms at the surface of the nanoparticles occurs, which leads to silver ion release and gradual erosion of the nanoparticles. Moreover, by vacuum depositing a second, ultrathin layer, the release rate can be controlled. In most reported studies, neither the amount of silver nor the release rate can readily be tailored to specific requirements. For a technology to be applicable to a range of biomedical devices, a smarter approach is needed which enables control over those parameters with minor process variation when using one basic technology. In particular, it is desirable to adjust the release rate of silver ions to comply with the different requirements of different biomedical devices, so that effective therapeutic dosage can be delivered over time periods as short as a few days or as long as a month or more. In most systems the loading and the release kinetics cannot be tailored independently.

The design of a system that enables independent control over both loaded amounts and delivery kinetics [50], while retaining the advantage of good coating adhesion

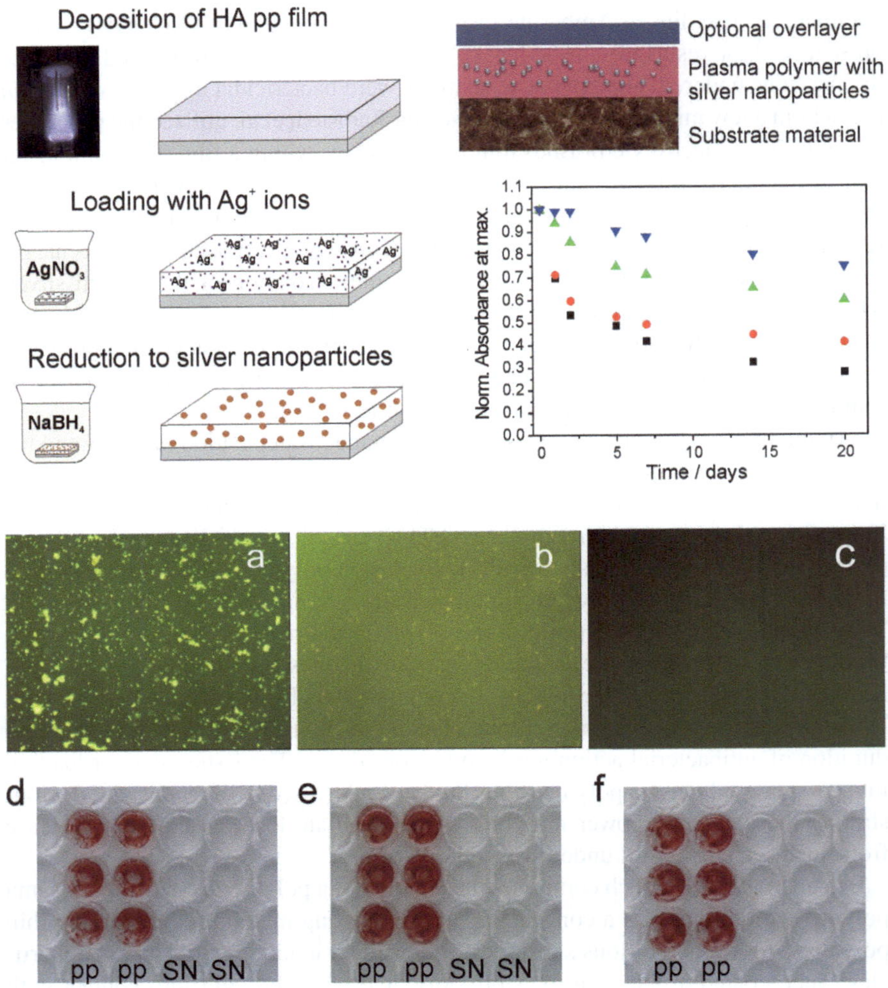

Fig. 14.3 Schematic diagram of the approach developed by Vasilev et al. [50] for the loading of silver nanoparticles into amine plasma polymer coatings, and the control of the release kinetics of silver ions via adjustable thickness of an additional plasma polymer overlayer. Images a to c show photomicrographs of bacterial colonization after 4 hrs onto surfaces of (**a**) control plasma polymer surface without silver, (**b**) silver-loaded, no overlayer, and (**c**) silver-loaded plus 6 nm heptylamine plasma polymer overlayer. Images d to f show photographs of multiwell plates stained for biofilm after bacterial growth for 24 hrs on samples coated with a silver-loaded plasma polymer layer followed by deposition of a diffusion-controlling overlayer of thickness (**d**) 6 nm; (**e**) 12 nm, and (**f**) 18 nm; pp denotes samples of plasma polymer controls without silver, and SN denotes samples containing silver nanoparticles

inherent in the plasma deposition of polymeric coatings, is shown in Fig. 14.3. Silver nanoparticles are loaded into an amine plasma polymer matrix by in-diffusion from a silver salt solution followed by reduction of the silver ions, which leads to their spontaneous agglomeration to nanoparticles within the polymeric matrix. A subsequently applied ultrathin barrier overlayer of adjustable thickness was used to control the rate of out-diffusion of silver ions. Plasma polymerization is a uniquely suitable approach for this as it enables the creation of ultrathin polymeric coatings with nanometer precision for the film thickness and excellent coating uniformity. The silver loading can be controlled either via the solution concentration of silver ions during loading or via the time of immersion for loading, leading to an approach with considerable flexibility and a wide range of controllable properties [50]. Plasma polymerization of *n*-heptylamine or allylamine was used to create amine-rich uniform polymeric coatings into which silver ions were diffused by placing samples in a solution of $AgNO_3$. The coalescence of silver ions to nanoparticles upon reaction with a reducing agent such as $NaCNBH_3$ was detected by a characteristic plasmon band in the UV/Vis absorption spectrum (Fig. 14.3). When placed in an aqueous solution, oxidation causes release of silver ions. The rate of release was shown to be adjustable by the application of a second plasma polymer layer (not containing silver ions and applied after loading the first coated layer with silver nanoparticles), with the thickness of this overlayer providing a convenient means of tailoring the release rate (Fig. 14.3). In this way, the loading and the release kinetics were adjustable independently over a considerable range. Even when slowing the rate of release such as to achieve sustained release over several weeks, antibacterial effectiveness was still obtained *in vitro* (Fig. 14.3).

An antibacterial coating can also be produced by the plasma polymerization of a phosphine-stabilized silver maleimide complex [51]. This method is attractive in terms of comprising a single step but requires synthesis of the precursor molecule used for the plasma deposition of the antibacterial coating and is largely contact-active due to the limited amount of silver within the film.

More commonly, conventional polymers and fabrication methods have been used. Eksik et al. [52] used electron transfer reaction and free radical polymerization processes to prepare triglyceride oil based polymer-silver nanocomposites. The antibacterial effect of the polymer nanocomposite was examined against Gram-positive, Gram-negative, and spore forming bacteria. Furno et al. [53] impregnated silicone with nanoparticulate silver metal using supercritical carbon dioxide and examined bacterial adherence and the rate of killing of planktonic bacteria by chemiluminescence and viable counts. The release rate of silver ions from the polymers in the presence and absence of blood was measured using inductively coupled plasma mass spectrometry (ICP-MS). It was found that the antibacterial efficiency of the coatings was dependent on the rate of release of silver ions. Kelly et al. [54] prepared polypyrrole and polyaniline conducting polymer composites with individual fibers of cellulose by direct polymerization of the respective monomers and used the redox active surface of the polymers to reduce silver ions to silver metal and thus producing cellulose/conducting polymer/silver composites. Kong et al. [55] synthesized poly(methyl methacrylate) nanofibers containing silver nanoparticles by radical-mediated dispersion polymerization and tested antibacterial properties against Gram-negative (*E. coli*) and Gram-positive (*S. aureus*) bacteria; MIC results showed that the silver loaded nanofibers had superior antimicrobial efficacy compared to that of silver sulfadiazine and silver nitrate at the same silver concentration. Liu et al. [56] developed a procedure for the synthesis of Ag nanoparticles in transparent PVA films via a two-step procedure consisting of ion incorporation in the matrix followed by thermal reduction. These composite films containing silver nanoparticles showed excellent antimicrobial performance toward bacteria such as *E. coli*. Silver nanoparticles were synthesized also in covalently linked layer-by-layer polymeric assemblies from methoxysilane polymers [57], in methoxy polyethylene glycol [58], an epoxy polymer matrix [59], poly(ether urethanes) [60], polyethylene [61], poly(vinyl alcohol) [62], diamond-like carbon [63], polyacrylate film [64], polytetrafluorethylene [47], and polyrhodanine nanofibers [65]. Vachon et al. [66] loaded silver ions into sulphonated polymer hydrogels. Kumar et al. [67] reported an approach to synthesize silver nanoparticle-embedded paint in a single step from common household paint, taking advantage of the naturally occurring oxidative drying process in oils, involving free-radical exchange, as a mechanism for reducing silver salts and dispersing silver nanoparticles in the oil media without the use of external reducing or stabilizing agents. Surfaces coated with silver-nanoparticle paint showed antimicrobial activity against Gram-positive human pathogens (*S. aureus*) and Gram-negative *E. coli.* A system devised by Sambhy et al. [68] uses a dual-action approach where a cationic polymer is infused with silver bromide. Silver is released from the film while the polymer, from the quaternized ammonium compound family, itself acts to disrupt bacteria cell membranes on the surface. Antibacterial properties were recorded both before and after exposure to mammalian fluids and the release rate of silver ions may be modulated by tailoring the nanoparticle size. Gordon et al. [69] prepared coordination polymers of ethanediyl bis(isonicotinate) and silver nitrate which can be grafted onto medical implants. Silver release was moderated by tuning the solubility of the compound. This method showed *in vitro* antimicrobial activity against *Staphylococcus epidermidis* over several days without significantly reducing the viability of eukaryotic cells.

Silver can not only be incorporated into polymeric carrier matrices/thin films, but also into some ceramics. Schneider et al. [70] described the use of silver on amorphous tricalcium phosphate (TCP) nanoparticles for electrospun, highly porous poly(lactide-co-glyclolide) (PLGA) fibrous composites. In vitro bioactivity tests of PLGA/Ag-TCP composite containing 0.5 wt% silver showed hydroxyapatite deposition on the nanocomposite within 2 days. Antibacterial tests using *E. coli* demonstrated a prolonged antibacterial effect of the scaffolds containing finely dispersed silver on TCP compared to current clinically used methods based on soaking the scaffolds with a tetracycline solution prior to implantation.

Even though only a small (but representative) part of the literature has been discussed in any detail, it is clearly evident that many approaches exist towards the same goal: elemental silver, nanoparticulate silver, ionic silver, complexed silver and nanocomposite silver. Each of these forms of silver have advantages and disadvantages according to (1) the desired release rate of silver, (2) the total required loading dose of silver, (3) the substrate to which the coating is to be applied, and (4) the considerations of the ultimate application of the material. It is also apparent that although much research is being carried out in the field of silver-containing biomaterials, there is little in the way of standardized testing to allow comparison of antibacterial properties, silver release and potentially adverse cytotoxic effects of the coatings.

Increasingly, due to the large array of antibacterial coatings which utilize silver, differentiation will be governed by the ability of a synthetic surface to resist the proliferation of bacteria whilst simultaneously providing a suitable environment for eukaryotic cells in the desired application. Thus, silver-containing coatings must be optimized for the specific requirements of a specific application, rather than the trend to date of claiming generic utility; clearly, the doses and rates of release, as well as the sensitivity of the host environment, might differ markedly for the wide range of biomedical devices that require antibacterial protection.

In addition, ease of upscaling, manufacturing costs and feasibility need to be taken into consideration when designing antimicrobial biomaterials. It is likely that future antimicrobial coatings will be derived from current iterations with advantages pertaining to speed and cost of manufacture, alongside application relevancy (release rate etc.) and ability to combat the evolutionary ability of bacteria to resist the effects of silver.

14.3 Bacterial Activity Versus Cytotoxicity

The antibacterial action of silver has been the subject of a number of studies and while opinions differ, there seems to be a likely scenario of silver ions binding to the bacterial cell membrane and damaging it by interfering with membrane receptors and interfering with bacterial electron transport (impeding the production of adenosine triphosphate, the cell's energy production). Gordon et al. [69] found that silver interfered with respiratory chain enzymes. Another theme is that silver binds to bacterial DNA and thus interferes with cell replication, and causes intracellular formation of insoluble compounds with nucleotides, proteins, and the amino acid histidine (making them unavailable as intracellular "building blocks").

Given similar structural elements of human cells, the question arises whether silver ions will also interfere with human cell and tissue functions in analogous ways. Indeed, Heidenau et al. [71] found an LD of 3.5×10^{-3} mmol/l for Ag^+ exposure of L929 fibroblasts, and Schrand et al. [72] reported that induction of reactive oxygen species (ROS), degradation of mitochondrial membrane integrity, disruption of the actin cytoskeleton, and reduction in proliferation after stimulation with nerve growth factor were observed upon exposing neuroblastoma cells to silver nanoparticles. Enzymes analogous to those found to be affected by silver [69] are also important in human cells, raising the question whether fundamentally silver ions can be applied without concomitant damage to nearby human cells and tissue. It may be a matter of what is the lesser evil—infection or damage to human tissue.

Perhaps relative cell size matters; bacteria are generally around 10–100 times smaller than eukaryotic cells, and therefore contain a much smaller surface/cell membrane area (but a higher surface area to volume ratio), which might mean that bacteria might be more easily affected by the toxic effects of silver per unit concentration. However, a larger number of bacterial cells per unit volume would reduce this effect.

There are many reports claiming that silver does not cause any significant damage and an equally large number of reports of damage to eukaryotic cells and tissue. This apparent controversy might result from various factors such as widely different tests, dosages, release rates, models, and techniques used to quantify effects at times perhaps not being sufficiently sensitive. We review here some representative reports. A thin coating of an antimicrobial composite of poly(4-vinylpyridine)-co-poly (vinyl-*N*-hexylpyridinium bromide) and AgBr nanoparticles on Tygon elastomer tubes caused disruption and activation of blood platelets, making this coating unsuitable for blood-contacting devices applications [52], but it was not clear whether it was the cationic copolymer or the Ag^+ ions, or both, that caused the adverse effects. It has also been reported that silver nanoparticles inhibited cell proliferation and migration in bovine retinal endothelial cells [73]. A megaprosthesis galvanically coated with elemental silver [74] showed antimicrobial efficacy against *S. aureus* in an *in vivo* rabbit model, but elevated silver levels were found in blood and organs, though without histological changes in organs. In a follow-up human trial, no systemic toxicity or side effects were found [75]. Chen et al. [76] created a co-sputtered silver-containing hypoxyapatite coating for implants which was found to be antimicrobial to both *S. aureus* and *S. epidermidis*, while also nontoxic to mammalian cells by comparison to hypoxyapatite alone.

Concerns have been raised by clinicians; for example, it was found that a person with burns who received local treatment with a commercial silver wound dressing showed hepatotoxicity and argyria-like symptoms and the silver levels in his plasma and urine were clearly elevated, as well as the liver enzymes, during the treatment period, raising concern about potential silver toxicity. It was suggested to monitor silver levels in plasma and/or urine during treatment [77]. Despite such concerns regarding toxicity to human cells and evidence of absence of antibacterial effectiveness *in vivo* [46, 78, 79], companies are building business on silver-releasing coatings, and research continues on variations of the theme. Perhaps more thought needs to be given where and when such coatings may be suitable or unsuitable; for example, does it matter if a layer of cells next to an implant is killed by antibacterial silver?

A key question is whether given the chemically rather indiscriminate nature of silver ions, can antibacterial effectiveness be balanced with minimal damage to mammalian cells? Agarwal et al. [80] reported that carefully controlled loading of silver nanoparticles within molecularly thin polymeric films can lead to antimicrobial activity without cytotoxicity. The system developed by Vasilev et al. [50], described above (Fig. 14.3), for obtaining control over loading and rate of release may, in addition to controlling release rates, also enable obtaining antibacterial effectiveness while reducing damage to mammalian cells and tissue, as suggested by data (Fig. 14.4) showing that osteoblast cells were able to attach and spread on multilayer coatings of the type shown in Fig. 14.3 despite silver ion release.

Yet, when analogous silver-loaded plasma polymer coatings were investigated with four other cell lines, a more complex picture emerged of differing sensitivities by different cell lines (S.S. Griesser et al., unpublished data). With the mouse fibroblast cell line NIH-3T3, the human fibroblast osteosarcoma line MG63, the human cornea epithelium line HCE-2, and the human retinal pigmented epithelium line ARPE-19 (all ATCC), cell attachment and spreading was investigated on samples with and without additional overlayers and over time periods up to 5 days. On heptylamine plasma polymer surfaces without silver, all cell lines attached and spread to confluence within 48 h, though ARPE-19 showed slower growth than the others. Optical micrographs showed that the responses of these four cell lines varied

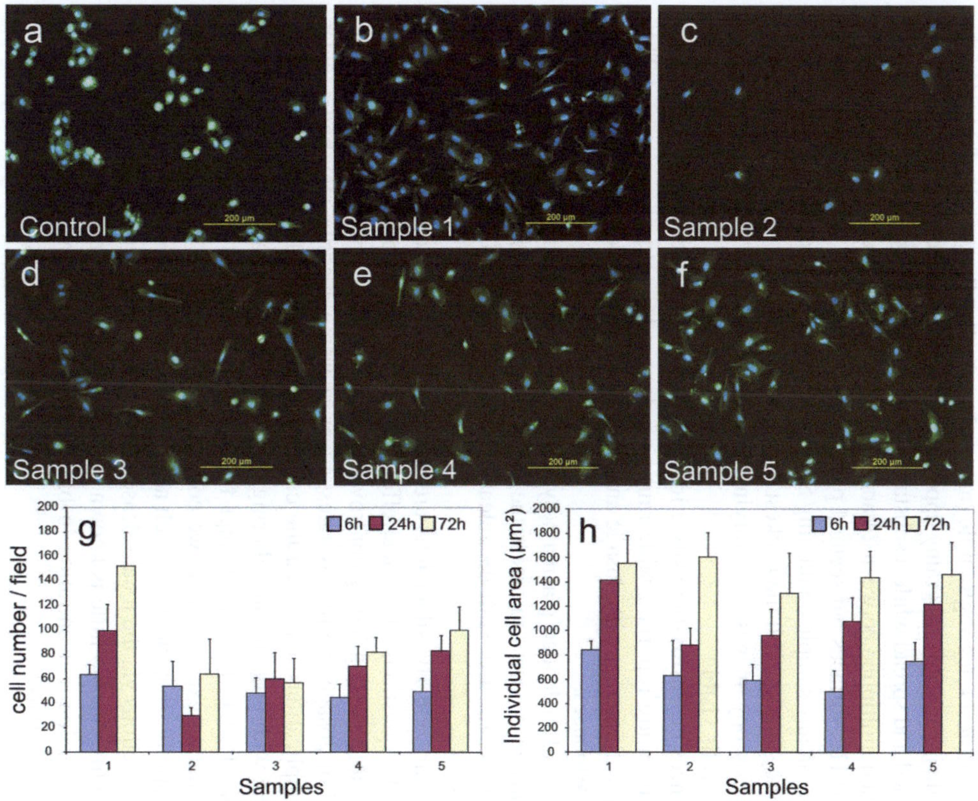

Fig. 14.4 Osteoblast cell adhesion and spreading after 24 h on (**a**) glass control; (**b**) heptylamine plasma polymer coating without silver; (**c**) HApp coating loaded with silver nanoparticles without an overlayer; (**d–f**) same as (**c**) but coated with an HApp overlayer of thickness 6 nm (**d**), 12 nm (**e**), and 18 nm (**f**); (**g**) cell numbers and (**h**) individual cell areas after 72 h on these samples [50]

considerably between cell lines and with silver release rates. HCE cells exposed to silver ions showed reduced growth rates and some cell death, but a significant number of cells still managed to attach and spread, followed by proliferation to areas of an almost confluent cell layer. Slower release of silver, via a thicker overlayer, increased the cell attachment and spreading. With ARPE-19 cells, release of silver ions similarly led to reduction in growth and a fraction of dead cells, but again the cell density increased from 24 to 48 h. 3T3 cells were much more affected by silver ions. On the fastest releasing sample (no overlayer), all cells were dead or dying, as shown by their rounded and grainy appearance. With slower release of silver, some cells attached and spread while a substantial percentage still was killed. The surviving cells then multiplied between 24 and 48 h. MG-63 cells grew extremely well on (non-silver) HA surfaces but seemed particularly sensitive to silver with no viable cell growth on any sample containing SNPs other than a few cells that managed to attach on the samples with the thickest overlayer. Figure 14.5 shows representative optical microscopy images, with varying proportions of round dead/dying cells and spread live cells visible.

Differences in sensitivity to silver ions of different cell lines may be one reason why the literature reports a wide range of observed cytotoxic responses. This sensitivity will interact with dosage and release rate to produce varying extents of cytotoxic responses.

For these silver-loaded plasma polymer coating systems, animal model studies are currently in progress to assess the extent of possible effects on tissue *in vivo* from silver ion release, with the release rate being varied via different overlayer thicknesses.

In summary, clearly, silver ions are effective against various bacteria *in vitro*. However, the questions of clinical effectiveness and damage to human cells and tissue are subject of continuing controversy. With improved understanding and better control of dosage and release rates, silver-releasing coatings may find a wider range of applications, but given the concern of potential side effects from silver complexing with proteins and altering their functions, we are of the opinion that application of silver release coatings for biomedical devices must be approached with due caution except where, we conjecture, much of the silver is likely to be removed rapidly from the body, such as with urinary catheters. The elimination of bacterial infections via the local release of silver has the potential to bring considerable benefits, but too little is known about downstream consequences such as a spike in silver ion concentration in blood and the excretion or accumulation in organs of silver ions.

14.4 Triggered Release

In addition to "passive" or "controlled passive" release systems, the approach of "active" silver release has been investigated. Raad et al. used electrochemical stimulation of metallic silver in a catheter to produce controlled levels of silver

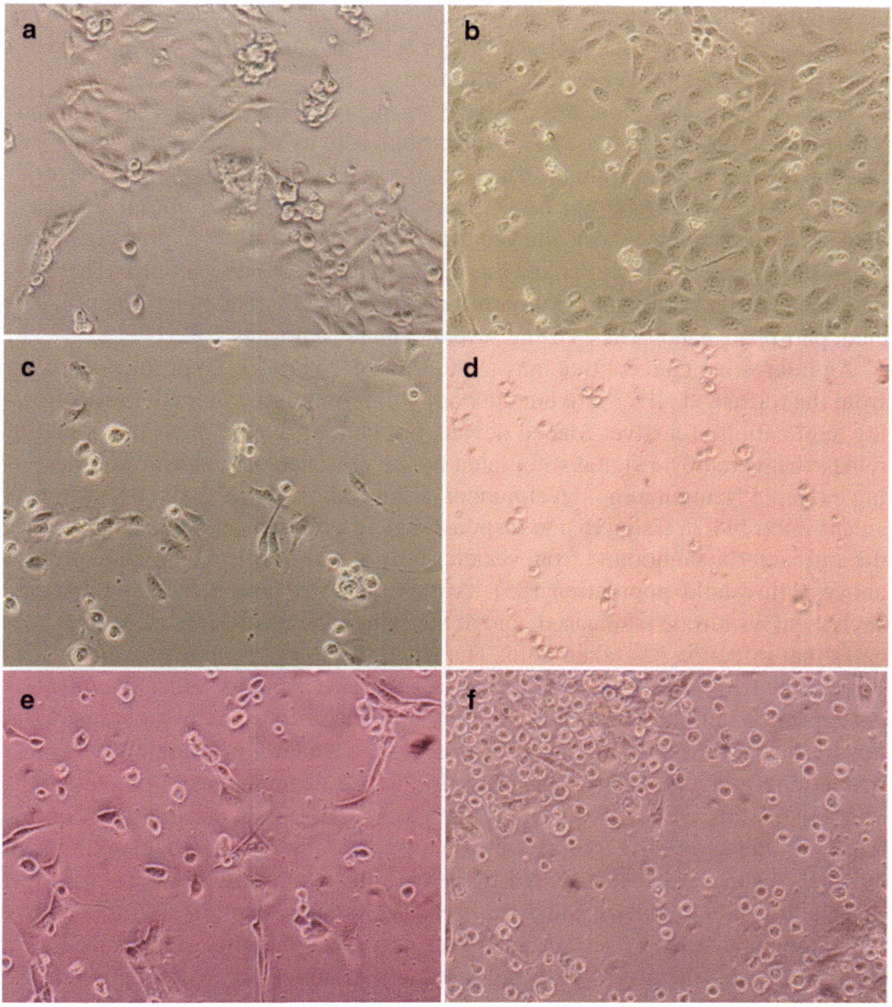

Fig. 14.5 Optical microscopy images of (**a**) HCE on HA+SNP/HA 12 nm, 24 h; (**b**) HCE on HA+SNP/HA 12 nm, 48 h; (**c**) ARPE on HA+SNP/HA 6 nm, 24 h; (**d**) 3T3 on HA+SNP/no overlayer, 24 h; (**e**) 3T3 on HA+SNP/HA 18 nm, 24 h; (**f**) MG-63 on HA+SNP/HA 18 nm, 48 h

ions [81]. This silver iontophoretic catheter, which consists of metallic silver wire wrapped around a silicone catheter and connected to an electric source, was shown to prevent *in vitro* migration of *S. epidermidis* from a contaminated hub to the tip over a 40-day period against a control of metallic silver which showed no effect, and also afforded protection from infection *in vivo* over a 7-day period in a rabbit model [82], whilst no evidence of silver toxicity was found in the rabbits [83]. Such an approach is by its very essence tuneable and long-lasting in terms of antibacterial effect, but requires a far greater quantity of silver than nanoparticulate, crystalline or purely ionic coatings, therefore potentially greatly increasing production costs, not to mention the battery setup and operation, which provides a practical obstacle in both use and application.

For potential application in bone tissue regeneration, Loher et al. [84] described a general approach to increase the antimicrobial activity of a silver-containing surface by 2–3 orders of magnitude. The authors used 1–2 nm silver particles decorating the surface of 20–50 nm carrier particles consisting of a phosphate-based, biodegradable ceramic, which allowed triggered release of silver as growing microorganisms dissolved the carrier and thus released the silver nanoparticles. They also described the rapid self-sterilization of polymer surfaces containing silver on calcium phosphate nanoparticles using several human pathogens.

An interesting concept that may avoid some of the potential problems arising from the release of silver into human body fluids and tissue immediately following application of a silver-loaded device, implant, or wound dressing, is that of release triggered by a signal stimulated by the presence of bacteria. An interesting example is an ongoing development at the University of Bath (UK) towards an antimicrobial dressing that is responsive to infection in a wound and releases the antibacterial compound from vesicles only if pathogenic bacteria are present above a threshold population [85]. An encapsulated antimicrobial substance such as silver nitrate is released, and at the same time a colorimetric signal indicates that infection has taken hold. The trigger in this system utilizes lytic toxins secreted by key pathogenic bacteria such as *S. aureus* and *Pseudomonas aeruginosa* as infection takes hold. Such bacteria secrete toxins to break down host tissue, providing nutrients and assisting the bacteria in colonizing the host wound environment. Many *S. aureus* strains, for example, secrete hemolysins and leukocidins which create pores in healthy cells, thus causing their death, while *P. aeruginosa* expresses phospholipases (amongst other toxins and enzymes) which catalytically hydrolyze cell phospholipids. It has been shown that pathogenic *S. aureus* and *P. aeruginosa* induce vesicle lysis [86], with a nonpathogenic strain of *E. coli* not affecting vesicle integrity. This is an excellent example of the use of bacterial metabolic products to switch on a defense response from a smart device.

14.5 Antimicrobial Resistance

The ability of microorganisms to develop resistance to antibiotics is a most serious issue for infection control. Bacteria are exposed to repeated cycles of treatment, which eliminates most but causes selection of resistant strains. It has been known for some time that some strains of bacterial species such as *Staphylococci*, which are responsible for the majority of the BAIs [1] have become resistant to β-lactam antibiotics, such as penicillin and methicillin. Continually exposing a bacterial population to an antimicrobial, especially at subcritical concentrations (below that required to kill all the population) is a highly effective method to accelerate the evolution of resistant species [36, 87]. It is therefore of crucial importance to consider and test whether medically relevant bacteria might evolve to adapt to silver ion release.

Silver appears to be the broadest spectrum antimicrobial agent available, and while concerns regarding the development of resistance have been expressed, so far there is no clear evidence of resistant strains. It has been demonstrated that bacteria such as *E. coli* may resist the bactericidal effects of silver by use of ion efflux pumps, which transport silver ions out of the bacterial cytoplasm and through the cell membrane [88]. Use of silver chaperones (ligands), such as amine functional groups or chitosan as in the approach by Shi et al., which provides a dual antimicrobial effect [89], may prevent the activation of such resistance mechanisms. Also, it is relevant to note that silver has been used as a biocide at subinhibitory concentrations for many years without widespread resistance of bacteria to silver becoming prevalent [40].

14.6 An Alternative Strategy: Surface-Immobilized Antibacterial Compounds

Given the potential drawbacks of silver/silver ions in regard to the mammalian cytotoxicity possibly being too high for some desired applications, and the limited lifetime of current release approaches, the alternative approach of covalent attachment of antibacterial compounds onto medical device surfaces may provide a route to obviating cytotoxicity issues and achieving permanent protection against biofilm formation. The covalent grafting of organic antibacterial compounds also avoids the issue of potential accumulation in organs.

Covalently immobilized antibacterial molecular layers have been the subject of rapidly increasing recent interest. It is beyond the scope of this chapter to review that literature; we merely wish to briefly mention a few representative possibilities. One approach comprises the covalent surface grafting of low molecular weight antimicrobial compounds. A number of established antibiotics, such as vancomycin and penicillin, have been immobilized onto the surface of polymers that are useful for biomedical device fabrication but by themselves are prone to bacterial biofilm colonization. A recent review [90] summarizes such studies.

Several studies have reported the grafting of biocides such as quaternized ammonium compounds (QACs) onto substrates [91, 92]. QACs act by lysing the membrane of bacteria, thus causing their rapid death, but they also have shown equally potent cytotoxic effects towards mammalian cells. Accordingly, grafted QAC coatings are better suited for applications where they do not contact human cells and tissue.

Alternatively, antibacterial synthetic polymers and antimicrobial peptides can be grafted onto polymer surfaces; this has also been the subject of considerable recent research in the search for antibacterial surfaces. Such approaches have also been reviewed in [90].

As for silver, a key issue is the question of differential toxicity. Some synthetic polymers of the Nylon-3 class have shown substantial differences between bacterial activity and damage to red blood cells [93]. Host defense peptides are innate to the human host body and as such good candidates for grafted antibacterial layers; their covalent immobilization has been reviewed [94] and their structure–activity relationships discussed [95].

Commercially available antibiotics can also affect human cells, though few data exist for surface-grafted layers. Some antibiotics may no longer be active when grafted onto a solid two-dimensional surface, as was observed for furanones due to the requirement for intracellular action [96]. Thus, there is ongoing interest in the identification of antibacterial compounds that exhibit a highly specific, selective mechanism of action with minimal biochemical activity in contact with human cells. Many existing antibiotics are derived from leads provided by compounds extracted from natural sources; plants and marine organisms continue to furnish interesting lead compounds as their defense mechanisms against microbial colonization commonly involve the biosynthesis of antimicrobial compounds. Antibiotics extracted from natural sources form the basis of many novel therapeutic agents as their chemical diversity provides wide opportunities for new drug leads [97–99]. Antibacterially active natural products offer structural variety and complexity, including distinctive ring architectures (e.g., vancomycin) or multifaceted supramolecular pharmacophores (e.g., erythromycin). The antibacterial penicillin-binding proteins (PCBs) and β-lactams are examples of naturally derived compounds with novel and even multiple modes of action [98].

Antibacterial coatings utilizing compounds from plant and marine life origins, or synthetic mimics, covalently grafted onto biomaterials surfaces, may thus be able to provide durable biofilm protection to biomaterials and biomedical devices. However, few detailed clinical studies exist and the issue of cytotoxicity remains to be studied for many of the antibacterial coatings that have shown promising *in vitro* antibacterial effects.

14.7 Conclusions

While a wide variety of strategies have been investigated for the fabrication of antibacterial coatings for biomedical devices, the use of silver occupies a prominent position with a large body of literature. Yet, it has gained only limited product acceptance apart from wound dressings. One reason is that it is relatively straightforward to perform antibacterial testing for attachment and biofilm formation on biomedical device surfaces *in vitro*, whereas clinical studies are much more complex and costly. Another reason is that *in vitro* tests utilize simplified environments that may neglect some key environmental factors that the coating needs to tolerate, such as acute inflammation following the wounding trauma to tissue. It is likely that there is no universal answer to the problem of biomedical device infection; coatings may have to be designed for specific applications taking into account the bacteria involved, their attachment and growth mechanisms, and the specifics of the biological milieu.

Clearly, silver ions are effective biocides and act against a wide range of bacteria *in vitro*. However, some studies have reported high levels of cell toxicity to mammalian cells; caution is indicated by, for example, an LD of 3.5×10^{-3} mmol/l for Ag^+ exposure of L929 fibroblasts [71] and interference with key enzymes common to bacteria and mammalian cells [69] by silver ions. The silver release approach may be suitable for some clinical applications, particularly wound dressings where side effects may be less important than the elimination of infection, but for other potential applications this high toxicity may cause serious problems. Input by clinicians and tissue biologists needs to be solicited to decide where the presence of killed cells adjacent to a biomedical device may or may not be a problem, or where silver ions may not accumulate to toxic levels, for example possibly with urethral catheters. In any case, the discrepancies in the scientific literature need to be resolved or at least rationalized in terms of different tests, dosages, release rates, and other possible factors, and a better understanding gained of how such factors might affect possible eukaryotic cell and tissue toxicity. Also, company literature on silver antibacterial approaches needs to be put in perspective and qualified in terms of possible uses; silver is not a panacea.

An issue common to all silver release approaches is the limited time over which the released concentration exceeds the therapeutic threshold. The effectiveness decays and usually vanishes within days or a few weeks. This is not a problem for applications such as many catheters, but for many biomedical implants, much longer durations are desired. For such long-term applications, silver release is not a viable option; instead, a suitable alternative strategy may be the covalent anchoring of antibacterial molecules onto biomedical device surfaces. Some studies have shown this approach to be successful *in vitro*, but very few *in vivo* data exist, and the potential of this alternative approach, and possible problems, cannot be assessed reliably at this point in time.

A number of issues need to be addressed in order to arrive at better-designed solutions for a range of biomedical devices and implants. A key issue is the question of cytotoxicity. There is continuing controversy in the literature about the cytotoxicity of silver. Is this a reflection of test systems that neglect key biological variables, of the use of different cells, of different ways in which silver was applied, or other factors? Firm cytotoxicity data for various cell lines are essential. Such data can then form a basis for informed decisions regarding where silver, or alternative, strategies may be suitable and where they should be avoided.

From a materials science perspective, a main issue is that many of the studies performed by clinicians contain little or no reliable information about the materials used. For example, concentrations and depth distributions of incorporated silver nanoparticles and silver ions within the materials are often not characterized. Even information on release kinetics are lacking in some studies, making it very difficult to understand scientifically the observations. In the absence of such information, it can be difficult to arrive at secured conclusions, let alone repeat such studies. It is highly recommended that materials scientists be included in such teams in order to ensure that the materials/devices used are well characterized, free of surface contaminants such as processing aids, and reproducible. In turn, materials scientists need to include biologists in order to ensure that their studies are not biologically naive.

Over a 5-year time frame, possibly the most exciting prospect is the development and design of smart coatings that offer triggered release of silver ions when demand is indicated by a signal such as a bacterial metabolite or enzyme. Such on-demand release, with no release unless bacteria are present, may circumvent the problem of limited duration offered by current release approaches in which release starts immediately upon contact with aqueous media. Advances in the understanding of bacterial biochemistry and advances in nanotechnology now offer the potential for the rational design of such triggered systems, and we expect that in the next few years there will be significant progress in this area.

Acknowledgments This work was supported in part by the Australian Government under NHMRC grant 1000737. KV acknowledges the Australian Research Council for support through fellowship FT100100292.

References

1. Harris LG, Richards RG. Staphylococci and implant surfaces: a review. Injury. 2006;37((2, Supplement 1)):S3–14.
2. Darouiche RO. Current concepts—treatment of infections associated with surgical implants. N Engl J Med. 2004;350(14):1422–9.
3. Costerton JW, Stewart PS, Greenberg EP. Bacterial biofilms: a common cause of persistent infections. Science. 1999;284(5418):1318–22.
4. Hoyle BD, Jass J, Costerton JW. The biofilm glycocalyx as a resistance factor. J Antimicrob Chemother. 1990;26(1):1–5.
5. Cheng G, Xite H, Zhang Z, Chen SF, Jiang SY. A switchable biocompatible polymer surface with self-sterilizing and nonfouling capabilities. Angew Chem Int Ed. 2008;47(46):8831–4.
6. Gottenbos B, Van der Mei HC, Busscher HJ, Grijpma DW, Feijen J. Initial adhesion and surface growth of *Pseudomonas aeruginosa* on negatively and positively charged poly(methacrylates). J Mater Sci Mater Med. 1999;10(12):853–5.
7. Martin TP, Kooi SE, Chang SH, Sedransk KL, Gleason KK. Initiated chemical vapor deposition of antimicrobial polymer coatings. Biomaterials. 2007;28(6):909–15.
8. Tiller JC, Liao CJ, Lewis K, Klibanov AM. Designing surfaces that kill bacteria on contact. Proc Natl Acad Sci USA. 2001;98(11):5981–5.
9. Fundeanu I, van der Mei HC, Schouten AJ, Busscher HJ. Polyacrylamide brush coatings preventing microbial adhesion to silicone rubber. Colloids Surf B Biointerfaces. 2008;64(2):297–301.
10. Harris LG, Tosatti S, Wieland M, Textor M, Richards RG. *Staphylococcus aureus* adhesion to titanium oxide surfaces coated with non-functionalized and peptide-functionalized poly(L-lysine)-grafted-poly(ethylene glycol) copolymers. Biomaterials. 2004;25(18):4135–48.
11. Kingshott P, Wei J, Bagge-Ravn D, Gadegaard N, Gram L. Covalent attachment of poly(ethylene glycol) to surfaces, critical for reducing bacterial adhesion. Langmuir. 2003;19(17):6912–21.
12. Ostuni E, Chapman RG, Liang MN, et al. Self-assembled monolayers that resist the adsorption of proteins and the adhesion of bacterial and mammalian cells. Langmuir. 2001;17(20):6336–43.
13. Schierholz JM, Lucas LJ, Rump A, Pulverer GJ. Efficacy of silver-coated medical devices. J Hosp Infect. 1998;40(4):257–62.
14. Hetrick EM, Schoenfisch MH. Reducing implant-related infections: active release strategies. Chem Soc Rev. 2006;35(9):780–9.
15. Zilberman M, Elsner JJ. Antibiotic-eluting medical devices for various applications. J Control Release. 2008;130(3):202–15.

16. Wu P, Grainger DW. Drug/device combinations for local drug therapies and infection prophylaxis. Biomaterials. 2006;27(11):2450–67.
17. Vasilev K, Cook J, Griesser HJ. Antibacterial surfaces for biomedical devices. Exp Rev Med Devices. 2009;6(5):553–67.
18. Schnieders J, Gbureck U, Thull R, Kissel T. Controlled release of gentamicin from calcium phosphate—poly(lactic acid-co-glycolic acid) composite bone cement. Biomaterials. 2006;27(23):4239–49.
19. Alt V, Bitschnau A, Osterling J, et al. The effects of combined gentamicin-hydroxyapatite coating for cementless joint prostheses on the reduction of infection rates in a rabbit infection prophylaxis model. Biomaterials. 2006;27(26):4627–34.
20. Rauschmann MA, Wichelhaus TA, Stirnal V, et al. Nanocrystalline hydroxyapatite and calcium sulphate as biodegradable composite carrier material for local delivery of antibiotics in bone infections. Biomaterials. 2005;26(15):2677–84.
21. Jones SA, Bowler PG, Walker M, Parsons D. Controlling wound bioburden with a novel silver-containing hydrofiber((R)) dressing. Wound Repair Regen. 2004;12(3):288–94.
22. Shanmugasundaram N, Sundaraseelan J, Uma S, Selvaraj D, Babu M. Design and delivery of silver sulfadiazine from alginate microspheres-impregnated collagen scaffold. J Biomed Mater Res B Appl Biomater. 2006;77B(2):378–88.
23. Kumar R, Munstedt H. Polyamide/silver antimicrobials: effect of crystallinity on the silver ion release. Polym Int. 2005;54(8):1180–6.
24. Nablo BJ, Rothrock AR, Schoenfisch MH. Nitric oxide-releasing sol-gels as antibacterial coatings for orthopedic implants. Biomaterials. 2005;26(8):917–24.
25. Poulter N, Donaldson M, Mulley G, Duque L, Waterfield N, Shard AG, Spencer S, Jenkins AT, Johnson AL. Plasma deposited metal Schiff-base compounds as antimicrobials. New J Chem. 2011;35(7):1477–84.
26. Zafar F, Ashraf SM, Ahmad S. In situ development of Zn/Cd-incorporated poly(esteramide-urethane) from sustainable resource. J Appl Polym Sci. 2008;110(1):584–94.
27. Coleman NJ. Aspects of the in vitro bioactivity and antimicrobial properties of Ag(+)- and Zn(2+)-exchanged 11 A tobermorites. J Mater Sci Mater Med. 2009;20(6):1347–55.
28. Duque L, Forch R. Plasma polymerization of zinc acetyl acetonate for the development of a polymer-based zinc release system. Plasma Process Polym. 2011;8(5):444–51.
29. Zhang W, Zhang YH, Ji JH, Zhao J, Yan Q, Chu PK. Antimicrobial properties of copper plasma-modified polyethylene. Polymer. 2006;47(21):7441–5.
30. Mary G, Bajpai SK, Chand NJ. Copper (II) Ions and copper nanoparticles-loaded chemically modified cotton cellulose fibers with fair antibacterial properties. J Appl Polym Sci. 2009;113(2):757–66.
31. Anyaogu KC, Fedorov AV, Neckers DC. Synthesis, characterization, and antifouling potential of functionalized copper nanoparticles. Langmuir. 2008;24(8):4340–6.
32. Mahony DE, Lim-Morrison S, Bryden L, Faulkner G, Hoffman PS, Agocs L, Briand GG, Burford N, Maguire H. Antimicrobial activities of synthetic bismuth compounds against *Clostridium difficile*. Antimicrob Agents Chemother. 1999;43(3):582–8.
33. Bland MV, Ismail S, Heinemann JA, Keenan JI. The action of bismuth against *Helicobacter pylori* mimics but is not caused by intracellular iron deprivation. Antimicrob Agents Chemother. 2004;48(6):1983–8.
34. Rai M, Yadav A, Gade A. Silver nanoparticles as a new generation of antimicrobials. Biotechnol Adv. 2009;27(1):76–83.
35. Poon VKM, Burd A. In vitro cytotoxity of silver: implication for clinical wound care. Burns. 2004;30:140–7.
36. Russell AD, Hugo WB. Antimicrobial activity and action of silver. In: Ellis GP, Luscombe DK, editors. Progress in medicinal chemistry. New York: Elsevier Science; 1994. p. 351–69.
37. Asharani PV, Wu YL, Gomg ZY, Valiyaveettil S. Toxicity of silver nanoparticles in zebrafish models. Nanotechnology. 2008;19:Article No:255102.
38. Liu JY, Sonshine DA, Shervani S, Hurt RH. Controlled release of biologically active silver from nanosilver surfaces. ACS Nano. 2010;4(11):6903–13.

39. Amberg M, Grieder K, Barbadoro P, Heuberger M, Hegemann D. Electromechanical behavior of nanoscale silver coatings on PET fibers. Plasma Processes Polym. 2008;5(9):874–80.
40. Morones JR, Elechiguerra JL, Camacho A, Holt K, Kouri JB, Ramirez JT, Yacaman MJ. The bactericidal effect of silver nanoparticles. Nanotechnology. 2005;16(10):2346–53.
41. Ho CH, Tobis J, Sprich C, Thomann R, Tiller JC. Nanoseparated polymeric networks with multiple antimicrobial properties. Adv Mater. 2004;16(12):957–61.
42. Vimala K, Sivudu KS, Mohan YM, Sreedhar B, Raju KM. Controlled silver nanoparticles synthesis in semi-hydrogel networks of poly(acrylamide) and carbohydrates: a rational methodology for antibacterial application. Carbohydr Polym. 2009;75(3):463–71.
43. Lee H, Lee Y, Statz AR, Rho J, Park TG, Messersmith PB. Substrate-independent layer-by-layer assembly by using mussel-adhesive-inspired polymers. Adv Mater. 2008;20(9):1619–23.
44. de Santa Maria LC, Souza JDC, Aguiar M, et al. Synthesis, characterization, and bactericidal properties of composites based on crosslinked resins containing silver. J Appl Polymer Sci. 2008;107(3):1879–86.
45. Ramstedt M, Cheng N, Azzaroni O, Mossialos D, Mathieu HJ, Huck WTS. Synthesis and characterization of poly(3-sulfopropylmethacrylate) brushes for potential antibacterial applications. Langmuir. 2007;23(6):3314–21.
46. Ramstedt M, Ekstrand-Hammarstrom B, Shchukarev AV, Bucht A, Osterlund L, Welch M, Huck WTS. Bacterial and mammalian cell response to poly(3-sulfopropyl methacrylate) brushes loaded with silver halide salts. Biomaterials. 2009;30(8):1524–31.
47. Zaporojtchenko V, Podschun R, Schurmann U, Kulkarni A, Faupel F. Physico-chemical and antimicrobial properties of co-sputtered Ag-Au/PTFE nanocomposite coatings. Nanotechnology. 2006;17(19):4904–8.
48. Despax B, Raynaud P. Deposition of "polysiloxane" thin films containing silver particles by an RF asymmetrical discharge. Plasma Processes Polym. 2007;4(2):127–34.
49. Korner E, Aguirre MH, Fortunato G, Ritter A, Ruhe J, Hegemann D. Formation and distribution of silver nanoparticles in a functional plasma polymer matrix and related Ag(+) release properties. Plasma Process Polym. 2010;7(7):619–25.
50. Vasilev K, Sah V, Anselme K, Ndi C, Mateescu M, Dollmann B, Martinek P, Ys H, Ploux L, Griesser HJ. Tunable antibacterial coatings that support mammalian cell growth. Nano Lett. 2010;10:202–7.
51. Poulter N, Munoz-Berbel X, Johnson AL, Dowling AJ, Waterfield N, Jenkins ATA. An organosilver compound that shows antimicrobial activity against *Pseudomonas aeruginosa* as a monomer and plasma deposited film. Chem Commun. 2009;47:7312–4.
52. Eksik O, Erciyes AT, Yagci Y. In situ synthesis of oil based polymer composites containing silver nanoparticles. J Macromol Sci A Pure Appl Chem. 2008;45(9):698–704.
53. Furno F, Morley KS, Wong B, et al. Silver nanoparticles and polymeric medical devices: a new approach to prevention of infection? J Antimicrob Chemother. 2004;54(6):1019–24.
54. Kelly FM, Johnston JH, Borrmann T, Richardson MJ. Functionalised hybrid materials of conducting polymers with individual fibres of cellulose. Eur J Inorg Chem. 2007;35:5571–7.
55. Kong H, Jang J. Antibacterial properties of novel poly(methyl methacrylate) nanofiber containing silver nanoparticles. Langmuir. 2008;24(5):2051–6.
56. Liu SX, He JH, Xue JF, Ding WJ. Efficient fabrication of transparent antimicrobial poly(vinyl alcohol) thin films. J Nanoparticle Res. 2009;11(3):553–60.
57. Sambhy V, Peterson BR, Sen A. Multifunctional silane polymers for persistent surface derivatization and their antimicrobial properties. Langmuir. 2008;24(14):7549–58.
58. Mallick K, Witcomb MJ, Scurrell MS. Self-assembly of silver nanoparticles: formation of a thin silver film in a polymer matrix. Mater Sci Eng C Biomimetic Supramolecular Syst. 2006;26(1):87–91.
59. Lu J, Moon KS, Wong CP. Silver/polymer nanocomposite as a high-k polymer matrix for dielectric composites with improved dielectric performance. J Mater Chem. 2008;18(40):4821–6.
60. Gray JE, Norton PR, Griffiths K. Mechanism of adhesion of electroless-deposited silver on poly(ether urethane). Thin Solid Films. 2005;484(1–2):196–207.

61. Sanchez-Valdes S, Ortega-Ortiz H, Valle L, Medellin-Rodriguez FJ, Guedea-Miranda R. Mechanical and antimicrobial properties of multilayer films with a polyethylene/silver nanocomposite layer. J Appl Polymer Sci. 2009;111(2):953–62.
62. Galya T, Sedlarik V, Kuritka I, Novotny R, Sedlarikova J, Saha P. Antibacterial poly(vinyl alcohol) film containing silver nanoparticles: preparation and characterization. J Appl Polym Sci. 2008;110(5):3178–85.
63. Schwarz F, Thorwarth G, Wehlus T, Stritzker B. Silver nanocluster containing diamond like carbon. Phys Status Solidi A Appl Mater Sci. 2008;205(4):976–9.
64. Voccia S, Ignatova M, Jerome R, Jerome C. Design of antibacterial surfaces by a combination of electrochemistry and controlled radical polymerization. Langmuir. 2006;22(20):8607–13.
65. Kong H, Jang J. Synthesis and antimicrobial properties of novel silver/polyrhodanine nanofibers. Biomacromolecules. 2008;9(10):2677–81.
66. Vachon DJ, Yager DR. Novel sulfonated hydrogel composite with the ability to inhibit proteases and bacterial growth. J Biomed Mater Res A. 2006;76A(1):35–43.
67. Kumar A, Vemula PK, Ajayan PM, John G. Silver-nanoparticle-embedded antimicrobial paints based on vegetable oil. Nat Mater. 2008;7(3):236–41.
68. Sambhy V, MacBride MM, Peterson BR, Sen A. Silver bromide nanoparticle/polymer composites: dual action tunable antimicrobial materials. J Am Chem Soc. 2006;128(30):9798–808.
69. Gordon O, Slenters TV, Brunetto PS, Villaruz AE, Sturdevant DE, Otto M, Landmann R, Fromm KM. Silver coordination polymers for prevention of implant infection: thiol interaction, impact on respiratory chain enzymes, and hydroxyl radical induction. Antimicrob Agents Chemother. 2010;54(10):4208–18.
70. Schneider OD, Loher S, Brunner TJ, Schmidlin P, Stark WJ. Flexible, silver containing nanocomposites for the repair of bone defects: antimicrobial effect against *E. coli* infection and comparison to tetracycline containing scaffolds. J Mater Chem. 2008;18(23):2679–84.
71. Heidenau F, Mittelmeier W, Detsch R, et al. A novel antibacterial titania coating: metal ion toxicity and in vitro surface colonization. J Mater Sci Mater Med. 2005;16(10):883–8.
72. Schrand AM, Braydich-Stolle LK, Schlager JJ, Dai LM, Hussain SM. Can silver nanoparticles be useful as potential biological labels? Nanotechnology. 2008; 19 (23):Article Number 235104.
73. Kalishwaralal K, Banumathi E, Pandian SRK, Deepak V, Muniyandi J, Eom SH, Gurunathan S. Silver nanoparticles inhibit VEGF induced cell proliferation and migration in bovine retinal endothelial cells. Colloids Surf B Biointerfaces. 2009;73(1):51–7.
74. Gosheger G, Hardes J, Ahrens H, Streitburger A, Buerger H, Erren M, Gunsel A, Kemper FH, Winkelmann W, von Eiff C. Silver-coated megaendoprostheses in a rabbit model—an analysis of the infection rate and toxicological side effects. Biomaterials. 2004;25(24):5547–56.
75. Hardes J, Ahrens H, Gebert C, Streitbuerger A, Buerger H, Erren M, Gunsel A, Wedemeyer C, Saxler G, Winkelmann W, Gosheger G. Lack of toxicological side-effects in silver-coated megaprostheses in humans. Biomaterials. 2007;28(18):2869–75.
76. Chen W, Liu Y, Courtney HS, Bettenga M, Agrawal CM, Bumgardner JD, Ong JL. In vitro anti-bacterial and biological properties of magnetron co-sputtered silver-containing hydroxyapatite coating. Biomaterials. 2006;27(32):5512–7.
77. Trop M, Novak M, Rodl S, Hellbom B, Kroell W, Goessler W. Silver coated dressing Acticoat caused raised liver enzymes and argyria-like symptoms in burn patient. J Trauma. 2006; 60(3):648–52.
78. Riley DM, Classen DC, Stevens LE, Burke JP. A large randomized clinical-trial of a silver-impregnated urinary catheter—lack of efficacy and Staphylococcal superinfection. Am J Med. 1995;98(4):349–56.
79. Srinivasan A, Karchmer T, Richards A, Song X, Perl TM. A prospective trial of a novel, silicone-based, silver-coated Foley catheter for the prevention of nosocomial urinary tract infections. Infect Control Hosp Epidemiol. 2006;27(1):38–43.
80. Agarwal A, Weis TL, Schurr MJ, Faith NG, Czuprynski CJ, McAnulty JF, Murphy CJ, Abbott NL. Surfaces modified with nanometer-thick silver-impregnated polymeric films that kill bacteria but support growth of mammalian cells. Biomaterials. 2010;31(4):680–90.

81. Raad I, Hachem R, Zermeno A, Dumo M, Bodey GP. In vitro antimicrobial efficacy of silver iontophoretic catheter. Biomaterials. 1996;17(11):1055–9.
82. Raad I, Hachem R, Zermeno A, Stephens LC, Bodey GP. Silver iontophoretic catheter: a prototype of a long-term antiinfective vascular access device. J Infect Dis. 1996;173(2):495–8.
83. Hachem RY, Wright KC, Zermeno A, Bodey GP, Raad II. Evaluation of the silver iontophoretic catheter in an animal model. Biomaterials. 2003;24(20):3619–22.
84. Loher S, Schneider OD, Maienfisch T, Bokorny S, Stark WJ. Micro-organism-triggered release of silver nanoparticles from biodegradable oxide carriers allows preparation of self-sterilizing polymer surfaces. Small. 2008;4(6):824–32.
85. Jenkins ATA, Young AER. Smart dressings for the prevention of infection in pediatric burns patients. Expert Rev Anti Infect Ther. 2010;8:1063–5.
86. Zhou J, Loftus AL, Mulley GJ, Jenkins ATA. A thin film detection/response system for pathogenic bacteria. J Am Chem Soc. 2010;132(18):6566–70.
87. Anwar H, Strap JL, Costerton JW. Establishment of aging biofilms: possible mechanism of bacterial resistance to antimicrobial therapy. Antimicrob Agents Chemother. 1992;36(7):1347–51.
88. Silver S. Bacterial silver resistance: molecular biology and uses and misuses of silver compounds. FEMS Microbiol Rev. 2003;27(2–3):341–53.
89. Shi ZL, Neoh KG, Kang ET, Wang W. Antibacterial and mechanical properties of bone cement impregnated with chitosan nanoparticles. Biomaterials. 2006;27(11):2440–9.
90. Munoz-Bonilla A, Fernandez-Garcia M. Polymeric materials with antimicrobial activity. Prog Polym Sci. 2012;37(1):281–339.
91. Dhende VP, Samanta S, Jones DM, Hardin IR, Locklin J. One-step photochemical synthesis of permanent, nonleaching, ultrathin antimicrobial coatings for textiles and plastics. ACS Appl Mater Interfaces. 2011;3(8):2830–7.
92. Jampala SN, Sarmadi M, Somers EB, Wong ACL, Denes FS. Plasma-enhanced synthesis of bactericidal quaternary ammonium thin layers on stainless steel and cellulose surfaces. Langmuir. 2008;24(16):8583–91.
93. Mowery BP, Lindner AH, Weisblum B, Stahl SS, Gellman SH. Structure-activity relationships among random nylon-3 copolymers that mimic antibacterial host-defense peptides. J Am Chem Soc. 2009;131:9735–45.
94. Costa F, Carvalho IF, Montelaro RC, Gomes P, Martins MCL. Covalent immobilization of antimicrobial peptides (AMPs) onto biomaterial surfaces. Acta Biomater. 2011;7:1431–40.
95. Palermo EF, Kuroda K. Structural determinants of antimicrobial activity in polymers which mimic host defense peptides. Appl Microbiol Biotechnol. 2010;87(5):1605–15.
96. Kuehl R, Al-Bataineh S, Gordon O, Luginbuehl R, Otto M, Textor M, Landmann R. Furanone at subinhibitory concentrations enhances staphylococcal biofilm formation by *luxS* repression. Antimicrob Agents Chemother. 2009;53(10):4159–66.
97. Cos P, Vlietinck AJ, Berghe DV, Maes L. Anti-infective potential of natural products: How to develop a stronger in vitro 'proof-of-concept'. J Ethnopharmacol. 2006;106(3):290–302.
98. von Nussbaum F, Brands M, Hinzen B, Weigand S, Habich D. Antibacterial natural products in medicinal chemistry—exodus or revival? Angew Chem Int Ed. 2006;45(31):5072–129.
99. Ndi CP, Semple SJ, Griesser HJ, Pyke SM, Barton MD. Antimicrobial compounds from the Australian desert plant Eremophila neglecta. J Nat Prod. 2007;70(9):1439–43.

Chapter 15
Quaternary Ammonium Compounds

J.A. Loontjens

Abstract The main cause of failure of biomedical implants is bacterial infections. Despite all efforts, it will never be possible to completely free operating theaters from bacteria, as human bodies contain already 10^{14} bacteria. Once on a surface, bacteria start proliferating, while protecting themselves with a slime layer against the immune system and administered antibiotics. A promising route to prevent or at least reduce bacterial infections caused by implants is by making surfaces of the devices antibacterial. One way to eradicate bacteria on implants is by contact killing.

Quaternary ammonium compounds (quats) are very potent biocides. The generally accepted mechanism is that quats destabilize the cytoplasmic membrane, which leads to leakage and eventually to cell death. Unfortunately, a similar mechanism provokes cytotoxicity. Fortunately, it is feasible to optimize the balance between biocidal activity and cytotoxicity by adapting the chemical structures.

Low molecular weight quats are effective, but leachable and thus only temporarily effective. Moreover, leachable quats will be transported throughout the whole body and may cause cell lysis. A more sustainable approach is to immobilize quats on surfaces. Although this development is still in an early stage in the last decade much progress has been made. There is a vast amount of literature describing successful *in vitro* experiments, and a number of papers have shown antibacterial activity in vivo.

15.1 Introduction

In 1847 I. Semmelweis proposed that physicians should wash their hand before medicating women during childbirth. He was fired proposing such a "senseless" procedure. Subsequently, he went to another hospital where the procedure was

J.A. Loontjens (✉)
Zernike Institute for Advanced materials, Polymer Chemistry Department,
University of Groningen, Nijenborgh 4, Groningen, 9747AG, The Netherlands
e-mail: J.A.Loontjens@rug.nl

T.F. Moriarty et al. (eds.), *Biomaterials Associated Infection: Immunological Aspects and Antimicrobial Strategies*, DOI 10.1007/978-1-4614-1031-7_15,

introduced. As a result the number of women dying during delivery decreased from 25 to 1%. It was unknown at that time that bacteria were responsible for many diseases. More than 30 years later, it was R. Koch who discovered that the evildoers were pathogenic bacteria, invisible to the naked eye.

Bacterial infections were a major cause of mortality in the history of mankind. The discovery of antibiotics in the last century resulted in a revolution in medication. Remarkably, it still took more than 10 years, after the discovery of penicillin by Fleming in 1928, before the world started to realize the importance of that finding. After the general acceptance of penicillin and the discovery of several other antibiotics it was thought, in the second half of last century, that mortality and morbidity due to infections were surmounted in the western world. Antibiotics interfere in specific growth steps of bacteria, enabling them to circumvent the treatment by using another pathway, thereby building up resistance. Antibiotics that are used in health care are also abundantly used in animal husbandries, in current veterinary practice. Via that route, the resistance of bacteria against certain antibiotics has increased enormously, which becomes a threat for mankind.

In contrast to antibiotics, biocides kill bacteria in a nonspecific manner. As a consequence the probability for bacteria to build up resistance is therefore much more difficult, although not completely absent [1]. Buildup of biocidal resistance after treatment is sometimes permanent, but often only temporary. After ceasing administration of the biocide the next generations of the same bacterial strain become susceptible again.

Biocidal compounds are able to kill a wide spectrum of bacterial strains in many applications. However, biocides are in principle not drugs, but meant to prevent microbial (cross) contamination. On November 4, 2003 The European Biocidal Product Directive was published [2], a list comprising about 1300 biocidal compounds that were commercially available on the market in Europe. Due to Reach (Registration, Evaluation, Authorization, and Restriction of Chemicals), which enforces companies to supply an extensive set of supporting data, it is expected that this list will be reduced. Nevertheless, there are still many low molecular weight biocidal compounds commercially available. As most of these compounds are already in use for decades in industrial and domestic applications, they have a proven track record. For instance, metal ions, such as tin and copper were widely used in marine paint. Due to environmental constrains tin is almost completely abandoned, while copper is under dispute. Silver is broadly used in domestic and in biomedical applications. Due to this wide usage, for instance in antibacterial socks, washing machines, and deodorants, silver will pollute our waste water. As a result, resistance against the use silver has become stronger in the last decade [3]. The same holds for halogen containing compounds and phenolic derivatives. Organic acids and alcohols are considered to be environmental friendly biocides, but unfortunately not very potent.

The list of the European Biocidal Product Directive offers an outstanding entry to select the most promising candidates for biomedical applications. From that compilation quaternary ammonium compounds (quats) are perhaps the most promising class of materials. There are, particularly in the last decade, quite some new approaches on quaternized materials, as described in a number of excellent reviews [4–10].

Quats cover a wide range of compounds with many different structural features. Their high potency has experimentally been proven against numerous bacterial strains. If a quaternary ammonium compound comprises, besides the positive charge, a hydrophobic moiety in close proximity biocidal properties can be expected. A large variety of structures and compounds is then conceivable. Another benefit is that quats are synthetically easily accessible, which opens a broad field of potential applications.

It is proposed that quaternary ammonium compounds interfere with the cytoplasmic membrane of bacteria, causing leakage by membrane destabilization [11]. This mechanism will occur in a lesser extent in case of mammalian cells, which have a different cytoplasmic membrane compared to bacteria (vide infra). As a result, quaternary ammonium compounds have a good balance between potency to kill bacteria and cytotoxicity for mammalian cells.

Phosphonium ions have rather similar properties as quaternary ammonium compounds, although some publications suggest a higher potency [12, 13]. Due to the similarity we restricted ourselves here to the synthetically more readily accessible quaternary ammonium compounds.

15.2 Antibacterial Mechanism of Quaternary Ammonium Compounds

There is quite some consensus on the concept that quaternary ammonium compounds disturb and eventually destroy the cytoplasmic membrane of bacteria. This disturbance causes leakage resulting finally in destruction of the organism (Fig. 15.1).

However, it is not fully understood how this process develops on a molecular level. Currently, researchers predominately support two mechanisms. Bacteria generally have a peptidoglycan outer layer (cell wall) and a phospholipid inner membrane (Fig. 15.2). Although gram-positive bacteria have a thicker (~80 nm) and more cross-linked outer layer than gram-negative bacteria (~20 nm), there is no evidence that these features determine the efficacy of quats. Both types of bacteria have a peptidoglycan envelope composed of a polysaccharide network and cross-linked by peptides. This network comprises glycerol phosphates (teichoic acids), in which the charge is neutralized by calcium or magnesium ions.

The phospholipid inner layer is a dynamic membrane that regulates the metabolic processes. Phospholipids comprise in general two long alkyl tails (C_{18}) connected via phosphoric acid. The remaining negative charge is neutralized by calcium or magnesium cations. This layer is a dynamic assembly of surfactant types of molecules (phospholipids), which is susceptible for damage. In contrast, the peptidoglycan layer is a cross-linked polysaccharide network, giving bacteria their shape stability. However, this network is permeable, enabling metabolites to cross the cell wall, and thus also penetrable for quats.

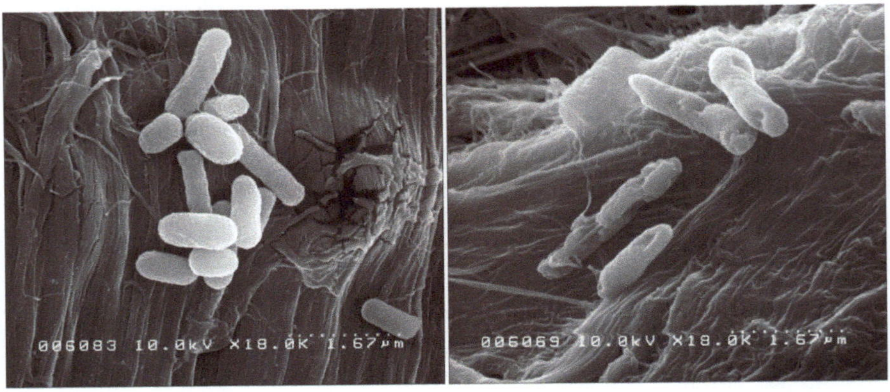

Fig. 15.1 Destruction of bacteria on surfaces comprising quaternary ammonium compounds (Quick med technologies) [47]

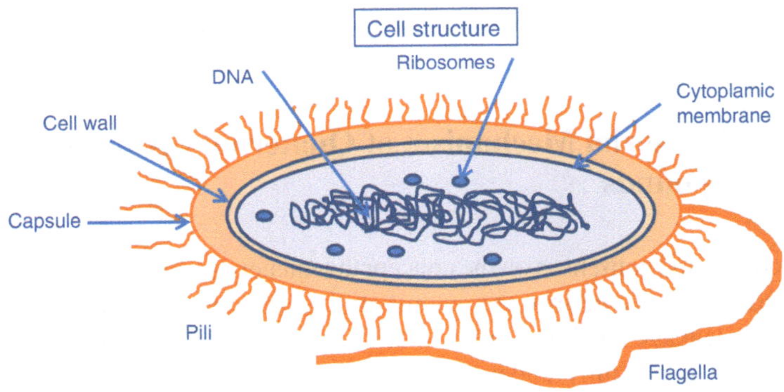

Fig. 15.2 Schematic representation of a bacterium

Tiller et al. [11] and others [14] proposed a mechanism by which the quaternary ammonium compounds penetrate through the peptidoglycan layer into the cytoplasmic membrane.

This seems a plausible mechanism as quaternary ammonium compounds are only effective if they have an amphiphilic structure, matching with phospholipids. It is thought that these foreign surfactant molecules disorder the inner phospholipid layer, thereby causing leakage of bacteria (Fig. 15.3). As a result, bacteria will lose essential compounds, indispensable for continuation of life.

In the second mechanism, it is proposed that quats penetrate into the peptidoglycan layer in the same way as discussed in the first mechanism. However, now

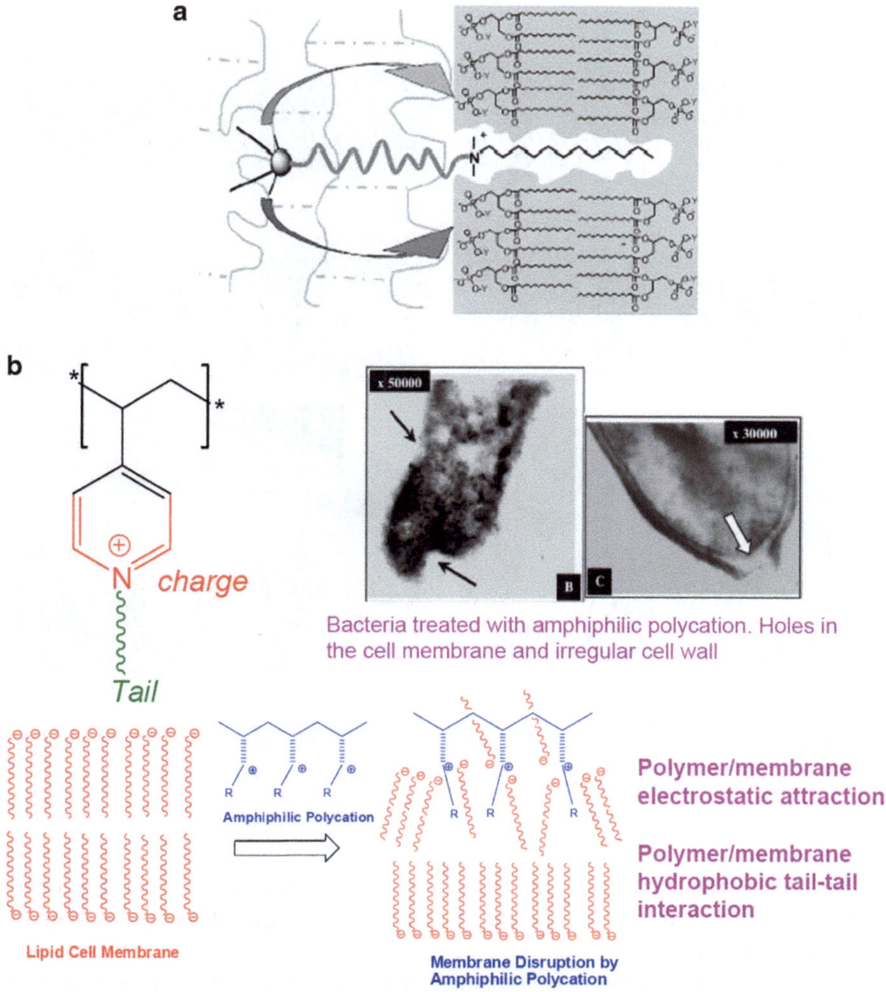

Fig. 15.3 (**a**) Schematic representation of disturbance of the cytoplasmic membrane by quats (scheme from Waschinski et al. [11] (**b**) Damaged bacteria. Kenawy et al. [5])

destabilization is caused by the exchange of cationic species [15–17]. If positive-charged species (quats) enter the cell membrane, other cationic species have to leave, to obey the principle of electrical neutrality. Calcium ions are not only the counter ions of the phospholipids for neutralizing charge but they are also responsible for the stability of phospholipid layer due to the ionic interactions (Scheme 15.1).

Destabilization of the cytoplasmic membrane by foreign cationic species can now take place at a "distance" as calcium ions are distributed throughout the peptidoglycan layer. Ionic exchange reactions can take place through equilibrium shifts

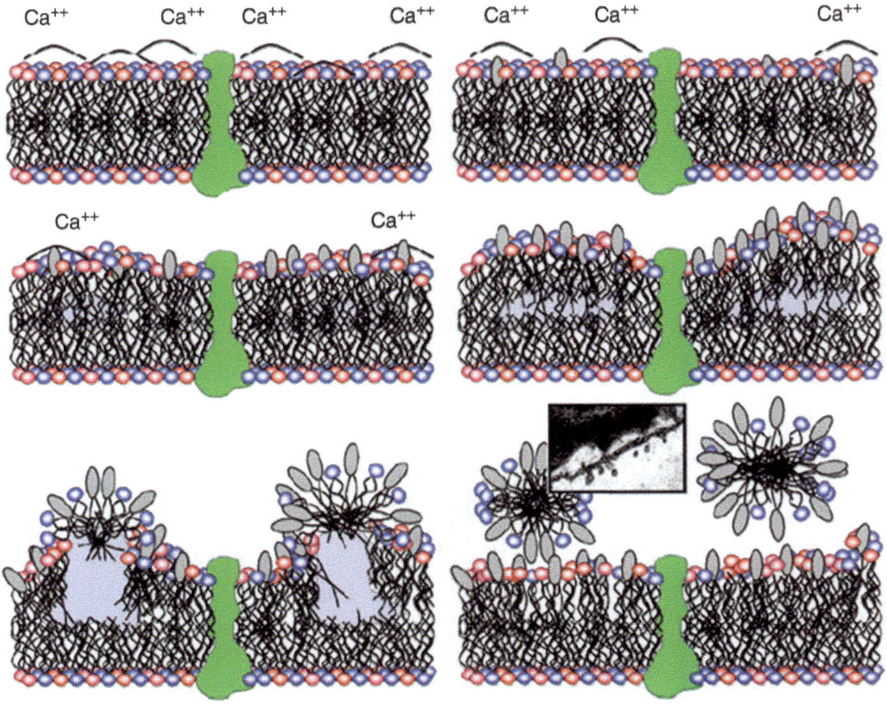

Scheme 15.1 Mechanism of disturbance of quats due to exchange of calcium ions (Gilbert et al. [29])

throughout the whole bacterial envelope, eventually reaching the cytoplasmic membrane. This second concept corresponds better with the results of immobilized quats with short spacers.

Immobilization of quats on biomedical devices is one of the most important study themes of the last decade. The advantage is that there is no leakage of biocidal substances into body fluids. Moreover, in principle, the efficacy could last longer since no exhausting of the biocide takes place, although fouling might interfere. Nevertheless, immobilization of quats is a very promising technique for eradicating bacteria on devices, while preventing biocides from contaminating the patient's body fluids.

One of the oldest and most successfully immobilized quats on the market is *N*-octadecyl, *N,N*-dimethyl [3-(trimethoxysilyl)propyl]chloride. Immobilization occurs via the condensation reaction of the siloxane groups with hydroxyl groups on the surface of a device. This immobilized compound is effective, although the length of the alkyl spacer group (octadecyl) between the tethered moiety and the quat is by far too short to penetrate through the peptidoglycan layer (20–80 nm) [18]. Similarly, cross-linked poly(ethylenimines) are very effective biocides. It is also in that system improbable that the dangling ends have sufficient length to reach the cytoplasmic membrane. Matyjaszewski et al. [19] came to the same conclusion with tethered polyacrylates.

Nevertheless, all these coatings are active biocides. Because of their efficacy, irrespective of the length of the spacers, the second mechanism seems more probable than the first one, at least for immobilized compounds. It can of course not be excluded and it is even probable that both mechanisms are operative, depending on the coating system.

15.3 Cytotoxicity

So far no conclusive model is proposed to explain why bacteria are more sensitive to quats than mammalian cells. Prokaryotic cells (bacteria) and eukaryote cells (human) display quite some structural resemblances. The active content (DNA, RNA, ribosomes, potassium ions, etc.) of both cells is enclosed by a phospholipid double layer, surrounded by a permeable shape-stable network (peptidoglycan and cytoskeleton respectively). The similarity between bacteria and mammalian cells is obvious, but there are also important differences, which might shed light on the difference in sensitivity for quats. One crucial distinction can be found in the structure of the phospholipid layers (Fig. 15.4).

The outer layer of mammalian cells contains mostly lipids with a zwitter ionic structure, i.e., each lipid molecule comprises a negative and a positive charge. Zwitter ions do not need an external counter ion, as they are internally neutralized. On the other hand, bacteria possess more an anionic outer layer (di-phosphate ester), of which the charge is neutralized by calcium or magnesium ions. As aforementioned, quats can replace calcium or magnesium ions, thereby destabilizing the cytoplasmic membrane. As a result, this distinction in structure of the lipids might be responsible for the difference in efficacy of quats on bacteria and mammalian cells. However, more investigations have to be done to fully prove this hypothesis.

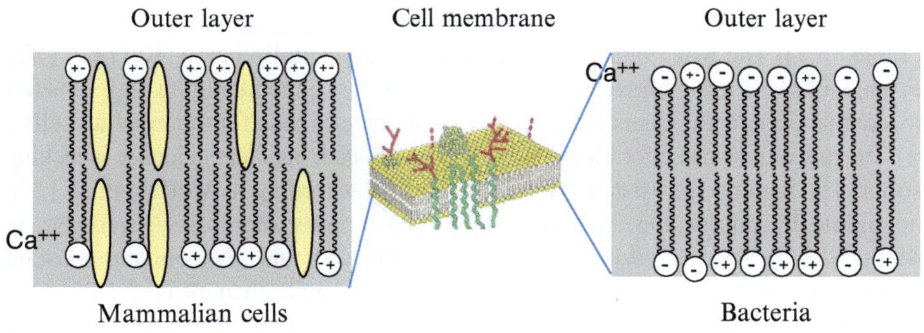

Fig. 15.4 Structure of phospholipid layers of bacteria and mammalian cells [108]

Table 15.1 Lethal dose of various quaternary ammonium compounds (Block et al. [21])

Compounds	LD_{50} (mg/kg)[a]
Cetylpyridinium chloride	200
Benzalkonium chloride	300
Domiphen bromide	320
Methylbenzethonium chloride	350
Benzethonium chloride	420
Didecyldimethyl ammonium chloride	530
Octadecyldimetylammonium chloride	720
BTC 2125M	750
Dowcil 200	2,200
6,10-ionene bromide	1,000
WSCP on bussan 77	2,770
Mirapol A-15	2,850
Onamer M or polyquat® or polyquaternium-1	4,470

[a]LD_{50} median lethal dose

Thus, quaternary ammonium compounds cause in general more damage to bacteria than to the membranes of mammalian cells. But, quats are not harmless for eukaryotic cells. Hence, it is crucial to find the optimal balance between the desired activity (killing pathogens) and the undesired action (cytotoxicity). This balance is dependent on the molecular structure of the quats and can be controlled by a proper design.

Some quats, such as cetylpyridinium chloride and benzalkonium chloride, are already used for many years in clinical applications, in food production and in health care. These compounds have a long practical history and thus apparently the right structural features. Kourai et al. [20] have measured the minimum inhibitor concentration (MIC) and the minimum biocidal concentration (MBC) for a number of bacteria, and the lethal dose (LD_{50}) of a number of quats for erythrocyte and fibroblast (NB1RGB) cells. In all cases the MIC was lower than the LD_{50} (lethal dose where 50% of the cells die), while in nearly all case the MBC was lower than the LD_{50}. Depending on the structure of the cationic species the balance can be optimized. An evidence-based choice for the best option can be made only after an experimental testing program.

Block [21] has summarized the LD_{50} values of a number of common quats (Table 15.1). The MIC values of potent quats are generally below 100 ppm [22]. From Table 15.1 it can be seen that the LD_{50} values are generally quite high. It is fair to say that, in general, quats are more toxic for bacteria than for mammalian cells.

The risk of resistance for biocides is often mentioned, but clear evidence is lacking so far. The buildup of resistance towards quats is, according to McBain et al. [23], negligible. In contrast, the risk of resistance of bacteria for antibiotics is well known. Methicillin-resistant *staphylococcus aureus* (MRSA) and Vancomycin-resistant *staphylococcus aureus* (VRSA) have become more regular than exceptional.

At this point it is also important to distinguish between intrinsic resistance and acquired resistance. Quaternary ammonium compounds have biocidal activity towards many bacterial stains. Meyer et al. [1] have shown that some strains acquire resistance during administration of quats, which was lost after administration had been ceased. So, after some time the next generations of the same bacterial strains become sensitive again for the same biocide.

Polymeric quaternary ammonium compounds are generally very potent biocides (vide infra). However, some of these polymeric cationic compounds are also quite toxic [24]. But, similarly as for low molecular weight quats, the toxicity of polymers is dependent on the molecular structure. For instance Polyquaternium® and, particularly, poly(hexamethylene biguanide) display a rather low cytotoxicity [25, 26]. These polymers are therefore well suited for biomedical applications.

Thus the balance between antibacterial properties and hemolysis depends on the polymeric structure and can be optimized. Kroda et al. [27] have demonstrated that the relation between antibacterial and hemolytic properties can be influenced by changing the balance between cationic and hydrophobic groups in copolymers. Also Sambhy et al. demonstrated that the antibacterial and hemolytic properties are influenced differently by changing the composition of copolymers [28]. These approaches offer tools to optimize the potency of polymeric quats, while limiting the cytotoxicity.

An alternative route to reduce toxicity is to use water insoluble or tethered cationic polymers, as they cannot contaminate body fluids (vide infra)

15.4 Low Molecular Weight Quaternary Ammonium Compounds

In clinical, domestic, and industrial applications low molecular weight quaternary ammonium compounds are widely used. Although many different structures exist, all effective compounds comprise a positive-charged species while bearing a hydrophobic moiety nearby. In Fig. 15.5 dodecyl trimethyl ammonium chloride (**a**), benzalkonium chloride (b), and chlorhexidine (**c**) are depicted as simple but very useful examples. Chlorhexidine is a strong basic compound that becomes a cationic species after protonation in an aqueous environment. Chlorhexidine is frequently used as an antiseptic agent in biomedical applications.

The aliphatic alkyl chain of low molecular quats consist in most cases of 6–18 CH_2-groups. The optimal length of the alkyl chains depends on the other structural elements of the molecules. These kinds of quats have been used since the 1930s, with no apparent reduction of efficacy [29]. These rather simple compounds are potent biocides, having a MIC generally below 100 ppm. Although, quaternary ammonium compounds with similar structural elements as the compounds in Fig. 15.5 are often biocidal, exceptions exist. As a result clinical evidence is mandatory to determine the balance between the biocidal efficacy and the cytotoxicity under appropriate conditions.

Low molecular weight quaternary ammonium compounds that do not fulfill the hydrophobic requirements, i.e., lacking long alkyl moieties, are not biocidal.

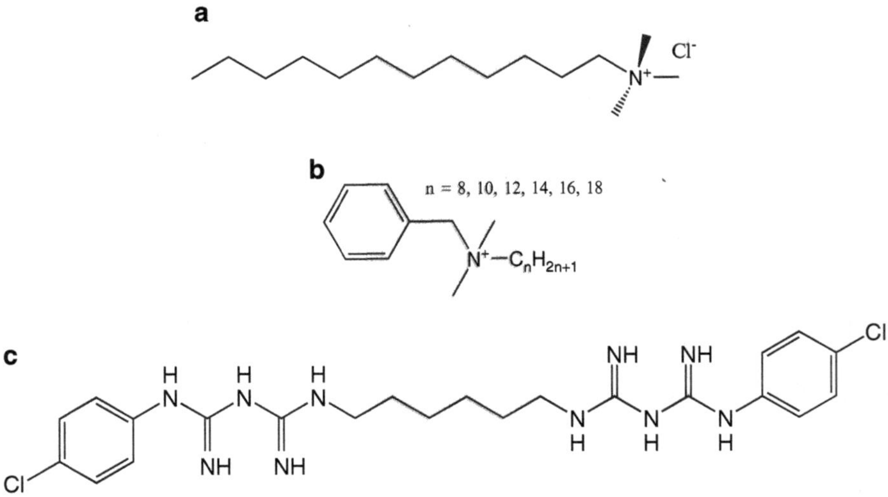

Fig. 15.5 Structure of (**a**) dodecyl trimethyl ammonium chloride, (**b**) benzalkonium chlorides and (**c**) chlorhexidine

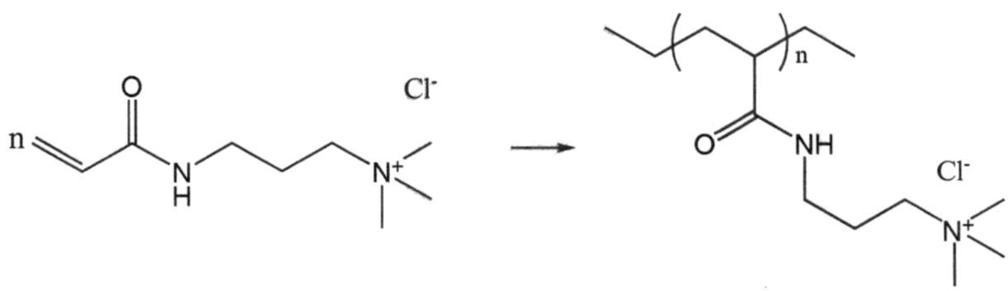

Scheme 15.2 Polymerization of (3-acrylamido propyl) trimethyl ammonium chloride [109]

Tetramethyl ammonium bromide for instance is not biocidal. However, if these low molecular weight quats comprise polymerizable groups than they may become antibacterial due to the hydrophobicity of the formed polymer backbone. The biocidal activity is then related to the molecular weight of the polymers [30]. In Scheme 15.2 an illustrative example of such system is given. The monomer, (3-acrylamido propyl) trimethyl ammonium chloride, does not display antibacterial properties, whereas the polymer is a potent biocide.

This result suggests that the hydrophobicity of the polymer backbone contributes to the hydrophobic requirements for biocidal behavior of quaternary ammonium compounds. A contribution of a cooperative effect of an assembly of positive charges in a polymer chain on the biocidal behavior, i.e., an attack of multiple cationic species simultaneously, cannot be excluded. It is even more probably that a number of cationic species has to act cooperatively to kill bacteria. With polymers such a cooperative action is assured.

15.5 Polymeric Quaternary Ammonium Compounds

Low molecular weight quaternary ammonium compounds are water soluble and therefore especially applicable in a wet environment. Wet conditions are obviously also ideal conditions for the growth of bacteria, fungi and algae. It is, therefore, not surprising that quats are beneficially applied in cosmetics, sanitation, swimming pools, etc.

Low molecular weight quats are less suited as a biocidal additive for biomedical devices because they are not only active on the devices but also contaminate body fluids, damaging human cells. By increasing the molecular weight, i.e., making polymers, the release profile can be controlled by reducing the solubility in body fluids. Moreover, polymeric quats can be compounded with the polymeric constituents of plastic devices, without deteriorating the mechanical properties. Besides that, polymeric materials offer the possibility to make antibacterial coatings. Coating formulations based on antibacterial polymers are very attractive, as it is an enabling technology to protect all existing devices and implants against bacterial colonization. Numerous excellent reviews on polymeric quats have been published in the last two decades [5–10].

Here we summarize briefly a number of cationic polymers with a high biocidal potency, which are intentionally designed for biomedical applications.

15.5.1 Cationic Polyacrylates

Polyacrylates comprising quaternary ammonium compounds are perhaps most often described in the literature, due to the importance of polyacrylates in coatings and because of the commercial availability of quat comprising monomers. In Fig. 15.6 a general structure of a polyacrylate copolymer is depicted with (3-methacrylamido alkyl) trimethyl ammonium chloride as antibacterial moiety.

The polymers can comprise, beside the cationic monomer, various comonomers allowing optimization of all other material properties. Poly(meth)acrylate polymers are used in many commercial coating formulations. This has obviously the advantage that the antibacterial coatings can be made in existing equipment, with common recipes. Creavis [31–37] has actively pushed polyacrylate quats in the beginning of this decade. They supplied these polymeric quats commercially as additives to add to other polymeric matrices. The slow leachability of these quats is due to the hydrophobic nature and the (co) polymeric structure.

Acrylate dispersion coatings are commonly made with anionic or nonionic surfactants, but have also been described with cationic dispersing agents. Ciba was one of the first to prepare cationic dispersions (in 1959) [38], although they were at that time not meant to be used in antibacterial coatings. Hoechst was one of the first (1991) to report on aqueous dispersions based on polyacrylate quats ((3-methacrylamido alkyl) trimethyl ammonium chloride (MAPTAC)) for making intentionally antibacterial coatings [39].

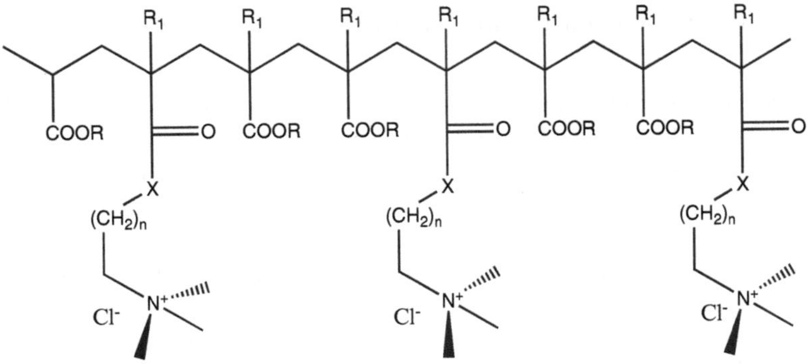

Fig. 15.6 Polyacrylate copolymers containing quaternary ammonium comprising monomers (MAPTAC, $n=2$, $X=\mathrm{NH}$, $R_1=\mathrm{CH_3}$)

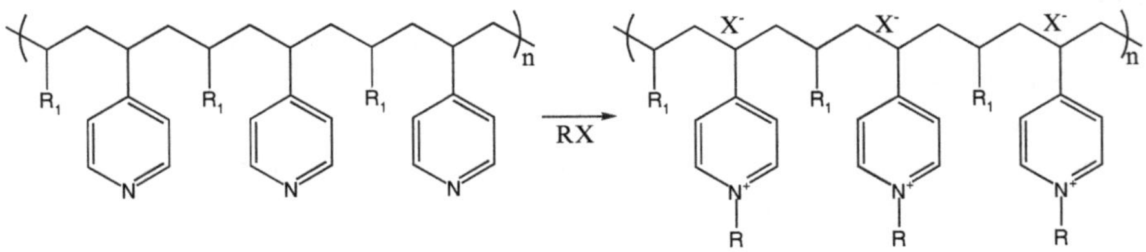

Scheme 15.3 Preparation of poly(vinyl pyridinium chloride) copolymers

15.5.2 Cationic Poly(Vinyl Pyridines)

Poly(vinyl pyridines) are particularly studied in the academia to demonstrate the applicability of polycations in antimicrobial coatings [40]. They are readily accessible via alkylation of commercially available poly(vinyl pyridine), forming poly(vinyl pyridinium chloride) (Scheme 15.3). The advantage of this system is that the type of R group and the number of quaternized pyridinium monomers can be chosen at will. Both factors have a profound influence on the antibacterial efficacy. The radical polymerization of vinylpyridine allows control over the synthesis, permitting to make well-defined structures [41].

15.5.3 Cationic Poly(Ethylene Imines)

Quaternized poly(ethylene imines), which are a very promising class of biocidal polymers, were pioneered by Klibanov et al. [42–46]. Poly(ethylene imines) (PEI) are probably one of the polymers with the highest density of secondary and tertiary amino groups, and can consequently afford a high charge density after quaternization

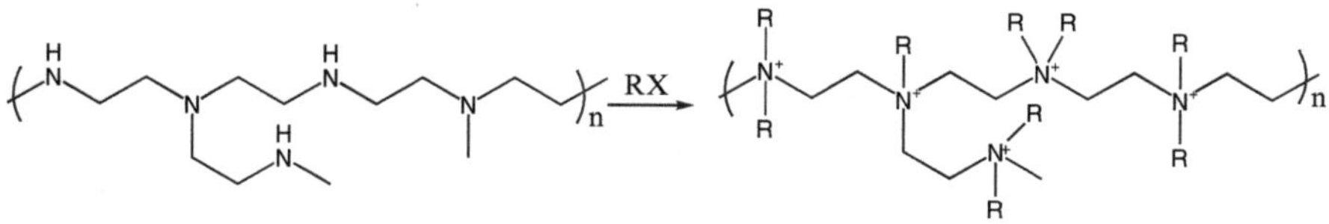

Scheme 15.4 Preparation of branched quaternized poly(ethylene imines)

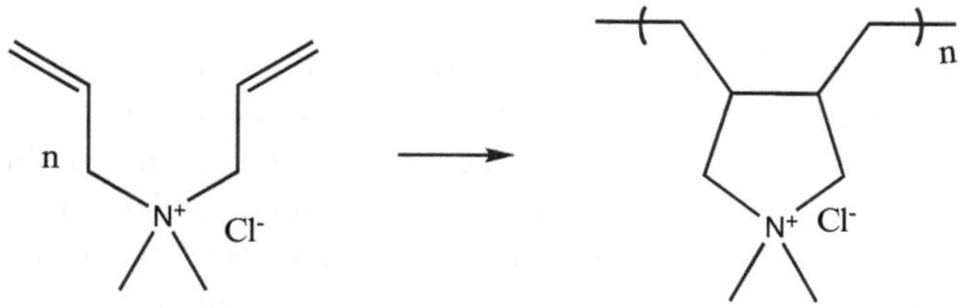

Scheme 15.5 Preparation poly(diallyl dimethylammonium chloride)

(Scheme 15.4). In contrast to most other polycations in this system the C_6 R-groups displayed the most potent biocidal efficacy. The commercial availability of various branched PEIs is another important advantage towards applications. Recently Klibanov et al. have shown that these hydrophobic polycationic coatings improved considerabe bone healing in sheep [46].

Soluble quaternized poly(ethylene imines) having a high charge density are cytotoxic. However, immobilization of quaternized PEIs shifts the balance between killing efficiency and toxicity in the desired direction, making these polymers promising candidates for antibacterial coatings (vide infra).

15.5.4 Cationic Poly(Diallyl Amines)

Poly(diallyl dimethylammonium chlorides) (PDADMAC) are used in commercial wound dressings [47], which demonstrates the efficacy of polymeric quats in biomedical applications. In Scheme 15.5 the structure of PDADMAC is depicted.

These polymers are barely water soluble due to the hydrophobic backbone. This is an advantage because these materials can be deposited on substrates, yielding non-leachable antibacterial surfaces in spite of their lack of cross-links or tethering units.

15.5.5 Cationic Poly(Hexamethylene Biguanides)

Poly(hexamethylene biguanides) (PHMB) are already for many decades commercially on the market and widely used in domestic as well as in clinical applications

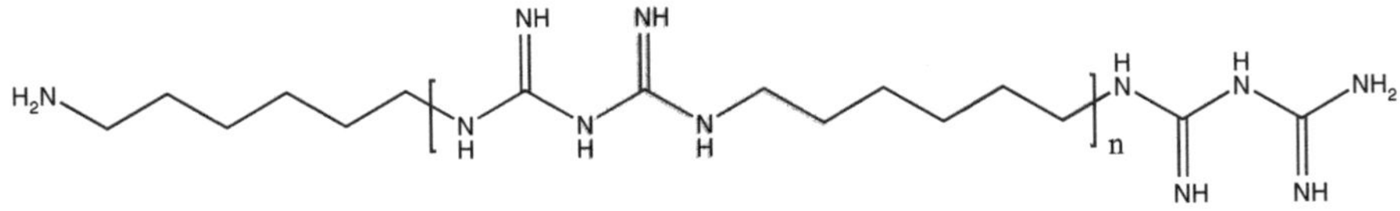

Fig. 15.7 Structure of poly(hexamethylene biguanides) ($n \approx 12$)

(Fig. 15.7). The synthesis was already described in 1949 by ICI [48]. PHMBs are, in contrast to the previous polymers, neutral species becoming polycations in an aqueous environment by protonation of their strong basic guanidine groups. Their high biocidal potency is remarkable, since protonated polyamines are usually only weak biocides. Their biocidal behavior depends on the pH of the medium. Because of their strong basicity PHMBs maintain their polycationic characteristics under physiological conditions. PHMBs are mostly used as additives, since they are not suited as construction material due to the low molecular weights of the commercial polymers (Mn ∪2,200 Da).

It is noteworthy that in this case the combination of an excellent potent biocidal behavior goes along with a low thrombogenic characteristic, making PHMB exceptional and, consequently, highly useful in medical applications.

15.5.6 Cationic Chitosan

Chitosan is an abundantly available natural polysaccharide, prepared from chitin, a main component of the shell of crustaceans (Fig. 15.8). The amide side group that is present in chitin is hydrolyzed to create a free amino group, which is responsible for the biocidal activity. Alkylation to make quaternary ammonium ions can be done, but it seems that protonation of the amino groups is sufficient to obtain biocidal activity.

Chitosan is a hydrophilic hydrogel if most of the acetyl groups of chitin are removed. These compounds are very attractive for lubricous antibacterial medical coatings [49], for instance on catheters. Conversely, hydrogels are not well suited if robust mechanical coating properties are required. This is frequently requested, because during surgery the forces on devices are quite often quite high.

15.6 Immobilized Quaternary Ammonium Compounds

The aforementioned soluble cationic polymers are very potent biocides, but are quite often cytotoxic, causing hemolysis. Kuroda et al. [50, 51], Ferruti et al. [52] and Sen et al. [28] have demonstrated a way to circumvent this by optimizing the hydrophobic/hydrophilic balance. After optimization of the structures, good compromises between both properties were found.

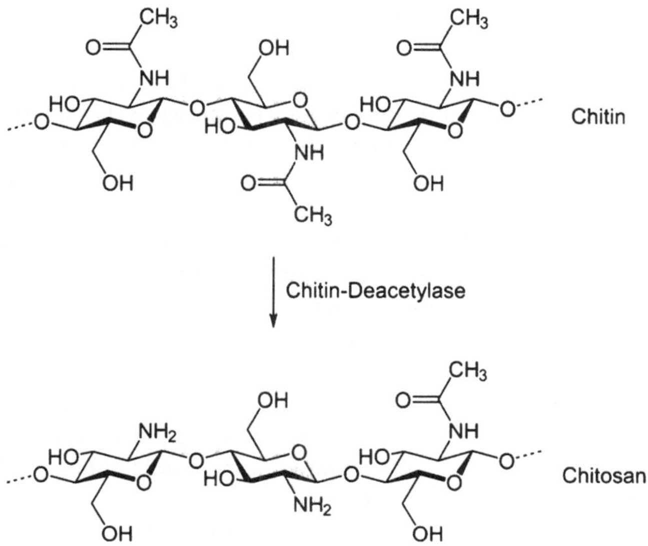

Fig. 15.8 Partial hydrolysis of the amide groups of chitin yields biocidal chitosan

Another way to circumvent health issues is to take advantage of the capability of (biocidal) polymers to immobilize them in coatings or to use them as construction material for biomedical devices and implants. Immobilized quats are not leachable and will therefore not contaminate body fluids. Because of that beneficial perspective numerous investigations on immobilized quats have been reported during the last decade. In most studies antibacterial polymers were applied as coatings. To secure immobilization the polymers have to be covalently tethered onto reactive groups on surfaces of substrates. Substrates comprising hydroxyl and amino groups are very favorable for coupling reactions. If these groups are absent they can be introduced on surfaces by plasma treatment procedures. If a direct tethering of a coating is not possible then coupling agents (primers) can be used to cover the surface with suitable anchor points.

It is impossible to make coatings having nil leachable biocides, since ppm amounts of leachables can be enough to kill bacteria. To assure the absence of any leachable component an intensive washing is needed. In order to detect leachables samples are often immersed in an agar gel loaded with bacteria cultures. The zone of inhibition of bacterial growth around the samples is considered as a good measure of leachability. However, the lack of an inhibition zone around a coated substrate can be interpreted falsely as absence of leachables. If the transfer of the biocide towards bacteria goes via a direct contact between bacteria and the biocide (if the water solubility is very low), no inhibition zone will be visible. Nevertheless, such a system will not contaminate in body fluids and is thus virtually a non-leaching coating system.

To ensure that no leachables are present it is necessary to remove all soluble residues of the coating by thoroughly washing. However, only in a few cases an exhaustive washing has been reported. Fortunately, in these few cases it was clearly demonstrated that the antibacterial properties were preserved (vide infra).

Immobilized quaternary ammonium compounds are obviously less accessible towards bacteria. In contact killing, bacteria should be able to approach the surface closely and there should be enough accessible quats to kill bacteria. Rondelez et al. [53] have shown, with poly(4-vinyl-*N*-butyl pyridinium) bromide, that the number of immobilized quats should surpass a threshold of about 10^{15} quaternary ammonium groups per cm^2 to be effective. Similar values were reported by Matyjaszewski et al. [54] for polyacrylates. Another interesting observation was made by Brizzolara et al. [18]. They found that silica particles were biocidal, whereas glass slides with the same concentration of quats were not. The explanation was that particles can surround bacteria, whereas flat glass slides cannot. So, with particles more quats were accessible for killing bacteria. Below a threshold little or no biocidal efficacy of immobilized quats can be expected.

In the next sections a brief overview of some of the most promising concepts is given.

15.6.1 Cationic Siloxanes

One of the oldest and perhaps most successful classes of immobilized cationic species is constituted by siloxanes, as depicted in Scheme 15.6. A compound in which R_1 is octadecyl is already for many decades on the market [55]. As a consequence, quite a large number of companies have filed patents on formulations comprising this compound [56–77], indicating the usefulness of the concept.

The alkoxy silane moiety is able to react with itself (cross-linking) as well as with hydroxyl groups on surfaces. This biocide forms a cross-linked coating, which is firmly immobilized due to a covalent coupling. Gottenbos et al. [78] used this technology to coat silicon rubbers. After a plasma treatment to obtain hydroxyl groups on the surface, samples were coated with 3-(trimethoxysilyl) *N*, *N*′-dimethyl, *N*-octadecyl propyl ammonium chloride. Before testing, the samples were washed

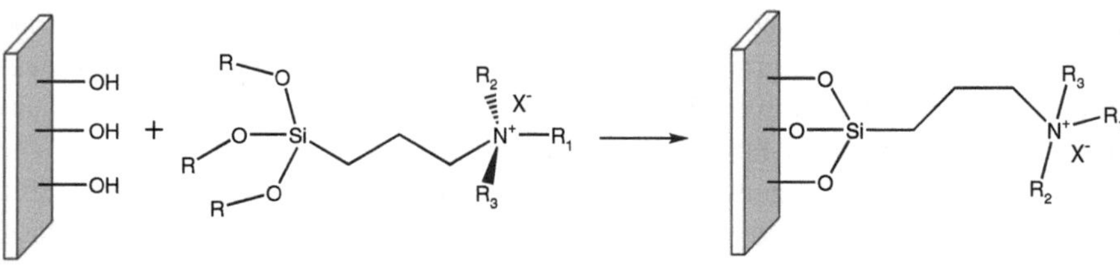

Scheme 15.6 Trialkyl aminopropyl trialkoxy silane (R2 and R3 are methyl, R1 is octadecyl)

with PBS for 30 min. This seems rather short to remove all leachables, but it might be sufficient for such a thin layer. It was demonstrated that these coatings were biocidal *in vitro*, even in the presence of proteins and, most importantly, also *in vivo* in rats. Remarkably, the *in vivo* experiments showed a higher antibacterial efficacy than the *in vitro* experiments.

A disadvantage of siloxane bonds could be that they can hydrolyze slowly under wet conditions, reducing perhaps the long lasting properties in biomedical applications. If during testing already some hydrolysis takes place, it could result in leachables. In that case the results could wrongly be interpreted that the biocidal compounds are tethered. But, Brizollara et al. showed convincingly that no leachables were released, while the surfaces displayed persuasively antibacterial properties [18]. Hence, this concept has proved its usefulness in many industrial applications.

15.6.2 Cationic Polyacrylates

Polyacrylates are abundantly used in coating formulations and, obviously, also studied for antibacterial coatings. The properties of copolymers of quaternary ammonium comprising (meth)acrylate monomers with a variety of other acrylate monomers can be tuned widely. The variation in the composition of these copolymers is nearly unlimited.

Particularly after the discovery of the living radical polymerizations in the 1990s grafting of polyacrylates on substrates became attractive. After grafting an initiator on a substrate the living polymerizations of numerous cationic and other monomers became feasible. Matyjaszewski et al. [19] reported the living polymerization of acrylates from a grafted initiator as depicted in Scheme 15.7. He demonstrated that these immobilized coatings displayed excellent antibacterial properties if the charge density surpassed a certain threshold (>10^{15} cations/cm^2).

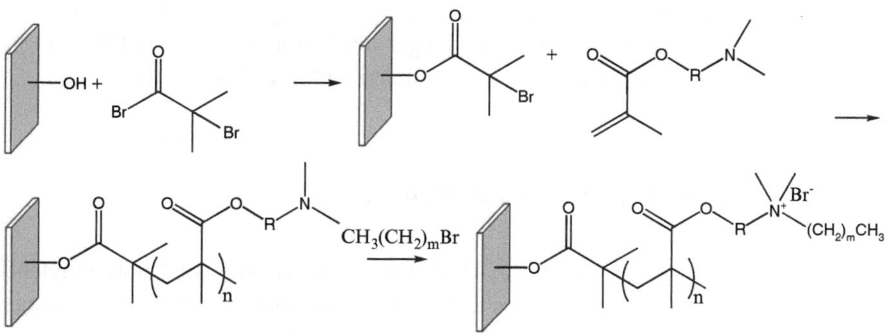

Scheme 15.7 Grafting poly(meth)acrylates comprising quaternary ammonium moieties from a hydroxyl functional surface ($R=(CH_2)x$, with $x=2$ or 3)

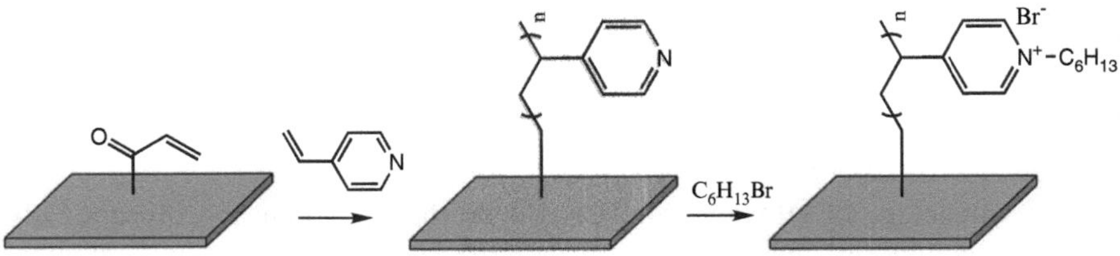

Scheme 15.8 Grafted poly(vinyl)pyridine on a substrate. *Left*: the copolymerization of vinyl pyridine with the tethered acrylate and subsequent alkylation

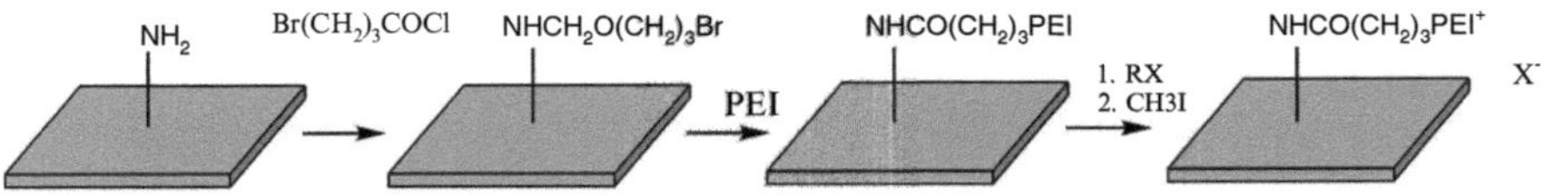

Scheme 15.9 Functionalizing substrates with quaternized poly(ethylene imine) (PEI)

The length of the polymer chains was varied, by taking advantage of a living polymerization technique. Antibacterial properties were obtained even when the length was too short (10 nm) to penetrate the peptidoglycan layer, but only if the charge density was above the aforementioned threshold. Hence, the length of the chains seems to play only a minor role in the killing mechanism.

15.6.3 Cationic Poly(Vinyl Pyridine)

Lewis et al. [79–81] have reported on the immobilization of alkylated poly(vinyl pyridine) (PVP). A copolymerization of vinyl pyridine with on glass slides grafted acrylate moieties yielded surfaces covered with covalently tethered PVP (Scheme 15.8). After alkylation biocidal coatings were obtained.

They found that pyridine alkylated with C_3–C_6 alkyl groups were most potent. In many other systems it is shown that longer alkyl tails are more potent ($>C_{12}$), demonstrating that each particular system requires optimized on his own.

15.6.4 Cationic Poly(Ethylene Imines)

Pioneering work of Klibanov et al. [43, 82] showed that alkylated poly(ethylene imines) are very promising candidates for preparing antibacterial materials. Coatings based on PEIs were applied via two routes. In one approach the surface was covered with alkylbromides, as depicted in Scheme 15.9. Fig. 15.3 Peptide tethered according to Ferreira et al. [91]

Polyethylene imines were immobilized by reacting the amino groups of PEI with the alkylbromides tethered to the surface of glass slides. After that the remaining amino groups were converted into quats by alkylation with alkyl halides, affording highly biocidal coatings.

Later on, the same group discovered a more simple approach in which quaternized poly(ethylene imines) were applied as coating without coupling agents and without cross-linking [83]. According to the authors, these hydrophobic antibacterial polymers were virtually insoluble in water after alkylation and could therefore be considered as immobilized. It is, however, in that cases difficult to unambiguously prove that no species were transferred directly during contact from the surface towards bacteria, without being transported via the water phase. Nonetheless, if no leachables elute in the aqueous phase then virtually non-eluting coatings are obtained. Interestingly, these coatings display not only excellent antibacterial properties but are also active against viruses [84, 85].

Non-leachable cross-linked cationic poly(ethylene imines) are successfully described for dental applications. In a series of publications Weiss and Domb et al. [86, 87] showed that 1 wt% of quaternized cross-linked PEI particles, imbedded in a commercial orthodontic matrix, offered a high killing efficiency. The non-leachability was convincingly demonstrated. For that the samples were washed for 3 months, while refreshing the PBS every 48 h. After this intensive washing the biocidal activity was still observed. In this case there is no doubt that all water soluble compounds were removed completely, while the remaining material was still active. These results are important as a solid proof that non-leachable quats are able to kill bacteria by a contact killing mode.

Another solid proof for contact killing is shown by Kang et al. [88] and Kenawy et al. [89]. They prepared cross-linked particles by the radical polymerization of vinyl benzyl chloride, 2-chloro ethyl vinyl ether and divinyl benzene. The polymer beads were converted into quaternary ammonium and phosphonium compounds by alkylating tertiary amines or phosphines. These beads were washed in a Soxhlet extraction until all un-reacted species were removed. The elegance of this method is that the beads do not contain any hydrolysable groups. Hence, after intensive washing no novel leachables can be released. Still antibacterial properties of these materials were clearly observed.

15.6.5 Cationic Peptides

In nature flora and fauna protect themselves against microbes with numerous antibacterial peptides. Also in the human body numerous antibacterial peptides exist. A general structural feature of antibacterial peptides is that they have amino acids with ammonium groups and hydrophobic moieties. In that respect the aforementioned antibacterial cationic compounds display a lot of similarities with these peptides. So far there is no convincing proof that the killing mechanism of cationic peptides differs from that with synthetic cationic species.

Hancock et al. [90] have summarized, in an excellent paper, the synthesis and tethering of a wide variety of biocidal peptides on surfaces. Further experimental evidence for the antibacterial properties of immobilized peptides is disclosed in a patent of Ferreira et al. [91]. To introduce acrylate groups they describe the transesterification of vinyl acrylate with the hydroxyl groups of polysaccharides. The acrylate group was used for cross-linking and/or to introduce amino functional substrates by adding diamines (known as the Michael addition). To optimize the biocidal properties of synthetic peptides it was emphasized to link them in the right way. So, it is important to select the right tethering terminus (C or N), to use appropriate type of (cationic) amino acids and to ensure that the folding of the peptides is aligned in the right positions. If the amino acids and the sequence order are correctly selected, the chains will fold in such a way that hydrophobic groups arrange at one side of the peptide creating hydrophobic areas. The hydrophilic moieties, comprising the cationic species, are grouped at the other side of the chain. This is conceptually the same approach as is found in nature. A spacer between the substrate and the cationic peptides is strongly recommended, sometimes even suggested to be indispensable.

Peptides of natural amino acids are highly susceptible for enzymatic hydrolysis. They usually degrade in the body within 1 h. Hence, peptides from natural amino acids are useless if long lasting properties are required. For that reason Messerschmith et al. described the immobilization of β-peptoids [92]. β-peptoids are much less inclined to degrade enzymatically, and hold, therefore, a promise for realizing long-lasting properties. Unfortunately, β-peptoids are missing the advantage of natural peptides.

15.6.6 Cationic Biguanides

Biguanides are cationic species that are well known for their favorable combination of being a high potent biocide, while having a low tendency to cause thrombosis. Chlorohexidine is a low molecular biguanide and is already on the market since 1954, mainly used for hygienic purposes, such as soaps, toothpastes, etc. It is also widely described as a coating additive [93–100]. Baxter [101], for instance, describes chlorohexidine (together with silver sulfadiazine) comprising coatings for anti-thrombogenic catheters.

Poly(hexamethylene biguanides) (PHMB) were developed as a slow leachable biocide that can be compounded with many other polymer matrices.

In order to create non-leachability, i.e., immobilization in coating systems, Sawan et al. [102, 103] described in a number of patents the cross-linking of poly(hexamethylene bi-guanides) (PHMB). In that case PHMB coating formulations were made with diepoxies as cross-linkers, forming thermally curable resin.

15.7 Dual Action

An important observation related to biomaterial-associated infections was that bacteria not only adhere onto surfaces of biomedical materials but also grow in the vicinity thereof. Zaat et al. [104, 105] showed in a series of *in vivo* experiments with mice that intentionally contaminated implants caused serious accumulation of bacteria in the soft tissue in the neighborhood of these implants. It is not yet known if under clinical conditions similar processes take place, but there is a risk that contact killing alone on a surface may not be sufficient, unless it works very fast. In that case a dual action system with a releasing and a non-eluting biocide may be more beneficial than a single action system. Releasing biocides can give a burst release, which can be an advantage in the early post-operative period, when the chance on infections is the highest. The immobilized biocides could guarantee long lasting performances.

Sawan et al. [106] combined cross-linked poly(hexamethylene biguanides) with silver ions. The silver ions are loaded in the coating, by dipping the coated sample in a diluted solution of AgI. The silver ions form a complex with the PHMB. The coating was washed to remove the excess of the silver salt that had not formed a complex. Complex formation is important to keep the eluting concentration of silver ions low to prevent too high contaminations levels of silver ions in body fluids (burst release). Moreover, this not only lowers the toxicity but also increases the life time of active silver ions. According to the patent bacteria are able to subtract silver ions directly from the polymer complex, without being transferred via the aqueous phase. In that case no silver ions will float around in the body fluids.

Sen et al. [107] has described similar synergistic effects of the combination quaternary poly(vinyl pyridines) and silver. The bromide ions formed during the quaternization reaction of poly(vinyl pyridines) with alkyl bromides were used to make the sparingly soluble silver bromide salt. Although these coatings contain silver ions, they were also considered as non-leachable. They explained as well that the biocidal properties were obtained by contact killing, meaning that the biocide (Ag^+) was transferred directly to bacteria and not via the aqueous medium.

15.8 Phosphonium Compounds

The structural features of quaternary ammonium compounds are not unique for killing bacteria. Other -onium compounds, like phosphonium ions, provided with a hydrophobic moiety are biocidal as well. Phosphonium compounds are, according to Endo et al. [12, 13] even ten or more times more potent that quaternary ammonium compounds. An explanation for this much better performance is not given. Although phosphonium compounds seem to perform even better, there is much more interest in quaternary ammonium compounds due to the availability of convenient synthetic routes.

15.9 Conclusions

Due to the aging of the world population and because of the unwillingness of mankind to further accept lack of function, the number of biomedical implants will substantially increase in the coming decades. In spite of all efforts in modern health care, a substantial further reduction of the incidence of bacterial infections during surgery seems to be impossible. Bacteria are literally everywhere, on the skin of patients and also in the air of operation theaters. It is simply unfeasible to eliminate them all during surgery. Hence, infections will continue to be the main cause of failure of biomedical implants. Moreover, implants hamper the immune system and also the ability of administered antibiotics to eliminate bacteria. The only way to eradicate bacteria properly is to kill them as soon as they contact surfaces of implants.

The use of antibacterial coatings is perhaps the best option to combat infections because they can cover, in principle, all conceivable biomedical materials. As a result, all existing implants could become antibacterial, without changing the remaining performance. Catheter coatings comprising leachable biocides, such as silver ions, are well known and commercially available. Although they are able to kill bacteria efficiently, they will contaminate body fluids when used on implants. Moreover, the release will stop in due time.

A more durable route is to immobilize biocides. Immobilized quaternary ammonium compounds are probably the best candidates to eradicate bacterial infections that arise from implants. In the last decade many excellent investigation have been reported on immobilized quats, showing good efficacies *in vitro* and *in vivo*. Unfortunately, only a limited number of *in vivo* studies have been published so far. Fortunately, those do confirm the *in vitro* results. The main challenge for the near future is to demonstrate more clearly that the many encouraging *in vitro* results are unambiguously confirmed *in vivo* experiments.

Acknowledgments The author wants to thank Dr. T. Dirks and Dr. A. Piermattei from DSM (in the Netherlands) for reading the manuscript carefully and for their comments.

Glossary

Coatings Are polymer compositions that cover surfaces that determine the interfacial properties.

Polymers (macromolecules, plastics) Are high molecular materials comprising numerous monomers units.

Quaternary ammonium compounds Are compounds that comprise positively charged nitrogen moieties ($R_4N^+X^-$).

Leachable antibacterial compounds Are low molecular compounds that are able to release from a surface into the surroundings.

Contact-killing Is the operating mechanism if bacteria are only killed upon direct contact with a surface.
Cytotoxicity Determines the damage on human cells.
Biocompatibility Is a measure for the acceptance of the body of foreign entities.

References

1. Meyer B, Cookson B. Does microbial resistance or adaptations to biocides create a hazard in infection prevention and control? J Hosp Infect. 2010;76:200–5.
2. Biocidal Product Directive. Commission Regulation (EC). 2003; No 2032/2003, November 4, 2003.
3. Friends of the earth, http://www.foe.org/sites/default/files/Nano-silverReport_US.pdf, 2009.
4. Al-bataineh S, Britcher I, Groesser H. XPS characterization of the surface of immobilization of antibacterial furanones. Surface Sci. 2006;600:952–62.
5. Kenawy E, Worley S, Broughton R. The chemistry and applications of antimicrobial polymers: a state-of-the-art review. Biomacromolecules. 2007;8(5):1359–84.
6. Ferreira L, Zumbuehl A. Non-leaching surfaces capable of killing microorganisms on contact. J Mater Chem. 2009;19:7796–806.
7. Timofeeva L, Kleshcheva N. Antimicrobial polymers: mechanism of action, factors of activity, and applications. Appl Microbiol Biotechnol. 2011;89:475–92.
8. Vasilev K, Cook J, Griesser H. Antibacterial surfaces for biomedical devices. Expert Rev Med Devices. 2009;6(5):553–67.
9. Tiller J. Antimicrobial surfaces. Adv Polym Sci. 2010. doi:10.1007/12_2010_101
10. Gabriel G, Som A, Madkour A, Eren T, Tew G. Infectious disease: connecting innate immunity to biocidal polymers. Mater Sci Eng. 2007;R57:28–64.
11. Waschinski C, Herdes V, Schueler F, Tiller J. Influence of satellite groups on telechelic antimicrobial functions of polyoxazolines. Macromol Biosci. 2005;5:149–56.
12. Kanazawa A, Ikeda T, Endo T. Novel polycationic biocides: synthesis and antibacterial activity of polymeric phosphonium salts. J Polym Sci A Polym Chem. 1993;31:335–43.
13. Kanazawa A, Ikeda T. Multifunctional tetracoordinate phosphorus species with high self-organizing ability. Coord Chem Rev. 2000;198:117–31.
14. Tossi A, Sandri L, Giangaspero A. Amphipathic, α-helical antimicrobial peptides. Biopolymers (Peptide Science). 2000;55:4–30.
15. Chen C, Cooper S. Interactions between dendrimer biocides and bacterial membranes. Biomaterials. 2002;23:3359–68.
16. Lenoir S, Pagnoulle C, Galleni M, Compère P, Jérôme R, Detrembleur C. Polyolefin matrixes with permanent antibacterial activity: preparation, antibacterial activity, and action mode of the active species. Biomacromolecules. 2006;7:2291–6.
17. Norris V, Chen M, Goldberg M, Voskuil J, McGurk G, Holland I. Calcium in bacteria: a solution to which problem? Mol Microbiol. 1991;5(4):775–8.
18. Brizzolara R, Stamper D. The effect of covalent surface immobilization on the bactericidal efficacy of a quaternary ammonium compound. Surf Interface Anal. 2007;39:559–66.
19. Lee S, Koepsel R, Morley S, Matyjaszewski K, Sun Y, Russell A. Permanent, nonleaching antibacterial surfaces. 1. Synthesis by atom transfer radical polymerization. Biomacromolecules. 2004;5:877–82.
20. Ohkura K, Sukeno A, Nagamunea H, Kourai H. Bridge-linked bis-quaternary ammonium anti-microbial agents: relationship between cytotoxicity and anti-bacterial activity of 5,50-[2,20-(tetramethylenedicarbonyldioxy)-diethyl]bis(3-alkyl-4-methylthiazonium iodide)s. Bioorg Med Chem. 2005;13:2579–87.

21. Block S. Disinfectants, sterilization and preservation. Philadelphia, PA: Lippincott Williams & Wilkins; 2001. ISBN ISBN 0-683-30749-1.
22. Mazzola P, Jozala A, de Lencastre Novaes L, Moriel P, Penna T. Brazilian Minimal inhibitory concentration (MIC) determination of disinfectant and/or sterilizing agents. J Pharm Sci. 2009;45(2):241–8.
23. McBain A, Ledder R, Moore L, Catrenich C, Gilbert P. Effects of quaternary-ammonium-based formulations on bacterial community dynamics and antimicrobial susceptibility. Appl Environ Microbiol. 2004;70:3449–56.
24. Etrych T, Leclercq L, Boustta M, Vert M. Polyelectrolyte complex formation and stability when mixing polyanions and polycations in salted media: a model study related to the case of body fluids. Eur J Pharm Sci. 2005;25:281–8.
25. Poly(bishexamethylene biguanides) Arch. http://www.archchemicals.com/Fed/Corporate/MyAcct/msdsloc.htm.
26. Polyquaternium BASF. http://worldaccount.basf.com/wa/NAFTA~en_US/Catalog/Cosmetics/pi/BASF/product_inci/polyquaternium_16.
27. Kroda K, Palermo E. Chemical structure of cationic groups in amphiphilic polymethacrylates modulates antimicrobial and hemolytic activities. Biomacromolecules. 2009;10:1416–28.
28. Sambhy V, Peterson B, Sen A. Antibacterial and hemolytic activities of pyridinium polymers as a function of the spatial relationship between the positive charge and the pendant alkyl tail. Angew Chem. 2008;120:1270–4.
29. Gilbert P, Moore L. Cationic antiseptics: diversity of action under a common epithet. J Appl Microbiol. 2005;99:703–15.
30. Ikeda T, Yamaguchi H, Tazuke S. Molecular weight dependence of antibacterial activity in cationic disinfectants. J Bioact Compat Polym. 1990;5:31–41.
31. Creavis WO200269709.
32. Creavis DE 10061251.
33. Creavis DE 10061250.
34. Creavis DE 10048614.
35. Creavis DE 10048613.
36. Creavis WO 200217724.
37. Creavis WO200119878.
38. Ciba DE1053783
39. Hoechst US5049383
40. Lewis G, Nguyen V, Cohen Y. Synthesis of poly(4-vinylpyridine) by reverse atom transfer radical polymerization. J Polym Sci A Polym Chem. 2007;45:5748–58.
41. Cheng Z, Zhu X, Shi L, Neoh K, Kang E. Polymer microspheres with permanent antibacterial surfaces from surface-initiated atom transfer radical polymerization of 4-vinylptridine and quaternization. Surf Rev Lett. 2006;13(2 & 3):313–8.
42. Lin J, Qiu S, Lewis K, Klibanov A. Mechanism of bactericidal and fungicidal activities of textiles covalently modified with alkylated polyethylenimine. Biotechnol Bioeng. 2003;83(2):168–72.
43. Milovic N, Wang J, Lewis K, Klibanov A. Immobilized N-alkylated polyethylenimine avidly kills bacteria by rupturing cell membranes with no resistance developed. Biotechnol Bioeng. 2005;90(6):715–22.
44. Klibanov A. Permanently microbiocidal materials coatings. J Mater Chem. 2007;17:2479–82.
45. Wong S, Li Q, Veselinovic J, Kim B, Klibanov A, Hammond P. Bactericidal and virucidal ultrathin films assembled layer by layer from polycationic N-alkylated polyethylenimines and polyanions. Biomaterials. 2010;31:4079–87.
46. Schaer T, Stuart S, Hsu B, Klibanov A. Hydrophobic polycationic coatings that inhibit biofilms and support bone healing during infection. Biomaterials. 2012;33:1245–54.
47. Mikhaylova A, Liesenfeld B, Moore D. Quick Med Technologies, Inc., Efficacy of BIOGUARD™ dressings utilizing advanced NIMBUS® technology. http://content.stockpr.com/qmdt/media/ea16ad90f18b558d673990c0c51b067e.pdf.
48. ICI GB 702,268, filed 22 August 1949.

49. Liu N, Chen X, Park H, Liu C, Liu C, Meng X, et al. Effect of MW and concentration of chitosan on antibacterial activity of Escherichia coli. Carbohydr Polym. 2006;64:60–5.
50. Kuroda K, Palermo E. Chemical structure of cationic groups in amphiphilic polymethacrylates modulates the antimicrobial and haemolytic activities. Biomacromolecules. 2009;10:1416–28.
51. Kuroda K, Caputo G, DeGrado W. The role of hydrophobicity in the antimicrobial and hemolytic activities of polymethacrylate derivatives. Chem Eur J. 2009;15:1123–33.
52. Marchisio M, Bianciardi P, Longo T, Ferruti P, Ranucci E, Neri M. Comparison between the hemolytic and antibacterial activities of new quaternary ammonium polymers. J Biomater Sci Polym Ed. 1994;6(6):533–9.
53. Kügler R, Bouloussa O, Rondelez F. Evidence of a charge-density threshold for optimum efficiency of biocidal cationic surfaces. Microbiology. 2005;151:1341–8.
54. Murata H, Koepsel R, Matyjaszewski K, Russell A. Permanent, non-leaching antibacterial surfaces—2: how high density cationic surfaces kill bacterial cells. Biomaterials. 2007;28:4870–9.
55. Aegis, http://www.aegismicrobeshield.com/
56. Chisholm US 2008/0181862.
57. Biosafe US2006223962.
58. Biosafe US65722926.
59. Biosafe WO9932157.
60. Dokuritsu Gyosei JP2005068191.
61. Bioshield WO9903866.
62. Bioshield WO9742200.
63. Univ Emory WO9741876.
64. Lamba-Kohlin US200618837.
65. Corpura EP1493452.
66. Burlington, US4414268.
67. Kimberly-Clark US6712121.
68. Kimberly-Clark US10429502.
69. Johnson & Son US6528472.
70. Invista North America US6790797.
71. Reckitt Benckiser Inc US 7304022.
72. Ciba WO2008006744, WO2008132045, WO2008006744.
73. NDSU foundations US2008181862.
74. Nano X DE 102007020404.
75. NDSU Res Found WO9903866.
76. Oniv Emory WO9741876.
77. Procter & Gamble WO99/32539.
78. Gottenbos B, van der Mei H, Klatter F, Nieuwenhuis P, Busscher H. In vitro and in vivo antimicrobial activity of covalent coupled quaternary ammonium silane coatings on silicone rubber. Biomaterials. 2002;23:1417–23.
79. Lewis K, Klibanov A. Surpassing nature: rational design of sterile-surface materials. Trends Biotechnol. 2005;23(7):343–8.
80. Klibanov A, Liao C, Tiller J, Lewis K. Antimicrobial polymeric surfaces. WO02/085542 (= US7151139)
81. Tiller J, Liao C, Lewis K, Klibanov A. Designing surfaces that kill bacteria on contact. Proc Natl Acad Sci USA. 2001;98(11):5981–5.
82. Lin J, Qiu S, Lewis K, Klibanov A. Bactericidal properties of flat surfaces and nanoparticles derivatized with alkylated poly(ethylene imines). Biotechnol Prog. 2002;18:1082–6.
83. Park D, Wang J, Klibanov A. One-step, painting-like coating procedures to make surfaces highly and permanently bactericidal. Biotechnol Prog. 2006;22:584–9.
84. Haldar J, Chen J, Tumpey T, Gubareva L, Klibanov A. Hydrophobic polycationic coatings inactivate wild-type and zanamivir- and/or oseltamivir-resistant human and avian influenza viruses. Biotechnol Lett. 2008;30:475–9.

85. Haldar J, An D, Alvarez de Cienfuegos L, Chen J, Klibanov A. Polymeric coatings that inactivate both influenza virus and pathogenic bacteria. Proc Natl Acad Sci USA. 2006;103(47): 17667–71.
86. Yudovin-Farber I, Beyth N, Weiss E, Domb A. Antibacterial effect of composite resins containing quaternary ammonium polyethyleneimine nanoparticles. J Nanopart Res. 2010; 12:591–603.
87. Yudovin-Farber I, Golenser J, Beyth N, Weiss E, Domb A. Quaternary ammonium polyethyleneimine: antibacterial activity. J Nanomater. 2010; Article ID 826343, 11, doi:10.1155/2010/826343
88. Cheng Z, Zhu X, Shi Z, Neoh K, Kang E. Polymer microspheres with permanent antibacterial surface from surface-initiated atom transfer radical polymerization. Ind Eng Chem Res. 2005;44:7098–104.
89. Kenawy E, Abdel-Hay F, El-Magd A, Mahmoud Y. Biologically active polymers: VII. Synthesis and antimicrobial activity of some crosslinked copolymers with quaternary ammonium and phosphonium groups. React Funct Polym. 2006;66:419–29.
90. Hilpert K, Elliott M, Jenssen H, Kindrachuk J, Fjell C, Körner J, et al. Screening and characterization of surface-tethered cationic peptides for antimicrobial activity. Chem Biol. 2009;16:58–69.
91. Ferreira L, Langer R, Loose C, O'shaughnessy W, Zumbuehl A, Stephanopous S. Medical devices and coatings with non-leachable antibacterial peptides. WO2007/095393.
92. Statz A, Kuang J, Ren C, Barron A, Szleifer I, Messersmith P. Experimental theoretical investigation of chain length and surface coverage on fouling of surface grafted polypeptoids. Biointerphases. 2009;4(2):22–32.
93. Tyco healthcare group US2007237812.
94. Nippon Soda WO20066118283.
95. Hodge US2003013794.
96. Arch WO2004100663.
97. Roby US2003236552.
98. Neste CHEM WO200112725.
99. Medtronic US 6033719, US 5928916, US5925552.
100. Becton Dickinson US2007202177.
101. Baxter WO200023124-A1
102. Sawan S, Shalon T, Subramanyam S, Yurmovetskiy A. Contact killing non-leaching antimicrobial materials. US5849311 (15 December 1998).
103. Sawan S, Shalon T, Subramanyam S, Yurmovetskiy A. Non-leaching antimicrobial films. US6030632 (29 February 2000).
104. Boelens J, Dankert J, Murk J, Weening J, van der Poll T, Dingemans K, et al. Biomaterial-associated persistence of staphylococcus epidermidis. J Infect Dis. 2000;181:1337–49.
105. Broekhuizen C, de Boer L, Schipper K, Jones C, Quadir S, Feldman R, et al. Peri-implant tissue is an important niche for Staphylococcus epidermidis in experimental biomaterial-associated infection in mice in pericatheter macrophages. Infect Immun. 2007;75(3):1129–36.
106. Sawan S, Shalon T, Subramanyam S, Yurmovetskiy A. Contact killing antimicrobial devices. US5817325 (6 October 1998).
107. Sambhy V, MacBride M, Peterson B, Sen A. Silver Bromide nanoparticles/polymer composites: dual action tunable antimicrobial materials. J Am Chem Soc. 2006;128:9798–808.
108. Zasloff M. Antimicrobial peptides of multicellular organisms. Nature. 2002;415:389–95.
109. Mei L, Loontjens T, Ren Y, van der Mei H, Busscher H. Contact-killing of adhering streptococci by a quaternary ammonium compound incorporated in an acrylic resin. Int J Artif Organs.

Chapter 16
Anti-adhesive and Antibacterial Polymer Brushes

K.G. Neoh, Z.L. Shi, and E.T. Kang

Abstract Bacterial infection associated with medical devices is a serious complication. One promising strategy to combat this problem is the functionalization of the device surface with a dense layer of polymer chains which can resist bacterial adhesion and colonization. These polymer chains may present large exclusion volumes to inhibit protein and bacterial adhesion and/or possess bactericidal functional groups. The coating of surfaces with these polymers may be carried out via a number of techniques such as self-assembly, grafting and surface-initiated polymerization. This article focuses mainly on polymer coatings which achieved the antibacterial effect without the leaching of the bactericidal components into the environment.

16.1 Introduction

The use of man-made materials and devices such as catheters, cardiac pacemakers, and prosthetic implants in the human body will continue to escalate. These materials present surfaces to opportunistic bacteria to attach, colonize, and form biofilms. The biofilm protects the colonizing microorganisms, and thus, bacteria in a biofilm can be several orders of magnitude more resistant to antibacterial agents than their planktonic counterparts, and biofilms once formed are very difficult to eradicate [1, 2]. The fate of the implant is determined by the race for the surface between the bacteria and the host tissue cells [3]. If the race is won by the bacteria, infection instead of tissue integration will result. It has been estimated that device-associated infections are

K.G. Neoh (✉) • Z.L. Shi • E.T. Kang
Department of Chemical and Biomolecular Engineering, National University of Singapore, Kent Ridge, Singapore 119260, Singapore
e-mail: chenkg@nus.edu.sg

T.F. Moriarty et al. (eds.), *Biomaterials Associated Infection: Immunological Aspects and Antimicrobial Strategies*, DOI 10.1007/978-1-4614-1031-7_16,

responsible for ~50% of all nosocomial infections [4]. The implications of this resistance and persistence are clearly manifested in biomaterials-associated infection which can potentially be one of the most devastating medical complications, since very often it becomes necessary to remove the infected device at considerable cost and suffering to the patient.

Since initial bacterial adhesion and colonization are the key steps to subsequent biofilm formation, strategies to prevent biomaterials-associated infection have focused on modifying the physicochemical surface properties of the biomaterials. One such strategy is the use of surface-confined macromolecular chains. Polymer chains tethered to the solid substrate via end groups are commonly known as polymer brushes, although polymers tethered to a surface via side groups can form loops and tails with behavior similar to a brush-like layer [5]. At sufficiently high density, tethered polymer chains are forced to stretch away from the surface. Polymer brushes prepared for antibacterial purposes may present large exclusion volumes making the surface difficult to approach by incoming proteins or bacteria [6]. Alternatively, the polymer brushes may possess functional groups which can kill bacteria on contact or they may be conjugated or complexed with bactericidal groups or agents. Tethered antibacterial brushes are advantageous in offering long lasting effects and minimizing environmental problems associated with the leaching of antibacterial agents. This chapter presents an overview of polymer brushes which confer anti-adhesive and/or bactericidal properties on surfaces.

16.2 Strategies for Surface Functionalization with Polymer Brushes

The tethering of polymer chains to a surface is usually carried out using either of the two techniques: physi-sorption or chemical bonding (Fig. 16.1) [7]. Polymer brushes synthesized via the physi-sorption method usually consist of two-component polymer chains (e.g., a diblock copolymer) where one component strongly adheres to the surface while the other extends away from the surface [8]. Due to the physical nature of the attachment, the polymer brushes formed by this approach may not be as stable as those from chemical bonding. Covalent bonding of polymer chains can be carried out by either the "grafting to" or "grafting from" method. In the "grafting to" technique, preformed polymer molecules with reactive end groups react with complementary reactive groups located on the surface to form tethered chains. This method can be easily carried out but it is difficult to achieve a high grafting density since the chains which are already attached present a steric barrier to the approaching polymer molecules. Furthermore, the film thickness is limited by the molecular weight of the polymer in solution [9]. To circumvent these limitations, the "grafting from" approach can be used. In this method, initiating groups are first attached to the substrate surface. These groups, upon exposure to a monomer solution, initiate polymerization with the chains propagating outwards from the substrate surface. Hence, this method is also referred to as surface-initiated polymerization.

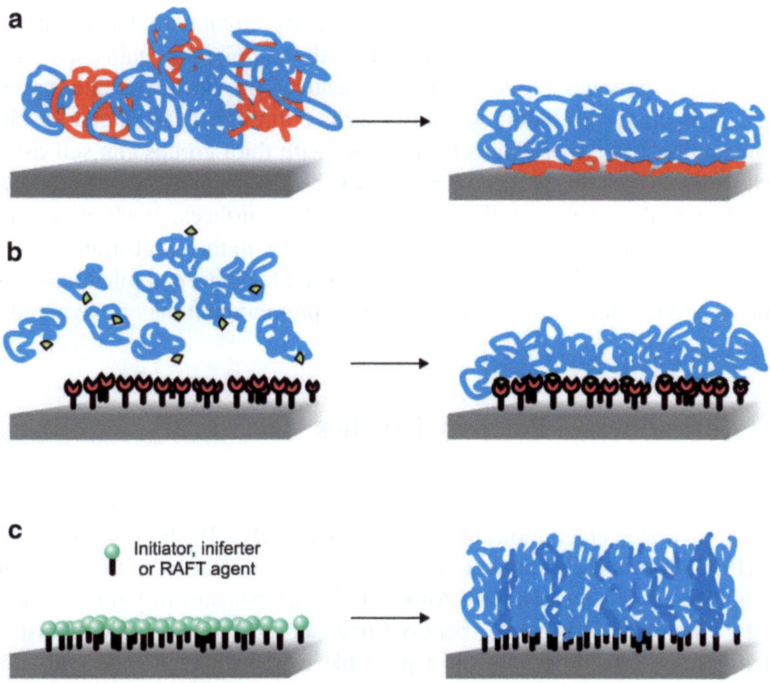

Fig. 16.1 Synthetic strategies for the preparation of polymer brushes: (**a**) physi-sorption of diblock copolymers via preferential adsorption of the *red blocks* to the surface (*grafting to* approach); (**b**) chemisorption via reaction of appropriately end-functionalized polymers with complementary functional groups at the substrate surface (*grafting to* approach); (**c**) polymer brushes grown via surface-initiated polymerization techniques (*grafting from* approach). Reprinted with permission from Ref. [7]. Copyright 2009 American Chemical Society

The polymerization can be carried out via conventional free radical polymerization or controlled/"living" polymerization techniques such as nitroxide-mediated polymerization (NMP), reversible addition–fragmentation chain transfer (RAFT) polymerization, and atom transfer radical polymerization (ATRP). Controlled/"living" polymerization techniques are considered particularly attractive for the "grafting from" strategy, as they allow better control over brush thickness, composition, and architecture [7]. It is beyond the scope of this review to discuss all the recent developments on the synthesis of polymer brushes, and the reader is referred to a number of excellent reviews which have been published on this topic (e.g., [7–9]). Among the different surface-initiated controlled polymerization techniques, ATRP has been the most extensively used to produce antibacterial polymer brushes [7]. ATRP relies on a redox process involving a transition metal complex in which a halogen atom is reversibly transferred between the transition metal and the growing polymer chain end. Its versatility and compatibility with a variety of functionalized monomers, and tolerance for impurities contribute to its popularity as a technique for synthesizing well-defined antibacterial coatings.

Depending on the nature of the substrate, surface pretreatment may or may not be necessary to achieve the desired interactions with the polymer chains. The stability of the attached polymer brush is often dependent on surface anchoring group. Polymer chains may self-assemble on the surface through hydrophobic or electrostatic interactions. Polymer chains or initiators with thiol groups can self-assemble on gold, while those with silane groups will self-assemble on metal and silicon/glass surfaces. Due to the ease of modifying gold and glass surfaces, much of the work on antibacterial polymer brushes have been carried out with these substrates. However, for other materials which are more likely to be used for *in vivo* applications, such as titanium or polyethylene, more involved surface pretreatment may be required.

16.3 Anti-adhesive Polymer Brushes

Anti-adhesive properties can be conferred on biomaterial surfaces by modification with highly hydrated chain molecules. Poly(ethylene glycol) (PEG) or poly(ethylene oxide) (PEO) forms the chemical basis of the most versatile approach to inhibit protein and bacterial adhesion [10]. Although the mechanisms of inhibition have not been fully elucidated, it generally believed that steric barrier, osmotic repulsion and excluded-volume effects are the most probable explanations for the anti-adhesive effects of polymer brushes. Attachment of anti-adhesive polymer chains to the surface via self-assembly through hydrophobic or electrostatic interactions, and covalent immobilization by means of either "grafting to" or "grafting from" approaches will be presented in the following sections.

16.3.1 Self-Assembled Polymer Brushes

Polyethylene oxide–polypropylene oxide–polyethylene oxide (PEO-PPO-PEO) amphiphilic triblock copolymers, commercially available as Pluronics®, are widely investigated as anti-adhesive coatings. Coating of hydrophobic surfaces with Pluronics® can be easily carry out since the polymer chains spontaneously assemble on these surfaces through the hydrophobic PPO domain, while the hydrophilic PEO blocks extend into the aqueous solvent phase providing a brush conformation at high packing density. A brush coating of PEO chains on a surface is expected to provide nonadhesive properties, and it has been determined that adsorption of Pluronic F127 ($PEO_{99}PPO_{65}PEO_{99}$ with a molecular weight of 12600) on hydrophobic surfaces with a water contact angle >80° yields a brush-like conformation with a thickness between 6 and 9 nm [11].

Pluronic coatings can be applied on different hydrophobic materials, including tissue culture polystyrene, methylated glass, silicone, and polylactic-co-glycolic acid. They have been shown to be stable under flow-induced shear stresses [12] and in physiological fluids [13]. Bridgett et al. used a series of 16 Pluronic block copolymers of different block lengths to study adhesion of *S. epidermidis* on polystyrene surfaces [14]. Substantial reductions (up to 97%) in bacterial adhesion levels were achieved with all copolymers tested, irrespective of the PPO or PEO block lengths. Wesenberg-Ward et al. reported that Pluronic F127 coating dramatically reduced the adhesion of *C. albicans* to polystyrene. The biofilm growth studies, analyzed by confocal scanning laser microscopy, revealed that Pluronic F127 decreased the biofilm surface coverage, cluster group size, thickness, and the presence of hyphal elements compared to the untreated polystyrene [15].

Busscher and coworkers have extensively studied the application of Pluronic coatings for preventing bacterial adhesion and biofilm formation with a parallel plate flow chamber *in vitro* and infected pockets in mice *in vivo* [16, 17]. The *in vitro* results showed that adhesion of *S. aureus* and *S. epidermidis* on Pluronic F-127-coated silicone rubber decreased tenfold compared to pristine silicone rubber. While the viability of biofilms on brush coatings was higher than that on pristine silicone rubber, the former developed more slowly and detached almost fully by high fluid shear. Brush coatings remained nonadhesive after *S. epidermidis* biofilm formation and subsequent removal whereas a part of its functionality was lost after removal of *S. aureus* biofilms. However, adhesion, growth, and detachment of *P. aeruginosa* were not significantly different on brush coatings compared to pristine silicone rubber [16]. Thus, the results demonstrated that the inhibition effect of the brush coating is strain dependent. In vivo experiments with Pluronic F-127 coated silicone rubber in infected pockets in mice demonstrated that the polymer brush coatings are less prone to reinfection than pristine silicone rubber when used in revision surgery [17].

PEG chains have been prepared with functional ends or side groups which act as anchors for self-assembly on surfaces. Peptide sequences with specific affinity for titanium [18] or polystyrene [19] have recently been identified using a combinatorial phage display technique and applied as anchors for PEG chains. The peptide conjugated to PEG assembled from dilute aqueous solution onto the selected surface through adsorptive mechanisms. The coated surface efficiently blocked the adsorption of fibronectin and significantly reduced the extent of *S. aureus* attachment and biofilm formation *in vitro*. Phosphorylated PEG derivatives have been investigated for use as an anti-adherent coating on hydroxyapatite (HA) which is widely used in orthopaedic and dental biomaterials [20]. Methacryloyloxydecyl phosphate (MDP)-PEG derivatives were prepared from the polymerization of PEG methacrylate, methacrylic acid, and MDP in aqueous solution containing 2-mercaptoethanol and ammonium persulfate. Pretreatment of HA surfaces with MDP-PEG prior to saliva incubation hydrophilized the surface and reduced salivary protein adsorption and saliva-promoted *S. mutans* attachment to HA. However, when MDP-PEG was added to the saliva-pretreated HA surface, its inhibitory effect on bacterial binding diminished since the adsorption of MDP-PEG was inhibited by salivary components already bound to the HA. On the other hand, treatment of saliva-pretreated HA surfaces with MDP-PEG plus pyrophosphate (which can desorb protein and acquired pellicles from HA surfaces) completely inhibited saliva-promoted *S. mutans* adhesion even when followed by additional saliva treatment.

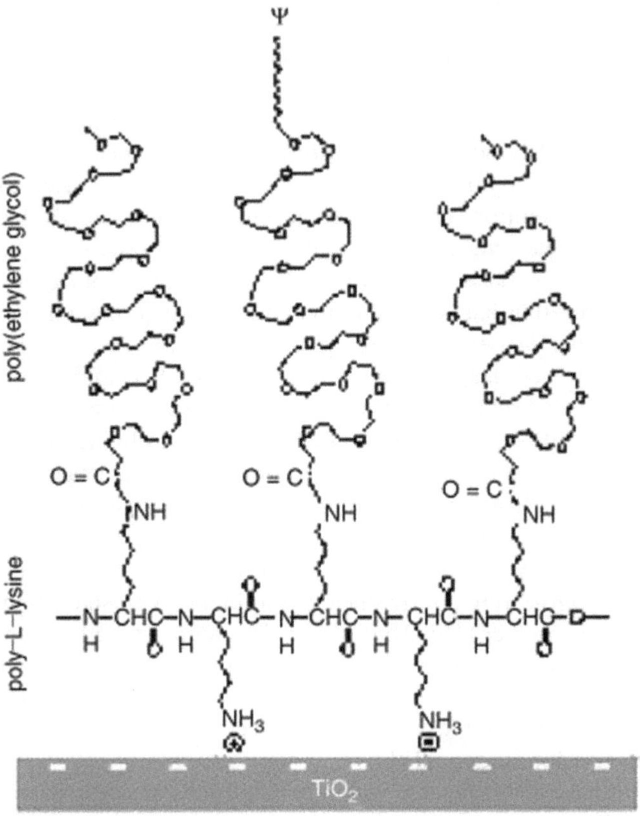

Fig. 16.2 Schematic diagram of the structure of PLL grafted with PEG side chains for imparting protein resistance to a titanium oxide surface. The PLL has positively charged amino-terminated side chains that bind to the negatively charged titanium oxide surface through multiple-site electrostatic interactions. Ψ = specific protein. Reprinted with permission from Ref. [23]. Copyright 2004 Elsevier

Textor and coworkers developed another type of polymer brush coating by applying a comb-like graft copolymer comprising a polycationic poly(L-lysine) (PLL) backbone and PEG side chains (PLL-g-PEG) [21–23]. The copolymer has approximately 120 L-lysine units, a PEG side chain of 47 ethylene glycol units and a grafting ratio (number of lysine monomers per PEG side chain) of between 3.3 and 4.5. This copolymer adsorbs rapidly and strongly through electrostatic interactions to negatively charged surfaces such as TiO_2 at physiological pH (Fig. 16.2), rendering them highly protein and bacterial resistant. The adhesion of *S. aureus* to smooth and rough (chemically etched) Ti surfaces coated with PLL-g-PEG decreased by 89–93% after exposure to *S. aureus* cultures for 1 to 24 h. PLL-g-PEG can be functionalized with the cell adhesive peptide, RGD (Arg–Asp–Gly), at the PEG side chain (PLL-g-PEG/PEG-RGD), and Ti coated with PLL-g-PEG/

PEG-RGD decreased bacterial adhesion by 69%. The PLL-g-PEG/PEG-RGD-coated surfaces exhibit selective biointeractiveness (adhesive to cells such as fibroblasts, osteoblasts, and endothelial cells, but repellent to extracellular matrix proteins and bacteria), and thus would be potentially useful for enhanced osseointegration required in osteosynthesis and orthopaedics. More recently, Busscher and coworkers have utilized PLL-g-PEG/PEG-RGD surfaces to test the concept of "race for the surface" between bacteria and mammalian cells [24]. *S. epidermidis* and U2OS osteoblast cells were grown simultaneously on the surfaces for 48 h under low shear ($0.14\ s^{-1}$). After 48 h, biofilms formed on all surfaces but the biofilms detached from the PLL-g-PEG and PLL-g-PEG/PEG-RGD coatings when exposed to an elevated shear ($5.6\ s^{-1}$). U2OS cells neither adhered nor spread on PLL-g-PEG brush coatings, regardless of the presence of biofilm. In contrast, in the presence of biofilm, U2OS cells adhered and spread on PLL-g-PEG/PEG-RGD coatings with a significantly higher surface coverage than on bare TiO_2. For implants, a post-implantation "decisive period" of 6 h has been identified to be critical for preventing bacterial adhesion as well as achieving long-term success of implantation [25]. Thus, even with a limited lifespan, a coating with bacteria-repellent properties and host-cell adhesive qualities may still serve a very useful function during this critical period.

An investigation of the long-term stability of the PLL-*g*-PEG coatings showed that the PLL-*g*-PEG-coated polystyrene and metal oxide surfaces remain protein- and cell-resistant in contact with full cell culture medium at 37°C over a timeframe of at least 8–12 days [26]. However, such adsorbed layers are unstable under extreme pH values (>10 or lower than isoelectric point of surface) or high ionic strength. Electrostatic interactions between the PLL-*g*-PEG and the surface will be compromised at high pH when the density of positive charges on the PLL backbone is reduced and also at low pH which results in a decrease in the negative surface charges. Similarly, in high ionic strength media, decreased electrical double layer interactions compromise adhesion of the PLL-g-PEG layer [27]. The use of covalent linkages for the attachment of the PLL-g-PEG has been carried out to enhance its stability and extend its range of applicability (discussed in the next section). There are also concerns that PEG, despite displaying excellent non-bioadhesive properties *in vitro*, may undergo oxidative degradation, and lose their function *in vivo* [24]. The use of poly(2-methyl-2-oxazoline) (PMOXA) as a potential alternative to PEG has been investigated by Textor and coworkers [28]. Comb copolymers consisting of a polycationic PLL backbone and PMOXA side chains (analogous to PLL-g-PEG) were anchored onto Nb_2O_5-coated silicon wafers via electrostatic interactions. With an optimal side chain grafting density, the resulting adsorbed monolayer of the copolymer could reduce protein adsorption from full human serum to below 2 ng/cm^2, which equaled the protein repellant properties of the best PEG-based coatings [28]. Poly(2-ethyl-2-oxazoline) end-functionalized with a quaternary ammonium salt has been prepared and shown to have higher antimicrobial activity against *S. aureus* than analogous end-functionalized PMOXA and PEO [29, 30]. Thus, poly(2-oxazoline)s may be potentially useful as a antibacterial coating for surfaces.

16.3.2 Grafting of Polymer Chains to Surfaces

One of the earliest works on the modification of a polymer surface with PEO to achieve bacterial repellence was via the surface-physical-interpenetrating-network (SPIN) method [31, 32]. This technique involves immersing a polymeric substrate in a liquid which is a mutual solvent for the polymer chains to be grafted and the substrate. As the surface of the substrate swells, the polymer chains will diffuse into the semi-dissolved interface. When the system is quenched with water, the interface collapses, resulting in the entrapment of the PEO chains. This method has been applied to modify polyethylene terephthalate (PET) films with PEO (M_w = 18,500) [32]. Significant reductions (between 70 and 95%) in adherent *S. epidermidis*, *S. aureus*, and *P. aeruginosa* cultured in different media (tryptic soya broth, human plasma, and human serum) were observed on the PEO-modified substrates compared to the untreated control. However, the SPIN method is limited by the choice of a suitable solvent and the molecular weight of the polymer chain to be grafted on the surface since a low molecular weight chain cannot be effectively entrapped while very large molecules face mass transfer limitations.

A number of groups have modified the surfaces of polymeric substrates via wet chemistry [33, 34] or plasma treatment [35] for subsequent grafting of PEG chains. Park et al. modified polyurethane (PU) surface with PEG1k (M_w = 1,000) carrying terminal hydroxyl, amino and sulfonate groups, PEG3.4k (M_w = 3,350) and PEG3.4k-heparin, respectively [33]. The PU surface was first grafted with hexamethylenediisocyanate through an allophanate reaction between the urethane proton and the isocyanate in the presence of a catalyst (di-*n*-butyl tin dilaurate). The free isocyanate groups on the surface then reacted with the terminal groups of the PEG (–OH and –NH_2). The PEG-grafted surfaces were investigated for bacterial adhesion using *S. epidermidis* and *E. coli* in broth and human plasma. The adhesion level was found to depend on the surfaces as well as the media. In the case of surfaces modified with PEG1k, *S. epidermidis* adhesion in plasma media was reduced but no reduction in adhesion was demonstrated in broth media regardless of terminal functional groups of PEG1k. However, for surfaces with the relatively longer PEG3.4k and PEG3.4k-heparin, bacterial adhesion was significantly reduced in both media. In the case of *E. coli* adhesion, significant reduction in adherent bacteria was observed on all PEG1k-, PEG3.4k-, and PEG-heparin-modified surfaces in both media compared to controls.

Kingshott et al. grafted PEG on PET via a multistep process: hydroxylation of PET surface by exposure to formaldehyde in acetic acid, carboxylation of hydroxyl groups by exposure to bromoacetic acid in sodium hydroxide, amidation by reaction with polyethyleneimine (PEI, M_w = 25,000) using carbodiimide chemistry, and finally reaction with linear methoxy-terminated PEG-aldehyde (M_w = 5,000) in the presence of sodium cyanoborohydride [34]. The last step was carried out at the lower critical solution temperature (LCST) of the PEG to ensure maximal surface coverage. The PEG layer was estimated to be 4–4.5 nm in thickness, and the coating reduced the level of adhesion of *Pseudomonas sp* by between 2 and 4 orders of magnitude for up to 5 h. Surface-sensitive techniques, such as X-ray photoelectron spectroscopy (XPS) and time-of-flight secondary ion mass spectrometry (ToF-SIMS), showed that low molecular weight proteins and peptides from the bacterial growth medium were not fully prevented from adsorbing on the surfaces, which may account for the residual bacterial adhesion.

Plasma techniques have been proposed as a simpler alternative to wet chemical modification of surfaces for reducing biofilm formation. A cold plasma technique was used to functionalize the surfaces of a polyamide and a polyester for the grafting of PEG [35]. The surfaces of the polymer substrates were first treated with silicon tetrachloride ($SiCl_4$) plasma, followed by the covalent immobilization of PEG via interaction of its OH end group with the active surface functional group. The polyamide and polyester surfaces modified with PEG with a molecular weight of 600 Da showed a 96% reduction in biofilm formation by *L. monocytogenes* compared with that of the corresponding unmodified substrates. Cold plasma enhanced processes were also employed for the deposition of PEG-like structures onto stainless steel surfaces [36]. Stainless steel samples were subjected to radio frequency plasma treatment in the presence of 1,4,7,10-tetraoxacyclododecane (12-crown-4)-ether and tri(ethylene glycol) dimethyl ether (triglyme). The plasma-modified stainless steel surfaces were more hydrophilic and had lower surface roughness values compared to those of unmodified substrates. Unmodified and plasma-modified substrates were subjected to bacterial attachment and biofilm development experiments using a mixed culture of *S. typhimurium*, *S. epidermidis*, and *P. fluorescens*. Compared to the unmodified surfaces, bacterial attachment after 1 h decreased by 56% and 82% on the 12-crown-4-plasma-modified and triglyme-plasma-modified surfaces, respectively. The corresponding decrease in biofilm formation after 1 day on the respective surface was 72% and 94%.

Covalent attachment of PLL-g-PEG copolymers has been carried out on aldehyde plasma-modified inorganic (silica wafers and optical waveguide chips) and polymeric (perfluorinated ethylene-co-propylene) substrates [27]. The substrates were first coated with a thin plasma polymer layer using a propionaldehyde plasma, followed by covalent immobilization of PLL-g-PEG via reductive amination between amine groups of the PLL backbone with aldehyde groups on the plasma deposited interlayer. The coatings produced via covalent immobilization proved to be stable when immersed in a 2.4 M salt solution for 24 h and maintained protein repellency, whereas PLL-g-PEG polymers that had been adsorbed electrostatically desorbed to substantial extents under such high salt conditions.

Another strategy which can greatly simplify the grafting of non-fouling polymers to biomaterial surfaces was developed by Messersmith and coworkers [37, 38]. This strategy was based on the adhesive characteristics of 3,4-dihydroxyphenylalanine (DOPA), an important component of mussel adhesive proteins. Linear monomethoxy-terminated PEGs were conjugated either to a single DOPA residue (mPEG-DOPA) or to the N-terminus of Ala-Lys-Pro-Ser-Tyr-Hyp-Hyp-Thr-DOPA-Lys (mPEG-MAPD), a decapeptide analogue of a protein found in *Mytilus edulis* adhesive plaques. Titanium surfaces can be readily modified by adsorption of mPEG-DOPA and mPEG-MAPD from solution. Although the exact mechanism of mussel adhesive proteins adhesion to surfaces is not fully understood, it has been widely speculated as being due in part to chemical interactions between the catechol side chain of DOPA residues and the surfaces to which the protein is adsorbed. The modified Ti surfaces exhibited low fibroblast adhesion for up to 2 weeks in culture, indicating that the non-fouling properties of mPEG-DOPA and mPEG-MAPD treated surfaces persist for extended periods of time [38]. Titanium modified with mPEG-DOPA was also shown to exhibit a substantial decrease in attachment of cells of *N. perminuta* and zoospores of *U. linza*, as well as the highest detachment of attached cells under flow compared to control surfaces [39]. Messersmith and coworkers also synthesized a class of antifouling and antimicrobial polymers composed of N-substituted glycine (peptoid) chains [40, 41]. These polymers were immobilized onto TiO_2 surfaces via an anchoring peptide consisting of alternating residues of DOPA and lysine. The polypeptoid-modified surfaces exhibited resistance to enzymatic degradation, significant reductions in adsorption of lysozyme, fibrinogen, and serum proteins, and reduced adhesion of fibroblast cells, *E. coli* and *S. epidermidis*. Besides mussel adhesive proteins, another biomimetic anchor based on the cyanobacterial iron chelator, anachelin, has been used to anchor PEG chains to TiO_2 surfaces [42]. While the PEG-anachelin conjugate has a catechol moiety, similar to PEG-DOPA, the former results in a more stable coating as a result of the interaction between the positive charge of anachelin and the negative charge of the surface and the anti-oxidative property of anachelin.

It can be expected that the different methods for grafting PEO or PEG to surfaces as described above will result in very different grafting densities and chain conformation. Furthermore, since different bacteria and test methods were used by different investigators, it would be difficult to make direct comparisons of the parameters which govern the anti-adhesive efficacy among the various studies. The effects of PEO chain length and temperature on bacterial and yeast adhesion have been studied by Busscher and coworkers [43, 44]. In their investigation, glass surfaces were modified by end-grafting methacryl-terminated PEO chains having molecular weights of 526, 2,000, or 9,800 Da with the surface silanol groups in a polymer melt (at 145°C). This resulted in PEO brushes on the surface with estimated lengths in water of 2.8, 7.5, and 23.7 nm, and chain grafting densities of 2.3, 1.0, and 0.2 chains/nm^2, respectively. With increasing brush length, there was a decrease in both the initial deposition rate and the number of adhering bacteria (*S. epidermidis* and *P. aeruginosa*) after 4 h (Fig. 16.3). This reduction was less for the larger yeast cells (*C. albicans* and *C. tropicalis*) as compared to the smaller bacteria, and the authors suggested that Lifshitz-van der Waals interactions play a crucial role in the residual adhesion of the microorganisms at the outer edge of the brush. The more hydrophobic organisms (*P. aeruginosa* and *C. tropicalis*) adhered in larger numbers to the PEO chains than the hydrophilic ones (*S. epidermidis* and *C. albicans*) and no significant differences were found in adhesion and retention between experiments conducted at 20°C and those conducted at 37°C. The adhesion of the microorganisms to the brush was weaker than that to bare glass since in the former the adhering microorganisms can be easily removed from the brush by a passing air bubble.

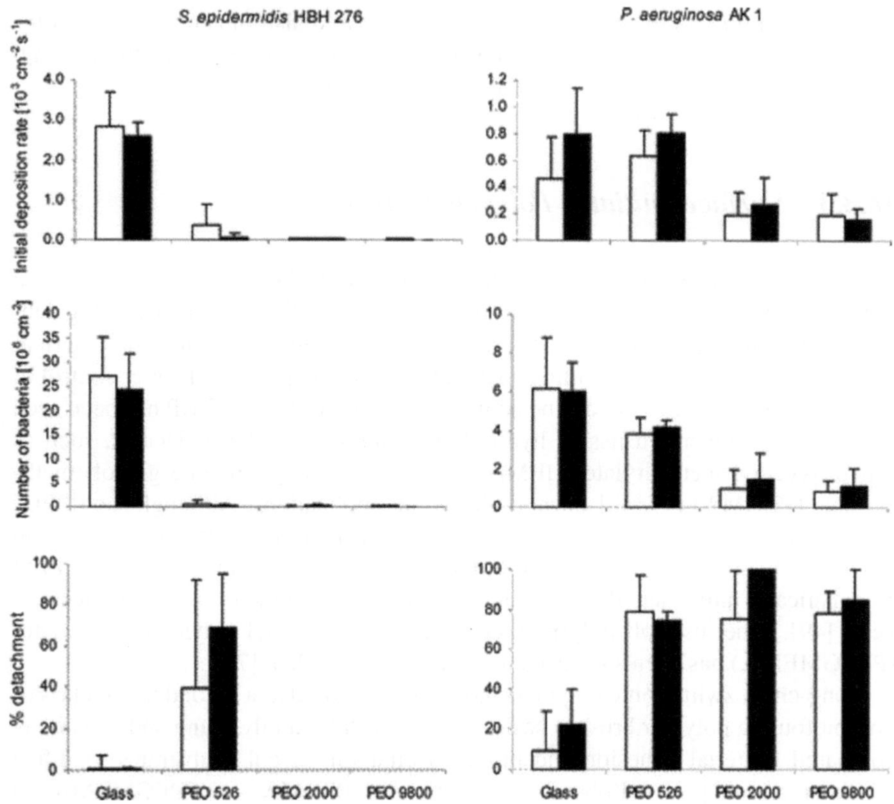

Fig. 16.3 Initial deposition rates, number of bacteria adhering after 4 h, and percent detachment of *S. epidermidis* HBH 276 and *P. aeruginosa* AK 1 at 20 (*white bars*) and 37°C (*black bars*) on glass and glass coated with the PEO brushes of M_w of 526, 2000, and 9800. For detachment of *S. epidermidis*, only data for glass and glass coated with PEO of M_w of 526 are shown. Error bars represent the average standard deviation over three separate experiments. Reprinted with permission from Ref. [44]. Copyright 2004 American Chemical Society

The stability and effectiveness of the PEO coating coupled to a glass surface via silyl ether bonds (as described above) was also investigated in saliva, urine, and PBS at 37°C [45]. The coating repressed *S. epidermidis* adhesion for 24, 48, and 0.5 h in PBS, urine, and saliva, respectively. The variation in effectiveness was attributed to conditioning film formation. Salivary proteins may adhere to and penetrate between the polymer chains, thereby covering the PEO coating and reducing its long-term effectiveness, whereas in urine, some factors can reduce overall bacterial adhesion by increasing the surface negative charge. The short period of effectiveness in PBS was attributed to degradation of the coating, and the grafting site of the PEO chain to the surface, rather than the PEO chain itself, is thought to be the weak link. In this study, the Si–O–C linkage to the glass and the ester group close to the anchoring site are susceptible to hydrolysis. The increasing amount of adherent bacteria with time as the PEO brush density decreased also illustrates the need for an optimum brush density to maintain the anti-adhesive property of the coating. This issue is discussed further in the following section.

16.3.3 Surface-Initiated Polymer Brushes

As indicated in the previous section, the grafting of PEO/PEG chains to surfaces can reduce bacterial adherence to various degrees, and brush thickness and grafting density are important parameters. With surface-initiated controlled radical polymerization, different types of non-fouling coating have been prepared and better control of brush thickness and density can be achieved. Surface-initiated ATRP has been used to produce polymer brushes of hydrophilic neutral PEG-like molecules, such as 2-hydroxyethyl methacrylate (HEMA) [46, 47] and poly(ethylene glycol) methacrylate (PEGMA) [47, 48] for non-biofouling applications. Although such PEG-like coatings show effectiveness under *in vitro* conditions, there is concern that *in vivo* the hydroxyl end groups at the side chains of these brushes may be oxidized enzymatically into aldehydes and acids, thereby reducing their long-term effectiveness [49]. The use of poly(poly(ethyleneglycol) methyl ether methacrylate) (PPEGMEMA) has been suggested to counter this problem [7].

Long-chain zwitterionic polymers have been proposed as a second major class of non-biofouling polymer brushes besides the PEG/PEO family. Jiang and coworkers compared bacterial adhesion and biofilm formation on gold substrates modified with zwitterionic poly(sulfobetaine methacrylate) (PSBMA) and PPEGMEMA via surface-initiated ATRP from a self-assembled monolayer of ω-mercaptoundecyl bromoisobutyrate on the gold surface [50, 51]. Both types of modified surfaces show short-term (3 h) and long-term (24 h) efficacy against *S. epidermidis* and *P. aeruginosa*. After 3 h of exposure, adhesion of *S. epidermidis* and *P. aeruginosa* on the PSBMA brush-coated substrates was reduced by 92% and 96%, respectively, compared to bare glass. The same group also tested another zwitterionic polymer, poly(carboxybetaine methacrylate) (PCBMA), grafted from glass surfaces via ATRP [52]. PCBMA was comparable to PSBMA in resisting biofilm formation; with reduction in biofilm formation of *P. aeruginosa* by 95% for up to 240 h at 25°C and by 93% for 64 h at 37°C, and suppressed *P. putida* biofilm accumulation by 95% for 192 h at 30°C, with respect to bare glass.

Fundeanu et al. prepared polyacrylamide (PAAM) brushes on silicon rubber by surface-initiated ATRP using a three-step reaction procedure comprising immobilization of a coupling agent, γ-aminopropyltriethoxysilane, anchoring of an ATRP initiator, 4-(chloromethyl) benzoyl chloride, and polymerization of acrylamide either in *N,N*-dimethylformamide (DMF) or in water [53]. PAAM brushes grown in water reduce the adhesion of *S. aureus* by 58%, *S. salivarius* by 52% and *C. albicans* by 77%. The anti-adhesive properties of brush coatings grown in DMF are slightly better due to the thicker polymer layer. No deterioration of the PAAM coating was detected after exposure to PBS and saliva for 1 month at 37°C. Another type of polymer brush, poly(methacrylic acid), has been formed on titanium surfaces via surface-initiated ATRP of methacrylic acid sodium salt with immobilized trichloro(4-(chloromethyl)-phenyl)silane. The functionalized surfaces are highly hydrophilic (water contact angle of 8°) and demonstrates inhibition of bacterial adhesion [54].

A number of studies on antibacterial brushes have relied on organosilanes for anchoring polymer chains to surfaces or to initiate polymerization from the surface. A study has shown that poly(PEGMA) or (PPEGMA) brushes, prepared by surface-initiated ATRP from glass or silicon oxide substrates modified with a trimethoxysilane-based ATRP initiator, detached rapidly from the substrate when high density brushes were incubated in cell culture medium but not in water [55]. Unlike the PPEGMA brushes, poly(HEMA) (or PHEMA) brushes are stable in cell culture medium. It was proposed that osmotic stresses acting on the PPEGMA brushes in the cell culture medium and steric crowding induce additional tension along the already stretched polymer brush backbones, which promotes hydrolysis of the Si–O bonds and detachment of the brush. This cleavage process may be further facilitated by the relatively ill-defined nature of the trialkoxysilane based initiator layer (as compared to alkyl thiol self-assembled monolayer and catechol-modified ATRP initiators). The stability of the PPEGMA brushes in cell culture medium could be improved from less than 1 to more than 7 days, without compromising the antifouling properties by appropriately decreasing of the grafting density. An interesting comparison was made between the degradation process of dense PPEGMA brushes and PEG monolayers grafted onto SiO_x surfaces using silane chemistry. The latter showed gradual detachment of the chains whereas PPEGMA brushes are released in the form of a continuous film. The free surface area that becomes exposed upon detachment of a single polymer chain in a tethered PEG monolayer may be accommodated for by neighboring chains, which will adopt a less stretched conformation. As a result, the non-fouling properties of these PEG brushes can still be preserved for a period of time. However, such a "repair mechanism" at the molecular level is not possible for the PPEGMA brushes due to their cross-linked nature. Hence, after cleavage of a sufficient number of Si–O bonds, a macroscopic film is released, which will adversely affect the non-fouling properties [55].

16.4 Bactericidal Polymer Brushes

The second class of antibacterial surfaces relies on immobilized bactericidal polymer chains to kill the bacteria upon contact. These polymers usually contain cationic groups, such as alkyl pyridinium or quaternary ammonium. While the exact bacteria-killing mechanism of these chains is still debatable, it is generally thought that the interaction of the cationic groups with the cell phospholipid membrane causes disruptions in its integrity and loss of intracellular components. Pioneering work on the covalent attachment of bactericidal polymers to surfaces was carried out by Klibanov and coworkers [56–59]. In their study, aminopropyltrimethoxysilane-coated glass slides acylated with acryloyl chloride, copolymerized with 4-vinylypyridine, and N-alkylated with hexylbromide were able to kill 94% of deposited *S. aureus* cells and more than 99% of deposited *S. epidermidis*, *P. aeruginosa* and *E. coli* [56]. The influence of the length of the alkyl group on the bactericidal activity was highlighted in this study. Pyridine groups N-alkylated with alkylbromide with 6 carbon atoms (C_6) showed the highest killing efficacy, followed by C_3 and C_4 chains, while the C_8–C_{16} chains are significantly less effective. However, it is not clear whether the concentrations of surface pyridinium groups on the substrates were the same when different alkyl bromides were used for N-alkylation. The group also developed a methodology for covalently derivatizing the surfaces of polymers with poly(vinyl-*N*-pyridinium bromide) by first coating the surface with a nanolayer of silica followed by treatment with 3-aminopropyltriethyoxysilane, alkylation with 1,4-bromobutane, and reaction with poly(4-vinylpyridine) and 1-bromohexane [57].

Our group demonstrated a simpler method for introducing pyridinium groups on PET films and cellulose by plasma pretreatment of the surfaces followed by UV-induced surface graft copolymerization with 4-vinylpyridine, and the subsequent alkylation of the grafted 4-vinylpyridine groups [60]. The bacteria-killing efficiency was found to be dependent on the surface pyridinium concentration and a surface concentration of 15 nmol/cm^2 on PET was highly effective against *E. coli*. A similar method has been used to confer pyridinium groups and antibacterial properties on electrospun polyurethane [61] and poly(vinylidene fluoride-co-hexafluoropropylene) fibrous membranes [62]. Besides the use of *N*-alkylated poly(vinylpyridine), the attachment of sufficiently long polyethylenimine chains on different substrates (glass, textile, iron oxide nanoparticles) followed by N-alkylation was shown to also result in bactericidal properties towards both Gram-positive and Gram-negative bacteria [58, 59].

Chitosan, a cationic polysaccharide with antibacterial activity [63, 64], has also been employed for the functionalization of surfaces [65–68]. An assembly of polyelectrolyte multilayers (PEMs) comprising hyaluronic acid and chitosan on titanium has been shown to result in an order of magnitude decrease in the number of adherent *S. aureus* and *E. coli* compared to pristine Ti [67]. Chemical cross-linking of the multilayers imparts structural stability, and the antibacterial properties were preserved without significant deterioration after 21-day immersion in PBS. Chitosan has also been grafted onto titanium alloy via a dopamine linker [68]. This substrate possesses antibacterial properties, and attachment of RGD peptide on the chitosan promotes osteoblast attachment, proliferation, and ALP activity. Thus, this substrate may be advantageous for combating biofilm-related infections and promoting tissue integration related to implants. The substrate remains bacteria-resistant after immersion in PBS for 14 days. These experiments illustrate the stability of the covalent linkages employed for surface functionalization and support the view that the antibacterial properties resulted from the surface-immobilized functional groups and not due to leaching of these groups into the media.

Another class of antimicrobial compounds is the *N*-halamines. This class of compounds contains nitrogen-halide covalent bonds, and unlike the cationic bactericidal polymers described above, the probable antimicrobial action involves the transfer of the oxidative halogen to the bacterial cell [69, 70]. Following the loss of the halogen, the *N*-halamines can be regenerated by exposure to additional free halogens. *N*-halamine moieties have been grafted onto the surfaces of polypropylene tubing by first hydroxylating the surfaces with potassium persulfate, followed by graft polymerization of methacrylamide using ceric(IV) ammonium nitrate as an initiator, and treatment with diluted chlorine bleach [71]. The last step converts some of the amide groups in the grafted poly(methacrylamide) side chains into acyclic *N*-halamines. When the *N*-halamine-functionalized tubing was challenged with *P. aeruginosa* in a continuous flowing model, no bacteria could be recovered from the surfaces after 1 week, and after 4 weeks the number of recovered bacteria was 2 orders of magnitude lower than on the unmodified tubing. Upon recharging with bleach, the antibacterial property was recovered and repeated recharging does not seem to significantly affect its efficacy. A similar method has been used for functionalization of cotton cellulose with acyclic *N*-halamines [72]. The functionalized cotton cellulose eliminated 10^8–10^9 CFU/mL of *E. coli*, *S. aureus*, and *C. tropicalis* in 3 min, 10^6–10^7 PFU/mL of MS2 virus in 5 min, and 10^6–10^7 spores/mL of *B. subtilis* in 4 h.

Better control of the composition, architecture, and functionalities of the bactericidal polymer coating can be achieved through surface-initiated ATRP as demonstrated by the work of Russell and coworkers [73–75]. In these works as well as in a number of others based on controlled polymerization techniques, the bactericidal polymer brushes were prepared by the polymerization and subsequent quaternization of 2-(dimethylamino)ethyl methacrylate (DMAEMA). This technique has been applied to filter paper and glass by first reacting 2-bromoisobutyryl bromide with the hydroxyl groups of cellulose in filter paper and free amine groups on amino glass slides via esterification or amidation, respectively, to produce the active ATRP initiator on the surfaces [73]. ATRP was then used to polymerize DMAEMA from the functionalized surfaces, and quaternization was carried out with ethylbromide. The modified surfaces had substantial antimicrobial capacity against *E. coli* and *B. subtilis*. The poly(DMAEMA) or (PDMAEMA) brush density can be controlled by varying the number of initiating sites through the use of a mixture of 2-bromoisobutyryl bromide and propionyl bromide. Propionyl bromide acts as a blocking agent since it produces a non-polymerizable site after reaction with the surface. The influence of polymer brush chain length and grafting density on the bacterial killing properties was investigated using a combinatorial screening method. Surface charge density was identified as a critical parameter which determines the bactericidal activity, and the most effective surfaces in killing *E. coli* had charge densities of >1–5×10^{15} accessible quaternary amine units/cm^2 on the glass slide, regardless of whether the dry brush layer thickness was 15 or 75 nm [74].

Surface-initiated ATRP of DMAEMA has been carried out on polypropylene using benzophenonyl 2-bromoisobutyrate as the initiator [75]. Immobilization of the initiator on polypropylene was achieved through two sequential steps: deposition of initiator on the polymer surface via spin coating followed by UV irradiation which promoted the reaction between the benzophenone group and the polypropylene chain. The *E. coli* killing efficacy was similar to that obtained in the earlier

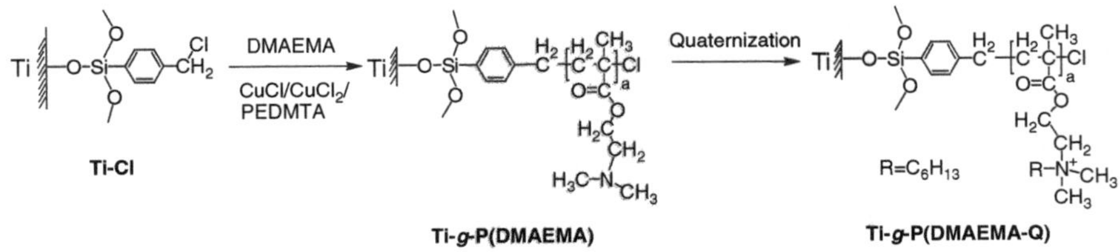

Fig. 16.4 Schematic diagram illustrating the processes of surface-initiated ATRP of DMAEMA from Ti-Cl surface and quaternization of grafted DMAEMA polymer chains. Reprinted with permission from Ref. [78]. Copyright 2007 American Chemical Society

work with glass slide [64], and essentially all bacteria were killed when the number of quaternary ammonium units exceeds $1.4 \times 10^{15}/cm^2$. Poly(vinylidene fluoride) (PVDF) with PDMAEMA side chains have been synthesized via vinylidene fluoride-initiated ATRP of DMAEMA [76]. The resulting PVDF-*g*-PDMAEMA copolymers were then cast into microfiltration membranes by phase inversion in an aqueous medium. The surfaces of the membranes are enriched with PDMAEMA chains which can be further quaternized with 1-bromohexane to result in an antibacterial membrane.

DMAEMA can also be polymerized from cellulosic filter paper via RAFT polymerization [77]. The hydroxyl groups of the cellulose fiber were first reacted with 2-chloro-2-phenylacetyl chloride. Further reaction with phenylmagnesium chloride and carbon disulfide resulted in cellulose-*S*-methoxycarbonylphenylmethyl dithiobenzoate RAFT agent. Surface-initiated RAFT polymerization of DMAEMA from the initiator-functionalized cellulose was carried out in the presence of the free chain transfer agent, *S*-methoxycarbonylphenyl dithiobenzoate and 2,2′-azobis(isobutyronitrile) (AIBN) in ethanol. Quaternization was carried out with 1-bromooctane, 1-bromododecane or 1-bromohexadecane. The PDMAEMA-grafted cellulose quaternized with the C_8 chain was found to exhibit particularly high activity against *E. coli*, while quaternization with the longer alkyl chain (C_{12} and C_{16}) resulted in little antibacterial effect.

Metals such as titanium and stainless steel are widely used as medical implants and hence an antibacterial surface would be highly beneficial. Titanium can be functionalized with bactericidal polymer brushes based on quaternized PDMAEMA using trichloro(4-(chloromethyl)phenyl) silane as the ATRP initiator as shown in Fig. 16.4 [78]. Alternatively, for electrically conductive substrates, a combination of cathodic electrografting and grafting of chains from the electrodeposited macroinitiator by ATRP was used by Jérôme and coworkers to prepare antibacterial surfaces [79, 80]. Stainless steel was first electrografted with 2-(2-chloropropionate)ethyl acrylate followed by ATRP of 2-(tert-butylamino)-ethyl methacrylate (TBAEMA) or copolymerization of TBAEMA with either monomethyl ether of poly(ethyleneoxide) methacrylate (PEOMA) or acrylic acid (AA) or styrene [79]. Compared to the bare stainless steel surface, brushes of poly(TBAEMA), poly(TBAEMA-co-PEOMA), and poly(TBAEMA-co-AA) decrease *S. aureus* adhesion by 3 to 4 orders of

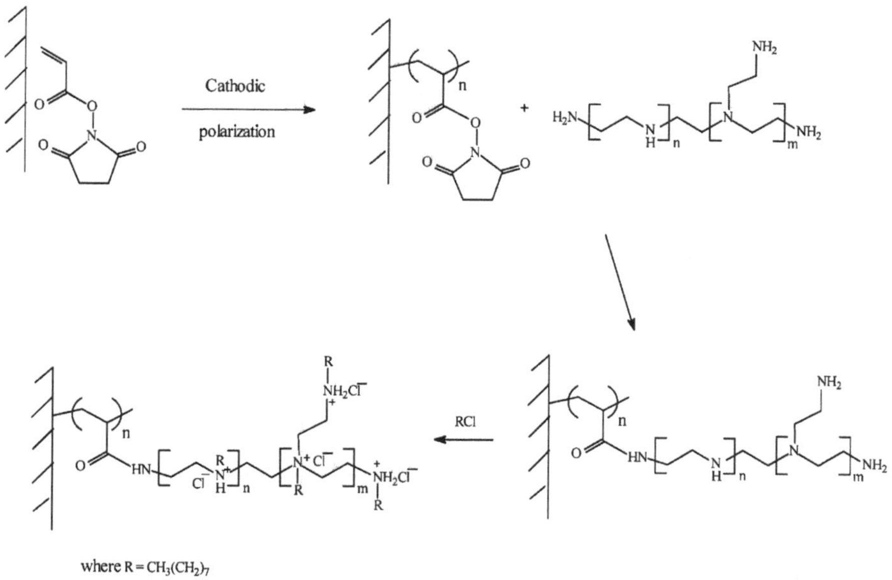

Fig. 16.5 Three-step strategy for the grafting of quaternized hyperbranched poly(ethyleneimine) onto stainless steel. Reprinted with permission from Ref. [80]. Copyright 2007 American Chemical Society

magnitude. In addition, poly(TBAEMA-co-PEOMA) and poly(TBAEMA-co-AA) are also effective in reducing fibrinogen adsorption by two- to threefold compared to poly(TBAEMA). A similar method was used to prepare hyperbranched quaternized polymer brushes on stainless steel [80]. ATRP of an inimer, 2-(2-bromopropionate) ethylacrylate, in the presence or absence of heptadecafluorodecyl acrylate (to modulate the surface properties) was carried out using electrografted poly[2-(2-chloropropionate)ethyl acrylate] as a macroinitiator. Quaternization was then carried out by reaction with pyridine. Compared to the untreated stainless steel, a 3 orders of magnitude decrease in the number of viable *S. aureus* cells was observed on the modified substrate. An even higher killing efficacy was observed with stainless steel electrografted with poly(*N*-succinimidyl acrylate) followed by grafting of preformed hyperbranched poly(ethyleneimine) (M_w = 50–60000, relative contents of primary, secondary, and tertiary amines = 1:1:1) and quaternization by 1-chlorooctane (Fig. 16.5). The higher efficacy may be attributed to the higher density of quaternary ammonium groups in the hyperbranched poly(ethyleneimine) in contact with the bacteria.

A combination of ATRP and "grafting to" technique has been employed to prepare antibacterial surfaces based on the following steps: (1) ATRP synthesis of well-defined block copolymers consisting of a PDMAEMA functional segment and a poly(3-(trimethoxysilyl)propyl methacrylate) (PTMSPMA) anchoring segment, and also DMAEMA/TMSPMA random copolymers; (2) immobilization of the copolymers onto glass slide surface via reaction between the anchoring

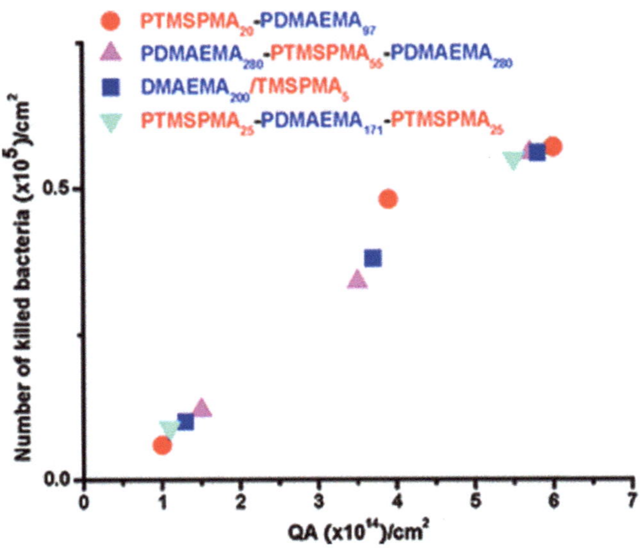

Fig. 16.6 Bactericidal activity of surfaces modified with diblock, triblock, and random copolymer versus density of quaternary ammonium units on surfaces (2.9×10^5 bacteria in control, surface area: 5 cm^2). Reprinted with permission from Ref. [81]. Copyright 2008 American Chemical Society

trimethoxysilyl groups in the copolymers and silanol groups from the surface; (3) conversion of the pendant amino groups in the surface-linked PDMAEMA to quaternary ammonium groups in the presence of ethyl bromide [81]. The grafting density increased with polymer concentration and immobilization time, and bactericidal activity increased with concentration of quaternary ammonium groups but was not strongly affected by polymer architecture as shown in Fig. 16.6. A comparison of the bactericidal activity of these surfaces with the surfaces prepared by the "grafting from" method in an earlier work by the same group [74] showed that the "grafting from" method provides a much higher density of quaternary ammonium groups than the "grafting to" method ($\sim 1 \times 10^{16}$ vs. $6 \times 10^{14}/cm^2$). As a result, the latter shows a lower overall bactericidal activity, but at the same density of quaternary ammonium groups, the bactericidal activity of surfaces prepared by "grafting to" was higher than for surfaces prepared by "grafting from." The authors proposed that the "grafting to" method results in an uneven distribution of quaternary ammonium groups, and the localized patches with a high concentration of these groups increases the killing efficiency as compared to surfaces with the same average polymer density but with a uniform distribution.

The plethora of work on polymer brushes with quaternary ammonium functionalities has confirmed their bactericidal effect. However, the bacteria which were killed may remain on the surfaces of the polymer brushes. The accumulation of dead bacteria on the polymer brushes may block the bactericidal groups, and may also trigger adverse reactions when used *in vivo*. Jiang and coworkers prepared a surface which can switch from being bactericidal to antifouling [82]. In this system, poly(N,N-dimethyl-*N*-(ethoxycarbonylmethyl)-*N*-[2′-methacryloyloxy)ethyl]-ammonium bromide) was grafted by surface-initiated ATRP on a gold surface covered with initiators. The modified surface can kill>99.9% of *E. coli* in 1 h. When this surface was then hydrolyzed to poly(*N*-(carboxymethyl)-*N*,*N*-dimethyl-2-[(2-methyl-1-oxo-2-propen-1-yl)-oxy]-ethanaminium) using *N*-cyclohexyl-3-aminopropanesulfonic acid buffer (10 m mM, pH 10.0) at 37°C for 8 days, 98% of the dead bacterial cells were released.

16.5 Polymer Brushes with Antimicrobial Agents

The previous section highlights the work carried out with polymer brushes possessing bactericidal properties as a result of their functional groups which are primarily quaternary ammonium or alkyl pyridinium. This section describes the use of polymer brushes which are either conjugated with agents such as antibiotics or antimicrobial peptides or complexed with silver for preparing an antimicrobial surface. The polymer brushes may fulfill a number of roles such as providing an anti-adhesive surface, serving as a spacer for the tethered antimicrobial agent or a reservoir for the agent.

16.5.1 Conjugation with Antibiotics

Since most systemically administered antibiotics lack efficacy against implant-associated infections, there have been a number of attempts to modify implant surfaces with antibiotics. The antibiotics are usually accommodated within a polymer matrix which allows for sustained release either by diffusion or degradation of the matrix [83, 84]. There are few reports on the conjugation of antibiotics to polymer brushes. Wach et al. synthesized a hybrid molecule comprising a PEG chain with an anachelin chromophore at one end and an antibiotic, vancomycin, at the other [85]. Vancomycin inhibits bacterial cell wall synthesis, and the antimicrobial activity of this hybrid molecule against *B. subtilis* is roughly a factor of two lower than vancomycin. When titanium substrates were treated with a solution of this hybrid molecule in (3-(*N*-morpholino)propanesulfonic acid) buffer, the anachelin groups bind to the surface oxide layer resulting in a surface which combines the anti-adhesive property of PEG with the action of vancomycin. The modified surface was shown to retain its activity after five cycles of incubation with bacteria followed by rinsing with a buffer.

Antibiotics have also been tethered to titanium surfaces via PHEMA chains [78]. This was accomplished by surface-initiated ATRP of HEMA using trichloro(4-(chloromethyl)phenyl) silane as the initiator, followed by the conversion of the pendant hydroxyl end groups of the grafted HEMA chains to carboxyl or amine groups to allow the coupling of the antibiotics, gentamicin and penicillin, respectively. The antibacterial activity of the antibiotics-modified surfaces against *S. aureus* was found to be comparable to that of quaternized PDMAEMA. Since the bactericidal action of gentamicin results from its binding to the bacterial small ribosomal subunit, it is believed that hydrolysis of the surface-attached antibiotics by bacterial proteases occurred resulting in the gradual release of surface-bound antibiotics.

Penicillin has also been attached to expanded poly (tetrafluoroethylene) (ePTFE) surfaces [86, 87]. The ePTFE surfaces were first treated with microwave plasma in the presence of maleic anhydride leading to the formation of acid groups, followed by esterification reactions with PEG. Penicillin was then attached to the PEG chains using 4-(dimethylamino)-pyridine and dicyclohexyl-carbodiimide. The penicillin-modified surfaces were effective against the Gram-positive *S. aureus* but not against Gram-negative bacteria *P. aeruginosa*. Hydrolytic stability of the penicillin-modified surfaces was investigated by immersing the substrates in PBS at 37°C. The results indicate that after 24 h, 32% of the tethered penicillin was lost due to the cleavage of the ester linkages between the antibiotic and the PEG linker but the surface still retained significant antimicrobial activity. In a later investigation, the group attached a broad spectrum antibiotic, ampicillin, to ePTFE using a slightly different method [88]. NH_2–PEG–COOH was attached to the maleic anhydride plasma-treated surface via amidation reactions, followed by attachment of antibiotic to the PEG spacer by amidation reactions between its NH_2 groups and COOH-terminated PEG chains. With this approach, hydrolysis was reduced and only 10% of the tethered antibiotic was lost after 24 h. The ampicillin-modified surface is effective against Gram-positive *S. aureus*, *B. thuringiensis* and *E. faecalis*, and Gram-negative *E. coli*, *P. putida*, and *S. enterica* bacteria.

16.5.2 Conjugation with Antibacterial Peptides

Antimicrobial peptides (AMP) are produced by many organisms such as plants, microorganisms, insects, and vertebrates as components of their innate immunity, and may be promising agents against antibiotic-resistant pathogens [89, 90]. The antimicrobial activity has been attributed to the interaction of the positively charged AMPs with the negatively charged bacterial cell membrane increasing its permeability and causing disruption, leading to cell death. The effect of immobilization of AMPs on their activity has been investigated by Bagheri et al. [91]. The peptides were immobilized at the C and N termini and via different side chain positions to the surface of resin beads bearing PEG spacers of different lengths. The results indicated that the peptide's antimicrobial activity decreases upon immobilization, and a shorter spacer will reduce the activity more. However, the activity is less dependent upon the position of immobilization. AMPs have also been immobilized on PEGylated polyethylene film [92] and titanium [93]. The polyethylene surface was first oxidized with chromic/sulfuric acid, then PEGylated by using carbodiimide chemistry to attach ω-amino-α-carboxyl-PEG (H_2N–PEG–COOH). The E14LKK peptide with side chain (lysine) amines protected with *tert*-butyloxycarbonyl was covalently coupled to the free COOH terminus of the PEG, again using carbodiimide coupling. Deprotection of the protected peptide was carried out after immobilization. Antibacterial activity against *E. coli* was observed when conjugation of the AMP was carried out as described, in contrast to the case when E14LKK with unprotected side chain amines was used.

Direct conjugation of the AMP, cathelin LL37, to titanium without a spacer also resulted in no antibacterial activity [93]. Two schemes of conjugating the peptide to the titanium surface via a PEG spacer were investigated. N-terminal coupling of the peptide to titanium was carried out by PEGylation of the silanized surface using α-*N*-hydroxysuccinimidyl-ω-maleimidyl-PEG followed by Michael-type coupling of the peptide exclusively at its N-terminus. For random coupling of the peptide to the spacer, the substrate was first treated with glycidyloxypropyl triethoxysilane followed by H_2N–PEG–COOH and coupling with the peptide using carbodiimide chemistry. The peptide randomly immobilized on the PEG spacer also showed no antibacterial activity, in contrast to peptide conjugated to the PEG spacer via its N-terminus. These investigations highlight the importance of a spacer which can provide lateral mobility and the appropriate orientation to the conjugated peptide for interacting with the bacterial membrane.

Huck and coworkers prepared antibacterial coatings which combined the bactericidal activity of an immobilized natural AMP, magainin I, with non-fouling polymer brushes based on 2-(2-methoxyethoxy)ethyl methacrylate (MEO2MA) and hydroxyl-terminated oligo(ethyleneglycol) methacrylate (HOEGMA) [94]. The copolymer coating, poly(MEO2MA-co-HOEGMA), was prepared by surface-initiated ATRP from silanized silicon wafers, and the relatively long water-soluble and flexible oligo(ethyleneglycol) side chains of the copolymer brushes were used for tethering the peptide in order to maintain its mobility and accessibility. A cysteine residue was attached to the C-terminus of magainin I, and since magainin I does not contain cysteine residues, the cysteine-derivatized magainin I could be exclusively attached to the brushes by its C-terminus via a *N*-(*p*-maleimidophenyl)isocyanate heterolinker (Fig. 16.7). The density of immobilized magainin I is dependent on the availability of the hydroxyl-reactive groups which can be changed by varying the ratio of HOEGMA to MEO2MA. The high antibacterial efficiency of the AMP-functionalized copolymer brushes against *L. ivanovii* and *B. cereus* was evident for a wide range of brush compositions.

16.5.3 Complexation with Silver

Silver-based compounds have been used in the treatment of wounds for centuries and the antibacterial properties of silver were investigated as early as the nineteenth century [95, 96]. The inhibitory action of silver ions on microorganisms is believed to be the result of the interaction of silver ions with thiol and other key functional groups in enzymes and proteins [97]. The actions of silver nanoparticles on microorganisms are not entirely understood. Silver nanoparticles have been shown to cause pit formation in bacteria cell wall and increased membrane permeability

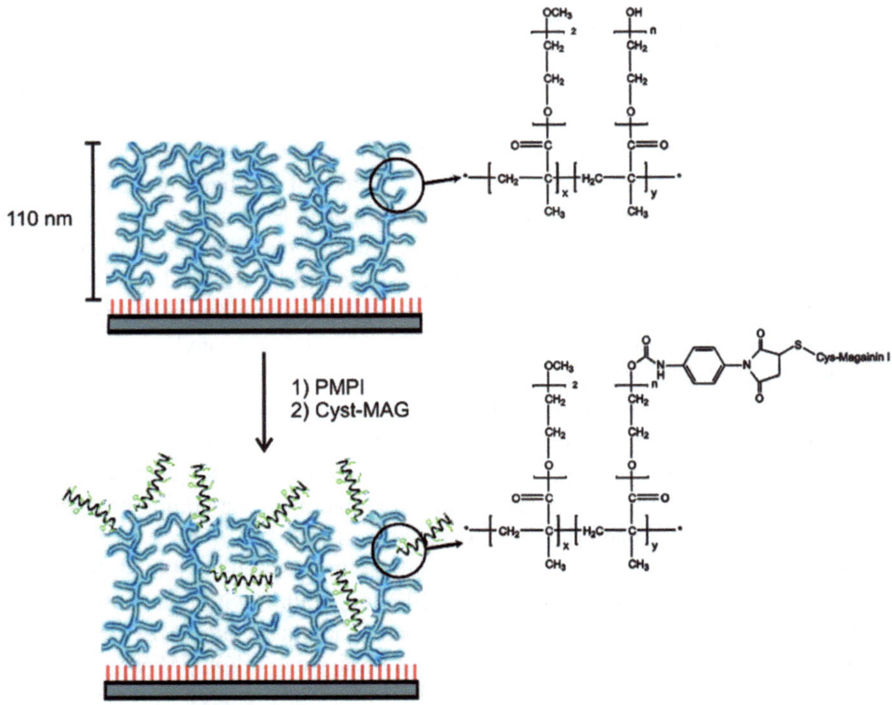

Fig. 16.7 Oriented grafting of Cyst-MAG derivative on poly(MOE2MA-co-HOEGMA) brushes via a *N*-(*p*-maleimidophenyl)isocyanate (PMPI) heterolinker. Reprinted with permission from Ref. [94]. Copyright 2009 American Chemical Society

[98]. In addition, silver nanomaterials may exert bactericidal activity through release of silver ions [99]. Anionic polyelectrolyte brushes bearing sulfonate groups have been used to trap silver ions [100]. The surface-initiated ATRP of 3-sulfopropylmethacrylate was carried on Si/SiO_2 and gold surfaces using 2-bromo-2-methyl-propionic acid 3-trichlorosilanyl-propyl ester and ω-mercaptoundecyl bromobutyrate as initiator respectively. Silver loading was carried out by immersing the polymer brush-functionalized substrate into a silver salt solution. The polyelectrolyte brushes "collapsed" in contact with the salt solutions, and the contraction in the brush thickness in the dry state is ~20–30% depending on the concentration of the salt solution. The silver-loaded brushes exhibited slow leaching of silver ions in water and in NaCl medium, and inhibited the growth of both Gram-positive (*S. aureus*) and Gram-negative (*P. aeruginosa*) bacteria.

Tang et al. prepared antibacterial filter paper using silver nanoparticles as the antibacterial agent via a four-step process: (1) immobilization of the ATRP initiator, 2-bromoisobuturyl bromide, by reaction with the OH groups of the filter paper, (2) surface-initiated ATRP of *tert*-butyl acrylate (tBA), (3) hydrolysis of the ester groups of the grafted poly(tBA) into carboxyl groups by trifluoroacetic acid for chelating Ag^+ and (4) reduction of chelated Ag^+ into silver nanoparticles using sodium borohydride [101]. The silver nanoparticle-decorated filter paper showed excellent antibacterial properties against *E. coli* compared to the original filter paper and the filter paper with grafted polymer brushes but without the silver nanoparticles.

While silver ions and nanoparticles have been demonstrated to be effective bactericidal agents, their impact on human health and the environment is not so clear [99, 102]. Unlike the bactericidal polymer brushes which are immobilized on the surface of substrates, the bactericidal action of silver relies to a certain extent on the leaching of the ions. In view of the increasing use of silver as an antibacterial agent, more studies on the toxicity of silver are warranted.

16.6 Conclusions and Outlook

The high human and economic costs associated with biomaterial-associated infections have spurred research on the development of anti-infective surfaces. One strategy is to functionalize surfaces with polymer brushes which are anti-adhesive and/or bactericidal. The advances in techniques for surface functionalization and tailoring of polymer composition and architecture in the past decade have resulted in many promising developments. A multitude of different techniques for immobilization of polymer brushes have been applied to substrates of different materials and shapes, and *in vitro* experiments have clearly demonstrated their antibacterial properties. Nevertheless, a number of issues remain to be resolved before such modified materials can be applied clinically. Substrates which showed high efficacy in preventing short-term bacterial adhesion *in vitro* (usually in PBS) may not function in the same manner and be as successful in preventing biofilm formation under conditions encountered *in vivo*. The long-term stability of the coating in the presence of body fluids is a concern. In addition, some applications may require sterilization of the substrates before use which may affect the stability of the coating. Potential cytotoxic effects of the coating and any degradation products on healthy tissues and cells need to be elucidated. Finally, when modified biomaterials for implant applications are applied clinically, they will be subjected to abrasion which may compromise the mechanical integrity of the coating. These challenges indicate that much more work remains to be done in order to successfully combat biomaterials-associated infections through material surface modification strategies.

Glossary

Bactericidal Capable of destroying bacteria
Biofilm An organized community of bacteria growing on a surface
Biomimetic Biologically inspired
Chelator A chemical that combines with metal ions and remove them from solution

Heterolinker A molecule with dissimilar functional groups used for coupling other molecules

Hyperbranched Multibranching structure

Inimer A compound which contains both a polymerizable group and a group able to initiate polymerization by forming active species

Isoelectric point The pH value at which a particular molecule or surface carries no net electrical charge

LCST Lower critical solution temperature is the temperature below which a mixture is miscible

Macroinitiator A polymer or oligomer which can initiate a polymerization reaction

PEGylation Chemical process of modifying a molecule or surface with polyethylene glycol (PEG)

Physi-sorption Process of adsorption where there is minimal perturbation of the electronic states of the adsorbed molecule and the surface. Also known as physical adsorption

Zwitterionic A molecule which carries both a positive and a negative charge

References

1. Davies D. Understanding biofilm resistance to antibacterial agents. Nat Rev Drug Discov. 2003;2:114–22.
2. Schachter B. Slimy business—the biotechnology of biofilms. Nat Biotech. 2003;21:361–5.
3. Gristina AG. Biomaterial-centered infection: microbial adhesion versus tissue integration. Science. 1987;237:1588–95.
4. Darouiche RO. Antimicrobial coating of devices for prevention of infection: principles and protection. Int J Artif Organs. 2007;30(9):820–7.
5. Minko S. Grafting on solid surfaces: "Grafting to" and "grafting from" methods. In: Stamm M, editor. Polymer surfaces and interfaces. Berlin, Heidelberg: Springer; 2008. p. 215–34.
6. Roosjen A, Norde W, van der Mei H, Busscher H. The use of positively charged or low surface free energy coatings versus polymer brushes in controlling biofilm formation. Prog Colloid Polym Sci. 2006;132:138–44.
7. Barbey R, Lavanant L, Paripovic D, et al. Polymer brushes via surface-initiated controlled radical polymerization: synthesis, characterization, properties, and applications. Chem Rev. 2009;109(11):5437–527.
8. Anthony MG, William JB. Recent advances in polymer brush synthesis. In: Advincula RC, Brittain WJ, Caster KC, Rühe J, editors. Polymer brushes. Weinheim: Wiley-VCH; 2005. p. 33–50.
9. Edmondson S, Osborne VL, Huck WT. Polymer brushes via surface-initiated polymerizations. Chem Soc Rev. 2004;33(1):14–22.
10. Harris JM, Zalipsky S. Poly(ethylene glycol): chemistry and biological applications. ACS symposium series, 680. Washington, DC: American Chemical Society; 1997.
11. Nejadnik MR, Olsson AL, Sharma PK, van der Mei HC, Norde W, Busscher HJ. Adsorption of Pluronic F-127 on surfaces with different hydrophobicities probed by quartz crystal microbalance with dissipation. Langmuir. 2009;25(11):6245–9.
12. McLean SC, Lioe H, Meagher L, Craig VSJ, Gee ML. Atomic force microscopy study of the interaction between adsorbed poly(ethylene oxide) layers: effects of surface modification and approach velocity. Langmuir. 2005;21(6):2199–208.
13. Liu VA, Jastromb WE, Bhatia SN. Engineering protein and cell adhesivity using peo-terminated triblock polymers. J Biomed Mater Res. 2002;60(1):126–34.

14. Bridgett MJ, Davies MC, Denyer SP. Control of staphylococcal adhesion to polystyrene surfaces by polymer surface modification with surfactants. Biomaterials. 1992;13(7):411–6.
15. Wesenberg-Ward KE, Tyler BJ, Sears JT. Adhesion and biofilm formation of *Candida albicans* on native and pluronic-treated polystyrene. Biofilms. 2005;2(01):63–71.
16. Nejadnik MR, van der Mei HC, Norde W, Busscher HJ. Bacterial adhesion and growth on a polymer brush-coating. Biomaterials. 2008;29(30):4117–21.
17. Nejadnik MR, Engelsman AF, Saldarriaga Fernandez IC, Busscher HJ, Norde W, van der Mei HC. Bacterial colonization of polymer brush-coated and pristine silicone rubber implanted in infected pockets in mice. J Antimicrob Chemother. 2008;62(6):1323–5.
18. Khoo XJ, Hamilton P, O'Toole GA, Snyder BD, Kenan DJ, Grinstaff MW. Directed assembly of PEGylated-peptide coatings for infection-resistant titanium metal. J Am Chem Soc. 2009;131(31):10992–7.
19. Kenan DJ, Walsh EB, Meyers SR, et al. Peptide-PEG amphiphiles as cytophobic coatings for mammalian and bacterial cells. Chem Biol. 2006;13(7):695–700.
20. Shimotoyodome A, Koudate T, Kobayashi H, et al. Reduction of *Streptococcus mutans* adherence and dental biofilm formation by surface treatment with phosphorylated polyethylene glycol. Antimicrob Agents Chemother. 2007;51(10):3634–41.
21. Kenausis GL, Voros J, Elbert DL, et al. Poly(L-lysine)-*g*-poly(ethylene glycol) layers on metal oxide surfaces: attachment mechanism and effects of polymer architecture on resistance to protein adsorption. J Phys Chem B. 2000;104(14):3298–309.
22. Tosatti S, De Paul SM, Askendal A, et al. Peptide functionalized poly(L-lysine)-*g*-poly(ethylene glycol) on titanium: resistance to protein adsorption in full heparinized human blood plasma. Biomaterials. 2003;24(27):4949–58.
23. Harris LG, Tosatti S, Wieland M, Textor M, Richards RG. *Staphylococcus aureus* adhesion to titanium oxide surfaces coated with non-functionalized and peptide-functionalized poly (L-lysine)-grafted-poly(ethylene glycol) copolymers. Biomaterials. 2004;25(18):4135–48.
24. Subbiahdoss G, Pidhatika B, Coullerez G, et al. Bacterial biofilm formation versus mammalian cell growth on titanium-based mono- and bi-functional coating. Eur Cell Mater. 2010;19:205–13.
25. Poelstra KA, Barekzi NA, Rediske AM, Felts AG, Slunt JB, Grainger DW. Prophylactic treatment of gram-positive and gram-negative abdominal implant infections using locally delivered polyclonal antibodies. J Biomed Mater Res. 2002;60:206–15.
26. VandeVondele S, Voros J, Hubbell JA. RGD-grafted poly-L-lysine-graft-(polyethylene glycol) copolymers block non-specific protein adsorption while promoting cell adhesion. Biotechnol Bioeng. 2003;82(7):784–90.
27. Blättler TM, Pasche S, Textor M, Griesser HJ. High salt stability and protein resistance of poly (L-lysine)-*g*-poly(ethylene glycol) copolymers covalently immobilized via aldehyde plasma polymer interlayers on inorganic and polymeric substrates. Langmuir. 2006;22(13):5760–9.
28. Konradi R, Pidhatika B, Muhlebach A, Textor M. Poly-2-methyl-2-oxazoline: a peptide-like polymer for protein-repellent surfaces. Langmuir. 2008;24(3):613–6.
29. Waschinski CJ, Tiller JC. Poly(oxazoline)s with telechelic antimicrobial functions. Biomacromolecules. 2005;6(1):235–43.
30. Hoogenboom R. Poly(2-oxazoline)s: a polymer class with numerous potential applications. Angew Chem Int Ed. 2009;48(43):7978–94.
31. Desai NP, Hubbell JA. Solution technique to incorporate polyethylene oxide and other water-soluble polymers into surfaces of polymeric biomaterials. Biomaterials. 1991;12(2):144–53.
32. Desai NP, Hossainy SF, Hubbell JA. Surface-immobilized polyethylene oxide for bacterial repellence. Biomaterials. 1992;13(7):417–20.
33. Park KD, Kim YS, Han DK, et al. Bacterial adhesion on PEG modified polyurethane surfaces. Biomaterials. 1998;19(7–9):851–9.
34. Kingshott P, Wei J, Bagge-Ravn D, Gadegaard N, Gram L. Covalent attachment of poly(ethylene glycol) to surfaces, critical for reducing bacterial adhesion. Langmuir. 2003;19(17):6912–21.
35. Dong BY, Jiang HQ, Manolache S, Wong ACL, Denes FS. Plasma-mediated grafting of poly(ethylene glycol) on polyamide and polyester surfaces and evaluation of antifouling ability of modified substrates. Langmuir. 2007;23(13):7306–13.

36. Denes AR, Somers EB, Wong ACL, Ferencz G. 12-Crown-4-ether and tri(ethylene glycol) dimethyl-ether plasma-coated stainless steel surfaces and their ability to reduce bacterial biofilm deposition. J Appl Polym Sci. 2001;81(14):3425–38.
37. Lee H, Dellatore SM, Miller WM, Messersmith PB. Mussel-inspired surface chemistry for multifunctional coatings. Science. 2007;318(5849):426–30.
38. Dalsin JL, Hu BH, Lee BP, Messersmith PB. Mussel adhesive protein mimetic polymers for the preparation of nonfouling surfaces. J Am Chem Soc. 2003;125(14):4253–8.
39. Statz AR, Finlay J, Dalsin J, Callow M, Callow JA, Messersmith PB. Algal antifouling and fouling-release properties of metal surfaces coated with a polymer inspired by marine mussels. Biofouling. 2006;22(5–6):391–9.
40. Statz AR, Barron AE, Messersmith PB. Protein, cell and bacterial fouling resistance of polypeptoid-modified surfaces: effect of side-chain chemistry. Soft Matter. 2008;4(1):131–9.
41. Statz AR, Park JP, Chongsiriwatana NP, Barron AE, Messersmith PB. Surface-immobilised antimicrobial peptoids. Biofouling. 2008;24(6):439–48.
42. Zurcher S, Wackerlin D, Bethuel Y, et al. Biomimetic surface modifications based on the cyanobacterial iron chelator anachelin. J Am Chem Soc. 2006;128(4):1064–5.
43. Roosjen A, Kaper HJ, van der Mei HC, Norde W, Busscher HJ. Inhibition of adhesion of yeasts and bacteria by poly(ethylene oxide)-brushes on glass in a parallel plate flow chamber. Microbiology. 2003;149(Pt 11):3239–46.
44. Roosjen A, van der Mei HC, Busscher HJ, Norde W. Microbial adhesion to poly(ethylene oxide) brushes: influence of polymer chain length and temperature. Langmuir. 2004;20(25):10949–55.
45. Roosjen A, de Vries J, van der Mei HC, Norde W, Busscher HJ. Stability and effectiveness against bacterial adhesion of poly(ethylene oxide) coatings in biological fluids. J Biomed Mater Res. 2005;73B(2):347–54.
46. Mei Y, Wu T, Xu C, et al. Tuning cell adhesion on gradient poly(2-hydroxyethyl methacrylate)-grafted surfaces. Langmuir. 2005;21(26):12309–14.
47. Tugulu S, Arnold A, Sielaff I, Johnsson K, Klok HA. Protein-functionalized polymer brushes. Biomacromolecules. 2005;6(3):1602–7.
48. Stefano T, Harm-Anton K. Surface modification of polydimethylsiloxane substrates with nonfouling poly(poly(ethylene glycol)methacrylate) brushes. Macromolecular Symposia. 2009;279(1):103–9.
49. Ostuni E, Chapman RG, Holmlin RE, Takayama S, Whitesides GM. A survey of structure–property relationships of surfaces that resist the adsorption of protein. Langmuir. 2001;17(18):5605–20.
50. Zhang Z, Chen S, Chang Y, Jiang S. Surface grafted sulfobetaine polymers via atom transfer radical polymerization as superlow fouling coatings. J Phys Chem B. 2006;110(22):10799–804.
51. Cheng G, Zhang Z, Chen S, Bryers JD, Jiang S. Inhibition of bacterial adhesion and biofilm formation on zwitterionic surfaces. Biomaterials. 2007;28(29):4192–9.
52. Cheng G, Li G, Xue H, Chen S, Bryers JD, Jiang S. Zwitterionic carboxybetaine polymer surfaces and their resistance to long-term biofilm formation. Biomaterials. 2009;30(28):5234–40.
53. Fundeanu I, van der Mei HC, Schouten AJ, Busscher HJ. Polyacrylamide brush coatings preventing microbial adhesion to silicone rubber. Colloids Surf B. 2008;64(2):297–301.
54. Zhang F, Zhang Z, Zhu X, Kang ET, Neoh KG. Silk-functionalized titanium surfaces for enhancing osteoblast functions and reducing bacterial adhesion. Biomaterials. 2008;29(36):4751–9.
55. Tugulu S, Klok HA. Stability and nonfouling properties of poly(poly(ethylene glycol) methacrylate) brushes under cell culture conditions. Biomacromolecules. 2008;9(3):906–12.
56. Tiller JC, Liao CJ, Lewis K, Klibanov AM. Designing surfaces that kill bacteria on contact. Proc Natl Acad Sci U S A. 2001;98(11):5981–5.
57. Tiller JC, Lee SB, Lewis K, Klibanov AM. Polymer surfaces derivatized with poly(vinyl-*N*-hexylpyridinium) kill airborne and waterborne bacteria. Biotechnol Bioeng. 2002;79(4): 465–71.
58. Lin J, Qiu S, Lewis K, Klibanov AM. Bactericidal properties of flat surfaces and nanoparticles vderivatized with alkylated polyethylenimines. Biotechnol Prog. 2002;18(5):1082–6.
59. Lin J, Qiu S, Lewis K, Klibanov AM. Mechanism of bactericidal and fungicidal activities of textiles covalently modified with alkylated polyethylenimine. Biotechnol Bioeng. 2003;83(2): 168–72.

60. Cen L, Neoh KG, Kang ET. Surface functionalization technique for conferring antibacterial properties to polymeric and cellulosic surfaces. Langmuir. 2003;19(24):10295–303.
61. Yao C, Li X, Neoh KG, Shi ZL, Kang ET. Surface modification and antibacterial activity of electrospun polyurethane fibrous membranes with quaternary ammonium moieties. J Membrane Sci. 2008;320(1–2):259–67.
62. Yao C, Li X, Neoh KG, Shi ZL, Kang ET. Antibacterial activities of surface modified electrospun poly(vinylidene fluoride-co-hexafluoropropylene) (PVDF-HFP) fibrous membranes. Appl Surf Sci. 2009;255(6):3854–8.
63. Rabea EI, Badawy MET, Stevens CV, Smagghe G, Steurbaut W. Chitosan as antimicrobial agent: applications and mode of action. Biomacromolecules. 2003;4:1457–65.
64. No HK, Park NY, Lee SH, Meyers SP. Antibacterial activity of chitosans and chitosan oligomers with different molecular weights. Int J Food Microbiol. 2002;74:65–72.
65. Fu J, Ji J, Yuan W, Shen J. Construction of anti-adhesive and antibacterial multilayer films via layer-by-layer assembly of heparin and chitosan. Biomaterials. 2005;26:6684–92.
66. Huh MW, Kang IK, Lee DH, et al. Surface characterization and antibacterial activity of chitosan-grafted poly(ethylene terephthalate) prepared by plasma glow discharge. J Appl Polym Sci. 2001;81:2769–78.
67. Chua PH, Neoh KG, Shi ZL, Kang ET. Structural stability and bioapplicability assessment of hyaluronic acid–chitosan polyelectrolyte multilayers on titanium substrates. J Biomed Mater Res. 2008;87A:1061–74.
68. Shi ZL, Neoh KG, Kang ET, Poh C, Wang W. Bacterial adhesion and osteoblast function on titanium with surface-grafted chitosan and immobilized RGD peptide. J Biomed Mater Res. 2007;86A:865–72.
69. Chen Y, Worley SD, Huang TS, et al. Biocidal polystyrene beads. III. Comparison of *N*-halamine and quat functional groups. J Appl Polym Sci. 2004;92(1):363–7.
70. Cao ZB, Sun YY. *N*-halamine-based chitosan: preparation, characterization, and antimicrobial function. J Biomed Mater Res. 2008;85A(1):99–107.
71. Luo J, Sun YY. Acyclic *N*-halamine-based biocidal tubing: preparation, characterization, and rechargeable biofilm-controlling functions. J Biomed Mater Res. 2008;84A(3):631–42.
72. Luo J, Sun YY. Acyclic *N*-halamine-based fibrous materials: preparation, characterization, and biocidal functions. J Polym Sci. 2006;44A(11):3588–600.
73. Lee SB, Koepsel RR, Morley SW, Matyjaszewski K, Sun Y, Russell AJ. Permanent, non-leaching antibacterial surfaces. 1. Synthesis by atom transfer radical polymerization. Biomacromolecules. 2004;5(3):877–82.
74. Murata H, Koepsel RR, Matyjaszewski K, Russell AJ. Permanent, non-leaching antibacterial surface–2: how high density cationic surfaces kill bacterial cells. Biomaterials. 2007;28(32):4870–9.
75. Huang J, Murata H, Koepsel RR, Russell AJ, Matyjaszewski K. Antibacterial polypropylene via surface-initiated atom transfer radical polymerization. Biomacromolecules. 2007;8(5):1396–9.
76. Xue J, Chen L, Wang HL, et al. Stimuli-responsive multifunctional membranes of controllable morphology from poly(vinylidene fluoride)-graft-poly[2-(*N*,*N*-dimethylamino)ethyl methacrylate] prepared via atom transfer radical polymerization. Langmuir. 2008;24(24):14151–8.
77. Roy D, Knapp JS, Guthrie JT, Perrier S. Antibacterial cellulose fiber via RAFT surface graft polymerization. Biomacromolecules. 2008;9(1):91–9.
78. Zhang F, Shi ZL, Chua PH, Kang ET, Neoh KG. Functionalization of titanium surfaces via controlled living radical polymerization: from antibacterial surface to surface for osteoblast adhesion. Ind Eng Chem Res. 2007;46(26):9077–86.
79. Ignatova M, Voccia S, Gilbert B, et al. Combination of electrografting and atom-transfer radical polymerization for making the stainless steel surface antibacterial and protein antiadhesive. Langmuir. 2006;22(1):255–62.
80. Ignatova M, Voccia S, Gabriel S, et al. Stainless steel grafting of hyperbranched polymer brushes with an antibacterial activity: synthesis, characterization, and properties. Langmuir. 2009;25(2):891–902.

81. Huang J, Koepsel RR, Murata H, et al. Nonleaching antibacterial glass surfaces via "grafting onto": the effect of the number of quaternary ammonium groups on biocidal activity. Langmuir. 2008;24(13):6785–95.
82. Cheng G, Xue H, Zhang Z, Chen S, Jiang S. A switchable biocompatible polymer surface with self-sterilizing and nonfouling capabilities. Angew Chem Int Ed. 2008;47(46):8831–4.
83. Wu P, Grainger DW. Drug/device combinations for local drug therapies and infection prophylaxis. Biomaterials. 2006;27(11):2450–67.
84. Hetrick EM, Schoenfisch MH. Reducing implant-related infections: active release strategies. Chem Soc Rev. 2006;35(9):780–9.
85. Wach JY, Bonazzi S, Gademann K. Antimicrobial surfaces through natural product hybrids. Angew Chem Int Ed. 2008;47(37):7123–6.
86. Aumsuwan N, Heinhorst S, Urban MW. Antibacterial surfaces on expanded polytetrafluoroethylene; penicillin attachment. Biomacromolecules. 2007;8(2):713–8.
87. Aumsuwan N, Heinhorst S, Urban MW. The effectiveness of antibiotic activity of penicillin attached to expanded poly(tetrafluoroethylene) (ePTFE) surfaces: a quantitative assessment. Biomacromolecules. 2007;8(11):3525–30.
88. Aumsuwan N, Danyus RC, Heinhorst S, Urban MW. Attachment of ampicillin to expanded poly(tetrafluoroethylene): surface reactions leading to inhibition of microbial growth. Biomacromolecules. 2008;9(7):1712–8.
89. Zasloff M. Antimicrobial peptides of multicellular organisms. Nature. 2002;415(6870):389–95.
90. Zaiou M. Multifunctional antimicrobial peptides: therapeutic targets in several human diseases. J Mol Med. 2007;85(4):317–29.
91. Bagheri M, Beyermann M, Dathe M. Immobilization reduces the activity of surface-bound cationic antimicrobial peptides with no influence upon the activity spectrum. Antimicrob Agents Chemother. 2009;53(3):1132–41.
92. Matthew DS, Joseph HH. Covalent immobilization of an antimicrobial peptide on poly(ethylene) film. J Appl Polym Sci. 2008;110(5):2665–70.
93. Gabriel M, Nazmi K, Veerman EC, Nieuw Amerongen AV, Zentner A. Preparation of LL-37-grafted titanium surfaces with bactericidal activity. Bioconjug Chem. 2006;17(2):548–50.
94. Glinel K, Jonas AM, Jouenne T, Leprince J, Galas L, Huck WT. Antibacterial and antifouling polymer brushes incorporating antimicrobial peptide. Bioconjug Chem. 2009;20(1):71–7.
95. Klasen HJ. Historical review of the use of silver in the treatment of burns. I. Early uses. Burns. 2000;26(2):117–30.
96. Klasen HJ. A historical review of the use of silver in the treatment of burns. II. Renewed interest for silver. Burns. 2000;26(2):131–8.
97. Jung WK, Koo HC, Kim KW, Shin S, Kim SH, Park YH. Antibacterial activity and mechanism of action of the silver ion in *Staphylococcus aureus* and *Escherichia coli*. Appl Environ Microbiol. 2008;74(7):2171–8.
98. Sondi I, Salopek-Sondi B. Silver nanoparticles as antimicrobial agent: a case study on *E. coli* as a model for Gram-negative bacteria. J Colloid Interf Sci. 2004;275(1):177–82.
99. Marambio-Jones C, Hoek EMV. A review of the antibacterial effects of silver nanomaterials and potential implications for human health and the environment. J Nanopart Res. 2010;12: 1531–51.
100. Ramstedt M, Cheng N, Azzaroni O, Mossialos D, Mathieu HJ, Huck WT. Synthesis and characterization of poly(3-sulfopropylmethacrylate) brushes for potential antibacterial applications. Langmuir. 2007;23(6):3314–21.
101. Tang F, Zhang LF, Zhang ZB, Cheng ZP, Zhu XL. Cellulose filter paper with antibacterial activity from surface-initiated ATRP. J Macromol Sci. 2009;46A(10):989–96.
102. Ramstedt M, Ekstrand-Hammarstrom B, Shchukarev AV, et al. Bacterial and mammalian cell response to poly(3-sulfopropyl methacrylate) brushes loaded with silver halide salts. Biomaterials. 2009;30(8):1524–31.

Part III
Clinical Applications, Case Studies

Chapter 17
Infection in Fracture Fixation: Device Design and Antibiotic Coatings Reduce Infection Rates

Gerhard Schmidmaier, Abhay D. Gahukamble, T. Fintan Moriarty, and R. Geoff Richards

Abstract Musculoskeletal infection is one of the most common complications associated with surgical fixation of bones fractured during trauma. Severe fractures with extensive tissue damage are particularly prone to infection due to the high risk of wound contamination and compromised vascularity in the affected tissues. An infection associated with a fracture fixation device can delay healing, greatly increase patient morbidity, require multiple surgeries for effective treatment outcomes, and may tremendously increase treatment costs. In the following chapter, two approaches to reduce the incidence of infection associated with fracture fixation devices will be described. The first is a passive approach involving aspects of implant design and application, whereby the implant used and the techniques used to place them can influence resistance to infection, at least in animal studies. The second approach involves antibiotic-coated intramedullary nails with a focus on two different gentamicin coatings.

17.1 An Introduction to Infection in Fracture Fixation

One of the most common complications associated with the surgical fixation of bones fractured during trauma is the development of infection [1]. The presence of an infection may lead to pain, implant loosening or nonunion of the fracture and require surgical removal of infected tissues and even the implant itself in an attempt to clear the infection. In common with many other implant-related infections, infections

G. Schmidmaier (✉)
Department of Orthopedics, Research Center for Experimental Orthopedics,
Trauma Surgery and Paraplegiology, Heidelberg University Hospital, Heidelberg, Germany
e-mail: gerhard.schmidmaier@med.uni-heidelberg.de

A.D. Gahukamble • T.F. Moriarty • R.G. Richards
AO Research Institute Davos, Clavadelerstrasse 9, Davos Platz, Switzerland

T.F. Moriarty et al. (eds.), *Biomaterials Associated Infection: Immunological Aspects and Antimicrobial Strategies*, DOI 10.1007/978-1-4614-1031-7_17,

associated with implanted fracture fixation devices result in significant patient morbidity, delayed healing, difficulty in achieving successful treatment outcomes, and great financial costs. In contrast with elective surgeries, however, there is a greater risk of infection in open trauma, where soft tissue damage is generally greater, and the barrier to contamination provided by the skin is broken, often outside of the operation room. These open wounds are almost certainly contaminated by bacteria. Furthermore, standard preoperative practices such as skin decolonization and timely prophylactic antibiotic administration may be either compromised or even impossible due to the unpredictable nature of trauma cases. Direct tissue destruction and the resultant edema can lead to local circulatory compromise which causes necrosis and prevents perfusion of systemic antibiotics and host immune factors. Considering this, it is not surprising that there are comparatively high infection rates in trauma cases in comparison with elective procedures such as total joint replacements.

Infections associated with fracture fixation devices are commonly classified according to the elapsed time between the surgery and the diagnosis or presentation of the infection. The defined thresholds vary within the literature but is often broken down to groups of early (<2 weeks), delayed (2–10 weeks) and late (more than 10 weeks) infections [2–5]. In terms of treatment protocols and outcomes, early infections are more likely to be treated successfully by aggressive antibiotic therapy, with additional debridement and/or drainage of infected or necrotic tissues. Delayed and late infections may, however, involve a chronic infection with an antibiotic-resistant biofilm present on the implanted device. These established or chronic infections can persist despite even aggressive parenteral antibiotic treatment which is primarily suppressive rather than therapeutic in nature. Definitive treatment of the infection will require extensive debridement and lavage and removal of the colonized implant. In cases where healing is nearly complete the implant may be retained with suppressive antibiotic therapy until fracture healing is complete; however, the infection is rarely eliminated as long as the implant is present and so the implant is removed upon fracture healing.

The socioeconomic impact of musculoskeletal infection is significant [6, 7] with one recent study showing that the cost associated with treating early infection of hip fractures quadrupled in comparison with noninfected cases [6]. The costs associated with delayed or late infections may be expected to be significantly more, as numerous surgical interventions may be required.

New therapeutic and preventative interventions are obviously required and, broadly speaking, the approaches taken within industrial and academic sectors may be broken down into either passive or active strategies. The passive strategies used to minimize the risk of developing fracture fixation implant-related infection include the following: the use of biocompatible implant materials, minimization of dead spaces in the patient or in the implant itself that preclude vascular access, careful surgical approach to minimize additional tissue destruction, and finally careful design and application of the fracture fixation implant itself to ensure protection of the viable tissues present [8]. In contrast to these passive strategies, which are designed with fracture healing as primary goal and infection prevention as a secondary goal, active strategies clearly focus on the problem of implant-related infection. Active strategies include any strategy whereby antimicrobial agents specifically targeting bacteria are delivered to the surgical site. The most common active strategy has involved the local delivery of antimicrobial agents within a carrier biomaterial to the site of implantation. This approach achieves far greater concentrations of antimicrobial at the implantation site than that which is achievable by systemic administration. The antimicrobials thus delivered are most often delivered by antibiotic-loaded bone cements, implant coatings, or degradable materials such as collagen fleeces.

In the present chapter, we shall focus upon both the passive and active strategies employed to minimize infection rate in fracture fixation.

17.2 Fracture Fixation Implants: An Overview

In order to facilitate rapid, pain-free healing, modern fracture fixation implants are designed to provide optimal fixation of the fractured bone with respect to both mechanical stability and protection of tissues such as the periosteum [9–11]. The use of these fixation implants also allows earlier mobility than what is otherwise achievable and thus potentiates improved functional outcomes for patients. Broadly speaking, the fracture fixation implants in current use may be placed into one or other of three implant categories: intramedullary nail, internal fixation plate and screws, or external fixation device (Fig. 17.1). Each implant type impacts upon the fractured bone and surrounding soft tissues in very different ways. Internal fixation plates, for example, are placed along the outer surfaces of the bone and interface with both the periosteum and the overlying soft tissues. Plates, therefore, typically require significant surgical incisions adjacent to the fracture site to allow placement of the plate and cause some damage to cortical bone through screw insertion.

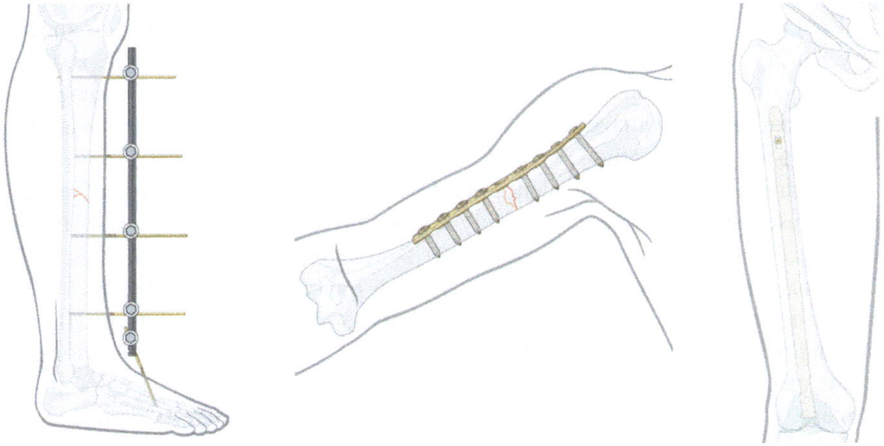

Fig. 17.1 Schematic drawings of (L-R): External fixation of a tibial fracture; Internal plate fixation of a humeral fracture; and Intramedullary nail in a femur. Copyright by AO Foundation, Switzerland. Source: AO Surgery Reference, www.aosurgery.org

Screws when used alone usually aim to provide inter-fragmentary compression and have the advantages of requiring smaller incisions. Intramedullary nails, in contrast, occupy a niche within the intramedullary canal and have a significant impact on intramedullary tissue structures and blood flow. The surgical incisions required for IM nail insertion are generally smaller than those required for plate procedures since the nail may be inserted through an opening at an adjacent joint rather than over the fracture site itself and the interlocking bolts can be placed using minimally invasive techniques. In the case of external fixation, multiple percutaneous pins connect the bone with external supports, creating minimal impact on the soft and hard tissues at a location removed from the fracture site. External fixators require only minimal skin incisions but there remains a possible route from the skin through to the cortical bone over the entire period of fixation leading to high risk of pin tract infection.

The stability of the fixed fractures will be somewhat different for each implant type used and there is a large amount of data available in the literature describing the influence of biomechanical properties such as stability, rigidity, or flexibility of the implant and the pronounced effects on the healing processes of bone [9–11]. Considering the great impact biomechanics plays on fracture inflammation and healing, it is also expected that biomechanics can also play a significant role in infection, although these factors remain to be fully elucidated.

17.2.1 The Role of Implant Type on Infection Susceptibility

The particular fixation technique chosen to fix any given fracture is dependent upon the type of fracture to be treated, the anatomical location of the fracture, functional goals, the extent of soft tissue injury, contamination present and in certain cases, surgeon preference [12]. In general, plate fixation is routinely applied for diaphyseal fractures of the upper extremity and periarticular fractures, whilst intramedullary nailing is often applied in diaphyseal fractures of the lower limbs. External fixation is most commonly used for cases with significant soft tissue damage and contamination or in damage control orthopedics where the external fixator is applied as a temporary stabilization.

In terms of the clinical prevalence of infection as it pertains to the type of implant system used, infection rates vary widely from report to report. Infection rates related to IM nail fixations have been reported at 2–3% for closed tibial fractures, up to 21% for open tibial fractures [13, 14], though infection rates of 50% have been reported for severe open fractures with extensive soft tissue damage (type III) [15, 16] and 5.6% for open femoral fractures [17]. Infection rates in plate fixation also show variability with infection rates most commonly cited at 2–4% for plated humeral fractures for example [18–20]. The infection rate after external fixation varies extremely widely with pin tract infection rates from 0% to 100% reported in human and veterinary medicine [21–24] and average values therefore extremely difficult to reliably compile. The clinical significance of pin tract infections lie in the fact that they preclude conversion to intramedullary nailing as a definite treatment option due to the high risk of infection [25].

In studies into open diaphyseal fractures of the tibia which may be fixed by either external fixation or IM nailing, infection rates seem to be equivalent between the two implant types [14, 26], though the clinical significance of these infections may differ. In diaphyseal fractures of the humerus, which may be fixed by internal fixation plates or IM nails, infection was found to be more common with plating in one study [27] though no differences were found in other studies [28, 29].

The wide variation in infection rates associated with each type of fracture fixation implant reflects the wide variation in fracture types and patient characteristics. This variation complicates the assessment of the susceptibility of a particular fixation implant type to infection. It is consistently apparent however, that the overriding influence on susceptibility to infection is the fracture type and extent of soft tissue damage. Although the clinical data does not allow us to tease out the role played by the type of implant used on the susceptibility to infection, this is believed to play a relatively minor role.

17.2.2 The Role of Implant Design and Surgical Technique on Infection Susceptibility

Considering the wide variation in the clinical presentations of infection associated with fracture fixation implants, animal models are required to gain scientific data on the resistance to infection of a particular fracture fixation device or procedure under controlled conditions and with a minimum of confounding factors. Animal models of infection have been used to compare infection susceptibility of numerous implants over the past decades as each generation of fracture fixation implants have been developed [10, 30–32].

The design of fracture fixation devices has evolved to allow a favorable compromise between protection of the local soft and hard tissues, commonly described as the biology, and correct anatomical and biomechanical fixation [9, 12]. One of the predecessors of the modern internal fixation plate was the dynamic compression plate (DCP, Fig. 17.2), which acted by compressing anatomically reduced fracture fragments together allowing rigid fixation upon application and direct bone healing. This technique relies on friction between plate and bone and therefore results in some compression necrosis of the interposed periosteum. With time, fixation theory progressed from compression plating to locked plates. These limited contact plates offer more protection to the periosteal tissues (Fig. 17.2) [9]. The infection resistance of these implants was also assessed in one of the first animal studies of its kind: a rabbit model of infection was used to compare the resistance to infection of the DCP with the newer point contact fixation plate (PC-Fix) [31]. Despite the fact that infection resistance was not a primary design goal of the PC-Fix, it was found to display

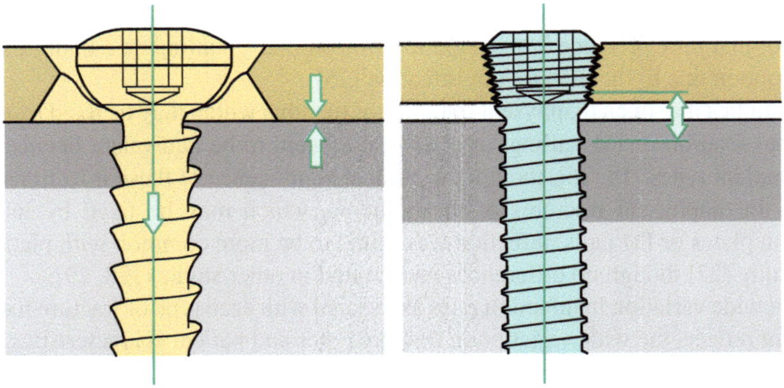

Fig. 17.2 Dynamic compression plate and screw (*Left*) acting to compress the plate across the fracture and (*Right*) a locking compression plate with fixation that protects the underlying bone. Copyright by AO Foundation, Switzerland

more than tenfold increase in resistance to infection in comparison with its DCP predecessor [31]. Clinical corroboration of this data is unfortunately unavailable; however, there is now widespread clinical uptake of the locking concept.

IM nail design characteristics have also been shown to influence infection rates in animal models of infection. For example, in a rabbit model of infection comparing solid, hollow, or cannulated nails the infection resistance of each was determined [33]. The hollow and slotted nails were more susceptible to infection than the solid nail, which was attributed to the presence of an avascular dead space in the hollow nails [33]. Such dead spaces where circulation does not reach seemingly provide a favorable niche for bacterial colonization and propagation of infection.

The intramedullary canal may also be reamed prior to nail insertion to provide optimal fit of the nail and optimal fixation of the fracture. Reaming of the cavity causes tissue destruction and may theoretically also influence infection rate. In an animal study, it was found that reamed nails had a small but significant increase in infection rate [34]. In the case of reamed versus unreamed IM nailing, there is some clinical data to compare with the animal model data, with clinical reports of infection after IM nailing of tibial fractures not significantly different between reamed and unreamed nails [35, 36]. The discrepancy between the outcomes from the animal model data and the clinical case reports may be due to small patient numbers or patient variability in the clinical studies, or the accepted limitations in extrapolating animal model data to the clinical situation.

More recent advances in implant design have focused upon minimally invasive surgical techniques such as minimally invasive osteosynthesis (MIO) whereby the surgical procedure limits the surgical wound as much as possible. Whilst this in theory reduces the open wound in size, this may be offset by increased blunt disruption of tissue while manipulating the implant into place. Further investigations into such matters are expected to be provided by clinical reports as minimally invasive procedures gain popularity. There are, however, some early indications that infection rate may be increased due to the minimal invasive approach for total joint replacement surgery [37], though further clinical observations are required for definitive assessment of this risk.

17.2.3 The Role of Implant Material on Infection Susceptibility

Metals, primarily titanium and its alloys and stainless steel, are the materials of choice for the fabrication of most fracture fixation devices. Metals display high stiffness, good strength, good ductility, and are usually well tolerated in terms of the host response to these foreign bodies [38]. Titanium and its alloys have a lower modulus of elasticity than steel, giving titanium greater flexibility and thus do not shield the bone from any stresses that may impact upon the biomechanics of the healing fracture [39, 40]. Furthermore, owing to this lower elastic modulus, titanium has increased fatigue resistance compared with electropolished stainless steel (EPSS) [38–40]. One additional important difference between EPSS and titanium is the tissue reaction at the implant–tissue interface. Standard titanium and its alloys are observed to support direct anchorage to bone, termed osseointegration, whilst a continuous one or two cell thick fibrous tissue layer is commonly found to separate EPSS devices from bone [41, 42]. This difference is believed to be due to a combination of the obvious differences in surface chemistry between the metals, but also the topographical differences between stainless steel which is usually electropolished to a very smooth surface, and titanium which has a microrough surface as standard. Preclinical animal studies have clearly shown that fibrous encapsulation of EPSS implants is more common than that of titanium implants Fig. 17.3 [41, 42]. The creation of a fibrous capsule surrounding an implant has important implications for the removal of the affected implant; however, considering the obvious and significant effect on cellular response, there is also obviously a potential impact on infection susceptibility.

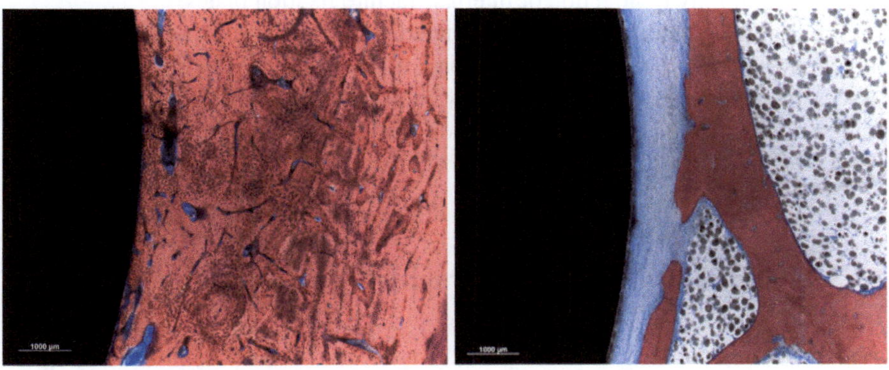

Fig. 17.3 Tissue contact adjacent to (L) standard titanium alloy IM nails after 12 months implantation in sheep tibia showing direct bone contact with the nail and (R) Electropolished stainless steel nail showing a fibrous tissue layer between the implant and the bone. Image courtesy of Jessica Hayes, AO Research Institute Davos

There is some preclinical [43] and clinical data [44, 45] comparing infection rates associated with implants composed of steel in comparison with implants composed of titanium. In general, there is a small trend towards reduced infection rates for titanium percutaneous pins [44] and spinal implants [45]. There is no clear evidence as to the reason for this difference, although it is commonly attributed to the superior biocompatibility of titanium over steel and may be due to the closer integration of bone to titanium metal. Preclinical studies have been performed to specifically determine the difference between stainless steel and titanium with regard to infection susceptibility. Hauke and colleagues tested the difference in infection susceptibility between IM nails composed of EPSS with those composed of titanium alloy. They showed in a rabbit study that titanium had a higher infection resistance compared with steel [32]. In a subsequent study Arens and colleagues also demonstrated that titanium DCPs had a higher infection resistance compared with steel DCPs in rabbits [30]. The outcome of these studies indicates superiority of titanium over steel; however, at this point it still remained unclear whether this is due to the chemical or topographical differences between microrough titanium and EPSS. We have more recently shown in a rabbit model using locked plates that infection susceptibility was equal between EPSS to titanium and titanium alloy as standard or as polished equivalents, indicating that neither implant material nor surface topography influence infection resistance in models with locked plates and minimal tissue trauma [46]. These results may appear to be in conflict with earlier study by Arens et al. who found that implant material does influence infection susceptibility [30], whilst in our more recent study, implant material was not found to impact upon infection susceptibility [46]. The crucial difference in the two studies and one which somewhat explains this apparent conflict surrounds the design of the implant. DCPs were used in the study of Arens et al. [30], and LCPs were used in the more recent study [46]. It had already been shown that the limited contact approach to fixation offered significantly greater protection from infection [31] and in the more recent LCP study it seems that this design improvement overrides the material influence observed to a point where implant material was no longer a factor in the susceptibility to infection of internal fracture fixation plates.

The data thus indicates that implant material may play a role in susceptibility to infection; however, the mechanical impact of the implant construct on the local tissues may override any comparatively minor impacts of implant material. To achieve a significant impact upon infection susceptibility and in particular in high-risk cases, active antimicrobial strategies are required.

17.3 Antibiotic Coatings on Fracture Fixation Devices

The tissue–implant interface is especially prone to contamination in all surgical procedures that require the implantation of a medical device [47–49]. Systemic prophylactic antibiotics are routinely given in orthopedic and trauma surgery to prevent implant-related infection, however, due to the disturbance of the bony structure and compromised local vasculature, appropriate local antibiotic concentrations might not be achieved by intravenously administered antibiotics. The use of locally applied antibiotic delivery vehicles is an accepted means to achieve the goal of killing contaminating bacteria before a biofilm has a chance to form. These biofilms are particularly important because once adherent on the surface of a biomaterial, bacteria produce an extracellular polymeric substance (EPS) [48, 50] culminating in an antibiotic- and host defense-resistant biofilm that makes eradication of the infection difficult [51]. Using a local delivery vehicle also achieves higher local concentrations and avoids the toxicity risk associated with high doses or prolonged exposure to systemic antibiotics. Several strategies for local antibiotic prophylaxis have been attempted in orthopedic surgery including antibiotic-loaded bone cement, polymethylmethacrylate (PMMA), as beads or as spacers, or antibiotic-impregnated collagen sponges and calcium sulfates [50, 52–54].

Certain aspects need to be considered if a successful local prophylaxis is to be achieved: Bactericidal antibiotics should be favored over bacteriostatic ones and the use of degradable drug carriers is of advantage as secondary surgery for removal is not necessary. The technique of delivery must guarantee a rapid release of the antibiotic from the carrier and local drug levels should be well above the minimal inhibitory concentration (MIC) of the most prevalent organisms. The drug release must also be restricted to a limited period of time to prevent development of antibiotic resistance.

Gentamicin-loaded PMMA beads are one of the most commonly used local antibiotic delivery vehicles in the treatment of infections associated with fracture fixation devices and prosthetic joint surgery. There is some clinical evidence that although loaded with gentamicin, PMMA beads used in septic joint prosthesis revision surgery can harbor bacterial colonies and that a high percentage of these may display increased resistance to gentamicin [55, 56]. This has raised concerns regarding the possibility of emergence of antibiotic-resistant bacteria due to use of antibiotic loaded cement beads. In vitro studies [57, 58] have also corroborated this data, for example one study has shown that PMMA loaded with gentamicin is susceptible to biofilm formation [58]. In prosthetic joint arthroplasty, the lack of widespread clinical reports of gentamicin-resistant bacteria colonizing PMMA spacers and the reduced revision rates when gentamicin-loaded spacers have been used has resulted in a favorable attitude to their clinical use in general [59], though conflicting opinions are also observed [60].

17.3.1 First Generation Gentamicin Eluting Devices and Their Drawbacks

The well-established use of antibiotic-loaded bone cement in infected fractures and prosthetic joint septic revision surgery prompted the off-label fabrication of antibiotic cement-coated intramedullary nails in operating rooms in a bid to provide a simple method of achieving high local concentrations of antibiotics [61–63]. This may be performed intra-operatively and involves coating conventional intramedullary nails with PMMA bone cement containing gentamicin or another heat-stable antibiotic. Once coated, the nail is passed through a sizer to achieve a smooth finish with a uniform diameter. In addition to the stabilization of the surgical site and local delivery of antibiotics, the coated nail occupies the dead space and facilitates ease of nail replacement after eradication of infection [62]. The indications for such devices have included infected nonunions, chronic osteomyelitis, intramedullary infections after primary nailing or bone transport, and as a temporary spacer for metaphyseodiaphyseal bone loss in the revision of a chronically infected total hip replacement [63]. Their use as a bridge to definitive intramedullary nailing after spanning external fixation has been suggested though not reported to date [61]. However, use of PMMA as a carrier has drawbacks. Being non-resorbable it requires a second procedure for implant removal, up to 70% of the antibiotic dose is unavailable as it is permanently sequestered in the PMMA and being a foreign body it can be colonized by bacteria after the elution is complete [56, 58, 64]. Despite the simplicity and ready availability, there has not been widespread clinical use of these devices.

17.3.2 Gentamicin-Coated Intramedullary Nail

Considering the problems with the off-label use of PMMA bone cement as a local delivery vehicle in intramedullary nails, it was aimed to develop a specific custom-made local drug delivery system suitable for trauma cases and the fixation of fractures. Gentamicin was chosen as the antibiotic of preference as it has been used successfully as a locally applied antibiotic in orthopedic surgery [65–68], it has a broad antimicrobial spectrum, covering most bacteria commonly involved in osteomyelitis, and displays bactericidal activity even on nonproliferating microorganisms [67, 69]. All these factors make gentamicin favorable for local application. The gentamicin is delivered from a biodegradable poly(D,L–Lactide) coating on titanium alloy implants [70]. Titanium alloy was chosen as it is the material of preference in the manufacture of intramedullary nails. The PDLLA coating degrades by hydrolysis within 3–6 months of implantation and the degradation products are metabolized in the citric acid cycle [71–75]. The coating provides mechanical, abrasion-resistant properties suitable for the needs of intramedullary nailing [74], i.e., the coating withstands the mechanical forces present when the implant is inserted. The total loss of the PDLLA coating during insertion of coated femoral stem prosthesis into human cadaveric femora was determined to be less than 5% [76].

17.3.2.1 In Vitro Characterization

The antibiotic release profile of the PDLLA coating with gentamicin was first investigated *in vitro* on titanium implants. For the *in vitro* antibiotic release experiments, titanium implants were coated with PDLLA/gentamicin and the antibiotic release in

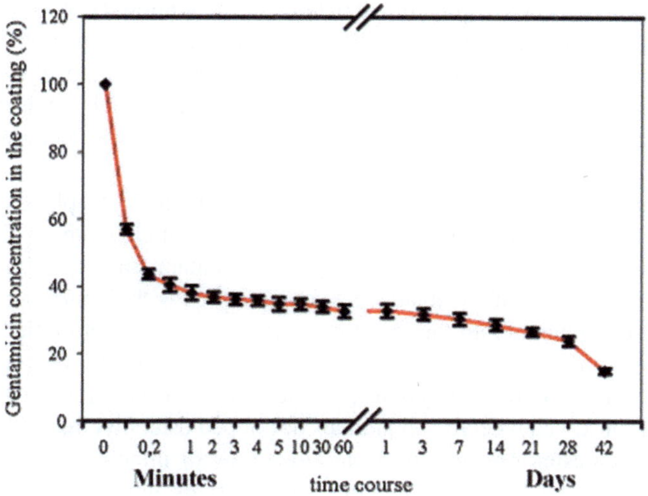

Fig. 17.4 In vitro gentamicin release over 42 days

aqueous solution was analyzed over time (up to 42 days). Bactericidal efficacy of the eluate was determined by biological and chemical assays (zone of inhibition and fluorescence polarization immuno-assay). Additional studies were also performed to determine bacterial adhesion and viability on the surface of the coating, and antibiotic resistance development upon exposure to the surface [77].

Gentamicin release from the coating displays a rapid initial burst release with 60% of the incorporated antibiotic released within the first minute (Fig. 17.4). A reduced yet constant release continued until approximately 3 weeks, at which time an additional 10% was eluted and after 6 weeks, a total of about 85% of the antibiotic was eluted from the coating. No resistance development was observed, and significantly fewer viable bacteria were attached to the implant in comparison with controls Fig. 17.4 [77].

In addition, in order to assess influence of the gentamicin coating on bone cells, alkaline phosphatase activity, protein expression, cell count, and Collagen Type I carboxyterminal propeptide (CICP) expression of osteoblasts were analyzed after culture in contact with the coated implants. Osteoblasts were not negatively affected by the gentamicin released out of the coating, with gentamicin-coated implants resulting in results that were similar to uncoated and empty vehicle-coated implants. The *in vitro* results indicated that this coating seemed to be a suitable option for local prophylaxis of implant-associated infections [77].

17.3.2.2 Preclinical In Vivo Efficacy

In a subsequent supportive preclinical *in vivo* study, Titanium K-wires coated with PDLLA containing 10% gentamicin were inserted as IM nails into rat tibia and *in vivo* pharmacokinetics of gentamicin in bone were monitored over 1 week. A peak gentamicin level of approximately 50 μg/g bone was achieved after 1 h, which reduced by 50% after 6 h and then remained more or less constant over the remainder of the first 24 h. The concentration of gentamicin within the rat bone decreased continuously and after 7 days the gentamicin concentration was determined to be 7.8 ± 5.04 μg/g [77].

In a subsequent infection model, the efficacy of the local prophylaxis provided by the PDLLA gentamicin coating was compared to a single systemic dose of gentamicin and to a combination of systemic plus the gentamicin-coated implants [78]. The medullary cavities of rat tibiae were contaminated with *S. aureus* and titanium K-wires were implanted into the medullary canals. For local antibiotic therapy, the implants were coated with PDLLA loaded with gentamicin. In the control group, which did not receive local or systemic application of the antibiotic, a 100% infection rate was observed whereby all animals developed osteomyelitis and all cultures of the implants tested positive for *S. aureus.* Onset of infection was prevented in 80% of animals treated with gentamicin-coated K-wires, with systemic prophylaxis and in 90% of animals without systemic prophylaxis Fig. 17.5 [78]. These *in vivo* studies show that under the conditions of the experiment, that local application of gentamicin delivered from a PDLLA coating on titanium K-wires was effective in prophylaxis of osteomyelitis.

17.3.2.3 Clinical Evaluation

Leading on from the promising results from the experimental *in vitro* and *in vivo* studies, a small early clinical trial was performed. Eight patients with open tibial fractures were treated with PDLLA and gentamicin-coated Universal Tibial Nail (UTNProtect, Synthes) in a first pilot study. Patients either received the coated nail as initial treatment ($n = 5$) or as a secondary stabilization after initial stabilization was provided by an external fixator ($n = 2$). One patient underwent a bone transport using the coated nail. All patients were followed up for at least 1 year, with X-rays taken for documentation of bone healing and signs of bone infection. Furthermore, soft tissue reactions were scored and serum infection parameters and systemic gentamicin levels analyzed. None of the treated patients developed any signs of infection on X-ray examination, soft-tissue reactions, or systemic infection parameters. All fractures consolidated within 6 months. None of the patients underwent a secondary intervention within the consolidation time. Gentamicin was not detected in any of the investigated patients' serum [70].

The UTN PROtect® IM nail was CE-certified in August 2005 and in a prospective, nonrandomized case series, the clinical, laboratory, and radiological outcomes of 21 patients who underwent surgical treatment for closed or open tibial fractures, as well as revisions with the UTN PROtect® gentamicin-coated intramedullary nail were investigated [79]. Nineteen of 21 patients (13 men, 8 women) completed the 6-month follow-up. Clinical outcomes measured adverse events, including infections and the short form 36 (SF-36) physical score. Laboratory outcomes, including

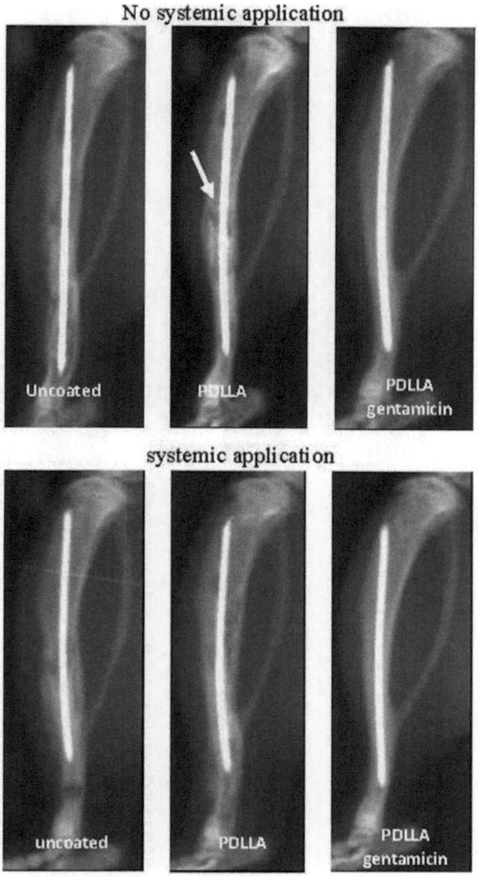

Fig. 17.5 Radiographic evaluation showing development and progression of bone infection in a rat infection model after 6 weeks. Clear radiographic signs of osteomyelitis were seen in all animals not treated with local or systemic application of gentamicin. Animals treated only with a systemic application of gentamicin showed nearly the same radiographic appearances. In animals with gentamicin-coated implants radiographic signs of infection were almost absent in all animals

C-reactive protein and leukocyte count, inflammatory markers, hemoglobin, and serum gentamicin, were measured at baseline and up to 6 months postoperatively. Radiographic assessments of fracture healing and weight-bearing capacity were determined at 5 weeks and 3 and 6 months after surgery.

In this study no implant-related infections occurred with one patient displaying superficial wound healing problems. Mean C-reactive protein levels remained below 5 mg/L throughout the study, with a peak at 4–7 days after surgery (4.4 mg/L; range 0.5–16.1 mg/L). Leukocyte counts and hemoglobin levels did not vary over time during the study. The mean SF-36 physical score at 6 months was 42.6 (range 19.4–56.7). Radiographic union, defined as three or four bridged cortices, was achieved in 11 patients (58%) after 6 months. The remaining eight patients showed partial fracture healing with one or two bridged cortices. Additionally, 13 patients (68%) demonstrated full weight-bearing capacity after 6 months. These first clinical uses of the gentamicin-coated intramedullary nail were associated with good clinical, laboratory, and radiological outcomes after 6

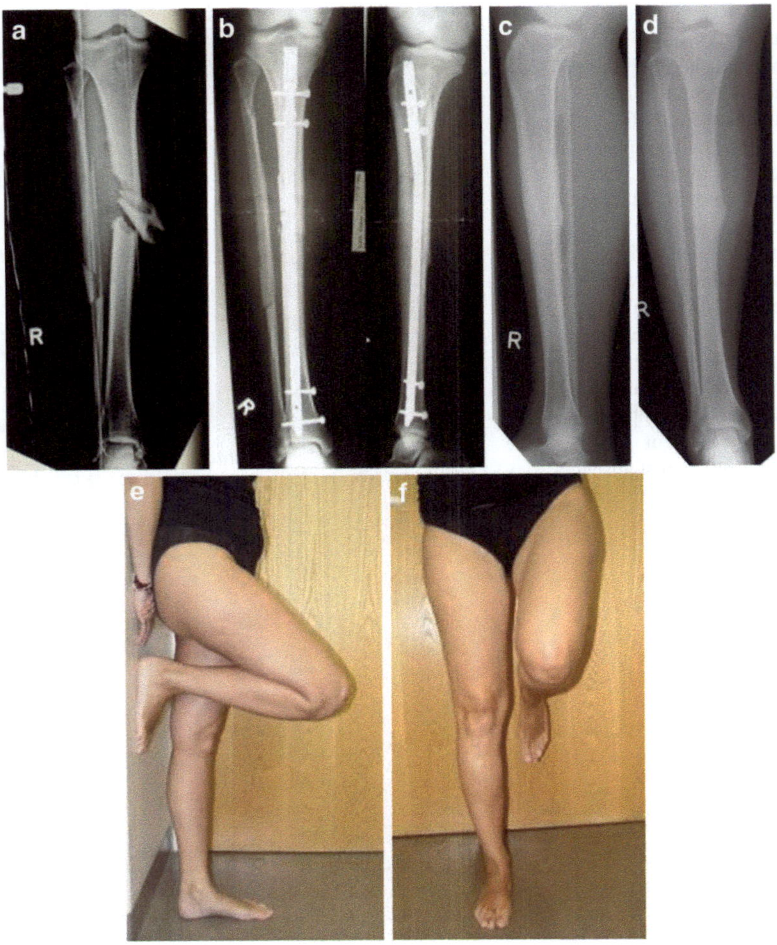

Fig. 17.6 Thirty-four-year-old woman with a 2° open lower leg fracture. (**a**) preoperative X-ray, (**b**) 4 weeks after stabilization with PDLLA-gentamicin-coated nail. (**c** and **d**) X-ray 1 year after fracture after implant removal and (**e** and **f**) Clinical outcome after implant removal

months. These preliminary results support the use of gentamicin-coated implants as a new potential option for the prevention of infection in trauma patients and in revision cases [79].

In January 2011 the CE-certified Expert Tibia Nail Protect (ETNProtect, Synthes), which is a newer generation IM nail, with the same coating technology, was introduced in the clinic. A prospective open-label post-market clinical follow-up multicenter study was started to evaluate the clinical usefulness of the operative stabilization of closed and open diaphyseal fractures of the tibia using the newer generation nail Fig. 17.6. This study is still in progress and results are as yet not available.

17.4 Conclusions

The combined impacts of a traumatic injury to soft and hard tissues, surgical fixation of a fracture with an implanted metal fixation device, the mechanical forces acting on the healing bone, and finally the presence of contaminating bacteria is a complex interaction that ultimately results in infection in a certain percentage of cases despite best clinical practice. The primary risk factors for the development of an infection are the extent of the initial trauma, anatomical location of the fracture, extent of soft and hard tissue damage, whether there is an open or a closed wound, and any patient comorbidities. Sufficient wound debridement and appropriate antibiotic prophylaxis are required to minimize the incidence of early onset perioperative infections. The surgical approach, the fixation device used and further, the design of the chosen fracture fixation device and its application can influence the susceptibility to infection in addition to the obvious crucial role in fracture healing. Passive strategies such as protection of periosteum and minimization of dead spaces have been shown in animal studies to impact upon infection resistance and should be considered important in routine cases. However, in high-risk cases, or for those with established infection, active strategies are required. Preclinical and clinical data have shown that local delivery of gentamicin from IM nails is a useful addition to the armamentarium of the practicing orthopedic surgeon in the treatment of cases at a high risk for infection.

Glossary

Biological internal fixation A technique of careful surgical exposure, fracture reduction, and fixation, which favors the preservation of the blood supply of the fracture site and, thereby, optimizes the healing potential of the bone and soft tissues.

Dynamic compression plate (DCP) A plate through which screws can be inserted to provide compression across a fracture site.

External fixation Skeletal stabilization using pins, wires, or screws that protrude through the skin and are linked externally by bars or other devices.

References

1. Boxma H, Broekhuizen T, Patka P, Oosting H. Randomised controlled trial of single-dose antibiotic prophylaxis in surgical treatment of closed fractures: the Dutch Trauma Trial. Lancet. 1996;347(9009):1133–7.
2. Gustilo RB, Gruninger RP, Davis T. Classification of type III (severe) open fractures relative to treatment and results. Orthopedics. 1987;10(12):1781–8.
3. Leutenegger AF. Acute infection following osteosynthesis. Ther Umsch. 1990;47(7):593–6.

4. Trampuz A, Zimmerli W. Diagnosis and treatment of infections associated with fracture-fixation devices. Injury. 2006;37 Suppl 2:S59–66.
5. Willenegger H, Roth B. Treatment tactics and late results in early infection following osteosynthesis. Unfallchirurgie. 1986;12(5):241–6.
6. Edwards C, Counsell A, Boulton C, Moran CG. Early infection after hip fracture surgery: risk factors, costs and outcome. J Bone Joint Surg Br. 2008;90(6):770–7.
7. Poultsides LA, Liaropoulos LL, Malizos KN. The socioeconomic impact of musculoskeletal infections. J Bone Joint Surg Am. 2010;92(11):e13.
8. Harris LG, Richards RG. Staphylococci and implant surfaces: a review. Injury. 2006;37 Suppl 2:S3–14.
9. Perren SM. Fracture healing. The evolution of our understanding. Acta Chir Orthop Traumatol Cech. 2008;75(4):241–6.
10. Schlegel U, Perren SM. Surgical aspects of infection involving osteosynthesis implants: implant design and resistance to local infection. Injury. 2006;37 Suppl 2:S67–73.
11. Stein H, Perren SM, Mosheiff R. New insights into the biology of fracture healing. Orthopedics. 2004;27(9):915–8.
12. Clifford RP. Open fractures. In: Rüedi TP MW editor. AO principals of fracture management. Stuttgart New York: AO Publishing Thieme-Verlag; 2000. p. 621–43.
13. Wiss DA, Stetson WB. Unstable fractures of the tibia treated with a reamed intramedullary interlocking nail. Clin Orthop Relat Res. 1995;315:56–63.
14. Henley MB, Chapman JR, Agel J, Harvey EJ, Whorton AM, Swiontkowski MF. Treatment of type II, IIIA, and IIIB open fractures of the tibial shaft: a prospective comparison of unreamed interlocking intramedullary nails and half-pin external fixators. J Orthop Trauma. 1998;12(1):1–7.
15. Esterhai Jr JL, Queenan J. Management of soft tissue wounds associated with type III open fractures. Orthop Clin North Am. 1991;22(3):427–32.
16. Gustilo RB, Mendoza RM, Williams DN. Problems in the management of type III (severe) open fractures: a new classification of type III open fractures. J Trauma. 1984;24(8):742–6.
17. Noumi T, Yokoyama K, Ohtsuka H, Nakamura K, Itoman M. Intramedullary nailing for open fractures of the femoral shaft: evaluation of contributing factors on deep infection and nonunion using multivariate analysis. Injury. 2005;36(9):1085–93.
18. Bell MJ, Beauchamp CG, Kellam JK, McMurtry RY. The results of plating humeral shaft fractures in patients with multiple injuries. The Sunnybrook experience. J Bone Joint Surg Br. 1985;67(2):293–6.
19. Dabezies EJ, Banta CJ, Murphy CP, D'Ambrosia RD. Plate fixation of the humeral shaft for acute fractures, with and without radial nerve injuries. J Orthop Trauma. 1992;6(1):10–3.
20. Heim D, Herkert F, Hess P, Regazzoni P. Surgical treatment of humeral shaft fractures—the Basel experience. J Trauma. 1993;35(2):226–32.
21. Antoci V, Ono CM, Antoci Jr V, Raney EM. Pin-tract infection during limb lengthening using external fixation. Am J Orthop (Belle Mead NJ). 2008;37(9):E150–4.
22. Egger EL. Complications of external fixation. A problem-oriented approach. Vet Clin North Am Small Anim Pract. 1991;21(4):705–33.
23. Green SA. Osteomyelitis. The Ilizarov perspective. Orthop Clin North Am. 1991;22(3):515–21.
24. Harari J. Complications of external skeletal fixation. Vet Clin North Am Small Anim Pract. 1992;22(1):99–107.
25. Della Rocca GJ, Crist BD. External fixation versus conversion to intramedullary nailing for definitive management of closed fractures of the femoral and tibial shaft. J Am Acad Orthop Surg. 2006;14:S131–5. 10 Spec No.
26. Tornetta III P, Bergman M, Watnik N, Berkowitz G, Steuer J. Treatment of grade-IIIb open tibial fractures. A prospective randomised comparison of external fixation and non-reamed locked nailing. J Bone Joint Surg Br. 1994;76(1):13–9.
27. Changulani M, Jain UK, Keswani T. Comparison of the use of the humerus intramedullary nail and dynamic compression plate for the management of diaphyseal fractures of the humerus. A randomised controlled study. Int Orthop. 2007;31(3):391–5.

28. Chapman JR, Henley MB, Agel J, Benca PJ. Randomized prospective study of humeral shaft fracture fixation: intramedullary nails versus plates. J Orthop Trauma. 2000;14(3):162–6.
29. Denies E, Nijs S, Sermon A, Broos P. Operative treatment of humeral shaft fractures. Comparison of plating and intramedullary nailing. Acta Orthop Belg. 2010;76(6):735–42.
30. Arens S, Schlegel U, Printzen G, Ziegler WJ, Perren SM, Hansis M. Influence of materials for fixation implants on local infection. An experimental study of steel versus titanium DCP in rabbits. J Bone Joint Surg Br. 1996;78(4):647–51.
31. Arens S, Eijer H, Schlegel U, Printzen G, Perren SM, Hansis M. Influence of the design for fixation implants on local infection: experimental study of dynamic compression plates versus point contact fixators in rabbits. J Orthop Trauma. 1999;13(7):470–6.
32. Hauke C, Schlegel U, Melcher GA, et al. Local infection in relation to different implant materials. An experimental study using stainless steel and titanium, unlocked, intramedullary nails in rabbits. Orthop Trans. 1997;21:835–6.
33. Horn J, Schlegel U, Krettek C, Ito K. Infection resistance of unreamed solid, hollow slotted and cannulated intramedullary nails: an in-vivo experimental comparison. J Orthop Res. 2005;23(4):810–5.
34. Melcher GA, Metzdorf A, Schlegel U, Ziegler WJ, Perren SM, Printzen G. Influence of reaming versus nonreaming in intramedullary nailing on local infection rate: experimental investigation in rabbits. J Trauma. 1995;39(6):1123–8.
35. Finkemeier CG, Schmidt AH, Kyle RF, Templeman DC, Varecka TF. A prospective, randomized study of intramedullary nails inserted with and without reaming for the treatment of open and closed fractures of the tibial shaft. J Orthop Trauma. 2000;14(3):187–93.
36. Keating JF, O'Brien PJ, Blachut PA, Meek RN, Broekhuyse HM. Locking intramedullary nailing with and without reaming for open fractures of the tibial shaft. A prospective, randomized study. J Bone Joint Surg Am. 1997;79(3):334–41.
37. Mahomed NN, Barrett JA, Katz JN, et al. Rates and outcomes of primary and revision total hip replacement in the United States medicare population. J Bone Joint Surg Am. 2003;85-A(1): 27–32.
38. Hayes JS, Richards RG. The use of titanium and stainless steel in fracture fixation. Expert Rev Med Devices. 2010;7(6):843–53.
39. Disegi JA. Titanium alloys for fracture fixation implants. Injury. 2000;31 Suppl 4:14–7.
40. Pohler OE. Unalloyed titanium for implants in bone surgery. Injury. 2000;31 Suppl 4:7–13.
41. Hayes JS, Vos DI, Hahn J, Pearce SG, Richards RG. An in vivo evaluation of surface polishing of TAN intermedullary nails for ease of removal. Eur Cell Mater. 2009;18:15–26.
42. Hayes JS, Seidenglanz U, Pearce AI, Pearce SG, Archer CW, Richards RG. Surface polishing positively influences ease of plate and screw removal. Eur Cell Mater. 2010;19:117–26.
43. Ganser A, Thompson R, Tami I, Neuhoff D, Steiner A, Ito K. An in vivo experimental comparison of stainless steel and titanium Schanz screws for external fixation. Eur J Trauma and Emerg Surg. 2007;33(1):59–68.
44. Pieske O, Geleng P, Zaspel J, Piltz S. Titanium alloy pins versus stainless steel pins in external fixation at the wrist: a randomized prospective study. J Trauma. 2008;64(5):1275–80.
45. Soultanis KC, Pyrovolou N, Zahos KA, et al. Late postoperative infection following spinal instrumentation: stainless steel versus titanium implants. J Surg Orthop Adv. 2008;17(3):193–9.
46. Moriarty TF, Debefve L, Boure L, Campoccia D, Schlegel U, Richards RG. Influence of material and microtopography on the development of local infection in vivo: experimental investigation in rabbits. Int J Artif Organs. 2009;32(9):663–70.
47. Gristina AG, Costerton JW. Bacterial adherence to biomaterials and tissue. The significance of its role in clinical sepsis. J Bone Joint Surg Am. 1985;67(2):264–73.
48. Gristina AG, Oga M, Webb LX, Hobgood CD. Adherent bacterial colonization in the pathogenesis of osteomyelitis. Science. 1985;228(4702):990–3.
49. Stemberger A, Grimm H, Bader F, Rahn HD, Ascherl R. Local treatment of bone and soft tissue infections with the collagen-gentamicin sponge. Eur J Surg. 1997;578(Suppl):17–26.
50. Hettfleisch J, Schottle H. Local preventive antibiotic treatment in intramedullary nailing with gentamycin impregnated biomaterials. Aktuelle Traumatol. 1993;23(2):68–71.

51. Jansen B, Peters G. Foreign body associated infection. J Antimicrob Chemother. 1993;32(Suppl A):69–75.
52. Blaha JD, Calhoun JH, Nelson CL, et al. Comparison of the clinical efficacy and tolerance of gentamicin PMMA beads on surgical wire versus combined and systemic therapy for osteomyelitis. Clin Orthop Relat Res. 1993;295:8–12.
53. Letsch R, Rosenthal E, Joka T. Local antibiotic administration in osteomyelitis treatment—a comparative study with two different carrier substances. Aktuelle Traumatol. 1993;23(7):324–9.
54. Nelson CL, Hickmon SG, Harrison BH. Elution characteristics of gentamicin-PMMA beads after implantation in humans. Orthopedics. 1994;17(5):415–6.
55. Kendall RW, Duncan CP, Smith JA, Ngui-Yen JH. Persistence of bacteria on antibiotic loaded acrylic depots. A reason for caution. Clin Orthop Relat Res. 1996;329:273–80.
56. Neut D, van de Belt H, Stokroos I, van Horn JR, van der Mei HC, Busscher HJ. Biomaterial-associated infection of gentamicin-loaded PMMA beads in orthopaedic revision surgery. J Antimicrob Chemother. 2001;47(6):885–91.
57. Arciola CR, Campoccia D, Montanaro L. Effects on antibiotic resistance of Staphylococcus epidermidis following adhesion to polymethylmethacrylate and to silicone surfaces. Biomaterials. 2002;23(6):1495–502.
58. van de Belt H, Neut D, Schenk W, van Horn JR, van der Mei HC, Busscher HJ. Staphylococcus aureus biofilm formation on different gentamicin-loaded polymethylmethacrylate bone cements. Biomaterials. 2001;22(12):1607–11.
59. Dunbar MJ. Antibiotic bone cements: their use in routine primary total joint arthroplasty is justified. Orthopedics 2009;32(9).
60. Hanssen AD. Prophylactic use of antibiotic bone cement: an emerging standard—in opposition. J Arthroplasty. 2004;19(4 Suppl 1):73–7.
61. Bhadra AK, Roberts CS. Indications for antibiotic cement nails. J Orthop Trauma. 2009;23 Suppl 5:S26–30.
62. Mendicino RW, Bowers CA, Catanzariti AR. Antibiotic-coated intramedullary rod. J Foot Ankle Surg. 2009;48(2):104–10.
63. Qiang Z, Jun PZ, Jie XJ, Hang L, Bing LJ, Cai LF. Use of antibiotic cement rod to treat intramedullary infection after nailing: preliminary study in 19 patients. Arch Orthop Trauma Surg. 2007;127(10):945–51.
64. Montali A. Antibacterial coating systems. Injury. 2006;37 Suppl 2:S81–6.
65. Bauer TW, Schils J. The pathology of total joint arthroplasty. II. Mechanisms of implant failure. Skeletal Radiol. 1999;28(9):483–97.
66. Kuhlback B, Falck H, Tornroth T, Wallenius M, Lindstrom BL, Pasternack A. Renal transplantation in amyloidosis. Acta Med Scand. 1979;205(3):169–72.
67. Southorn PA, Plevak DJ, Wright AJ, Wilson WR. Adverse effects of vancomycin administered in the perioperative period. Mayo Clin Proc. 1986;61(9):721–4.
68. Taylor GJ, Bannister GC, Calder S. Perioperative wound infection in elective orthopaedic surgery. J Hosp Infect. 1990;16(3):241–7.
69. Eron LJ. Prevention of infection following orthopedic surgery. Antibiot Chemother. 1985;33: 140–64.
70. Schmidmaier G, Lucke M, Wildemann B, Haas NP, Raschke M. Prophylaxis and treatment of implant-related infections by antibiotic-coated implants: a review. Injury. 2006;37 Suppl 2:S105–12.
71. Hutmacher D, Hurzeler MB, Schliephake H. A review of material properties of biodegradable and bioresorbable polymers and devices for GTR and GBR applications. Int J Oral Maxillofac Implants. 1996;11(5):667–78.
72. Kulkarni RK, Pani KC, Neuman C, Leonard F. Polylactic acid for surgical implants. Arch Surg. 1966;93(5):839–43.
73. Kulkarni RK, Moore EG, Hegyeli AF, Leonard F. Biodegradable poly(lactic acid) polymers. J Biomed Mater Res. 1971;5(3):169–81.
74. Schmidmaier G, Wildemann B, Stemberger A, Haas NP, Raschke M. Biodegradable poly(D, L-lactide) coating of implants for continuous release of growth factors. J Biomed Mater Res. 2001;58(4):449–55.

75. Schmidmaier G, Wildemann B, Bail H, et al. Local application of growth factors (insulin-like growth factor-1 and transforming growth factor-beta1) from a biodegradable poly(D, L-lactide) coating of osteosynthetic implants accelerates fracture healing in rats. Bone. 2001;28(4):341–50.
76. Förster C, Schmidmaier G, Gollwitzer H, Stemberger A, Raschke M, Haas NP. Eine biodegradierbare (Poly(d;l-Laktid)-Beschichtung für Endoprothesen zur lokalen Applikation von Wirksubstanzen. Hefte z d Unfallchirug. 1999;275:219.
77. Vester H, Wildemann B, Schmidmaier G, Stockle U, Lucke M. Gentamycin delivered from a PDLLA coating of metallic implants: in vivo and in vitro characterisation for local prophylaxis of implant-related osteomyelitis. Injury. 2010;41(10):1053–9.
78. Lucke M, Wildemann B, Sadoni S, et al. Systemic versus local application of gentamicin in prophylaxis of implant-related osteomyelitis in a rat model. Bone. 2005;36(5):770–8.
79. Fuchs T, Stange R, Schmidmaier G, Raschke MJ. The use of gentamicin-coated nails in the tibia: preliminary results of a prospective study. Arch Orthop Trauma Surg. 2011;131(10): 1419–25. PMID: 21617934.

Chapter 18
The Indwelling Bladder Catheter: Attempts to Prevent Infection and the Development of Bacterial Biofilms

David Stickler and Roger Feneley

Abstract The Foley indwelling bladder catheter is the most commonly deployed prosthetic medical device. While this catheter provides a convenient way to drain urine from the bladder, it also provides easy access to the bladder for bacteria contaminating the skin insertion site. In addition the catheter undermines the basic antibacterial defenses within the lower urinary tract. As a result, catheter-associated urinary tract infections are the most common infections acquired by patients in healthcare facilities. Bacterial biofilms form readily on these catheters and play important roles in the pathogenesis of the conditions that complicate the care and seriously threaten the health of catheterized patients. This chapter reviews the many attempts that have been made to prevent infection and biofilm formation by incorporating antimicrobial agents such as silver, nitrofurazone, minocycline, rifampicin, other antibiotics and biocides into catheters. The failure of antimicrobial catheters to prevent the development of the particularly troublesome crystalline biofilms is discussed and the need explained for fundamental changes in the design of catheters.

18.1 Introduction

The urinary catheter has been an essential item of medical equipment for centuries, challenging the ingenuity of successive generations of clinicians in their search for an ideal tube in size, shape, and composition to drain the bladder through the urethra. Catheters were discovered in the ruins of the surgeon's

D. Stickler (✉)
Cardiff School of Biosciences, Cardiff University, Cardiff, Wales CF10 3AX, UK
e-mail: stickler@cardiff.ac.uk

R. Feneley
North Bristol NHS Trust, Southmead Hospital, Bristol, UK
e-mail: rogerfeneley@dial.pipex.com

T.F. Moriarty et al. (eds.), *Biomaterials Associated Infection: Immunological Aspects and Antimicrobial Strategies*, DOI 10.1007/978-1-4614-1031-7_18,

house at Pompeii (79 AD). Historic accounts record the use of catheters made from hard metals such as gold, silver, and bronze or soft materials in the form of hollow leaves from the onion plant, reeds or skins from animals. In 1937 Dr Robert Foley described a self-retaining indwelling bladder catheter manufactured from latex claiming it to be the “finally perfected catheter” which left “little or nothing to be desired.” Some 70 years later, the basic design of the catheter remains virtually unchanged and it is the most commonly deployed prosthetic medical device [1]. It is estimated that worldwide around 100 million are used annually.

Foley catheters are flexible tubes generally manufactured from either latex or a silicone polymer with shaft lengths for adults of either about 38 cm for use in males or about 22 cm for females and diameters ranging from 4 to 8 mm. Catheter sizes are expressed in terms of the Charrière (Ch) or French Gauge (FG) which is three times the external diameter measured in mm. The tip of the device, which is located in the bladder, has a pair of drainage eyes and the other end has a drainage funnel for connection to a urine collection bag. Alongside the drainage funnel is the inflation funnel fitted with a valve for injection of sterile water (5–30 ml) through a channel in the catheter wall, filling a balloon which is located just below the eye-holes. The inflated balloon ensures that the catheter is retained in the bladder. Cross sections of a catheter reveal the central lumen through which the urine flows and the inflation channel in the thick catheter wall (Fig. 18.1).

Indwelling Foley catheters are used for short-term (1–7 days), medium-term (7–28 days) or long-term (>28 days) drainage of the bladder and have a wide range of applications in modern medicine. In acute care they have important roles draining the obstructed bladder, facilitating drainage during postoperative recovery and monitoring urinary output from unconscious patients. In chronic care they are used for the long-term management of urinary retention and incontinence in the elderly and in patients disabled by strokes, spinal injury, or neuropathies such as multiple sclerosis. Prevalence studies indicate that around 10% of patients in hospitals and 5% of those in nursing homes or home care will be undergoing indwelling bladder catheterization [2–5]. Unfortunately, catheters pose serious threats to the health and welfare of many of these patients. As long ago as 1958 Paul Beeson was warning clinicians of the dangers of indwelling bladder catheterization. In an editorial entitled “The case against the catheter” he concluded that “… the decision to use this instrument should be made with the knowledge that it involves the risk of producing a serious disease which is often difficult to treat” [6].

In this chapter we review (a) the complications encountered in the care of catheterized patients, (b) the infections that result from catheterization, (c) the development of bacterial biofilms on the catheters and their role in pathogenesis, (d) attempts to prevent infections and biofilms by incorporating antimicrobial agents into catheters, and (e) the major problem of catheter encrustation by crystalline biofilms. In conclusion the need to fundamentally change catheter design is explained.

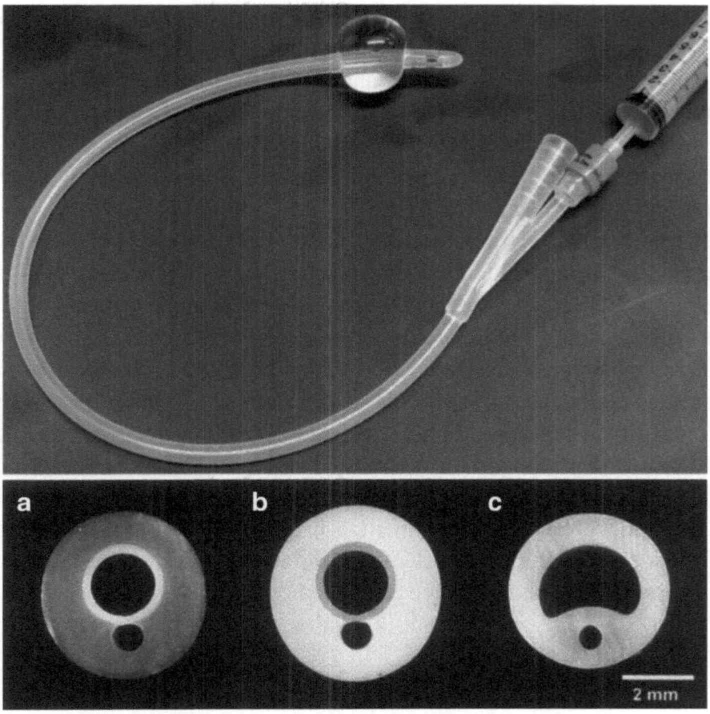

Fig. 18.1 An all-silicone Foley catheter with its retention balloon inflated (from Stickler DJ, Surface coatings in Urology, in *Coatings for Biomedical Applications* ed M. Driver 2012 ISBN 1845695682, Woodhead Publishing Cambridge, with permission). The bottom panel shows cross-sections of (**a**) a silicone-coated latex catheter, (**b**) a hydrogel-coated latex catheter and (**c**) an all-silicone catheter (from Stickler DJ et al. (2003) *Urological Research*, 31:306–311 with permission)

18.1.1 Complications in the Care of Patients Enduring Catheterization

The urethra provides the most common route for introducing a catheter into the bladder, being less invasive than the alternative suprapubic approach which requires a surgically created tract. Traumatic urethral catheterization however is a major cause of discomfort, pain, and urethral stricture [7, 8]. A review of short-term bladder drainage showed that the suprapubic route reduced the risk of infection, but this has to be set against the risks of suprapubic insertion [9]. In the longer term suprapubic catheterization offers advantages in terms of patient comfort and nursing care [10].

The continued presence of the catheter in the urethra can cause ulceration and septic lesions such as periurethral abscesses. Infection of the urinary tract is a common consequence of catheterization, the incidence depending on the duration of catheterization. Around 10–25% of patients undergoing short-term catheterization will acquire infection. For those enduring long-term catheterization infection is inevitable [11]. For most of the time the majority of these patients do not exhibit symptoms of infection and the condition is termed asymptomatic bacteriuria. They are at risk however from a range of infection-associated complications which can seriously compromise their health. A prospective study in nursing homes demonstrated significantly higher morbidity and mortality rates in patients undergoing long-term catheterization than in matched noncatheterized controls. Over a study period of 12 months catheterized patients were significantly more likely to have received antibiotics and to have spent more time in hospital. They were also three times more likely to be dead [12].

Catheter-associated infection becomes particularly life-threatening when bacteria in the bladder urine invade the kidneys and bloodstream causing pyelonephritis and septicemia. Surveys have revealed that 1–4% of hospital patients with bladder catheters will develop symptomatic bloodstream infections [13, 14]. Although this is a small percentage, catheterized patients are so numerous that these urinary tract infections are a major cause of nosocomial bacteremia [15]. In the case of long-term patients many of the serious complications are triggered by encrustation of the catheters. Crystalline deposits can form on the retention balloon, obstruct the drainage eyes and block the lumen. As a result urine either leaks around the outside of the catheter and patients become incontinent or urine is retained in the bladder causing its painful distension. Reflux of infected urine to the kidneys can then precipitate serious symptomatic episodes such as pyelonephritis, septicemia and endotoxic shock [16].

18.1.2 Catheter-Associated Urinary Tract Infections

While the Foley catheter provides a convenient way to drain urine from the bladder, it also gives the bacterial flora colonizing the skin insertion site easy access to the bladder. Unfortunately the catheter also undermines an important defense mechanism the bladder has against infection. The cycle of bladder filling and emptying normally ensures that bacterial cells contaminating the urethra and any managing to gain access to the bladder are flushed out of the lower urinary tract. This does not occur in the catheterized patient. The bladder does not fill, instead a sump of residual urine forms below the drainage eyes. Urine then trickles through the catheter into the drainage bag rather than flushing the urethra. This facilitates the migration of bacteria through the urethra and on arrival in the bladder they find themselves provided with a continually replenished reservoir of an excellent growth medium. Rapid bacterial multiplication results in urinary populations that commonly reach 10^7 cfu/ml [17, 18]. The numbers of patients being catheterized is so large that catheter-associated urinary tract infection is the most common of the infections acquired by patients in healthcare facilities [3, 19, 20].

In most cases infection results when organisms colonizing the periurethral skin migrate along the outside of the catheter into the bladder. Organisms can also pass into the bladder through the lumen of the catheter. This happens because the regular opening of the tap to drain urine from the bag can result in its contamination. Large urinary populations of bacteria can then develop in the bag and spread through the drainage tube and catheter to the bladder [11, 21]. Strategies for the development of new technologies to prevent these infections thus need to block both these major routes.

Single species such as *Staphylococcus epidermidis*, *Enterococcus faecalis*, or *Escherichia coli* usually produce the infections in the early stages of catheterization [22]. As time goes by however, a variety of species accumulate in the bladder and patients undergoing long-term catheterization become infected by complex mixed communities. Organisms such as *Klebsiella*, *Pseudomonas aeruginosa*, *Providencia stuartii*, *Morganella morganii*, and *Proteus mirabilis* become particularly common and stable additions to the urinary flora [17, 23]. These polymicrobial communities are very difficult to eliminate by antibiotic therapy while the catheter remains in place. The mere colonization of the bladder urine is not generally considered to be clinically significant. Because of this and the danger of promoting antibiotic resistance, it is common practice not to treat these infections with antibiotics unless clinical symptoms develop indicating that the kidneys or bloodstream have become involved [10]. In patients undergoing long-term catheterization, contaminated urine can thus be flowing through catheters for long periods (catheter changes commonly being scheduled at up to 10–12 week intervals).

Catheters provide attractive sites for bacterial colonization, cells adhere to the surfaces and biofilms form readily. The bacterial cells thrive embedded in their matrix gel and bathed in a gentle flow of warm nutritious urine. In patients who develop bacteriuria during short-term catheterization, colonization of the catheter does occur but the biofilms that form are generally sparse [24]. As the devices are removed within a day or so, they cause few problems. In contrast, on long-term catheters huge populations develop, often greatly outnumbering those present in the urine. These extensive biofilms become visible as thick coatings on the catheter surfaces and can have profound effects on the health of patients [25].

While catheterized patients with bacteriuria are generally asymptomatic for most of the time, they are at risk of a range of infection-associated complications which can severely threaten their health [26]. Many attempts have thus been made to prevent catheter-associated infections with apparently rational strategies to block the routes of infection with antibacterial agents [16]. The lesson has been that the more comprehensive the strategy to erect antibacterial barriers, the less effective the strategies have been in preventing infection. For example, a policy was devised which involved preparing the periurethral skin with chlorhexidine, instilling a gel containing the antiseptic into the urethra to lubricate the passage of the catheter into the bladder, daily cleaning of the catheter-metal junction with chlorhexidine solution, applying cream containing the antiseptic to the periurethral skin and instilling chlorhexidine into the urine drainage bags every time they were emptied. The result was that with all the conceivable routes of infection blocked, an extensive outbreak occurred with a chlorhexidine resistant multidrug resistant strain of *Pr. mirabilis* involving at least 90 patients. The problem was only resolved when the policy was abandoned [27].

In view of all the difficulties involved in preventing these infections using antibacterial agents in the daily care of catheterized patients and the complications induced by the formation of biofilms on catheters, it is not surprising that attempts have been made to prevent bacterial colonization and infection by incorporating an antibacterial into the catheters themselves. The first attempt at such a strategy was made by Butler and Kunin in 1968 [28]. These workers conducted a clinical trial to examine the efficacy of catheters impregnated with either tetra-methyl-thiuramdisulphide or with a proprietary agent of the cyclic thiohydroxamic group. The criterion of infection used in the study was bacteriuria (>10^5 cfu/ml). Over a 7 day observational period no significant differences were observed in the rates of bacteriuria in patients with control or the two antibacterial catheters. They also found that catheters removed from patients after 24–48 h had lost their antibacterial activity. This early attempt at infection control with antibacterial catheters thus highlighted one of the central problems of the strategy, i.e., maintaining antibacterial activity over the lifetime of the catheter.

The essential requirements for an infection resistant antimicrobial catheter material were listed by Guggenbichler et al. [29]. Such a material should possess a broad spectrum of antimicrobial activity and be resistant to colonization by microbial biofilms. The activity should endure for the lifetime of the device and should not be reduced by contact with body fluids. In addition it should not select for and spread resistance to antibiotics and other antimicrobials. The failure to produce catheters that achieve these aims in the years since Butler and Kunin's early attempts is testimony to the intrinsic difficulties involved achieving these aspirations. As a partially implanted device, the urinary catheter is constantly under threat from cross-contamination of its drainage system and from the rich microbial skin flora at the insertion site [30, 31]. Antimicrobial catheters thus have to deal with a constant threat from a diverse range of organisms including many multidrug resistant nosocomial pathogens.

In recent years it has become clear that the catheter most likely to succeed in preventing infection is one from which a broad-spectrum agent elutes into the surrounding fluid. The agent can then attack any contaminating organisms before they can reach the bladder, attach to the catheter surfaces and assume the antimicrobial resistant biofilm phenotype [32, 33]. To block the various routes of infection, the active agent should elute from both the external and internal surfaces of the catheter. It is difficult to conceive how an agent that is firmly incorporated into the catheter material and does not elute, could prevent organisms ascending the urethra in a film of mucus between the catheter and the urethral mucosa or in the columns of urine that form in the catheter and its drainage system. In addition, the rapid coating of implanted devices with conditioning films of urinary proteins and other organic molecules [24, 34] can mask the properties of a nondiffusing antibacterial material. Despite considerable research activity to try and produce an infection resistant antimicrobial catheter, few such devices have been developed to the stage at which they have been tested in clinical trials. Of these only two have become available for clinical use.

18.1.3 Silver Catheters

Silver is an attractive candidate for use in antimicrobial catheters because silver-coated or impregnated catheters satisfy many of the specifications of the "gold standard"

[35]. Silver has a low toxicity profile and has activity against a wide range of nosocomial organisms including antibiotic resistant staphylococci, enterococci, Gram-negative enteric bacilli, *Ps. aeruginosa*, and *Candida albicans* [36]. However, while silver-containing materials have been exploited successfully in several other prosthetic devices, their use in urinary catheters remains contentious [37]. In the case of Foley catheters the prerequisite for activity is the release of free silver ions at antibacterial concentrations from both the inner and outer surfaces. While doubts have been expressed about the activity of silver ions in urine due to neutralization by chloride ions and proteins [38] silver has been presented in several formats on prototype urinary catheters.

Several types of technology designed to deliver silver ions from catheters have been tested under laboratory conditions. Kumon et al. [39] for example, developed a silver coating for silicone catheters which contained a mixture of lecithin, silver citrate, and liquid silicone. The lecithin was designed to produce a hydrophilic surface and to modulate the release of silver ions. Experiments in laboratory models showed that this material resisted colonization by *P. aeruginosa* and inhibited the migration of *E. coli* over catheters. Guggenbichler et al. [29] and Samuel and Guggenbichler [40] described a technology which achieved the distribution of billions of submicron particles of metallic silver evenly through matrices of polyurethane and silicone. Interaction of electrolytes with the silver particles throughout the matrix resulted in the slow release of silver ions into an aqueous film close to the surface of the material. Catheters manufactured from this material released 0.5 μg of silver/24 h and it was calculated that continuous elution of silver from these materials could be expected for several months. An alternative way to deliver silver ions to urine is by applying direct electric currents to catheters that incorporate silver electrodes. This iontophoresis of silver has been shown not only to generate antimicrobial activity in urine [41] but also to inhibit crystalline biofilm formation on catheters over short periods in bladder models infected with *Pr. mirabilis* [42]. Despite these encouraging *in vitro* results none of these catheters seem to have undergone further development.

A silicone catheter coated with a thin layer of silicone elastomer containing micronized silver oxide was developed as far as clinical trials. Johnson et al. [43] reported the results of a trial on a catheter that had been coated on both internal and external surfaces with the silver oxide. A total of 482 unselected patients undergoing short-term (3–4 days) bladder catheterization in a general hospital, were involved. The incidence of urinary tract infection defined as two consecutive urine cultures with $>10^2$ cfu/ml of the same organism, or a single culture with $>10^5$ cfu/ml, was 10% in both the control and test groups. A later study [44] also demonstrated that silver oxide catheters had little if any protective effect.

The antimicrobial catheter that has undergone the most extensive testing in clinical trials is the hydrogel/silver alloy latex-based device that has been marketed by C.R. Bard as their Bardex IC catheter. In this device metallic silver is chemically anchored to the latex surface in a coating of gold and palladium, which apparently allows an extended slow release of silver ions from both the internal and external surfaces. The external hydrogel layer gives the catheter its lubricity and is claimed to facilitate the controlled release of the silver ions to exert their antibacterial activity in the surrounding fluids [45].

The initial clinical trials on these devices claimed dramatic reductions in the incidence of bacteriuria in patients undergoing short-term catheterization. In a letter to the *The Lancet* Lundeberg [46] reported that the incidence of bacteriuria (defined as >100 cfu/ml) was reduced from 34% in patients using a "standard catheter" to 12% in patients using the silver-coated catheters ($P<0.001$). A subsequent paper [47] reported the incidence of bacteriuria (this time defined as $>10^5$ cfu/ml) in patients randomized to receive either a Teflon-coated latex catheter or a silver alloy catheter. After 6 days, bacteriuria was reported in six of the sixty (10%) patients who had the silver catheter and 22 (37%) of the sixty who had the Teflon catheter ($P<0.01$). The third trial conducted by this group [48] investigated the incidence of bacteriuria ($>10^5$ cfu/ml) in patients randomly assigned to receive either a noncoated latex (NC) catheter, a hydrogel-coated latex (HC) catheter or a silver alloy hydrogel (SHC) catheter. Of the 30 patients in each group, 50% of the NC group, 33% of the HC group and 10% of those in the SHC group were bacteriuric after 5 days. Statistical analysis revealed a significant difference in these rates between the NC and SHC groups ($P<0.002$) but not between the SHC and HC groups ($P<0.06$).

Nine years passed before the next reports of clinical trials appeared as full papers in peer reviewed journals. Verleyen et al. [49] reported two studies on the efficacy of these catheters in which bacteriuria was defined as $>10^5$ cfu/ml. In the first of these, 27 male patients who had undergone prostatectomy were fitted with either a Bard IC catheter or a silicone control catheter. After 14 days there was no significant difference in the number of patients with bacteriuria in the two sets of patients (50.0% with the silver group and 53.3% in the silicone group). In the second trial, patients in a urology ward randomly received either a "classical" latex catheter (101 patients) or the Bard IC catheter (79 patients). It was reported that on day 5, only 6.3% of patients in the silver catheter group had bacteriuria compared to 11.9% in the latex catheter group ($P<0.003$). The choice of the control catheters in these two trials is curious. The hydrogel-coated latex catheter (the Bard Biocath) would have made a more rational control. Nevertheless, the authors' concluded that in the short-term, the silver-coated catheter significantly delayed the onset of bacteriuria compared to the basic latex catheter. Karchmer et al. [50] compared the infection rates in hospitalized patients given the hydrogel/silver alloy-coated or silicone-coated latex catheters. Wards rather than patients were randomized into two groups. One group of wards was supplied with the silver-coated catheters, the other with the silicone-coated catheters. After 6 months cross-over was instituted. Surveillance for nosocomial infections was performed using the Center for Disease Control criteria [51] which register both symptomatic and asymptomatic bacteriuria. The infection rates were 1.1% in the 13,945 patients fitted with the silver catheter, compared to 1.36% in the 13,933 having the silicone-coated catheter ($P=0.07$). When the data was expressed as infection rates per patient days, infections on the wards using silver catheters were calculated to be 2.66 infections per 1,000 patients days, compared to 3.35 per 1,000 days in the control wards ($P=0.04$). Secondary bloodstream infections complicated the care of 14 of the 343 infected patients. There was no significant difference in the rates of these infections in the two groups of patients ($P=0.42$).

Several groups have reported data indicating that the hydrogel/silver-coated latex catheter does not give protection against infection. Thibon et al. [52] for example, came to this conclusion from a randomized, prospective, double-blind, multicenter trial to compare the silver catheters with all-silicone standard catheters. Unlike many other studies, the two groups were comparable with no significant difference regarding sex, age, weight, antibiotic treatment at time of catheter insertion and duration of catheterization. Infection was defined as bacteriuria $>10^5$ cfu/ml with $>10^5$ leucocytes/ml. There was no significant difference in the infection rates in those using the silver catheter (10%) and the control catheter (11.9%).

Saint et al. [53] conducted a meta-analysis on the data from the trials on the silver catheters. They acknowledged that the studies based on poor methodology generally showed a greater benefit to an intervention than the more rigorous studies. They proceeded however to analyze the pooled data from all the published studies. They came to the tentative conclusion that the silver alloy catheters were significantly more effective than the silver oxide-coated devices and may be worth the extra cost in terms of preventing a common nosocomial infection. The Working Party [54] that provided guidelines for infection prevention in healthcare facilities in the Netherlands, were critical of this analysis, expressing the opinion that meta-analysis should be based on clinically homogeneous studies of high methodological quality. They assessed each study for five basic requirements of a good quality controlled clinical trial, awarding each a score of from zero to five. The two trials by Verleyen et al. [49] scored zero and those by Liedberg et al. [47, 48] scored 1, all being designated of poor quality. Only the trial conducted by Thibon et al. [52] was considered to be of good quality (scoring five). The fundamental problems with these trials led the working party to conclude that there was insufficient evidence to recommend the use of the silver alloy catheters. The subsequent literature has not resolved the controversy over the efficacy of the Bard IC catheter [55, 56].

In response to concerns over latex allergy, the silver alloy coating has also been applied to all-silicone catheters. The technical difficulties of binding the silver alloy to silicone surfaces were overcome by improved pretreatment and primer coating of the all-silicone catheters. A trial [57] of these silver/silicone catheters involved a 14 month baseline period, during which all-silicone catheters (Lubri-Sil, CR Bard) were used throughout the hospital. In a 10 month intervention period, silver alloy hydrogel-coated silicone catheters (Lubri-Sil IC, CR Bard) were substituted. The criterion of infection was bacteriuria $>10^5$ cfu/ml urine with no more than two bacterial species. In the silver group 116 cases of significant bacturia were recorded (rate of 9.6 per 100 catheters) and 218 in the nonsilver group (rate of 11.65 per 100 catheters). There was no significant difference in these rates ($P=0.15$). The median time to onset of UTI was 4 days in each group and the authors concluded that the silver-coated catheters did not prevent infection. In response Crnich and Drinka [58] have pointed out that the nature of the control catheters used in these studies might profoundly affect the conclusions. They observed that many studies showing a beneficial effect of the silver alloy catheters used latex based rather than silicone catheters as controls.

18.1.4 Nitrofurazone Catheters

Nitrofurazone is related to nitrofurantoin, a drug that has been used as a prophylactic and a treatment for catheter-associated urinary infections [59]. The chemical nature of nitrofurazone allows its combination with silicone to form a stable biomaterial which has been used by Rochester Medical Corporation (Stewartsville, Minn.) to coat silicone Foley catheters. Laboratory tests on these catheters were encouraging. Release of nitrofurazone was reported to last for at least 30 days. The activity of nitrofurazone was tested against isolates of a range of species that are commonly responsible for catheter-associated urinary tract infection [60]. Staphylococci, *E. coli*, enterococci and some isolates of *Citrobacter* and *Klebsiella* sp. were sensitive to nitrofurazone (MICs of 2–16 mg/L). Isolates of *Enterobacter*, *Serratia* sp., and *P. mirabilis* exhibited a degree of resistance (MICs 32–64 mg/L). *Pseudomonas* however was resistant with MICs >128 mg/L. This pattern of activity was also revealed when sections of the impregnated catheters were incubated on lawns of test species on agar. Large inhibition zones were produced around the catheter segments against most of the isolates. The pseudomonads and many of the strains of *Serratia*, *Proteus*, and *Enterobacter* were not inhibited by the catheter segments. As these resistant Gram-negative organisms are much more commonly associated with infections in long-term catheters, it was suggested that these nitrofurazone catheters would be more appropriate for short-term use. A subsequent laboratory study [61] compared the antibacterial activity of the nitrofurazone containing silicone catheter with that of the silver/hydrogel-coated latex catheter. The six species selected for study were *E. coli*, *K. pneumoniae*, *Citrobacter freundii*, *Staphylococcus aureus*, coagulase-negative staphylococci, and *Enterococcus faecium.* Recent multidrug resistant isolates of these species were tested against segments of the two types of catheter in zone of inhibition tests on agar. The nitrofurazone segments produced inhibition zones against all of the multidrug resistant isolates of the test species, with the exception of vancomycin-resistant *E. faecium.* A subset of strains were then tested against sections of both catheter types. Inhibition zones were visible around the silver catheters on lawns of *S. aureus* but not of *E. coli*, *K. pneumonia*, *C. freundii*, or *E. faecium.* The durability of the *in vitro* activity of the catheters was also assessed against the test organisms by daily serial transfer of catheter segments to fresh agar plates containing the test organisms. The silver catheters lost the little activity they had shown by day 2 whereas the nitrofurazone catheter material retained activity until day 5. Other workers have claimed that nitrofurazone is released from these catheters at greater than 50 mg/l for 44 days [62].

Johnson et al. [61] suggested that multidrug resistance among the range of important uropathogens was not likely to undermine the efficacy of the nitrofurazone catheter. They also argued that nitrofurazone resistance was unlikely to develop as a result of the use of these catheters but omitted to consider the possibility that these catheters would select for the intrinsically resistant *Pseudomonas, Proteus, Serratia,* and *Candida.* In conclusion they considered that nitrofurazone catheters were substantially more effective than the silver catheters at providing antibacterial activity. These claims for superior performance of the nitrofurazone catheters in laboratory tests have yet to be examined in clinical trials [63].

The efficacy of the nitrofurazone catheters compared to control silicone catheters has been studied in four clinical trials. Maki et al. [64] in a report which only appeared as a conference abstract, claimed that over periods of up to 7 days the nitrofurazone catheter reduced the incidence of bacteriuria (>10^3 cfu/ml) threefold. There was no significant reduction however in infections by *Candida.* Al-Habdan et al. [65] tested the ability of the nitrofurazone catheters to prevent bacteriuria (>10^5 cfu/ml) in postoperative orthopedic and trauma patients. In this case silicone-coated latex catheters were used for the control group. There were 50 patients in each group, six cases of bacteriuria were reported in the control group compared to none in the nitrofurazone group ($P=0.028$). A third trial [62] reported that over periods of up to 7 days, 16/38 patients in a control group acquired bacteriuria (>10^3 cfu/ml) compared to 12/58 in the nitrofurazone group, a difference that was significant ($P=0.026$). Scanning electron micrographs of catheters removed from patients after 7 days appeared to show less dense bacterial colonization on the nitrofurazone catheters. It was concluded that the nitrofurazone coating prevented infection by inhibiting the formation of biofilm on the catheter.

More recently Stensballe et al. [66] reported a randomized, double-blind controlled trial to determine the ability of the nitrofurazone catheters to prevent catheter-associated bacteriuria and funguria in short-term catheterization of trauma patients. The incidence of bacteruria and funguria (>10^3 cfu/ml) was less in the nitrofurazone group than in a silicone catheter group, the infection rates expressed as incidence per 1,000 catheter days, were 13.8 vs. 38.6 ($P=0.005$). The authors pointed out that a limitation of this study was that the clinical significance of asymptomatic bacteriuria and funguria is unclear. They also reported that the incidence of clinical signs of infection did not differ between groups and no cases of catheter related bacteremia were observed during the trial.

18.1.5 Minocycline and Rifampicin Impregnated Catheters

There are strong arguments in favor of using a combination of antimicrobial agents in the development of anti-infective catheters. Darouiche et al. [67] suggested that a combination of minocycline and rifampin could be ideal for this purpose. This combination has synergistic and broad-spectrum activity against almost all potential urinary pathogens. Neither of these antibiotics is generally used to treat urinary tract infections. Another factor that may be useful in preventing the development of resistance is that they have different modes of action, minocycline inhibiting bacterial protein synthesis and rifampicin inhibiting DNA-dependent RNA polymerase. These authors impregnated all-silicone Foley catheters with this antibiotic combination and conducted a clinical study to determine the durability of the antimicrobial activity. Impregnated catheters were passed via the urethra into the bladders of patients with spinal cord injury. Catheters were removed at various times up to 21 days and the residual antimicrobial activity was assessed by measuring the zones of inhibition produced by 1 cm segments of the catheters against lawns of test bacteria. The activity was shown to persist for 7 days against all of an impressive list of urinary tract pathogens. By day 14 however, there was little activity against *C. albicans* and by day 21 the segments had also lost activity against *E. coli*, *K. pneumoniae*, *E. cloacae*, and *Citrobacter diversus*. Activity against important species such as *P. aeruginosa*, *P. mirabilis*, and *E. faecalis,* however, was retained for the full 21 days.

Experiments in a laboratory model of the catheterized bladder showed that the migration of urinary pathogens along catheters was inhibited by the presence of these antibiotics [68]. A subsequent clinical trial of their ability to prevent bacteriuria (>10^4 cfu/ml) in 14 day test periods demonstrated significantly lower rates of infection at day 7 (15.2% vs. 39.7%) and day 14 (58.5% vs. 83.5%) [69]. This difference was due to a reduction of Gram-positive bacteriuria; there was no effect on Gram-negatives or yeasts. The authors concluded that while these catheters might be useful for the control of infection over the short-term, they would be unlikely to be effective for patients undergoing long-term drainage such as spinal cord injured patients and elderly nursing home residents because of the heavy Gram-negative skin contamination in these groups. These catheters have not been developed further and were not marketed by the manufacturers.

18.1.6 What Can We Conclude from the Clinical Trials of Antimicrobial Catheters?

Some authors consider that coating catheters with the antibacterials silver or nitrofurazone have "engineered out" the risk of infection in catheterized patients [18, 45]. Others are less enthusiastic, for example in their comprehensive Cochrane review Brosnahan et al. [70] concluded that while the data from clinical trials on hospitalized adults undergoing short-term catheterization suggest that the use of silver alloy catheters might reduce the risk of infection, the evidence is not strong and the trials are generally of poor quality.

Trautner et al. [71] were even more critical stating that clinical trials simply using bacteriuria, at whatever level, as the criterion for infection are fundamentally flawed. They argued that in attempting a rational assessment of the evidence in the literature concerning the prevention of catheter-associated infection, it is important to appreciate the distinction between infection and asymptomatic bacteriuria. The mere presence of bacteria in the urine of asymptomatic patients is distinct from the situation in which bacteria are present in the urine and the patient is exhibiting symptoms indicating invasion of the tissues of the urinary tract. More than 90% of cases of catheter-associated bacteriuria are asymptomatic. It would clearly be more impressive if clinical trials showed that the various antimicrobial catheters could significantly reduce the incidence of pyelonephritis or bloodstream infections. The difficulty here is that the diagnosis of symptomatic infection in catheterized patients is problematic [72, 73]. When a catheterized patient has a fever even in the absence of localizing urinary tract symptoms, because the urine culture is invariably positive, the episodes are usually described as symptomatic urinary tract infection. A prospective study using serological criteria identified only one-third of such episodes in long-term patients as originating from a urinary source [74]. There are other difficulties of course, in performing such trials on symptomatic UTI because of the lower incidence of these conditions [75]; they would have to be on a much larger scale than those performed to date.

Johnson et al. [63] pointed out that while clinical trials suggest that both nitrofurazone and silver alloy-coated catheters seem to reduce the development of asymptomatic bacteriuria during short-term use, they provide little or no data on the effect of these devices on symptomatic urinary tract infection, morbidity, secondary bloodstream infection, or mortality rates. In view of the lack of data on these clinically meaningful end-points, they concluded that it is not possible to make definitive recommendations about the use of these catheters.

These trials on antimicrobial catheters raise other important issues. Little attention has been paid to the bacteriology and lessons should be learned from a study of the nature of the organisms that cause bacteriuria in patients fitted with the antimicrobial catheters. How did these organisms manage to infect and colonize the urinary tract? Were they colonizing the catheters? Were they resistant to the concentrations of the antibiotics that had eluted into the urine? An understanding of how they manage to survive would surely give valuable insights into ways of improving the efficacy of the antimicrobial catheter strategy.

A technical issue relating to these studies is that in none of them were steps taken to neutralize the activity of the antimicrobials in the urine once the samples had been collected. In the cases of minocycline, rifampicin, and nitrofurazone (but not silver) substantial concentrations of drugs elute from the catheter into the urine. While the drainage tube is clamped to allow aspiration of a urine sample from the catheter, further leaching of the drugs can occur. What then happens to bacterial populations that might be present in the bladder urine on exposure to these drugs during the time between sampling and analysis? So are these studies recording the effect of the antibacterial agents in the urine in the sample pot rather than the prevention of bacteriuria in the patient? These doubts must be settled by the use of suitable controls.

So what can we realistically conclude from the clinical trials? Rational examination of the data and the methodologies inevitably leads to the conclusion that even with hospitalized patients undergoing short-term catheterization, the evidence that antimicrobial catheters reduce the incidence of urinary tract infection is inconclusive [76, 77]. In view of the uncertainty of the efficacy of the silver catheters, the UK National Institute for Health Research (NIHR) has commissioned a large scale, properly controlled clinical trial to try and settle the issue. The report of this study is due to be published by the end of 2012. An expert panel of the Infectious Diseases Society of America recently concluded that the most effective way to reduce the incidence of catheter-associated urinary tract infection was (a) to restrict the use of catheters and (b) to remove indwelling catheters as soon as they are no longer needed [77].

Further research is needed in this area to answer several important questions. For example, to what extent does asymptomatic bacteriuria lead to symptomatic infection? Does asymptomatic bacteriuria resolve spontaneously when short-term catheters are removed? Is there a distinction in these respects between bacteriuria with mixed populations of skin organisms and the presence of pyelonephritic organisms such as *E. coli* and *P. mirabilis*? Answers to these questions might help to resolve the issues. In the meantime before firm conclusions can be made about the efficacy of the antimicrobial catheters and their cost-effectiveness, we must await for more reliable, quality data on reduction in real infection rates from properly designed and controlled trials [77, 78].

18.1.7 Antimicrobial Catheters Not Tested in Clinical Trials

The literature records many suggestions for the development of antimicrobial Foley catheters that have only been examined in laboratory or animal models. Richards et al. [79] for example, suggested that chlorhexidine could be incorporated into silicone catheters. Laboratory tests however, revealed that the antiseptic eluted rapidly from this material within 24 h. Chlorhexidine is probably not a good choice for preventing infection as it has limited activity against *P. mirabilis,* an organism that causes serious complications in catheterized urinary tracts [80, 81]. Goankar et al. [82, 83] produced catheter materials loaded with combinations of silver sulfadiazine, triclosan, and chlorhexidine. In a laboratory model designed to simulate the migration of urinary pathogens over the surface of catheters from the urethral meatus to the bladder, they found that their catheters significantly delayed the migration of test bacteria compared to control catheters, silver/hydrogel and nitrofurazone catheters.

Polymer coatings for catheters containing gentamicin, norfloxacin, or ciprofloxacin have been developed and reported to perform well in laboratory or animal models [84–87]. Pugnash et al. [88] coated silicone catheters with a gelatin hydrogel containing ciprofloxacin encapsulated in liposomes. Gentamicin and fluoroquinolones are important broad-spectrum drugs that are of particular value in the treatment of the life-threatening septicemias that can develop from catheter-associated urinary tract infections. Because of the danger of selecting for antibiotic resistance, they are not appropriate candidates for incorporation into polymer coatings for catheters.

Liposome technology is interesting however as it can be engineered to release active agents under specific conditions such as at specific temperatures or pHs [89, 90]. Enzyme-activated liposomes can also be produced which release their contents in response to phospholipase, inflammatory enzymes, hemolysins, and other bacterial enzymes [91]. These types of coatings thus offer the prospect that active agents could be retained on the catheter, only being released when an infection threatens.

Woo et al. [92] developed an infection-responsive drug release system in which ciprofloxacin was incorporated as a monomer into a polyurethane backbone. The enzyme cholesterol esterase which is specifically released during the inflammatory process was shown to trigger degradation of the polymer and release of the antibiotic. Tanahara et al. [93] had produced a similar system by conjugating gentamicin to a carboxylated poly (vinyl alcohol) hydrogel via a thrombin-sensitive peptide linker. Exposure of this polymer to fluid from a wound infected with *Staphylococcus aureus* resulted in drug release. The authors suggested that combinations of drugs could be released in this way. Coating catheters with such bioresponsive polymers may have a future role in urology.

An ultrasound responsive drug release system was developed by Kwok et al. [94]. It exploited a drug-loaded poly (2-hydroxyethyl) methacrylate (pHEMA) hydrogel which was covered by layers of ordered methylene chains that self-assemble to create a membrane. This layer formed a barrier which inhibited passive release of the drug from the hydrogel matrix. Exposure to ultrasound disturbed the methylene chain membrane and released a pulse of drug. The barrier reassembled when the ultrasound was turned off and drug release ceased. Such a system that delivered ciprofloxacin was developed and tested by Norris et al. [95]. Experiments in flow cells demonstrated that the ability of *Ps. aeruginosa* to colonize glass slides coated with the hydrogel was reduced on exposure to 20 min daily bursts of ultrasound. Coating catheters with such a material might be feasible as ultrasound at the low power levels necessary to release the drug is noninvasive.

The silicones used in catheter manufacture have high coefficients of friction and their resulting lack of lubricity can cause discomfort and even trauma on catheter insertion or removal. Malcolm et al. [96] developed an ingenious method of producing self-lubricating silicones. A cross-linker derived from the lubricious oil oleyl alcohol was used to cure condense a poly dimethylsiloxane. Oleyl alcohol, the product of the cross-linking reaction in the polymer synthesis is thus distributed throughout the silicone elastomer matrix. The slow diffusion of the oleyl alcohol and its subsequent persistence at the material surface produces a highly lubricious biomaterial that could be used in the manufacture of catheters to facilitate their insertion and removal. The controlled release of drugs can also be achieved from these polymers by varying the concentration of the cross-linker. It is possible therefore that these self-lubricating silicones loaded with appropriate agents could well find applications in the manufacture of urinary catheters.

Dariouche et al. [97] explored the efficacy of silicone catheters that had been coated with a combination of chlorhexidine and protamine sulfate. Laboratory tests demonstrated that the activity of this catheter material against a panel of urinary tract pathogens persisted on agar for at least 10 days. Sections of control silicone catheters, silver/hydrogel-coated latex catheters, and chlorhexidine + protamine coated catheters were inoculated with *E. coli* and then implanted subcutaneously into rabbits. On removal from the animals, bacterial colonization was found to be significantly less on the chlorhexidine + protamine catheters than on sections of the other two catheters. Hachem et al. [98] produced silicone catheters coated with Gendine a novel dye composed of a mixture of chlorhexidine and gentian violet. When sections of test catheters were incubated for 24 h in broth cultures of *E. coli*, vancomycin-resistant enterococci, *P. aeruginosa*, and *Candida* sp., biofilms were formed on sections of control silicone catheters and silver/hydrogel-coated latex catheters, but not on Gendine-coated catheters. Experiments were also performed in rabbits in which catheter insertion sites were inoculated daily with *E. coli*. After 4 days, invasive bladder infection (defined as >100 cfu/g of bladder wall tissue and histological evidence of cystitis) was recorded in 56% of rabbits fitted with control silicone catheters, 43% of those with silver/hydrogel catheters but in none of those animals fitted with Gendine-coated devices. Recently Regev-Shoshani et al. [99] reported the results of laboratory tests suggesting that impregnating silicone catheters with gaseous nitric oxide prevented their colonization by *E. coli*. These three novel approaches warrant further evaluation.

18.1.8 The Catheter Encrustation Problem

About half of patients undergoing long-term catheterization will experience the complication of catheter encrustation and blockage [100, 101]. All currently available catheters are vulnerable to encrustation and there are no effective methods available to nursing staff for preventing the process [102, 103]. If we are to succeed in controlling this complication we need a clear understanding of the mechanisms which deposit crystalline material on catheter surfaces.

In recent years it has become obvious that catheter encrustation is a result of colonization by crystalline bacterial biofilm (Fig. 18.2). The principal crystalline components of these biofilms are struvite (magnesium ammonium phosphate) and a poorly crystalline form of apatite (a hydroxylated calcium phosphate in which a variable proportion of the phosphate groups are replaced by carbonate). Large numbers of bacilli, especially species capable of producing the enzyme urease are associated with these crystals. Urease is the driving force for the generation of the crystals. In the pool of residual bladder urine it hydrolyses urea to produce ammonia causing a rise in pH. As the urine becomes alkaline, crystallization of the magnesium and calcium phosphates is induced. Aggregation of the crystalline material then occurs in the urine and in the biofilm developing on the catheter surfaces [104]. This process continues until the accumulating crystalline biofilm blocks the flow of urine through the catheter with potentially serious consequences for the patient.

Several bacterial species that infect the catheterized tract can produce urease. Of these, *P. mirabilis* is the most commonly found in the urine of patients suffering

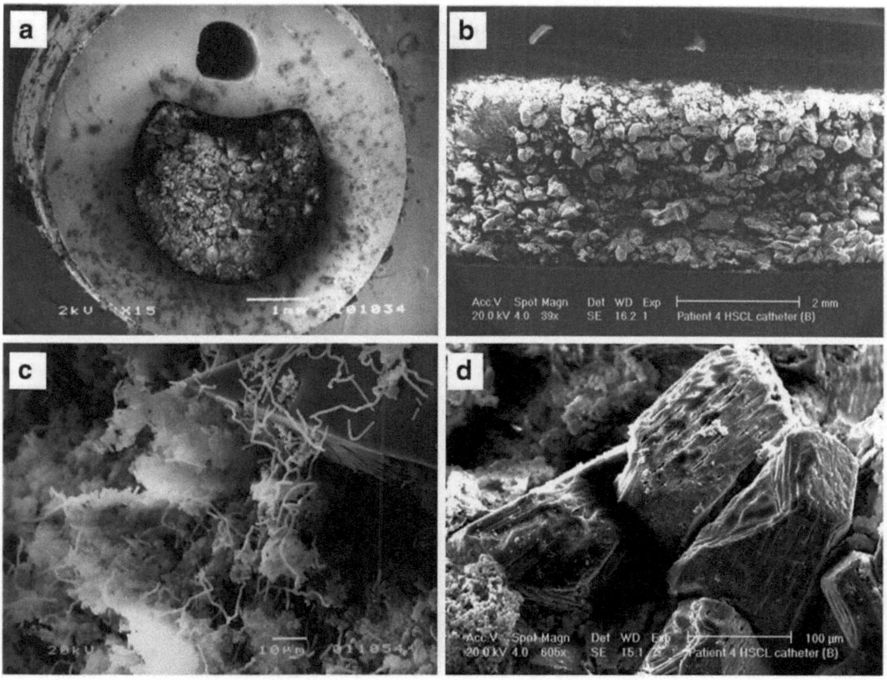

Fig. 18.2 Scanning electron micrographs of crystalline biofilms encrusting catheters. (**a**) shows a cross-section of an all-silicone catheter removed from a patient after 8 weeks in situ; (from Stickler, 1996, *European Urology Update* 5: 1–8, with permission), (**b**) shows a longitudinal-section of a silver/hydrogel-coated latex catheter removed from a patient after 11 days (from Morgan, Rigby and Stickler 2009, Urological Research, 37:88–93, with permission.). In both these cases extensive crystalline material can be seen occluding the catheter lumen. The presence of crystals and bacilli in the material encrusting catheters is shown in (**c**) (from Stickler et al. 1998 *European J Clin Microbiol and Infect Dis* 17:649–652, with permission) and (**d**)

from recurrent catheter encrustation [80, 105] and also the predominant organism on encrusted catheters [106]. The urease it produces can hydrolyze urea much faster than those of other species [107]. In laboratory models of the catheterized bladder, other urease producers such as *Morganella morganii*, *K. pneumoniae*, and *P. aeruginosa* fail to generate alkaline urine and do not produce appreciable crystalline biofilm on catheters [108, 109]. In these experiments only *P. mirabilis*, *Proteus vulgaris*, and *Providencia rettgeri* were capable of producing alkaline urine and extensive catheter encrustation. The latter two species are only found in 5–10% of catheter biofilms [110]. Epidemiological and experimental evidence thus points to *P. mirabilis* as the prime cause of crystalline biofilm formation on Foley catheters.

P. mirabilis is an ingenious organism that can initiate the formation of crystalline biofilms in several ways [25]. The primary stage in the formation of biofilms on implanted medical devices usually involves their rapid coating by a conditioning film of proteins from body fluids. These proteins provide receptor sites for bacterial attachment via the fine hair-like fimbriae (adhesins) that protrude from their cell

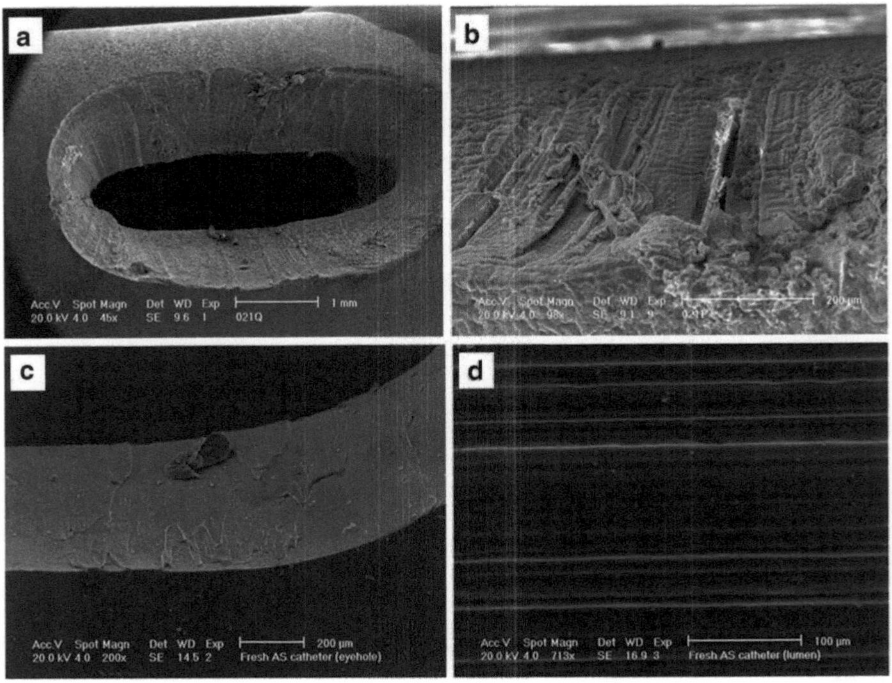

Fig. 18.3 Scanning electron micrographs of the surfaces of fresh un-used catheters. (**a**) and (**b**) show the nature of the surface around the drainage eye of a latex-based catheter (from Stickler DJ et al. 2003, *Urological Research*, 31:306–311, with permission). The irregularities of the surfaces around the eye-hole and of the central lumen of an all-silicone catheter are shown in (**c**) and (**d**) respectively (from Stickler DJ, *Proteus mirabilis* biofilm formation and catheter design, in *Biomaterials and Tissue Engineering in Urology* 2009, ed Denstedt J and Atala A , ISBN 978-1-84569-402-9, Woodhead Publishing, Cambridge with permission)

walls [111]. Several such adhesins have been identified on *P. mirabilis* cells [112] and protein coatings have been found on catheters removed from patients after short periods [24]. While *P. mirabilis* can probably attach to conditioning films in this way they can also bind directly to silicone surfaces [113].

Powerful physical forces can also initiate crystalline biofilm formation. The uneven nature of catheter surfaces, especially latex-based catheters, has been revealed by scanning electron microscopy [114, 115] (Fig. 18.3). The techniques used to produce the eye-holes, tear through the latex producing surfaces that appear in electron micrographs like rocky landscapes of craters and crevices. Silicone-based catheters have rather smoother surfaces but irregularities commonly occur around the eye-holes and where the extrusion production techniques have generated striations on the lumenal surfaces. As contaminated urine flows over these surfaces, bacteria become trapped in the irregularities. The drainage eyes are particularly vulnerable to bacterial colonization. Scanning electron microscopy on catheters removed from bladder models at various intervals after inoculation of the urine with *P. mirabilis* showed that within 2 h bacterial cells were trapped in the surface crevices and depressions (Fig. 18.4). Microcolonies of cells developed from these

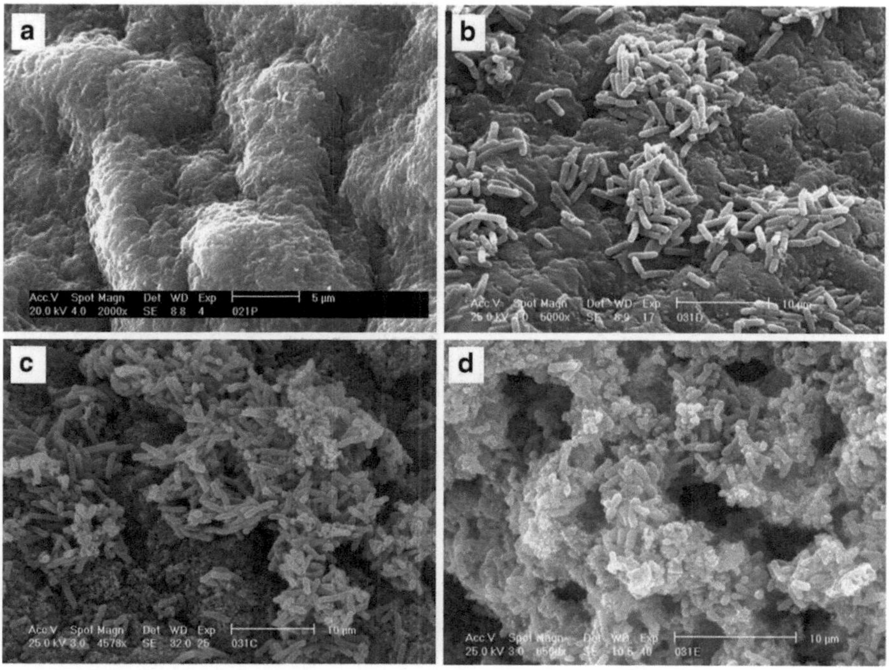

Fig. 18.4 Micrographs illustrating the stages in the colonization of a hydrogel-coated latex catheter by *Proteus mirabilis* in a laboratory model under conditions in which the urine is initially acid. (**a**) the irregular surface around the drainage eyes of an unused catheter; (**b**) microcolonies of cells colonizing the surface after 4h incubation; (**c**) by 6h the microcolonies have spread over the surface, the urine has become alkaline and microcrystals of apatite are forming on the surface and in the biofilm; (**d**) after 20h incubation a mature crystalline biofilm has formed

pioneer cells after 4 h and by 6 h the pH of the urine had risen and crystals appeared in the developing biofilm. Extensive crystalline biofilms then spread over the catheter surfaces [115]. Continued development of the crystalline biofilm usually blocked the catheter eye-hole or the central lumen in the balloon region.

Experiments in parallel-plate flow cells have demonstrated that when urine flows over smooth polymer films, the pH of the urine can be a major factor in determining bacterial adhesion. Some polymers with strongly electron donating hydrophilic surfaces will resist colonization at pH 6. If the pH rises however, calcium and magnesium phosphates come out of solution, macroscopic aggregates of crystals and cells form in the urine, settle on the surface and initiate crystalline biofilm formation [116].

In the many patients enduring recurrent catheter encrustation, the normal practice is to simply replace the blocked catheter. In these cases the new catheters are thus placed directly into urine cultures of *P. mirabilis* at alkaline pHs containing aggregates of microcrystals of apatite and struvite. An investigation of the early stages of biofilm formation under these conditions in a laboratory model, revealed a common sequence in development on all-silicone, silicone-coated latex, hydrogel-coated latex and silver/hydrogel-coated latex catheters [117]. After just 1 h in the models the catheter surfaces were covered by a microcrystalline layer which X-ray

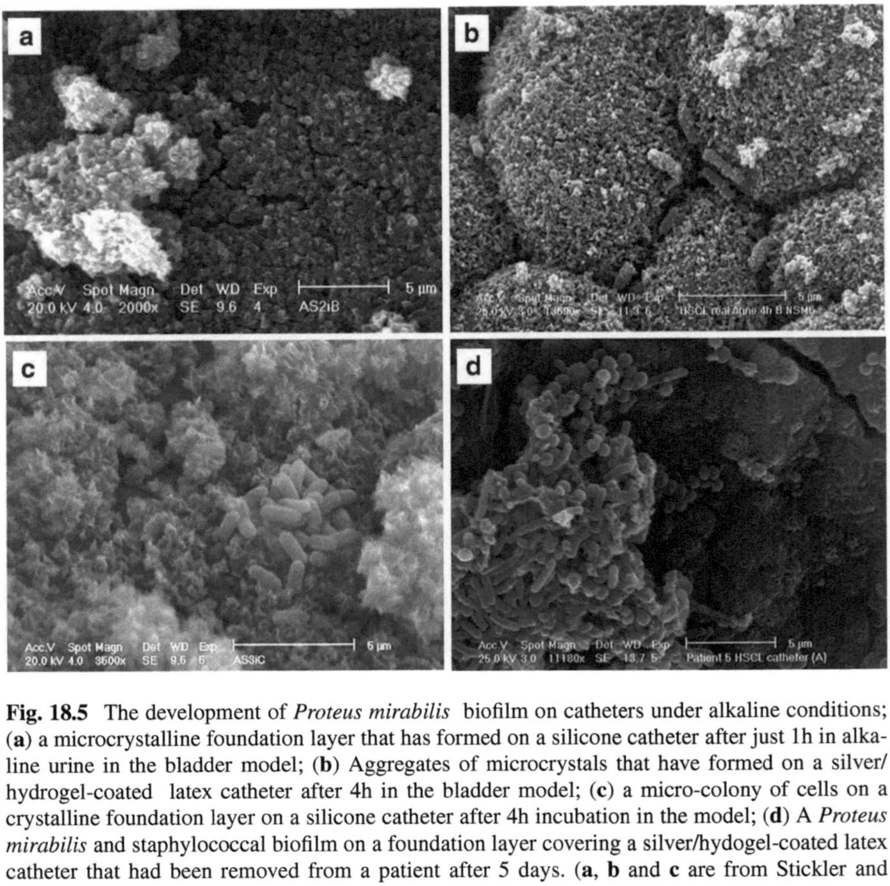

Fig. 18.5 The development of *Proteus mirabilis* biofilm on catheters under alkaline conditions; (**a**) a microcrystalline foundation layer that has formed on a silicone catheter after just 1h in alkaline urine in the bladder model; (**b**) Aggregates of microcrystals that have formed on a silver/hydrogel-coated latex catheter after 4h in the bladder model; (**c**) a micro-colony of cells on a crystalline foundation layer on a silicone catheter after 4h incubation in the model; (**d**) A *Proteus mirabilis* and staphylococcal biofilm on a foundation layer covering a silver/hydogel-coated latex catheter that had been removed from a patient after 5 days. (**a**, **b** and **c** are from Stickler and Morgan, 2008 *Journal of Hospital Infection*, 69:350–360, with permission); d is from Morgan, Rigby and Stickler 2009, *Urological Research*, 37:88–93, with permission)

microanalysis confirmed as calcium phosphate. Bacterial colonization of this foundation layer followed with microcolonies of cells developing on the crystalline surface. At 18 h the eye-holes and the lumenal surfaces were covered extensively by crystalline *P. mirabilis* biofilms (Fig. 18.5). Examination of silver/hydrogel latex catheters taken from patients who were infected with *P. mirabilis* and were recurrent blockers confirmed that microcrystalline foundation layers form on these antimicrobial catheters *in vivo* [118].

In the experimental models, little encrustation occurred on nitrofurazone catheters in the first 12 h but by 29 h the catheters had blocked. After just 4 h the urinary pH dropped from 8.6 to 6.4 and the bacterial population had fallen. By 18 h however the pH had risen to 8.3 and the viable cell count had recovered. Color changes occurred in the urine which suggested that the yellow nitrofurazone eluted rapidly from the catheters in the few first hours. It is possible that high concentrations of the drug were initially present in the urine and had an antibacterial effect. The cultures recovered, alkaline conditions were reestablished and catheter encrustation resumed as the concentration of nitrofurazone reduced with time.

These observations that crystalline biofilms of *P. mirabilis* can form in several distinct ways under various conditions have important implications for the development of encrustation-resistant catheters. It is clear that trying to inhibit bacterial attachment and crystalline biofilm formation by immobilizing an antibacterial in the catheter is unlikely to prevent the problem in patients infected with *P. mirabilis*. For example, in the case of the silver-coated catheters, the deposition of the crystalline foundation layer allows cells to attach and multiply protected from contact with the underlying silver. Thus, if antibacterials are to be incorporated into catheters to stop encrustation, they must diffuse out from the catheter surface and prevent bacteria from elevating the urinary pH over prolonged periods. Unfortunately silver does not elute from the silver/hydrogel catheters in sufficient concentrations to inhibit the activity of *P. mirabilis* [117].

Phosphoryl-choline (PC) the major polar head group on the outer surfaces of erythrocytes has been incorporated into synthetic polymers to produce biocompatible coatings for medical devices [119]. These polymers produced stable extremely hydrophilic surface coatings on a variety of substrates [120]. PC-coated surfaces were reported to be resistant to protein binding [121] and bacterial colonization [122]. PC-coatings did not however, inhibit the formation of crystalline *P. mirabilis* biofilm on silicone or latex Foley catheters [123].

Diamond-like carbon coatings have been used to control encrustation on ureteral stents [124]. Although the encrustation process on these devices differs from that on bladder catheters, not involving crystalline biofilm formation by urease producing bacteria [123], it would be interesting to test whether these coatings could prevent catheter encrustation.

18.1.9 Basic Design Problems of the Foley Catheter

There are intrinsic design problems with the Foley catheter which induce vulnerability to infection [125] and facilitate its colonization by the crystalline biofilms that compromise patient care [115]. It has become obvious that impregnating catheters with antibacterials fails to solve these problems. Despite the urgent need, manufacturers have been reluctant to invest in research to improve catheter design [125]. It is quite extraordinary that a device which so comprehensively undermines the defenses of the bladder against infection is still in such large scale use. A major issue is that the regular filling and emptying of the bladder which flushes any bacterial contamination from the lower tract, does not occur in the presence of the catheter. Instead, a stagnant sump of residual urine forms which can commonly be around 100 ml in volume [126]. Fresh urine is of course supplied from the kidneys and the overflow dribbles down the catheter. A continuous flow system into a reservoir of nutrient medium is thus established in which bacterial communities can flourish. In the case of infection by urease producing organisms, it results in the development of a pool of alkaline urine in which the crystals that encrust catheters and form bladder stones are precipitated. The problems are compounded because the balloon and tip of the catheter can erode and damage the smooth mucosal surface of the bladder. The pressure exerted by the catheter on the walls of the urethra attenuates its blood supply and can block the lubricating periurethral glands. The stressed urethral surface and the outer surface of the catheter provide convenient substrates for bacterial colonization and ascending migration into the bladder.

The Foley catheter thus constitutes a persistent threat to the integrity of the bladder and upper urinary tract. The challenge is to produce a device that allows the bladder to retain its normal voiding system and does not damage the mucosal surfaces of the lower urinary tract. Kunin [125] suggested that to maintain the integrity of the urethra a thin-walled, continuously lubricated, collapsible catheter will be required. The retention balloon should also be replaced by a device which does not result in a residual sump of urine and will permit regular complete emptying of the bladder. Such catheters would of course be more difficult to manufacture and more expensive, but surely would be well worth the investment. Currently catheters themselves are not costly, but the eventual price we have to pay for managing the complications they cause is enormous.

It is possible to partially reinstate the periodic filling and emptying of the catheterized bladder by replacing the catheter drainage tube and bag with a simple manual valve [127, 128]. When the valve is closed the bladder fills with urine. Regular opening of the valve then allows the periodic release of urine from the bladder. It has been suggested that this helps to maintain bladder tone and capacity. It has also been claimed that the valve regulated periodic flushing of the catheter might decrease infection and catheter encrustation [129, 130]. Experiments in laboratory models of the catheterized bladder fitted with manual valves have demonstrated that opening the valve to drain urine from the bladder at 2 h intervals four times during the day, significantly delayed the blockage of the catheter by *P. mirabilis* crystalline biofilm. An even greater beneficial effect was recorded when an automatic valve was used to achieve release of urine at 4 h intervals throughout the day and night [131]. As manual valves present difficulties to many people with poor manual dexterity, electronically controlled automatic valves are being developed. One such compact, reliable prototype shows considerable promise [132]. Other valves have been developed that respond to pressure changes in the bladder (e.g., The Urocycler, UroSolutions Inc. Orlando, FL). Even with the application of a valve, the retention balloon of the Foley catheter still ensures that a sump of residual urine will remain in the bladder. What is really needed is a design which allows the complete emptying of the bladder.

A catheter with an intraurethral section that was so soft and compliant that it conformed to the shape of the urethra was described over 20 years ago [133]. Not only did patients find these catheters more comfortable but also significantly fewer ($P < 0.001$) of the conformable catheters had intraluminal obstruction with struvite than silicone-coated latex control catheters. Despite these encouraging results this catheter has not been developed further.

Another early idea which has not been taken up is that of self-lubricating catheters. Kunin and Finkelberg [134] suggested that as in many cases the urethra becomes dry, irritated and inflamed on long-term catheterization, it would be beneficial if the catheter was repeatedly lubricated. They developed a prototype in which lubricant was extruded intermittently through an additional channel within the catheter. A controlled trial showed a significant reduction in bacteriuria in females who received daily methyl cellulose or polymyxin B lubricant compared to those with catheters that were not lubricated. They concluded that if a less cumbersome delivery system could be developed it might have considerable potential. The recent ingenious development of self-lubricating silicones is thus of great interest [96].

If the medical device industry cannot be persuaded to exploit some of these ideas, then at the least we should encourage them to make relatively simple improvements in the current Foley catheters. The narrow internal diameters of the current catheters induce acute vulnerability to blockage. Alternatives to latex and silicone as the base materials for manufacture might be considered. For example prototypes have been made from a nylon braid encased in a polyurethane matrix which has excellent flow properties and resistance to kinking. They have substantially thinner walls giving much larger internal diameters which are not easily occluded [135]. Surely it would also be relatively simple to produce catheters with larger eye-holes and smoother surfaces.

18.1.10 Conclusions

It is difficult to understand why in the twenty-first century we are unable to drain urine from a dysfunctional bladder without producing infection and a range of associated complications. The morbidity caused and the costs to health services are surely no longer acceptable. We need to realize that simply by impregnating catheters with antibacterials or coating them with surfaces that in clean laboratory conditions appear to prevent their colonization by bacterial biofilms is not the way forward. These approaches show a lack of understanding of how biofilms develop in the catheterized bladder and underestimate the ability of bacteria to overcome attempts at control using chemical agents. It is important to realize that we need to make fundamental changes in catheter design [136]. Better eye-holes, wider luminal internal diameters and smother surfaces should be relatively simple to achieve. To obtain significant progress however we need designs that do not produce a residual sump of urine in the bladder and do not undermine innate defense mechanisms against infection by trauma, irritation and inflammation of urethral and bladder epithelia. We need to encourage the medical device industry to take up these challenges. If we are to achieve these advances it is high time that we took the advice given by Calvin Kunin in the conclusion to his editorial in the *New England Journal of Medicine* in 1988 [125].

> We must therefore consider initiating a national program that will combine the efforts of government, investment by industry and the talent in academic institutions to develop better indwelling catheters and bring them to clinical trial. The catheters currently available are inexpensive instruments but the eventual price we have to pay is great indeed. Catheters of the future may well be more expensive, but should be well worth the investment.

References

1. Darouiche RO. Device-associated infections: a macroproblem that starts with microadherence. Clin Infect Dis. 2001;33:1567–72.
2. Jepsen OB, Larsen SO, Dankert J, et al. Urinary tract infection and bacteraemia in hospitalized medical patients—a European multicentre prevalence survey on nosocomial infection. J Hosp Infect. 1982;3:241–52.
3. Warren JW, Steinberg L, Hebel JR, Tenney JH. The prevalence of urethral catheterization in Maryland nursing homes. Arch Intern Med. 1989;149:1535–7.
4. Zimakoff J, Pontoppidan B, Larsen SO, Stickler DJ. Management of urinary bladder function in Danish hospitals, nursing homes and home care. J Hosp Infect. 1993;24:183–99.
5. Sørbye LW, Finne-Soveri H, Ljunggren G, Topinkova E, Bernabei R. Indwelling catheter use in home care: elderly aged +65 in 11 different countries in Europe. Age Ageing. 2005;34:377–81.
6. Beeson P. The case against the catheter. Am J Med. 1958;24:1–3.
7. Hendlin K, Meyers J, Monga M. Foley catheter characteristics: predicting problems. J Endourol. 2009;23:169–71s.
8. Tan GW, Chan SP, Ho CK. Is transurethral catheterization the ideal method of bladder drainage? A survey of patient satisfaction with indwelling transurethral urinary catheters. Asian J Surg. 2010;33:31–6.
9. Niel-Weise BS, van den Broek PJ. Urinary catheter policies for short-term bladder drainage in adults. Cochrane Database Sys Revs. 2005;3:1–39.
10. Tenke P, Kovacs B, Bjerklund, Johansen T, Matsumoto T. European and Asian guidelines on the management and prevention of catheter-associated urinary tract infections. Int J Antimicrob Agents. 2008; 31 Suppl 1:S68–S78.
11. Stamm WE. Catheter-associated urinary tract infections: epidemiology, pathogenesis and prevention. Am J Med. 1991;91(Suppl 3B):65s–71s.
12. Kunin CM, Douthitt S, Dancing J, Anderson J, Moeschberger M. The association between the use of urinary catheters and morbidity and mortality among elderly patients in nursing homes. Am J Epidemiol. 1992;135:291–301.
13. Krieger JN, Kaiser DL, Wenzel RP. Urinary tract etiology of bloodstream infections in hospitalized patients. J Infect Dis. 1983;148:57–62.
14. Bryan CS, Reynolds KL. Hospital-acquired bacteremic urinary tract infection: epidemiology and outcome. J Urol. 1984;132:484–98.
15. Saint SJ. Clinical and economic consequences of nosocomial catheter-related bacteriuria. Am J Infect Control. 2000;28:68–75.
16. Kunin CM. Care of the urinary catheter. In: Kunin CM, editor. Detection, prevention and management of urinary tract infections. 5th ed. Baltimore: Williams and Wilkins; 1997. p. 226–78.
17. Clayton CL, Chawla JC, Stickler DJ. Some observations on urinary tract infections in patients undergoing long-term bladder catheterization. J Hosp Infect. 1982;3:39–47.
18. Maki DG, Tambyah PA. Engineering out the risk for infection with urinary catheters. Emerging Infect Dis. 2001;7:342–7.
19. Meers PD, Ayliffe GAJ, Emmerson AM, Leigh DA, Mayon-White RT, Mackintosh CA et al. National survey of infections in hospitals 1980. Part 2, Urinary tract infection. J Hosp Infect. 1981;Suppl 2:23–28.
20. Hayley RW, Culver DH, White JW, Morgan WM, Emori TG. The nationwide nosocomial infection rate. Am J Epidemiol. 1985;121:159–67.
21. Tambyah PA, Halvorson KT, Maki DG. A prospective study of pathogenesis of catheter-associated urinary tract infections. Mayo Clin Proc. 1999;74:131–6.
22. Matsukawa M, Kunishima Y, Takahashi S, Takeyama K, Tsukamoto T. Bacterial colonization on intraluminal surface of urethral catheters. Urology. 2005;65:440–4.
23. Warren JW, Tenney JH, Hoopes JM, Muncie HL, Anthony WC. A prospective microbiologic study of bacteriuria in patients with chronic indwelling urethral catheters. J Infect Dis. 1982; 146:719–23.

24. Ohkawa M, Sugata T, Sawaki M, Nakashima T, Fuse H, Hisazumi H. Bacterial and crystal adherence to the surfaces of indwelling urethral catheters. J Urol. 1990;143:717–21.
25. Stickler DJ. Bacterial biofilms and catheter-associated urinary tract infection. Nature Clin Prac Urol. 2008;5:598–608.
26. Stickler DJ, Zimakoff J. Complications of urinary tract infections associated with devices used for long-term bladder management. J Hosp Infect. 1994;28:177–94.
27. Dance DA, Pearson AD, Seal DV, Lowes JA. A hospital outbreak caused by a chlorhexidine and antibiotic resistant *Proteus mirabilis*. J Hosp Infect. 1987;10:10–6.
28. Butler HK, Kunin CM. Evaluation of polymyxin catheter lubricant and impregnated catheters. J Urol. 1968;100:560–6.
29. Guggenbichler JP, Boswald M, Lugauer S, Krall T. A new technology of microdispersed silver in polyurethane induces antimicrobial activity in central venous catheters. Infection. 1999;27 Suppl 1:s16–23.
30. Kunin CM, Steele C. Culture of the surfaces of urinary catheters to sample urethral flora and study the effect of antibacterial therapy. J Clin Microbiol. 1985;21:902–8.
31. Fawcett C, Chawla JC, Quoraishi A, Stickler DJ. A study of the skin flora of spinal cord injured patients. J Hosp Infect. 1986;8:149–58.
32. Danese PN. Antibiotic approaches: prevention of catheter colonization. Chem Biol. 2002;9: 873–80.
33. Knudsen BE, Chew BH, Denstedt JD. Drug-eluting biomaterials in urology: the time is ripe. Brit J Urol Int. 2005;95:726–7.
34. Santin M, Motta A, Denyer SP, Cannas M. Effect of the urine conditioning film on ureteral stent encrustation and characterization of its protein composition. Biomaterials. 1999;20:1245–51.
35. Stickler DJ. Biomaterials to prevent nosocomial infections: is silver the gold standard? Curr Opin Infect Dis. 2000;13:389–93.
36. Russell AD, Hugo WB. Antimicrobial activity and mechanism of action of silver. Prog Med Chem. 1994;39:351–70.
37. Darouiche RO. Anti-infective efficacy of silver-coated medical prostheses. Clin Infect Dis. 1999;29:1371–7.
38. Schierholz JM, Yücel N, Rump AF, Beuth J, Pulverer G. Anti-infective and encrustation-inhibiting materials: myths and facts. Int J Antimicrobial Ag. 2002;19:511–6.
39. Kumon H, Hashimoto H, Nishimura M, Monden K, Ono N. Catheter-associated urinary tract infections: impact of catheter materials on their management. Int J Antimicrob Ag. 2001;17:311–6.
40. Samuel U, Guggenbichler JP. Prevention of catheter-related infections: the potential of a new nano-silver impregnated catheter. Int J Antimicrob Ag. 2004;23 Suppl 1:S75–8.
41. Davis CP, Anderson MD, Hoskins S, Warren MM. Electrode and bacterial survival with iontophoresis in synthetic urine. J Urol. 1992;147:1310–3.
42. Chakravati A, Gangodawila S, Long ML, Morris NS, Blacklock AR, Stickler DJ. An electrified catheter to resist encrustation by *Proteus mirabilis* biofilm. J Urol. 2005;174:1129–32.
43. Johnson JR, Roberts PL, Olsen RJ, Moyer KL, Stamm WE. Prevention of catheter-associated urinary tract infection with a silver oxide urinary catheter: clinical and microbiologic correlates. J Infect Dis. 1990;162:1145–50.
44. Riley DK, Classen DC, Stevens LE, Burke JP. A large randomized clinical trial of a silver-impregnated catheter: lack of efficacy and staphylococcal superinfection. Am J Med. 1995;98:349–56.
45. Davenport K, Keeley FX. Evidence for the use of silver alloy-coated urethral catheters. J Hosp Infect. 2005;60:298–303.
46. Lundeberg T. Prevention of catheter-associated urinary tract infections by the use of silver-impregnated catheters. Lancet. 1986;2:1031.
47. Leidberg H, Lundeberg T. Silver alloy-coated catheters reduce-catheter-associated bacteriuria. Br J Urol. 1990;65:379–81.
48. Leidberg H, Lundeberg EP. Refinements in the coating of urethral catheters reduces the incidence of catheter-associated bacteriuria. Eur Urol. 1990;17:236–40.
49. Verleyen P, De Ridder D, Van Poppel H, Baert L. Clinical applications of the Bardex IC Foley catheter. Eur Urol. 1999;36:240–6.

50. Karchmer TB, Giannetta ET, Muto CA, Strain BA, Farr BM. A randomized cross-over study of silver-coated urinary catheters in hospitalized patients. Arch Int Med. 2000;160:3294–8.
51. Garner JS, Jarvis WR, Emori TG, Horan TC, Hughes JM. CDC definitions for nosocomial infections. Am J Infect Control. 1988;16:128–40.
52. Thibon P, Le Coutour X, Leroyer R, Fabry J. Randomized multi-centre trial of the effects of a catheter coated with hydrogel and silver salts on the incidence of hospital-acquired urinary tract infection. J Hosp Infect. 2000;45:117–24.
53. Saint SJ, Elmore JG, Sullivan SD, Emerson SS, Koepsell TD. The efficacy of silver alloy-coated urinary catheters in preventing urinary tract infection: a meta-analysis. Am J Med. 1988;105:236–41.
54. Niel-Weise BS, Arend SM, van den Broek PJ. Is there evidence for recommending silver-coated urinary catheters in guidelines? J Hosp Infect. 2002;52:81–7.
55. Lai KK, Fontecchio SA. Use of silver hydrogel urinary catheters on incidence of catheter-associated urinary tract infections in hospitalized patients. Am J Infect Control. 2002;30:221–5.
56. Rupp ME, Fitzgerald T, Marion N, et al. Effect of silver-coated urinary catheters: efficacy, cost effectiveness and antimicrobial resistance. Am J Infect Control. 2004;32:445–50.
57. Srinivasan A, Karchmer T, Richards A, Song X, Perl TM. A prospective trial of a novel silicone-based silver-coated Foley catheters for the prevention of nosocomial urinary tract infections. Infect Control Hosp Epidemiol. 2006;27:38–43.
58. Crnish CJ, Drinka PJ. Does the composition of urinary catheters influence clinical outcomes and the results of research studies? Infect Control Hosp Epidemiol. 2007;28:102–3.
59. Lambert HP, O'Grady FW. Antibiotic and chemotherapy. 6th ed. Edinburgh: Churchill Livingstone; 1992.
60. Johnson JR, Berggren T, Conway AJ. Activity of a nitrofurazone matrix urinary catheter against catheter-associated uropathogens. Antimicrob Ag Chemother. 1993;37:2033–6.
61. Johnson JR, Delavari P, Azar M. Activities of a nitrofurazone containing catheter and a silver hydrogel catheter against multi-drug resistant bacteria characteristic of catheter-associated urinary tract infection. Antimicrobial Ag Chemother. 1999;43:2990–5.
62. Lee SM, Short TD, Unsworth A. Design and development of a novel automatic valve system for long-term catheterized urinary incontinence patients. Proc Inst Mech Eng [H]. 2007;221:665–75.
63. Johnson JR, Kuskowski MA, Wilt TJ. Systematic review: antimicrobial urinary catheters to prevent catheter-associated urinary tract infection in hospitalized patients. Ann Int Med. 2006; 144:116–26.
64. Maki DG, Knasinski V, Tambyah PA. A prospective investigatory blinded trial of a novel nitrofurazone-impregnated urinary catheter. Infect Control Hosp Epidemiol. 1997;18:P50.
65. Al-Habdan I, Sadat-Ali M, Corea JR, Al-Othman A, Kamal BA, Shriyan DS. Assessment of nosocomial urinary tract infections in orthopaedic patients: a prospective and comparative study using two different catheters'. Int Surg. 2003;88:152–4.
66. Stensballe J, Tvede M, Looms D, et al. Infection risk with nitrofurazone-impregnated urinary catheters in trauma patients. Ann Int Med. 2007;147:285–93.
67. Darouiche RO, Hampel TB, Boone TB, Raad II. Antimicrobial activity and durability of a novel antimicrobial-impregnated bladder catheter. Int J Antimicrob Ag. 1997;8:243–7.
68. Darouiche RO, Safar H, Raad II. In vitro efficacy of antimicrobial-coated bladder catheters in inhibiting bacterial migration along catheter surface. J Infect Dis. 1997;176:1109–12.
69. Darouiche RO, Smith JA, Hanna H, et al. Efficacy of antimicrobial-impregnated bladder catheters in reducing catheter-associated bacteriuria: a prospective, randomized multi-center clinical trial. Urology. 1999;54:976–81.
70. Brosnahan J, Jull A, Tracey C. Types of urethral catheters for management of short-term voiding problems in hospitalized adults'. Cochrane Database Syst Rev. 2004;1:1–29.
71. Trautner BW, Hull RA, Darouiche RO. Prevention of catheter-associated urinary tract infection. Curr Opin Infect Dis. 2005;18:37–41.
72. Galloway A, Green HT, Windsor JJ, Menon KK, Gardner BD. Serial concentrations of C-reactive protein as an indicator of urinary tract infection in patients with spinal injury. J Clin Path. 1986;39:851–5.

73. Nicolle LE. Catheter-related urinary tract infection. Drugs Ageing. 2005;22:627–39.
74. Orr PH, Nicolle LE, Duckworth H, et al. Febrile urinary tract infection in the institutionalized elderly. Am J Med. 1996;100:71–7.
75. Tambyah PA, Maki DG. Catheter-associated urinary tract infection is rarely symptomatic: a prospective study of 1497 catheterized patients. Arch Intern Med. 2000;160:678–82.
76. Pratt RJ, Pellowe CM, Wilson JA, et al. Epic2: National evidence-based guidelines for preventing healthcare-associated infections in NHS hospitals in England. J Hosp Infect. 2007;65 Suppl 1:S1–64.
77. Hooton TM, Bradley SF, Cardenas DD, et al. Diagnosis, prevention, and treatment of catheter-associated urinary tract infection in adults: 2009 international clinical practice guidelines from the Infectious Diseases Society of America. Clin Infect Dis. 2010;50:625–63.
78. Schumm K, Lam TB. Types of urethral catheters for management of short-term voiding problems in hospitalized patients: a short version Cochrane review. Neurourol Urodyn. 2008;27:738–46.
79. Richards CL, Hoffman KC, Bernard JM, Winslow SD, Norman JC, Whalen RL. Development and characterization of an infection inhibiting urinary catheter. ASAIO J. 2003;49:449–53.
80. Mobley HTL, Warren JW. Urease-positive bacteriuria and obstruction of long-term urinary catheters. J Clin Microbiol. 1987;25:2216–7.
81. Stickler DJ. Susceptibility of antibiotic-resistant Gram-negative bacteria to biocides: a perspective from the study of catheter biofilms. J Appl Microbiol. 2002;92(suppl):163S–70.
82. Gaonakar TA, Sampath LA, Modak SM. Evaluation of the antimicrobial efficacy of urinary catheters impregnated with antiseptics in an invitro urinary tract model. Infect Cont Hosp Epidemiol. 2003;24:506–13.
83. Goanakar TA, Careos L, Modak SM. Efficacy of a silicone urinary catheter impregnated with chlorhexidine and triclosan against colonization with *Proteus mirabilis* and other urinary pathogens. Infect Control Hosp Epidemiol. 2007;28:596–8.
84. Kwok CS, Horbett TA, Ratner BD. Design of infection resistant antibiotic-releasing polymers. II Controlled release of antibiotics through a plasma-deposited thin film barrier. J Control Release. 1999;62:301–11.
85. Cho YH, Lee SJ, Lee JY, et al. Prophylactic efficacy of a new gentamicin-releasing urethral catheter in short-term catheterized rabbits. Br J Urol Int. 2001;87:104–9.
86. Cho YW, Park JH, Kim SH, et al. Gentamicin-releasing urethral catheter for short-term catheterization. J Biomater Sci Polym Ed. 2003;14:962–72.
87. Park JH, Cho YW, Cho YH, et al. Norfloxacin-releasing urethral catheter for long-term catheterization. J Biomater Sci Polym Ed. 2003;14:951–62.
88. Pugnash JL, DiTizio V, Mittleman MW, Bruce AW, DiCosmo F, Khoury AE. Antibiotic hydrogel-coated Foley catheters for prevention of urinary tract infection in a rabbit model. J Urol. 1999;162:883–7.
89. Maeda M, Kumano A, Tirrell DA. H^+-induced release of contentsof phosphatidylcholine vesicles bearing surface-bound polyelectrolyte chains. J Am Chem Soc. 1988;110:7455–9.
90. Kono K. Thermosensitive polymer-modified liposomes. Adv Drug Rev. 2001;53:307–19.
91. Meers P. Enzyme-activated targeting of liposomes. Adv Drug Delivery Rev. 2001;53:265–72.
92. Woo GL, Mittleman MW, Santerre JP. Synthesis and characterization of a novel biodegradable antimicrobial polymer. Biomaterials. 2000;21:1235–46.
93. Tanahara M, Suzuki Y, Nishimura Y, Suzuki K, Kakimaru Y, Fukunishi Y. A novel microbial infection-responsive drug release system. J Pharm Sci. 1999;88:510–4.
94. Kwok CS, Mourad PD, Crum LA, Ratner BD. Self-assembled molecular structures as ultrasonically responsive barrier membranes for pulsatile drug delivery. J Biomed Mater Res. 2001; 57:151–64.
95. Norris P, Noble M, Francolini I, et al. Ultrasonically controlled release of ciprofloxacin from self-assembled coatings on poly(2-hydroxyethyl methacrylate) hydrogels for *Pseudomonas aeruginosa* biofilm prevention. Antimicrob Ag Chemother. 2005;49:4272–9.
96. Malcolm RK, McCullagh SD, Woolfson AD, Gorman SP, Jones DS, Cuddy J. Controlled release of a model antibacterial drug from a novel self-lubricating silicone biomaterial. J Control Rel. 2004;97:313–20.

97. Darouiche RO, Mansouri MD, Gawande PV, Madhyastha S. Efficacy of combination of chlorhexidine and protamine sulphate against device-associated pathogens. J Antimicrob Chemother. 2008;61:651–7.
98. Hachem R, Reitzel R, Borne A, et al. Novel antiseptic urinary catheters for prevention of urinary tract infections: correlation of in vivo and in vitro test results. Antimicrob Ag Chemother. 2009;53:5145–9.
99. Regev-Shoshani G, Ko M, Miller C, Av-Gay Y. The slow release of nitric oxide from charged catheters and its effect on biofilm formation by *Escherichia coli.* Antimicrobial Agents Chemother. 2010;54:273–9.
100. Getliffe K. The characteristics and management of patients with recurrent blockage of long-term urinary catheters. J Adv Nursing. 1994;20:140–9.
101. Kohler-Ockmore J, Feneley RCL. Long-term catheterization of the bladder, prevalence and morbidity. Brit J Urol. 1996;77:347–51.
102. Capewell AE, Morris SL. Audit of catheter management provided by district nurses and continence advisors. Brit J Urol. 1993;71:259–64.
103. Morris NS, Stickler DJ, Winters C. Which indwelling urethral catheters resist encrustation by *Proteus mirabilis* biofilms? Brit J Urol. 1997;80:58–63.
104. Morris NS, Stickler DJ, McLean RJ. The development of bacterial biofilms on indwelling catheters. World J Urol. 1999;17:345–50.
105. Kunin CM. Blockage of urinary catheters: role of microorganisms and constituents of the urine on formation of encrustations. J Clin Epidemiol. 1989;42:835–42.
106. Stickler DJ, Ganderton L, King J, Nettleton J, Morris NS. *Proteus mirabilis* biofilms and the encrustation of urethral catheters. Urol Res. 1993;21:407–11.
107. Jones BD, Mobley HLT. Genetic and biochemical diversity of ureases of *Proteus, Providencia* and *Morganella* species isolated from urinary tract infection. Infect Immun. 1987;55:2198–203.
108. Stickler DJ, Morris N, Moreno MC, Sabbuba NA. Studies on the formation of crystalline bacterial biofilms on urethral catheters. Eur J Clin Microbial Infect Dis. 1998;17:649–52.
109. Broomfield RJ, Morgan SD, Khan A, Stickler DJ. Crystalline bacterial biofilm formation on urinary catheters by urease producing urinary tract pathogens: a simple method of control. J Med Microbiol. 2009;58:1367–75.
110. MacLeod SM, Stickler DJ. Species interaction in mixed community biofilm on urinary catheters. J Med Microbiol. 2007;56:1549–57.
111. Donlan RM. Biofilms: microbial life on surfaces. Emerg Infect Dis. 2002;8:881–90.
112. Jacobsen SM, Stickler DJ, Mobley HTL, Shirtliff ME. Complicated catheter-associated urinary tract infections due to *Escherichia coli* and *Proteus mirabilis.* Clin Microbiol Revs. 2008;21:26–59.
113. Downer A, Morris NS, Feast WJ, Stickler DJ. Polymer surface properties and their effect on adhesion of *Proteus mirabilis.* Proc Inst Mech Eng. 2003;217:279–89.
114. Cox AJ. Comparison of catheter surface morphologies. Brit J Urol. 1990;65:55–60.
115. Stickler DJ, Young R, Jones G, Sabbuba NS, Morris NS. Why are Foley catheters so vulnerable to encrustation and blockage by crystalline bacterial biofilm? Urol Res. 2003;31:306–11.
116. Stickler DJ, Lear JC, Morris NS, et al. Observations on the adherence of *Proteus mirabilis* onto polymer surfaces. J Appl Microbiol. 2006;100:1028–33.
117. Stickler DJ, Morgan SD. Observations on the development of the crystalline bacterial biofilms that encrust and block Foley catheters. J Hosp Infect. 2008;69:350–60.
118. Morgan SD, Rigby D, Stickler DJ. A study of the structure of the crystalline biofilms that encrust and block silver Foley catheters. Urol Res. 2009;37:88–93.
119. Hayward JA, Chapman D. Biomembrane surfaces as models for polymer design: the potential for haemocompatibility. Biomaterials. 1984;5:135–42.
120. Campbell EJ, O'Byrne V, Stratford PW, et al. Biocompatible surfaces using methacroylphosphorylcholine laurylmethacrylate co-polymers. ASIO J. 1994;40:853–7.
121. Ishihara K, Iwasaki Y. Reduced protein adsorption on novel phospholipids polymers. J Biomater Appl. 1998;13:111–27.

122. Andrews CS, Denyer SP, Hall B, Hanlon GW, Lloyd AW. A comparison of the use of an ATP-based bioluminescent assay and image analysis for the assessment of bacterial adhesion to standard HEMA and biomimetic soft contact lenses. Biomaterials. 2001;22:3225–33.
123. Stickler DJ, Evans A, Morris NS, Hughes G. Strategies for the control of catheter encrustation. Int J Antimicrob Ag. 2002;19:499–506.
124. Laube N, Kleinen L, Bradenahl J, Meissner A. Diamond-like carbon coatings on ureteral stents—a new strategy for decreasing the formation of crystalline bacterial biofilms? J Urol. 2007;177:1923–7.
125. Kunin CM. Can we build a better urinary catheter? New Eng J Med. 1988;319:365–6.
126. Garcia MM, Gulati S, Liepman D, Stackhouse GB, Greene K, Stoller ML. Traditional Foley drainage systems—Do they drain the bladder? J Urol. 2006;177:203–7.
127. German K, Rowley P, Stone D, Kumar U, Blackford HN. A ramdomized cross-over study comparing the use of a catheter valve and a leg bag in urethrally catheterized male patients. Br J Urol. 1997;79:96–8.
128. Fader M, Petterssen L, Brooks R, et al. A multi-centre comparative evaluation of catheter valves. Br J Nurs. 1997;6:359–67.
129. Addison R. Catheter valves: a special focus on the Bard Flip-Flo catheter. Br J Nurs. 1999; 8:576–80.
130. Doherty W. The Sims Portex catheter valve: an alternative to the leg bag. Br J Nurs. 1999;8: 459–62.
131. Sabbuba NA, Stickler DJ, Long MJ, Dong Z, Short TD, Feneley RJC. Does valve regulated release of urine from the bladder decrease encrustation and blockage of indwelling catheters by *Proteus mirabilis* biofilms? J Urol. 2005;173:262–6.
132. Lee SJ, Kim SW, Cho YH, et al. A comparative multi-centre study on the incidence of catheter-associated urinary tract infection between nitrofurazone-coated and silicone catheters. Int J Antimicrobial Ag. 2004;24 Suppl 1:S65–9.
133. Brocklehurst JC, Hickey DS, Davies I, Kennedy AP, Morris JA. A new urethral catheter. Br Med J. 1988;296:1691–3.
134. Kunin CM, Finkelberg Z. Evaluation of an intraurethral lubricating catheter in preventing catheter-induced urinary tract infections. J Urol. 1971;106:928–30.
135. Lawrence EL, Turner IG. Kink flow and retention properties of urinary catheters part 2: novel design. J Mater Sci: Mater Med. 2006;17:153–9.
136. Feneley RCL, Kunin CM, Stickler DJ. An indwelling catheter for the 21st century. Br J Urol Int. 2012;109:1746–9.

Chapter 19
Antimicrobial-Modified Vascular Catheters

Rabih O. Darouiche

Abstract Vascular catheters constitute an essential component of modern health care. The escalating use of vascular catheters has highlighted the need to optimize prevention of infectious complications. Bloodstream infection is the most common serious complication of indwelling vascular catheters. Although strict implementation of traditional infection control measures continues to be the primary measure for preventing infection, the level of adherence by medical staff to such measures varies over time, between different units, and across various medical centers. This limitation underscores the need to assess the potential clinical impact of surface modification of vascular catheters. Since catheter colonization can be a prelude to infection, antimicrobial modification of the surfaces of catheters has the potential of not only inhibiting bacterial colonization of the catheter surfaces but also reducing the incidence of catheter-related bloodstream infection. A number of antimicrobial-modified vascular catheters are currently available for patient care, but they differ with regard to the type of antimicrobials, application on the external versus internal catheter surfaces, spectrum and durability of antimicrobial activity, and the ability to clinically protect against catheter-related bloodstream infection. Scientific evidence should guide the present and future applications of antimicrobial-modified catheters.

19.1 Introduction

The escalating use of advanced technology has contributed to the great strides in Medicine over the last decade. However, the more extensive use of vascular catheters and longer survival of patients at an inherently higher risk for infection have illuminated the increasing importance of catheter-related infection. Although

R.O. Darouiche, M.D. (✉)
Center for Prostheses Infection and Infectious Disease Section, Michael E. DeBakey Veterans Affairs Medical Center and Baylor College of Medicine, Houston, TX, USA
e-mail: rdarouiche@aol.com

T.F. Moriarty et al. (eds.), *Biomaterials Associated Infection: Immunological Aspects and Antimicrobial Strategies*, DOI 10.1007/978-1-4614-1031-7_19,

infection of vascular catheters can clinically manifest as exit-site infection or thrombophlebitis, the most common serious complication of vascular catheters is bloodstream infection. Since cure of catheter-related bloodstream infection usually requires removal of the infected catheter along with occasionally prolonged antibiotic therapy, we have witnessed over the past decade a soaring global interest by healthcare providers, scientific organizations, and regulatory agencies in preventing infection [1–3].

The objectives of this chapter are to: (1) address the need to prevent infection of vascular catheters; (2) discuss the pathogenesis of catheter-related bloodstream infection; (3) explore the optimal properties of antimicrobial-modified catheters; (4) analyze the protective efficacy of various types of antimicrobial-modified catheters; and (5) provide future directions for surface modifications of vascular catheters.

19.1.1 Need to Prevent Infection

The escalating drive to prevent catheter-related bloodstream infection is fueled by the need to achieve the following five goals:

(a) Decrease the unacceptably high incidence of infection: Most cases of catheter-related bloodstream infection arise from the almost 6 million central venous catheters that are inserted each year in the USA, including about 4.5 million short-term catheters (with a mean duration of placement of 7–10 days) and 1.5 million long-term catheters, mainly peripherally inserted central catheters (PICC, over a million) and tunneled catheters [4]. The vast majority of the 175 million vascular catheters that are inserted annually in the USA are peripherally placed and are very unlikely (<0.1–0.2%) to cause bloodstream infection [5]. Although the overall rate of catheter-related bloodstream infection has decreased over the past decade, still more than 200,000 cases of infection occur each year in the USA [1, 4, 6], including at least 80,000 cases among critically ill patients [7].
(b) Avoid serious infectious complications: These include septic shock, multiorgan dysfunction, and death, with an attributable mortality as high as 23% [8] to 25% in critically ill patients [9–11].
(c) Limit the economic impact of infection: it additionally costs as much as 29 [9–11] to 56 [12] thousand dollars to treat a bout of catheter-related bloodstream infection in critically ill patients. Furthermore, the surviving patients are hospitalized for a mean of 6.5 [9–11] to 22 [12] days longer than those who do not develop such an infection. The overall annual cost of management of catheter-related bloodstream infection in the USA is estimated to be about $2.3 billion.
(d) Limit the reservoir of antibiotic resistance: Bacteria embedded within the biofilm surrounding the indwelling vascular catheter are typically much less responsive to commonly used antibiotics than planktonic bacteria [13]. For biofilm-embedded organisms to respond to antibiotics, they require 100–1,000-fold higher concentrations as compared with free-floating organisms. Potential mechanisms of poor response of biofilm-embedded bacteria to antibiotics

include relatively slow growth of bacteria within the biofilm (thereby limiting the antimicrobial activity of cell-wall active antibiotics like glycopeptides and beta-lactams), limited penetration of antibiotics into the biofilm, and the presence of substances within the biofilm that could inhibit the antimicrobial activity of antibiotics. As a result, the biofilm around vascular catheters can act as a large reservoir of antibiotic-resistant bacteria, including vancomycin-intermediate or-resistant staphylococci [14, 15].

(e) Reduce the likelihood of device dysfunction: Bacterial colonization of the catheter, even in the absence of clinical manifestations of infection, can be associated with catheter dysfunction [13], including catheter occlusion and vascular thrombosis [16]. A similar relationship exists with other devices as loosening of joint prostheses [17], capsular contracture of mammary implants [18], blockage of biliary stents [19], and failure of dental implants [20] have all been associated with asymptomatic bacterial colonization of the device.

19.1.2 Pathogenesis of Vascular Catheter Infection

Optimal prevention should be based on comprehensive understanding of the pathogenesis of catheter-related bloodstream infection [21]. In that regard, there are three important pathogen-related factors that need to be addressed.

19.1.2.1 Source of Pathogens

The four potential sources of pathogens are the patient's skin around the site of catheter insertion, a contaminated catheter hub or needleless access device, hematogenous seeding from a distant site of infection, and an infected infusate. The specific source of infecting organisms eludes determination in about one-quarter of cases of catheter-related bloodstream infection. The first two sources of infecting pathogens illustrate the unparalleled importance of the skin (of patients and healthcare providers, respectively) as they collectively are responsible for the vast majority of cases of catheter-associated infections. The patient's skin around the catheter insertion site is the most common source of organisms that colonize short-term (mean duration, <7–10 days) central venous catheters [22]. The concentration of bacteria on the skin differs between various bodily sites with the highest concentration generally present in the femoral > jugular > subclavian area. Factors that lead to higher bacterial concentration on the skin at the site of catheter insertion include presence of hair follicles, high temperature, moist environment, and soiling by bacteria-containing bodily fluids and secretions. After colonizing the external surface of the catheter, skin-derived bacteria may migrate along the subcutaneous segment into the distal intravascular segment and cause bloodstream infection. Long-term vascular catheters that are subjected to frequent manipulation of the catheter hub and/or needleless access devices become contaminated by organisms that could originate from the hands of medical personnel then migrate along the internal surface of the catheter into the intravascular segment, thereby resulting in bloodstream infection [23]. The difference in the pathogenesis of infections associated with short-term versus long-term catheters helps explain why surface-modified vascular catheters with antimicrobial activity only along the external surface have the potential to protect against infection associated with short-term but not long-term catheters.

19.1.2.2 Type of Pathogens

The sources of infection of vascular catheters dictate the microbiologic etiology [24]. Since the patient's skin around the catheter insertion site and the skin of medical personnel hands constitute the two most common sources of infecting pathogens, at least two-thirds of cases of catheter-related bloodstream infection are caused by staphylococcal organisms (including coagulase-negative staphylococci and *Staphylococcus aureus*). Less common pathogens, including gram-negative bacteria and *Candida* spp., collectively cause about one quarter of cases of infection and are particularly instrumental in causing infection of long-term catheters and in immunocompromised subjects. Taking into consideration this wide array of potential pathogens, surface-modified catheters that are active only against gram-positive bacteria may not significantly reduce the overall rate of infection and, in some instances, may even predispose to superinfection by less common but more opportunistic pathogens.

19.1.2.3 Environment of Pathogens

As is the case with other medical devices, infection of vascular catheters is initiated by the formation of a layer of biofilm around the indwelling catheter [13, 25]. The biofilm comprises both bacterial products (fibroglycocalyx in the case of coagulase-negative staphylococci) and host elements (blood cells, platelets, and tissue ligands, including fibronectin, fibrinogen, and fibrin that variably adhere to well-described receptors on the surface of certain organisms, including staphylococci and *Candida* spp.). Not only does the biofilm act as a barrier that protects embedded organisms from host immune defenses like phagocytosis and opsonization [25], it also can impair the activity [26] and, possibly, the penetration [13, 26, 27] of antibiotics within the biofilm. This unique biofilm environment may help explain why surface-modified vascular catheters that contain antimicrobial agents that retain their activity within the biofilm and leach off the catheter surface to produce a zone of inhibition against organisms that are embedded deep within the biofilm tend to be clinically protective [28]. Since host tissue can act as a niche for catheter-related infection, it is possible that leaching of antimicrobial agents off the surface of coated vascular catheters could also kill bacteria present in the peri-implant tissues [29].

19.1.3 Optimal Properties of Antimicrobial-Modified Catheters

Since bacterial colonization of vascular catheters is a prelude to clinical infection, antimicrobial modification of the surface of catheters has the potential of protecting against the evolution of catheter-related bloodstream infection. Dipping the catheter in antimicrobial solutions is considered a form of antimicrobial modification of the catheter surfaces, but the vast majority of antimicrobial-modified vascular catheters entail coating/impregnation. Although the term "impregnation" technically reflects the presence of antimicrobial agents within the material of the catheter whereas the term "coating" indicates bonding of the antimicrobial agents to only the surface(s) of the catheter, impregnated catheters tend to have a longer durability of antimicrobial activity than coated catheters. However, these two terms are often used interchangeably in both publications and scientific presentations. To avoid confusion, this chapter will consistently utilize the word "coated" rather than "impregnated."

The presence, or lack thereof, of certain characteristics of antimicrobial modification of vascular catheters can help predict its ability to resist clinical infection. The following optimal characteristics are generally present in clinically protective antimicrobial-modified catheters, but are partially or completely absent in antimicrobial-modified catheters that have no proof of clinical efficacy.

19.1.3.1 Antimicrobial Activity Against Most Potential Pathogens

Although staphylococci cause most infections of vascular catheters, it is imperative that antimicrobial-modified catheters are also active against other potential pathogens so that superinfection by gram-negative bacteria and *Candida* spp. does not ensue. This adverse consequence has been observed, for example, with vancomycin-dipped vascular catheters which led to a higher incidence of Candida growth as compared with control catheters [30].

19.1.3.2 Antimicrobial Activity During the Time at Highest Risk for Inoculating Pathogens

Since organisms can adhere to the catheter surface both at the time of and subsequent to placement, surface-modified catheters that possess antimicrobial activity for only hours or few days may not be clinically ineffective.

19.1.3.3 Antimicrobial Agents Leach Off the Catheter to Produce Zones of Inhibition

This property not only could help ensure activity against organisms that adhere directly to the catheter surface and those that are embedded deep within the biofilm [28], but also could kill bacteria present in peri-implant tissues [29]. Since silver molecules are usually tightly bound to the modified surface of the device, the lack of zones of inhibition explains the discordant results of silver-coated devices in terms of observed efficacy *in vitro* (under circumstances that do not account for the biofilm) versus largely non-demonstrable efficacy *in vivo* (as silver molecules have no access to organisms deeply embedded within the biofilm or present in peri-implant tissues) [28, 29]. The size of the zone of inhibition against *S. aureus* in *in vitro* assays can also help predict the degree of *in vivo* efficacy and clinical protection [31–35].

19.1.3.4 Antimicrobial Activity Along Both Catheter Surfaces

Although most infections of short-term central venous catheters are caused by organisms that originate from the patients' skin and colonize the external surface of the catheter, prolonged duration of catheter placement favors colonization of the internal catheter surface. This helps explain why external coating with chlorhexidine and silver sulfadiazine generally is more clinically protective against infection of short-term than long-term vascular catheters. The value of using antimicrobial-modified vascular catheters that afford luminal protection against infection is becoming even more important because most currently used infection control practices (optimal skin antisepsis in patients, use of maximal barriers during catheter insertion, avoidance of femoral catheterization, etc.) primarily reduce contamination of the external catheter surface by the patients' skin flora, thereby potentially favoring a switch from external to internal surface colonization.

19.1.4 Protective Efficacy of Antimicrobial-Modified Catheters

Dozens of antimicrobial-modified vascular catheters have been found to reduce bacterial colonization of the catheter *in vitro* and several of those have been examined in animal models, but only a few have been clinically assessed and will be addressed in this chapter. Since the clinical efficacy of preventive approaches is best examined in randomized clinical trials, the protective efficacy of antimicrobial-modified catheters will be addressed by relying on the results of *only* such designed studies that have been reported in peer-reviewed journals. Clinical trials with less desirable designs, including non-randomized, retrospective, and cross-over studies will not be relied upon because confounding variables may lead to scientifically invalid conclusions. Likewise, the results of studies that have been reported in an abstract form but have not yet been subjected to the peer review process will be excluded.

Although catheter colonization is a prelude to catheter-related bloodstream infection, most colonized catheters do not become clinically infected. Therefore, a significant reduction in the rate of catheter colonization does not, in and by itself, constitute a proof of clinical efficacy. The ultimate proof of clinical efficacy is a significant reduction in the rate of catheter-related bloodstream infection in a sufficiently powered, randomized controlled trial. As a corollary, if an inadequately powered clinical trial that fails to demonstrate a significant reduction in the rate of

Table 19.1 Clinical efficacy of antimicrobial-modified vascular catheters

Catheters with no proof of significant decrease in catheter colonization or CRBSI[a]
Catheters coated with silver
Catheters coated with benzalkonium chloride
Catheters with proof of significant decrease in catheter colonization but not CRBSI
Catheters dipped in antibiotics
Catheters coated with silver–platinum–carbon
Catheters coated with rifampin–miconazole
Catheters with proof of significant decrease in CRBSI
Catheters coated with chlorhexidine–silver sulfadiazine
Catheters coated with minocycline–rifampin

[a] *CRBSI* catheter-related bloodstream infection

catheter-related bloodstream infection despite a significantly lower rate of catheter colonization in the antimicrobial-modified versus control group, it is implied that that particular antimicrobial-modified catheter experimental strategy either does not clinically protect against bloodstream infection or needs to be further examined in a larger-sized clinical trial. Although the most desirable impact of potentially preventive measure of using an antimicrobial-modified catheter would be a reduction in mortality, it would be impractical to conduct a several thousand-patient trial that would be sufficiently powered to assess this outcome which has a relatively very low incidence (equivalent to the incidence of catheter-related bloodstream infection multiplied by the risk of dying from the infection).

A critical analysis of the peer-reviewed literature allows the categorization of antimicrobial-modified vascular catheters into 3 groups based on how they compare with *control uncoated* catheters: (1) approaches with no proof of significant decrease in catheter colonization or catheter-related bloodstream infection; (2) approaches with proof of significant decrease in catheter colonization but not catheter-related bloodstream infection; and (3) approaches with proof of significant decrease in catheter-related bloodstream infection (Table 19.1). Not unexpectedly, not all studies of the same type of antimicrobial-modified catheter always yield the similar results; in that case, the quality and quantity of the available evidence from randomized controlled trials is assessed to accurately assign catheters to the most proper category. The findings of properly conducted meta-analyses and which correctly adjust for confounding variables can be relied upon to help with category assignment of antimicrobial-modified catheters.

19.1.4.1 Antimicrobial-Modified Catheters with No Proof of Significant Decrease in Catheter Colonization or Catheter-Related Bloodstream Infection

Catheters Coated with Silver

This represents the most investigated approach in this category and focuses on surface modification of the catheter surface with different silver-containing moieties. Not only did *in vitro* studies yield conflicting findings with regard to efficacy as some studies showed reduced bacterial adherence to the surfaces of polyurethane silver-coated catheters [36] and others indicated that silicone silver-coated catheters is ineffective [37], but also the results of animal models yielded inconclusive results [38]. Although an earlier randomized controlled trial reported that silver-coated central venous catheters are protective [39], several subsequent randomized trials failed to show that silver-coated short-term central venous catheters [40–42] and long-term hemodialysis catheters [43] significantly reduce the incidence of catheter colonization and catheter-related bloodstream infection. Not only is the silver application to the surface of short-term catheters generally ineffective, but its incorporation onto the surface of long-term hemodialysis catheters can negatively impact the incidence of infection and cause adverse skin changes [43]. In some instances, discrepancies in the efficacy between various types of silver-containing catheters could be attributed, in part, to the use of different polymers and chemical substances. In general, potential explanations for the poor clinical efficacy of silver-coated catheters include the following: (a) poor leaching of silver molecules off the surface of most coated catheters and, therefore, lack of effective zones of inhibition around the catheter surface limits the access of the silver molecules to biofilm-embedded organisms and bacteria present in peri-implant tissues; (b) silver tends to bind to host proteins, thereby resulting in lower concentration of free active silver molecules; and (c) the antimicrobial activity of silver can be impaired in the absence of bodily fluids [30, 44].

Catheters Coated with Benzalkonium Chloride

In vitro studies showed that heparin-coated catheters possess some antimicrobial activity, possibly attributable to the weak antiseptic benzalkonium chloride which is applied to the catheter surface primarily for its surfactant activity to allow bonding with heparin [45]. A randomized clinical trial reported significant reduction in colonization of benzalkonium chloride/heparin-coated catheters but did not assess the incidence of infection [46]. However, a small randomized controlled trial failed to show a significant decrease in the rate of colonization and bloodstream infection associated with benzalkonium chloride-coated versus uncoated catheters [47].

19.1.4.2 Antimicrobial-Modified Catheters with Proof of Significant Decrease in Catheter Colonization But Not Catheter-Related Bloodstream Infection

Catheters Dipped in Antibiotics

This bed-side approach was studied by dipping catheters that were pretreated with a positively charged surfactant (usually tridodecyl methyl ammonium chloride: TDMAC) in a solution of negatively charged antibiotics (such as cephalosporins and glycopeptides) just prior to catheter insertion. Although prospective, randomized trials showed that short-term central venous or arterial catheters that were pretreated with the surfactant TDMAC and dipped at the bedside in cefazolin [48] or vancomycin [49] just prior to insertion was associated with reduced rate of catheter colonization as compared with non-immersed catheters, there was no reported impact on the incidence of catheter-related bloodstream infection.

The common objective of both antimicrobial coating and bed-side dipping of the vascular catheter in an antimicrobial solution is to inhibit bacterial colonization of the device and, hopefully, protect against catheter-related bloodstream infection. However, the approach of antimicrobial coating is more advantageous than the strategy of dipping devices in antimicrobial solution. It is certainly more practical and quicker to insert an antimicrobial-coated catheter without having to prepare the antimicrobial solution and dip the catheter in it before insertion. Although it is a manufacturing requirement that a homogeneous antimicrobial film surrounds different parts of an antimicrobial-coated catheter, different segments of the antimicrobial-dipped catheter may have different antimicrobial exposure during the process of dipping. Whereas the amounts of antimicrobials that are bound to and leach from antimicrobial-coated catheters are generally known, that is not the case with antimicrobial-dipped catheters. As a corollary, the durability of antimicrobial activity can be manipulated when manufacturing antimicrobial-coated catheters to provide different duration of antimicrobial activity for short-term versus long-term catheters, but that goal is hard to achieve with antimicrobial-dipped catheters. In general, dipping catheters in antibiotic solutions incorporates relatively smaller amounts of antibiotics onto the surface of the catheter and results in a shorter duration of antimicrobial activity than coated catheters (could be hours vs. days or weeks). Finally, whereas catheters may be dipped in various types of antimicrobial solutions, there exist regulatory limitations on the choice of antimicrobials that could be approved for coating. For example, the American and European regulatory authorities have never approved vancomycin-coated catheters because this antibiotic is the drug of choice for treatment of established catheter-related bloodstream infections, with the understanding that should resistance or adverse reaction to this antibiotic occur it could affect the patient's care.

Catheters Coated with Silver, Platinum and Carbon

A randomized controlled trial showed that short-term central venous catheters coated with silver–platinum–carbon (so-called Oligon) were significantly less likely to become colonized than conventional uncoated catheters, but there was no significant reduction in the rate of catheter-related bloodstream infection [50]. A more recent randomized clinical trial that compared two types of antimicrobial-modifications demonstrated that catheters coated with silver–platinum–carbon were significantly more likely to be colonized than catheters coated with minocycline and rifampin, but there were no significant differences in the incidence of catheter-related bloodstream infection [51].

Catheters Coated with Rifampin and Miconazole

This antimicrobial-modified catheter which is currently not available in the USA is intended to render activity against both bacterial and fungal organisms. Since miconazole is primarily active against fungi and possesses no or minimal activity against bacteria, there is a concern that catheters coated with miconazole and rifampin could increase the risk of bacteria developing rifampin resistance. It has been shown *in vitro* to reduce catheter colonization by staphylococcal and enterococcal organisms [52]. Although a randomized clinical trial reported significantly lower rates of colonization of rifampin/miconazole-coated central venous catheters as compared with standard uncoated catheters, there was no significant impact on catheter-related bloodstream infection [53].

19.1.4.3 Antimicrobial-Modified Catheters with Proof of Significant Decrease in Catheter-Related Bloodstream Infection

Catheters Coated with Chlorhexidine and Silver Sulfadiazine

There exist two generations of catheters coated with the combination of chlorhexidine and silver sulfadiazine. The first-generation and most clinically studied catheter [54–63] has antimicrobial agents incorporated only along the external surface of the catheter. The second-generation catheter differs in two ways from the first-generation catheter: (1) both chlorhexidine and silver sulfadiazine are incorporated onto the external and internal surfaces of the catheter as well as the hub, and (2) it contains three times higher amount of chlorhexidine [64].

The randomized controlled trials of first-generation central venous catheters coated with chlorhexidine and silver sulfadiazine versus control uncoated catheters yielded variable results. For instance, some studies showed impact of this particular coating approach on neither catheter colonization nor catheter-related bloodstream infection [54–56], whereas others demonstrated significant reduction in the rates of catheter colonization but not catheter-related bloodstream infection [57–61]. Yet a third group of trials reported significant protection against catheter colonization and catheter-related bloodstream infection afforded by the chlorhexidine/silver sulfadiazine-coated catheters [62, 63]. It is important to note that some of the above-discussed trials were relatively small studies. However, a number of meta-analyses have concluded that chlorhexidine/silver sulfadiazine-coated central venous catheters significantly protect against catheter-related bloodstream infection [65–67].

Since this first-generation chlorhexidine/silver sulfadiazine-coated catheter provided short-lived (about 1 week) antimicrobial activity only along the external surface of the catheter [68], it expectedly did not protect against infection of catheters that remained in place for a mean of 20 days in patients with hematological malignancy as compared with uncoated catheters [56]. The second-generation chlorhexidine/silver suldfadiazine-coated catheters have a longer durability of antimicrobial activity than first-generation coated catheters. A large randomized controlled trial demonstrated that second-generation chlorhexidine/silver suldfadiazine-coated, polyurethane short-term, central venous catheters are less likely to become colonized than uncoated catheters, but it showed only a statistically insignificant trend for a lower rate of catheter-related bloodstream infection [64]. Since the incidence of catheter-related bloodstream infection in the uncoated catheter group was lower than usual, this study may not have had sufficient power to assess the outcome of bloodstream infection. As both the first- and second-generation catheters coated with chlorhexidine and silver sulfadiazine generally reduced catheter colonization by a comparable degree, it is reasonable to regard these two catheters as being equally protective against infection of short-term catheters. Although chlorhexidine-related events of anaphylaxis or anaphylactoid reactions have been reported primarily in Japan and less so in Europe and Australia, this issue has not been problematic in the USA [69].

It is worth noting that a recent randomized controlled trial demonstrated that first-generation central venous catheters coated with chlorhexidine-silver sulfadiazine and catheters externally coated with 5-fluorouracil had comparable rates of catheter colonization and catheter-related bloodstream infection [70]. However, since the latter catheters have not been clinically compared to uncoated catheters, the 5-flurouracil-coated catheters were not included in Table 19.1.

Catheters Coated with Minocycline and Rifampin

This combination of antibiotics is unique because (a) both agents are active against the vast majority of staphylococcal isolates, including methicillin-resistant *S. aureus* (MRSA) and methicillin-resistant *S. epidermidis* (MRSE) [71], (b) the combination of agents provides broad-spectrum antimicrobial activity against most pathogens that can cause catheter-related bloodstream infection, thereby reducing the likelihood of developing superinfection by gram-negative bacteria or *Candida* spp. [72], (c) since minocycline and rifampin have different mechanisms of activity (minocycline retards protein synthesis and rifampin inhibits DNA-dependent RNA polymerase), it is unlikely that a bacterial strain will become concomitantly resistant to both agents [71]; and (d) unlike many other antibiotics (such as glycopeptides, quinolones, and aminoglycosides) that are much less active against biofilm bacteria than planktonic bacteria, minocycline [73] and rifampin [74] are particularly active against biofilm-embedded bacteria.

The clinical efficacy of this antimicrobial-modified catheter was demonstrated in a large randomized controlled clinical trials which showed that polyurethane short-term central venous catheters coated with minocycline and rifampin are significantly less likely than uncoated catheters to become colonized and cause bloodstream infection [75]. An insufficiently powered trial did not yield significant protection against infection as compared with uncoated short-term central venous catheters [76]. In a large randomized controlled trial, polyurethane short-term central venous catheters coated with minocycline and rifampin were significantly more protective against catheter colonization and catheter-related bloodstream infection than first-generation chlorhexidine/silver sulfadiazine-coated catheters [77]. The superior clinical protection afforded by minocycline/rifampin-coated catheters could be predicted since these catheters (a) were shown in an animal study to prevent *S. aureus* infection of percutaneously placed catheter segments more than first-generation catheters coated with chlorhexidine and silver sulfadiazine [72] and (b) have a longer in-vivo durability of antimicrobial activity, as determined by the residual zones of inhibition generated by catheters removed from patients [78].

The clinical benefit achieved by incorporating the combination of minocycline and rifampin onto the surfaces of short-term polyurethane central venous catheters prompted the assessment of this approach in long-term silicone catheters. Indeed, a randomized controlled trial demonstrated that similarly coated, silicone long-term central venous catheters were significantly less likely than uncoated catheters to result in bloodstream infection [79]. Another randomized controlled trial was conducted to determine whether antimicrobial coating can obviate the need for the impractical, time-consuming, and expensive practice of tunneling long-term central venous catheters, and it demonstrated that non-tunneled minocycline/rifampin-coated catheters are significantly less likely than tunneled uncoated catheters to result in bloodstream infection [4]. Likewise, polyurethane hemodialysis catheters coated with minocycline and rifampin significantly protected against catheter-related bloodstream infection as compared with uncoated catheters in patients with acute renal failure [80]. A number of meta-analyses demonstrated that vascular catheters coated with minocycline and rifampin are clinically beneficial when compared to uncoated catheters [66, 67, 81–83] and are more protective against infection than catheters coated with chlorhexidine and silver sulfadiazine [67, 82, 83].

Although a single study had suggested that prolonged exposure of bacteria to rifampin *in vitro* may result in antibiotic resistance [84], two other *in vitro* studies showed no evidence for emergence of antibiotic resistance [85, 86]. More importantly, clinical trials have shown no evidence of developing resistance to the antimicrobials incorporated onto short-term [77] or long-term [79, 87] central venous catheters coated with minocycline and rifampin, as well as short-term chlorhexidine/silver sulfadiazine-coated catheters [62].

A scientific dialogue took place, on and off, during the past decade between advocates and doubters of the efficacy of antimicrobial-modified vascular catheters [88–90]. The preponderance of evidence and the most recent Centers for Disease Control and Prevention (CDC) guidelines for the prevention of intravascular catheter-related infections support the benefits and use of certain clinically protective vascular catheters (namely, catheters coated with chlorhexidine and silver sulfadiazine and catheters coated with minocycline and rifampin) [1].

19.1.5 Future Directions

As management of catheter-related bloodstream infection can be both difficult and expensive, prevention of such infections remains a priority. Since almost 90% of cases of bloodstream infection in critically ill patients arise from an indwelling infected vascular catheter [6, 91], it is essential that we optimize existing practices and explore new approaches for surface modification of vascular catheters.

19.1.5.1 Combining Basic Infection Control Measures with Existing Protective Catheters

Recent evidence indicated that strict implementation of basic infection control measures can significantly protect against catheter-related bloodstream infection in critically ill patients [92] and laid the ground for the just-released updated CDC guidelines to universally apply the so-called "aseptic-technique bundle" [1]. This bundle comprises hand washing with soap/water or alcohol-based hand rub, avoidance of femoral catheterization, use of >0.5% chlorhexidine preparations with alcohol to cleanse the patient's skin at the insertion site, utilization of maximal barriers during catheter insertion, and prompt removal of vascular catheters when no longer needed. Although institutional implementation of quality improvement programs to educate healthcare providers and improve their compliance with basic infection control measures when inserting or maintaining a vascular catheter are generally beneficial, the level of adherence and duration of benefit may not be permanent [93, 94]. Moreover, some aseptic measures have probably reached a point of limited return as they have reduced but did not eliminate the occurrence of catheter-related bloodstream infection [95]. In that regard, basic infection control measures and clinically protective antimicrobial-modified catheters can complement one another in achieving the desired goal of zero tolerance for catheter-related bloodstream infection. It is important to note that the updated CDC guidelines have issued the highest recommendation (category IA: Strongly recommended for implementation and strongly supported by well-designed experimental, clinical, and epidemiologic studies) for the use of "antiseptic/antibiotic impregnated short-term central venous catheters and chlorhexidine impregnated sponge dressings if the rate of infection is not decreasing despite adherence to other strategies (i.e., education and training, maximal sterile barrier precautions, and >0.5% chlorhexidine preparations with alcohol for skin antisepsis"). This represents an escalation of the strength of the recommendation for using antimicrobial-modified vascular catheters as compared with the last CDC guidelines that were issued in 2002 and had only a category 1B recommendation for using antimicrobial-modified vascular catheters [96]. Although optimal infection control measures save lives, more lives can potentially be saved by combining infection control measures and clinically protective technology.

19.1.5.2 Guidance for Using Currently Available Protective Catheters

Use of both patient-specific and institution-based factors should guide the utilization of the currently available antimicrobial-coated catheters that have been shown to be clinically protective against infection. The clinical protection afforded by antimicrobial-coated catheters has been documented primarily in populations of patients at high risk for infection, including, but not limited to, critically ill patients, immunocompromised subjects, and recipients of total parenteral nutrition. These catheters are not intended to be used in patients at low risk of infection such as those embarking on elective surgery procedure and expected to have the central venous catheter in place for only 1–2 days. An institutional incidence of catheter-related bloodstream infection of >3.3/1,000 catheter-days (equivalent to about 2%) could be considered high enough to consider the use of antimicrobial-coated vascular catheters [62, 97, 98]. Although clinically protective antimicrobial-coated catheters generally cost about 20% more than uncoated prototypes, tremendous savings can be incurred when using chlorhexidine/silver sulfadiazine [99] and, to a larger extent, the slightly more expensive but significantly more clinically protective minocycline/rifampin-coated catheters [100]. Thoughtful analysis can help streamline the proper selection of antimicrobial-modified catheters [101, 102].

19.1.5.3 Exploring New Technologic Advances

Like with the catheter itself, there is a need to examine the value of surface modification of other components of the vascular access system, including the needleless access devices, tubings, catheter hub, and catheter cap. Since all cases of catheter-related bloodstream infection are biofilm-based, incorporation of antibiofilm/antimicrobial agents into the catheter surface or catheter lock solutions may constitute the ultimate generation of surface modification.

Acknowledgments Dr. Darouiche is a coinventor of the impregnation of vascular catheters with the combination of minocycline and rifampin. Baylor College of Medicine, the employer of Dr. Darouiche, owns the corresponding patents and has executed a licensing agreement with Cook Inc. for manufacturing of vascular catheters impregnated with minocycline and rifampin. Baylor College of Medicine provides Dr. Darouiche with a certain portion of the royalties received from Cook Inc.

References

1. O'Grady NP, Alexander M, Burns LA, et al. Summary of recommendations: guidelines for the prevention of intravascular catheter-related infections. Clin Infect Dis. 2011;52:1087–99.
2. Demetriades D, May A, Gamble H. When does a Centers for Disease Control and Prevention recommendation become standard of care? Perhaps in the courtroom. American College of Surgeons mock trial: line sepsis liability. J Am Coll Surg. 2008;206:370–5.
3. McGee D, Gould M. Preventing complications of central venous catheterization. N Engl J Med. 2003;348:1123–33.
4. Darouiche RO, Berger DH, Khardori N, et al. Comparison of antimicrobial impregnation with tunneling of long-term central venous catheters: a randomized, controlled trial. Ann Surg. 2005;242:193–200.
5. Cicalini S, Palmieri F, Petrosillo N. Clinical review: new technologies for prevention of intravascular catheter-related infections. Crit Care. 2004;8:157–62.

6. Donelli G, Francolini I. Efficacy of antiadhesive, antibiotic and antiseptic coatings in preventing catheter-related infections: review. J Chemother. 2001;13:595–606.
7. Mermel LA. Prevention of intravascular catheter-related infections. Ann Intern Med. 2000;132: 391–402.
8. Warren DK, Quadir WW, Hollenbeak CS, et al. Attributable cost of catheter-associated bloodstream infections among intensive care patients in a nonteaching hospital. Crit Care Med. 2006;34:2084–9.
9. Pittet D, Tarara D, Wenzel RP. Nosocomial bloodstream infection in critically ill patients: excess length of stay, extra costs, and attributable mortality. JAMA. 1994;271:1598–601.
10. Heiselman D. Nosocomial bloodstream infections in the critically ill [Letter]. JAMA. 1994; 272: 1819–20.
11. Pittet D, Wenzel RP. Nosocomial bloodstream infections in the critically ill [Letter]. JAMA. 1994;272:1820–1.
12. Dimick JB, Pelz RK, Consunji R, et al. Increased resource use associated with catheter-related bloodstream infection in the surgical intensive care unit. Arch Surg. 2001;136:229–34.
13. Darouiche RO. Device-associated infections: a macroproblem that starts with microadherence. Clin Infect Dis. 2001;33:1567–72.
14. Smith TL, Pearson ML, Wilcox KR, et al. Emergence of vancomycin resistance in *Staphylococcus aureus*. N Engl J Med. 1999;340:493–501.
15. Sieradzki K, Roberts RB, Haber SW, Tomasz A. The development of vancomycin resistance in a patient with methicillin-resistant Staphylococcus aureus infection. N Engl J Med. 1999;340:517–23.
16. Raad II, Luna M, Khalil SM, Costerton JW, Lam C, Bodey GP. The relationship between the thombotic and infectious complications of central venous catheters. JAMA. 1994;271: 1014–6.
17. Darouiche RO, Landon GC, Patti JM, et al. Role of *Staphylococcus aureus* surface adhesions in orthopaedic device infections. J Med Microbiol. 1997;46:75–9.
18. Virden CP, Dobke MP, Stein P, Parsons CJ, Frank DH. Subclinical infection of the silicone breast implant surface as a possible cause of capsular contracture. Aesthetic Plast Surg. 1992;16:173–9.
19. Leung JW, Liu YL, Desta T, Libby E, Inciardi JF, Lam K. Is there a synergistic effect between mixed bacterial infection in biofilm formation in biliary stents? Gastrointest Endosc. 1998;48:250–7.
20. Esposito M, Hirsch J, Lekholm U, Thomsen P. Differential diagnosis and treatment strategies for biologic complications and failing oral implants: a review of the literature. Int J Oral Maxillofac Implants. 1999;14:473–90.
21. Safdar N, Kluger DM, Maki DG. A review of risk factors for catheter-related bloodstream infection caused by percutaneoulsy inserted, non-cuffed central venous catheters: implications for preventive strategies. Medicine (Baltimore). 2002;81:466–79.
22. Maki DG, Cobb L, Garman JK, et al. An attachable silver-impregnated cuff for prevention of infection with central venous catheters: a prospective randomized multicenter trial. Am J Med. 1988;85:307–14.
23. Salzman MB, Isenberg HD, Shapiro JF, et al. A prospective study of the catheter hub as the portal of entry for organisms causing catheter-related sepsis in neonates. J Infect Dis. 1993;167:487–90.
24. Raad II, Bodey GP. Infectious complications of indwelling vascular catheters. Clin Infect Dis. 1992;15:197–208.
25. Costerton JW, Stewart PS, Greenberg EP. Bacterial biofilms: a common cause of persistent infections. Science. 1999;284:1318–22.
26. Darouiche RO, Dhir A, Miller AJ, Landon GC, Raad II, Musher DM. Vancomycin penetration into biofilm covering infected prostheses and effect on bacteria. J Infect Dis. 1994;170:720–3.
27. Kumon H, Tomochika K, Matunaga T, et al. A sandwich cup method for the penetration assay of antimicrobial agents through *Pseudomonas* exopolysaccharides. Microbiol Immunol. 1994;38:615–9.
28. Darouiche RO. Anti-infective efficacy of silver-coated medical prostheses. Clin Infect Dis. 1999;29:1371–7.

29. Zaat SAJ, Broekhuizen CAN, Riool M. Future Microbiol. 2010;5:1149–51.
30. Thornton J, Todd NJ, Webster NR. Central venous line sepsis in the intensive care unit: a study comparing antibiotic coated catheters with plain catheters. Anaesthesia. 1996;51:1018–20.
31. Bassetti S, Hu J, D'Agostino Jr RB, Sherertz RJ. In vitro zones of inhibition of vascular catheters predict efficacy in preventing catheter infection with *Staphylococcus aureus* in vivo. Eur J Clin Microbiol Infect Dis. 2000;19:612–7.
32. Raad I, Darouiche R, Hachem R, Sacilowski M, Bodey GP. Antibiotics and prevention of microbial colonization of catheters. Antimicrob Agents Chemother. 1995;39:2397–400.
33. Raad I, Darouiche R, Hachem R, Mansouri M, Bodey GP. The broad spectrum activity and efficacy of catheters coated with minocycline and rifampin. J Infect Dis. 1996;173:418–24.
34. Darouiche RO, Raad II, Heard SO, et al. A comparison of two antimicrobial-impregnated central venous catheters. N Engl J Med. 1999;340:1–8.
35. Sherertz RJ, Heard SO, Raad II, et al. Gamma radiation-sterilized, triple-lumen catheters with a low concentration of chlorhexidine were not efficacious at preventing catheter infections in intensive care unit patients. Antimicrob Agents Chemother. 1996;40:1995–7.
36. Jansen B, Rinck M, Wolbring P, Strohmeier A, Jahns T. In vitro evaluation of the antimicrobial efficacy and biocompatibility of a silver-coated central venous catheter. J Biomat App. 1994;9: 55–70.
37. Kampf G, Dietze B, Grobe-Siestrup C, Wendt C, Martiny H. Microbicidal activity of a new silver-containing polymer, SPI-ARGENT II. Antimicrob Agents Chemother. 1998;42:2440–2.
38. Gilbert JA, Cooper RC, Puryear HA, et al. A swine model for the evaluation of efficacy of anti-microbial catheter coatings. J Biomater Sci Polym Ed. 1998;9:931–42.
39. Goldschmidt H, Hahn U, Satwender HJ, et al. Prevention of catheter-related infections by silver-coated central venous catheters in oncological patients. Zentralbl Bakteriol. 1995;283:215–23.
40. Bach A, Eberhardt H, Frick A, et al. Efficacy of silver-coating central venous catheters in reducing bacterial colonization. Crit Care Med. 1999;27:515–21.
41. Dunser MW, Mayr AJ, Hinterberger G, et al. Central venous catheter colonization in critically ill patients: a prospective, randomized, controlled study comparing standard with two antiseptic-impregnated catheters. Anesth Analg. 2005;101:1778–84.
42. Kalfon P, de Vaumus C, Samba D, et al. Comparison of silver-impregnated with standard multilumen central venous catheters in critically ill patients. Crit Care Med. 2007;35:1032–9.
43. Trerotola S, Johnson M, Shah H, et al. Tunneled hemodialysis catheters: use of a silver-coated catheter for prevention of infection-a randomized study. Radiology. 1998;207:491–6.
44. Schierholz JM, Wachol-Drewek Z, Lucas LJ, Pulverer G. Activity of silver ions in different media. Zentralbl Bakteriol. 1998;287:411–20.
45. Mermel LA, Stolz SM, Maki DG. Surface antimicrobial activity of heparin-bonded and antiseptic-impregnated vascular catheters. J Infect Dis. 1993;167:920–4.
46. Moss HA, Se T, Faroqui MH, et al. A central venous catheter coated with benzalkonium chloride for the prevention of catheter-related microbial colonization. Eur J Anesthesiol. 2000;17:680–7.
47. Jaeger K, Osthaus A, Heine J, et al. Efficacy of a benzalkonium chloride-impregnated central venous catheter to prevent catheter-associated infection in cancer patients. Eur J Chemother. 2001;47:50–5.
48. Kamal GD, Pfaller MA, Rempe LE, et al. Reduced intravascular catheter infection by antibiotic bonding. JAMA. 1991;265:2364–8.
49. Thornton J, Todd NJ, Webster NR. Central venous line sepsis in the intensive care unit: a study comparing antibiotic coated catheters with plain catheters. Anesthesia. 1996;51:1018–20.
50. Ranucci M, Isgro G, Giomarelli PP, et al. Impact of oligon central venous catheters on catheter colonization and catheter-related bloodstream infection. Crit Care Med. 2003;31:52–9.
51. Fraenkel D, Rickard C, Thomas P, Faogali J, George N, Ware R. A prospective, randomized trial of rifampicin-minocycline-coated and silver-platinum-carbon-impregnated central venous catheters. Crit Care Med. 2006;34:668–75.
52. Schierholz JM, Fleck C, Beuth J, Pulverer G. The antimicrobial efficacy of a new central venous catheter with long-term broad-spectrum activity. J Antimicrob Chemother. 2000;46:45–50.

53. Yucel N, Lefering R, Meagele M, et al. Reduced colonization and infection with miconazole-rifampicin modified central venous catheters: a randomized controlled clinical trial. J Antimicrob Chemother. 2004;54:1109–15.
54. Ciresi DL, Albrecht RM, Volkers PA, Scholten DJ. Failure of antiseptic bonding to prevent central venous catheter-related infection and sepsis. Am Surg. 1996;62:641–6.
55. Osma S, Kahveci SF, Kaya FN, et al. Efficacy of antiseptic-impregnated catheters on catheter colonization and catheter-related bloodstream infections in patients in an intensive care unit. J Hosp Infect. 2006;62:156–62.
56. Logghe C, Van Ossel C, D'Hoore W, Ezzedine H, Wauters G, Haxhe JJ. Evaluation of chlorhexidine and silver sulfadiazine impregnated central venous catheters for the prevention of bloodstream infection in leukaemic patients: a randomized controlled trial. J Hosp Infect. 1997;37:145–56.
57. Sheng WH, Ko WJ, Wang JT, Chang SC, Hsueh PR, Luk KT. Evaluation of antiseptic-impregnated central venous catheters for prevention of catheter-related infection in intensive care unit patients. Diagn Microbiol Infect Dis. 2000;38:1–5.
58. Carrasco MN, Bueno A, de las Cuevas C, et al. Evaluation of a triple-lumen central venous heparin heparin-coated catheter versus a catheter coated with chlorhexidine and silver sulfadiazine in critically ill patients. Intensive Care Med. 2004;30:633–8.
59. Heard SO, Wagle M, Vijayakumar E, et al. The influence of triple-lumen central venous catheters coated with chlorhexidine/silversulfadiazine on the incidence of catheter-related bacteremia: a randomized, controlled clinical trial. Arch Intern Med. 1998;158:81–7.
60. Tennenberg S, Lieser M, McCurdy B, et al. A prospective randomized clinical trial of an antibiotic- and antiseptic-coated central venous catheter in the prevention of catheter-related infections. Arch Surg. 1997;132:1348–51.
61. Brun-Buisson C, Doyon F, Sollet LP, Cochard JF, Cohen Y, Nitenberg G. Prevention of intravascular catheter-related infection with newer chlorhexidine-silver-sulfadiazine-coated catheters: a randomized controlled trial. Intensive care Med. 2004;30:837–43.
62. Maki DG, Stolz SM, Wheeler S, Mermel LA. Prevention of central venous catheter-related bloodstream infection by use of an antiseptic-impregnated catheter: a randomized, controlled study. Ann Intern Med. 1997;127:257–66.
63. Jeager K, Zenz S, Juttner B, et al. Reduction of catheter-related infections in neutropenic patients: a prospective controlled trial using chlorhexidine and silver sulfadiazine-impregnated central venous catheter. Ann Hematol. 2005;84:258–62.
64. Rupp ME, Lisco SJ, Lipsett PA, et al. Effect of a second-generation venous catheter impregnated with chlorhexidine and silver sulfadiazine on central catheter-related infections: a randomized, controlled trial. Ann Intern Med. 2005;143:570–80.
65. Veenstra DL, Saint S, Saha S, Lumley T, Sullivan SD. Efficacy of antiseptic-impregnated central venous catheters in preventing catheter-related bloodstream infection. JAMA. 1999;281:261–7.
66. Hockenhull JC, Dwan KM, Smith GW, et al. The clinical effectiveness of central venous catheters treated with anti-infective agents in preventing catheter-related bloodstream infections: a systematic review. Crit Care Med. 2009;37:702–12.
67. Ramritu P, Halton K, Colligan P, et al. A systematic review comparing the relative effectiveness of antimicrobial-coated catheters in intensive care units. Am J Infect Control. 2008;36:104–17.
68. Bassetti S, Hu J, D'Agastino Jr et al. Prolonged antimicrobial activity of a catheter containing chlorhexidine-silver sulfadiazine extends protection against catheter infection. Antimicrob Agents Chemother. 2001;45:1535–8
69. Raad I, Hana H. Intravascular catheters impregnated with antimicrobial agents: a milestone in the prevention of bloodstream infections. Support Care Cancer. 1999;7:386–90.
70. Walz JM, Avelar RL, Longtine KJ, et al. Anti-infective external coating of central venous catheters: a randomized, noninferiority trial comparing 5-flurouracil with chlorhexidine/silver sulfadiazine in preventing catheter colonization. Crit Care Med. 2010;38:2095–102.
71. Darouiche RO, Raad II, Bodey GP, Musher DM. Antibiotic susceptibility of staphylococcal isolates from patients with vascular catheter-related bacteremia: potential role of the combination of minocycline and rifampin. Int J Antimicrob Agents. 1995;6:31–6.

72. Raad I, Darouiche R, Hachem R, Mansouri M, Bodey GP. The broad spectrum activity and efficacy of catheters coated with minocycline and rifampin. J Infect Dis. 1996;173:418–24.
73. Raad I, Hanna H, Jiang Y, et al. Comparative activities of daptomycin, linezolid, and tigecycline against catheter-related methicillin-resistant *Staphylococcus* bacteremic isolates embedded in biofilm. Antimicrob Agents Chemother. 2007;51:1656–60.
74. Widmer AF, Frei R, Rajacic Z, et al. Correlation between in vivo and in vitro efficacy of antimicrobial agents against foreign body infections. J Infect Dis. 1990;162:96–102.
75. Raad I, Darouiche R, Dupuis J, et al. Central venous catheters coated with minocycline and rifampin for the prevention of catheter-related colonization and bloodstream infections: a randomized, double-blind trial. Ann Intern Med. 1997;127:267–74.
76. Leon C, Ruiz-Santana S, Rello J, et al. Benefits of minocycline and rifampin-impregnated central venous catheters. A prospective, randomized, double-blind, controlled, multicenter trial. Intensive Care Med. 2004;30:1891–9.
77. Darouiche RO, Raad II, Heard SO, et al. A comparison of two antimicrobial-impregnated central venous catheters. N Engl J Med. 1999;340:1–8.
78. Marick PE, Abraham G, Careau P, Varon J, Fromm Jr RE. The ex vivo antibacterial activity and colonization rate of two antimicrobial-coated central venous catheters. Crit Care Med. 1999;27: 1128–31.
79. Hanna H, Benjamin R, Chatzinikolau I, et al. Long-term silicone central venous catheters impregnated with minocycline and rifampin decrease rates of catheter-related bloodstream infection in cancer patients: a prospective, randomized clinical trial. J Clin Oncol. 2004; 22:3163–71.
80. Chatzinikolau I, Finkel K, Hanna H, et al. Antibiotic-coated hemodialysis catheters for the prevention of vascular-catheter related infections: a prospective, randomized study. Am J Med. 2003;115:352–7.
81. Falagas ME, Fragoulis K, Bliziotis IA, Chatzinikolau I. Rifampicin-impregnated central venous catheters: a meta-analysis of randomized controlled trials. J Antimicrob Chemother. 2007;59:359–69.
82. Pai MP, Pendland SL, Danziger LH. Antimicrobial-coated/bonded and –impregnated intravascular catheters. Ann Pharmacother. 2001;35:1255–63.
83. Gilbert RE, Harden M. Effectiveness of impregnated central venous catheters for catheter related blood stream infection: a systematic review. Curr Opin Infect Dis. 2008;21:235–45.
84. Sampath LA, Tambe SM. In vitro and in vivo efficacy of catheters impregnated with antiseptics or antibiotics: evaluation of risk of bacterial resistance to the antimicrobials in the catheters. Infect Control Hosp Epidemiol. 2001;22:640–6.
85. Aslam S, Darouiche RO. Prolonged bacterial exposure to minocycline/rifampicin-impregnated vascular catheters does not affect antimicrobial activity of catheters. J Antimicrob Chemother. 2007;60:148–51.
86. Munson EL, Heard SO, Doern GV. In vitro exposure of bacteria to antimicrobial impregnated-central venous catheters does not directly lead to the emergence of antimicrobial resistance. Chest. 2004;126:1628–35.
87. Chatzinikolaou I, Hanna H, Graviss L, et al. Clinical experience with minocycline and rifampin-impregnated central venous catheters in bone marrow transplantation recipients: efficacy and low risk of developing staphylococcal resistance. Infect Control Hosp Epidemiol. 2003;24:961–3.
88. McConnell SA, Gubbins PO, Anaissie EJ. Do antimicrobial-impregnated central venous catheters prevent catheter-related bloodstream infection? Clin Infect Dis. 2003;37:65–72.
89. Crnich CJ, Maki DG. Are antimicrobial-impregnated catheters effective? Don't throw out the baby with the bathwater. Clin Infect Dis. 2004;38:1287–92.
90. McConnell SA, Gubbins PO, Anaissie EJ. Are antimicrobial-impregnated catheters effective? Replace the water and grab your washcloth, because we have a baby to wash. Clin Infect Dis. 2004;39:1829–33.
91. Richards MJ, Edwards JR, Culver DH, Gaines RP. Nosocomial infections in medical intensive care units in the United States. National Nosocomial Infections Surveillance System. Crit Care Med. 1999;27:887–92.

92. Pronovost P, Needham D, Berenholtz S, et al. An intervention to decrease catheter-related bloodstream infections in the ICU. N Engl J Med. 2007;355:2725–32.
93. Warren DK, Zack JE, Mayfield JL, et al. The effect of an education program on the incidence of central venous catheter-associated bloodstream infection in a medical ICU. Chest. 2004;126:1612–8.
94. Sherertz RJ, Ely EW, Westbrook DM, et al. Education of physicians-in-training can decrease the risk for vascular catheter infection. Ann Intern Med. 2000;132:641–8.
95. Bijma R, Girbes AR, Klejer DJ, et al. Preventing central venous catheter-related infection in a surgical intensive-care unit. Infect Control Hosp Epidemiol. 1999;20:618–20.
96. O'Grady NP, Alexander M, Dellinger EP et al. Guidelines for the prevention of intravascular catheter-related infections. Centers for Disease Control and Prevention. MMWR Morb Mortal Wkly Rep. 2002;51(RR-10):1–29.
97. Crnich CJ, Maki DG. The promise of novel technology for the prevention of intravascular device-related bloodstream infection. I. Pathogenesis and short-term devices. Clin Infect Dis. 2002;34:1232–42.
98. Crnich CJ, Maki DG. The promise of novel technology for the prevention of intravascular device-related bloodstream infection. II. Long-term devices. Clin Infect Dis. 2002;34:1362–8.
99. Veenstra DL, Saint S, Sullivan SD. Cost-effectiveness of antiseptic-impregnated central venous catheters for the prevention of catheter-related bloodstream infection. JAMA. 1999;282:554–60.
100. Shorr AF, Humphreys CW, Helman DL. New choices for central venous catheters: potential financial implications. Chest. 2003;124:275–84.
101. Bouza E, Guembe M, Munoz P. Selection of the vascular catheter: can it minimise the risk of infection. Int J Antimicrob Agents. 2010;36 Suppl 2:S22–5.
102. Raad I, Hana H, Maki D. Intravascular catheter-related infections: advances in diagnosis, prevention, and management. Lancet Infect Dis. 2007;7:645–57.

Chapter 20
Topical Antimicrobial-Containing Biomaterials for Peri-Implant Infections in the Oral Cavity

Stefan Renvert and G. Rutger Persson

Abstract Implant mucositis and peri-implantitis have an infectious etiology. If left untreated the inflammatory responses and the infection will result in loss of alveolar bone around dental implants. Clinically, this is combined with edema and redness of the soft tissues surrounding the implant. The ultimate consequences of implant mucositis and peri-implantitis may be loss of the implant.

Dental implant manufacturers have focused on implant design, ease of placement, esthetics, and new prosthetic components. Enhancement of osseo-integration has been attempted by changing titanium surface structures, titanium purity, and various coatings.

The biofilm structures at titanium dental implants comprise of bacteria that co-aggregate with each other and are difficult to eliminate mainly due to the rough threaded surfaces of oral implants making them difficult to debride. The microbiota initiates a host inflammatory response that in many aspects are similar to what can be seen in periodontitis. The prevalence of peri-implantitis ranges between 15 and 25%. Patients who have lost teeth due to periodontitis may be more prone to peri-implantitis. The susceptibility may be directly associated with a genetically complex predisposing risk.

Recent data suggest that professional mechanical debridement of pockets around implants and the patients' own ability to keep implants free from bacteria is unpredictable. Professional debridement may be efficacious over shorter time periods and

S. Renvert (✉)
University of Kristianstad, Kristianstad, Sweden

Trinity College, Dublin, Ireland

Blekinge Tekniska Högskola, Karlskrona, Sweden
e-mail: stefan.renvert@hkr.se

G.R. Persson
University of Kristianstad, Kristianstad, Sweden

University of Washington,Seattle, WA, USA

University of Bern, Bern, Switzerland
e-mail: rutger.persson@hkr.se

T.F. Moriarty et al. (eds.), *Biomaterials Associated Infection: Immunological Aspects and Antimicrobial Strategies*, DOI 10.1007/978-1-4614-1031-7_20,

would most likely require intensive supportive therapy. In the event of limited response to nonsurgical mechanical supportive therapy, the adjunct use of local application of a slow-release antimicrobial agent could be considered. Surgical treatment to eliminate soft tissue pocket and attempts to establish hygienic conditions may not be acceptable to the patient in the esthetic areas. In these situations regenerative therapy may be considered.

Antibiotics are commonly used in medicine to treat infections. Restrictive use of antibiotics is a necessity to prevent the occurrence of wide-spread antibiotic resistance. Peri-implantitis is a local infection confined to the area around the affected implant. Therefore, local administration of an antibiotic drug in a high concentration is logical. Several local antibiotic products have been developed and studied extensively. Such local antibiotic drugs have been on the market for some years but due to a variety of factors some of these local antibiotics have been withdrawn from the market or may only be available in some countries. There is a need to develop new methods allowing effective delivery of local antibacterial and anti-inflammatory agents for the treatment of implant infections.

20.1 Introduction

Infectious complications in implant dentistry are common. The bacterial infection at dental implants can be described as a complex biofilm. Treatment of infection and inflammation associated with dental implants is technically difficult and not predictable.

Treatments with antibacterial agents are common both in medicine and dentistry. Usually the infection at dental implants is localized to a single or few implants and systemic administration of antibiotics may seem irrational. Nevertheless, it can be assumed that local infections and inflammatory processes can have systemic effects and should therefore not be ignored.

Local administration of an antibiotic drug in a high concentration provides an option in the management of infections around implants. Local administration of an antibiotic with a sustained drug release using a biomaterial that allows dispersion of the antibiotic drug over a period of time long enough to have bacteriostatic or bacteriocidal effects against the target microorganism(s) would be preferred to generalized antibiotic administration. A comprehensive review of the current understanding of peri-implantitis is provided.

20.2 Introduction to Titanium Implants

The implementation of titanium dental implants in clinical practice has made it possible to replace teeth that have been lost due to trauma, tooth decay, periodontitis or to compensate for congenitally missing teeth. Treatment with titanium dental implants as replacement for lost teeth has been available for approximately 40 years. The local oral environment where the implant is placed harbors a high density of commensal bacteria that easily colonize teeth, soft tissues, and any foreign material in the oral cavity. Early efforts with dental implants, using non-titanium alloys, were not successful and it was common with infectious complications. In addition to infection, poor tissue integration and adverse tissue reactions to the used foreign materials could also contribute to failures. In the 1970s, animal electron microscopy studies demonstrated that bone could become closely anchored to titanium plasma coated surfaces [1]. The authors described this event as an ancylotic contact between bone and the titanium surface. In these studies the investigators observed that the connective tissue fibers were functionally oriented and appeared to be inserted into the titanium-sprayed surface. They also noticed that "if the implant was placed in an area with an immobile keratinized mucosa signs of adhesion of the epithelial cells to the titanium sprayed surface became permanent" [2]. Following these early studies in a nonhuman primate animal model, the same research group proceeded with titanium sprayed hollow screw implants in humans. They observed that the clinical success of hollow screw implants with a plasma-sprayed titanium surface was clearly dependent on the presence of a tethered, immobile mucosa ("gingiva") around the post of the implant, and with a prerequisite of perfect oral hygiene [2]. In other words, the presence of keratinized attached tissues facilitated oral hygiene efforts. This observation was the first report on the success of using titanium dental implants and, furthermore, identifying the needs for excellent oral hygiene. At this time, dental plaque was merely considered as a composite of debris including bacteria.

These early studies also pointed at the necessity to control for accumulation of plaque (biofilm), which was in concordance with the principles for successful periodontal therapy. Excellent oral hygiene had previously been demonstrated as a prerequisite for control of gingivitis and periodontitis in patients undergoing periodontal therapy [3]. The ability to maintain an excellent level of oral hygiene at dental implant sites may not be easy. It appears that patients who have received treatment for peri-implantitis by nonsurgical methods often have a very high level of bacteria at the neck of implants in spite of extensive motivation and instructions in oral hygiene (Fig. 20.1). One explanation why patients fail to maintain an adequate local oral hygiene can be the design of the supra-structures which may make it difficult to adequately clean the implants with various types of oral hygiene aids.

Further studies on the interface between titanium implants and alveolar bone could demonstrate that a very close spatial relationship existed between bone and titanium. The soft tissues also adhered closely to the titanium implant. The interface between bone and titanium was considered to be a chemical bonding (osseo-integration) [4]. This perception was interpreted such that infectious inflammatory responses in the soft tissues and bone surrounding dental implants were unlikely. If an infectious inflammatory response occurred, it was usually attributed to a technical problem or contamination of the implant at the time of insertion of the implant.

During the early healing phase after implantation, histological findings have revealed the establishment of a long junctional epithelium with an underlying loose

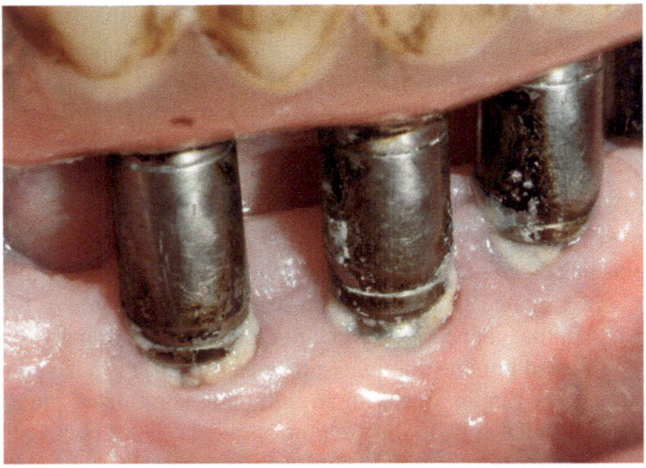

Fig. 20.1 Clinical picture of dental titanium implants demonstrating large amounts of bacterial plaque at the margin between the gingival soft tissues and implants. Notice that plaque can also be seen at least halfway up on the supra-gingival implant neck. Aerobic streptococci and anerobic bacteria are found in such biofilm structures. Proper oral hygiene measures can remove much of the extracellular matrix

connective tissue rich in inflammatory cells. Within 2–8 weeks of healing, signs of epithelial proliferation and a more organized and mature connective tissue without evidence of inflammation can be seen [5].

There may be several causes for implant failure including: mechanical failure, adverse loading, corrosion, or the effects of the infection—inflammation in implant surrounding tissues. Infections at implants may also be dependent on the synergy of several conditions and independent of what might have been the first factor in a chain reaction leading up to clinical symptoms of inflammation.

Implant therapy is considered to be highly effective, reliable, and safe. For many years, the so called survival rate (the implant still being present in the oral cavity) of implant therapy has been reported to be within the range of 95–99% (i.e., [6–10]). Nevertheless, there is evidence that the survival rate of dental implants in patients with a history of periodontitis is lower [11].

During the initial years of oral implant therapy, failures with titanium dental implants focused on technical or anatomical issues as being the reason for bone and implant loss [12–14]. Technical issues such as the retention of the superstructures have also been discussed [15].

20.2.1 Titanium Dental Implant Complications

Due to great variation in the characteristics of study populations and varying criteria for the diagnosis of infectious and inflammatory conditions at dental titanium implants, there is heterogeneity in the data on the prevalence of dental implant complications. One study has shown that biological complications occur in 17%, and technical failures in 30% of patients who have received dental implants [16]. Other studies have shown that the prevalence of peri-implantitis is in the range of 16–25% [17–19]. In contrast to such reports showing relatively high complication rates, others have identified that the prevalence of peri-implantitis is only approximately 7% [20,21]. World-wide, it has been estimated that millions of dental titanium implants are being placed every year. With an increasing population having dental implants, the prevalence of implant related infections will most likely increase and may cause major challenges to clinical therapy [17,18].

Early horizontal bone loss adjacent to an osseo-integrated implant is a result of bone remodeling due to the surgical procedure itself. Horizontal bone loss that continues after the initial healing phase has been associated with microbial plaques being present at dental implants adjacent to clinically inflamed tissues. This plaque (biofilm) has been reported to differ from what had been described for dental plaque around teeth [22]. The microbiota at titanium dental implants includes predominantly coccoid cells and nonmotile rods, and 65% of human biopsies from soft tissues surrounding oral implants present with evidence of inflammation [23]. These findings are consistent with the early findings by Schroeder et al. [1], and with case reports on the loss of implant integration and subsequent implant being associated with poor oral hygiene [24]. The development of a biofilm occurs both around titanium implants and around natural teeth. The implant surface provides a nonbiological structure which may allow the development of a characteristic biofilm. Certain bacteria may have a greater affinity to an implant surface than to tooth surfaces. Staphylococcus aureus has been suggested as one of those pathogens frequently found at infected and inflamed tissues surrounding titanium implants [25–30]. Periodontal infections around teeth are dominated by Gram-negative anaerobes [31].

Such concerns about inflammation and infection associated with titanium dental implants resulted in a consensus report suggesting that peri-implant mucositis could be diagnosed and defined as a reversible inflammatory condition of the soft tissues at dental implants [32]. If the inflammation was combined with loss of alveolar bone surrounding the implant, the condition would be referred to as peri-implantitis and a nonreversible condition [33]. Within this concept it is understood that during the first year after implant insertion a certain amount of crestal bone may be lost during the first phase of bone remodeling and this would equal approximately the length of 1–1.5 mm. In order to define peri-implantitis a radiographic analysis is necessary. Examples of extensive bone loss and a diagnosis of severe peri-implantitis are illustrated (Figs. 20.2 and 20.3).

An example of a titanium dental implant that has been implanted without complication for more than 5 years is illustrated (Fig. 20.4) showing no evidence of bone loss around implant threads.

It is now established that biomaterial implants are surrounded by an fibro-inflammatory, integration-deficient zone within which stimulation of cellular immune responses occur [34]. Thus, superoxide radical and cytokine-mediated tissue damage with increased susceptibility to infection has been demonstrated [35].

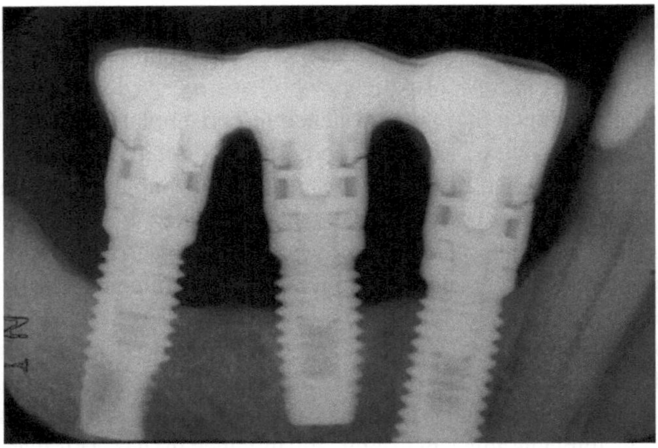

Fig. 20.2 An example of an intra-oral radiograph from the lower jaw identifying three titanium dental implants with a horizontal pattern of bone loss. Two of these implants (to the *left*) present with typical radiographic evidence of bone loss with exposure of three implant threads or more depending on how the implants were inserted. The implant to the right shows radiographic evidence of no significant loss of implant surrounding bone

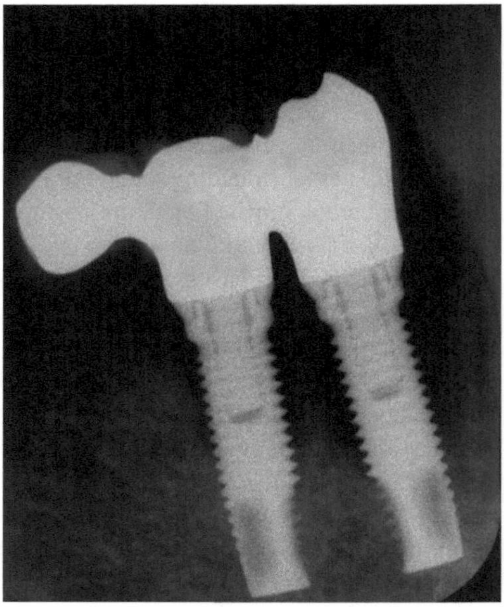

Fig. 20.3 An example of an intra-oral radiograph from the lower jaw identifying two titanium dental implants with extensive bone loss. The pattern of bone loss is both horizontal and vertical and with radiographic evidence of crater defects at both implants

From an immuno-histological perspective, peri-implantitis lesions seem to comprise of larger inflammatory infiltrates than what can be seen at peri-implant mucositis lesions and with significantly greater proportions of B cells (CD19+) and elastase-positive cells [34]. In addition, the loss of osseo-integration of dental titanium implants appears to be associated with an inflammatory process similar to that observed in aggressive periodontitis [36].

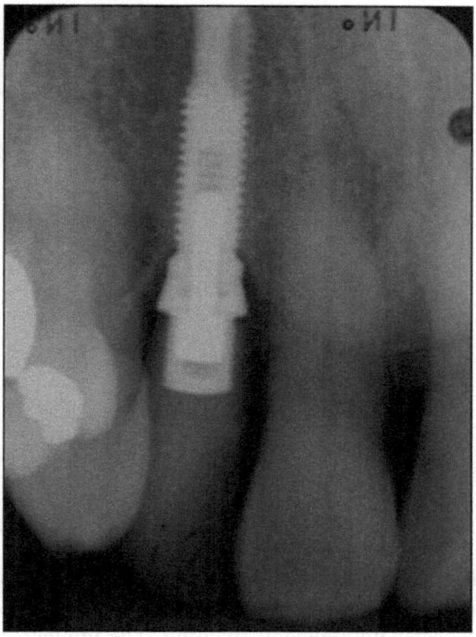

Fig. 20.4 An example of a single titanium dental implant at the former location of tooth (22, [FDI system] or 10 [international system]). This implant has been in function >5 years and with no evidence of bone loss. It should be noticed that some remodeling of the bone architecture has taken place and this change is within normal limits

20.3 The Infectious Etiology of Oral Implant Mucositis, and Peri-Implantitis

Subjects who have had a history of periodontitis may be at a greater risk for complications with titanium dental implants [21]. Although the microorganisms are the initiating factor, periodontitis is considered to be multi-factorial [37]. Genetic predisposition are reported as an important factor for the susceptibility of periodontitis and would most likely also be a risk factor for peri-implantitis [38]. There are new data to suggest that the microbial content of the periodontal pocket is a determinant of gene expression in the gingival tissues [39]. This is supported by immunological data demonstrating that patients who have experienced titanium dental implant loss have low serum IgG antibody titers and affinity (function) to *Staphylococcus aureus* and *Tannerella forsythia*. This suggests that the immune system of such susceptible subjects cannot control the infections that occur in tissues surrounding dental implants [40].

The oral cavity is populated by a very large number and diverse range of bacteria [41,42]. Approximately 70% of these bacteria are non culturable phylotypes [43]. Several studies have confirmed that microorganisms are essential for the development of infections around dental implants (i.e., [44–49]). Some studies have suggested that the microbiota associated with peri-implantitis is similar to the microbiota associated with periodontitis (for a review, see [48,50]).

Studies using routine microbiological assays such as checkerboard DNA–DNA hybridization have demonstrated that the infection at dental titanium implants appears to be similar to what can be found in infections (periodontitis) of the tooth [51]. However, other bacteria such as *Escherichia coli, Helicobacter pylori*, and *S. aureus* may also be present at oral implant infections [25,28–30]. This is not

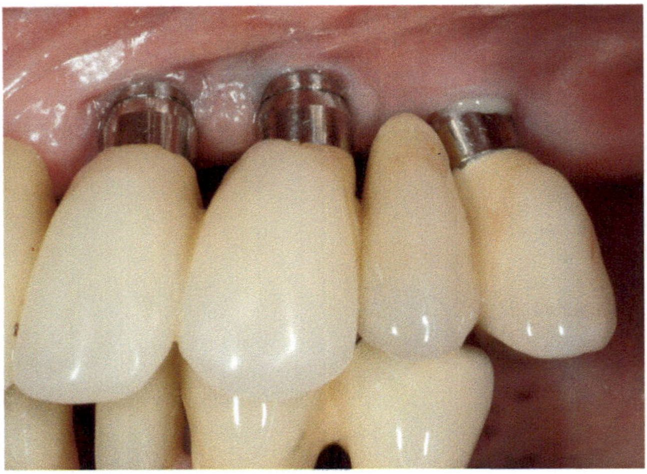

Fig. 20.5 A clinical picture illustrating titanium dental implants with evidence of soft tissue inflammation. The implant to the right in the picture has evidence of suppuration at the mid-buccal aspects and a general swelling. The implant to the left shows evidence of severe marginal tissue inflammation and swelling with tissue redness suggesting an early acute lesion

surprising as foreign bodies are commonly colonized by *S. aureus*. It has been reported that titanium favors colonization by *S. aureus* [26,27]. Infectious complications of hip joint titanium implants by *S. aureus* have also been reported [52]. Data have demonstrated that pathogens associated with periodontitis can also be identified at infected dental implant sites [53].

An example of severe peri-implantitis with presence of suppuration, which is a common feature in advanced peri-implantitis, is illustrated in Fig. 20.5.

There are several studies suggesting that the colonization of bacteria around dental implants occurs rapidly from the first minutes after insertion and a complex biofilm is formed within the first few weeks [49,54,55]. Data suggest that if *S. aureus* can be identified around titanium dental implants shortly after insertion, *S. aureus* can most likely also be identified 1 year later, thereby posing a risk for future implant failure due to infection [56]. Staphylococcus species appears to have a high affinity to titanium implant surfaces [26,57]. It appears that using various peptide coatings on titanium surfaces colonization of *S. aureus* may be prevented [58].

Intra-coronal compartments of screw-retained fixed restorations retained on titanium dental implants are heavily contaminated with pathogens and the restorative margin may serve as the principal pathway for bacterial leakage [59]. This suggests that in order to control for the infection in soft tissues around dental implants, the design and later the retention of the superstructure is an important factor. The design of implants usually makes it difficult for mechanical debridement. Data suggest that whereas the supragingival component can be populated by a dense biofilm the subgingival implant area is usually not covered by a dense biofilm [60]. Although most studies are focused on bacterial counts or total bacterial load the inflammatory response is driven by the pathogeniety of the bacterial species that have infected the subgingival niche around implants or teeth.

20.4 Treatment of Infectious Titanium Dental Implant Conditions

There are several nonsurgical and surgical treatment options available for the management of infectious dental implant complications. The adjunct use of antibiotics and antiseptics in combination with debridement has been considered. If the infection is not successfully treated, peri-implantitis may lead to the complete loss of integration and implant loss [25,48,61].

Based on meta-analysis of the available literature, the conclusion is that there is weak evidence that antibacterial mouth-rinses are effective in reducing plaque and marginal bleeding around implants [62]. Recent studies have demonstrated that mechanical debridement of tissues and dental implants diagnosed with peri-implantitis only results in a temporary and limited improvement [63,64]. An alternative treatment option for the treatment of peri-implantitis using an Erbium Ytrium-Aluminum-Granat Laser (Er:YAG) may also not be effective in controlling the infection [65].

Antimicrobial adjunct therapy of peri-implantitis lesions may therefore be required to control implant mucositis and peri-implantitis. Due to the fact, that it is difficult to treat peri-implantitis the use of prophylactic systemic antibiotics have been suggested in many publications. Currently a Cochrane review has identified four randomized controlled clinical trials (RCTs) that have assessed the benefits of prophylactic antibiotics at the time of implant placement. The analysis identified that it would be necessary to treat 33 patients with prophylactic antibiotics to prevent one case with early implant loss [66]. In the case of prophylaxis there is some evidence suggesting that 2 g of amoxicillin given orally 1 h preoperatively significantly reduce failures of dental implants placed [66,67]. There are, however, also several reports suggesting that the preoperative use of antibiotics does not provide significant benefits and that pre-medication at the time of implant insertion should not be used on a routine basis [68]. Thus there is currently no consensus whether prophylactic antibiotics should be given routinely or not prior to placement of dental titanium implants.

Chemical agents have been tested to decontaminate the implant surface during surgery, including 3% H_2O_2, saturated citric acid (pH 1), or chlorhexidine. Such treatments of the implant surface do not seem to alter the titanium surface [69]. No single method of surface decontamination has been found to be superior [70].

20.5 Modifying Implant Surfaces to Prevent Bacterial Colonization and the Development of a Pathogenic Biofilm

In regards to titanium dental implants, manufacturers have focused on implant surface structures to enhance biocompatibility, which may, if exposed to the oral environment, favor bacterial colonization and the development of a pathogenic biofilm.

At the atomic or molecular levels it is possible to change the surface properties of the implant using nanotechnology. Numerous studies have been performed to assess whether conditioning of a titanium implant surface by various nanotechnologies alters the osseo-integration of the implant. When titanium dioxide (TiO_2) is irradiated with near-UV light, this semiconductor exhibits strong bactericidal activity. Attempts are currently underway to assess if a new interfacial bond-promoting concept for an on demand long term bacterial inhibition can be achieved by using available TiO_2 nanoparticles as photocatalytic bacteriocidal agents when exposed to UV light sources [71–73]. The *in vitro* results are encouraging and it may be possible to develop a TiO_2 coating of dental titanium implants that can be activated as needed to control for severe infections. There are currently no clinical studies that have assessed whether such TiO_2 coating of dental titanium implants differ in terms of osseo-integration as traditional standard implants.

In vitro experiments have shown that a silver-nanoparticle modified titanium surface is effective against *S. aureus* and *E. coli* and also shows cytocompatibility [74].

Nanocrystalline anatase coated implants enhance osseo-integration [75]. Nanoparticles are not antimicrobial agents in the strictest sense, but such a coating would represent a modulation of a surface interacting with human cells, thus providing a biomaterial as a barrier against bacterial infections. Irradiation of nanocrystalline anatase coated implants with ultraviolet light changes the implant surface such that it can be characterized as a superhydrophilic surface. The ability of bacteria to attach to such a surface may decrease significantly.

Through specific binding of antimicrobial peptides to titanium, biofilm development can reduce the binding of *P. gingivalis* to conditioned titanium surfaces [76]. There are recent efforts to fast-load biomimetic hydroxyapatite coatings on implants with various antibiotics such as Amoxicillin, Gentamicin sulfate, and Tobramycin to develop implant devices providing a sustained release of antibiotics at the time of implantation. During the process the antibiotic is soaked on to the titanium HA coating during 15 min. Coating of titanium surfaces with an adhesion enhancing gradient layer of titanium oxide may provide a drug delivery vehicle through loading into hydroxy-apatite surfaces. In vitro studies of such antibiotic coatings have identified antibacterial effects against *S. aureus* with duration of at least 24 h [77]. Studies have also shown that biodegradable polypeptide multilayer nanofilms can act as antibiotic carriers at the implant/tissue interface and that they are compatible with human osteoblasts while being bactericidal to bacteria such as *S. aureus* [78,79].

Initial bacterial adhesion can also be prevented by impregnating a titanium implant surface with an antimicrobial coating (poly-D,L-lactide, or politerefate), in combination with chlorhexidine diacetate. In vitro studies have shown that such titanium implant surface modulation can prevent colonization with *S. aureus* while maintaining cytocompatibility [57]. Furthermore, deposition of silver or strontium apatite on titanium surfaces appears to outperform hydroxyl apatite and produce more osteoblast proliferation and providing antibacterial capacity over at least 30 days [80].

The implant industry has dedicated much time and effort to the enhancement of the titanium surface in regards to osseo-integration. One of the concerns has been changes of the surface characteristics may enhance the biocompatibility allowing osteoblast activities. According to the above cited studies it appears that, using nanotechnology or other recent surface coating methodologies, it may be possible to develop titanium implants with both enhanced biocompatibility while also providing a protective barrier against bacterial colonization of the implant surface. This may, at least, be possible under the early phase of wound healing following implant insertion. This may be an advantage as early bacterial colonization of titanium dental implants has been documented [49,54,55]. Data suggest that if *S. aureus* can be identified in samples taken from implants shortly after implantation *S. aureus* will with great likelihood also be present in samples taken 1 year later and although no clinical evidence of peri-implantitis exists [56].

20.6 Topical Antibiotic Releasing Biomaterials as an Adjunct to the Treatment of Peri-Implantitis

Both systemic and local modes of administration of antibiotics have been proposed to treat or control periodontitis and peri-implantitis. The literature on the efficacy of systemic antibiotics in the treatment of chronic periodontitis is extensive. Data suggest that systemic antibiotics contribute to clinically excellent results [81]. A difference between treatment of chronic periodontitis and implant mucositis and peri-implantitis is that periodontitis usually has a more generalized pattern whereas infection at implants is usually localized to a single or few implants. Therefore localized treatment with antibiotics which can be administered at high concentrations is a particularly attractive concept in the treatment of peri-implantitis.

It is important that the administration of antibiotics to the pockets around an implant is simple to perform and accurate in the delivery of the drug at a concentration that is adequate over a time span necessary to have an anti-bacterial impact. One of the problems with local administration of antibiotics is that there is a constant flow of excudate in an outwards direction from inflamed sites. If the pocket is treated with mechanical debridement immediately before administration of local antibiotics it is also likely that there is a profound bleeding from the pocket. Bleeding and exudates can easily either dilute the concentration of the antibiotic or actually eliminate the antibiotic drug administered. It is therefore important that the delivery system provides a vehicle that effectively controls the fluid flow and distributes the drug over a sufficient period of time. The experience from periodontal therapy using local antibiotics has been that it is difficult to retain the drug in the periodontal pocket. Bacteria in biofilm structures also exchange antibiotic resistance capabilities through plasmids transfer. This adds to the challenge in designing antibiotic drugs that effectively can control the infection.

Since the late 1990s several different modes of administration of local antibiotics in the treatment of peri-implantitis have been suggested (i.e., loading polymers, functionalizing fibers, or mixing with bone fragments). Some of the devices and methods used to administer local antibiotics may no longer be available on the market, or are only available in some countries, as a result of either regulatory or financial reasons. Different antibiotics have also been tested but the primary outcome measures have in most studies been clinical changes. Few studies have actually assessed changes in bacterial presence in pockets that have been treated.

20.6.1 Tetracycline and Tetracycline Derivates in Slow Release

Tetracycline was introduced as a broad-spectrum antibiotic in the late 1940s, and is effective against a broad spectrum of pathogens including Gram-positive and Gram-negative bacteria. The mechanism of action of tetracycline is related to the inhibition of protein synthesis by binding to the 30S bacterial ribosome. In addition to the antibacterial effect, tetracycline also inhibits matrix metallo-proteinases. Thus, local treatment of peri-implantitis with tetracycline may also have an anti-inflammatory effect.

Resistance to tetracycline is common in Bacteroides and Parabacteroides strains [82]. A high resistance rate to tetracycline has also been reported for Staphylococcus spp. [83]. In regards to the common appearance of resistance to tetracycline among bacteria that have been associated with peri-implantitis, it can be questioned if tetracycline is the drug of choice for local administration. If, however, tetracycline can be administered in high doses and during a prolonged period of time, tetracycline may be efficacious due to the combined antimicrobial and anti-inflammatory properties.

There is a substantial literature on the local administration of tetracycline for the treatment of periodontitis. A polymeric tetracycline HCl-containing fiber (Actisite® ALZA Corporation Pal Alto CA, and Procter & Gamble Cincinatti OH) can be packed into periodontal pockets of infected teeth (Fig. 20.6). Actisite® periodontal fiber consists of a 23 cm (9 in.) monofilament of ethylene/vinyl acetate copolymer, 0.5 mm in diameter, containing 12.7 mg of evenly dispersed tetracycline hydrochloride (25%). The release of tetracycline appears to be limited to approximately 5 days [84].

Unfortunately, locally administered tetracycline usually results in a transitory selection of sub-gingival bacterial species intrinsically resistant to tetracycline [85]. This resistance appears to disappear over time (6 months) and did not seem to have a negative clinical impact on the treatment of periodontitis.

The literature on the efficacy of local administration of tetracycline for the treatment of peri-implantitis is limited. In comparison to local debridement alone, the adjunct use of the tetracycline fiber did not result in better clinical results [85]. An early study including eight subjects with implant mucositis who were treated with sustained release tetracycline HCl (Actisite®) demonstrated that this adjunct treatment to nonsurgical debridement can reduce the extent of hyperplasia [86]. Another study has been published using the tetracycline

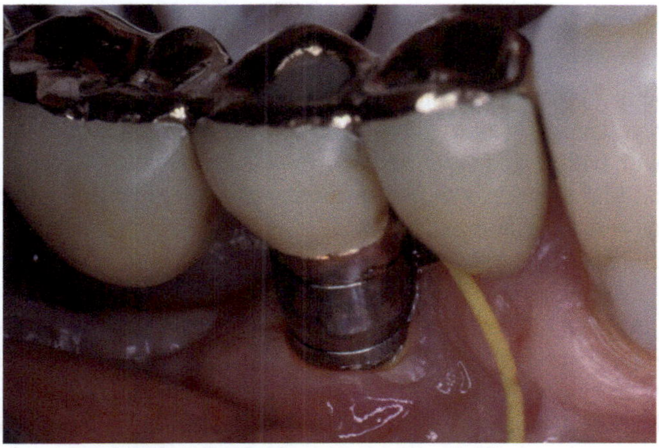

Fig. 20.6 Clinical situation with the application of the Actisite fiber®

HCl-containing fiber in a case series including 25 subjects who received such adjunct tetracycline treatment immediately after routine local debridement [87]. The fiber was left in situ for approximately 10 days. In combination with prior mechanical debridement and post-administration with chlorhexidine rinse, data suggest that clinical conditions improve with an average reduction of probing depths of 2 mm (from on average 6 mm before to 4 mm after treatment. Significant decreases in the frequency of detection of *Campylobacter rectus*, *Fusobacterium nucleatum. P. intermedia,* and *T. forsythia* was reported [87]. This study was limited to standard bacterial culture methods. Bacterial recolonization approaching approximately the same level as before was reported to occur at 12 months after therapy. It appears that the administration of a fiber for the release of an antibiotic drug was not further pursued, since a literature search using PubMed Medline failed to identify additional studies in regards to treatment of peri-implantitis and there are no data reported for the treatment of periodontitis after 2004. The product is currently not available on the market.

A combination of tetracycline HCl and a hydroxyapatite porous bone scaffold coated with poly(epsilon-)caprolactone and hydroxyapatite powder hybrids, has been studied but is currently not available on the market. It was developed for tissue-regeneration therapy with a sustained release of tetracycline [88]. Recent *in vitro* studies have shown that tetracycline HCl dispersed and absorbed onto Bio-Oss® granules (Geistlich Pharma AG, Wolhusen, Switzerland) effectively reduces levels of *Aggregatibacter actinomycetemcomitans*, *P. gingivalis*, and *S. aureus* [89]. Such an administration route in the surgical management of peri-implantitis may prove to be efficacious. There is also evidence that a mixture containing autologous ground calvaric bone, human recombinant tissue factor (rhTF), platelet-rich plasma (PRP), and tetracycline as a osteogenic bone paste in maxillary sinus floor augmentation has a high (90%) success rate and may facilitate a controlled healing and bone regeneration with limited risk for infection [90].

20.6.2 Doxycycline in Sustained Release

Doxycycline is a broad-spectrum antibiotic synthetically derived from oxytetracycline. Doxycycline is effective against several of the bacterial species that have been associated with both periodontitis and peri-implantitis. Doxycycline can also inhibit matrix metallo-proteinases, which contribute to tissue destruction activities in diseases such as peri-implantitis.

Doxycycline has been incorporated in a slow release system designed for use in the treatment of periodontal diseases and also applicable for the treatment of peri-implantitis. ATRIDOX® (Atridox, Block Drug, Jersey City, NJ) is a subgingival controlled-release product with doxycycline hyclate 10% distributed through a 2 syringe mixing system. Syringe A contains 450 mg of the ATRIGEL® Delivery System, which is a bioabsorbable, viscous polymeric formulation composed of 36.7% poly (DL-lactide) (PLA) dissolved in 63.3% *N*-ethyl-2-pyrrolidone (NMP). Syringe B contains doxycycline hyclate (equivalent to 42.5 mg doxycycline). The constituted product is a yellow viscous liquid with a concentration of 10% of doxycycline hyclate. In contact with crevicular fluid, the liquid product solidifies and doxycycline is released during a 1 week time period.

Comparing the clinical outcomes in patients with a diagnosis of periodontitis with deep periodontal pockets between local administration of three slow-release products (a metronidazole gel (Elyzole Dental Gel®), a chlorhexidine chip (PerioChip®), or doxycycline (ATRIDOX®)) the best clinical effects were found using the doxycycline treatment [91]. However, none of the antibiotic drugs tested resulted in a decrease in bacterial load. Thus the somewhat better effect by doxycycline treatment may be explained by anti-inflammatory properties of the doxycycline.

In one longitudinal randomized clinical trial over 3 years, adjunct application of Atrodix® resulted short-term improvements in the group receiving the adjunct doxycycline gel in comparison to mechanical debridement alone, but only at 3 months after therapy [92]. The 3 year follow up results were similar with an average reduction in pocket depth of approximately 1 mm in both study groups. The impact on the subgingival microbiota was greater in the group receiving debridement and the doxycycline gel but only for 3 of the 40 species studied by the checkerboard DNA–DNA hybridization method at different time points during the study duration of 3 years. This included *Fusobacterium periodonticum*, *P. gingivalis* and *Prevotella nigrecsens*. At year three no differences in the counts of the sub-gingival bacteria assessed could be found [92]. One explanation why the microbiological impact of therapy was restricted can most likely be explained by the marginal improvement in oral hygiene and gingival inflammation. A prerequisite for successful antibacterial treatment is clearly that excellent oral hygiene must be maintained. Otherwise, short term treatment gain in clinical conditions and microbiological counts may not be sustainable. This would apply not only to periodontal therapy but also to the treatment of peri-implantitis.

Case-control studies in assessing the efficacy of the administration of doxycycline (Atridox) in cases with peri-implantitis have shown that such administration of the antibiotic releasing material, as an adjunct to mechanical debridement, provides significant clinical improvements in comparison to what might be achieved by debridement alone [93]. The combined use of an autogenous bone graft and a biodegradable polymer barrier that delivered sustained release of Atridox™ appeared to be effective in a case with severe peri-implantitis [94].

20.6.3 Minocycline in Sustained Release

Minocycline hydrochloride microspheres (ARESTIN® Orapharma Inc Warminster PA) is a sub-gingival sustained-release product containing minocycline hydrochloride incorporated into a bioadhesive bioresorbable polymer (glycolide-co-dl-lactide) with a unit-dose cartridge that delivers minocycline hydrochloride equivalent to 1 mg of minocycline. The mode of action is that crevicular fluid hydrolyzes the polymer resulting in a sustained release. The drug concentration is retained and released over a time period of 21 days. No changes in the presence of minocycline-resistant bacteria or *Candida albicans* or *S. aureus* have been documented and the local minocycline concentration may be up to 1,000 times higher than when systemically administered. Several studies have assessed the clinical benefits of local administration of minocycline applied to periodontal pockets in patients with periodontitis. Such studies have shown that adjunct administration of minocycline microspheres is efficacious in terms of reducing probing depth [95].

The efficacy of debridement and adjunct administration with either minocycline microspheres or a chlorhexidine gel at baseline was evaluated in a randomized clinical trial in peri-implantitis patients. A greater reduction in probing pocket depth at the deepest site for implants treated with minocycline (ARESTIN®) was found at days 30, 90 and 180 but not at the study endpoint after 1 year. Bleeding on probing significantly decreased and decreased to a greater extent at implants treated with the active drug. In both study groups, treatment reduced the levels of bacteria studied (including: total bacterial count, *A. actinomycetemcomitans*, *P. gingivalis*, *P. intermedia*, *T. forsythia*, or *Treponema denticola*) [28,29,96]. Thus, both the minocycline microsphere treatment and the chlorhexidine treatment as adjunct therapy to local mechanical debridement were effective in controlling for the infection in peri-implantitis for at least 1 year after the first intervention. The question that can be raised is if debridement alone would have yielded the same results; however, this was not determined in the study [96].

In a case series including subjects with peri-implantitis, local debridement and administration of minocycline microspheres resulted in bacterial decreases identified as early as 10 days after intervention for 6 of the 40 species studied [97]. At day 180 significantly lower total bacterial counts were found in comparison to baseline specifically for *A. actionomycetemcomitans*, *P. intermedia*, *P. gingivalis*, *T. forsythia*, and *Streptococcus anginosus*. At the study endpoint only the levels of A. *actinomycetemcomitans* were lower. Clinically significant improvements were obtained when comparing baseline and 1 year data [98].

20.6.4 Gel for Sustained Release of Metronidazole

Metronidazole is an oral synthetic antibacteriozidal agent 1-β hydroxyeth-2-methyl-5-nitriomidazole readily taken up by Gram—negative anaerobic bacteria disrupting bacterial DNA. Elyzol 25% Dental Gel (Colgate-Palmolive, Surrey, UK) distributed in periodontal pockets provides metronidazole concentrations >100 mg/ml as measured locally in the crevicular fluid for at least 8 h. At 36 h, concentrations above 1mg/ml can still be measured. Metronidazole is released slowly from the dental gel with a bioavailability of about 70%. The maximal plasma concentration is reached at 4 h but serum concentrations >1.3 mg/ml have not been measured. In general, metronidazole is an antibiotic active against most of the bacteria in the subgingival flora in periodontitis and with bactericidal effect against Bacteroides spp., Fusobacterium, Selemonas, Campylobacter, Spirochetes, and other obligate anaerobic organisms, but is also effective against *A. actinomycetemcomitans*.

The information on the efficacy of a metronidazole gel in the treatment of peri-implantitis is limited. One study has shown that peri-implantitis lesions can be positively resolved by using the metronidazole dental gel (25%). This drug treatment led to a 60–70% decrease of Gram negative, and 40–50% increase for Gram positive microorganisms, bringing these back to normal values in almost all the patients. Moreover, a good recovery of the peri-implant soft tissues was observed [99]. In a pilot study including treatment of implants with experimentally induced peri-implantitis in dogs, treatment with a metronidazole gel and local debridement resulted in more new bone to implant contact than in placebo treated animals [65].

20.6.4.1 Moxifloxacin Gel in Sustained Release

Moxifloxacin is a fluoroquinolone antibiotic with broad antibacterial activity against both aerobic and anaerobic bacteria and therefore suitable also for the treatment of peri-implantitis lesions. In a phase II clinical trial, the local administration of 0.4% moxifloxacin gel in combination with nonsurgical periodontal therapy has demonstrated a greater reduction in probing depth following an one-time administration in comparison to debridement alone [100]. This regimen has not been tested for the treatment of peri-implantitis. There is currently no product on the market that is available that includes this drug formulation.

20.6.5 Antiseptics in Sustained Release

Therapeutical concentrations of antibacterial such as chlorhexidine, and cetylpyridiniumchloride may inhibit the enzymatic breakdown of membranes used to enhance regeneration of tissues at teeth and implants [101]. Thus, it is possible that such local administration of antibacterial agents may reduce bacterial enzymatic effects during regenerative procedures. Both local antibiotics and administration of local antiseptic intervention may effectively enhance infection control and reduction/elimination of inflammation at dental implants. Antiseptics in solution may not have sustained effects and must first be incorporated in slow release agents. There are currently no such release systems available with the exception of a chlorhexidine chip (PerioChip® Dexcel Pharma Technologies Ltd, Jerusalem, Israel). PerioChip® contains 2.5 mg chlorhexidine gluconate provided in a biodegradable matrix of hydrolyzed gelatin (cross-linked with glutaraldehyde). PerioChip® also contains glycerin and purified water. The chip is placed in a periodontal pocket and will provide a slow release of chlorhexidine. Clinical data suggest that the treatment of patients with severe and aggressive periodontitis results in a greater reduction of pro-inflammatory prostaglandin E2 in crevicular fluid than in the control receiving mechanical debridement alone. In addition, a decrease in counts of spirochetes was greater in the PerioChip® treated group than in the control group [102]. This chlorhexidine containing chip incorporated 2.5 mg chlorhexidine gluconate in a biodegradeable matrix of cross-linked hydrolyzed gelatin.

In one study assessing the effects of mechanical debridement with or without the local application of a chlorhexidine gelatum (Plaque Out Gel, Hawe-Neos Dental, Weissenfluh, Switzerland) at one time in subjects with periodontitis the results demonstrated that both modalities of treatment were effective in reducing peri-implant mucositis and probing depths. The addition of chlorhexidine gel to mechanical debridement did not enhance the results as compared to mechanical debridement alone [103].

20.7 Summary and Conclusions

It is currently well established that both implant mucositis and peri-implantitis have an infectious etiology. Nevertheless, there is insufficient evidence to identify which specific pathogens that are the key bacteria causing implant mucositis or peri-implantitis. Given the different characteristics of titanium and root cementum it seems reasonable to believe that the biofilm development may differ between the two environments. This would then have consequences and may require the use of alternative anti-infective strategies which may also require alternative antibiotic and antiseptic approaches. To begin with, it seems reasonable to use the same types of biomaterials that have successfully been used to control for periodontitis at natural teeth also when treating infected tissues around dental titanium implants.

If left untreated the inflammatory responses and the infection result in loss of alveolar bone around dental implants and edema and redness of the soft tissues surrounding the implant. The consequences of this condition may be loss of the implant. It is also possible that such infections may have other systemic consequences and similar to what has been reported for associations between periodontitis and diseases such as diabetes mellitus, cardiovascular diseases, rheumatoid arthritis. Due to the fact that the placement of dental titanium implant is a highly reliable and effective method to replace missing teeth, the medical/dental community focused on technologies aiming at faster osseo-integration of the implants. Dental implant manufacturers and distributors have focused on design, ease of placement of implants, esthetics, and new prosthetic components. Enhancement of osseo-integration has been attempted by changing titanium surface structures, titanium purity, and various coatings.

The diagnostic criteria for implant mucositis and peri-implantitis were developed recently [33]. It is therefore understandable that the scientific evidence about the prevalence of implant complications beyond technical failures is under debate. Even if the prevalence of peri-implantitis approaches 20–30% of all dental implants placed, the success rate would remain at 70–80% and this should be considered as a remarkable success for an organ replacement device that functions in an environment with a more complex microbiota than anywhere else in the human body. There is, in fact, emerging data suggesting that the infection in cases with peri-implantitis may also include pathogens that are less commonly recognized in periodontitis. Findings of Staphylococci, Pseudomonas, Bacteroides, Enterococci, Fusobacterium species and currently less well defined microorganisms in a complex biofilm together with species traditionally associated with periodontitis, would make therapy to eradicate the infection a major challenge to the clinician. The biofilm structure at titanium dental implants comprise of bacteria that co-aggregate with each other, that compete over space and nutrition and arrange themselves within layers of aerobic anaerobic species with protective layers of surface extracellular matrix containing proteins, long chain polysaccharides and lipids.

The attachment to the titanium surface is characterized by chemical and, electrostatic binding forces making mechanical debridement very difficult, if not impossible. In addition the threaded surfaces of oral implants are also difficult to debride. The location of the bacteria at the external side of the epithelial soft tissue lining with limited or no intra- and intercellular invasion adds to the complexity of treatment. Management of the implant surface characteristics may be the most advantageous method to control or guide the development of a biofilm of bacteria.

The host inflammatory responses to infection at dental titanium implants with a biocompatible but yet foreign body characteristics may appear similar to what can be seen in periodontitis. The scientific evidence is, however, not comprehensive and almost all publications on the host responses have been focused on traditional histological analysis suggesting similarities between periodontitis and peri-implantitis lesions. There are few studies assessing humoral immunity, and secretion of pro-inflammatory cytokines in peri-implantitis. Furthermore, studies assessing the microbiota from lesions associated with peri-implantitis have focused on the same array of bacteria as are commonly studied in periodontitis. It may be a misconception that the infectious etiology in periodontitis and peri-implantitis is similar. The development of a biofilm is not only dependent on similarities in aqueous solution (i.e., crevicular fluid) but is also dependent on surface characteristics. In this regard the surface of a natural tooth and that of the titanium surface differ greatly. As for many other foreign bodies/implants of various types that have and are being used in medicine Staphylococci spp. are of primary concerns. This appears to be the same for peri-implantitis lesions.

During the earlier period in implant dentistry most implants were placed in subjects who lost teeth due to trauma, tooth decay, or congenitally missing teeth. More recently, a large number of implants are being and have been placed in subjects who lost teeth due to periodontitis, and in older patients. Patients who lost teeth due to periodontitis may be more prone to oral infections resulting in bone resorption. The susceptibility may be directly associated with a genetically complex predisposing risk, or changes in immunity as a result of aging. Currently neither of these two factors can be controlled for.

Recent data also suggest that the professional mechanical debridement of pockets around implants and the patients' own ability to keep implants free from bacteria is not predictable. Professional debridement may be efficacious over shorter time periods and would most likely require rather intensive supportive therapy, In the event of limited response to nonsurgical mechanical supportive therapy, the adjunct use of local application of a slow-release antimicrobial agent should be considered. The anatomical characteristics of implant design including the superstructure to reestablish esthetics and function make hygiene difficult. Surgical treatment to eliminate soft tissue pocket and attempts to establish hygienic conditions would in many instances compromise esthetics and would probably not be acceptable to the patient.

Antibiotics are commonly used in medicine to treat infections. Restrictive use of antibiotics is a necessity to prevent the occurrence of wide-spread antibiotic resistance. Clearly, peri-implantitis is a local infection confined to the area around the affected implant. Therefore, local administration of an antibiotic drug in a high concentration is logical. Furthermore there are few studies that have identified pathogens within soft or hard tissues in periodontitis and there are to our knowledge no such confirmed reports for peri-implantitis infections. It is, therefore, important that the antibiotic drug is deposited at the site of infection. A systemic route of administration may not be efficacious. In contrast the local administration of an active depot of antibiotic with a sustained release of drug, over a period of time long enough may have bacteriostatic or bacteriocidal effects against the target microorganism(s).

Several local antibiotic products have been developed and studied extensively. Such local antibiotic drugs have been on the market for some years but due to a variety of factors some of these local antibiotics have been withdrawn from the market or may only be available in some countries. There is clearly a need to develop cost-effective methods to distribute antibiotics locally at infected implant sites especially as the infection usually is located to a single implant. Once, the specific infectious etiology of peri-implantitis has been identified it may be advantageous to locally administer a targeted antibiotic drug for the treatment of this local infections. The same could be said about the application of a local antiinflammatory drug to control the inflammatory process.

Glossary

Autogenous bonegraft Involves utilizing bone obtained from the same individual receiving the graft

Bactericidal A substance that kills bacteria

Bacteriostatic A substance that inhibit growth and reproduction of bacteria without killing them

Biofilm A biofilm is an aggregate of microorganisms in which cells adhere to each other and/or to a surface

Gingivitis A reversible inflammation of the soft tissue surrounding the teeth

Implant Mucositis An inflammatory lesion that resides in the mucosa. This is clinically identified by redness, swelling, and bleeding on probing

Nanotechnology Nanotechnology deals with structures sized between 1 and 100 nm in at least one dimension and involves developing materials or devices within that size

Osseo-integration A direct structural and functional connection between living bone and the surface of a load-bearing artificial implant

Peri-implantitis An inflammatory lesion in the mucosa also affecting the supporting bone. This is clinically identified by redness, swelling bleeding on probing and bone loss

Periodontitis An irreversible inflammation of the soft and hard tissues surrounding the teeth leading to a loss of bone support

References

1. Schroeder A, Pohler O, Sutter F. Tissue reaction to an implant of a titanium hollow cylinder with a titanium surface spray layer SSO. Schweiz Monatsschr Zahnheilkd. 1976;86: 713–27.
2. Schroeder A, van der Zypen E, Stich H, et al. The reactions of bone, connective tissue, and epithelium to endosteal implants with titanium-sprayed surfaces. J Maxillofac Surg. 1981;9:15–25.
3. Axelsson P, Lindhe J. Effect of controlled oral hygiene procedures on caries and periodontal disease in adults. J Clin Periodontol. 1978;5:133–51.
4. Albrektsson T, Brånemark PI, Hansson HA, et al. Osseointegrated titanium implants. Requirements for ensuring a long-lasting, direct bone-to-implant anchorage in man. Acta Orthop Scand. 1981;52:155–70.
5. Vignoletti F, de Sanctis M, Berglundh T, et al. Early healing of implants placed into fresh extraction sockets: an experimental study in the beagle dog. III: soft tissue findings. J Clin Periodontol. 2009;36:1059–66.
6. Babbush CA. Titanium plasma spray screw implant system for reconstruction of the edentulous mandible. Dent Clin North Am. 1986;30:117–31.
7. Buser D, Weber HP, Brägger U. The treatment of partially edentulous patients with ITI hollow-screw implants: presurgical evaluation and surgical procedures. Int J Oral Maxillofac Impl. 1990;5:165–75.
8. Wedgwood D, Jennings KJ, Critchlow HA. Experience with ITI osseointegrated implants at five centres in the UK. Br J Oral Maxillofac Surg. 1992;30:377–81.
9. Laney WR, Jemt T, Harris D, et al. Osseointegrated implants for single-tooth replacement: progress report from a multicenter prospective study after 3 years. Int J Oral Maxillofac Impl. 1994;9:49–54.
10. Aglietta M, Siciliano VI, Zwahlen M, et al. A systematic review of the survival and complication rates of implant supported fixed dental prostheses with cantilever extensions after an observation period of at least 5 years. Clin Oral Implants Res. 2009;20:441–51.
11. Ong CT, Ivanovski S, Needleman IG, et al. Systematic review of implant outcomes in treated periodontitis subjects. J Clin Periodontol. 2008;35:438–62.

12. Bosker H, Jordan RD, Sindet-Pedersen S, Koole R. The transmandibular implant: a 13-year survey of its use. J Oral Maxillofac Surg. 1991;49:482–92.
13. Ekfeldt A, Johansson LA, Isaksson S. Implant-supported overdenture therapy: a retrospective study. Int J Prosthodont. 1997;10:366–74.
14. Brägger U, Aeschlimann S, Bürgin W, et al. Biological and technical complications and failures with fixed partial dentures (FPD) on implants and teeth after four to five years of function. Clin Oral Implants Res. 2001;12:26–34.
15. Henry PJ, Laney WR, Jemt T. Osseointegrated implants for single-tooth replacement: a prospective 5-year multicenter study. Int J Oral Maxillofac Impl. 1996;11:450–5.
16. Simonis P, Dufour T, Tenenbaum H. Long-term implant survival and success: a 10–16-year follow-up of non-submerged dental implants. Clin Oral Impl. 2010;21:772–7.
17. Fransson C, Lekholm U, Jemt T, Berglundh T. Prevalence of subjects with progressive bone loss at implants. Clin Oral Impl Res. 2005;16:440–6.
18. Roos-Jansåker AM, Lindahl C, Renvert H, et al. Nine- to fourteen-year follow-up of implant treatment. Part II: presence of peri-implant lesions. J Clin Periodontol. 2006;33:290–5.
19. Koldsland OC, Scheie AA, Aass AM. Prevalence of peri-implantitis related to severity of the disease with different degrees of bone loss. J Periodontol. 2010;81:231–8.
20. Esposito M, Hirsch JM, Lekholm U, Thomsen P. Biological factors contributing to failures of osseointegrated oral implants. I. Success criteria and epidemiology. Eur J Oral Sci. 1998;106:527–51.
21. Máximo MB, de Mendonça AC, Alves JF, et al. Peri-implant diseases may be associated with increased time loading and generalized periodontal bone loss: preliminary results. J Oral Implantol. 2008;34:268–73.
22. Bower RC, Radny NR, Wall CD, Henry PJ. Clinical and microscopic findings in edentulous patients 3 years after incorporation of osseointegrated implant-supported bridgework. J Clin Periodontol. 1989;16:580–7.
23. Adell R, Lekholm U, Rockler B, et al. Marginal tissue reactions at osseointegrated titanium fixtures (I). A 3-year longitudinal prospective study. Int J Oral Maxillofac Surg. 1986;15:39–52.
24. van Steenberghe D, Klinge B, Lindén U. Periodontal indices around natural and titanium abutments: a longitudinal multicenter study. J Periodontol. 1993;64:538–41.
25. Leonhardt Å, Dahlén G, Renvert S. Five year clinical, microbiological, and radiological outcome following treatment of peri-implantitis in man. J Periodontol. 2003;74:1415–22.
26. Harris LG, Richards RG. Staphylococcus aureus adhesion to different treated titanium surfaces. J Mater Sci Mater Med. 2004;15:311–4.
27. Antoci Jr V, Adams CS, Parvizi J, et al. Covalently attached vancomycin provides a nanoscale antibacterial surface. Clin Orthop Relat Res. 2007;461:81–7.
28. Renvert S, Lindahl C, Renvert H, et al. Clinical and microbiological analysis of subjects treated with Brånemark or AstraTech implants: a 7-year follow-up study. Clin Oral Implants Res. 2008;19:342–7.
29. Renvert S, Lessem J, Dahlén G, et al. Mechanical and repeated antimicrobial therapy using a local drug delivery system in the treatment of peri-implantitis: a randomized clinical trial. J Periodontol. 2008;79:836–44.
30. Persson GR, Samuelsson E, Lindahl C, et al. Mechanical non-surgical treatment of peri-implantitis: a single-blinded randomized longitudinal clinical study. II. Microbiological results. J Clin Periodontol. 2010;37:563–73.
31. Haffajee AD, Socransky SS. Introduction to microbial aspects of periodontal biofilm communities, development and treatment. Periodontol. 2006;42:7–12.
32. Albrektsson T, Isidor F. Consensus report of session IV. In: Lang NP, Karring T, editors. Proceedings of the First European Workshop on Periodontology. London: Quintessence; 1994. p. 365–9.
33. Lindhe J, Meyle J, Group D of European Workshop on Periodontology. Peri-implant diseases: Consensus Report of the Sixth European Workshop on Periodontology. J Clin Periodontol. 2008;35 Suppl 8:282–5.
34. Gualini F, Berglundh T. Immunohistochemical characteristics of inflammatory lesions at implants. J Clin Periodontol. 2003;30:14–8.

35. Gristina AG. Implant failure and the immuno-incompetent fibro-inflammatory zone. Clin Orthop Relat Res. 1994;298:106–18.
36. Bullon P, Fioroni M, Goteri G, et al. Immunohistochemical analysis of soft tissues in implants with healthy and peri-implantitis condition, and aggressive periodontitis. Clin Oral Implants Res. 2004;15:553–9.
37. Page RC, Beck JD. Risk assessment for periodontal diseases. Int Dent J. 1997;47:61–87.
38. Michalowicz BS, Diehl SR, Gunsolley JC, et al. Evidence of a substantial genetic basis for risk of adult periodontitis. J Periodontol. 2000;71:1699–707.
39. Papapanou PN, Behle JH, Kebschull M. Subgingival bacterial colonization profiles correlate with gingival tissue gene expression. BMC Microbiol. 2009;18(9):221.
40. Kronström M, Svensson B, Erickson E, et al. Humoral immunity host factors in subjects with failing or successful titanium dental implants. J Clin Periodontol. 2000;27:875–82.
41. Aas JA, Paster BJ, Stokes LN, et al. Defining the normal bacterial flora of the oral cavity. J Clin Microbiol. 2005;43:5721–32.
42. Paster BJ, Olsen I, Aas JA, et al. The breadth of bacterial diversity in the human periodontal pocket and other oral sites. Periodontol. 2006;42:80–7.
43. Dewhirst FE, Chen T, Izard J, et al. The human oral microbiome. J Bacteriol. 2010;192: 5002–17.
44. Pontoriero R, Tonelli MP, Carnevale G. Experimentally induced peri-implant mucositis. A clinical study in humans. Clin Oral Impl Res. 1994;5:254–9.
45. Augthun M, Conrads B. Microbial findings of deep peri-implant bone defects. Int J Oral and Maxillofac Impl. 1997;12:106–12.
46. Salcetti JM, Moriarty JD, Cooper LF. The clinical, microbial and host response characteristics of the failing implant. Int J Oral Maxillofac Impl. 1997;12:32–42.
47. Leonhardt A, Renvert S, Dahlén G. Microbial findings at failing implants. Clin Oral Implants Res. 1999;10:339–45.
48. Quirynen M, De Soete M, van Steenberghe D. Infectious risks for oral implants: a review of the literature. Clin Oral Impl Res. 2002;13:1–19.
49. Quirynen M, Vogels R, Peeters W, et al. Dynamics of initial subgingival colonization of "pristine" peri-implant pockets. Clin Oral Implants Res. 2006;17:25–37.
50. Mombelli A. Microbiology and antimicrobial therapy of peri-implantitis. Periodontol. 2002;28:177–89.
51. Renvert S, Roos-Jansåker AM, Lindahl C, et al. Infection at titanium implants with or without a clinical diagnosis of inflammation. Clin Oral Implants Res. 2007;18:509–16.
52. Stoodley P, Kathju S, Hu FZ, et al. Molecular and imaging techniques for bacterial biofilms in joint arthroplasty infections. Clin Orthop Relat Res. 2005;437:31–40.
53. Shibli JA, Melo L, Ferrari DS, et al. Composition of supra- and subgingival biofilm of subjects with healthy and diseased implants. Clin Oral Impl. 2008;19:975–82.
54. De Boever AL, De Boever JA. Early colonization of non-submerged dental implants in patients with a history of advanced aggressive periodontitis. Clin Oral Implants Res. 2006;17:8–17.
55. Fürst MM, Salvi GE, Lang NP, Persson GR. Bacterial colonization immediately after installation on oral titanium implants. Clin Oral Implants Res. 2007;18:501–8.
56. Salvi GE, Fürst MM, Lang NP, et al. One-year bacterial colonization patterns of Staphylococcus aureus and other bacteria at implants and adjacent teeth. Clin Oral Implants Res. 2008; 19:242–8.
57. Harris LG, Mead L, Müller-Oberländer E, Richards RG. Bacteria and cell cytocompatibility studies on coated medical grade titanium surfaces. J Biomed Mater Res A. 2006;78:50–8.
58. Khoo X, O'Toole GA, Nair SA et al. Staphylococcus aureus resistance on titanium coated with multivalent PEGylated-peptides. Biomaterials. 2010;31:9285–92.
59. Cosyn J, Van Aelst L, Collaert B, Persson GR, De Bruyn H. The peri-implant sulcus compared with internal implant and suprastructure components: a microbiological analysis. Clin Implant Dent Relat Res. 2011;13:286–95.
60. Elter C, Heuer W, Demling A, et al. Supra- and subgingival biofilm formation on implant abutments with different surface characteristics. Int J Oral Maxillofac Impl. 2008;23:327–34.

61. Esposito M, Hirsch J, Lekholm U, Thomsen P. Differential diagnosis and treatment strategies for biologic complications and failing oral implants: a review of the literature. Int J Oral Maxillofac Impl. 1999;14:473–90.
62. Grusovin MG, Coulthard P, Worthington HV et al. Interventions for replacing missing teeth: maintaining and recovering soft tissue health around dental implants. Cochrane Database Syst Rev. 2010;8:CD003069.
63. Máximo MB, de Mendonça AC, Renata Santos V, et al. Short-term clinical and microbiological evaluations of peri-implant diseases before and after mechanical anti-infective therapies. Clin Oral Implants Res. 2009;20:99–108.
64. Renvert S, Samuelsson E, Lindahl C. Mechanical non-surgical treatment of peri-implantitis: a double-blind randomized longitudinal clinical study. I: clinical results. J Clin Periodontol. 2009;36:604–9.
65. Schwarz F, Bieling K, Nuesry E, et al. Clinical and histological healing pattern of peri-implantitis lesions following non-surgical treatment with an Er:YAG laser. Lasers Surg Med. 2006;38:663–71.
66. Esposito M, Cannizarro G, Bozzoli P, et al. Effectiveness of prophylactic antibiotics at placement of dental implants: a pragmatic multicentre placebo-controlled randomised clinical trial. Eur J Oral Implantol. 2010;3:135–43.
67. Esposito M, Worthington HV, Loli V, Grusovin P. Interventions for replacing missing teeth: antibiotics at dental implant placement to prevent complications. Cochrane Database Syst Rev. 2010;7:CD004152.
68. Abu-Ta'a M, Quirynen M, Teughels W, van Steenberghe D. Asepsis during periodontal surgery involving oral implants and the usefulness of peri-operative antibiotics: a prospective, randomized, controlled clinical trial. J Clin Periodontol. 2008;35:58–63.
69. Ungvári K, Pelsöczi IK, Kormos B, et al. Effects on titanium implant surfaces of chemical agents used for the treatment of peri-implantitis. J Biomed Mater Res B Appl Biomater. 2010;94:222–9.
70. Claffey N, Clarke E, Polyzois I, Renvert S. Surgical treatment of peri-implantitis. J Clin Periodontol. 2008;35 Suppl 8:316–32.
71. Maness PC, Smolinski S, Blake DM, et al. Bactericidal activity of photocatalytic TiO(2) reaction: toward an understanding of its killing mechanism. Appl Environ Microbiol. 1999;65:4094–8.
72. Tsuang YH, Sun JS, Huang YC, et al. Studies of photokilling of bacteria using titanium dioxide nanoparticles. Artif Organs. 2008;32:167–74.
73. Necula BS, Fratila-Apachitei LE, Zaat SA, et al. In vitro antibacterial activity of porous TiO_2-Ag composite layers against methicillin-resistant Staphylococcus aureus. Acta Biomater. 2009;5:3573–80.
74. Liao J, Anchun M, Zhu Z, Quan Y. Antibacterial titanium plate deposited by silver nanoparticles exhibits cell compatibility. Int J Nanomedicine. 2010;5:337–42.
75. Rupp F, Haupt M, Klostermann H, Kim HS et al. Multifunctional nature of UV-irradiated nanocrystalline anatase thin films for biomedical applications. Acta Biomater. 2010 [Epub ahead of print]
76. Yoshinari M, Kato T, Matsuzaka K, et al. Prevention of biofilm formation on titanium surfaces modified with conjugated molecules comprised of antimicrobial and titanium-binding peptides. Biofouling. 2010;26:103–10.
77. Brohede U, Forsgren J, Roos S, et al. Multifunctional implant coatings providing possibilities for fast antibiotics loading with subsequent slow release. J Mater Sci Mater Med. 2009;20:1859–67.
78. Jiang B, Li B. Polypeptide nanocoatings for preventing dental and orthopaedic device-associated infection: pH-induced antibiotic capture, release, and antibiotic efficacy. J Biomed Mater Res B Appl Biomater. 2009;88:332–8.
79. Li H, Ogle H, Jiang B, Hagar M, Li B. Cefazolin embedded biodegradable polypeptide nanofilms promising for infection prevention: a preliminary study on cell responses. J Orthop Res. 2010;28:992–9.
80. O'Sullivan C, O'Hare P, O'Leary ND, et al. Deposition of substituted apatites with anticolonizing properties onto titanium surfaces using a novel blasting process. J Biomed Mater Res B Appl Biomater. 2010;95:141–9.

81. Cionca N, Giannopoulou C, Ugolotti G, Mombelli A. Microbiologic testing and outcomes of full-mouth scaling and root planing with or without amoxicillin/metronidazole in chronic periodontitis. J Periodontol. 2010;81:15–23.
82. Boente RF, Ferreira LQ, Falcão LS, et al. Detection of resistance genes and susceptibility patterns in Bacteroides and Parabacteroides strains. Anaerobe. 2010;16:190–4.
83. Perreten V, Kadlec K, Schwarz S, et al. Clonal spread of methicillin-resistant Staphylococcus pseudintermedius in Europe and North America: an international multicentre study. J Antimicrob Chemother. 2010;65:1145–55.
84. Kenawya ER, Bowlin GL, Kevin Mansfield K, et al. Release of tetracycline hydrochloride from electrospun poly(ethylene-co-vinylacetate), poly(lactic acid), and a blend. J Control Rel. 2002;81:57–64.
85. Rodrigues RM, Gonçalves C, Souto R, Feres-Filho EJ, Uzeda M, Colombo AP. Antibiotic resistance profile of the subgingival microbiota following systemic or local tetracycline therapy. J Clin Periodontol. 2004;31:420–7.
86. Schenk G, Flemmig TF, Betz T, et al. Controlled local delivery of tetracycline HCl in the treatment of periimplant mucosal hyperplasia and mucositis. A controlled case series. Clin Oral Implants Res. 1997;8:427–33.
87. Mombelli A, Feloutzis A, Brägger U, Lang NP. Treatment of peri-implantitis by local delivery of tetracycline. Clinical, microbiological and radiological results. Clin Oral Implants Res. 2001;12:287–94.
88. Kim HW, Knowles JC, Kim HE. Development of hydroxyapatite bone scaffold for controlled drug release via poly(epsilon-caprolactone) and hydroxyapatite hybrid coatings. J Biomed Mater Res B Appl Biomater. 2004;15:240–9.
89. Dashti A, Ready D, Salih V, et al. In vitro antibacterial efficacy of tetracycline hydrochloride adsorbed onto Bio-Oss bone graft. J Biomed Mater Res B Appl Biomater. 2010;93: 394–400.
90. Philippart P, Brasseur M, Hoyaux D. Human recombinant tissue factor, platelet-rich plasma, and tetracycilne induce a high-quality human bone graft: a 5-year survey. Int J Oral Maxillofac Impl. 2003;18:411–6.
91. Salvi GE, Mombelli A, Mayfield L, et al. Local antimicrobial therapy after initial periodontal treatment. J Clin Periodontol. 2002;29:540–50.
92. Bogren A, Teles RP, Torresyap G, et al. Locally delivered doxycycline during supportive periodontal therapy: a 3-year study. J Periodontol. 2008;79:827–35.
93. Büchter A, Meyer U, Kruse-Lösler B, Joos U, Kleinheinz J. Sustained release of doxycycline for the treatment of peri-implantitis: randomised controlled trial. Br J Oral Maxillofac Surg. 2004;42:439–44.
94. Büchter A, Kleinheinz J, Meyer U, Joos U. Treatment of severe peri-implant bone loss using autogenous bone and a bioabsorbable polymer that delivered doxycycline (Atridox). Br J Oral Maxillofac Surg. 2004;42:454–6.
95. Williams RC, Paquette DW, Offenbacher S, et al. Treatment of periodontitis by local administration of minocycline microspheres: a controlled trial. J Periodontol. 2001;72:1535–44.
96. Renvert S, Lessem J, Dahlén G, et al. Topical minocycline microspheres versus topical chlorhexidine gel as an adjunct to mechanical debridement of incipient peri-implant infections: a randomized clinical trial. J Clin Periodontol. 2006;33:362–9.
97. Persson GR, Salvi GE, Heitz-Mayfield LJ, et al. Antimicrobial therapy using a local drug delivery system (Arestin) in the treatment of peri-implantitis. I: Microbiological outcomes. Clin Oral Implants Res. 2006;17:386–93.
98. Salvi GE, Persson GR, Heitz-Mayfield LJ, et al. Adjunctive local antibiotic therapy in the treatment of peri-implantitis II: clinical and radiographic outcomes. Clin Oral Implants Res. 2007;18:281–5.
99. Stellini E, Migliorato A, Mazzoleni S, et al. Topical treatment of peri-implantitis with metronidazole dental gel 25%. Clinical analysis and microbiological control. Minerva Stomatol. 2000;49:59–67.

100. Flemmig TF, Gregor Petersilka G, Völp A, et al. Efficacy and safety of adjunctive local moxifloxacin delivery in the treatment of periodontitis. J Periodontol. 2011;82:96–105.
101. Sela MN, Babitski E, Steinberg D, et al. Degradation of collagen-guided tissue regeneration membranes by proteolytic enzymes of Porphyromonas gingivalis and its inhibition by antibacterial agents. Clin Oral Implants Res. 2009;20:496–502.
102. Mizrak T, Güncü GN, Caglayan F, et al. Effect of a controlled-release chlorhexidine chip on clinical and microbiological parameters and prostaglandin E2 levels in gingival crevicular fluid. J Periodontol. 2006;77:437–43.
103. Porras R, Anderson GB, Caffesse R, et al. Clinical response to 2 different therapeutic regimens to treat peri-implant mucositis. J Periodontol. 2002;73:1118–25.
104. Christenson EM, Anseth KS, van den Beucken JJ, et al. Nanobiomaterial applications in orthopedics. J Orthop Res. 2007;25:11–22.
105. Duarte Fde F, Lotufo RF, Pannuti CM. Local delivery of chlorhexidine gluconate in patients with aggressive periodontitis. J Int Acad Periodontol. 2008;10:31–5.
106. Emrani J, Chee W, Slots J. Bacterial colonization of oral implants from nondental sources. Clin Implant Dent Relat Res. 2009;11:106–12.
107. Flores CY, Diaz C, Rubert A, Benítez GA, Moreno MS, Fernández Lorenzo deMele MA. Spontaneous adsorption of silver nanoparticles on Ti/TiO2 surfaces. Antibacterial effect on Pseudomonas aeruginosa. J Colloid Interface Sci. 2010;350:402–8.

Chapter 21
Preventive Strategies in VAP: Focus on Silver-Coated Endotracheal Tubes

Marin Kollef

Abstract *Introduction*: Ventilator-associated pneumonia (VAP) is problematic because of its frequency, morbidity, and mortality. Many strategies have been studied in an attempt to reduce the rate of VAP. This chapter provides a comprehensive review of silver-coated endotracheal tubes in the prevention of VAP.

Methods: Two silver-coated endotracheal tubes have been studied in comparison with similar, uncoated tubes—each beginning with preclinical models and progressing to clinical studies. One is commercially available (*Ag*ento® IC, C. R. Bard, Covington, Georgia, United States) and is coated with silver ions micro-dispersed in a proprietary hydrophilic polymer. The other remains investigational and is prepared by submerging a standard endotracheal tube into silver sulfadiazine (with or without chlorhexidine) and polyurethane.

Results: Both silver-coated tubes were active in preclinical models designed to mimic surrogate endpoints for VAP, such as *in vitro* bacterial adherence, biofilm formation, and bacterial burden in animal models. Both tubes were active in phase 2 studies of patients requiring mechanical ventilation. The commercially available tube was active in a randomized, phase 3 study and reduced the incidence of microbiologically confirmed VAP at any time after intubation (silver vs. uncoated, 37/766 [4.8 %] vs. 56/743 [7.5 %]; $P=0.03$; relative risk reduction, 35.9 %) and within 10 days of intubation (27/766 [3.5 %] vs. 50/743 [6.7 %]; $P=0.005$, relative risk reduction, 47.6 %).

Conclusions: The silver-coated endotracheal tube is a promising strategy for preventing VAP. The commercially available tube meets the standards for evidence-based guidelines because effectiveness is supported by the results of a well-designed,

M. Kollef (✉)
Washington University School of Medicine,
660 South Euclid Avenue, St. Louis, MO 63110, USA
e-mail: mkollef@dom.wustl.edu

T.F. Moriarty et al. (eds.), *Biomaterials Associated Infection: Immunological Aspects and Antimicrobial Strategies*, DOI 10.1007/978-1-4614-1031-7_21,

phase 3 study. In addition, the device is suitable for use in the intensive care unit because it becomes user independent after intubation and does not add to the burden of busy healthcare providers.

Abbreviations

AMMI Canada	Association of Medical Microbiology and Infectious Disease Canada
ATS/IDSA	American Thoracic Society/Infectious Diseases Society of America
BSAC	British Society for Antimicrobial Chemotherapy
CCCTG	Canadian Critical Care Trials Group
CPIS	Clinical pulmonary infection score
GPP	Good practice point
HAP	Healthcare-associated pneumonia
ICU	Intensive care unit
IHI	Institute for Healthcare Improvement
NASCENT study	North American Silver-Coated Endotracheal Tube study
NHSN	National Healthcare Safety Network
SHEA/IDSA	Society for Healthcare Epidemiology of America/IDSA
SMART	Specific, measurable, achievable, relevant, and time bound
VAP	Ventilator-associated pneumonia

21.1 Introduction

Ventilator-associated pneumonia (VAP) is problematic in the intensive care unit (ICU) because of its frequency, morbidity, and mortality. The mean pooled VAP rate per 1,000 ventilator days was recently estimated to be 24.1 in developing countries [1]; 13.6 in Europe, Asia, Africa, and Asia [2]; and 3.3 in the United States [2]. This variability is attributable partly to location and also to type of ICU, year of survey, and other factors. VAP prolongs length of stay in the ICU and hospital, resulting in a mean attributable cost to the hospital of $25,000 per episode based on a meta-analysis that adjusted costs to 2005 levels [3]. The crude mortality rate is 44 %, resulting in an excess mortality rate of 29.3 % [2]. These cost and mortality estimates are higher than corresponding estimates for other types of nosocomial infection, such as bloodstream infection and catheter-associated urinary tract infection [2, 3].

Failure to promptly initiate appropriate antimicrobial therapy for VAP further increases mortality [4]. Selection of initial therapy is based on the microbiology of VAP, which is often polymicrobial including a variety of bacterial pathogens, and difficult to treat [5]. The most common are aerobic gram-negative bacilli, such as

Pseudomonas aeruginosa, *Klebsiella pneumoniae*, and *Acinetobacter* species; and gram-positive cocci, such as *Staphylococcus aureus*, which is often resistant to methicillin. The presence of multidrug-resistant pathogens varies depending on the hospital, patient population, exposure to antibiotics, type of ICU, and change over time, which underscores the need for local surveillance [5]. Treatment guidelines for VAP emphasize the need for early, appropriate antimicrobial therapy with risk stratification based on the potential for multidrug-resistant pathogens and other variables [5, 6]. Patients without risk factors are candidates for monotherapy, whereas patients with risk factors require combination therapy. To avoid excessive use, antimicrobial therapy should be deescalated on the basis of culture results and clinical response [5, 6].

The purpose of this chapter is to provide a comprehensive review of silver-coated endotracheal tubes in the prevention of VAP, including preclinical and clinical studies on available and investigational devices. To place this review into perspective, this chapter begins with a brief overview of other preventive strategies and lessons learned from studies of those strategies.

21.2 Advances in the Prevention of VAP

Many strategies have been studied in an attempt to reduce the rate of VAP and have led to the development of evidence-based guidelines by organizations around the world (Table 21.1). Guidelines from the American Thoracic Society/Infectious Diseases Society of America (ATS/IDSA) [5] and Society for Healthcare Epidemiology of America/IDSA (SHEA/IDSA) [7] are combined for the purpose of this review. Guidelines from the Canadian Critical Care Trials Group (CCCTG) [8] and Association of Medical Microbiology and Infectious Disease Canada (AMMI Canada)/Canadian Thoracic Society [9] are also combined. Guidelines from the European HAP (hospital-acquired pneumonia) Working Group (European Group) [10] and Working Party on HAP of the British Society for Antimicrobial Chemotherapy (BSAC) [11] are also considered.

To prepare guidelines, organizations review relevant studies and sometimes consider clinical issues such as safety, feasibility, and cost. Although organizations use different scales to rank supporting evidence, level I or A generally represents stronger evidence than level III or C. For example, ATS/IDSA [5] defines level I evidence as based on well-conducted, randomized, controlled studies, whereas level III evidence is based on case studies and expert opinions. SHEA/IDSA [7] uses an analogous scale to grade quality of evidence and adds letters to grade strength of recommendation; grade A is defined as good evidence to support a recommendation, whereas grade C is defined as poor evidence. AMMI Canada [9] uses an analogous approach to grade both strength of recommendation and quality of evidence. The BSAC [11] uses a 4-letter scale in which grade A is based on at least one meta-analysis, systematic review, or well-designed (level 1) randomized controlled study; whereas grade D is, for example, extrapolated evidence from level II studies. Good

Table 21.1 Evidence-based guidelines for preventing ventilator-associated pneumonia (VAP)

Strategy	American Groups [5, 7]	Canadian Groups [8, 9]	European Group [10]	British Group [11]
General infection control strategies		A-1		
Active surveillance for VAP	A-II			GPP
Hand hygiene	I		Recommend	GPP
Minimum ventilation duration and noninvasive ventilation	B-III		Recommend	A
Daily sedation interruption and assessment of readiness to wean	I		Recommend	A
Education program	A-II			B
Prevention of aspiration				
Semirecumbent positioning	B-II	A-2	Recommend	B
Avoidance of gastric overdistention	I			GPP
Avoidance of unplanned extubation and reintubation	I		Recommend	
Cuffed endotracheal tube with subglottic suctioning	B-II	A-2	Recommend	B
Cuff pressure ≥20 cm water	II		Recommend	GPP
Reduction of colonization				
Orotracheal (not nasotracheal) intubation	II	B-2	Recommend	C
Avoidance of antiulcer prophylaxis for low-risk patients	Unresolved		Recommend	A
Regular antiseptic oral care	A-I	Consider	Recommend	
Silver-coated endotracheal tube			Consider[b]	
Minimization of equipment contamination				
Sterile water for rinsing equipment	IB			
Removal of condensate from ventilatory circuits	IB			GPP
Change of circuit only when soiled or malfunctioning	IA	A-1		A
Proper storage and disinfection of equipment	Recommend			GPP

American groups, American Thoracic Society/Infectious Diseases Society of America (IDSA) and Society for Healthcare Epidemiology of America/IDSA; British group, Working Party on Hospital-Acquired Pneumonia (HAP) of the British Society for Antimicrobial Chemotherapy; Canadian groups, Association of Medical Microbiology and Infectious Disease Canada/Canadian Thoracic Society and Canadian Critical Care Trials Group; Euro Group, European HAP Working Group

[a] Each organization used a different scale for grading recommendations, with A or I representing-stronger recommendations than C or III. See text for additional details

[b] Consider in distinct settings and populations

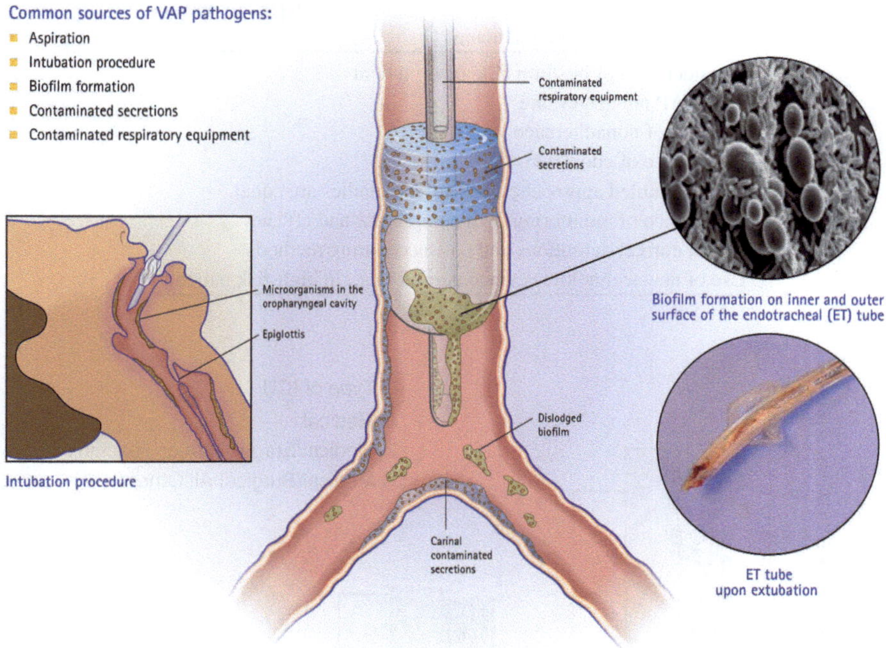

Fig. 21.1 Pathogenesis of ventilator-associated pneumonia (VAP). © 2009, C. R. Bard, Inc. Reproduced with permission from C. R. Bard, Inc., Covington, Georgia, USA

Practice Point (GPP) refers to a best practice point based on BSAC clinical experience. The CCCTG [8] and European Group [10] consider both strength of evidence and clinical issues before determining whether guidelines should be recommended, considered, or not recommended.

Exact strategies differ among organizations, but trends are evident. Preventive strategies should begin with general infection control measures, such as routine hand hygiene between patient contacts and educational programs for staff members. Next, strategies are based on the pathogenesis of VAP (Fig. 21.1) and therefore aim to prevent aspiration of secretions (e.g., semirecumbent positioning of the patient) and bacterial colonization of aerodigestive tract (e.g., orotracheal intubation instead of nasotracheal intubation). Lastly, strategies include measures to minimize the risk of contaminated equipment, such as properly storing and disinfecting equipment.

Important lessons can be learned from studies of these preventive strategies (Table 21.2) and have implications for evaluating new strategies. The first lesson is to consider the bustling environment within the ICU. Critically ill patients require care from a multitude of physicians, nurses, respiratory therapists, and other specialists as well as multiple therapeutic interventions that often involve complicated equipment. To increase the likelihood of successful outcomes in the ICU, healthcare providers should follow the specific, measurable, achievable, relevant, and time bound (SMART) approach [12]. Whenever possible, they should choose

Table 21.2 Lessons learned from studies of strategies to prevent ventilator-associated pneumonia

Consideration of bustling ICU environment
Zero VAP not achievable
Problem of nonadherence
Importance of education
Use of bundled approaches, but not all bundles are equal
Importance of monitoring both compliance and efficacy
Use of correct definitions and cost-accounting methods
Use of new technologies when needed (e.g., in high-risk patients)

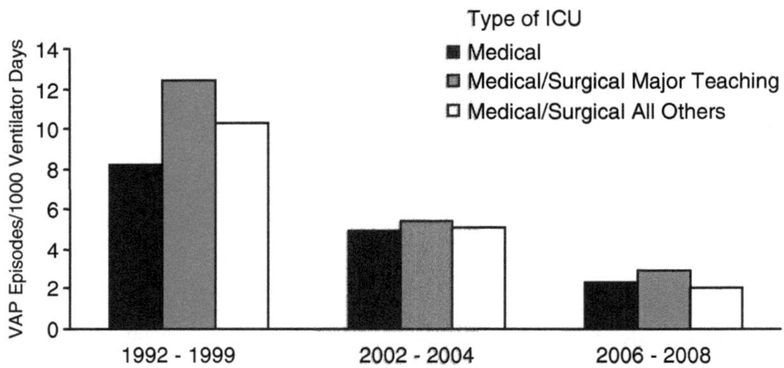

Fig. 21.2 Decreasing VAP rate in intensive care units in the United States [14–16]

simple tactics over complex, multifaceted strategies. All relevant stakeholders should be engaged in planning and execution. Nurses and other stakeholders are in the best position to identify the tactics that will be feasible and achievable within their busy ICUs as well as any barriers. A team leader should be appointed to champion the intervention and collaborate with administrators and ICU staff [13].

The second lesson is that a zero VAP rate is not achievable. Individual institutions may claim that VAP has been eliminated and national surveys may show progress over time, but VAP persists. For example, surveillance data from the United States demonstrate remarkable progress since the 1990s (Fig. 21.2) [14–16]. Between the 1990s and early part of this decade, the VAP rate decreased by half in medical/surgical ICUs at nonteaching hospitals—from 10 to 5 episodes per 1,000 ventilator days. By 2008, the rate had plummeted to 2 episodes per 1,000 ventilator days, but this low rate is not zero. Other types of ICUs have demonstrated analogous progress; however, VAP continues to occur.

The third lesson is the problem of failure to adhere to guidelines, which has been documented among physicians [17, 18], nurses [19–21], and respiratory therapists [21]. Adherence is variable depending on the guideline, method of assessment, and other factors. For example, nurses and respiratory therapists reported adherence rates of 83 % for 11 effective strategies; however, adherence to individual strategies ranged from 27 % to 97 % [21]. Furthermore, this aggregate was only marginally higher than that for all guidelines, including those considered to be ineffective or of undetermined effectiveness. The lack of correlation between adherence and level of evidence in this and other surveys [18, 19, 21, 22] suggests that evidence has not yet impacted upon clinical decision making. Another important consideration is that self-reported rates are often inflated and therefore better than those measured by direct observation.

Potential barriers to adherence to various types of guidelines include lack of awareness, familiarity, agreement, self-efficacy (e.g., lack of belief that the individual can perform the behavior), and outcome expectancy (e.g., lack of belief that the behavior will yield the desired outcome)—as well as guideline-, patient-, and environment-related barriers [23]. Surveys confirm that these barriers also diminish adherence to preventive guidelines for VAP. Among physicians, the most common barriers are disagreement with interpretation of clinical studies (35 %), lack of resources (31 %), and costs (17 %) [18]. Among nurses, the most common barriers are lack of resources (37 %), miscellaneous (overwork, lack of time for hand washing; 22 %), patient discomfort (8 %), disagreement with reported study results (8 %), and fear of potential adverse events (6 %) [19].

The fourth lesson is the importance of education as demonstrated by a series of pre- and postintervention observational studies at Barnes-Jewish Hospital [24, 25]. To improve awareness of preventive strategies, a multidisciplinary task force developed an educational program for ICU nurses and respiratory care practitioners [24]. Participants took a preintervention test, reviewed a study module, and then took an identical postintervention test. The VAP rate decreased from 12.6 to 5.7 episodes per 1,000 ventilator days ($P<0.001$) and was associated with an estimated annual cost savings of more than $400,000 [24]. After this educational program was expanded to the four largest hospitals in the local healthcare system, the aggregate VAP rate decreased from 8.8 to 4.7 episodes per 1,000 ventilator days ($P<0.001$). Improvements were statistically significant at three hospitals, but not at the one where respiratory therapists had the lowest rate of study module completion, which underscores the importance of adhering to the educational component of the protocol. When this educational program was implemented at a hospital in Thailand, the VAP rate decreased from 20.6 to 8.5 episodes per 1,000 ventilator days in the medical ICU ($P=0.001$) and was associated with shorter duration of ICU stay ($P<0.001$) and cost savings ($P<0.001$) [26].

The fifth lesson is bundling strategies to prevent VAP, but not all bundles are equal. The Institute for Healthcare Improvement (IHI) launched a highly publicized campaign to save lives and improve the quality of healthcare in the United States by implementing feasible, evidence-based interventions [27]. To prevent VAP and other adverse events, IHI recommends the following elements: (1) elevation of head of bed, (2) daily interruption of sedation (aka, "sedation vacation") and assessment of readiness to extubate, (3) peptic ulcer prophylaxis, and (4) deep venous thrombosis prophylaxis. To evaluate the relationship between VAP bundles and outcomes, we performed a systematic review of the literature and found only four studies that met our criteria for review and data extraction [28]. Bundles in these studies had at least three of the IHI elements and also had adjunct elements such as dedicated teams, hand washing, and daily goals. Although not always reported, adherence was at least 95 % in most centers in one study and 100 % in another, which is consistent with IHI's target of 95 %. The four studies yielded favorable outcomes; the VAP rate decreased from a range of 2.7–13.3 to a range of 0.0–9.3 episodes per 1,000 ventilator days. Each study, however, had major flaws in design, reporting, and results; including bias, confounding, and lack of generalizability. For example, selection bias may have occurred because observers were not blinded and may not have been consistent in their efforts to detect VAP, especially if different team members were involved and if diagnostic criteria were not rigorously defined. More recently, the IHI bundle was evaluated at Boston Medical Center [29]. Compliance was assessed twice daily with monthly feedback to the multidisciplinary team. Total bundle adherence increased from 53 % to 91 % in the trauma ICU and from 63 % to 81 % in the surgical ICU. At the same time, the VAP rate decreased from 10.2 to 3.4 episodes per 1,000 ventilator days ($P=0.004$) [29]. This study design is noteworthy because infection control team members did not change during the study period; however, they were not blinded and therefore may have been biased if they were aware of the overarching purpose of the intervention [29]. Collectively, these flaws preclude definitive statements about the effectiveness of the IHI bundle to prevent VAP.

The sixth lesson is the importance of monitoring both adherence and efficacy. Surprisingly, 38 % of nurses and respiratory therapists are not aware of VAP rates in their institutions [21]. Institution-specific VAP rates should be used to create awareness of the problem, whereas serial rates should be used as feedback after implementation of preventive strategies. The benefit of monitoring and providing feedback was demonstrated in recent studies [30–32] published after our review [28]. For example, passive education was not as effective as active strategies at Stirling Royal Infirmary in the United Kingdom [30]. In addition to staff education, active strategies included feedback of process measurement, feedback of outcome measurement, and removing barriers to delivery of care. After implementing these active strategies, the adherence rate to all six bundle elements increased from 48 % to 54 % ($P<0.0001$) and the VAP rate decreased from 19.2 to 7.5 episodes per 1,000 ventilator days [30] Progressive improvement occurred during stepwise implementation of a multifaceted program at Bichat-Claude-Bernard Hospital in Paris [31]. The program included a multidisciplinary task force, educational sessions, direct observations and performance feedback, and technical improvements and reminders. After implementation, median composite adherence scores increased over time ($P<0.0001$) and the VAP rate decreased from 23.5 to 11.5 episodes per 1,000 ventilator days ($P<0.0001$) [31]. Lastly, a real-time compliance dashboard was used to monitor real-time compliance with each bundle element and to provide continuous feedback to staff at Vanderbilt University Medical Center in Nashville [32]. Adherence improved from 39 % to 89 % ($P<0.001$) and the VAP rate decreased from 15.2 to 9.3 episodes per 1,000 ventilator days ($P=0.01$) [32].

The seventh lesson is the importance of using correct definitions of VAP and cost-accounting methods. We recently reviewed the diagnostic conundrum and concluded that there is no absolute gold standard for diagnosing VAP [33]. The most popular diagnostic criteria for VAP are the presence of a new or progressive consolidation on chest radiology plus at least two of the following clinical criteria: fever >38 °C, leukocytosis or leukopenia, and purulent secretions. Unfortunately, these clinical criteria are nonspecific in critically ill patients in the ICU because of the presence of underlying cardiopulmonary disorders. The chest radiograph can also be nonspecific in this setting. Another diagnostic tool is the Clinical Pulmonary Infection Score (CPIS), which is based on fever, leukocytosis, tracheal aspirates, oxygenation, radiographic infiltrates, and semiquantitative cultures of tracheal aspirates with gram-negative stain. CPIS, however, is not as accurate as quantitative bacterial cultures of the lower respiratory tract. Biomarkers, such as procalcitonin, are probably not accurate for diagnostic purposes; however, serial procalcitonin measurements may be useful markers for detecting sepsis and for facilitating decisions about when to stop antibiotic therapy. In view of the lack of accuracy of individual tests, we recommend a multimodal approach. When VAP is suspected, patients should undergo evaluation that is supported by local expertise and should include imaging procedures (chest radiograph, computed tomography), bacteriologic cultures from the lower respiratory tract, and possibly biomarkers [33].

Appropriate cost-accounting methods should be used to estimate the costs and associated savings of preventive strategies. Accounting methods should be both rigorous and precise as demonstrated by an economic analysis of an evidence-based sepsis protocol [34]. Ideally, inputs should be based on actual costs to the healthcare system rather than on charges to the patient. If costs are not available, charges can be converted on the basis of cost-to-charge ratios—preferably using ratios that are specific for each department. Estimated costs should be as comprehensive as possible and should include direct costs (e.g., labor and materials) and indirect costs (e.g., administrative overhead). Data should be collected prospectively to reduce the risk of recall, coding, or other bias. If previously published values are used to estimate inputs, they should be adjusted for inflation. Control groups should be appropriate (e.g., at risk for VAP) and should be studied concurrently to limit bias due to changes in procedure and costs. Published studies should carefully define all model inputs and include sensitivity analyses to demonstrate the effect of changing the most influential inputs. In our study [34], implementation of the evidence-based sepsis protocol was associated with lower median total costs (pre- vs. post-implementation, \$21,985 vs. \$16,103; $P<0.01$). Importantly, use of appropriate cost-accounting methods revealed that savings were primarily attributable to 35 % lower ICU costs ($P=0.03$) and 30 % lower ward costs ($P=0.03$). Implementation of the sepsis protocol did *not* increase component costs, such as pharmacy, laboratory, radiology, and respiratory therapy [34].

The eighth lesson is to take advantage of new technologies, especially in patients at high risk for VAP. New technologies should be considered if they have been shown to prevent VAP in well-designed clinical studies and to be cost-effective, especially if the new technologies are easy to integrate into the ICU environment and do not further burden the healthcare provider [12].

21.3 Silver-Coated Endotracheal Tubes

Placing an antibacterial coating on an endotracheal tube is a theoretically attractive method for preventing VAP in view of lessons learned from other preventive strategies for VAP and experience with other devices. Silver has broad-spectrum antibacterial activity *in vitro* [35] and, when placed on urinary catheters, prevented urinary tract infections in a meta-analysis of clinical studies [36]. Silver sulfadiazine is a sulfa derivative, which also has antibacterial properties, that is applied topically to prevent infection in burn victims and that is generally considered to be safe despite widespread use.

Two silver-coated endotracheal tubes have been evaluated in two series of studies—each beginning with preclinical models and progressing to clinical studies. Findings from these independent series should not be compared directly because the tubes differ substantially in preparation and evaluation methods. Differences in preparation methods may have implications for the release of silver and resultant silver concentrations. One of these tubes is commercially available (*Ag*ento® IC, C. R. Bard, Covington, Georgia, United States; hereafter, silver-coated tube). The other has not been as thoroughly evaluated and remains investigational (hereafter, silver-sulfadiazine-coated tube).

The commercially available endotracheal tube is coated with silver chloride colloidal particles micro-dispersed in a proprietary hydrophilic polymer. Both the inner and outer lumens are coated, permitting silver ions to migrate to the tube surface where they can provide a sustained antimicrobial effect. In addition, silver [37, 38] and the hydrophilic polymer [39–42] both block bacterial adherence to devices. Thus, the silver-coated endotracheal tube is thought to reduce the risk of VAP by multiple mechanisms, including creating an environment that facilitates interaction of silver salt with bacterial contaminants and by discouraging bacterial adherence (Fig. 21.3).

The investigational silver-sulfadiazine-coated endotracheal tube is prepared by submerging the tip of a standard endotracheal tube (Lo-Contour; Mallinckrodt Inc., St. Louis, Missouri, United States) into a mixture of silver sulfadiazine (with or without chlorhexidine) and polyurethane in *N*, *N*-dimethylacetamide; rapidly aspirating the mixture to the level of the connector piece; draining the tube; air drying the tube by placing the plastic tube in a rotational device for 12 h; and sterilizing the tube with ethylene oxide gas [43–45]. This process results in a coating on the inner, but not the outer, lumen of the endotracheal tube. The silver sulfadiazine in polyurethane coating was chosen for several reasons. First, it is smooth and resistant to torque. Second, it does not require ultraviolet light or another type of special care. Third, silver sulfadiazine is widely and safely used as topical preparations for burn patients and on intravenous and urinary catheters [45].

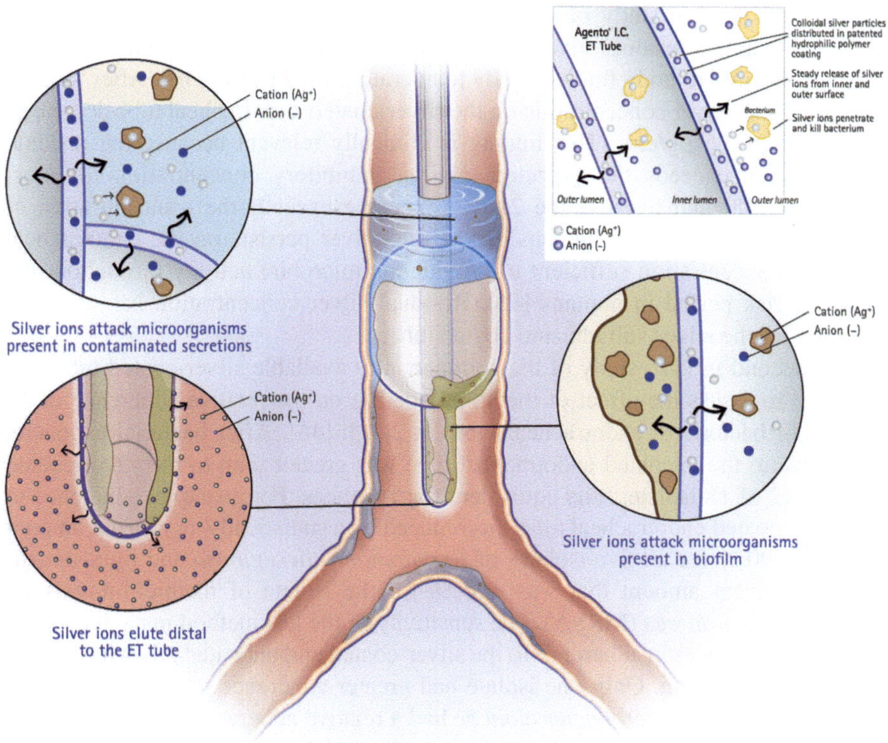

Fig. 21.3 Multiple mechanisms of action of silver-coated endotracheal tube, including contributions of hydrophilic polymer. © 2009, C. R. Bard, Inc. Reproduced with permission from C. R. Bard, Inc., Covington, Georgia, USA

21.3.1 Preclinical Studies

Both endotracheal tubes have been investigated in *in vitro* models [44, 46]. The commercially available silver-coated tube has been investigated in rabbits [46] and dogs [47], whereas the investigational silver-sulfadiazine-coated tube has been investigated in sheep [43, 44]. Each study is described separately, beginning with studies on the commercially available silver-coated tube.

The commercially available silver-coated tube has been evaluated in a series of sequentially conducted preclinical studies [46] that were designed to elucidate the mechanism of action. Specifically, these studies were designed to sequentially evaluate surrogate endpoints in the early pathogenesis of VAP, beginning with 2- to 4-h exposure *in vitro* [46] and progressing to 96-h exposure in dogs [47]—thereby providing a rationale for further evaluation in patients requiring mechanical ventilation. Each preclinical study was controlled; the control tube was similar to the silver-coated tube except that it was not coated.

The first *in vitro* study of the commercially available silver-coated tube was designed to assess residual silver concentration in a model simulating and approximating exposure to body fluids at body temperature for 21 days [46]. After incubation, the mean silver concentration on the silver-coated endotracheal tube decreased from 13.3 to 2.6 μg/cm². This finding is clinically relevant because the residual concentration exceeds the expected minimal inhibitory concentration for most organisms [48] and because the 21-day exposure exceeds the usual duration of mechanical ventilation in humans. Therefore, silver persists on the endotracheal tube at a concentration sufficient to provide antimicrobial activity throughout the expected risk period in humans [46]. Residual silver concentration has not been reported for the silver-sulfadiazine-coated tube.

The second *in vitro* study of the commercially available silver-coated tube was designed to assess the effect of the silver coating on the relative adherence of 18 isolates of bacteria to endotracheal tubes at 2–4 h [46]. After bacterial exposure, adherence to the uncoated endotracheal tube was greater than to the silver-coated tube for 12 of 18 isolates, and equivalent for 5 isolates. For example, adherence to the silver-coated endotracheal tube was reduced by a statistically significant amount that was >90 % for all five isolates of *Pseudomonas aeruginosa* and by a statistically significant amount that was <90 % for the isolate of methicillin-resistant *Staphylococcus aureus* (MRSA). The sensitivity of the test method made it difficult to detect differences in adherence to the silver-coated and uncoated tubes for low- or nonadhering bacteria. Only one isolate had greater adherence to the silver-coated endotracheal tube; *Enterobacter cloacae* had a relative adherence score of −1 [46]. These findings are clinically relevant because bacterial adherence is the first step in the formation of biofilm by bacteria that frequently cause VAP, such as *P. aeruginosa* and MRSA.

The *in vitro* study of the investigational silver-sulfadiazine-coated tube was designed to assess the effect of the silver coating on bacterial adhesion and biofilm formation [44]. Tubes were examined 6, 24, and 72 h after initial exposure to *P. aeruginosa*, which was repeated every 24 h. No bacteria or secretions were detectable by scanning electron and confocal laser scanning microscopy on coated tubes at any time. Similarly, no bacterial colonization was detectable by microbiologic studies on the four coated tubes, except for two tubes at 24 h. In contrast, bacterial adhesion at 6 h, microcolony formation at 24 h, and uniform protein-like deposits at 72 h were detectable on all four uncoated tubes. Similarly, colonization was detectable on all four uncoated tubes at all observation periods. The bacterial burden in broth cultures was lower for coated tubes at 6 and 72 h ($P<0.01$) [44]. These findings are clinically relevant because scanning electron and confocal laser scanning microscopy provide visual confirmation that silver coating blocks biofilm formation.

The first animal study of the commercially available silver-coated tube was designed to evaluate the effect of the silver coating on bacterial burden in rabbits [46]. Rabbits were exposed to a respiratory isolate of *P. aeruginosa* at 0 and 6 h and were intubated for 16 h. Uncuffed tubes were used to accommodate the small diameter of the rabbit trachea. Colonization was detectable on three sampling sites of the tube

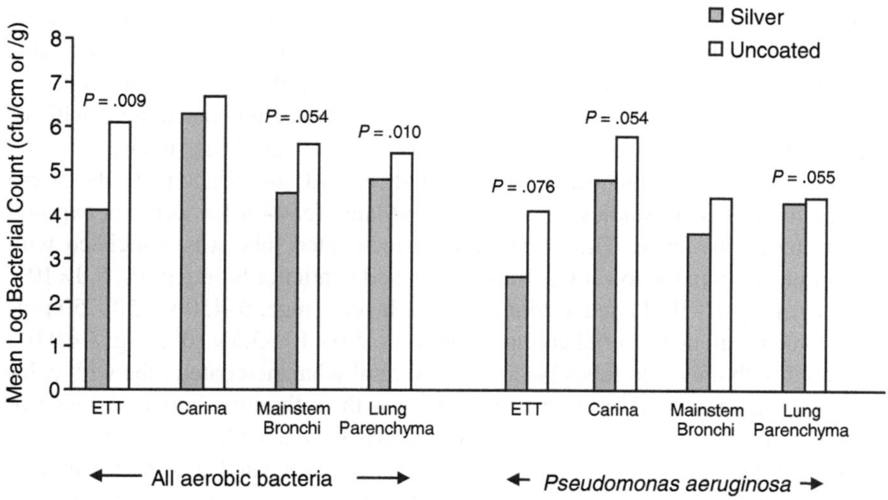

Fig. 21.4 Decreased bacterial burden in dogs intubated with silver-coated tubes in a randomized preclinical study [47]

and adjacent trachea in only two of six rabbits that were intubated with the silver-coated tube. Similarly, *P. aeruginosa* was isolated from the lungs of none of the rabbits intubated with the silver-coated tube. In contrast, the tube samples and adjacent trachea were heavily colonized in all rabbits that were intubated with the uncoated tube. *P. aeruginosa* was isolated from the lungs of four of six rabbits intubated with the uncoated tube. The silver coating was associated with at least a 99 % reduction in the mean log concentrations of *P. aeruginosa* on the combined results from the three tube sampling sites (silver vs. uncoated, 1.05 vs. 3.84 colony-forming units [cfu]/cm; $P<0.0001$), adjacent trachea (1.54 vs. 3.83 cfu/g; $P<0.05$), and lung tissue (0.00 vs. 2.48 cfu/g; $P<0.05$) [46]. These findings are clinically relevant because they extend 4-h *in vitro* findings to rabbits that were intubated (but not mechanically ventilated) for 16 h.

The second animal study of the commercially available silver-coated tube was designed to evaluate the effect of the silver coating on bacterial burden in dogs that were mechanically ventilated [47]. Dogs were intubated with cuffed tubes, exposed to *P. aeruginosa* at 1 and 8 h, and ventilated for 96 h. The silver-coated tube was associated with reductions in the bacterial burden (Fig. 21.4). The reductions for all aerobic bacteria were statistically significant on the tube (silver vs. uncoated, 4.1 vs. 6.1 log cfu/cm; $P=0.009$) and in the lung parenchyma (4.8 vs. 5.4 log cfu/g; $P=0.010$); the reduction in the mainstem bronchi approached statistical significance (4.5 vs. 5.6 log cfu/g; $P=0.054$). The reductions for *P. aeruginosa* approached statistical significance on the tube (2.6 vs. 4.1 log cfu/cm; $P=0.076$), in the carina (4.8 vs. 5.8; $P=0.054$), and in multiple sites in the lung parenchyma (4.3 vs. 4.4 log cfu/g; $P=0.055$). In addition to reducing the bacterial burden, the silver-coated tube was associated with less pulmonary damage as demonstrated by lower histologic scores (2.8 vs. 7.1; $P < 0.001$) [47]. These findings are clinically relevant because they extend 16-h rabbit findings to dogs that were not only intubated but also mechanically ventilated for 96 h—thereby better simulating clinical conditions.

Animal studies of the investigational silver-sulfadiazine-coated tube with or without chlorhexidine were designed to evaluate the effect of the silver coating on bacterial burden in mechanically ventilated sheep [43, 44]. Uncoated tubes were used as controls in all studies. Sheep were ventilated for 24 h but were not exposed to a bacterial challenge. The silver-sulfadiazine-coated tube was associated with less colonization in the lower respiratory tract and ventilator tubing (0 vs. 5.0×10^5–3.5×10^8 cfu/g; $P < 0.01$) and a thinner mucus layer (range, 0–450 vs. 50–750 μm) in the lumen with no bacterial colonization (0 vs. 5.0×10^6–3.5×10^8 cfu/g; $P < 0.01$) compared with uncoated tubes [44]. The silver-sulfadiazine-coated tube with chlorhexidine was associated with no bacterial growth in the tube lumen in seven of eight sheep, whereas the uncoated tube was associated with colonization in all animals (median, 7.1×10^6 cfu/g; $P = 0.001$) [43]. The coated tube had a thinner bacterial biofilm (32 vs. 236 μm; $P < 0.001$), and was also associated with less bacterial colonization in bronchial samples ($P = 0.004$) and lung parenchyma ($P = 0.01$) [43]. Collectively, these findings are clinically relevant because they provide visual evidence on scanning electron microscopy that silver blocks biofilm formation in an animal model. The silver-sulfadiazine-coated tube (without chlorhexidine) was chosen for further evaluation because of safety considerations and the potential for systemic hypersensitivity reaction with chlorhexidine.

21.3.2 Phase 2 Clinical Studies

The commercially available silver-coated [49] and investigational silver-sulfadiazine-coated [45] endotracheal tubes have been evaluated in separate feasibility studies that were conducted as randomized, phase 2 clinical studies. A single-blind design was used for the study of the silver-coated tube with investigators blinded to randomization block length and microbiology laboratory personnel blinded to group assignments [49]; blinding was not indicated in the other study [45]. The silver-coated tube was evaluated in 149 adults in the ICU who were expected to require intubation for at least 24 h [49]; the silver-sulfadiazine-coated tube was evaluated in 46 post-cardiac surgery patients who were intubated for up to 24 h (mean, 16 h) [45].

The commercially available silver-coated tube appeared to be safe as demonstrated by the lack of overt safety issues [49]; the distribution and pattern of adverse events were similar for patients intubated with the silver-coated or uncoated tubes. Evaluation of silver concentrations in the subset of patients with paired values revealed that blood concentrations were higher after extubation than at baseline ($P < 0.0001$), but the median difference of 0.8 ng/mL [49] was not clinically significant compared with findings from normal volunteers [50] and patients with burns treated with silver-containing preparations [51]. These findings are clinically relevant because of implications for both safety and clinical activity as silver release is required for the antibacterial activity of the silver-coated tube.

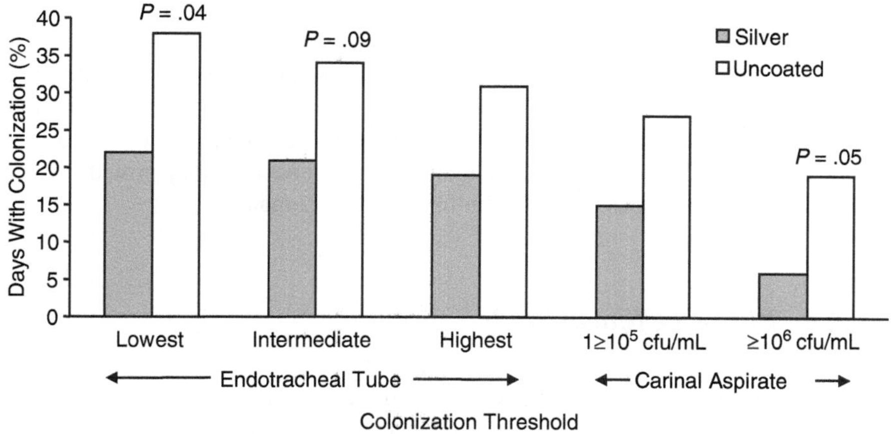

Fig. 21.5 Decreased colonization in patients intubated with silver-coated tubes in a randomized feasibility study [49]. Lowest threshold, ++, +++, or $\geq 10^4$ cfu/mL; intermediate threshold, ++, +++, or $\geq 10^5$ cfu/mL; highest threshold, ++, +++, or $\geq 10^6$ cfu/mL

In addition to feasibility, the phase 2 study of the silver-coated tube was also designed to assess the effect on bacterial burden at two sites [49]. First, bacterial burden on the inner surface of the endotracheal tube was assessed by swab cultures. In this multicenter study, some centers used qualitative methods and others used quantitative methods. Moderate growth (++) and abundant growth (+++) were combined because of lack of quantitative differentiation between these qualitative results. Results from different centers were combined into three colonization thresholds (lowest threshold, ++, +++, or $\geq 10^4$ cfu/mL; intermediate threshold, ++, +++, or $\geq 10^5$ cfu/mL; highest threshold, ++, +++, or $\geq 10^6$ cfu/mL) Second, bacterial burden in the tracheal aspirate was assessed quantitatively by all centers at two thresholds ($\geq 10^5$ and $\geq 10^6$ cfu/mL). The silver-coated tube was associated with reduced colonization as measured by the percentage of days with colonization (Fig. 21.5). The reductions were statistically significant on the tube at the lowest threshold ($P=0.04$) and in the carinal aspirate at $\geq 10^6$ cfu/mL ($P=0.05$). In addition, the silver-coated tube was associated with delayed colonization on the tube (lowest threshold, $P=0.02$, log-rank test; $P=0.10$, Wilcoxon's test) and in carinal aspirates (highest threshold, $P=0.08$, log-rank test; $P=0.05$, Wilcoxon's test) [49], and lower maximal bacterial burden in carinal aspirates for 7 days (mean log-transformed burden, 4.2 vs. 5.5 log cfu/mL; $P=0.02$, Wilcoxon's test) [49]. These findings are clinically relevant because they extended preclinical findings by demonstrating that the silver-coated tube reduced colonization in carinal aspirates, a precursor for VAP [52].

The investigational silver-sulfadiazine-coated tube appeared to be safe as demonstrated by the lack of adverse events related to the use of the coated tube during intubation or at any time during hospitalization [45]. This phase 2 study was also designed to assess the effect on bacterial burden as measured by microbiology, confocal scanning laser microscopy, and scanning electron microscopy. Upon extubation, no evidence of bacterial colonization was evident on the inner lumen of the silver-sulfadiazine-coated tube (frequency, 0/23 vs. 8/23; $P<0.01$). In addition, the silver-sulfadiazine-coated tube prevented organized bacterial biofilm formation in the tube lumen and was associated with a thinner mucus layer (range, 0–200 vs. 50–700 μm) [45]. These findings are clinically relevant because they provide visual evidence that silver blocks early biofilm formation in humans.

21.3.3 Phase 3 Clinical Study

The commercially available silver-coated endotracheal tube was evaluated in a phase 3 study known as the North American Silver-Coated Endotracheal Tube (NASCENT) study [53]. This prospective, randomized, single-blind, controlled study was conducted at 54 centers throughout North America. We had to screen more than 9,000 patients in order to identify 2003 patients expected to be intubated for ≥24 h and therefore suitable for randomization. The primary endpoint was VAP incidence defined by rigorous microbiologic criteria (i.e., quantitative bronchoalveolar lavage [BAL] fluid culture with $\geq 10^4$ cfu/mL). The trigger point for collecting BAL fluid was the most popular diagnostic criteria for VAP (i.e., radiographic evidence of VAP combined with at least two clinical criteria) [53]. Therefore, this endpoint is consistent with the recommended multimodal diagnostic approach [33]. Investigators followed institutional strategies for preventing VAP and were encouraged to avoid changes during the study.

The silver-coated tube was associated with a reduced incidence of microbiologically confirmed VAP (Fig. 21.6) [53]. This benefit was statistically significant at any time after intubation (silver vs. uncoated, 37/766 [4.8 %] vs. 56/743 [7.5 %]; $P=0.03$), for a relative risk reduction of 35.9 %. The benefit was even greater within 10 days of intubation as demonstrated by the relative risk reduction of 47.6 % (27/766 [3.5 %] vs. 50/743 [6.7 %]; $P=0.005$), which is clinically relevant because the duration of intubation is usually less than 10 days. Multivariate logistic regression analysis provided additional evidence that the observed benefits were attributable to the silver-coated tube. After adjusting for other risk factors in the multivariate analysis, treatment group was associated with reduced risk of developing VAP at any time (odds ratio for silver-coated vs. uncoated, 0.52; 95 % CI, 0.33–0.82; $P=0.005$) and within 10 days of intubation (odds ratio, 0.44; 95 % CI, 0.26–0.73; $P=0.001$). In addition, the silver-coated tube delayed the onset of microbiologically confirmed VAP ($P=0.005$). After adjusting for other risk factors in the Cox regression analysis, treatment group was associated with delayed time to occurrence of VAP (hazard ratio, 0.55; 95 % CI, 0.37–0.84; $P=0.006$) [53]. These findings are clinically relevant because they extend findings from preclinical studies [46, 47] and the feasibility study [49], which had demonstrated significant benefits as measured by surrogate endpoints for VAP. The NASCENT study demonstrated a statistically significant impact on VAP as defined by a multimodal approach based on the

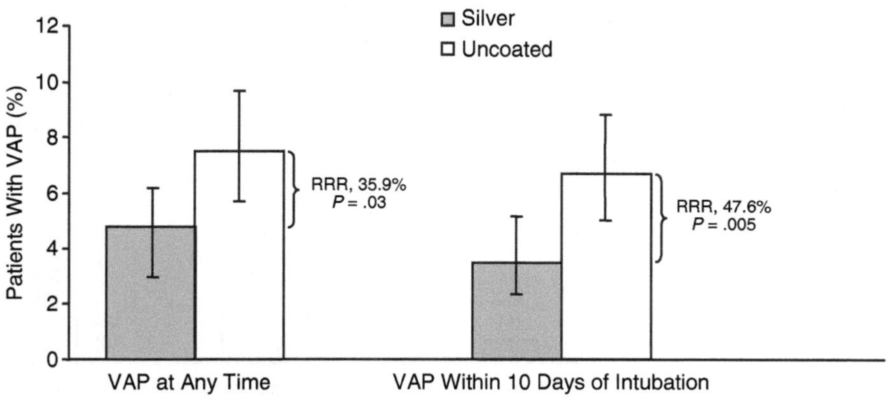

Fig. 21.6 Decreased ventilator-associated pneumonia (VAP) in patients intubated ≥24 h with silver-coated tubes in the NASCENT randomized trial [53]. RRR, relative risk reduction. Lines above and below each column represent 95 % confidence intervals

popular diagnostic criteria as the trigger for performing quantitative microbiologic studies.

The silver-coated tube appeared to be safe as evidenced by the similar distribution and pattern of adverse events between treatment groups [53]. Of 59 adverse events reported as definitely related to the endotracheal tube, 21 occurred in patients intubated with the silver-coated tube and 38 occurred in patients intubated with the uncoated tube. Of 74 adverse events reported as definitely related to the intubation procedure, 39 occurred in patients intubated with the silver-coated tube and 35 occurred in patients intubated with the uncoated tube [53].

One unexpected finding was the occurrence of microbiologically confirmed VAP in only 7.5 % of patients intubated for ≥24 h with the uncoated tube. This low incidence of VAP may have confounded the ability to detect between-group differences in other outcomes, such as length of stay and mortality.

To further investigate the potential impact of the silver-coated tube on mortality, we performed a retrospective cohort analysis [54] of patients who developed VAP in the NASCENT study [53]. The silver-coated tube was associated with reduced mortality in this subset (5/37 [14 %] vs. 20/56 [36 %]; $P=0.03$), for a relative risk reduction of 62.2 % (Fig. 21.7) [54]. The multivariate analysis supported these findings as being assigned to the uncoated treatment group was a predictor of mortality (odds ratio, silver vs. uncoated group, 0.28; 95 % CI, 0.09–0.89; $P=0.03$); APACHE II ≥20 and inappropriate empiric antibiotic therapy also remained in the model ($P<0.1$). Infection-related deaths occurred in only 1 patient (3 %) who was intubated with the silver-coated tube and in 9 patients (15 %) who were intubated with the uncoated tube. Next we focused on potentially multidrug-resistant bacteria, because they are associated with higher mortality compared with other bacteria [55–58], and determined the cause of death. For the purpose of this study [54], we defined potentially multidrug-resistant bacteria as MRSA, *P. aeruginosa*,

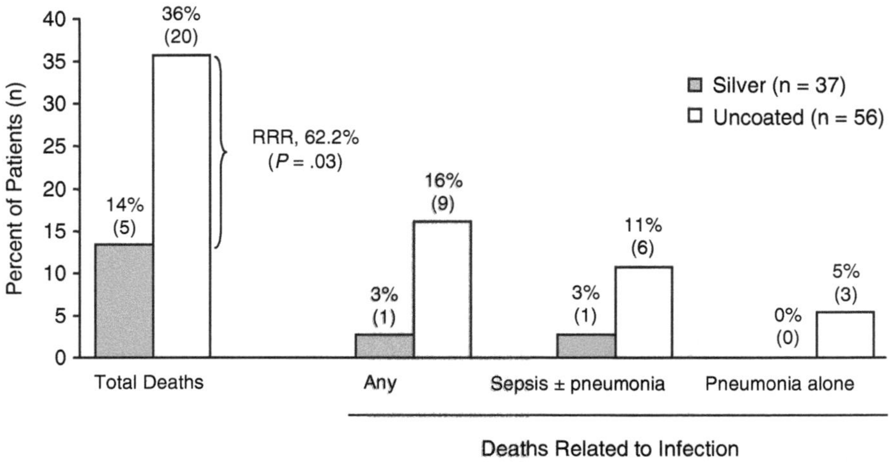

Fig. 21.7 Decreased mortality in a cohort analysis [54] of patients with VAP intubated with silver-coated tubes in the NASCENT randomized trial [53]. RRR, relative risk reduction

Acinetobacter baumannii, *Stenotrophomonas maltophilia*, and *Burkholderia cepacia*. These potentially multidrug-resistant bacteria contributed to the only death due to sepsis or pneumonia in patients intubated with the silver-coated tube and to 6 of 9 such deaths in patients intubated with the uncoated tube [54]. Furthermore, the rate of VAP due to potentially multidrug-resistant bacteria was 50 % lower in patients intubated with the silver-coated tube than in those intubated with the uncoated tube, a difference that was statistically significant when all patients in the NASCENT study [53] were considered (13/766 [1.7 %] vs. 25/743 [3.4 %]; $P<0.05$) [54]. These findings provide further evidence that the silver coating prevents VAP by reducing the bacterial burden, including potentially multidrug-resistant bacteria.

The investigational silver-sulfadiazine-coated tube has not been evaluated in a phase 3 study.

21.3.4 Pharmacoeconomic Studies

The potential economic impact of the commercially available silver-coated tube has been evaluated in a hypothetical cohort [59] and in a series of observational studies [60–62]. The silver-sulfadiazine-coated tube has not been evaluated in a pharmacoeconomic study.

The first study was a cost-effectiveness analysis of the economic outcomes of VAP avoidance based on a simple decision model using a hypothetical 1,000-patient cohort intubated with silver-coated or uncoated tubes, which was followed to VAP or no VAP [59]. The design is noteworthy because the inputs were intentionally biased to favor uncoated tubes. For example, the relative risk reduction in incidence

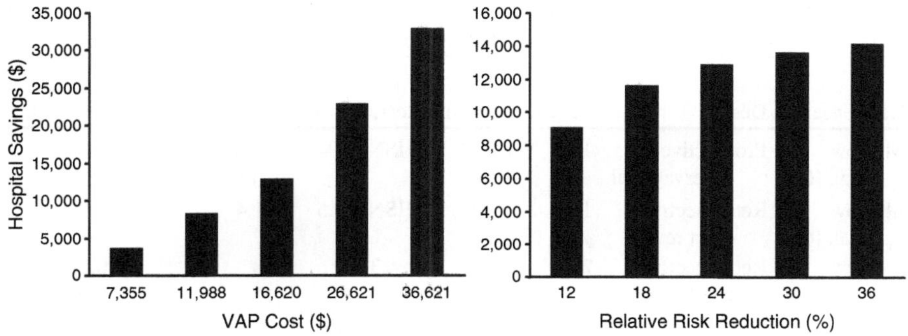

Fig. 21.8 Marginal Hospital savings for each case of VAP prevented with use of silver-coated endotracheal tube in a cost-effectiveness analysis [59]. Sensitivity analysis for various VAP costs (shown on the *left*) and for various relative risk reductions (shown on the *right*)

of microbiologically confirmed VAP in patients intubated ≥24 h was lowered from 35.9 % as demonstrated in the NASCENT study [53] to 24 % in the base-case analysis [59]. Cost inputs were a conservative estimate of $16,620 for the marginal VAP cost, $90.00 for coated tubes, and $2.00 for uncoated tubes. The last input was a pooled cumulative VAP risk of 9.7 % based on a systematic review [63]. With these inputs, 23 of 97 expected cases of VAP could be avoided with the silver-coated ETT in the hypothetical cohort, resulting in a savings to the hospital of $12,840 per case of VAP avoided in the base case [59]. Extensive sensitivity testing was performed to simulate uncertainty about model inputs, and the unique costs and epidemiologic situations of different hospitals. The most influential inputs were VAP cost and relative risk reduction with the silver-coated tube. When these inputs were evaluated in a two-way sensitivity analysis, hospital savings persisted across a range of VAP costs (from $7355 to $36,621) and relative risk reductions (from 12 % to 36 %; Fig. 21.8). Moreover, hospital savings persisted in multivariate Monte Carlo simulations of 10,000 for each outcome with a 95 % CI of $9,630 to $16,356 per case of VAP avoided with the silver-coated tube. The break-even cost for the silver-coated tube was $388 with other base-case inputs held constant [59]. These findings are clinically relevant because they suggest that the silver-coated tube may represent a strategy for avoiding VAP that will be cost-effective across the wide range of inputs likely to occur in different hospitals around the world.

A series of observational studies was conducted at a university teaching hospital in Omaha, Nebraska, United States, where systematic use of the silver-coated endotracheal tube was implemented in April 2008 [60–62]. The silver-coated tube was to be used in all high-risk patients requiring intubation for mechanical ventilation in the emergency department, in ICUs, or on general wards; the silver-coated tube was also to be used in at-risk patients potentially requiring postoperative mechanical ventilation (otherwise, conventional tubes were used in the operating room). The diagnosis of VAP was based on National Healthcare Safety Network (NHSN) criteria in all three studies [60–62], which are frequently used for surveillance purposes in

Table 21.3 Observational studies of the effect of the silver-coated on ventilator-associated pneumonia (VAP) at Creighton University Medical Center

Reference	Design	Observation period (months)	VAP[a] rate/1,000 days	*P* value	
				Preintervention	Intervention
Morrow et al. [60]	Prospective, observational	12	NHSN, 9.04	2.64	0.002
				0.56	<0.001
Morrow et al. [61]	Retrospective chart review	12	NHSN, 19.5	11.4	0.07
Morrow et al. [62]	Retrospective chart review	24	Any, 33	18	0.0002
			NHSN, 18	10	0.01
			Clinical, 15	8	0.04

[a]VAP was diagnosed by National Healthcare Safety Network (NHSN) unless otherwise specified; see text for explanation of clinical VAP

the United States. The third study included two additional definitions [62]. Clinically significant VAP was defined as abnormal chest x-ray suggestive of VAP, CPIS ≥7, and ≥3 days of antimicrobial therapy appropriate for respiratory source. Any VAP was defined as either NHSN VAP or clinically significant VAP. The first study [60] was a prospective, observational study of nearly 1,000 patients evaluated over two consecutive 6-month periods immediately before and after implementation of the silver-coated tube; the same infection control nurse diagnosed VAP throughout the study. The second and third studies were retrospective chart reviews; eligibility criteria were VAP based on diagnostic codes and intubation for ≥2 days [61, 62]. The second study was conducted over identical 6-month periods in consecutive years before and after implementation of the silver-coated tube, and included a total of 275 evaluable patients [61]. The third study was conducted over two consecutive 12-month periods immediately before and after implementation of the silver-coated tube, and included a total of 546 evaluable patients [62]. These differences in study design and diagnostic criteria have implications for VAP rates and preclude direct comparison of rates from different studies.

Implementation of systematic use of the silver-coated tube was associated with statistically significant decreases in VAP rates in the first and third studies (Table 21.3) [60, 62]. For example, the VAP rate per 1,000 ventilator days fell from 9.04 to 2.64 episodes in all patients ($P=0.002$) and to 0.56 episodes in the subset of postintervention patients who were actually intubated with the silver-coated tube ($P<0.001$) in the first study [60]. In addition, the rate fell progressively as use of the silver-coated tube increased from 42 % to 77 % over the 6-month intervention period. The benefit approached, but did not achieve, statistical significance in the second study ($P=0.07$) [61], presumably because of the smaller sample size. These findings are clinically relevant because they expand on previous well-designed studies [46, 47, 49, 53] by demonstrating that systematic use of the silver-coated tube is associated with statistically significant benefits in clinical practice. In addition, these findings confirm that the benefit is detectable for a variety of definitions of VAP, which had ranged from surrogate endpoints [46, 47, 49] to microbiologically confirmed VAP and now include both NHSN criteria [60–62] and clinically diagnosed VAP [62].

Implementation of systematic use of the silver-coated tube was associated with decreased medical resource utilization in the second and third studies [61, 62]. For example, both antibiotic use and the duration of hospitalization were lower during the intervention period in the second study ($P<0.05$ for both endpoints) [61]. Similarly, antibiotic use was lower during the intervention period in all patients and in the subset with NHSN VAP in the third study ($P=0.003$ for both groups) [62]. These findings are clinically relevant because they expand on the findings from the cost-effectiveness analysis of the hypothetical cohort [59], which had predicted hospital cost savings. Although we did not measure savings per se, they are inferred by shorter duration of hospitalization because it is one of the main determinants of costs. In addition, our findings suggest that the lower VAP rate may translate to tangible changes in physician behavior, such as decreasing antibiotic prescriptions, which in turn may have implications for antibiotic resistance.

21.4 Conclusions

The silver-coated endotracheal tube is a promising strategy for preventing VAP based on lessons learned from studies of other preventive strategies (Table 21.2). First, the device meets the high standards for most evidence-based guidelines because effectiveness is supported by the results of a well-designed, phase 3 study [53]. The silver-coated tube is included in recent guidelines from the European HAP Working Group [10], but not in guidelines that preceded publication of the phase 3 study [53]. Second, the device had statistically significant effects on VAP as defined by a wide variety of diagnostic criteria, including rigorous microbiologic criteria in the phase 3 study [53] as well as surrogate endpoints in earlier studies [46, 47, 49] and NHSN surveillance criteria in clinical practice [60–62]. Third, the device reduced the relative risk of VAP by approximately 36 % in the phase 3 study [53], an amount that is both statistically and clinically significant. The relative risk reduction was nearly 50 % within 10 days of intubation, which is clinically relevant because the duration of intubation is usually less than 10 days. Fourth, the device is suitable for use with other preventive strategies because the relative risk reduction occurred while adding the silver-coated tube to a wide variety of existing bundles at 54 institutions throughout the United States in the phase 3 study [53]. Fifth, the device is suitable for use in the ICU because it becomes user independent after intubation and does not add to the burden of busy healthcare providers. This has implications for addressing the problem of nonadherence. Of course, education is critical to the success of any preventive strategy, and systematic use of the silver-coated endotracheal tube did increase throughout the 6-month intervention period at a university teaching hospital [60]. Sixth, the silver-coated endotracheal tube is an example of a new technology that is suitable for use in patients at risk of developing VAP.

Lastly, careful and accurate analyses of infection control practices should be conducted at the hospital level. These analyses should include appropriate cost-accounting methods and assessments of all costs associated with VAP. Only in this way can ICU directors and infection control practitioners truly evaluate the impact of implementing new technologies such as the silver-coated ETT in their ICU settings.

References

1. Rosenthal VD, Maki DG, Salomao R, et al. Device-associated nosocomial infections in 55 intensive care units of 8 developing countries. Ann Intern Med. 2006;145:582–91.
2. Rosenthal VD, Maki DG, Jamulitrat S et al. International Nosocomial Infection Control Consortium (INICC) report, data summary for 2003–2008, issued June 2009. Am J Infect Control. 2010;38:95–104 e102.
3. Anderson DJ, Kirkland KB, Kaye KS, et al. Underresourced hospital infection control and prevention programs: penny wise, pound foolish? Infect Control Hosp Epidemiol. 2007;28: 767–73.
4. Kuti EL, Patel AA, Coleman CI. Impact of inappropriate antibiotic therapy on mortality in patients with ventilator-associated pneumonia and blood stream infection: a meta-analysis. J Crit Care. 2008;23:91–100.
5. Michael S. Niederman, Donald E, et al. Guidelines for the management of adults with hospital-acquired, ventilator-associated, and healthcare-associated pneumonia. Am J Respir Crit Care Med. 2005;171:388–416.
6. File Jr TM. Recommendations for treatment of hospital-acquired and ventilator-associated pneumonia: review of recent international guidelines. Clin Infect Dis. 2010;51 Suppl 1: S42–7.
7. Coffin SE, Klompas M, Classen D, et al. Strategies to prevent ventilator-associated pneumonia in acute care hospitals. Infect Control Hosp Epidemiol. 2008;29 Suppl 1:S31–40.
8. Muscedere J, Dodek P, Keenan S, Fowler R, Cook D, Heyland D. Comprehensive evidence-based clinical practice guidelines for ventilator-associated pneumonia: diagnosis and treatment. J Crit Care. 2008;23:138–47.
9. Rotstein C, Evans G, Born A, et al. Clinical practice guidelines for hospital-acquired pneumonia and ventilator-associated pneumonia in adults. Can J Infect Dis Med Microbiol. 2008;19: 19–53.
10. Torres A, Ewig S, Lode H, Carlet J. Defining, treating and preventing hospital acquired pneumonia: European perspective. Intensive Care Med. 2009;35:9–29.
11. Masterton RG, Galloway A, French G, et al. Guidelines for the management of hospital-acquired pneumonia in the UK: report of the working party on hospital-acquired pneumonia of the British Society for Antimicrobial Chemotherapy. J Antimicrob Chemother. 2008;62: 5–34.
12. Kollef M. SMART approaches for reducing nosocomial infections in the intensive care unit. Chest. 2008;134:447–56.
13. Craven DE. Preventing ventilator-associated pneumonia in adults: sowing seeds of change. Chest. 2006;130:251–60.
14. National Nosocomial Infections Surveillance (NNIS) System report, data summary from January 1990–May 1999, issued June 1999. Am J Infect Control. 1999;27:520–532.
15. National Nosocomial Infections Surveillance (NNIS) System Report, data summary from January 1992 through June 2004, issued October 2004. Am J Infect Control. 2004;32:470–485.

16. Edwards JR, Peterson KD, Mu Y et al. National Healthcare Safety Network (NHSN) report: data summary for 2006 through 2008, issued December 2009. Am J Infect Control. 2009;37:783–805.
17. Cook D. Ventilator associated pneumonia: perspectives on the burden of illness. Intensive Care Med. 2000;26 Suppl 1:S31–7.
18. Rello J, Lorente C, Bodi M, Diaz E, Ricart M, Kollef MH. Why do physicians not follow evidence-based guidelines for preventing ventilator-associated pneumonia?: a survey based on the opinions of an international panel of intensivists. Chest. 2002;122:656–61.
19. Ricart M, Lorente C, Diaz E, Kollef MH, Rello J. Nursing adherence with evidence-based guidelines for preventing ventilator-associated pneumonia. Crit Care Med. 2003;31:2693–6.
20. Cason CL, Tyner T, Saunders S, Broome L. Nurses' implementation of guidelines for ventilator-associated pneumonia from the Centers for Disease Control and Prevention. Am J Crit Care. 2007;16:28–36. discussion 37; quiz 38.
21. Kaynar AM, Mathew JJ, Hudlin MM, et al. Attitudes of respiratory therapists and nurses about measures to prevent ventilator-associated pneumonia: a multicenter, cross-sectional survey study. Respir Care. 2007;52:1687–94.
22. Cook D, Ricard JD, Reeve B, et al. Ventilator circuit and secretion management strategies: a Franco-Canadian survey. Crit Care Med. 2000;28:3547–54.
23. Cabana MD, Rand CS, Powe NR, et al. Why don't physicians follow clinical practice guidelines? A framework for improvement. JAMA. 1999;282:1458–65.
24. Zack JE, Garrison T, Trovillion E, et al. Effect of an education program aimed at reducing the occurrence of ventilator-associated pneumonia. Crit Care Med. 2002;30:2407–12.
25. Babcock HM, Zack JE, Garrison T, et al. An educational intervention to reduce ventilator-associated pneumonia in an integrated health system: a comparison of effects. Chest. 2004;125:2224–31.
26. Apisarnthanarak A, Pinitchai U, Thongphubeth K, et al. Effectiveness of an educational program to reduce ventilator-associated pneumonia in a tertiary care center in Thailand: a 4-year study. Clin Infect Dis. 2007;45:704–11.
27. Berwick DM, Calkins DR, McCannon CJ, Hackbarth AD. The 100,000 lives campaign: setting a goal and a deadline for improving health care quality. JAMA. 2006;295:324–7.
28. Zilberberg MD, Shorr AF, Kollef MH. Implementing quality improvements in the intensive care unit: ventilator bundle as an example. Crit Care Med. 2009;37:305–9.
29. Bird D, Zambuto A, O'Donnell C, et al. Adherence to ventilator-associated pneumonia bundle and incidence of ventilator-associated pneumonia in the surgical intensive care unit. Arch Surg. 2010;145:465–70.
30. Hawe CS, Ellis KS, Cairns CJ, Longmate A. Reduction of ventilator-associated pneumonia: active versus passive guideline implementation. Intensive Care Med. 2009;35:1180–6.
31. Bouadma L, Mourvillier B, Deiler V, et al. A multifaceted program to prevent ventilator-associated pneumonia: impact on compliance with preventive measures. Crit Care Med. 2010;38:789–96.
32. Zaydfudim V, Dossett LA, Starmer JM, et al. Implementation of a real-time compliance dashboard to help reduce SICU ventilator-associated pneumonia with the ventilator bundle. Arch Surg. 2009;144:656–62.
33. Morrow LE, Kollef MH. Recognition and prevention of nosocomial pneumonia in the intensive care unit and infection control in mechanical ventilation. Crit Care Med. 2010;38: S352–62.
34. Shorr AF, Micek ST, Jackson Jr WL, Kollef MH. Economic implications of an evidence-based sepsis protocol: can we improve outcomes and lower costs? Crit Care Med. 2007;35: 1257–62.
35. Petering HG. Pharmacology and toxicology of heavy metals: silver. Pharmacol Ther. 1976;1: 127–30.
36. Saint S, Elmore JG, Sullivan SD, Emerson SS, Koepsell TD. The efficacy of silver alloy-coated urinary catheters in preventing urinary tract infection: a meta-analysis. Am J Med. 1998;105:236–41.

37. Ahearn DG, Grace DT, Jennings MJ, et al. Effects of hydrogel/silver coatings on in vitro adhesion to catheters of bacteria associated with urinary tract infections. Curr Microbiol. 2000;41:120–5.
38. Gabriel MM, Sawant AD, Simmons RB, Ahearn DG. Effects of silver on adherence of bacteria to urinary catheters: in vitro studies. Curr Microbiol. 1995;30:17–22.
39. Chandra J, Patel JD, Li J, et al. Modification of surface properties of biomaterials influences the ability of *Candida albicans* to form biofilms. Appl Environ Microbiol. 2005;71: 8795–801.
40. Cunliffe D, Smart CA, Alexander C, Vulfson EN. Bacterial adhesion at synthetic surfaces. Appl Environ Microbiol. 1999;65:4995–5002.
41. Patel JD, Ebert M, Ward R, Anderson JM. *S. epidermidis* biofilm formation: effects of biomaterial surface chemistry and serum proteins. J Biomed Mater Res A. 2007;80:742–51.
42. Tebbs SE, Elliott TS. Modification of central venous catheter polymers to prevent in vitro microbial colonisation. Eur J Clin Microbiol Infect Dis. 1994;13:111–7.
43. Berra L, De Marchi L, Yu ZX, Laquerriere P, Baccarelli A, Kolobow T. Endotracheal tubes coated with antiseptics decrease bacterial colonization of the ventilator circuits, lungs, and endotracheal tube. Anesthesiology. 2004;100:1446–56.
44. Berra L, Curto F, Li Bassi G, et al. Antimicrobial-coated endotracheal tubes: an experimental study. Intensive Care Med. 2008;34:1020–9.
45. Berra L, Kolobow T, Laquerriere P, et al. Internally coated endotracheal tubes with silver sulfadiazine in polyurethane to prevent bacterial colonization: a clinical trial. Intensive Care Med. 2008;34:1030–7.
46. Rello J, Afessa B, Anzueto A, et al. Activity of a silver-coated endotracheal tube in preclinical models of ventilator-associated pneumonia and a study after extubation. Crit Care Med. 2010;38:1135–40.
47. Olson ME, Harmon BG, Kollef MH. Silver-coated endotracheal tubes associated with reduced bacterial burden in the lungs of mechanically ventilated dogs. Chest. 2002;121:863–70.
48. Grier N. Silver and its compounds. In: Block SS, editor. Disinfection, sterilization, and preservation. 2nd ed. Philadelphia: Lea & Febiger; 1997. p. 395–407.
49. Rello J, Kollef M, Diaz E, et al. Reduced burden of bacterial airway colonization with a novel silver-coated endotracheal tube in a randomized multiple-center feasibility study. Crit Care Med. 2006;34:2766–72.
50. Perrelli G, Piolatto G. Tentative reference values for gold, silver and platinum: literature data analysis. Sci Total Environ. 1992;120:93–6.
51. Wan AT, Conyers RA, Coombs CJ, Masterton JP. Determination of silver in blood, urine, and tissues of volunteers and burn patients. Clin Chem. 1991;37:1683–7.
52. Feldman C, Kassel M, Cantrell J, et al. The presence and sequence of endotracheal tube colonization in patients undergoing mechanical ventilation. Eur Respir J. 1999;13:546–51.
53. Kollef MH, Afessa B, Anzueto A, et al. Silver-coated endotracheal tubes and incidence of ventilator-associated pneumonia. The NASCENT Randomized Trial. JAMA. 2008;300: 805–14.
54. Afessa B, Shorr AF, Anzueto AR, Craven DE, Schinner R, Kollef MH. Association between a silver-coated endotracheal tube and reduced mortality in patients with ventilator-associated pneumonia. Chest. 2010;137:1015–21.
55. Fagon JY, Chastre J, Domart Y, et al. Nosocomial pneumonia in patients receiving continuous mechanical ventilation. Prospective analysis of 52 episodes with use of a protected specimen brush and quantitative culture techniques. Am Rev Respir Dis. 1989;139:877–84.
56. Fagon JY, Chastre J, Hance AJ, Montravers P, Novara A, Gibert C. Nosocomial pneumonia in ventilated patients: a cohort study evaluating attributable mortality and hospital stay. Am J Med. 1993;94:281–8.
57. Kollef MH, Silver P, Murphy DM, Trovillion E. The effect of late-onset ventilator-associated pneumonia in determining patient mortality. Chest. 1995;108:1655–62.
58. Rello J, Jubert P, Valles J, Artigas A, Rue M, Niederman MS. Evaluation of outcome for intubated patients with pneumonia due to *Pseudomonas aeruginosa*. Clin Infect Dis. 1996;23:973–8.

59. Shorr AF, Zilberberg MD, Kollef M. Cost-effectiveness analysis of a silver-coated endotracheal tube to reduce the incidence of ventilator-associated pneumonia. Infect Control Hosp Epidemiol. 2009;30:759–63.
60. Morrow LE, Jarrett JE, Malesker MA, Kollef MH. Systematic use of a silver-coated endotracheal tube reduces rates of venitlator-associated pneumnia (poster 2146). San Diego, CA: American Thoracic Society International Conference; 2009.
61. Morrow LE, Jarrett JE, Malesker MA, Kollef MK. Effect of a silver-coated endotracheal tube on ventilator-associated pneumonia and medical resource utilization in clinical practice (poster 388). Society of Critical Care Medicine's 39th Critical Care Congress. Miami Beach, FL, 2010
62. Morrow LE, Mintz EP, Malesker MA. Effect of a silver-coated endotracheal tube on ventilator-associated pneumonia and medical resource utilization in clinical practice (poster 983). Fifth Decennial International Conference on Healthcare-Associated Infections. Atlanta, GA, 2010
63. Safdar N, Dezfulian C, Collard HR, Saint S. Clinical and economic consequences of ventilator-associated pneumonia: a systematic review. Crit Care Med. 2005;33:2184–93.

Index

T.F. Moriarty et al. (eds.), *Biomaterials Associated Infection: Immunological Aspects and Antimicrobial Strategies*, DOI 10.1007/978-1-4614-1031-7,

T

U

MIX
Papier aus verantwortungsvollen Quellen
Paper from responsible sources
FSC® C105338

If you have any concerns about our products,
you can contact us on
ProductSafety@springernature.com

In case Publisher is established outside the EU,
the EU authorized representative is:
Springer Nature Customer Service Center GmbH
Europaplatz 3, 69115 Heidelberg, Germany

Printed by Libri Plureos GmbH
in Hamburg, Germany